Electricity for Heating, Ventilation, Air Conditioning, and Related Areas

Electricity for Heating, Ventilation, Air Conditioning, and Related Areas

TIMOTHY L. ALTON
Palliser Institute

Prentice Hall

Upper Saddle River, New Jersey   Columbus, Ohio

Library of Congress Cataloging-in-Publication Data

Alton, Timothy L.
Electricity for heating, ventilation, air conditioning, and related areas / Timothy L. Alton.
p. cm.
Includes index.
ISBN 0-13-524893-0
1. Air conditioning—Electric equipment. 2. Heating—Equipment and supplies. 3. Ventilation—Equipment and supplies. 4. Electric engineering. I. Title.
TK4035.A35A48 1998
621/3′024′697—dc21 97-24649
CIP

Editor: Charles E. Stewart, Jr.
Production Editor: Mary Harlan
Design Coordinator: Julia Zonneveld Van Hook
Text Designer: Custom Editorial Productions, Inc.
Cover Designer: Proof Positive/Farrowlyne Assoc., Inc.
Cover Photo: International Stock
Production Manager: Deidra M. Schwartz
Illustrations: Timothy L. Alton and Custom Editorial Productions, Inc.
Marketing Manager: Debbie Yarnell

This book was set in Times Roman by Custom Editorial Productions, Inc., and was printed and bound by Courier/Westford, Inc. The cover was printed by Phoenix Cover Corp.

Printed in the United States of America

10 9 8 7 6 5 4 3 2 1

ISBN: 0-13-524893-0

Prentice-Hall International (UK) Limited, *London*
Prentice-Hall of Australia Pty. Limited, *Sydney*
Prentice-Hall Canada Inc., *Toronto*
Prentice-Hall Hispanoamericana, S. A., *Mexico*
Prentice-Hall of India Private Limited, *New Delhi*
Prentice-Hall of Japan, Inc., *Tokyo*
Simon & Schuster Asia Pte. Ltd., *Singapore*
Editora Prentice-Hall do Brasil, Ltda., *Rio de Janeiro*

To the memory of my father

Joseph H. Alton

Preface

Electricity for Heating, Ventilation, Air Conditioning and Related Areas was written for persons involved in careers—such as heating, ventilation and air conditioning—who have their roots in mechanical technology, but also have a significant electrical component. The text introduces the tools, instruments, and practices of electrical work. It explores electrical theory from the basics of direct current circuit operation through the complexities of single- and three-phase alternating current. It examines a broad array of electrical control devices and all types of electric motors, and provides detailed descriptions and explanations of typical HVAC circuits.

The focus of the text is on the practical realities of electricity in real world applications. Theory is covered with a minimum of mathematics. Step-by-step examples—most taken directly from the HVAC field—are used to tie the theory lessons to reality. Troubleshooting tips are integrated into these descriptions where they are appropriate, and safety is stressed throughout the book. Warnings of potential hazards are provided, particularly for situations where the dangers are not obvious. Good work habits are promoted through practical advice on testing procedures that are not only safe but also efficient and effective.

It is my intention and hope that this book will serve the beginning student as a learning tool, and the practicing tradesperson or professional as a valued reference.

Acknowledgments

I would like to express my thanks to the many individuals without whom this book would not have become a reality, in particular, my wife, Wonda, for her assistance and patience, and my children, Michael, Harley, and Joel, for their tolerance and encouragement. I would also like to give special thanks to the faculty of the Refrigeration and Air Conditioning Department, the Plumbing and Gas Fitting Department, and the Electronics Department of Kelsey Institute for their cooperation and for access to their labs and workshops when I needed to take photographs. My gratitude is also extended to my copy editor, Amy Lemmon-Bowen, for her professionalism, and Jim Reidel of Custom Editorial Productions, Inc. for his cooperation. Lastly, I want to thank the reviewers, Fred R. Burgess, CHI Institute; Gregory E. Jourdan, Wenatchee Valley College; Tim Losey, Milwaukee Area Technical College; and Bob Rogers, Oregon Institute of Technology, for their comments and suggestions.

Tim Alton

Contents

Part I

Tools and Instruments

Chapter

The first section of this book is intended primarily as a reference. The first portion of Chapter 1, "The Electrician's Toolkit," provides practical information on the selection and use of tools related to the installation, modification, and maintenance of electrical equipment. "Understanding Installations" provides some insight into the parts and pieces that make up the electrical system, and the terminology that is used to describe them. "Tips and Techniques" describes the performance of basic electrical tasks. This information should be of particular interest to tradespersons and technicians who are required to perform these tasks as a function of their primary occupation.

The second and final chapter of this section, "Meter Selection and Use," is a critical one for those actively involved in "hands-on" electrical testing and troubleshooting. It describes the various electrical instruments and their functions, as well as the variations in different makes and models available in the marketplace. The nature of this chapter makes it difficult to fully appreciate some of the details without prior knowledge of electrical theory and practice.

It may be appropriate for most students to consult this section at various points in their training.

1 Hand Tools and Techniques

Objectives

This chapter is a general reference on the tools, terminology, and techniques of electrical installations. When study of this chapter has been completed, you should be able to:

- Describe tool features that are unique to the performance of electrical tasks
- Use correct terminology when referring to specific parts of an electrical installation
- Describe the purpose of the electrical ground system
- Describe or demonstrate the correct methods for stripping and terminating conductors to each other, and to various types of terminals
- Describe some of the techniques and pitfalls of retrofitting conductors into existing buildings

1.1 The Electrician's Toolkit

This chapter is intended as a guide for those who are not trained as electricians, but are required, as part of other responsibilities, to work on equipment that is electrically powered or controlled. The tools listed and discussed are far from a full range of electrical tools. They are only examples of the everyday toolkit items carried by many electricians. Some are unique to electrical work. Others are common to many occupations, but may be somewhat different in size or style when selected for this purpose. As with many things, good quality is usually worth the extra cost. Good tools provide years of useful service and are a pleasure to use. Poor quality tools are a constant frustration.

A common, but sometimes questionable, feature of tools designed specifically for electrical work is insulated handles and driver shanks. Insulated handle grips on pliers and cutters are usually desirable, for the added comfort and improved grip, if nothing else. This feature increases the electrical safety of the tool, but only if the insulation is in good condition. Insulated handle grips on good quality pliers and cutters are either very well made or are replaceable.

Insulated grips on adjustable wrenches may be of less value, depending on the situation. The insulation must be applied by dipping or similar process, so the handles cannot be replaced if they become damaged. Insulation on screwdriver and nut driver shanks must

FIGURE 1–1 From left to right, these tools are a needlenose pliers, wire stripper and crimping tool, cabinet point screwdriver, pump pliers, adjustable wrench, Phillips screwdriver, larger cabinet point screwdriver, nut driver, multi-tap tool, Robertson screwdriver, and 9″ linesman pliers.

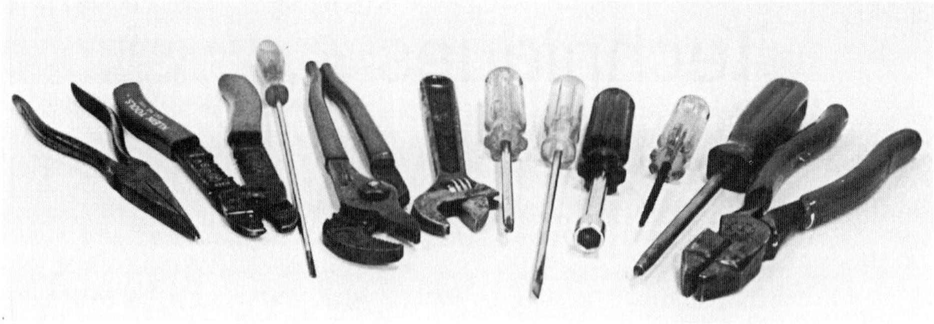

be evaluated on the basis of its electrical safety only, for it adds nothing to the normal functionality of the tool. Low quality tools with thin coatings of plastic added to the shanks should be avoided completely. The plastic may do little more than hide the finish on the metal shank.

When evaluating a tool, try to imagine its condition after several years of use. A primary concern with all insulated tools is that the wear and tear from normal usage, and the unavoidable accumulation of sweat and grime, may render the insulation ineffective. Unless the insulation is thick, free of deep cuts or cracks, and reasonably clean and dry, it is unwise to trust your life to its reliability. Contact between live circuits and tools is not always accidental. When dealing with electricity, there are occasions when there is no practical alternative to intentionally engaging energized metal with the business end of a tool. Operations such as tightening a live terminal screw or intentionally cutting an energized wire are part of normal electrical work. Such situations call for understanding and caution, and for reliable insulation on the tool handles. For this reason it is important that insulated parts are not abused, and that damaged handles are replaced if necessary. When an occasion arises where the insulation is all that is between you and a live circuit, make sure it is in good condition. Some common tools are shown in Figure 1–1.

Drivers

Electrical work requires an assortment of screwdrivers. The most commonly used tips are flat blade, Robertson, and Phillips. Most electricians prefer drivers with large, comfortable handles and moderately long shafts. The extra length is handy for reaching into electrical boxes, and is generally more convenient in a toolkit. For those occasional crowded locations, a couple of stubby or offset screwdrivers can be tossed into the bottom of the kit

The most common screw head, for terminals, is a straight slot. Three or four *cabinet* or *parallel point* flat blade screwdrivers, ranging from approximately $^{1}/_{8}''$ to $^{3}/_{8}''$ wide, are generally adequate. Cabinet and parallel tips do not flare out above the blade end. The very end of the blade is as wide as, or slightly wider than, the rest of the tip or screwdriver shank. This small detail is very important. It is often necessary to reach into relatively deep, narrow recesses to engage the heads of terminal screws. The material around the terminal is plastic insulation, which is usually quite brittle. When the driver used has a

FIGURE 1–2 Flared and parallel blade tips.

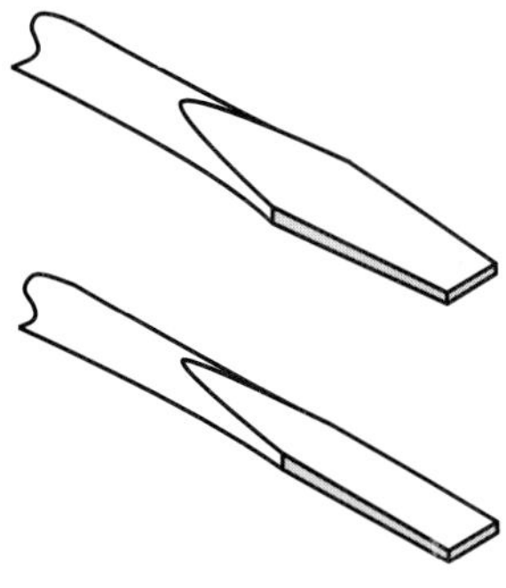

conventional flared tip (also called a keystone tip), the usual result is damage to the surrounding material. Circuit breakers are a common victim of such encounters.

For all but the hardest blade materials, the difference between a parallel tip and a flared one (see Figure 1–2) is 10 minutes of work with a sharp, flat metal file. Only moderate amounts of patience and skill are required to produce a smooth, accurate finish. A grinder will work, but is definitely not recommended. It is difficult to get an accurate cut with a grinder, and all too easy to burn the tip and ruin the temper. The same logic applies to dressing worn tips. Use a file, not a grinder.

A screw starter can be a lifesaver, particularly when the screws have slotted or Phillips heads. There are several different types on the market. The best way to choose is to try out several different ones, to see how easy they are to use and how well they hold the screw. Regardless of the type or style, this is one tool that should be insulated. It often requires both hands to align the screw properly, and to steady it as it is started into the threads. This operation often involves an energized circuit, where it is not possible to reach in and start the screw with your fingers.

For the machine bolts used in electrical boxes, and for sheet metal and wood screws used in field installations, the square-cut Robertson is very common. Two drivers, size 1 and 2 (usually equipped with color-coded green and red handles), will fit most situations. The smaller size 0 and larger size 3 are used occasionally. These are coded yellow for the small one, black for the larger one.

The X-shaped Phillips head is also common for terminal screws, and for general use in manufactured equipment. Sizes 1 and 2 are the most common. Note that the Phillips is easily confused with the similar Posidrive, and Reed and Prince, tips. This can easily lead to a poor fit and screw head damage. The need for these other types depends on the equipment encountered. If the drivers you have do not fit the screw heads correctly, find out what type does fit and get one. Damaged screw heads are a sign of poor workmanship and may make the screw very difficult to remove and replace.

There are other tips as well, such as the six-pointed Torx or the bowtie-shaped clutch head. These styles have not been particularly popular in electrical equipment in the past, but they are becoming more common. When they are encountered, there is no alternative but to have the appropriate drivers on hand.

In addition to an assortment of screwdrivers, most electricians also carry at least a few nut drivers. The correct range of sizes will vary somewhat with the circumstances, but the most common ones include $^{1}/_{4}''$, $^{5}/_{16}''$, $^{11}/_{32}''$, $^{3}/_{8}''$, and $^{7}/_{16}''$. These tools are particularly convenient for reaching into junction boxes, such as those on small motors, where the connections are made by looping the conductor and placing it between washers on a captive stud. To finish the connection, a retaining nut must be tightened down to secure the wire. A small socket set is a workable alternative, but nut drivers are much handier to use. They also reduce the risk of stripping the threads on these small brass studs.

Most nut drivers have a hollow shaft, allowing them to tighten a nut several inches onto a threaded rod or long bolt.

Pliers and Cutters

Many electricians carry 9″ side-cutting lineman's pliers, medium-size pump pliers, relatively large needlenose pliers, and a 7″ diagonal cutter. Personal preferences vary, of course, but these are very workable tools. The 9″ lineman's pliers is an excellent general-purpose cutter and pliers. A smaller one may be better for typical service work, but there are few hand tools that can match a good quality full-size lineman's pliers for power and durability. Choose a model that has a high-leverage hinge point for maximum cutting power. The hinge pin on pliers of this type is placed as close as possible to the cutting blades.

Pump pliers are generally used as a light-duty pipe wrench. They can be adjusted very quickly to accommodate a wide range of sizes. They should not be used for nuts and bolts, because they can easily damage the machined corners and flat surfaces of the nut or bolt head. They are very good for gripping pipe and pipe fittings. In electrical work, they are used primarily for tightening conduit fittings and threaded hub cable connectors. Lower quality versions of this tool have a multiposition slip joint arrangement at the hinge point. This is similar to two-position hinge joints found on common slip joint pliers. This style of pump pliers is best avoided. The far superior groove or channel style hinge joint is easy to adjust, does not bind, and is very strong.

Diagonal cutters are subject to more personal preference than lineman's pliers, because they are made in a greater variety of styles and sizes. Variations include different lengths and different angles to the blades. Cutters designed specifically for electrical work are available with optional stripping holes drilled through the blades. This is one feature that the tool user will either love or hate. It can be very convenient to have a combined stripping and cutting tool when you are making hundreds of connections a day. However, for general service work, the notches in the blade are more likely to be cursed than appreciated.

A relatively inexpensive cutting tool that is highly recommended for both service and construction work is a small diagonal cutter designed to cut flush with the surface. The cutting edges are sharpened by grinding on the inside of the jaw. This is the opposite of typical cutters, which have the cutting blades formed, at least partially, by grinding the edges from the outside. When plastic cable ties are clipped with standard cutters, the cut end is usually sharp enough to cause a cut or deep scratch. While this is not a serious safety hazard, it can make it difficult and very uncomfortable to work, particularly in crowded locations. Clipping the ends flush to the base material virtually eliminates the sharp ends. This tool is worth owning for that reason alone.

The variety of long-nose pliers available can fill a small tool catalog. Thankfully, most of them are designed for specific applications in electronics, jewelry making, or similar fine work. Such small tools are generally of little value when working at the larger scale of typical electrical equipment. A relatively large, straight-tipped pliers is usually the best choice. It should be strong enough that gripping an object in the tip and applying full power with one hand will not distort the tip, or give any other indication that it might damage the tool. Long-nose pliers are most prized for their ability to reach into awkward places, such as when manipulating wires in the back of an electrical box.

The hardworking pliers and cutters typically found in an electrician's toolkit are expensive tools that see constant use. For service work, lesser tools may be suitable. A smaller pair of pliers and a lighter cutter will handle control wiring with ease. Pump pliers may not be necessary at all, but there is nothing better for tightening threaded hub cable connectors. Hammering the locknut with a hammer and punch (or, as often happens, with lineman's pliers and a screwdriver) is the usual alternative, but it is not always a good one. For long-nose pliers, almost any good quality pair will do. Flush-cutting pliers are generally inexpensive and should be considered a must for any area where plastic cable ties are used. They are also very handy for many small tasks, but do not expect these small cutters to withstand general-purpose use, as they are not intended for heavy work.

Knives and Strippers

A knife is an essential tool for electrical work. It should have a short, sharp blade that either is rigid or locks securely in place. Retractable blade utility knives, or medium-duty adjustable blade types, are often entirely suitable for light-duty jacket stripping and insulation removal. Both of these knives use inexpensive replaceable blades, so keeping them sharp is not a problem. The adjustable blade type can be set to any depth, a definite advantage for slitting the jackets on round cables.

Specialized jacket removal knives are also available, but rarely justified for service work. Heavy-duty hook-type electrical knives are designed for removing the thick, very tough jackets used on rough-usage cables, and the equally thick, even tougher conductor insulation used on high-voltage cables. These knives are ideal for their intended task, but are clumsy for less demanding chores. Knives with long, narrow blades or those that taper to a sharp point are generally unsuitable for electrical work. Knives are subject to a lot of personal preference. Follow the general guidelines for blade length and rigidity, and keep them sharp.

There are many different wire stripping tools available, but the best general-purpose wire stripper is a knife. The thin conductor insulation typically found on wires used for controls and low-voltage power circuits is relatively easy to cut. The problem is to avoid cutting rings or nicks in the conductor below (see Figure 1–3). A common method, which almost always causes conductor damage, is holding the knife blade perpendicular to the conductor, and then cutting a slit all the way around the wire. This is very difficult to do without cutting into the conductor inside. A much better method is to make three or four small cuts around the wire with the knife blade held at a low angle so it will not dig into the conductor. The final cut can be pulled to the end of the wire, usually pulling the insulation off at the same time. Alternatively, a single low-angle cut can be used to remove a strip of insulation right to the end, and then the remaining insulation can be folded back and cut off.

Not ringing or notching the wire is much more important than is generally believed. When conductors break off at a connection point, the break is almost always at a ring or nick caused by careless stripping. Copper hardens when it is flexed. Any movement, including the mechanical vibration of equipment, can cause the conductor to become brittle. The flexing will be concentrated at the weakest point, and that is where it will eventually break.

FIGURE 1–3 Correct and incorrect methods of stripping wire.

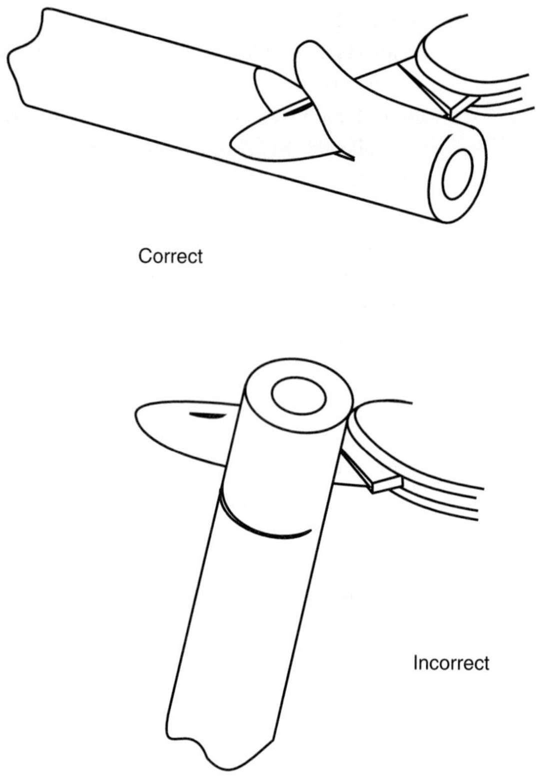

Stranded wire presents different problems. Care must be taken to avoid cutting off some of the strands. If the insulation is relatively tough, the best approach is to take extra care not to cut completely through to the conductor. If the insulation is soft, such as the rubber coverings used on many portable equipment cables, it can often be removed simply by pinching the insulation and pulling it straight off the end of the wire. If this proves too difficult, a single slit at the desired strip point may be enough to allow the rubber to be pulled free.

With a little practice, almost any type of conductor insulation can be removed properly and efficiently with a knife. However, there are also a wide variety of wire stripping tools available. A common design combines several different sizes of stripping notches, a crimper for compression-type wire ends, a wire cutter, and tapped holes for cutting small bolts without destroying the threads. A good one can be indispensable, not necessarily as a wire stripper, but for its bolt-cutting features, and possibly as a crimping tool. Beware of combination tools that are poorly made or clumsy to use.

When control panels are installed or extensively modified, hundreds of wires must be stripped. With such a large volume of work, an automatic stripper is definitely worth the time and effort it will save. The most common style has an action similar to that of a pair of pliers. The wire is placed in the appropriate stripping notch and, when the handles are squeezed, the tool grips the wire, closes the stripping jaws, and pulls the insulation off in one quick movement. Used properly, this is a fast, accurate method for stripping wire. Automatic strippers are a bit bulky in a toolkit, and more expensive than other types of strippers, but are worth it in certain situations.

Many wire strippers, even good quality ones, can easily damage the conductor if they are not used properly. The best way to judge the tool and the stripping technique is to experiment. Examine the conductor closely after each try. If it is nicked or ringed, consider changing tools, techniques, or both. When using notch-type strippers, try the next larger notch, rather than using the one that matches the conductor size. Do not rotate the stripper around the wire. This almost always damages the conductor inside. Expect to have to pull the stripper with some force. Pull straight off the end, with the tool held perpendicular to the wire. Tilting or angling the tool usually causes it to dig into the conductor. If you find it is almost impossible to remove the insulation without some damage to the conductor, try using a knife.

Crimpers

Wire crimpers are now used almost exclusively to compress the preinsulated sleeves of wire ends such as ring or spade lugs, push-on "quick connect" terminals, or in-line splicing sleeves. The crimp tool used must be suitable for the brand of devices being crimped. Some manufacturers are very insistent on specific, often expensive, custom crimping tools. Unless there is a good reason to change, it is usually good practice to purchase wire ends that match the available crimping tools. Samples should be compared for mechanical strength and the general quality of the connection. A good quality combination tool, as mentioned above, may be entirely suitable, if it is matched to an appropriate make of wire ends.

Older compression-style wire connectors are uninsulated metal sleeves. The wire ends are twisted together, and then the sleeve is slipped over the end and compressed with a crimping tool. After the sleeve is secure, it is covered with a matching insulating cap or with electrical tape. Like most compression connectors, these may have to be cut off if they must be removed, making them unsuitable for use where occasional disconnection is required. The indent-type crimp tools used for these uninsulated crimp connectors are not suitable for crimping the preinsulated types. The correct tool will compress the internal metal sleeve without destroying the outer layer of insulation. This caution applies to the entire range of preinsulated wire terminal ends.

Note that some manufacturers do use compression connectors inside their equipment; if changes must be made to the wiring, they may have to be cut off and replaced. Try loosening them by compressing them with a pliers first; this will often open them enough so that the wires will slip out. If further wiring changes are likely, consider using a removable connector, such as a wire nut, instead of a compression type. Twist-on wire connectors (wire nuts) have largely replaced the crimp-on types for wire-to-wire connections in field-installed electrical circuits.

Wrenches

In addition to the most common nut drivers or a small socket set, one or two adjustable wrenches will round out the basic set of electrical tools. Adjustable wrenches in the 6″ to 10″ range are the most suitable. Note that no adjustable wrench is ever as good as a proper fitting socket or flat wrench. Although electrical work is rarely as "nuts and bolts" intensive as mechanical installation or maintenance, there are times when a full set of flat wrenches and sockets is required. A pump pliers is never a suitable substitute for anything but a small pipe wrench. It should never be used on a machined part such as a bolt or nut.

Miscellaneous Tools

Other items that find their way into many toolkits include Allen wrenches, small taps, a drywall saw, and a punch awl. Sets of folding Allen wrenches are particularly convenient. A multitap tool is much handier than a tap handle and a few easy to lose taps, although it is also more likely to be broken. Multitap tools are shaped like a small screwdriver, with a tip that has a combination of three different-size taps. The smallest tap is at the very end, with progressively larger ones further up the shaft. One common example can rethread #6-32, 8-32, and 10-24 with the same tool. While these tools are not intended for tapping large numbers of holes in heavy metal, they are very convenient for cleaning up threaded holes in sheet metal enclosures such as electrical boxes.

A punch awl is excellent for making starting holes for wood and sheet metal screws. A small, straight-handle drywall saw takes up little more space than a large screwdriver, and can be very convenient when dealing with plasterboard.

1.2 Understanding Installations

Introduction

The purpose of the electrical system in any building is to distribute electrical energy obtained from some source—generally an electrical utility—to all the electrical equipment and outlets in the building. This must be done in such a way that the wiring system does not expose the building occupants to danger from electrical shock, or create a fire hazard. All of the electrical energy, for every electrically operated device in the building, must pass through the point where the utility supply connects to the consumer's equipment. This point of connection is generally referred to as the building's electrical service. The power capacity of the service—the conductors, fuses or breakers, and the associated equipment—is greater than at any other point in the building.

It is, of course, impractical to maintain this high capacity throughout the building. Instead, the electrical system is broken into segments, each requiring only a portion of the total energy available. The size of the conductors is reduced to match the load being served, rather than maintained at the size of the overall building supply. In large buildings there may be several steps, or stages, of reduction. This way the system is kept as compact and cost effective as possible.

Each time the wire size is reduced, lower rated fuses or circuit breakers are installed to reduce the maximum possible current. This protects the lighter wiring and equipment from the damage that could otherwise occur during a fault. Without these reductions in fuse size, a wiring or equipment fault, or an usually heavy load, could overload the wiring system so severely that the building would be set on fire. Although the electrical energy supplied to every load in the building comes from the electrical utility through the service connection, it passes through at least two, and often many more, levels of fuse or circuit breaker protection.

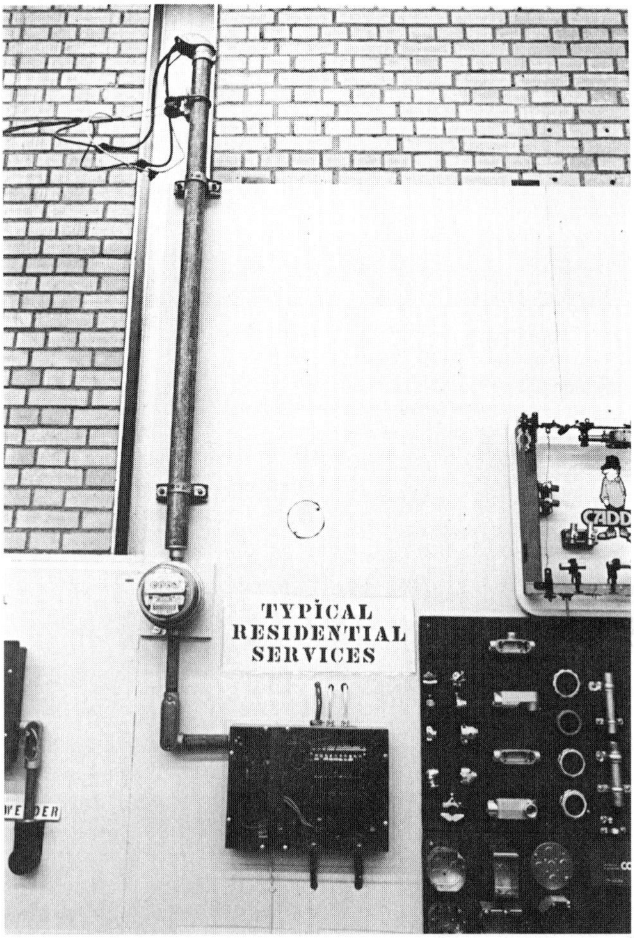

FIGURE 1–4 A school display service, with a meter and combination panelboard.

In most systems, electrical circuits come from panelboards (See Figures 1–4 and 1–5). These function as distribution points where a high-current feeder can be split off to supply other loads. The panelboard is arranged so that each circuit connects to the high-current supply through a fuse or circuit breaker. This protects each lower current circuit, and provides a point where it can be disconnected for service or revisions.

Main feeders supply panelboards; branch circuits connect from the panelboard to the individual loads. The conductors may be individual wires or cables inside a protective conduit, or groups of wires in a multiconductor cable. Single-conductor wire and cable is available in a wide variety of sizes, with many different insulation coverings. Multiconductor cables can be protected with covering jackets, flexible metal armor, or both. Electrical boxes provide termination points to interconnect the conductors of the system, and to provide a mounting for devices such as switches and receptacles. The box protects the wiring from damage, and the building and its occupants from the hazards of fire or electrocution.

Grounding is a critical part of all electrical systems, and one that is often misunderstood. In virtually all electrical installations, all of the metal boxes and conduits, and all equipment

FIGURE 1–5 This display panelboard is typical of the type found in residential services.

frames and enclosures, are electrically interconnected with a system of protective ground conductors. This system is in turn connected to all of the metal framework and piping in the building and to the earth itself. The entire ground system plays no active role in normal electrical operation, but is critical to the safety of the electrical system, and to the building and its occupants.

The Electrical Utility

For most buildings, the electrical energy originates with the supply utility and is delivered through their distribution system. This may be an exposed aerial system mounted overhead on poles, or a much less noticeable underground system. It may also be a mixture of aerial and underground equipment. Aerial conductors terminate in the open, often insulated only by their separation from other conductors and grounded structures or equipment. Underground cables terminate inside pad-mounted transformers, connection kiosks, or underground equipment vaults. Kiosks—the small metal enclosures that serve as connection points for feeders that fan out to various buildings—are often shared with other utilities such as telephone and cable television companies.

High-voltage cables link the transformers to the main supply lines, and low-voltage cables extend to individual buildings. In residential areas, the utility usually supplies a group of houses from each distribution transformer More demanding loads, such as larger

FIGURE 1–6 This pole-mounted three-phase utility transformer supplies this commercial service through overhead conductors. The utility equipment ends, and the customer's portion of the service begins, where the cables connect below the weatherproof service head on the building.

commercial or industrial buildings, may have dedicated transformers (see Figure 1–6) or transformer banks. A transformer bank is a group of transformers operating as a unit.

The Service

The connection between the incoming utility supply conductors and the building electrical equipment is called the *service entrance.* The exact equipment used for the service entrance varies with the type of supply—underground or overhead—and the total electrical demand.

Regardless of the size of the service, the incoming energy must be measured by the *utility metering* equipment. The service conductors must pass through the metering equipment before they connect to the electrical system inside the building. On smaller buildings this may be an external meter socket mounted in the service conduit or cable run. The service conductors connect to jaws inside the meter socket. The upper set of jaws are the line side, meaning they connect to the utility supply, and the lower set are the load side, meaning they

FIGURE 1–7 A typical overhead service originates at the utility pole line, where a transformer reduces the voltage to the appropriate level. Aerial conductors connect the low-voltage output of the transformer to the consumers' service conductors at the service head. The power is routed through the utility meter and into the service disconnect at the service panel. A ground connection is established at the utility pole and at the service panel.

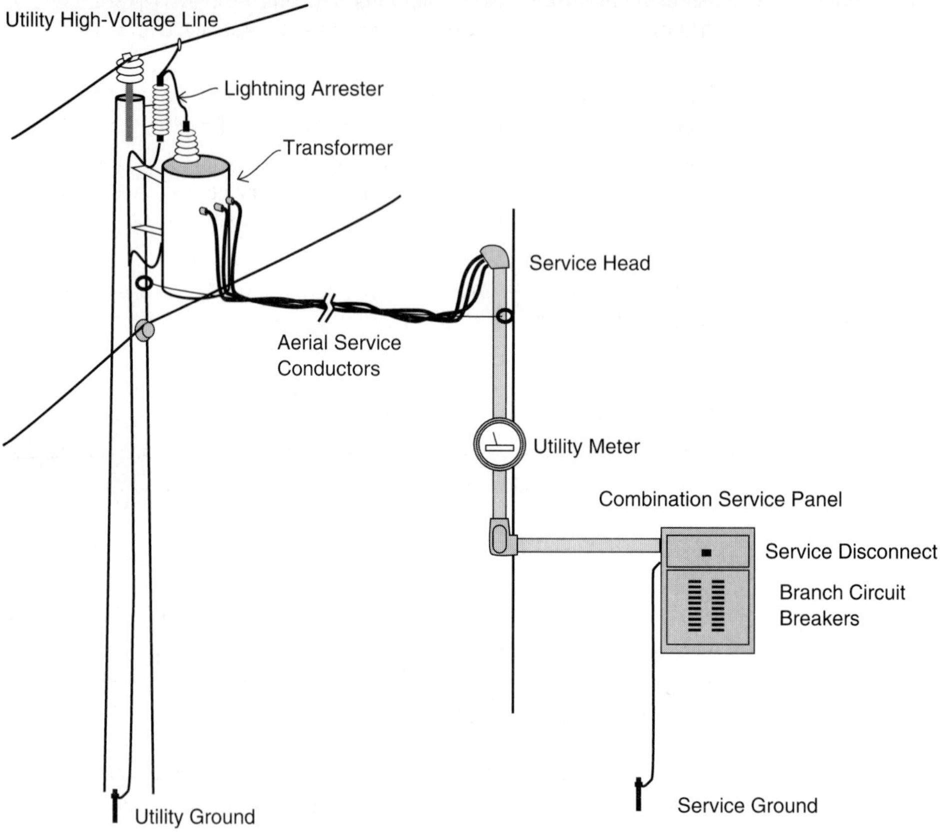

connect to the building load. When the meter is installed—it plugs into the jaws of the socket—it completes the circuit from the utility supply into the building. On services supplied from aerial conductors, the service cable or conduit must be cut so the meter can be installed. A typical overhead service is shown in Figure 1–7. On underground services, the meter socket serves as the junction point between the underground supply conductors and the conductors that carry power through the wall of the building to the service equipment inside.

Larger services are more elaborate. The demand is too great for an in-line meter, so the metering equipment is connected through instrument transformers. Instead of a self-contained meter located outside the building, there will usually be a meter and instrument transformer combination—possibly integrated into the service equipment or installed in a metering enclosure.

Apartments and strip malls, where there are several tenants in the same building, may have a multiple meter service to allow for individual billing. These divided sevices are usually connected much as if the individual customers were in separate buildings. The individual service feeders are tapped off the main service conductors inside the service equipment. Each of these subservices has an individual service switch and meter socket.

Note

As a general rule, each building is supplied by only one electrical service. There are exceptions, but only when warranted by specific circumstances.

The *main service disconnect* is the fused switch or circuit breaker that can shut off the utility power to the entire building or, in the case of multimetered services, the entire serviced area. Electrical codes demand that this disconnect be located as close as practical considerations allow to the point where the service entrance conductors come into the building. One reason for the strict enforcement of this rule is that the incoming feeders have no overcurrent protection, except the utility supply fuses, until they pass through the main service disconnect. The utility fusing may be set high enough to supply several buildings, and is often far too high to safely protect the service conductors. An electrical fault in these feeders, such as a short circuit or ground, can cause a lot of damage. Once through the main fuses or breaker, the conductors are protected. Locating the service as close as possible to the supply keeps the unprotected portion as short as possible, reducing the chances of a problem. The current rating of the main service fuses or circuit breaker must be adequate to supply the entire building, or the area being served.

Panelboards

Following the main service, the power must be distributed to the various building loads. In a small building, everything is supplied from one distribution *panel* or *panelboard.* In many cases the main service breaker and the panel are combined in a single *combination service panel.* This is a convenient and cost-effective method to provide both functions in one compact package.

Panelboards have a system of internal *busbars,* or *buses,* that allow the individual distribution breakers mounted in the panel to make the connection to the incoming supply. These are sometimes referred to as the panel *mains*. When the panel is equipped with a main breaker, the breaker completes the connection between the supply and the busbars. When the panel has no main breaker, the incoming supply conductors are connected directly to the busbars through lugs mounted on each bus. This type of panelboard is sometimes referred to as a mains-only type.

The busbars are usually made of copper or aluminum alloy. They are, of course, insulated from the panelboard and each other. The busbars are designed so the individual distribution circuit breakers, installed in the panelboard, will connect directly to the busbars. The breaker to bus bar connection may be a set screw or bolt-on type, but for smaller panels some type of spring clip or friction fit is much more common. One common design slips over an exposed busbar tab; another has connection points that insert into punched holes or slots cut into the busbar. Circuit breakers are not standardized; most will only fit into a particular make of panelboard. The circuit breakers feed the individual loads, or other circuits, that the panelboard supplies.

There are a variety of breakers available for all panelboards. These may have different current ratings, different numbers of poles—one or two poles for single phase, three poles for three phase—or be a special-purpose breaker such as a GFCI (Ground Fault Circuit

Interrupter). GFCIs are required by electrical codes to reduce the shock hazard on certain circuits, such as bathroom receptacles. Most manufacturers have several different lines of panels and breakers, designed for different types of applications. The breakers from one product line usually will not fit into the panels of another.

Fused type panels are similar in design to breaker panels. The fuse panel is equipped with busbars, just as in the breaker panel, but the connection between the bus and the circuit conductor is made through fuses rather than through a breaker. The panel may be equipped with switches for each fuse, or the fuses may connect directly to the bus. Lower ampacity fuses are usually screw-in *plug fuses* with bases similar to a standard lamp socket. Those designed for higher currents are usually cylindrical *cartridge fuses*. Panels with built-in main fuses may use a switch or a *pullout* or *draw-out* fuse holder to allow the fuses to be safely removed without powering down the supply to the panel.

Draw-out fuse holders are insulated containers for the fuses. They have handles on the front for easy removal and the fuses are mounted inside, out of sight. The connections between the panel and the ends of the fuses are made through copper stabs that fit into matching jaws inside the panel. Removing the draw-out fuse holder disconnects the circuit. In some designs the holder can then be turned over and inserted back into the panel. In the inverted position, the fuse holder covers the exposed conductors but does not reconnect the circuit. There will be some indication on the holder to show whether it is in the connected or disconnected position. Use caution: Some types of removable fuse holders will make the connection with the fuse holder inserted in either direction.

In a fuse panel installation, the main fuses may be in the panel as draw fuses, or in a separate fused switch. The main fuses connect the incoming supply and the bus bars, just like the main breaker in the combination breaker panel. In some panels, draw fuses are also supplied for higher current loads such as electric ranges and dryers. The fuses in the panel make the connection between the main buses and the individual circuits, much like the breakers in a breaker panel. Fused panels are rarely seen in newer installations.

Panelboards vary in current rating and rated voltage. Combination panelboards, for example, are rated for a specific voltage and a maximum main breaker ampere rating. Panels that have no main breaker are also rated at a certain maximum current, but they depend on external breakers or fuses to protect the panel from overload. Panelboards also vary in the number of breaker spaces provided, with the maximum generally limited to 42. The number of spaces is not necessarily the maximum number of breakers because, although single-pole breakers do take up only one space, most two-pole breakers take up two spaces and three-pole types take up three. When additional breaker space is required—but not additional current capacity—a second panelboard can be installed, and fed from the first. It will normally feed from a breaker in the main panel, but may also come from a separate main breaker or a fused switch.

As the size of the building and the total current demand increases, the interior electrical distribution system becomes more complex. The service to the building not only becomes larger, it is also almost always three-phase, instead of single-phase, power. Smaller three-phase services can be installed in much the same way as a single-phase system. A single combination panelboard may be used, with breakers for both single-phase and three-phase loads in the same panel. The single-phase loads are supplied from single- or two-pole breakers, depending on the circuit voltage, and the three-phase loads come from three-pole breakers. When more breaker spaces are required—three-pole breakers take up three panel spaces each—this basic system can be expanded with additional panels.

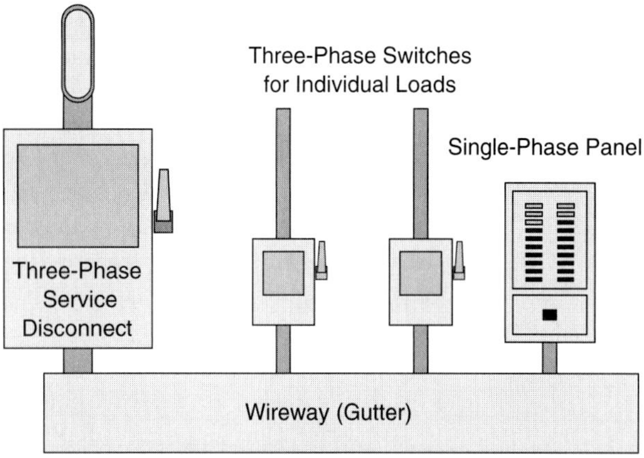

FIGURE 1–8 Some three-phase services must supply the three-phase and single-phase loads separately. This one uses individual switches for the three-phase loads, and a panelboard for the single-phase loads.

Loads are often grouped. For example, power loads, such as motors, may be supplied from one panel, while lighting and appliance circuits are supplied from one or more different panels. Building layout permitting, panelboards may be located at some distance from the service, in an area that is close to the loads they supply. This can greatly reduce the length of the conductors required to power up the individual loads. They can be fed from a panelboard close by, rather than each circuit having to originate in the main electrical service area.

One type of three-phase service requires the single- and three-phase panels to be physically separated (see Figure 1–8). In these cases, the main switch is usually a fused disconnect or an enclosed circuit breaker with no attached panelboard. The main switch supplies power to two or more separate panels through an enclosure called a *gutter,* or *wireway.* The gutter contains insulated sections of busbar, multiconnection terminal blocks, or similar devices that provide for interconnections between the heavy feeders coming from the main switch and the smaller subfeeders that extend to the other switches or distribution panelboards. The gutter provides some flexibility in connections, because it can accommodate power loads supplied through individual fused switches or enclosed breakers, or complete panels that in turn supply other individual circuits.

Feeders, Branch Circuits, and Loads

The overcurrent devices—the fuses or circuit breakers that protect the system from excessive current—are generally the dividing points between various defined functions of an electrical system. For example, from the utility connection to the supply side of the main service overcurrent device—the main breaker or fuses—is the *service.* On the supply side of the service switch or breaker, the conductors are protected from short-circuit damage by the utility overcurrent protection. On the load side, they are protected first by the main breaker, and then, as the system fans out, by additional breakers or fuses. All the circuits in the building extend from the service, but the current for each circuit must first pass through a fuse or breaker with ratings that are appropriate for the circuit capacity. These overcurrent devices also separate the system by function.

When the connected conductors supply an additional distribution point, such as another panel or a gutter, they are called *feeders.* If they supply a mix of individual loads and panels, they may be called *mains*. Circuits that supply power directly to the various electrical loads and outlets, not other panelboards, are called *branch circuits*. Branch circuits extend from the last overcurrent out to the loads.

Instead of a single distribution panel, larger buildings may have many panels, often serving different functions. *Feeder distribution panels* are used to supply other panels. These panels have smaller numbers of breakers, with higher ratings to accommodate the heavy load. *Power distribution panels* may be used to supply loads that are too demanding to be serviced effectively from a single service panel. Additional distribution panels, sometimes called *subpanels,* may be added to any system. The subpanels are supplied from a main panel with a *feeder* heavy enough to supply all the equipment connected to the subpanel. Subpanelboards may supply branch circuits to a particular area, a specific type of load, or even to a separate building.

Larger services are often supplied at a high enough voltage that it is necessary to install one or more *transformers* within the building to supply the lower voltage loads, such as standard 120 V outlets that all electrified buildings require. The higher service voltage is used directly to supply the major power loads, such as large motors, high-capacity water heaters, or high-current process equipment. The transformers are used to reduce the service voltage to 120 V (usually a combination of 120/240 V single-phase or 120/208 V three-phase) to supply low-voltage panelboards that feed the 120 V outlets and other low-voltage loads. There are definite savings involved when the low-voltage demand is kept to a minimum. For this reason, ballasted lights, such as fluorescent, mercury vapor, metal halide, or high-pressure sodium, are often supplied at higher voltages than 120 or 240 V.

Very large buildings and demanding industrial complexes may have a much higher service voltage, and the associated switchgear to control and manage such a system. Installations of this type may have three or more different operating voltages in the same building.

In any building, the electrical equipment fans out from the service, where the utility supply enters, to every piece of electrical equipment and electrical outlet in the area the service supplies. All of the circuits may originate at a single panelboard, with separate runs of cable for each circuit. An example of this type of system can be found in a typical house. In contrast, a large electrical installation may contain a vast network of switches, fuses, breakers, panelboards, transformers, conduits, wires and cables, and control equipment.

Attempting to comprehend a complete system, even a relatively small one, is almost impossible. Even the relatively small electrical installation found in a modest home, small business, or apartment block can be overwhelming when viewed all at once. However, the individual component parts are not particularly difficult to understand as separate units or in the context of a larger application. Reducing the complex whole down to its component parts makes a system much easier to understand.

Wire and Cable

The terms "wire" and "cable," as they apply to conductors, can be confusing (Figure 1–9 shows wire and cable samples). This is particularly true if you are fussy about using proper terminology. Used correctly, the term "wire" means "a slender round rod or filament of drawn metal." In common usage it generally means individual conductors of modest size. These wires may be solid or stranded, which means that the "wire" is made up of many strands of much finer wire. Stranded wire is slightly twisted to hold the thin strands together. The wire may be bare, or it may be covered with a coating of electrical insulation.

FIGURE 1–9 Wire and cable samples. From bottom to top, 2- and 3-conductor #14 AWG nonmetallic sheathed cable, 2-conductor #14 AWG armored cable (spiral aluminum armor), and 3-conductor armored cable with outer jacket. This industrial cable is shown with a gland-type connector designed specifically for this cable. *Note:* The unjacketed armored cable shown here is a relatively old type with a flat aluminum ground strip. Newer versions contain a separate copper ground wire.

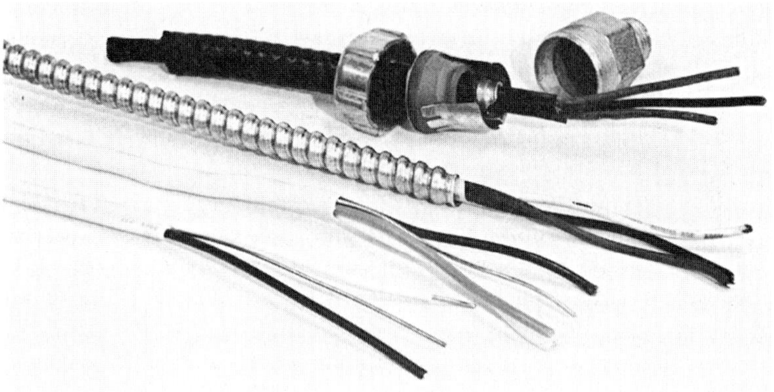

Technically speaking, all stranded conductors are cables; however, in common practice the term "cable," as it applies to electrical conductors, is used to describe large individual conductors, or any size of multiconductor assembly. Multiconductor cables are made up of two or more wires grouped together with the intention that the complete cable will be installed as a unit. The wires may be simply twisted together or, more often, covered with an outer jacket of protective plastic or rubber. Cables intended for physically demanding applications may have more durable coverings, such as a continuous aluminum, copper, or stainless steel outer sheath, or a spiral armor of aluminum or steel. Metal sheathed cables and armored cables may also have an additional outer jacket of plastic material. Armored cables are common in commercial and industrial applications.

In addition to two or more individually insulated conductors, multiconductor cables may also include a bare or insulated ground conductor. The ground is often not counted in common reference. For example, an electrician may refer to a cable containing two insulated conductors and a bare ground wire as a two-conductor cable. Multiconductor cables contain all the wires needed for one or more circuits. They are generally convenient to install, particularly inside wall framing or in industrial environments where cables are usually laid in cable trays specifically designed for that purpose.

Conductor size in North America is based on the American Wire Gauge (AWG), which extends, in reverse numerical order, from a very fine, 0.0031″ in diameter 40 gauge, up to a "four ought"—written 4/0 or 0000—which is about 0.46″ in diameter. Beyond the largest gauge size, the unit of measure is the circular mil—which is the area of a circle 1 mil (0.001″) in diameter. The circular mil is a small unit, so electrical conductors above 4/0 are always in thousands of circular mils. This unit is abbreviated as kcmil or MCM. Standard conductor sizes, listed in circular mils, start at 250,000 CM (250 kcmil), the first size above 4/0 AWG, and continue upward to 2000 kcmil. A 2000 kcmil conductor is about 1.63″ in diameter.

Stranded wires have the same total cross-sectional conductor area as an equivalent size of solid wire. With a few exceptions, conductors #8 AWG and larger are stranded to improve flexibility. Different strandings are used for various applications. Flexible power cables used for portable equipment have a much finer strand than would be used in a permanently installed power cable.

Virtually all single-family dwellings, and some multi-family buildings, are wired with multiconductor nonmetallic sheathed cables. The term "nonmetallic sheathed cable" is commonly used in the industry. These cables used to be jacketed with a paper and fabric material soaked in a tarlike substance. They are now jacketed with some type of plastic or nylon. Nonmetallic sheathed cable is generally available with two and three conductors—plus a bare ground—in standard sizes from 14 to 6 AWG or larger. It is well suited to the protected installation for which it is approved. It may be concealed inside framed walls or fastened to the sides of structural members. It is neither designed nor approved for installations where it could be exposed to damage. Although all modern cables of this type have plastic jackets, the general-purpose types found inside a home are not approved for damp locations or for underground installation. There are other nonmetallic sheathed cables with different cable designations that are approved for such situations.

Most buildings designed for multi-family occupancy or for commercial use are wired with a combination of metal conduit and armored cable. Conduits are used for main runs and often carry multiple circuits in a single pipe. The conduits terminate at outlets or junction boxes, and cable is used to extend the wiring system to lighting fixtures, switches, and additional outlets.

A typical armored cable, for dry location wiring, has a spiral-wrapped interlocking aluminum outer sheath. The conductors inside may be wrapped with plastic or paper, but are not jacketed. There is no overall outer jacket covering the armor. This cable is used in place of nonmetallic sheathed cables, primarily for its superior mechanical protection. It is available in a range of standard sizes with two, three, or more conductors. This is a common general-purpose cable that is often misunderstood and misused. It is not indestructible. Electrical codes generally require it be protected from damage, almost as strictly as a nonmetallic sheathed cable.

Note

This type of armored cable also requires special termination procedures. The insulation on the internal conductors must be protected from possible damage on the sharp edges of the cutoff armor. A small plastic bushing may have to be inserted around the conductors inside the armor at every cable termination.

When terminating this cable, it is generally not recommended to cut the armor with a side cutter. This results in a crude, distorted cut that is difficult to fit into a connector. It is also likely to have many sharp edges. Instead, the armor should be cut with a hacksaw or a tool designed specifically for cutting such cables. A shallow slit across the armor will allow it to be separated cleanly.

Cables always terminate in boxes or at similar fittings. The cable must be held securely to the box, without damage to the sheath or the wires inside. For plastic jacketed, nonmetallic sheathed cables, the electrical box must have a built-in clamp, or a suitable cable connector must be installed in one of the "knockout" holes. To install the cable, the sheath is first stripped back and removed to expose the conductors inside. The cable jacket should protrude into the box just far enough to protect the exposed wires from possible damage on the edges of the clamp or cable connector. The cable must be clamped on the sheath, not on the unprotected conductors.

Armored cable connectors are different from those typically used with nonmetallic sheathed cable. The metal armor must be prevented not only from pulling free of the box, but also from being pushed into the box. Approved connectors are designed with a reduced exit hole, or other provision, to prevent the armor from passing through the cable connector or clamp into the box. If the metal armor is forced through the connector, it can damage the conductors inside, or cause a short. Both straight and right-angle armored cable connectors are readily available. Interlocking spiral armor can come apart quite easily if the cable is bent at a sharp angle. For this reason, right-angle connectors should be installed wherever a straight connector might unnecessarily expose the cable to a sharp bend or other damage.

Typical industrial armored cables are more complex, and more expensive, than the cables used in most commercial applications. These cables have a full internal jacket and spiral armor, plus a full external jacket. They require special cable connectors that not only secure the armor, but also provide a seal to the outer jacket. This helps keeps moisture and other contaminants out of the cable system, electrical boxes, and connected equipment.

There are many code rules governing the installation of cables, but all are based on the need to provide a safe, reliable electrical system. The cable must be suitable for the physical environment in which it is installed. It must also be located and secured so that it is protected from damage. The cable must enter the electrical boxes or equipment in such a way that the conductors inside the cable are protected from damage. The cable must be secured to the box by an approved clamp or fitting. For armored cable, the armor is secured independent of the internal conductors. Nonmetallic sheathed cables are secured by clamping onto the external jacket. Wiring inside the box must not be expected to support or secure the cable.

Conduit

Conduit provides the best possible protection for the wiring system. It also allows for greater flexibility in the number, color, and size of conductors installed in each run. The conduit system—the pipe, the boxes, and all the necessary fittings—provides a complete protective enclosure for the circuit wiring. It is installed independent of the wiring it will contain—the conductors are pulled in after everything is in place.

Conduit is also called "raceway." One common type is a lightweight metal tubing—called electrical metallic tubing, or EMT—which is interconnected with slip-on fittings secured by set screws, or threaded hubs with internal compression rings. Heavier metal conduits, usually referred to as "rigid conduit," are thick enough for threading, and interconnections are made with threaded couplers and other fittings. There are also nonmetallic conduits, made from materials such as PVC or fiberglass. A complete range of fittings is available for these conduits, just as for the metallic types. Fittings may be secured with an appropriate solvent cement, or they may be a press-on type with internal seals or some other system specifically approved for that conduit. All metal- and plastic-type conduits can be bent in the field. This allows them to be installed with far fewer fittings than would otherwise be required. Extra fittings add cost and can make it harder to install the conductors.

The type of conduit used is generally dependent on the conditions into which the wiring system is to be installed. Metal conduits may be required in commercial buildings to comply with the fire rating of the structure as a whole. Conduits are used for building services to protect the service feeders from damage where they enter the building. Nonmetallic systems may be preferred in areas where corrosive chemicals could destroy metals.

A major role of the conduit system is to protect the wiring it contains. However, it also protects the building structure from fire and, in certain cases, from the danger of explosion. For example, the conduit system in a fuel handling depot or natural gas plant must prevent explosions that could otherwise occur from normal operation of the electrical equipment. It is virtually impossible to completely eliminate arcs from normal switching, or heat from a fault in the wiring or equipment. However, these occurrences must not ignite the explosive mix of air and flammable gases that may surround the equipment. The problem is complicated by the fact that the conduit system is likely to contain the same explosive mix.

The solution to this problem is conduits and fittings that mate with broad machined flanges, or threads of sufficient length. When an internal arc or explosion occurs, the heat is dissipated as the gases pass through between the flanges or around the threaded connection. The gases are cooled to below the ignition temperature of the external gas–air mix, and the hazard is eliminated. In addition to the specialized fittings required—such as the boxes and box covers, switch plates, lighting fixtures, etc.—the conduit system must also have explosion-proof seals installed in the conduit runs. These prevent a small explosion in one box from spreading to other parts of the system.

Note

The rules of installation for conduit are quite complex, but there are some basics that all service personnel should understand. Electrical conduits are not standard tubing or pipe, and no such substitutions are allowed. This also applies to standard pipe fittings of all types. Conduits may use standard pipe thread, but standard fittings cannot be used.

Metal conduits almost always provide the interconnecting ground. This means that loose or disconnected fittings can leave equipment ungrounded. This is a critical consideration in any system using metal conduit. All conduit couplings and box connections must be electrically and mechanically secure, and the conduit system must be continuous from the source to the frame of the equipment it supplies.

Although metal conduits are generally suitable as ground conductors, flexible portions of the conduit system are not. Flexible conduits almost always require an additional ground conductor to carry the ground system from the box or conduit fitting, where they connect, to the equipment to which the flexible section extends. For example, in a conduit-based electrical system, it is usually not practical to connect standard conduit to motors or other equipment that may require occasional disconnection for service or replacement. Instead, the conduit system extends to within a short distance of the connection point, and then flexible conduit is used from there to the motor or other equipment. Proper grounding of the equipment depends on an adequate ground connection though the flexible portion to the standard conduit. This typically requires a ground wire supplementing the flexible conduit section.

Note

There are times, such as during the process of minor wiring alterations, when it may be tempting to add wiring to the existing conduits. This may or may not be acceptable, depending on the operating voltage of the conductors in the conduit, and on the allowable conduit fill. The allowable fill is based on the conduit and conductor size, the number of circuits, the electrical load, and ambient conditions such as the maximum temperature of the surrounding air. It is not simply a question of how much room there is left inside the pipe. Do not assume that conductors can be added to existing conduits just because they can be made to fit.

Electrical Boxes

The U.S. (NEC—National Electrical Code) and Canadian (CEC—Canadian Electrical Code) electrical codes require most electrical connections be made inside an approved box or fitting. This includes virtually all interior wiring operating at more than 24 V. The box may be made of metal or nonmetallic material, but must be approved by the appropriate authority, normally Underwriters Laboratories (UL) or the Canadian Standards Association (CSA). Interestingly, not all countries have this provision in their codes. It is, however, a normal requirement in Canada and the United States. For this reason, only in very rare cases do cables and conduits not terminate at an electrical enclosure of some kind.

Sometimes the box serves only as a junction point for two or more cables or conduits. In other cases, the box also serves as a mounting location for a device such as a switch, receptacle, light fixture, or other equipment. The box then serves double duty as a device box and a junction box for continuation of the circuit. This is common in lighting and receptacle circuits, where more than one outlet is supplied from a single branch circuit. The usual practice is to run the wiring from one box to the next. Power is transferred through the cables that loop from box to box, with connections made to each device in the circuit as it goes along. This greatly reduces the amount of cable that must be run from the breaker panel or fuse box.

Like conduits, boxes have specific limitations regarding the number of wires allowed inside. Because there are so many possible combinations, the rules for calculating box fill can be quite confusing. Electrical codes take into account such things as box size, wire size, any mounted devices (such as switches or receptacles), and the type and number of wire connectors that are used. As with conduits, it is never a simple case of how many wires will fit.

Factory-built electrical equipment does not escape the rules about all connections being in a box. Boxes are built into the equipment where the supply conductors enter and connect. There may also be boxes or other enclosures inside the equipment at points where electrical devices are installed or wiring must be interconnected.

Note

There are many code restrictions regarding the installation of wires into boxes and conduits. These rules set the maximum number of conductors allowed and the length of free conductor required for connections, and also set limitations about mixing wires from different circuits. The number of conductors allowed depends on the volume of the box or cross section of the conduit, the size of the conductors, whether there are devices mounted in the box, and certain conditions of use. Control wires, for example, carry less current and therefore produce less heat than those of power circuits. For this reason, installation codes may permit more control conductors than power conductors in a given run of conduit. The allowable conduit fill may also have to be reduced when the temperature of the surrounding environment is unusually high.

For service personnel who may become involved in alterations to a system, but not in a major installation, there are a few main points to remember.

- No line voltage connections can be made outside an approved electrical box.
- Junction boxes must not be hidden in a wall or other inaccessible location. For example, a damaged cable cannot be spliced together and buried in a wall cavity.

- Wiring systems often loop along from one box to the next. A circuit may also be fed from some intermediate point in the system, rather than by a direct run to the main supply panel.
- All the wires in the box must have long enough "free" ends to allow the connections to be made conveniently outside the box. Electrical codes state a specific length, and maintaining this length requires some attention to details about how the wires will fit into the box when it is closed up and the cover is installed. When the conductors are just stripped and terminated, with no thought as to their final positions inside the box, the result is usually a mess.
- There are code restrictions on conductors from different circuits, or operating at different voltages, occupying the same conduit or box. This is part of a general attempt to keep different systems separated from each other, particularly if the voltages are also different. Sometimes barriers are required inside a box, between the two different wiring systems. They may also have to be placed in separate cables or conduit runs.

Grounding

Grounding is a critical part of all electrical systems, and one of the most likely to be misunderstood. It is deserving of special attention, because it is not part of the normal operation of the electrical system, and therefore is untested until something goes wrong. When things do go wrong, the ground system must be up to the task, or personnel and equipment will be in jeopardy. When a serious fault occurs, and the ground system proves to be inadequate, the results can be very serious. A problem like this may be due to faulty installation, but is far more likely to be the result of changes that have occurred over the life of the building.

An electrical ground system interconnects part of the electrical system and virtually everything in the building that conducts electricity. The grounded portion of the electrical system typically includes one of the incoming supply conductors, all metal enclosures, equipment frames, and metal conduits, plus the wires and cables of the ground system itself. A connection is also made to the earth on which the building rests, the building foundation, all structural steel, metal piping systems such as water lines, fire protection sprinkler systems, and gas lines, and all metal ductwork.

The necessity for a ground system starts with the utility system outside the building. Grounds help to protect the building and its occupants from the high voltages of the utility distribution system. The electrical energy supplied to a typical building is delivered through a voltage-reducing transformer that links the high utility voltage to the low voltage of the building electrical system. Inside the transformer case, the high- and low-voltage conductors are very close together. A transformer failure can connect the high utility voltage to the low-voltage system of the building. Such a situation endangers everyone and everything in the vicinity. With proper grounding, the damage is restricted to the transformer.

To be effective, the transformer ground connection must involve not only the metal frame and case of the transformer, but also one of the low-voltage conductors that supplies power to the building. This conductor must be connected directly to the earth, and to the electrical ground system of the building. With the ground in place, any faults that occur in the transformer will be directed through the earth back into the utility system. Because of the way the utility ground system is connected, this fault will cause a short that blows the protective fuses that supply the transformer. The building will be left without power, but the danger of a high-voltage disaster is eliminated.

The concern about high voltage entering the building electrical system is further increased by the presence of lightning. Overhead utility lines are common targets for lightning strikes. The only way to protect against lightning damage is to supply a path to ground that is as direct as possible.

Grounding is a very effective and essential part of protecting the building from high utility voltages and lightning. Unfortunately there is a catch. Once one conductor of the building electrical system is grounded, everything in the building that is conductive, or is in contact with the earth, is also connected to one line of the electrical system. This includes the ground outside the building, concrete floors, and all conductive materials with which the building occupants may come in contact. The only way to make the system safe is to connect all of the conductive parts together. With everything interconnected, any contact between the ungrounded "live" wires of the supply and the grounded material of the building will cause a short circuit and trip the circuit at its protective circuit breaker or fuse. An adequate ground system is essential for all single-phase and three-phase systems. Although some three-phase systems do not have a direct connection between the supply conductors and ground, even this does not reduce the need for grounding all the metal parts of the building and its equipment.

There may be two, three, or four conductors from the utility transformer to the electrical equipment inside the building. The exact number depends on whether the service is single or three phase, and how it is connected. In most cases, one of these incoming conductors must be grounded. **The grounded conductor is a circuit common, and is usually referred to as the *neutral*.** The neutral carries circuit current and is part of the normal system of conductors. It is unique only because it is connected to the ground system.

The neutral by itself is not a shock hazard. It is already connected to ground; connecting it a second time, through your body for example, does not cause any current to flow. However, making contact between another line conductor and anything that is grounded has the same effect as making contact between a line conductor and the neutral. This always results in a shock, possibly a serious or even fatal one, if conditions stack the odds against the victim.

The only safe alternative to adequate grounding is to isolate the electrical system from all conductive materials within the building, and from the earth itself. Such complete isolation is usually impractical, or impossible to achieve. To be safe it must be so effective that it is impossible for electrically live parts to come in contact with anyone, directly or through contact with other conductive materials. This can be done on a very small scale but is a practical impossibility on a large system.

Isolation, as an option to grounding, is usually restricted to individual items such as double-insulated tools, kitchen appliances, and home electronic equipment. These items have no ground connection. The electrical supply cord used has only two conductors, and there are only two blades on the attachment cap. They are safe to use because they have a nonconductive frame or case, or because their internal construction makes it virtually impossible for the case to become energized by contact with the internal wiring.

The intent of grounding is to shield the building and its occupants from the possibility of exposure to live electrical parts. It is done by interconnecting the earth, common (neutral) conductor of the electrical supply, all the "non-current carrying" metal parts of the electrical system (the metal boxes, conduits, and equipment enclosures), and all the metal parts of the building as a whole.

With everything grounded, including the supply common, any contact between the dangerous live conductors and virtually anything installed in the building that is made of metal will result in a short circuit. Current will flow through the metal to its ground connection, and back to the circuit common. This trips the breaker or blows the circuit's protective fuse. Grounding does not prevent such faults, it just ensures that the circuit fails to a safe condition.

Warning

Failure to ground metal parts that may be exposed to electrical contact creates a very dangerous situation. For example, assume the frame of an electrically powered tool or other piece of equipment has no direct connection to the ground system. The exposed metal of the device can easily become energized through contact with a live conductor inside. This may be a result of faulty wiring, insulation failure, or mechanical damage. With no ground conductor to act as a circuit return, the metal frame can remain energized indefinitely. All exposed metal of the device becomes a "live" conductor, a serious shock hazard waiting for an unsuspecting victim. Contact with the device and any grounded surface is exactly the same as contact with the live conductor to which the frame has become connected.

When the metal frame has a path to ground, but it is not a very good one, not only is there a shock hazard, but a fire hazard as well. With a poor connection back to ground, the current flow may not be high enough to trip the circuit. Instead, the current may arc, or build up heat across a high-resistance contact area, possibly setting fire to the building.

The ground system is obviously a very necessary part of the electrical system. However, there is no way to tell if it will provide the safety function it is intended for until a fault occurs. Disconnected grounds go unnoticed because the system still works. The integrity of the ground system depends on the integrity and workmanship of those who install it, and of anyone who alters it after the initial installation is done. This is why it is so important for all installers and service personnel to understand what the ground system is and how it is connected.

Warning

There are many ways to break the ground connection, rendering part of the ground system ineffective. The following are examples of broken ground connections that are easily ignored but potentially lethal.

Poor conduit connections, **such as loose locknuts or disconnected couplings.**

Disconnected bonding jumpers **to water lines and around water meters.**

Ground wires left unconnected, **or improperly connected inside boxes. Ground terminals are provided specifically for that purpose; using other connections such as the head of a wood screw, etc., is not acceptable.**

Ground wires left off motors. **If the motor is mounted on rubber—a common method of reducing sound transmission—a fault in the motor energizes the motor frame. With no path to ground, the motor remains energized, a deadly trap for the next service worker.**

Ground pins cut off extension cords, or two-wire cords used with three-wire appliances or tools. **The third wire in the cord is intended to carry the protective ground right to the frame of the tool. Without it, a faulty tool becomes a live wire, a common cause of electrical fatalities—and one of the easiest to avoid.**

1.3 Tips and Techniques

Introduction

There are many simple procedures to make the common tasks of electrical work quick, neat, and effective. Good workmanship is important, but it depends not only on good intentions, but also on the understanding and skill of the person doing the job. Good workmanship arises from good habits and well-organized approaches to tasks, and requires both a good attitude and experience. For experienced electricians, procedures such as those described below are a matter of habit. For service workers who complete such tasks only rarely, they may be awkward to perform. Unless the desired result is fully understood, the end product may be below acceptable standards.

Connecting Conductors

Wire connections are where most electrical work comes undone. Good connections are neither difficult nor time consuming, but a lot of things can and do go wrong. To understand what makes a good connection, it helps to know the reasons why connections sometimes fail.

Electricity travels through solid materials. At a connection point the current must pass from one conductor to the other, usually through surface-to-surface contact only, without the benefit of a bonding material such as solder. The size and quality of this contact area is critical. If the connection is not done properly, the contact area may be too small or too loose. When the conductor is made of aluminum, an insulating oxide layer may form between the two surfaces. When two different metals are involved, electrolytic action may attack the materials and destroy the connection. If the conductor was nicked or ringed during installation, vibration may anneal the metal at the weakened point, causing it to crack and break away.

Some of these problems can cause a failure all by themselves. Others simply increase the resistance of the connection and cause it to heat. Heat invariably leads to further deterioration, and eventually the connection fails completely.

Good installation practices eliminate problems with connections. The end result must be good physical contact between the two conductive surfaces. The contact must be over the widest practical area, and the surfaces must be clean. The conductor materials must also be compatible; copper and aluminum together will cause electrolytic action. The connection must be mechanically secure—tight, but not too tight. Experience has shown that overtightening also causes failures. A good connection is kept under tension by the natural "springiness" of the materials involved. Overtightening causes the materials to extrude out of the contact area, and the natural tension is lost. This phenomenon is sometimes called "cold flow." Much of the difficulty with aluminum house wiring has been attributed to cold flow.

Wire-to-Wire Connection Methods. Figure 1–10 shows the necessary ingredients for wire-to-wire connection. For two or more small gauge wires, the usual method of connection is to twist the conductors together and apply a wire nut, compression sleeve, or a set screw connector. A much less popular but generally superior method is solder.

A wire nut is an insulated cap with a tapered internal thread. They are probably the most popular of all the electrical connectors available for small conductors. They can be applied quickly and easily with no special tools, and the connection is secured and insulated in one step. They are easily removed without excessive damage to the conductors, and

FIGURE 1–10 The tool in the background is a combination wire stripper and crimper. In the back row, from left to right, items 1 to 5 are different makes and sizes of wire nuts. Items 6 and 7 are the cap and set screw sleeve of a set screw connector, item 8 is a set screw connector with the sleeve in place (inverted for shipping), and item 9 is another wire nut. In the front row are various preinsulated spade and ring lugs, male and female push-on "quick connect" terminals, and three sizes of in-line splice sleeves.

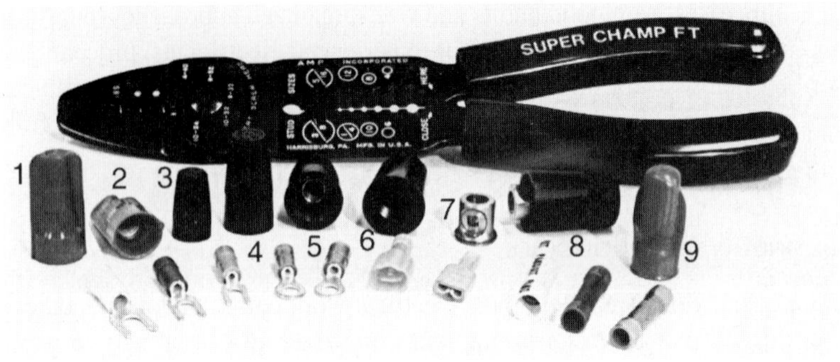

can usually be reused. The internal thread is typically a spiral spring that helps ensure good performance by allowing a small amount of expansion and contraction without a loss of tension. They are excellent for solid conductors, but can be difficult to apply properly if the conductors are stranded.

Set screw connectors are also relatively simple to apply. The wires are twisted together and the metal sleeve of the connector is placed over the twisted ends. A set screw in the sleeve is tightened, and the connection is covered with a separate insulating cover that screws on over the sleeve. The result is a secure, insulated connection that is easily removed. Set screw connectors can almost always be reused. They are difficult to tighten fully unless the sleeve is held with pliers, or with the special tool available for this purpose.

A common error is to put the sleeve on upside down, which happens because the connectors are shipped with the sleeves inside the insulating cover cap. For ease of removal, they are inserted upside down. When they are put on that way, the cover will not come down over the connection as far as it should. This may leave some of the stripped conductor exposed below the edge of the cover.

Wire connections may also be secured with a preinsulated compression connector or with an uninsulated compression sleeve, which is then covered with an insulating cap or electrical tape. Compression sleeves have the advantage of making a connection that is mechanically very strong. They are quick and simple to apply, if the appropriate tool is at hand. They have the disadvantage of being difficult to remove. Some brands are virtually impossible to take off without cutting the conductor. Uninsulated sleeves must be covered with the correct insulating cap or with a layer of electrical tape. Taping connections provides very good cover if it is done correctly, but it takes extra time.

Wires may also be soldered together, and then the joint insulated with electrical tape. This is probably the best possible connection, both electrically and mechanically, but it is also the most time consuming and difficult to perform. It also will not work on aluminum conductors. Solder is used extensively in electronic circuits. It is highly recommended anywhere that a very reliable, low-resistance connection is essential. In the heating, ventilation, and air conditioning industry, field-soldered electrical connections are specifically recommended for wire-to-wire connections in millivolt control systems. These systems operate from a supply of less than 1 V. Any resistance in the control wiring can reduce the voltage at the gas valve or other control device, and cause the system to fail.

For larger conductors, split bolt connectors are one option. These are available in brass—for copper conductors only—or in an alloy suitable for contact with both copper and aluminum. The alloy types come with a divider used to separate the copper conductor from the aluminum conductor. This provision makes the split bolt connector one of the only removable connectors that can contain both copper and aluminum conductors, keeping them physically separated to avoid electrolytic action between the two metals. Split bolts are most often used in overhead wiring, where the connections may not even be taped. When they are located out of reach, and positioned so that physical contact with other conductors or grounded metal is impossible, leaving the energized connector exposed does not cause a problem.

Split bolts may also be used to connect large conductors in electrical boxes. Here the connection must, of course, be insulated with suitable electrical tape. Another option, sometimes used on motor terminals, is to install a ring-type terminal lug on each conductor. The rings are then stacked as required and secured with a small bolt. Washers are usually installed on each side of the rings to make the connection easier to tighten properly, and to apply even pressure across the contact surfaces. This connection is then covered with tape.

Large compression connectors are also very common in outdoor wiring. They provide a very secure connection, are usually suitable for a mix of aluminum and copper, and are less expensive than split bolts. They do, however, require an expensive, heavy-duty compression tool.

Taping. Tape is used to insulate exposed electrical parts, such as connections made with solder, or mechanical connectors that are not supplied with a suitable insulating cover. Low-voltage connections, in dry locations, require only a few half lapped layers of any approved vinyl electrical tape. There are several methods of applying tape; the one described here (and illustrated in Figure 1–11) is only an example. With a minimum of practice it provides a very secure tape cover that is quick to apply and easy to remove.

The tape is first secured by turning a small tab over one wire, just below the connection, and sticking the tape to itself. Electrical tape has very little adhesive; this tape-to-tape contact holds better than from the tape to the surface of the insulated wire. Once the tape is secured, tension can be applied to stretch the tape as it is wrapped up the joint, toward the exposed end. Each layer should lap over the one below by at least half the tape width. The spiral wrap should be continued until it extends past the end of the connection by a little more than the width of the tape. This loose end is then folded back to provide a relatively thick pad of tape over the cut ends of the connected wires. The tape is then applied over the folded end, wrapping the opposite way around, and spiraled back down over rest of the joint. The tape is snapped off and the connection is done.

To remove the tape, loosen the exposed end and pull straight out—away from the connection. The tape will spiral off until the original tape-to-tape starting point is all that is left; this is pulled free and the tape is discarded. The method above works very well with all connections of this type. Line splices or taps are more difficult to cover neatly, but the approach is similar. Attaching the starting end to the conductor, as suggested above, makes it easier to get the tape on tight. The tape should be lapped at least half its width. Two or three layers is usually enough, depending on the voltage and the location.

FIGURE 1–11 Taping a wire connection.

There are many types of electrical tape and taping materials. Having a few of these on hand can be very helpful in certain situations. Self-amalgamating rubber tapes can provide a waterproof cover if they are applied properly. Shortly after application, the layers of tape become a seamless mass. If the connection is completely covered and the tape makes a good bond to clean wire insulation, the result is waterproof. Rubber tapes are relatively soft and must be protected with two or three layers of a suitable cover tape.

There are also pliable insulating compounds that can be used to fill in the gaps and holes and to pad sharp corners before taping begins. This is rarely required for simple wire joints, but may be necessary for bulky connections such as split bolts or bolted ring lugs. These compounds are usually self-amalgamating. Properly applied, they provide a smooth base for the layers of tape that follow. This can greatly simplify the final tape job.

Some of the most specialized tapes are used to cover high-voltage connections—2300 V and above. High-voltage conductors are not constructed the same way as those designed for lower voltages. Most high-voltage cables have a layer of semiconductive material around the conductor, underneath the main insulation layer. There is a second layer on the outer surface of the insulation that is followed by a layer of grounded metal tape or braid, and then the covering jackets and, possibly, armor. Terminating and insulating these conductors is a job for certified electricians. It should not be attempted by anyone without the necessary experience and training.

Low-temperature tape is a must when making connections in very cold conditions. Ordinary general-purpose tape is extremely frustrating in cold weather. Not only is the tape very stiff and brittle, the adhesive freezes, leaving the tape with no "stick" at all. Color tapes can be very convenient for identification purposes.

One option to tape that is particularly handy for straight connections such as in-line splices is heat-shrinkable tubing. This material can be slipped over one of the wires before the splice is made, and then repositioned over the splice. It is shrunk into place by heat from a lighter, match, heat gun, or soldering iron. The result is a neat, seamless cover over the connection.

Connections at Terminals. Small conductors terminating at devices or on terminal strips may be wrapped under a terminal screw or placed into a clamp- or set screw-type lug. For a plain screw terminal, such as found on a typical 120 V duplex receptacle or switch, the wire should be formed into a loop that will pass about three-fourths of the way around the screw before it is put in place. The loop should be sized to fit neatly under the edge of the screw head when it is tightened down (see Figure 1–12). It is usually helpful to give the loop a slight backward tilt. On most devices, this makes it easier to slip the loop under the terminal screw. It takes practice to form a good loop, but it is not difficult.

For the best connection, stranded wire should be tinned—filled with solder—before it is terminated under a standard screw terminal. Another option, if space allows, is to apply a suitable ring or spade lug to the stranded wire, and then put the lug under the screw. Do not use ring lugs if the terminals have captive screws, because these screws are very difficult to remove and replace. Other types of terminals are designed to contain any loose strands, making it possible to get a good connection directly on the untinned wire.

Some terminal screws have special washers, or either grooves or ridges in the base material, that are designed to hold the conductor in position. These can be convenient, because the conductor does not have to be formed into a loop, but extra attention must be paid to getting the conductor in the right place. When there are ridges, the wire must be trapped inside the ridges, but not caught between them and the edge of the screw. If that happens, tightening the screw will cut into the conductor material. Parallel grooves in the

FIGURE 1–12 Wire-to-terminal connections.

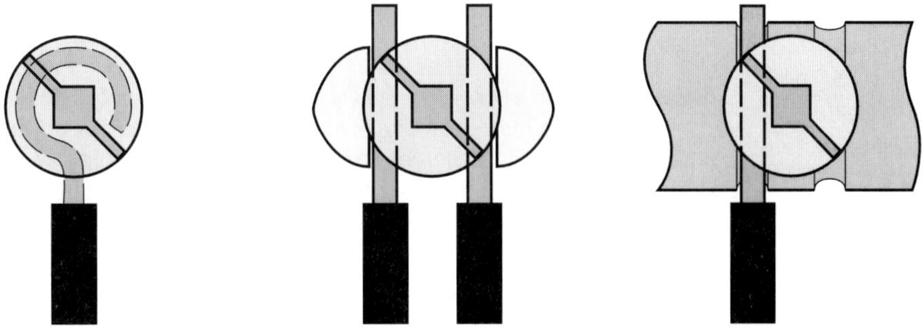

Wire must wrap around a standard screw terminal.

If there are ridges under the screw head, the wires must be trapped against them. A maximum of two can fit under the same screw head.

If there are grooves in the base material, the wire must lay in the groove.

base material indicate the wire may be left straight, and simply laid into the groove. It is important that the wire is still in the groove after the screw has been tightened.

Where larger conductors must be accommodated, the terminal point will almost always be equipped with a suitable lug. These are usually some type of set screw or clamp arrangement. Set screws that tighten directly on the wire have a greater tendency to cut into the conductor, especially if they are overtightened. Clamp types generally have a larger compression area. In both cases the termination should be tightened, the conductor should be shaken in an attempt to loosen it, and the terminal tightened a second time. This is particularly important on stranded conductors. It allows the strands to settle into position, and helps ensure the connection will stay tight through years of vibration and cycles of temperature change.

Terminals must be approved for use with aluminum conductors if these are being used. For large conductors, one option is a special pin lug that can be compressed onto the aluminum conductor before it is terminated. This allows a large aluminum conductor to be connected into a copper-rated lug.

Note

The following is a collection of rules that address the wire connection errors most commonly made by novices.

No insulation is allowed in the connection itself. For example, when wires are twisted together in preparation for a wire nut or other connector, there must be no insulation twisted up into the group. Following proper procedures will avoid this problem. The cut edges of the insulation—where the insulation ends and the conductor is exposed—should have no ragged ends that could twist into the connection. The conductors should be stripped a little longer than necessary, and the strip length should be the same for each wire. The cut ends should be lined up prior to twisting. This helps ensure all the conductors will twist equally. Although the insulated portions of the conductors may twist slightly—it is difficult to hold them—there should be no insulation in the actual connection. After the conductors are twisted, the ends can be cut to a length that is appropriate for the connector or connection method used.

There must be no unnecessary exposure of uninsulated conductor. For wire nuts, after fully tightening the connector, there should be no exposed conductor below the skirt, and no insulation up into the metal spiral inside. This same guideline applies to all connectors with insulated cover caps. The bare conductor should be just visible under the skirt of the connector, but should not be exposed below the bottom edge.

Sleeve-type connectors with a set screw must be put on right side up. This error, noted earlier, is common because the connectors are shipped with the sleeve inverted in the screw-on cap. The threaded rim of the sleeve must be toward the insulation, so that the cap can thread on fully. Another common problem with these connectors is allowing the cut end of the twisted conductors to extend too far out of the sleeve. This not only prevents the cap from covering the sleeve properly, it may also cause it to jam on the sharp ends of the conductor, making it difficult to remove.

Strip length is particularly important for preinsulated compression-type connectors, wire ends, and in-line splice sleeves. The position of the stripped end of the conductor cannot be seen after installation, so problem connections are very difficult to spot. The conductor must be fully inserted into the internal compression sleeve, with the conductor insulation just inside the plastic skirt. This provides full contact between the connection surfaces, and leaves no exposed conductor. Compression sleeves must be selected to match the conductor size. The sleeve will not compress properly on a conductor that is too small.

Connections must be mechanically secure but not overtightened. All connections must be able to withstand some physical movement and vibration without coming loose. The amount of movement varies with the situation, but remember the connection is supposed to stay tight forever. Terminations that are subject to constant movement or vibration deserve a little extra care and attention.

When tightening connections, remember that mechanically secure does not necessarily mean as tight as possible. Overtightening can cause problems. Some manufacturers specify tightening torque for set screw and bolted connections. Compression-type connectors must be done up with a tool that is approved for that make of connector. A correctly matched tool will not overcompress the connector.

Aluminum conductors require special attention. Metals exposed to air tend to oxidize. This is generally not a problem with copper conductors, because copper oxide is a relatively good conductor. Unfortunately, aluminum oxide is a poor conductor, and it can create a high-resistance barrier on the surface of the metal. To ensure a proper connection, this oxide must be removed and not allowed to re-form. An antioxidant compound is readily available from electrical equipment suppliers, usually packaged in easy-to-use plastic squeeze containers. This compound, with a consistency of a medium-weight lubricating grease, is applied to the exposed metal before the connection is made. A stiff brush should be used, particularly when the paste is applied to larger stranded conductors.

When aluminum conductors are used, the connectors must also be specifically approved for use with aluminum. Markings for approved connectors are something like "Al" or "Cu/Al." Split bolt connectors designed to interconnect copper and aluminum have a divider to separate the two conductors. The two conductors must be placed on opposite sides of the divider for this feature to be of benefit. In-line compression sleeves provide separation between copper and aluminum conductors by accepting one type of conductor in each end. Large compression connectors for aluminum wires are usually prefilled with antioxidant.

Attend to wire and equipment layout before making connections. A common error made by beginners is to connect the conductors too soon. The normal order of installation is boxes, then cables, then cable terminations, followed by wire connections and the installation of devices and equipment. Arranging the conductors in the box before any trimming, stripping, or terminating will save time overall. It also makes the final job much neater, and much easier to work on at a later date.

Cable and Conduit Installation

The first step in a cable installation, in wood or steel stud construction, is to install the boxes. These are placed as required for junctions and box-mounted devices. The cable route is planned, and the necessary holes are drilled or punched in the building framework. Then the cable is installed and secured in the building frame. Each cable is then prefitted to the point of entry into the box, and trimmed to length if necessary. There must be sufficient allowance for the free conductor ends that will be needed inside the box. Cables must not be cut too short—a few inches of excess conductor is a very minor waste compared to the time lost trying to terminate conductors that are too short.

With the cable approximately the right length, the jacket is stripped back far enough to provide the necessary free conductor in the box. Boxes may have built in cable clamps or, where no clamps are provided, an appropriately sized knockout is removed and a connector installed. The conductors are then inserted through the clamp or connector with the jacket extending through just far enough to protect the conductors from any sharp edges. The clamp or connector is tightened onto the cable jacket to secure the cable. Cables should not pull tight against the connector. Nonmetallic sheathed cables should have a loop of slack conductor between the box and the first staple.

All armored cables require special connectors, and in some cases a protective bushing must be installed inside the end of the armor to protect the conductors from the sharp edges where it has been cut. In other respects, armored cable is not much different to install than nonmetallic sheath cable. Some slack should be provided at the box, if possible, but care must be taken to avoid sharp bends in the cable, because this can cause the armor to separate. Ninety-degree connectors should always be used where they will help protect the cable: Common examples are where the cable enters into an equipment enclosure. The 90° connector will allow the cable to lay flat against the surface, whereas a standard connector would cause it to loop out, inviting damage.

Connectors that have internal glands to seal out moisture and contaminants are more difficult to install. Such connectors are available for nonmetallic sheathed cable and jacketed armored cables. Typical examples of this type of connector have a threaded retaining nut, a gland washer, and a soft rubber gland that must be fitted onto the cable first. When the cable has a jacket over the armor, a special part of the fitting will be used to ground the armor. This varies with the connector design, but a popular one is a thin metal ferrule that threads into the inside of the spiral armor and totally encloses the cut end. The grounding device will mate with the armor and the inside of the connector hub.

In most cases, connectors—at least the part that fastens to the box—should be installed on the box before the cable is placed into the connector. The threaded fitting can then be tightened into the box without twisting the cable. The hub should be turned into the locknut, not the other way around. The best tool for the job is a waterpump pliers. The sharp teeth of the locknut will bite into the panel metal to keep it from turning and to provide a good ground. Unless the connector has an insulated throat where the conductors pass through into the box, an appropriate conduit bushing may be required to protect the conductors.

When conduit is used, the complete conduit system is installed prior to pulling in the conductors. Boxes within the conduit runs are installed as the conduit is placed. This simplifies the problem of fitting the conduits accurately to the box connectors. Once the conduit is installed, a fish tape or pull line can be used to pull the conductors into place. Lines can be blown in with compressed air, or pulled in with a vacuum. There are several specialized systems used for this purpose. All fittings must, of course, be open when fishing or blowing in a line.

Conductors are pulled into each conduit as a group, not one at a time. This may require multiple spools of appropriately sized and colored wires. Alternately, the length of the run can be measured, and then multiple conductors can be spooled out and cut to length. Conductors must be fed in straight and be free of twists or crossings for a smooth trouble-free cable pull. Once the conductors are all in place, connections can be done just as they would be for a cable system. Conductors that pass through a box with no connection are left as a continuous run where possible, but some slack must be pulled into the box and left to allow for future connections.

Adding conductors to a conduit run after the initial installation can be very difficult. Unless there is a lot of extra space inside the pipe, it may be easier to install a second conduit run. In short runs, a wire can sometimes be added by simply sliding it in. Another option is to use one existing conductor to pull in two or more replacements. This will fail if the conductors were allowed to twist and cross over each other during the initial installation.

Organizing the Box. Once the conductors are all in the box and all cable clamps or connectors are secured, the box should be organized before any wires are trimmed off or stripped. Ground wires should be shaped and placed at the bottom of the box, where they can be easily attached to the ground screw. When a ground wire is required for a mounted device, such as a receptacle, one ground wire should be left long enough to extend from the ground screw to the device. Two ground wires may be terminated under a single approved ground screw. (This may be different in some code jurisdictions.) More than two may require a wire connector. Extra ground wires cannot be connected just by twisting them together: They must be connected with an approved wire connector.

After the grounds are in place, the neutrals should be organized. Each conductor should be laid, at least roughly, into its final resting place before trimming. Conductors that will be interconnected should be grouped so that after they have been trimmed and connected, they will fit into their intended positions without unnecessary bends and folds. Finally, all the neutral wires are trimmed to length, stripped, and terminated as required. Following the installation of the neutrals, the hot feeds and switch loops are organized and terminated. When a short connecting wire is needed to connect a device mounted in the box, these pieces should be cut, stripped, and installed at the same time as the other connections are made.

When power is going to be applied to a circuit without all the devices in place, all unterminated wires should be capped with a wire nut or insulated with tape. Devices such as receptacles, switches, and lights are not usually installed until after the interior finish is complete. All wiring should be pushed back into the boxes to avoid damage.

When wires are numbered or otherwise identified, make sure the markings are retained when the wire is terminated. When tape markers must be moved to a different position on the wire, use new markers, and put them on before the original ones are trimmed off. This will avoid confusion later. It can be very time consuming to identify unmarked conductors. Do not reuse tape-type wire markers, as the adhesive will almost certainly not hold the second time around. Loose-fitting markers, such as plastic ferrules or heat shrink that has not yet been heated, can easily be slid into position before the wire is trimmed.

Retrofitting

One of the most challenging and potentially time-consuming jobs in all electrical work is installing wiring after construction is complete. Retrofitting a fully finished building is always difficult. Wire can often be "fished" into wall cavities and carried from point to point through accessible areas, such as cavities above suspended ceilings, in attics, or unfinished crawl spaces and basements. An understanding of the structure of the building is essential for successful retrofitting. Outside walls are usually far more difficult than interior walls, because most buildings have thermal insulation in all outer walls. These walls should be avoided if alternatives are available. For truly impossible situations, surface-mountable raceways and boxes are available, but these generally should be regarded as a last resort.

There are a number of techniques and procedures that can greatly improve retrofitting efficiency and the quality of the finished product. The following tips are mostly applicable to residential construction, but the same basic principles apply to many commercial installations.

- Always predetermine the **complete** cable path before any cutting or drilling is done.
- Be on the lookout for alternatives in case an initial route is blocked by unexpected obstacles.
- Check the position of interior wall framing as accurately as possible. A stud finder can be very helpful. Beware of cross blocking at midwall.
- When working from an attic, locate the wall headers and determine distances from landmarks that are identifiable from both above and below. Good references are intersecting walls and ceiling fixtures. Measure carefully—the correct wall cavity must be identified. With a single top plate, the studs can be located from the nail heads. When the top plate is doubled, the studs will have to be located by measurement.
- To locate walls from below the floor, it may be necessary to drive a long nail or slender tool, such as a screw driver, through the floor. If the floor is carpeted, it may be acceptable to punch through the floor close to the baseboard. With a floor finish that would be damaged by this small hole, it may be possible to pull the baseboard away from the wall far enough to go through between the baseboard and wall finish. Be very careful not to damage the baseboard or the wall. In some cases it may be better to drill with a long, narrow bit.
- When drilling into a carpeted floor, the carpet thread will almost certainly wind onto the drill bit as it turns. If this happens, and is not corrected immediately, the carpet will unravel around the hole, causing permanent damage.
- Once a suitable cable path has been determined, holes can be drilled into the wall cavities from the attic or under the floor. When working from a driven marker, make sure the hole is drilled the correct distance, **in the correct direction** from the marker, so that the hole is inside the wall. It is depressingly easy to measure the wrong way and drill up through the finished floor—an expensive and embarrassing error.
- Wall boxes can be mounted on the stud or in the wall finish, but this decision must be made before the hole is cut. The best choice depends on the finish, and what the box must support. Fastening to the stud is the strongest, but is generally more difficult. Fastening to the wall finish should only be attempted if it is strong enough and the appropriate anchoring devices are available. Special mounts and suitable boxes are required if the stud is not going to be used.
- When the box is to be stud mounted, stay slightly clear of the stud on the initial cut. This will allow the edge of the stud to be located precisely. The box can then be used as a template to mark out the hole. Do not try to cut for a perfect fit. Allow a little clearance around the box. If the hole is too small, the finish may be damaged when the box is forced into place. This is particularly important in plasterboard, because it is easy to cut but difficult to trim.
- Once the hole is cut, check the box for fit, but do not mount it in the hole. The box is not mounted until the cables are pulled into place and terminated into the box clamps or connectors. The hole is used as the access for installing the cables.
- For a device such as a low-voltage thermostat that does not require a box, the hole in the wall may have to be much smaller. This is a bit of a disadvantage, but usually the

installation can still be done. If the thermostat is designed to be mounted on a box, a plaster ring may be installed instead. The ring will fit in a full-size hole, but has no back or cable clamps to complicate installation. Plaster rings can be difficult to mount after the wall finish is in place.

- Thermostats, in particular, should not be subject to drafts or other avoidable temperature influences from inside the wall cavity. Some provision may have to be made to seal the hole behind the thermostat, or it may never work properly.

Once the path for the cable run has been established, and all the necessary holes drilled or cut, cable installation can begin. The best starting point is usually, but not always, near one end of the run. Think through how the cable will be located and handled at each point. For wall outlets that are near the point of entry into the cavity, relatively stiff power cables can often be pushed into place. The cable can be located by reaching into the hole that was cut for the box. This is a two-person job. One can watch the destination hole while the other manipulates the cable from the attic or below the floor. A stiff wire hook can be helpful to grab the cable from the hole.

When installing light cables, such as thermostat wire, a fish tape or chain may have to be used. The cable is too soft to be manipulated in the wall. Both the chain and tape have advantages. A lightweight chain can be dropped from the attic down to the box hole, or from the hole down to the bottom of the cavity. A stiff wire hook can then be used to grab the chain and pull it through. Alternatively, a fish tape can be slid into place from above or below. Once the fish tape or chain is in place, it may be helpful to install a pull wire first, instead of the actual cable. A pull wire can be used to draw the cable in from either direction, while the fish tape can only be used to pull from the point of entry. When more than one cable must be installed in the same space, a pull wire can be used over again to the pull in several cables in turn.

It can be very difficult to fish a cable through from floor to ceiling without an access hole in the wall. Tricks that help include drilling at least one of the holes substantially larger than required for the cable. This makes it much easier to manipulate the catch hook or a second fish tape. A fish chain can be dropped from the top as before, or a fish tape can be run from either end. With two tapes, the usual method is to run both into the cavity and attempt to hook one with the other. Hooking two fish tapes inside a wall cavity without a hand-sized access hole can be very difficult.

Remember that all holes in areas that are subject to air circulation from unconditioned spaces must be properly sealed off after the installation is complete. This may mean replacing the insulation and installing a vapor barrier to seal around the hole edges and the cables.

Questions

1. Why is it so important to have cabinet point screw drivers for electrical work, rather than the more common keystone tip styles?
2. What are the handle colors of size 1 and 2 Robertson screwdrivers?
3. What three screwdriver tips are often confused, and what may result if the incorrect type is used?
4. What is the purpose of a panelboard, and how is it arranged to accomplish that task?
5. What is the purpose of a typical electrical box?
6. What components of a building are connected to the electrical ground system?

7. What is the electrical service entrance, and what is its purpose?
8. What is shut off when the main service disconnect is opened?
9. The individual circuit breakers of a panelboard connect to metal bars called _______.
10. What is the purpose of a ground fault circuit interrupter (GFCI)?
11. Large buildings have panelboards mounted at some distance from the main service. Why is this generally more cost effective than powering all loads from the main service area?
12. Explain the difference in the terms "service," "feeder," and "branch circuit" as they apply to the electrical system in a building.
13. When the building service is at a high voltage, what is required to supply energy to low-voltage loads such as 120 V outlets?
14. Explain why the circuits feeding fluorescent fixtures or other ballasted lights in large buildings are often much more dangerous than other electrical outlets.
15. What is the meaning of the term "cable" as it is commonly used to describe electrical conductors?
16. A cable commonly referred to as having two, three, or more conductors usually contains one extra conductor. What is the purpose of the extra wire?
17. What is the largest AWG conductor size?
18. What unit is used to measure conductors larger than those measured in AWG?
19. What is the correct method for cutting spiral cable armor?
20. What extra precaution may be required to protect conductors at the terminations of armored cables?
21. On what part of the cable should the cable clamp be tightened?
22. What is another term used to describe electrical conduit?
23. How could a loose electrical box cover be dangerous in an environment containing explosive gases?
24. Can standard pipe and fittings be used as electrical conduit?
25. What hazard may be associated with a loose conduit fitting?
26. Why is it that a conduit may be be filled to full legal capacity and still have room for more conductors?
27. List three general code requirements that apply specifically to electrical boxes.
28. What connects to the electrical ground system?
29. What high-voltage hazards are reduced or eliminated by the ground system?
30. The ground conductor and grounded circuit conductor are different. What name is usually used for the grounded circuit conductor?
31. Describe a situation where it would be dangerous to leave the ground wire disconnected.
32. List at least four ways that the ground connection may be broken.
33. List at least three problems that can cause connection failures.
34. Compare wire nut, set screw, and compression connectors by listing at least one advantage and disadvantage of each.
35. What is the purpose of the divider in a split bolt connector?
36. Describe the correct method of connecting a solid conductor under a terminal screw.
37. What is the purpose of raised ridges in the base material under a terminal screw? Explain how they can contribute to bad connections.
38. When a large stranded conductor is terminated in a set screw connector, the connection should be tightened down twice. Why is once not usually enough?

2 Meter Selection and Use

Objectives

When the study of this chapter has been completed, you should be able to:

- Identify typical electrical test instruments
- Explain the functions and features of typical analog and digital multimeters
- Explain how to use a multimeter safely and efficiently to perform typical electrical tests
- Describe the operation and use of clamp-on ammeters
- Describe the use of current shunts for reading very low or very high values of current
- Describe and explain various special-purpose meter features and functions
- Describe the purpose and use of megohmmeters
- Describe the purpose and use of wattmeters
- Identify the hazards associated with various meters and meter functions
- Describe and explain the safety habits and procedures that will protect against personal injury and meter damage

2.1 Introduction

A good quality, general-purpose multimeter is an absolute necessity for service work on electrical circuits of any kind. It is essential for ensuring personal safety, for routine testing, and for locating circuit problems. After the multimeter, the most popular electrical instrument is the clamp-on ammeter. This is the standard instrument for measuring current in power circuits. Another, less well-known instrument is the megohmmeter. This instrument is a diagnostic tool for testing the condition of insulation, particularly in motors. Megohmmeter tests are a key element in all preventive maintenance programs. A long-term record of periodic test results is a chart of the electrical health of a piece of equipment. Such records can identify problems and predict failures. The megohmmeter can also be used to check equipment that has been out of service for a lengthy period of time, to see if it is safe to use.

2.2 Analog vs. Digital Meters

Multimeters and clamp-on ammeters are available with either an analog (needle and scale) indication or a digital readout. New, high-quality analog multimeters are rare and relatively expensive. They have been almost entirely replaced by digital types. Digital meters seem to be following the present trends of electronics products. Every year new models come out that are better designed and have more features than those of the previous year—for the same price, or even less.

Quality analog meters, already owned, are usually worth keeping. With care and the occasional minor repair, they can last a lifetime. For a new purchase, a good quality digital meter is the best. Modern digital meters have features that are unavailable on analog versions. They are simpler to use, more rugged, and less expensive for equivalent capabilities. Low-quality analog or digital meters are not recommended, except for occasional use on 120 V or less.

For personnel who routinely work on equipment operating at higher voltages, an industrial-type 600 V rated test light or similar device is highly recommended. This is in addition to a good quality multimeter, not as a replacement. Test lights are ideal for safety-related checks. They fit in a pocket, are very reliable, and are always ready for immediate use. Carrying a test light encourages a "test first/touch later" attitude that has saved many service workers from serious injury.

2.3 Meter Functions and Features

Multimeters can measure a variety of electrical quantities. What can be measured, the range of values covered, and the accuracy of the readings all vary with the make and model of the meter. An inexpensive meter may be accurate enough when new, but is less likely to remain so through years of use and abuse. Digital meters are easy to read and generally provide three- or four-digit results compared to no more than two or three for an analog. However, this can provide an illusion of accuracy that is not always warranted. Good quality analog meters are as accurate as many digital versions. Typical analog and digital meters are shown in Figure 2–1.

Typical multimeter functions include measuring resistance, voltage, and current. Resistance ranges extend from less than one ohm to several megohms. Reading resistances below $^1/_{10}$ of an ohm requires a specialized meter and very good test leads. Voltage measurements, both AC and DC, are standard functions of virtually all multimeters. The range of voltages is typically from $^1/_{1000}$ V (a millivolt) or less to 600 V or more. Typical multimeter current ranges extend from low milliamps, or even microamps ($^1/_{1,000,000}$ of an ampere), to a maximum of about 10 amps. Some multimeters have very limited current capabilities; others have no current functions at all. Meters designed primarily for electronic work are more likely to have very low voltage and current capabilities.

Newer meters are more durable and less dangerous to use than some of the older models. They are less likely to be damaged by careless use, and far less likely to cause serious injury if they are incorrectly applied.

FIGURE 2–1 The analog meter on the left uses a needle-and-scale indication. The digital meter provides measurement information in a digital display.

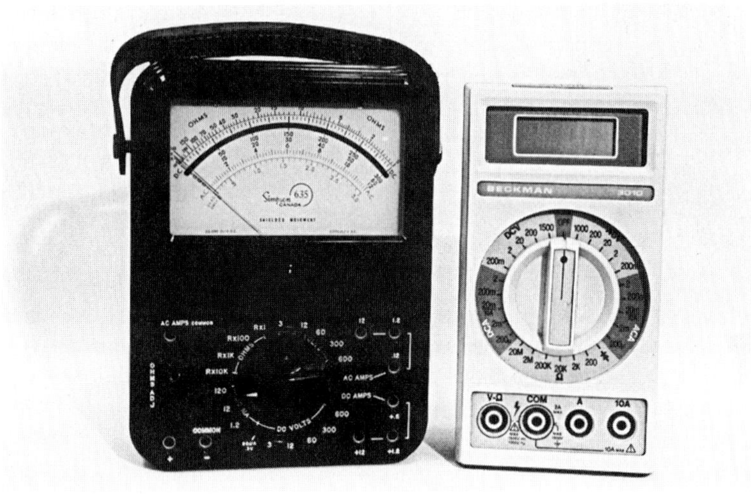

Resistance Measurement

Electrical resistance is "electrical friction" and is measured in ohms (Ω). A resistance meter is generally referred to as an ohmmeter. Resistance measurement is a function provided on all multimeters. It is intended for use in de-energized circuits only. The resistance of an individual component usually cannot be measured unless it is first disconnected on at least one end.

Typical analog meter resistance scales (shown in Figure 2–2) can be difficult to read. On most analog meters there is one set of ohms scale markings, with a single number set. The numbers are in reverse order, starting with 0 on the right and ending with an infinity symbol (∞) on the left. The scale is not linear. The lower values are well separated and easy to read. The higher values (at the left end of the scale) are very compressed. Each scale marking on the left side may represent many thousands of ohms, while each mark on the right may be only a fraction of an ohm. The meter usually has several resistance ranges, each with a different multiplier. Typical examples are RX1, RX100, RX1K, and RX10K. The actual resistance reading is the scale value multiplied by the value of the

FIGURE 2–2 Note the arrangement of the resistance scale of this analog meter face.

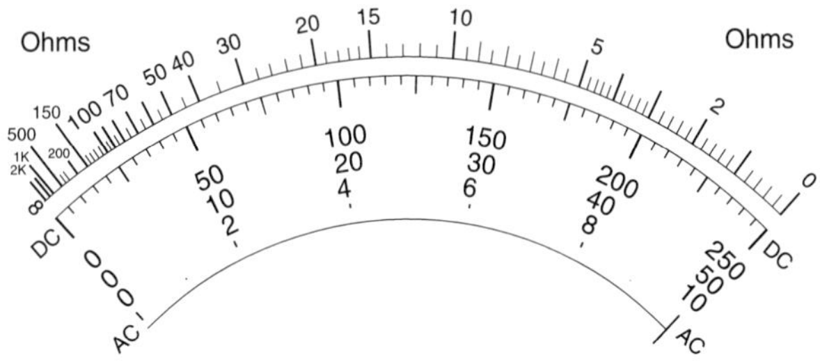

range multiplier. Using the ranges listed above, the actual meter reading would be multiplied by 1, 100, 1000, or 10,000, depending on the range selected.

Note

The infinity symbol (∞) appears on the extreme left end of the resistance scale of virtually all analog multimeters. An infinite resistance is the resistance of an open circuit. This symbol is not used on digital meters. Instead, the meter has an "out-of-range" or "over range" indication. This indication is not standardized, so it can vary with the make and model of the meter. Common examples include one or more flashing digits or the letters "OL."

Digital meters also provide several resistance ranges, but the range markings are not multipliers. On a digital meter, each resistance range can provide a reading only up to its listed maximum. For example, with a 200 Ω range selected, the meter cannot measure any resistance above 200 Ω—actually 199.9 Ω on most meters. When the resistance being read is greater than the selected range, the meter will display its "out-of-range" indication. On older digital meters, the ranges have to be manually selected. Many of the newer models select the appropriate range automatically. This feature is called "autoranging."

Many digital meter readouts use what is called a $3^1/_2$-digit display. This means the last three digits can be any value from 0 to 9, but the first digit cannot be more than 1. The four digits can therefore display a maximum value of 1999. The full range of resistance values that the meter can read is much wider than the total available digits. To cover such a range of values, instead of using multipliers like the analog meter, digital meters use different base units, and reposition the decimal place as required.

The lowest range is displayed in ohms. The decimal is placed one position from the right, making the minimum display value 0.1 Ω and the maximum 199.9—essentially 200 Ω. The 200 Ω range is followed by a series of ranges that read in thousands of ohms. To indicate the unit is in thousands, the prefix "kilo" is used, abbreviated to a lowercase "k." One kilohm (1 kΩ) equals 1000 Ω. The meter ranges are separated by a factor of 10, for four kilohm ranges—2 kΩ, 20 kΩ, 200 kΩ, and 2000 kΩ. The final resistance range on most multimeters is in millions of ohms. "Mega," which is abbreviated to an uppercase "M," is the unit prefix for 1,000,000. The megohm range is typically 20 MΩ, ten times greater than the next lower range.

Although the numeric value given by digital meters is much easier to read than an analog scale, errors in readings are surprisingly common. One thing that leads to confusion is that typical digital meters read all values above 200 Ω in thousands of ohms (kΩ), not in ohms. This means that values between 200 and 999 will be displayed with a leading decimal. For example, a reading of .535 (the meters may not show a zero before the decimal point) is actually 535 Ω because 0.535 kΩ is 535 Ω. Many meters indicate the unit (Ω, kΩ, or MΩ as appropriate) in the readout, but it is often overlooked.

Note

A variation on the $3^1/_2$-digit readout is a 4000 count display. In this system the first digit can be any value from 0 to 3, for a maximum count of 0–3999. The logic of this system is the same as for the 0–1999 displays, but each range can display a wider range of values.

Many meters require manual selection of both the function and the range. This may involve pushbuttons or a rotary selector switch. For the most accurate results, the range selected should be the lowest one available that can provide a reading. Ranges always represent maximums, so this means the best range is the closest one above the actual value. Selecting a range that is too high will cause inaccurate readings because the actual value will be rounded. In extreme cases the meter will round to zero. For example, attempting to read the value of a resistance that is less than 100 Ω on a range of 200 kΩ or higher will result in a reading of 0 Ω.

Autoranging meters automatically select the range to match the value being read. This feature is convenient, but requires attention to readout details to avoid misinterpretation. To help the user correctly interpret the measured value, the unit of measurement (Ω, kΩ or MΩ) and the range are provided in the meter display. Audible range-change warnings may be provided to remind the user that a different range is in use. Autoranging meters also incorporate a manual range selection, such as a pushbutton, that allows the user to cycle through the ranges. Once a range is manually selected, the meter will not automatically range up or down, even if the value is too high or too low. This can lead to errors, because the user may forget the autoranging feature is locked out. "Memory" and "hold" functions can also lead to errors for the same reason. On some meters, a memory feature is provided to deduct the value of an initial reading from all the readings that follow. This allows lead resistance to be deducted automatically when reading very low values of resistance. Hold buttons are used to lock the display value so that the meter can be read after the test probes have been disconnected. This can be very convenient when working in dark or crowded conditions.

Continuity Testing

The term "continuity," as it applies to electrical circuits and devices, means a continuous conductive path or complete circuit. For example, a copper wire that has no breaks would have continuity. So would a closed switch, a good fuse, or a solenoid coil that is in working order. When a device is checked for continuity, the test is intended to determine if it is capable of passing current. In most cases, a continuity check can determine if the device is defective or in working order.

Any ohmmeter can be used as a continuity tester. A low value of resistance indicates a complete circuit, and an infinite value indicates the circuit is open. The exact value of resistance is usually of less interest than the fact that the circuit is complete or open. Speed and convenience are more important than the exact resistance in ohms, so many meters have an audible continuity signal that "beeps" when the resistance is below a certain value. The audible tone allows continuity tests to be conducted without glancing at the meter. This is a nice hands-free feature for quick tests of cables, switches, and similar equipment. However, when many accurate, low-resistance readings have to be made, the continuity test signal can become very annoying. Better meters separate the continuity and resistance functions, or have a provision for turning off the tone.

Note

Many experienced installers will not accept circuit continuity tests done with ohmmeters, insisting instead on a test device, such as a battery and buzzer, that passes a higher current through the circuit. Occasionally a circuit will pass an ohmmeter test, but fail to work when it is put into service.

Warning

Forgetting to check the function selection, and applying voltage to a multimeter that is set to measure resistance, is a common cause of meter damage and personal injury. Analog meters, particularly the less expensive models, are very unforgiving of such mistakes. Digital meters are generally much less likely to be damaged by this simple error, but they are not indestructible.

Voltage Measurement, AC and DC

AC and DC voltage measurement functions are provided on all multimeters. AC and DC are different, so there are separate selections for each of them. In most cases, the ranges available for these two voltage functions will be similar, although the DC ranges may be more extensive. The ranges may have to be manually selected or—if the meter is a digital type—it may autorange just as it does on the ohmmeter function.

For typical analog meters, the voltage and current scale arrangements are somewhat different than for the resistance readings. Usually there are only one or two scale markings, but, unlike the resistance scale, they are linear. The divisions between the units are of equal length across the full length of the scale. The scale is also in "normal" order: zero on the left and the maximum values on the right.

Instead of one number set and several range multipliers, the voltage scales have several number sets that are matched to the ranges. For each range there is a number set provided that corresponds either to the actual range values or to some multiple of 10. For example, a number set that extends from 0 to 300 would be used with all ranges based on 3—such as 0.03, 0.3, 3, 30, or 300. The range value on the selector is always the maximum for that setting, so it corresponds to the last marking and digit on the right-hand end of the scale.

Analog meters require correct polarity for all DC functions. If the polarity is reversed, the meter needle will move slightly down scale. Interchanging the leads will correct the problem. This behavior can be used to identify the polarity of the circuit.

Note

Always watch the meter closely when taking DC readings. The downscale movement of the meter needle that occurs if the polarity is wrong is quite slight. The meter has needle stops at both ends of the scale, but they are not very far past the last scale markings. When the needle dips below zero, the leads have to be reversed and the reading taken again.

AC voltage ranges are not polarity sensitive for either an analog or a digital meter. The same reading will result with the leads reversed. The AC functions are usually very similar in layout to the DC functions, though there may be fewer ranges. Analog meters may have a separate AC scale. A digital meter may have an autoranging feature or require manual selection.

Digital meters provide three- or four-digit numeric voltage values. The range of available digits is identical to those provided for the resistance measurement functions. Typically this means a maximum display value of 1999, with the decimal positioned to match the range. Ranges usually start at 200 millivolts or less and go up past 600 V. The lowest range on a typical meter will display in millivolts. One millivolt (mV) is $^1/_{1000}$, or 0.001, of a volt.

A digital meter set to read DC volts is not polarity sensitive. The meter will read equally well with either polarity at the probe connections. However, the meter readout includes a polarity indication. This is typically a + or – sign, indicating the polarity at the positive input jack.

Some meters show no sign if the polarity is "normal"—positive to the positive (red) lead, negative to the common (black) lead—but will indicate a negative if the leads are reversed.

Low-voltage ranges on digital meters are very sensitive. It is not unusual for the meter to read some voltage value with the meter leads just in the open air, connected to nothing at all. This is particularly noticeable on autoranging meters, because they will automatically go to the lowest range when the meter is disconnected from a voltage source. These voltage readings are a result of electrostatic and electromagnetic fields that radiate through the air from all types of electrical or electronic equipment.

Note

A weakness of early digital meters was the difficulty of reading rapidly changing values. Many now include a bar graph that operates much like the needle of an analog meter. It is much easier to observe changes on a graph than as a series of numbers.

Voltage tests are one of the most common of all multimeter tests, and are usually the most critical. Resistance tests must be done with the circuit de-energized; therefore (assuming this critical precaution has been followed), such tests can be taken with virtually no risk. For voltage readings, the circuit must, of course, be energized, so the risks are much higher. If the voltage level is at all dangerous, taking readings in energized circuits requires a great deal more care and closer attention to details than readings taken with the power disconnected.

Warning

In addition to other safety precautions, safety glasses should be mandatory when working on energized electrical equipment. A mistake usually results in an electrical arc that projects hot particles of metal some distance from the point of contact. Higher voltages and high-current circuits can produce faults that are best described as explosive, but even 120 V arcs can be dangerous.

Before making contact with a live circuit, it should be an instinctive habit to check the meter function and range settings and the meter lead connections. The probes must be plugged into the correct jacks, the function must be AC or DC volts as appropriate, and, if the meter is not autoranging, the range selected must be above the expected value. When the circuit voltage is unknown, the highest available range is a good starting point. The meter can be ranged downward as required after the initial test.

The voltmeter reading must be taken across two points. If one meter lead is going to be connected to a circuit common or to ground, that connection should be made first. This permits full attention to be focused on the most critical contact point. Whenever possible, make contact with the live part of the circuit at a point that is easily accessible. The point of contact should be separated from other components and grounded surfaces, and should be secure enough to allow a solid connection.

The meter probes should be applied in a firm but controlled manner. Care must be taken to not slip, particularly if the circuit is operating at line voltage. The meter user should also pay attention to his own stability and position. It is much better to have no unnecessary contact with grounded metal or other conductive surfaces. This greatly reduces the chances of injury if something goes wrong.

Avoid contact between the exposed metal of the probe and any circuit components other than the ones under test. There can be serious consequences if a meter probe bridges the gap between an energized part of the circuit and another conductor, component lead, or grounded metal. The probe will short out that part of the circuit, usually causing at least minor damage. Burned meter probes and arcing damage to the area of contact are typical. Much more extensive damage is very likely, particularly in circuits with fragile control contacts or electronic components. In power circuits, the risk of personal injury can be very high.

Warning

There are a number of circumstances where a perfectly good meter will fail to read a potentially dangerous voltage. A typical multimeter set to read DC voltage will indicate 0 V when it is connected to an energized AC circuit. A meter with a hold or memory function can give the same results if that feature has been activated before the test. The meter is usually unharmed, but the zero reading can lead the unwary user into a deadly trap.

These simple errors can be extremely dangerous. Very often, service workers must trust their multimeter to confirm whether a circuit is live or is safe to touch. Obviously, it is very important that the meter be set to an appropriate function, and any "hold" function released, for such a test. When making safety-related checks it is a good habit to touch the meter to a live part of the circuit first (if one is conveniently available) before checking the part that is supposed to be de-energized. This pretests the meter to confirm it is in working order and properly set before the safety test is done.

Another good habit, particularly with analog meters that have no "off" position on the selector, is to always leave the meter set to read a relatively high AC voltage. That way, if the meter is connected to a live circuit without checking the function selections first, there will be no danger to the meter or the user. With the meter on a high AC range, it will provide a reading if the circuit is energized, even if it is energized with DC. Unlike DC ranges on AC, an AC range will read a DC voltage. The reading will not be accurate, but it will at least indicate the hazard.

AC and DC Current Measurement

Warning

Typical analog meters have unfused current inputs. Used carelessly, a meter without fuse protection may be damaged or destroyed. Connected to an industrial voltage, it can turn into a bomb. Typical digital meters have two dedicated current inputs. One is for very low current readings, in milliamps ($^1/_{1000}$ A) or even microamps ($^1/_{1,000,000}$ A). The second input is for ampere readings. The low-current jack is usually protected by a low-value internal fuse. The high-current jack may not be fused at all, depending on the make and model of the meter. On meters with no fuse protection, when the leads are inserted into the high current jacks, they are connected together inside the meter by a solid metal bar or strap. This short-circuits the leads, regardless of the position of the meter's function switches. Under these conditions, placing the probes across a voltage can be extremely hazardous to the user and the equipment under test.

As these warnings may indicate, current measurement is the most troublesome of all multimeter functions. Electrical current is a flow of electrons inside the circuit conductors. An ampere, or amp, is a measure of the electron flow rate. To detect electron flow, the meter must be connected into the circuit in such a way that the full circuit current passes through the meter. To be a part of the circuit, but have a minimal effect on normal circuit operation, the ammeter must have a very low internal resistance.

In terms of its behavior and its connection in the circuit, an ammeter is essentially a jumper wire. Just as an improperly placed jumper is a short circuit, so is an improperly placed ammeter. When the leads are connected to the current jacks, or the meter is otherwise set up to read current, the meter is a potential short circuit. The ammeter function of a typical multimeter can be a very useful feature, but it must be used with care.

Separate meter functions are required for AC and DC current measurements, just as they are for voltage measurements. An AC current value also may not register with the meter set for DC. A DC current may register on an AC function setting, but the reading will be inaccurate.

Typical analog meters provide several current ranges, accessed with a selector switch, extra lead jacks, or a combination of these. The lowest current ranges on the meter are usually part of the switch-selected functions. When the meter has high current ranges—up to several amperes—they are usually provided through separate dedicated current input jacks. These current inputs are usually unfused, and therefore—as mentioned above—must be used with extreme care when connecting to circuits that are fed from high-powered sources. DC current ranges often start at very low values. Many analog meters can read DC currents as low as a few microamps. AC current ranges are usually less sensitive. The lowest AC range usually starts at a somewhat higher current level than the lowest DC range.

Typical digital meters have two dedicated lead jacks provided specifically for measuring current. One is for low current ranges—micro- and milliamperes—and is usually rated to a maximum of only a few hundred milliamps. The second input is for high currents, typically up to 10 A. Fuse protection is normally provided on the low-current input. Better meters, particularly those that are intended for electrical service work rather than for electronics, have fuses on both low- and high-current inputs. These two fuses are very different. The high current range is typically protected by a compact 600 V industrial type cartridge fuse, while the low current range has a small glass fuse rated for 250 V or less. The high-voltage fuse adds significantly to the cost of manufacturing the meter, so the meter price may be higher.

On typical handheld meters, the meter case must be opened to change the fuse. A spare may be provided in a storage space inside the meter. Two different spares should be provided if the meter has fuse protection on both the low- and high-current inputs, but don't count on them being there. Extra fuses should be a toolbox item. Never replace the high-current fuse with one of a lower voltage rating.

Bench-style meters often have a fuse that is accessible from the front panel of the meter. One method used to provide this convenience is a current jack that can be removed to reveal the fuse. If the meter has this feature, the current jack will have small notches on each side to accept a thumbnail or coin. The jack can be removed by pushing it in and turning it one quarter of a turn.

The hazards of using ammeters can never be taken lightly; however, they can be greatly reduced. Some analog meters have little if any built-in protection: Poorly designed models may literally explode when they are improperly connected to an industrial-level

FIGURE 2–3 From left to right in the back row, an analog multimeter, a manually ranged digital multimeter, an autoranging digital meter without current ranges, a hand-cranked megohmmeter, and a digital clamp-on ammeter. In the front is an analog clamp-on multimeter that provides the basic multimeter functions plus clamp-on current measurement.

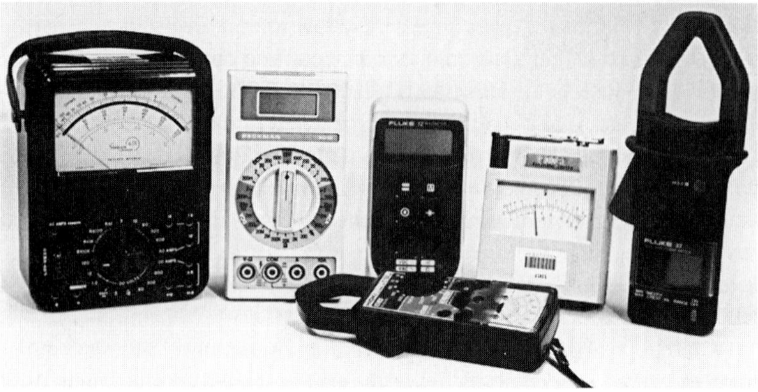

voltage. Such meters have been at the center of some tragic accidents. Digital meters without internal fusing are almost as dangerous. Accidental contact across a powerful circuit can blow the end off the meter probes, burn out the meter circuit board, and, in severe cases, set off more destructive and dangerous arcing. Newer multimeters, designed for industrial electrical use, have internal fuses on all current ranges. These fuses virtually eliminate the possibility of personal injury. It is still possible to damage the meter leads and destroy circuitry, but there is much less chance of being injured. Meters that are designed primarily for use in electronics are less likely to have this additional fuse protection. A selection of meters is shown in Figure 2–3.

Multimeters Without Current Ranges

At least one meter manufacturer now offers an industrial multimeter with no current functions. This meter is designed for service personnel in the power electrical field, where a separate, noncontact current meter is normally available. This meter, called a clamp-on ammeter, is described below. If there is no need for the milliammeter or ammeter functions found on a typical multimeter, a meter without them may have some advantages. These meters are less expensive than a conventional meter of similar quality, and are safe to the point of being almost foolproof. One popular model provides excellent voltage, resistance and continuity features, plus a built in capacitance testing function.

Clamp-On Ammeters

Clamp-on ammeters read current without a direct connection to the circuit. The meter has an open loop at one end that is sectioned to form a jaw. A simple trigger mechanism opens the jaw to accept a conductor into the loop. This loop can range in size from less than an inch to several inches in diameter, depending on the ratings of the meter. It must be large enough to easily accommodate the largest conductor that has to be monitored. To measure circuit current with this meter, the current loop is placed around a current-carrying conductor. Electrical current creates a magnetic field around the conductor that is proportional to the rate of current flow. The meter loop links with this magnetic field and reads the current with no electrical contact.

For added safety, the jaws of the meter are covered with insulating material. There is no need to shut off the circuit or change the circuit connections. The meter can be applied to any conductor, even while it is energized. The only complication is that the meter jaw must contain only one conductor of the circuit at a time. When the meter is clamped around a two-conductor cable, for example, conductors from both sides of the circuit pass through the meter loop. The current in the conductors is moving in opposite directions, so the magnetic fields cancel each other and the meter reads zero. To read the current with a clamp-on meter, a point must be found where the conductors are separated. The meter can then be hooked around only one of the wires instead of the entire cable. This often means opening junction boxes or panelboards; still, it is a lot easier and a lot safer than wiring a directly connected ammeter into the circuit.

Clamp-on ammeters have much higher current capacity than can be provided with a direct connect meter. Typical industrial versions can read several hundred amperes. Higher capacity meters, or standard meters with high-current adapters, can be used to measure currents of several thousand amperes.

Clamp-on ammeters are available in a variety of styles and prices, with either analog meter movements or digital readouts. Some of these meters are designed as general-purpose multimeters, with lead jacks, AC and DC voltage ranges, and resistance functions. They are slightly bulkier than many multimeters, and may lack advanced features, but a clamp-on meter with multimeter functions can perform virtually all the electrical power-related tasks encountered in heating, ventilation, and air conditioning service work.

Most clamp-on meters are designed to read only AC current, but there are some that will read either AC or DC. These two meters use very different methods to measure the current. Both detect the magnetic field around a current-carrying conductor, but the circuits used to convert field strength into a current value, are very different. In the AC-only type, the iron meter loop that encircles the conductor passes through an internal transformer coil. A small current, proportional to the main conductor current, is induced directly into this coil. The internal meter circuits measure the induced current, either directly, through an analog meter movement, or though electronic conversion to a digital display. The AC/DC type of clamp-on meter also uses an iron core in the loop, but the sensing element is a solid-state device that measures the strength of the magnetic field. This is then translated into a current value through the internal electronic circuits of the meter. DC currents in the ranges that can be measured with clamp-on meters are not particularly common in equipment driven by the usual utility power connection. DC power circuits are used in specialized equipment such as paper machines, printing presses, rolling mills, and electroplating. They are rare in heating, ventilation, and air conditioning equipment.

The ranges available on clamp-on meters start at a much higher level than the current ranges of a typical multimeter. This limitation can be at least partially overcome by passing multiple wraps of the conductor through the meter jaws. The actual current is the indicated value divided by the number of times the conductor passes through the meter's sensing loop. This can provide improved low-current capabilities, but unless there is enough slack in the circuit conductor to allow all the extra wraps, the circuit may have to be disconnected and a jumper installed. Coils with a convenient number of turns can be made up, or bought as an accessory. A 10-loop coil with alligator clips on the ends can be a handy accessory.

An attractive feature of most digital clamp-on meters is the ability to store and display current peaks. This can be very informative in circuits with loads that draw a high initial current, and then settle down to a much lower value. Most coil-operated equipment, such as motors and transformers, draws a high inrush current when it is first energized. It is impossible to measure these values unless the meter has some means of recording the peak. Analog meters are limited by their construction, making accurate peak-current readings difficult, if not impossible, to obtain. Digital meters, equipped with the necessary internal memory, can provide peak readings quickly and conveniently.

Some high-quality clamp-on ammeters feature true RMS readings. RMS stands for "root mean square," which is a mathematical method of determining the average value for a quantity that is constantly changing—in this case, the AC current. Some meters assume that the AC current wave is undistorted and provide a current reading based on that assumption. When the wave is distorted, these meters will be in error. When the amount of distortion is significant, the error is also significant.

True RMS meters provide a more accurate value, because they attempt to include all the variations in the current rather than making assumptions that are not necessarily true. Either meter will read undistorted currents equally well, but a true RMS meter provides a much more accurate value with a distorted signal. The ability to read a true RMS value is critical for tests involving equipment with electronic power supplies. Examples include variable-speed motor drives, large battery chargers or UPS (uninterruptable power supply) systems, and feeders to extensive computer installations. Standard digital meters cannot accurately read the current to such loads because they cannot properly interpret the distorted currents they draw.

As an option to a separate clamp-on ammeter, clamp-on current accessories are available for use with most multimeters. The accessory-current jaws connect to the meter through a small cable, allowing the readout to be remote from the conductor under test. This can be convenient in some situations. Note, however, that a good clamp-on accessory may cost as much as a separate clamp-on ammeter, and is unlikely to provide the same features.

Reading Current Without an Ammeter

Current can be read by using a voltmeter and a precision resistance wired into the circuit. The voltage drop across the resistance is proportional to the current flow. The resistance required for a given current and voltage drop can be calculated with the Ohm's Law formula, $R = E \div I$. If the resistance value is chosen wisely, the voltage readings can be interpreted as current with little or no calculation. For example, a resistance of 1 Ω would cause a voltage drop of 1 V/A. The voltage reading across the resistor can be interpreted directly as amps. With a resistance of 1 kΩ (1000 ohms), and the meter set to read millivolts, the reading is directly convertible to microamperes. A reading of 1 mV would be equal to 1 μA of current. This technique can be very useful for reading current values that are either very high or very low.

High-current DC circuits are usually equipped with precision resistors called "shunts." These have extremely low but very precise resistance values. They are rated in volts or millivolts as a proportion of the current. An example could be a 100 mV output for a current of 1000 A. A millivolt meter across the shunt would provide an accurate current reading up to the 1000 A limit of the shunt. The shunt resistance would be 0.0001 Ω, a very low value. The millivolt meter reading would have to be multiplied by 10 to convert it to amps. If the meter is permanently installed, a suitably scaled analog face plate or an internal digital scale multiplier is used to convert the actual value directly into an ampere display.

Shunts are standard equipment on most DC power equipment because, until recently, there was no other way to measure large values of DC currents. Shunts work on AC or DC, but AC circuits normally do not have them installed. Current transformers are used instead because they provide a similar ratio output, plus electrical isolation from the circuit. This is much safer than a shunt, which is energized at full circuit voltage.

At the other extreme, very low current values are also difficult to measure. Typical multimeters provide current ranges for most situations, but few will read down into the low microamp (μA) range. Unfortunately, such low current tests are sometimes required. For example, the flame sensor circuits used to monitor the pilot or main flame in many furnaces, and in commercial and industrial burners, do operate at only a few microamperes. This can make it difficult to troubleshoot an electronic burner control without a microammeter.

One option is to buy a multimeter that has a microammeter range. These tend to be expensive, electronics-oriented meters that may not be well matched to other aspects of electrical service work. A second option is to purchase a separate microammeter. This is also a relatively expensive meter with no other uses. Dedicated microammeters are normally analog types and are quite delicate, a potentially serious problem for mobile service personnel. A third option is to purchase or build a suitable current shunt for use with a standard multimeter. Commercial DC current adapters are little more than a series resistor with connections for the meter and the meter leads. A small value capacitor may be added across the resistor to filter out some of the voltage spikes. A similar adapter can be built from a few dollars worth of parts.

While high-current shunts have very low resistances, low-current shunts typically have much higher resistance values. Commercial adapters for testing microamp flame sensor current may have an internal resistance as high as one megohm (1 MΩ). This high value does not cause a problem because the operating voltage of the circuit is relatively high, and the normal circuit resistance is many megohms. As long as the shunt resistance is only a small fraction of the total circuit resistance, it will provide a satisfactory signal without interfering with the operation of the circuit.

With a 1 MΩ resistance, microamperes of current can be read directly on a voltage range. The meter is connected across the resistor and set to read in volts (see Figure 2–4). One volt indicates a current flow of one microampere (1 μA). A similar effect can be obtained with a one kilohm (1 kΩ) resistor and a millivolt meter range. A 1 kΩ resistor converts 1 μA of current flow into a reading of 1 mV. The 1 kΩ resistor has less effect on circuit operation, and the 1 mV per μA output is well within the capabilities of most multimeters. Other values can also be used. When adapters of this type are homebuilt, it is a good idea to compare the results to a genuine microammeter reading.

FIGURE 2–4 Reading current directly on a voltage range.

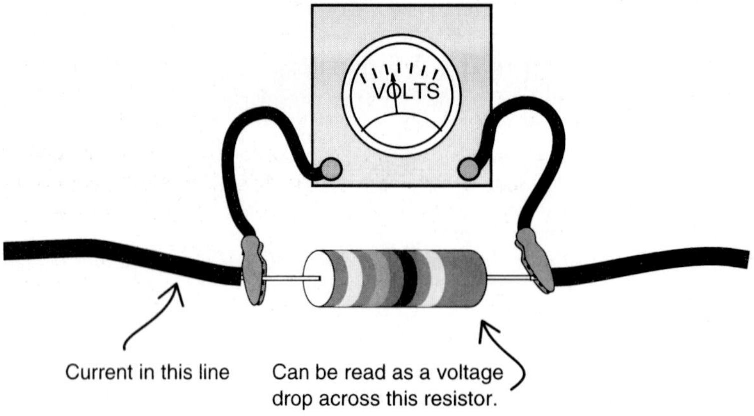

Other Multimeter Features and Accessories

The rapid developments in electronics that are so much a part of modern times have had a powerful effect on measuring instruments such as multimeters. Each year, more features become available, often at lower cost than the year before. Many of these features have been available in one form or another for many years, but never before have so many been available in a single, compact, and relatively inexpensive meter.

Although it will inevitably get even better and less expensive, the modern multimeter is a wonderful tool—smaller, safer, and with more useful features than ever before. The following list of options and features should provide some insight into the range of possibilities and may be helpful to anyone interested in purchasing a new meter. The final choice is a personal one, and should be made after careful consideration of at least some of the many excellent quality meters that are presently available.

The basic features of a multimeter, as listed previously, are: AC and DC voltage, resistance, and, usually, AC and DC current measurement. A common addition to the resistance function is an audible tone for continuity testing, although this feature may be combined with a different function, such as the diode test mentioned in the following paragraph. Some digital meters also have a memory feature to subtract an initial resistance reading from subsequent readings. This allows the user to automatically correct for meter lead resistance, for example. This is particularly convenient for measuring a series of very low resistance values, but can lead to mistakes, or even be a safety hazard, if it is engaged accidentally.

Resistance tests can be used to check diodes and other solid-state components, such as conventional transistors. However, a much better test is to measure the voltage drop across the diode or transistor terminals while a small current source is connected. This function has been a relatively standard feature on digital multimeters for many years. It is marked on the meter function selector with a standard diode symbol: ⊣◀⊢. On some meters, the audible continuity feature is combined with the diode test instead of the resistance function.

Other meter features include a "hold" function or a more automated "touch and hold." A hold feature locks the display so that it shows the last reading, allowing measurements to be taken without having to look at the meter while the probes are still in contact with the circuit. This can be convenient, and much safer, when taking voltage readings in awkward locations. The meter user can concentrate on the most dangerous area—the point of contact with the live circuit—and not have to look at the meter display until afterward. Note, however, that hold features are not without problems or hazards. The hold button can lock the display on zero, resulting in a zero reading on an energized circuit. This can lead to accidental contact with a live circuit—an extremely dangerous result for such a simple error.

Some digital meters are also equipped with internal memory to retain minimum and maximum readings. This feature may be combined with an internal time clock to "time stamp" the recorded values. With this capability, the meter can record both the minimum and maximum values and the actual times when they occurred. This can be handy for tracking suspected voltage-variation problems. It also allows a circuit to be monitored for suspected intermittent faults. Problems that occur at random and then disappear are notoriously difficult to locate, because the cause cannot be detected after the failure. With a recording meter, a suspected portion of the circuit can be monitored for an extended period of time. The recorded values and times can be used to narrow the possibilities until the problem has been isolated.

Testing capacitors using only standard ohmmeter functions has traditionally been one of the few areas where digital meters are less convenient than the older analog types. However, although the response of the analog ohmmeter is much easier to interpret than that of a digital meter, neither meter can measure the actual capacitance value. A true capacitance meter is far superior to any type of ohmmeter test. Until recently, true capacitance testing functions were only available on dedicated capacitance meters. They are now a common digital multimeter feature, providing actual capacitance values in microfarads (μf) or picofarads (pf). A picofarad is $^{1}/_{1,000,000,000,000}$ of a farad. Such small value capacitors are common in electronics, but rare in power circuits.

When there is even an occasional need to test capacitors, a capacitance measurement function can be very desirable. Be aware, however, that there is considerable variation in the ranges provided by different meters. Those that are designed primarily for electronic use tend to cover only very small capacitance values, while meters designed for use in electrical power applications tend to cover much larger values.

Temperature measurement is an obvious priority in heating, ventilation, and air conditioning applications, but it is also an important diagnostic function in other types of service work. There are dedicated thermocouple meters, multimeters with thermocouple inputs, and several types of temperature measurement accessories that work with most multimeters. These include various contact and noncontact probes, and adapters with inputs for one or two thermocouples. Thermal probes can be expensive, and some are quite fragile. Examine your needs carefully before you buy.

A new generation of multimeters is emerging, with graphical displays and many advanced features. At the present time they are too expensive to attract much attention as general-purpose service meters. However, it is going to be interesting in the future as the price declines and these sophisticated instruments become more popular. These meters put test equipment, superior in capabilities to a small testing laboratory of a few years ago, into the palm of one hand. It will take time for these instruments to become affordable. It will also take time for present-day technicians to learn to fully utilize the information they can provide.

FIGURE 2–5 This megohmmeter can read resistance values above 100 MΩ.

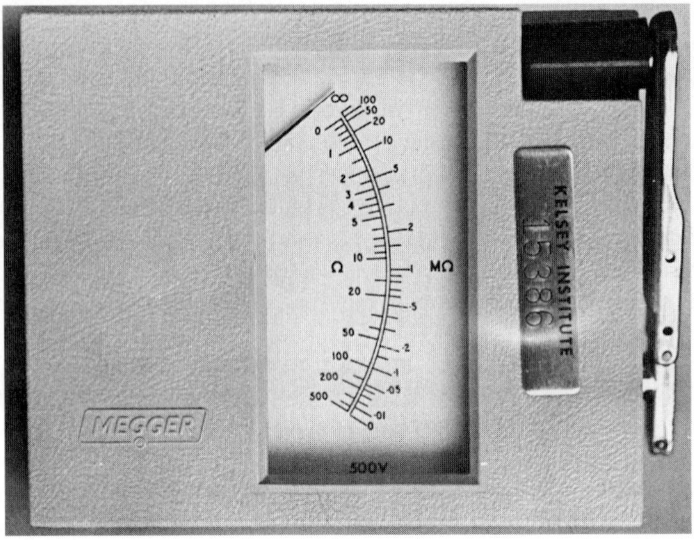

2.4 Other Electrical Test Instruments

Megohmmeters

Contrary to popular belief, electrical insulation is not perfect. It always leaks, just a little, allowing some of the energized electrons to pass through to other conductors or to ground. This leakage is very small, and normally occurs over a wide area. It can be measured with a megohmmeter (shown in Figure 2–5) to find an overall resistance value. This value is an important indicator of the condition of the insulation.

Consider, for example, the insulation of an electric motor winding. In a typical motor, the conductor used to make the coils is insulated with a thin coating of varnish. This is more than adequate for the small voltage difference between different turns of the coil, but is not enough to safely separate the conductor voltage from the grounded iron core. For this reason, the coils are further separated from the core iron by layers of heavy insulation paper, fitted into the core slots before the coils are installed. Extra insulation may also be installed between various coil groups. The exposed ends of the coils are secured to prevent movement that could wear through the insulation. After the winding is complete, it may be dipped or sprayed to saturate the exposed portions with additional varnish. This final coating is cured to a hard finish. The additional varnish further solidifies the windings to prevent movement and to protect the relatively thin conductor varnish from damage.

The finished motor winding is fairly complex and labor-intensive to install or replace. It represents a very high percentage of the total cost of the motor. Because of the way it is constructed, it is nearly impossible to locate and repair a single point of failure, so almost any insulation breakdown will result in replacement of the motor or a complete rewind. This is why adequate test procedures with appropriate test instruments are so important.

When the motor winding is tested with an ohmmeter, the reading will normally be an infinite resistance between the winding and the motor frame. This type of test is acceptable as a quick check to detect burned-out windings. However, if the motor winding insulation is degraded by an accumulation of dirt or moisture, the ohmmeter test will give it a pass mark. The ohmmeter applies only a few volts between the winding and the frame. This test voltage is far too low for the instrument to detect insulation leakage.

A megohmmeter applies a much higher test voltage, typically above the normal operating voltage of the motor. The voltage is DC, just as it is for the ohmmeter. Megohmmeters are available with several standard output voltages ranging from 100 to 5000 V or more. Some have switch-selectable voltages so they can be matched to the ratings of the equipment under test. All manufacturers of motors and similar equipment have recommendations regarding megohmmeter testing of their products.

A megohmmeter can also accurately measure much higher resistances than an ohmmeter. A very good digital ohmmeter cannot measure resistance values above 20 MΩ. Anything above that is interpreted as infinity. A typical megohmmeter scale extends to 100 MΩ or more.

When a megohmmeter is used to test a motor winding, it is connected with one lead to the windings and one to the iron frame. The instrument applies the DC test voltage between these connections, and the internal meter responds to any current flow in the circuit. The current through the meter is the total leakage from the winding through the insulation to the iron core. Because the test voltage is much higher than for an ohmmeter, the leakage current is also much higher. The meter responds to current but is scaled to indicate resistance: The higher the current, the lower the insulation resistance, and vise versa. For a healthy motor, the insulation resistance value is typically high, often into the hundreds of megohms. Lower than normal readings indicate the winding insulation is damp or otherwise contaminated. Below a certain level, which varies with the motor type and voltage ratings, the motor should not be put back into service until it has been properly cleaned and dried.

A megohmmeter test is intended to be nondestructive. Because of the relatively high voltage at which the test is done, there is always a possibility of damage. For this reason, it is important to follow the equipment manufacturer's recommendations. Used properly, a megohmmeter can provide an accurate measure of the condition of insulation, without the risks associated with energizing the motor at its normal operating voltage. For example, a megohmmeter test can be used to check motor windings that have just been cleaned, or to determine the condition of a motor that has been out of service for some time.

Motor windings that have been cleaned have to be properly dried before use. Stored motors often take on moisture and other contaminants. If the test results show a low insulation resistance, the motor can be dried out and retested. The drying process can be tracked with additional testing until an acceptable reading is obtained. The motor can then be safely put back into normal service. This way the motor can be salvaged and put to normal use. There was nothing wrong other than the moisture level of the insulation, but if it had been connected and energized without drying it out first, the motor would have been destroyed.

Megohmmeter tests are more difficult to perform and to interpret than ohmmeter tests. They are also more informative. Suitable test procedures have been part of preventive maintenance recommendations for many years, and should be a part of any preventive maintenance program. The degree to which such testing is going to be used can affect the choice of the make and model of instrument. Small megohmmeters, typically used for "go/no go" types of tests, may be powered by an internal hand-cranked generator. This is fine for short periodic tests, but the tests required for a full-blown preventive maintenance program may take many minutes to complete. It takes a seriously dedicated and physically fit service technician to keep up the required cranking speed for the duration of such a long test. Battery-operated or motor-driven megohmmeters are much more suitable for such demanding chores. They are also much easier to read than a small hand-cranked unit, which must be read while it is cranked at full speed.

Note

Megohmmeters apply a voltage high enough to damage sensitive electronic components. If the equipment has built-in electronic devices or circuit components, follow the manufacturer's recommendations. In most cases, the electronic components must be disconnected before any testing begins.

Wattmeters

Of the common electrical values, AC power (wattage) is the most elusive and the most difficult to measure. However, since operating costs are based on power, power measurement is important for comparing operating efficiencies and projecting energy costs. In a DC circuit, power can be found from voltage and current measurements. In an AC circuit, power cannot be found from voltage and current measurements alone, unless the load contains no coils or capacitors.

When the circuit load is made up exclusively of resistors, such as heating elements or incandescent lightbulbs, no special meter is necessary. Voltage and current readings, and a little calculation, are the only requirements. (Power in DC and AC resistive circuits is volts × amps). However, typical resistive loads such as lightbulbs and heating elements are usually labeled in watts, so there is rarely any need to measure their power input. Where a power measurement might be of more interest, such as finding the operating cost of a motor, it cannot be done without a wattmeter.

Because of the problems associated with measuring power, very few service technicians have access to the necessary instruments. Electrical power (watts) cannot be read with any of the common multimeters. Bench-type wattmeters are bulky, expensive, relatively fragile, and may require accessory transformers to connect them to the system under test. These complications are greatly increased when the circuit is three-phase. Both current and voltage must be read simultaneously for the meter to make a power measurement. In a three-phase circuit, for continuous measurement, two meters are required for a three-wire circuit, and three meters are required for a four-wire circuit. Each requires voltage and current connections. When the circuit is operating at more than 120 V, potential transformers may be required. When the current is above 5 A, current transformers may also have to be used. Properly connecting the instruments and transformers requires a considerable amount of knowledge and a fair amount of time. The number of connections that appear to work, but are inaccurate, greatly outnumber the few that are correct.

Convenient handheld single-phase wattmeters are available and are much easier to use than bench types. They look much like a clamp-on ammeter, and connect in a similar manner. The current reading is through the clamp-on current loop, and the voltage reading is through separate voltage leads that must also be connected to the circuit. Like other wattmeters, these meters tend to be expensive. Some are fully functional multimeters, reading all the usual quantities plus power. Power-reading multimeter accessories with similar capabilities are also available and similar in price. Some of the newer graphical multimeters include power-measurement features, in addition to a host of other capabilities.

Reading power on a three-phase circuit with a single-phase meter involves multiple tests and some elementary calculations. The exact combination of readings and calculations depends on the three-phase connection in use. For three-phase four-wire systems, the total three-phase power is the sum of the power delivered by all three phases individually. Three readings are taken with the voltage referenced to the neutral. For three-phase three-wire systems, the connections and calculations are slightly different and may vary with the set-up of the meter. Usually two tests are required, consisting of current in two phases referenced to voltage in the third. The power readings are summed as for three-phase four-wire measurements, but one of the readings may be negative.

Portable three-phase power meters can be connected to read all the necessary currents and voltages simultaneously. They are special-purpose instruments that cost thousands of dollars.

2.5 Meter Safety

Safety is a constant concern for electrical equipment installers and service personnel. Almost all electrical systems operate at voltages high enough to be dangerous. Standard voltages of 120 V or more are potentially lethal. Even experienced electricians have died from contact with 120 V AC circuits. In most cases, sustained contact with AC voltages up to about 480 V causes a loss of muscular control. A shock victim who has grasped an energized object is unable to release his grip. Death is usually due to loss of blood circulation caused by ventricular fibrillation, a random uncoordinated heart flutter that replaces normal heart function. Accidental contact with slightly higher voltages, such as 600 V, can produce violent muscular convulsions. Severe burns are also very common.

The most dangerous low-voltage shocks occur when the current passes through the chest cavity or the head. Contact from arm to arm, arm to leg, or directly to the chest or head area are all more dangerous than contact across one hand or some similar situation. The usual result of these “less dangerous” shocks is a painful reminder that the circuit is still energized. At higher voltages these minor annoyances can become severe burns and permanent damage.

The risks of electrical work do not end with the dangers of electrocution or electrical burns caused by direct contact. There are many situations where severe injury or even death can result with no actual contact between the victim and the energized parts of the circuit. Electricity at line voltages always arcs when a connection is broken. Normally this arcing is contained within the switch or other equipment. Equipment failures, mistakes, and accidental contact with metal tools or other equipment can cause explosive arcs that cannot be contained. With the right conditions, this can produce a devastating chain reaction.

An electrical arc is a conductive path. With an industrial voltage and a powerful supply, the arc, once started, does not burn itself out. Instead, it expands until enough equipment fails and the power supply is finally cut off. The overall result is an explosion of electrical energy and vaporized material that causes extensive damage and severe burns.

Warning

For the service technician, the single most important piece of safety equipment may be a handheld multimeter. One of its primary functions is to protect the user from injury or death. Any service person or equipment installer involved with electrical powered equipment must have a reliable meter, the knowledge and skill to use it safely, and the good sense to maintain it in good working order.

The following is a list of safety habits and procedures that greatly reduce the chances of an electrical accident.

- Always wear safety glasses.
- Make a habit of looking at the meter before applying the leads, to be sure the range and function selected are suitable and that the leads are plugged into the correct jacks.
- When testing voltage for safety reasons, if at all possible test a live circuit first, and then test the circuit that is supposed to be de-energized. This will pretest the meter and the meter leads before you put your life on the line. For example, when testing at a switch, check the "hot" (or line) side first, and then the load side.
- Avoid unnecessary contact with grounded surfaces. For example, when taking voltage readings, don't lean on the frame of the equipment. Stand clear, and you will be much less likely to become part of the circuit.
- Removable clip ends can be a handy accessory for maintaining contact during current or voltage measurements, or for multiple readings. A hand-to-hand shock is one of the most dangerous. The risk can be reduced by clipping one lead in place and then measuring with one hand kept away from any contact. This is particularly good for multiple readings from a neutral or ground conductor, but can also be applied to multiple-line conductors.
- Always use good quality meter leads. Poor quality leads are prone to cuts and cracks that can be deadly in wet hands.
- Check the leads regularly for damage and for increased lead resistance. An increase in lead resistance is a sure sign that some strands of the conductor inside are broken. This usually occurs at the flex points near the probes and jack connectors. Broken strands soon become broken conductors. The leads must be repaired or, preferably, replaced.
- Sharpening meter probes for piercing insulation is convenient for some types of work, **but caution is advised.** The sharp ends are dangerous to handle, and poking holes in insulation is not always such a good idea.
- If available, a protective slip case should be purchased with the meter. This provides excellent protection, because it can be left on the meter at all times. For storage in a toolbox or similar location, a full-coverage case is better because it protects both the meter and the leads.
- When the meter can be shut off with a function and range selected—there is no off position on the selector switch—make a habit of leaving it set for a moderately high AC voltage function. This will help prevent placing the meter on voltage with it set for a potentially dangerous selection (i.e., resistance or current).

One final, rather sobering, note of caution. It is often assumed the greatest danger of electrical work is electrical shock. However, around industrial voltages, you are more likely to be maimed, blinded, or burned to death by an electrical arc than to die by electrocution. The risks can be greatly reduced by good working habits.

Questions

1. What are the basic multimeter functions?
2. What unit is used to measure electrical resistance? What is its symbol?
3. What value appears on the extreme left end of an analog meter's resistance scale?
4. Analog meter resistance ranges are multipliers; digital meter ranges are maximums. Explain the difference.
5. What are the numeric values of the prefixes represented by μ, m, k, and M?
6. On a manually ranged digital ohmmeter, is the most accurate range selection the one above, or the one below, the actual resistance value?
7. What may occur if a multimeter set for resistance measurement is connected to an energized circuit?
8. What happens when an analog DC voltmeter or ammeter is connected with the wrong polarity?
9. What is the result if the same mistake is made with a digital meter?
10. When the circuit voltage is unknown, what range should be selected for the initial test?
11. Before touching the meter probes to a live circuit, the meter should be checked to ensure it is properly set up. What are the critical settings and selections?
12. List the key points of conducting safe voltage tests.
13. What circumstances can result in a zero voltage reading on a live circuit?
14. Why touch the meter to a live circuit before checking one that is supposed to be de-energized?
15. Explain why built-in fuses are an important part of ammeter safety.
16. Why do ammeters require extra caution when they are connected into a circuit?
17. Why does a clamp-on ammeter read zero when it is clamped around a multiconductor cable that is carrying current?
18. How can a clamp-on ammeter be used to measure values that are well below its lowest range?
19. If a 1 kΩ resistor is used as a current shunt, what is the current when the voltage drop across the resistor is 2.75 mV?
20. For what reasons may an megohmmeter test of a motor winding be more accurate than a similar test done with an ohmmeter?
21. What precautions should be taken into account when considering the use of a megohmmeter?
22. In what types of circuits can power be calculated from voltage and ampere readings?

Part II

Electrical Fundamentals

Chapter

The Electrical Fundamentals section contains five chapters covering the essential elements of electrical theory. With the exception of electrical power, which has been integrated into the section on motors, this section contains virtually all the electrical theory that will be presented in this book. Chapter 3, "The Nature of Electricity," explores what electricity is and where it comes from. Chapter 4, "Circuit Fundamentals," explains the variations in circuit arrangement, and circuit operation with DC sources. Chapter 5, "Alternating Current," examines the operation of circuits powered from AC supplies. Chapter 6, "Single-Phase and Three-Phase Systems," explains the difference between these two AC systems, starting at the supply generator where they are produced. The final chapter of this section, "Transformers and Services," explains the operation of transformers and their role in our electrical systems. Chapter 7 also examines the differences in various consumer service voltages and configurations, and the supply transformer connections by which they are created.

For students of electrical theory, this section and the electrical power portion of the section that follows are the critical parts of this text. For students more involved with electrical devices and applications, this section is primarily a reference. However, students from these two pursuits should note that it is best to integrate theory and practice whenever possible. Theory provides the answers to "how" and "why" electrical circuits and devices behave they way they do. Practice explores the rich variety of ways in which that knowledge has been applied.

3 The Nature of Electricity

Objectives

This chapter is a general reference on what electricity is and where it comes from. After the study of this chapter is completed, you should be able to:

- Describe static and current electricity, and explain the differences between them
- Define the term "conductor" and explain how electrical current can flow through solid materials
- Define the term "insulator" and explain why electrical insulators do not conduct electricity
- List the primary types of electrical sources and provide examples of each
- Explain the relationship between electricity and magnetism
- Describe how various sources of energy are used by electrical utilities to generate electricity
- Explain the limitations of sources and how they can affect circuit operation

3.1 Introduction

Electricity is a form of energy so commonplace in industrialized nations that a conscious effort is required to avoid it. However, even though the average person uses many different electrically operated devices every day, only a curious few understand how they operate.

There are good reasons for this general lack of understanding. Although the results of electrical activity are usually obvious, electrical current cannot be directly observed. Electricity is generally undetectable without the risk of a painful shock, unless a measuring instrument is used. And even with the appropriate instruments, the marvels of electricity remain hidden within the solid materials of its circuits. It should be no surprise that it remains a mystery even to those who grow up surrounded by electrical devices and equipment.

FIGURE 3–1 The orbiting electrons carry a negative charge. This makes them repel each other, but causes a mutual attraction between them and the positive protons in the nucleus.

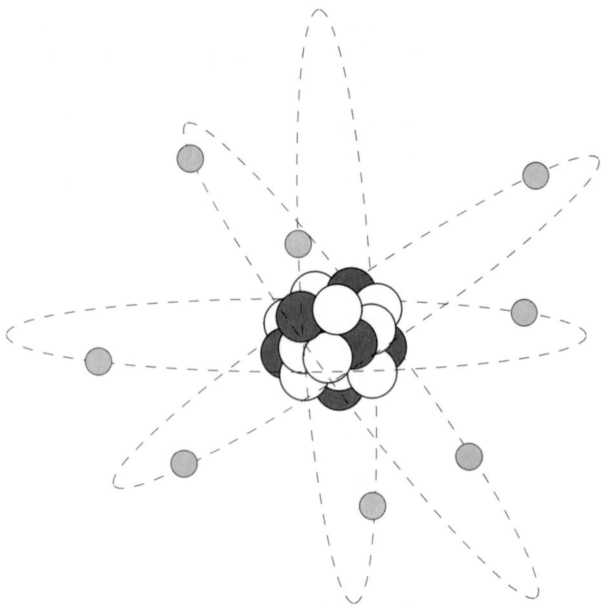

3.2 Electrons and Atomic Structure

Electrical energy is dependent on the characteristics and behavior of a minute atomic particle called the electron. In the simple terms of grade school science, atoms are the smallest particles of an element that retain the properties of that element, and electrons are the smallest particles of the atom. This definition is, of course, both incomplete and inaccurate, but it is adequate for the purpose of this text.

An atom consists of a nucleus containing one or more protons and, sometimes, one or more neutrons as well. Orbiting at high speed around the nucleus are the electrons, equal in number to the protons within. The nucleus is held together by the powerful natural force responsible for nuclear energy. The electron is kept in orbit around the nucleus by the attraction and repulsion of electrical charges. The protons have positive charges, the electrons negative charges. These two different charges are mutually attractive. This attraction holds the electrons in orbit around the nucleus similar to the way the moon is trapped in its orbit by the mutual gravity between it and the Earth.

Opposite charges attract, but like charges repel. Electrons repel electrons, and protons repel protons. For the orbiting electrons, this repelling force acts to hold the electrons separate from one another. In the nucleus, the repulsion between the protons is overcome by the powerful nuclear forces that bind the nucleus together (see Figure 3–1).

3.3 Static Electricity

A static electrical charge is an excess of electrons or protons on an insulated body. A wide variety of situations can transfer electrons from one material to another, resulting in a charge difference. Most of them occur when there is a mechanical interaction between certain types of materials. Rubbing a glass rod with silk or fur, or sliding certain types of shoe soles across a nylon carpet, are common examples. In each case electrons are stripped away from one material and deposited on the other. Where the electrons are removed, the net charge will be positive, because the number of protons is greater than the number of electrons. Where the electrons are deposited, the net charge will be negative. For a charge to accumulate, the two bodies of material must be insulated from one another.

When conditions are right, and charges do accumulate, the charged objects will follow much the same rules of behavior as the individual charges they carry. Objects with unlike charges attract one another; those carrying like charges repel. When two objects with different charges are brought close together, the objects themselves will initially be pulled toward each other. However, the charge difference between them can only be maintained while the objects are electrically insulated from one another. When the objects get close enough together, the high-voltage difference created by the two charges will break down the air molecules, and the excess electrons of the negatively charged object will jump across the gap. The air that once kept the charges insulated from one another suddenly becomes ionized and conductive, and the excess electrons rush across the space. The accumulated charges equalize in an instant, restoring the balance between the number of protons and electrons.

The energy stored by the charges is discharged, and the result is often a visible spark of static electricity. The size of the spark and the energy it carries depend on the total accumulated charge. A harmless, though annoying, static shock has the same root cause as a lightning stroke. The driving force is an accumulated charge difference; the particles in motion are the electrons and the charged ions of the air through which the arc passes. The difference is in the magnitude of the charge. The energy carried by the arc dissipates in the air and on the target where the charges are equalized.

It is interesting to note that the identification of these charges is a cause for some confusion. The electron is the mobile particle, and an accumulation of electrons is the usual cause of a charge buildup. During a static discharge, the electrons leap across the gap from the negatively charged body to the positive one. This is backward to normal logic. The negative identifier for the electrons implies that less moves toward more—negative moves toward positive, instead of the more logical positive to negative.

This confusion extends into the description of electrical circuits in general. The actual direction of electron flow in a circuit is not under debate—it is from negative to positive, just as in a static discharge. Unfortunately, many electrical conventions and a lot of the symbols used in electronics are based on the assumption that flow is positive to negative—backward to reality. These opposing views have caused so much debate that textbooks are sometimes published in two different versions, just to satisfy the opposing camps. Over the years I have encountered two different explanations of how this came about. One put the blame on Ben Franklin; the other on another experimenter named Alessandro Volta. Regardless of where fiction ends and fact begins, both stories are interesting and at least plausible. They also can help to make some sense of it all.

The story about Franklin is that he identified the charges as positive and negative when he recorded the details of his experiments with static electricity. His intent was that positive would describe a quantity of charge that would move naturally toward a deficiency—positive moving to negative, more moving to balance out less. Unfortunately, he made a bad guess about which way the static sparks jumped, and it took about a hundred years to prove him wrong. Due to this mistake, the electron was dubbed negative, the proton positive.

Volta's deductions were drawn from experiments with current electricity, not static. Volta did a lot of pioneering work with cells and batteries. However, one of the fundamental difficulties confronting him was a lack of workable instruments. Electricity was little understood, and metering equipment was almost nonexistent. Volta needed some way to measure and analyze the electrical phenomena he was discovering. One of his solutions was to experiment with his own senses, something he recorded in considerable detail. Among other things, he concluded a direction of flow, based on a sense of taste related to current through the tip of his tongue. This sounds quite bizarre, but reading his letters and attempting to imagine his situation makes it much easier to appreciate his efforts.

Regardless of the historical reasons for this confusion, every student of electricity must cope with the results. In the end, it is easy to forgive the mistakes of earlier explorers equipped with little more than an insatiable curiosity.

Electricity moves from negative to positive. Unfortunately, most equipment is built as if it goes the other way, and most drawings—and the drawing symbols—are also made the "wrong" way around. The "right" way is called electron flow; the "wrong" way is called conventional current flow. It is usually much easier to deal with both ways than to join one camp or the other. It may also help to realize that in many circuits the actual direction of flow is irrelevant: The circuit will operate the same with either polarity of connection.

3.4 Current Electricity

The term "static electricity" is associated with the sudden release of a built-up charge. This behavior makes it difficult to harness static electricity for any useful purposes. Although there is a very close relationship between static and current electricity, the two are usually treated separately for purposes of study.

Current electricity is the orderly flow of electrons through a conductive material. Like static electricity, a potential difference must exist for the electron movement to occur. Unlike static electricity, the electron flow is routed by the circuit conductors to a load device, where the energy is put to work. Figure 3–2 compares static and current electricity.

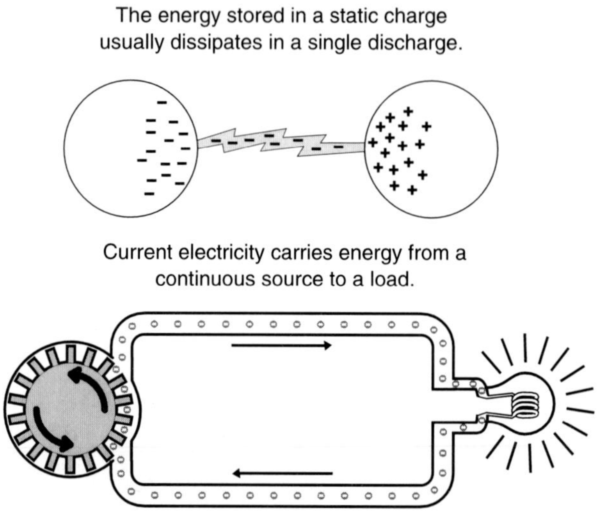

FIGURE 3–2 Although there is no such thing as the "electron pump" shown in this drawing, it illustrates how the source must constantly force the electrons through the circuit.

In such a circuit, electrical energy passes from the supply to the load, where the energy is converted to another form and released in an orderly, controlled manner. A sudden discharge of static electricity dissipates all the stored energy in a single stroke. The flow of energy associated with current electricity can continue indefinitely: As long as a potential difference and a conductive path are maintained, the electrons and the potential energy they carry continue to flow through the circuit. The energy the electrons carry is dissipated in the load, where it is converted to some other energy form. The electrons are not used up by the load. They continue on their way around the circuit—back to the source from which they came.

The terms "charge" and "potential difference" mean much the same thing. An electrical charge exists on an object when the number of protons and electrons does not match. A charge difference exists between two objects when one has a net negative charge and the other a net positive charge. Bringing these objects in contact, either directly or by connecting them with a conductor, results in a flow of electrons from the negatively charged body to the positively charged one. The difference in charges is called a "potential difference" because it is a form of potential energy.

Potential difference is the difference in electrical charge: voltage is the unit of measure. However, in common usage, the terms "potential difference" and "voltage" are essentially interchangeable. Both terms are also as applicable to static electricity as they are to current electricity. It does not matter whether a potential difference is a result of static charge or produced by some other means—the end result is much the same.

3.5 Conductors and Insulators

In most elements, the structure of the atom is symmetrical, so the electron-proton attraction stops the electrons from participating in any activity that would distance them from the parent atom. Materials like this are electrical *insulators*; current cannot flow through the material because the electrons cannot move away from their normal orbits. In a few elements, primarily metals, the structure is such that a few outer electrons are able to move from atom to atom with relative ease. These electrons are in constant motion, not only around their parent atoms, but also drifting freely from one atom to another. Under normal circumstances, this motion is completely random. When the energy level of the material rises, the rate of this random movement increases. If this energy is directionless—heat energy, for example—the movement is still random. However, when the energy has direction—such as an electrical potential difference—the electrons can move much like fluid in a pipe.

Materials that have these properties are called electrical *conductors*. It should be noted here that anytime an electrical charge is able to move, it can carry electrical energy and be classed as an electrical current. In solid materials the moving charge is the electron, but in fluids—either gases or liquids—it can be ions with either a positive or a negative charge. It does not matter what carries the electrical charge: The end result is electrical current flow and a transfer of energy.

Some fluids are good conductors; others are insulators. Many become conductive only when they contain dissolved salts or other chemicals that produce ions. Water, for example, is an excellent insulator if it is pure, but even minute amounts of salt or other additives make it conductive.

Conductors

Of all the common conductive materials, silver is the best and copper second. Gold, lead, aluminum, steel, and others are quite good. Excellent conductivity, attractive physical characteristics, and reasonable cost make copper and aluminum the two most widely used conductors. Copper is generally preferred for small-diameter conductors because it is stronger and harder than an aluminum wire of similar size. This makes it easier to install and connect properly. Aluminum is generally preferred for overhead lines, due to its light weight and lower cost. It is combined with a high-strength steel core to allow long spans between support structures. Aluminum is also very successfully applied in larger insulated conductors. It is not as good a conductor as copper. To compensate for this slightly higher resistance, aluminum conductors are typically one trade size larger than would be used if the conductor was of copper.

Silver is only rarely used as a conductor; it is too expensive for general use. However, silver is used extensively in alloys plated onto switch contacts, and is also used in some types of fuses. Gold wire is used in miniature electronic components because it can be drawn to extremely small sizes. These hairlike wires can be connected efficiently, under a microscope, to provide highly reliable interconnections between the microscopic electronic components and the connecting pins on the case of a device. Gold plating is used on contact surfaces of high-quality electronic components. It ensures long-term reliability at connection points because gold is not only an excellent conductor, but is also very resistant to oxidation.

Insulators

A conductive path is essential to a flow of electrical current; however, given a chance, the current will not only flow in the conductors provided, but also in every other conductive path that is available. For this reason, the intended path must be the only one to which the current has access. The electricity must be confined to this path by a barrier of insulating material. The insulation of one conductor from another can take several forms. For example, the conductors can be separated by suspending them in the air, which is a very good insulator. The conductors must be supported on insulators, but they do not have to be covered in insulation where they are out of reach of, and out of contact with, other materials.

When supporting the conductors in the air is not practical, such as for wire and cable installed in buildings, or buried in the ground, they must be covered with insulation. Most nonmetallic materials are insulators, but some are much more suited to this task than others. In modern electrical equipment, the insulating materials most commonly used include electrical varnishes, a variety of plastics and rubber compounds, paper that has been varnished or saturated with insulating oil, glass, ceramic, and epoxy. High-voltage systems also make wide use of insulating oils, sulfur hexafluoride (SF_6) gas, and even vacuum.

3.6 Sources of Electricity

Electrical devices require a continuous supply of electrical energy. Unfortunately, electricity cannot be found in nature in a readily usable form: It must be generated by some device or process that converts some other form of energy into electrical energy. A variety of ways have been found to do this, all of which are used where they best fit the situation. Because electricity is so much in demand, research continues even today in an attempt to find better energy converters and generators, and more efficient ways of putting them to work. Figure 3–3 illustrates five common sources of electricity.

FIGURE 3–3 These five examples represent the five common sources of electrical energy.

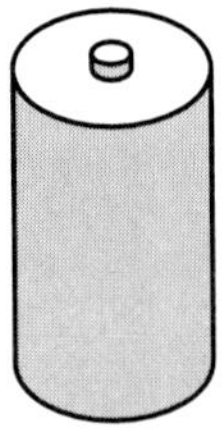

A cell or battery generates electricity through a chemical reaction inside the cell.

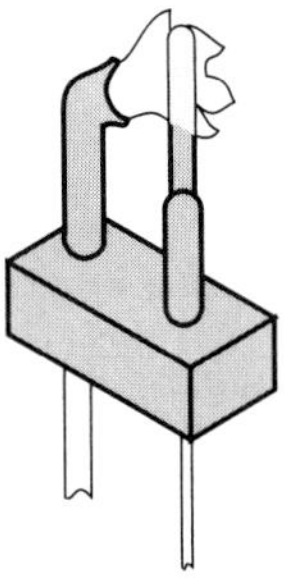

A thermocouple, such as the one in a furnace pilot light, generates electricity directly from heat.

A solar or photovoltaic cell generates electricity directly from light.

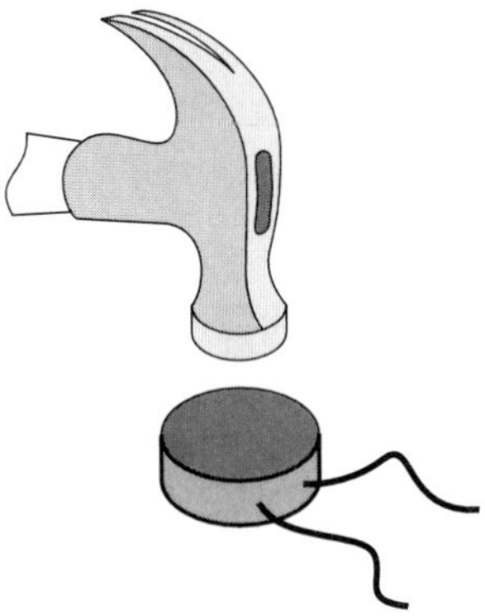

A piezo crystal generates electricity from mechanical stress.

Electromagnetic generators produce electricity from the mechanical energy provided by the driving engine or turbine.

Direct Conversions

The most convenient sources of electrical energy are the product of direct conversions. The other energy form—typically heat, light, or mechanical force—is changed into electrical energy without any electromechanical apparatus. Examples of direct conversions include some chemical reactions, a heat-operated device called a thermocouple, photovoltaic cells—commonly called solar cells—and piezo crystals, which generate electrical energy directly from mechanical stress.

Chemical Energy to Electrical Energy. All chemical reactions result in a change in the way atoms and molecules are combined. The chemical makeup of the materials involved is changed by the reaction, and in some cases an electrical charge or potential difference is also produced. When a chemical reaction produces opposite electrical charges at two physically separate points, the ingredients for an electrical cell, or battery, have been created. Note that the two terms—"cell" and "battery"—are often used interchangeably, but they are not necessarily the same thing. A cell is a single unit of chemically reactive material and its associated container. A battery is a group of interconnected cells, usually combined into a single case. Inside the battery, the cells are separated from each other, either by divisions in the larger case—such as in automobile batteries—or by their individual containers and, if necessary, by separators made of insulating material. The common AAA, AA, C, and D size "batteries" are really individual cells. The 6 V and 9 V versions contain multiple cells and are true batteries.

In practical cells and batteries, the chemical reaction not only produces an electrical potential—voltage—but is also controlled by that potential. The reaction continues until a specific voltage is reached, and then it stops. When an electrical circuit is connected to the cell, current begins to flow, the voltage drops slightly, and the reaction immediately restarts. This behavior is very important: The cell or battery can supply power only until the reacting chemicals are exhausted. Because the reaction stops each time the power flow stops, the chemical materials are preserved when the electrical device sits idle. This is why batteries can be left unused for long periods of time, yet provide power instantly when it is needed. There is a limit to this shelf life, because a small amount of chemical activity is inevitable, but modern dry cells can sit for many years with very little loss.

Rechargeable cells are based on reversible chemical reactions. When energy is supplied to the cell instead of drawn from it, the chemical reaction is reversed. The reactive materials inside the cell are actually returned to their original state. This cycle of charge and discharge can be repeated many times before the cell finally fails. The most common rechargeable devices are lead acid batteries used in automotive-type applications, and nickel cadmium (nicad) cells used as replacements for single-use alkaline dry cells, or in battery packs for cordless equipment such as tools, radios, and telephones. There are several other less common battery technologies available and in use.

Fuel cells are essentially batteries that can be supplied continuously with a chemical fuel. As long as the required fuel is available, energy can be obtained from the cell. Its useful life as an energy source is not dependent on the amount of chemical that can be packaged inside the cell. Fuel cell technologies have been used for years, but have not been able to compete economically with other energy systems, except in very specialized applications.

Thermal Energy to Electrical Energy. Thermal energy from various sources is used to create steam for large turbine-driven generators (see the information on electromagnetic generators that follows); however, direct conversion from heat to electricity is also possible and much simpler. Unfortunately, the output is very small. The device used is called a thermocouple.

A thermocouple is made from two dissimilar metals bonded together at two separate junction points. When there is a difference in the temperature at these two junctions, the result is a potential difference between them. This voltage can be used to supply electrical energy to a device wired into the circuit. The amount of energy produced is small, but the thermocouple is simple, reliable, and inexpensive. It is well suited to remote temperature measurement because the voltage produced is proportional to the temperature difference across the thermocouple circuit.

Thermocouples are also a critical part of the safety shutoff system used to monitor the pilot flame in gas-fired burners, such as furnaces and domestic hot water heaters. The thermocouple is mounted so that it is constantly heated by the pilot flame. The current from the thermocouple is used to hold open a safety valve in the main gas supply. If the pilot light goes out for any reason, the thermocouple quickly cools down, the current stops, and the spring-operated safety valve closes.

Multiple thermocouples are sometimes combined into a device called a thermopile. Thermopiles behave the same as a single thermocouple, but the voltage output is higher. Enough energy can be taken from a thermopile to operate a specially designed gas burner control. It has the advantage of being able to operate safely with no other power source. This allows the burner to be electrically controlled—by a wall-mounted thermostat, for example—without the need for an external source of electricity.

Light Energy to Electrical Energy. A photovoltaic cell, commonly called a solar cell, contains a material that converts light energy directly into electrical energy. These cells are relatively expensive and not very efficient; however, they have the tremendous advantage of having no fuel cost. They are well suited to applications where only very small quantities of energy are required, such as powering electronic calculators, or where the energy demands are modest and the situation makes it difficult, or undesirable, to supply the electricity in other ways.

The initial costs are high, but solar cells can be a reliable, cost-effective source of electrical energy for isolated installations. They are usually combined with a high-quality battery pack so power is available 24 hours a day. Steady improvements in efficiency and reductions in cost are making solar cells increasingly attractive.

Mechanical Stress to Electrical Energy. Piezo crystals are materials that produce electrical energy when they are mechanically stressed. This is the source of energy for spark igniters on barbecues and in many "electronic" cigarette lighters. The piezo crystal does not produce continuous power, just a brief burst of high voltage. However, the materials and mechanism are simple and inexpensive to manufacture. Piezo crystals are also used in low-cost microphones and similar electronic applications where a mechanical vibration must be converted into an electrical signal.

The action of the crystal is also reversible. When a voltage signal is applied, it vibrates in response to changes in the signal. This technique is used in some high-frequency loudspeakers.

Electromagnetic Generators

Modern industrial societies have a staggering appetite for electrical energy. To date, no direct conversion device exists that can produce electricity economically on such a large scale. Therefore, virtually all commercial-scale electrical energy comes from electromechanical generators. There are great differences among the appearances of various types of generating stations, but this has little to do with the generators themselves. The real differences are in the energy source and the prime movers—the engines and turbines—used to drive them. Regardless of their size or appearance, all generators work the same way.

Electromagnetism and Generation. Magnetism and electricity are very much interrelated. When electrical current flows, one result is a magnetizing force around the conductor. Likewise, when a conductor is placed within a changing magnetic field, an electrical force is generated that affects the electrons in the conductor. This force appears as a voltage across the conductor ends. These two relationships are exploited in electromagnets, generators, transformers, motors, and a diverse array of electronic devices. Electromagnetism is fundamental to the operation of much of the electrical equipment on which modern societies depend.

Generators Used by Electrical Utilities. Among the largest machines in the world, utility generators not only produce enormous quantities of electrical energy, they require even more energy to drive them. For example, a 300 MW (300,000,000 watt) generator—a very modest machine by present standards—requires an input power of over 400,000 Hp when it is operating at full load. To provide so much mechanical power requires convertible energy on an enormous scale. Typical sources are massive hydro dams, huge steam boilers, or, for smaller units, giant piston engines or gas turbines.

Hydro Power. Hydroelectric power generation is the most effective way to produce electrical energy. It is also the most attractive because the resource is renewable and pollution free. Unfortunately, an effective hydro plant requires a massive volume of water at the highest possible head. ("Head" is the height difference between the water level above and below the dam. The higher the head, the more energy that can be extracted from the water that flows through the turbine.) It is also greatly preferred to have the largest possible water storage area. This allows the utility greater flexibility in scheduling generation to match the daily and seasonal fluctuations in electrical demand. Smaller reservoirs on reliable rivers can provide a lot of energy, but are more difficult to use effectively because the power generation must closely match the rate at which the river flows.

Few places in the world are blessed with acceptable hydro sites that have not already been developed. The environmental impact of such large dams and reservoirs can cause irreversible damage. Increased awareness of these problems has made new projects very difficult to justify. For electrical utilities fortunate enough to have hydro stations, they are invaluable, low-maintenance workhorses. Unlike steam turbines, they can sit idle without consuming any of their energy source and be up and running in a matter of minutes.

Steam Turbines. The steam turbine is one of the most effective methods available to convert heat into electrical energy. Unfortunately, the power output needed means an enormous volume of steam at very high temperatures. These requirements usually dictate that the heat source be either a nuclear reactor or a massive power boiler with a shocking appetite for fossil fuel. Using the 300 MW example discussed above, a coal-fired boiler for a turbine of this size consumes several thousand tons of lignite coal per day.

One drawback of steam turbines is that it is virtually impossible to recover all the heat energy as mechanical power. A sizable portion must be lost in condensing the steam back to water. The overall efficiency of a typical boiler and steam turbine generating plant is less than 50%. Fossil-fuel plants are expensive to operate and maintain. Fuel consumption is enormous, the fuel is nonrenewable, and the problems of pollution and pollution control go on without end. However, it is socially unthinkable to turn off the power, and it has to come from somewhere.

Despite its limitations, for high-output horsepower at sustained load there are few machines, other than a water turbine, that can rival the steam turbine. Theoretically, any heat source can produce a steam supply. In practice, because of the extremely high demands, the most common heat sources are coal and nuclear reactors. Less common sources include gas, fuel oil, wood, incinerated garbage, geothermal heat, and solar collectors.

Steam turbines are not suited to intermittent operation. When load scheduling will allow it, steam turbines may run continuously, often at full load. They are only shut down for scheduled maintenance or emergency repairs. When they are taken off line but may be needed on short notice, they are kept on hot standby, with the boiler up to temperature and the machine turning slowly on an auxiliary drive. When they must be shut down completely, such as for a major mechanical problem or a boiler leak, it can take days to bring the boiler up to temperature and get the machine back on line.

Internal Combustion Engines. In some cases internal combustion engines burning diesel fuel, natural gas, or methane are also used to drive generators, but they are usually small in comparison with steam turbines. A stationary diesel plant is rarely more than a few megawatts. Piston engines of that size are impressive machines that dwarf the generators they drive.

Gas Turbines. Where natural gas is the fuel available, gas turbines can be used to drive the generator directly, eliminating the need for a boiler. Gas turbines cannot be built to match a large steam turbine, but they have an important advantage as reserve units because they can be brought into operation much faster. The gas turbine doesn't require the long heat-up time that a boiler does, so it can be totally shut down for idle periods.

Summary. Electrical utilities maintain massive generating stations and power distribution systems. These are generally taken for granted, and would be forgotten by many if there were no utility bills. It would be impossible to maintain anything close to our present standard of living without that constant flow of energy that our electrified lifestyles demand.

3.7 All Sources Have Limits

A brief tour of electrical sources easily becomes diverted toward the operation of electrical utilities. This is inevitable, given the fact that they supply almost all the electrical energy in everyday use. Considering how dependent our modern societies are on the continued supply of electricity, a little appreciation for the magnitude of this situation is probably a good thing. However, the focus of this discussion is on sources of electrical power, and how they relate to the operation of electrically powered equipment.

All electrical equipment is totally dependent on the quality and continuity of the power supply that provides the energy it consumes. When equipment malfunctions, the supply should not be taken for granted. When a piece of electrical equipment quits operating, the first check should be to confirm the power source. If that checks out and the equipment restarts, the supply should be retested with the equipment running. This is particularly important for less regulated sources such as batteries, solar cells, and thermocouples. Utility power is usually present at full rating or not at all. There are exceptions to this, but these are usually short-lived disturbances that may disappear before any testing can be done.

The output of a battery, solar cell, thermocouple, or other such device is affected by its condition, the condition of the circuit wiring and connections, the amount of load on the device, and the input energy to the conversion process. A solar cell, for example, provides proper output voltage only if all the following conditions are met. The cell itself must be in good working order. There must be no high resistances in the circuit connections or at switch contact points. The cell must not be overloaded—trying to draw too much current will cause the voltage to sag, and a short circuit can drop it to nothing. Last but not least, the cell's source of energy must be present—it must be receiving adequate light.

These types of considerations apply to many small power sources. All of the relevant conditions must be met or the circuit will not operate correctly. The supply may test perfectly with no load connected, and fail due to low voltage when the load is connected and operating. Battery voltage will drop off when the chemicals are nearly exhausted, but they will test close to normal with no load attached.

The output voltage of a thermocouple or thermopile varies with temperature and with the amount of electrical load. Circuits operated from such limited supplies are sensitive to every kind of problem. The normal operating voltage is so low that only a very slight reduction can cause the circuit to fail. For example, in standing pilot furnaces where the thermocouple keeps the safety valve open, only a small change in thermocouple output will shut down the furnace. The cause may be a loss of heat due to a weak or wavering pilot flame. It can also be a poor connection at the valve, or a defective thermocouple. As with most electrical supplies, the only reliable test is one that is done under load. A small testing accessory is available that allows the thermocouple output current to be checked while it is in operation. This is a much more conclusive than any other type of test.

Thermopile circuits in furnaces are generally more complex than thermocouples, because they involve external controls and a more elaborate gas valve. These systems are also subject to all kinds of problems because of the low operating voltage. Measuring the current with the thermopile connected to its normal load is the only conclusive test of the circuit and the condition of the thermopile. Weak pilots, bad connections, or extra resistance anywhere in the circuit can cause problems.

Supply problems are not always caused by the power source. They also may originate in the circuits that carry the current from the supply panel to the equipment that has failed. Faulty connections, long wiring runs that are overloaded, loss of one fuse on a three-phase system, bad contacts in motor control equipment, and transformer failures are just some of the many supply-related difficulties that can result in equipment shutdown.

Such failures can cause reduced voltages at the load, which do not occur unless it is drawing high current. A voltage test with the equipment shut off may show relatively

FIGURE 3–4 Overloads are one cause of low-voltage output from weak sources. A short circuit between the conductors is an extreme overload that drops the voltage of this cell nearly to zero. This also discharges the cell very rapidly.

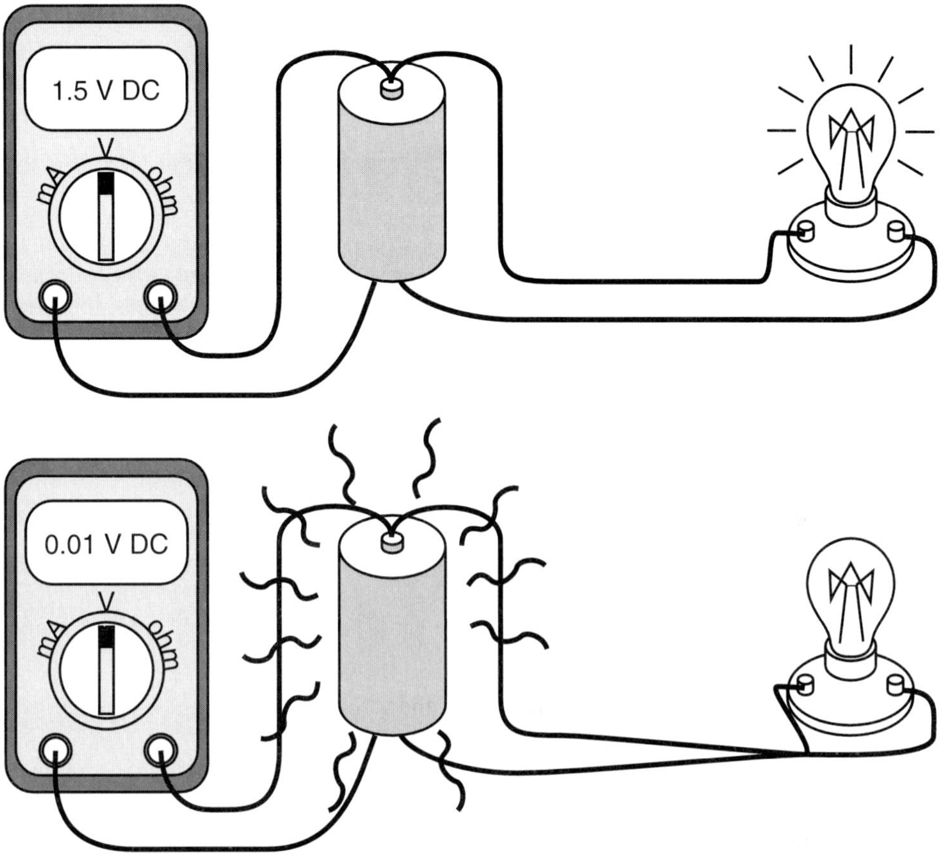

normal levels. However, if the test is repeated with the equipment in full operation, the voltage at that end of the circuit will sag as the load current rises.

Another power source problem that is easily overlooked is multiple sources in the same piece of equipment. A good example is electronic thermostats that fail when the batteries are dead. This simple problem is easily missed, because some thermostats operate without battery power—the batteries are only there to save the program from loss when the main power supply is interrupted. Other thermostats fail totally when the batteries go dead. A test at the wiring terminals of the thermostat shows normal voltages, but the thermostat will not function until the batteries are replaced.

When troubleshooting any electrical system, always confirm the source first before checking further. If possible, check with the load connected; this is particularly important when the source is a relatively weak one, such as battery, thermocouple, or solar cell (see Figure 3–4).

Questions

1. Define "electron" and "proton" in general terms, and explain the relationship between them.
2. What is a static electrical charge?
3. What is the cause of a static shock or spark?
4. What is the difference between "electron flow" and "conventional current flow," and which best represents reality?
5. Define current electricity.
6. Explain the difference between the terms "potential difference" and "voltage."
7. Explain why some materials are conductors and some are insulators.
8. Why are copper and aluminum the most widely used electrical conductors?
9. List the five common energy conversions used to produce electricity, and give an example of each.
10. Explain why the chemical reaction in a battery stops when the battery is not supplying any current.
11. In what way is the chemical reaction of a rechargeable cell different from the reaction in a single-use cell?
12. Describe two common uses for thermocouples.
13. What is the relationship between magnetism and electricity?
14. Whenever an electrical device fails to operate properly, or will not operate at all, what should be checked first?
15. Why is it best to test low-powered sources while they are supplying power to a load?

4 Circuit Fundamentals

Objectives

This chapter examines how electrical circuits are constructed and how they operate with a DC power supply. Detailed descriptions of multimeter use explain testing and troubleshooting procedures. The fundamental units of electricity are defined, and circuit calculations are introduced. (Much of what is learned in this chapter also applies to AC circuits, but no assumptions should be made about AC operation until after a thorough examination of Chapter 5, "Alternating Current.") When the study of this chapter has been completed, you should be able to:

- Identify the elements of a practical circuit, and explain the role of each in circuit operation
- Explain the difference between power and control circuits
- Construct a practical circuit
- Name and define the units of resistance, potential difference (voltage), and electrical current
- Demonstrate the proper use of a multimeter to measure resistance, voltage, and current
- Calculate electrical quantities using Ohm's Law
- Identify series circuit connections and calculate unknown values
- Identify parallel circuit connections and calculate unknown values
- Simplify combination circuits and perform and calculate unknown values
- Identify various types of resistors
- Correctly interpret resistor color codes
- Explain the purpose of overcurrent protection
- Define the terms "time delay," "current limiting," and "interrupting capacity" as they apply to fuses
- Explain the difference between thermal and magnetic trip elements in circuit breakers

4.1 Elements of a Circuit

All practical circuits, no matter how simple or complex, must have four basic components (see Figure 4–1):

1. A *source* of electrical energy;
2. *Conductors* to carry the electrical energy around the circuit;
3. A *control device* to allow the circuit to be regulated, or turned on and off; and
4. A *load* to convert the electrical energy into some other energy form and put it to some useful purpose.

The Source

The power source supplies electrical energy to the circuit. It may be attached directly, such as in many battery- or solar cell-operated devices, or it may be as remote as the distant generators of a power utility. The utility system is a vast, elaborate network of interconnecting conductors, transformers, and control equipment. Although the distances may be great and the system complex, all the energy delivered to every outlet in every building originates at the generators.

Every electrical circuit must have a source of electrical energy, and all electrical energy is obtained through conversion of some other energy form. While this statement is very true, it can lead to confusion in the use of terms. For convenience, the point of closest connection is often referred to as the "power source," or "power supply," for a circuit. This may be many different things, but common examples include a battery, a 120 V receptacle, or a DC power supply that is powered from a 120 V AC outlet. It could also be a transformer that receives power from a utility connection but changes the voltage to some different value.

Of all these examples, only the battery is truly a power source. All the rest are points of connection to the power utility and, ultimately, its generators. The DC power supply and the transformer both receive energy from the utility. The DC supply changes the voltage and converts it from AC to DC. The transformer changes the voltage, but leaves it as an AC output.

As important as it is to recognize the original source, it is foolish to let minor distinctions in terminology interfere with the normal flow of ideas. The points to be understood are, first, that each circuit must have a source of voltage and current to power its load devices. This is generally referred to as the power source or supply for that circuit,

FIGURE 4–1 The four basic elements of a circuit.

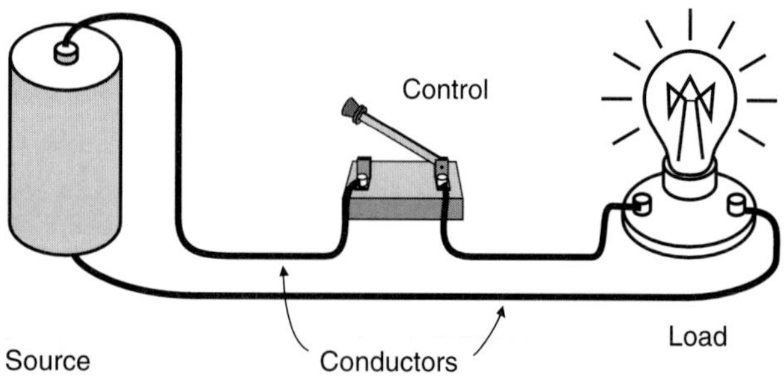

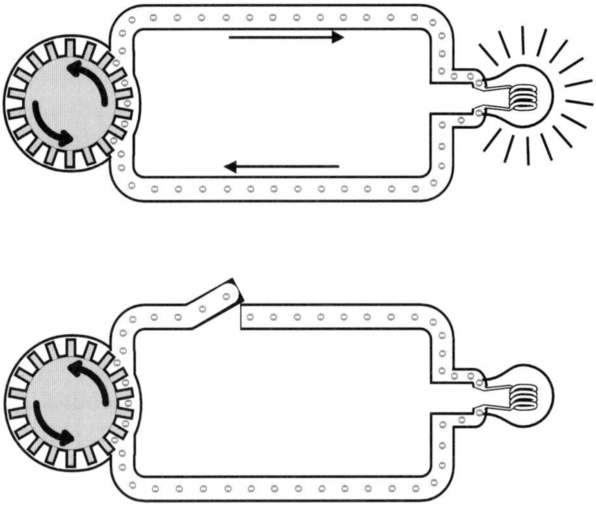

FIGURE 4–2 The energy of the source is carried to the load by the electrons that fill the conductive circuit. The same electrons are constantly recycled, gaining energy at the supply and giving it up at the load. If the conductive path is broken at any point, the source still supplies pressure, but all flow stops.

regardless of the real source of electrical energy. Second, because electricity does not exist in nature in a usable form, it must be created by converting some other type of energy. This point of conversion is ultimately the true source of electrical energy for the circuit, even though a remote source, such as a power utility generator, is rarely given credit for supplying a circuit immediately at hand.

The Conductors

One of the great strengths of electricity is its ability to move through solid conductors from the power source to the driven loads. Conductors usually take the form of round wire or cable covered in insulation, which may be used individually or in multiconductor cables. Other examples of conductors include uninsulated overhead wires, solid metal pipes and bars, thin metal coatings on printed circuit boards, and the metal frame of an automobile, which is used as a common return.

No matter what forms of conductors are used, all electrical circuits must have a complete conductive path from one side of the supply to the load, and from the load back to the other side of the supply. Electrical current is not a fuel—it is an energy carrier. It is not poured into the load—it is forced to pass through it, giving up energy in the process. For this to occur, the load must connect to the supply with two conductors, not one. The current flows from the supply to the load, and back, always in a complete loop. If either side of the conductive loop is broken at any point, the current flow stops, and the load is shut off (see Figure 4–2).

The Control Device

Although not absolutely required for circuit operation, for practical reasons almost all circuits include some kind of control device. The simplest is an on/off switch, but a variety of more complex devices are also used. The control may be manually operated or automatic, depending on the application. For example, a room light is usually operated by a wall switch, while a heating or air conditioning system is normally controlled by a thermostat. Both the switch and thermostat provide on/off control, but the lamp must be manually controlled by a person in the room, while the thermostat turns the heating or air conditioning system on and off automatically, maintaining the room temperature without constant attention.

A simple switch is an all-or-nothing type of control. The circuit is either fully on or it is off. A switch cannot regulate the flow at any in-between value. In contrast, there are other types of devices that can provide control over a range of different values. Common examples include volume controls on radios and stereos, and electronic controls such as lamp dimmers and variable-speed trigger controls on power tools. These devices can vary the voltage applied to the load they supply, giving the user full control of the volume of the sound output, the brightness of the light fixture, or the output power and speed of the tool.

The Load

The load device is usually the reason for the circuit, and all circuits include at least one load. Common examples include heating elements, lightbulbs, and motors. In each case the load device converts electrical energy into some other energy form. The heating element changes electrical energy to heat, the lightbulb produces light (and heat), and the motor changes the energy into mechanical rotation and torque.

The load is usually the reason the circuit exists. There would be no point to the circuit if there were no lightbulb, heater, motor, stereo receiver, etc., for it to power. Of course many loads are made up of lots of additional electrical or electronic components, and circuits within circuits, all containing control devices and loads of their own.

Power or Control? Circuits With Different Purposes

Most electrical circuits fall into two general categories, based on the type and purpose of loads they operate. Power circuits drive loads such as lighting, heating elements, and motors, where the electrical power is used primarily for the energy it provides. These circuits tend to have heavy conductors and high operating voltages and currents. They also do the most work and, not surprisingly, cost the most to operate. By comparison, control circuits use electricity mainly for the flexibility it provides to a control situation. Control circuits have loads, but they are usually low-power devices such as valves, relay coils, and indication lights. The control circuit typically operates at a lower voltage and much lower current than its associated power circuit.

Many types of equipment contain circuits of both kinds. The power circuit carries the bulk energy that is needed to drive the equipment. The control circuit extends and enhances the controllability of the system (see Figure 4–3). Simple on/off control is replaced with some degree of automation, operation from remote locations, or features such as indicating lights to keep the operator informed about how the equipment is performing. All of the control functions are handled by the control circuit.

FIGURE 4–3 This motor installation has a power circuit that provides electrical energy to the motor, and a control circuit that connects the operator's control station to the motor control device.

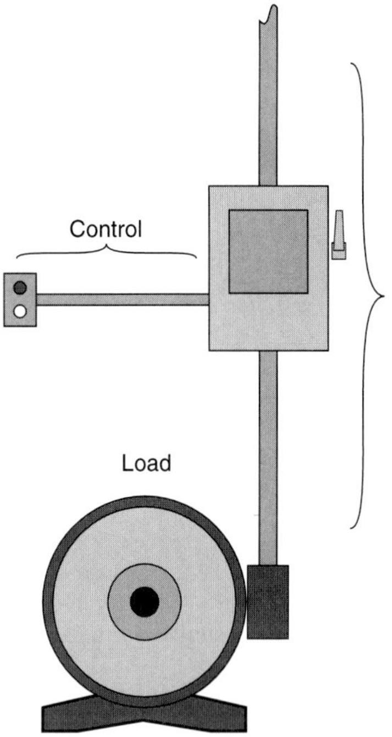

4.2 The Basic Circuit

Electrical circuits are often presented in the form of a diagram of some type. In the diagram, the circuit devices may be shown in a pictorial or symbolic manner, depending on the circuit complexity and the audience the drawing is intended to serve. Figure 4–4 is an example of a simple circuit shown in a pictorial representation and, below it, in a more symbolic diagram form.

The circuit contains the four components of a practical circuit. The battery is the power source; the wires form a continuous conductive loop from the battery through the switch to the lamp and back to the supply. The open-knife-blade switch provides the convenience of being able to turn the lamp on and off without disconnecting the wiring or removing the bulb from the socket. The lamp is the load and the only reason for the circuit to exist. The circuit is arranged in a sensible layout. The conductors run directly between their connection points in a neat and orderly manner. The basic operation of the circuit is immediately obvious.

The accompanying symbolic drawing shows the same circuit. The pictorial view of the battery is replaced by the standard battery symbol, representing two plates separated in a container of chemicals. The knife-blade switch is replaced by a standard switch symbol, shown in the open position. The lamp is replaced by a circle with a "filament" and two leads, one of several common lamp symbols. The circuit conductors are shown as simple lines. The symbolic drawing is also laid out in a sensible manner and, again, the circuit operation is quite obvious.

The lines representing the conductors show how the terminals of each of the other devices are interconnected. A wire connects one terminal of the battery to the switch. A

FIGURE 4–4 A simple circuit and its corresponding symbolic diagram.

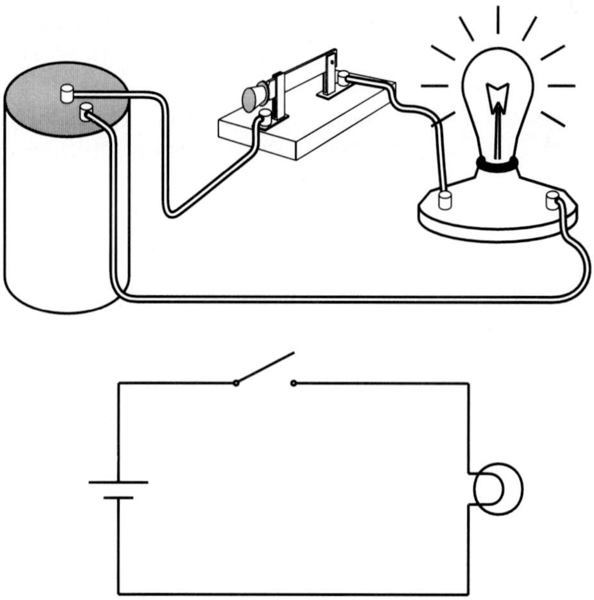

second wire connects from the opposite switch terminal to the lamp, and the final wire connects the other lamp terminal back to the supply. The drawing attempts to show the devices and the connections as clearly as possible, but it does not necessarily show them in their exact physical locations. This is a general rule with symbolic drawings of electrical equipment. The drawing is intended to show the equipment in a manner that makes the operation of the circuit as clear as possible: The physical location of the circuit components and wiring should be of secondary concern. For this reason, the location of devices on the drawing may not be a very close match to their actual location in the circuit.

Practical Example: A Battery-Powered Light (Part 1)

The information that follows describes a simple electrical experiment. The equipment is very basic, and the various tests are covered in a very detailed way. The intent is that readers can follow the example and realize the learning objectives, much as they would from doing the experiments under the author's supervision. The point of the exercise goes far beyond the performance of the experiment itself. Very simple electrical experiments can demonstrate key points that are crucial to understanding and working with electrical equipment.

This example can also be done as a simple experiment if suitable equipment is at hand. The materials include four 1.5 V nonrechargeable dry cells with holders (size C or D cells are the most suitable, but other sizes can be used); short wires with some type of clip end to interconnect the various parts; a single-pole single-throw switch, such as a simple on/off toggle switch, to be used for the control; and a small 6 V lightbulb with a suitable socket to make a convenient load. There is no need to actually perform the experiments, but the experience may be worthwhile.

In this circuit, the four cells are used to power a single 6 V lightbulb. The cells are the circuit power source—a true source in this case, because the electrical energy the bulb uses is generated entirely within the cells. The energy conversion is from chemical energy to electrical energy, the result of a reaction between the "plates" and the electrolyte in the cell. The conductors are small, insulated wires used to interconnect each of the circuit parts. The control device is a toggle switch, and the load is the lightbulb. Each of the circuit components will be examined in turn. A meter is used to provide additional information; it can be analog or digital, both of which are used in the description.

Checking the Cells. A close look at one of the cells should reveal polarity markings and, possibly, the rated output voltage and the ampere-hour rating. Voltage is the potential difference developed by the chemical reaction in the cell; the ampere-hour rating (probably in milliamp-hours) is a measure of the cell's capacity. This rating can be used to compare the useful life of different cells.

The post, or button end of the cell, is the positive terminal. On most cells there is a "+" sign marking on the side of the case at that end. The flat metal bottom is the negative terminal. Batteries, like all direct-conversion devices, produce direct current and have a specific output polarity. The cell's rated voltage may not be marked, but a typical alkaline cell is rated at approximately 1.5 V.

Checking the cell with a multimeter will reveal the actual voltage being generated. It is also an interesting exercise in meter use. The exact procedure varies with the meter available: Two different types are described below. Battery voltage provides some insight into the battery's condition, but it is not particularly conclusive. The output voltage of a typical cell or battery tapers off only slightly as the chemicals are used up. The voltage of a cell near the end of its useful life may be only a little lower than it was when it was new. The voltage remains relatively high until almost no usable chemicals are left, and then it drops off sharply and the cell becomes useless.

Testing With an Analog Meter. With an analog meter, the function selected must be DC volts, and the range at least slightly above the 1.5 V rating of the cell. A test will show the voltage rating is an approximate value. A new cell will have a somewhat higher output; an old one will be less.

The meter leads must be inserted into the correct meter jacks. The red lead goes to the positive jack (marked "+" on most analog meters) and the black lead goes to the negative jack (usually marked with a "–" sign or with the letters "COM" or "COMMON"). With the meter properly set, the red lead is put in contact with the positive battery terminal, and the black lead is connected to the negative terminal. The meter needle should move up the scale to the right and settle at some point along the scale.

To read the indicated value, the appropriate scale must first be identified. It is usually marked "DC" at either end. The next step is to locate the correct number set. Analog multimeters have multiple number sets to accommodate the many ranges provided. To determine which set to use, match the number at the end of the scale to the range selected. Range numbers always represent the maximum, or end-of-scale, value. For example, if a meter is set to a 10 V range, and exactly 10 V is applied to the leads, the meter needle will come to rest at the very end of the scale (see Figure 4–5).

The number set may not be an exact match for the range value. To reduce the amount of information printed on the meter face, the numbers supplied are either equal to, or some multiple of 10 of, each range on the meter. Obtaining an actual reading may

FIGURE 4–5 With the meter set for DC volts, 2.5 V range, this needle position represents 1.5 V DC. The DC scale is used with the number set 0 to 250. The numbers are interpreted as 0–2.5 to match the range selected.

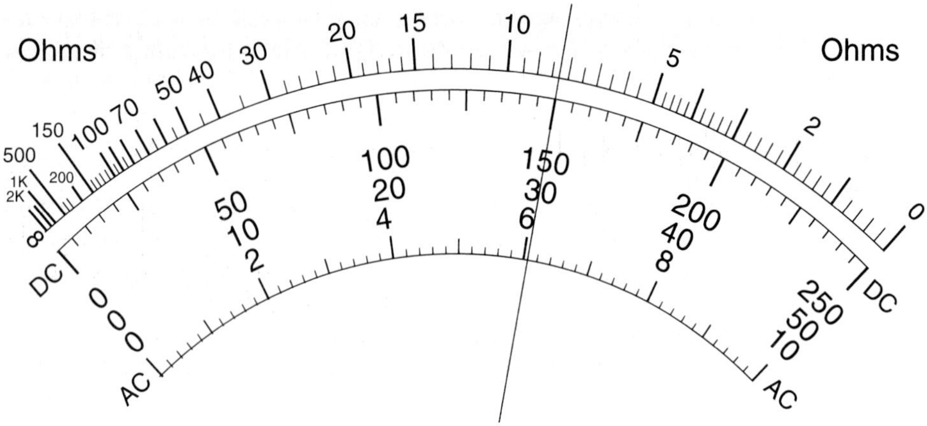

require a decimal place shift. For example, a 2.5 V range may make use of a set of numbers that ends in 25 or 250.

To determine the needle position on the scale as accurately as possible, look straight at the needle, perpendicular to the meter face. Good quality analog meters have a narrow mirror along the length of the scale. When the needle is being viewed properly, its reflection is hidden behind it. The viewer is then directly above that point of the scale, and the reading is as accurate as possible. Note the needle position on the DC scale and convert this into a numeric value. Estimate intermediate values when the needle does not fall directly on a scale marking.

This process may seem rather tedious for a simple voltage reading—and to some extent it is. This is part of the reason why most beginners prefer digital meters over analog types. However, with a little practice, analog meter readings become much easier. It should also be noted that, for accurate readings, similar attention to detail is also required when using a digital meter.

Tips for Using Analog Meters to Read DC Voltage

Analog meters are polarity sensitive when reading DC. This means care must be taken when connecting the meter and interpreting the result. If the leads are inserted incorrectly or the probe tips are connected to the "wrong" polarity (reversed—negative to positive and vice versa), the meter needle will move slightly down scale, stopping just below the 0 marking on the left-hand side of the meter scale.

The meter behaves this way because of the way it works. The needle is driven by a "meter movement," which is essentially a tiny electric motor. This motor works against a spring that acts to hold the needle at the left-hand 0 mark. When the meter is connected to a voltage with the correct polarity, the torque produced by the meter movement forces the needle to move up the scale, against the tension of the spring. When the polarity is reversed, the meter movement turns the wrong way, driving the needle down scale against the stop. This characteristic can be used to identify the polarity of the source voltage.

When no voltage is applied to the meter, the needle should sit exactly at the last mark on the left-hand end of the scale. This zero position depends on an accurate setting of the meter spring inside. When it gets out of adjustment, this can be corrected by turning a small screw head located at the bottom center of the meter face. This will bring the meter needle back into alignment and improve the accuracy of any readings.

Before touching the meter to a live circuit, always take care to select an appropriate range. It must be high enough to withstand the applied voltage without driving the needle past the end of the scale, but it should not be too high. For best accuracy, the needle should be as far as possible up the scale, without going over the end. Set the range just above the anticipated value or, if the voltage is unknown, select a high range first, and then reset the range as required.

Always glance at the meter just after making contact with the test point. This will quickly alert you to potential problems. If the range selected is too low, the needle will hit the end stop. If that happens, lift the probes away from the voltage source and select a higher range. If the polarity is wrong, the needle will drop to just below the zero mark: Watch the needle as the probes are lifted to confirm the direction of movement, and then reverse the probes and test again. If the range is too high, the needle will move up the scale, but not far enough to be accurate. Do not lift the probes immediately; take a reading on the range selected, and then lift the probes and select an appropriate range based on that value.

Testing With a Digital Meter. When the meter is a digital type and autoranging, all that must be selected is the DC voltage function. (A few new meters even do this automatically.) If the meter is not an autoranging type, an appropriate range must be selected, just as for the analog meter. Note that all autoranging meters can also be ranged manually. This can be confusing at first, because if the range button is pressed, the meter becomes locked into a range and will not automatically reset itself. When the locked range is too high, the reading will be rounded, possibly even to 0. If it is too low, the meter will indicate over range and will probably "beep" to attract attention to the problem. The digital meter may be easier to read, because the value is given directly as a number. However, its operation must be studied and understood, or simple errors like this will lead to frustration and misinterpretation.

On most digital meters the lead jacks are marked differently than on a typical analog type. The usual markings are "COMMON," or just "COM," on one jack, and "V" for volts on the other. Insert the black lead into the common jack. The one marked V will usually have other markings as well, such as Ω, kΩ, or mA. The red lead should be inserted into this jack.

Warning

Do not connect to a separate current input, particularly the high-current jack found on most digital meters. It will be marked something like "A" or "10A," usually in red, and will often have some kind of warning written on the meter face. Proper use of this input is described in a subsequent section.

Once the digital meter is correctly set up, a voltage reading can be taken without regard to the polarity of the probe connections. The meter will read the voltage value with the leads connected either way, and provide an appropriate polarity indication. Assuming the leads are inserted correctly—the negative black lead in the common, and the red lead in the positive V terminal—and the polarity is also correct, the meter readout will have either no polarity indication, or a "+" sign ahead of the voltage value. If the leads are reversed—red to negative, black to positive—the voltage value will be the same, but the polarity sign will be a minus "–". Do not misinterpret this indication: It is not +1.5 V or –1.5 V, it is just 1.5 V. The DC voltage has a positive and negative side—the polarity indication simply identifies whether the red meter lead is on the positive or the negative side. In most cases, the voltage value should be recorded without the sign.

Tips for Using Digital Meters to Read DC Voltage

Digital meters give the appearance of being simpler to use than the analog types, and in some ways this is true. However, a digital multimeter is a sophisticated, complex instrument that often has features not found on the simpler analog types. Spending a little time reading the meter manual can be very worthwhile. Learn the purpose and operation of the meter function selection switches or pushbuttons, and the meanings of all the jack markings.

Note the information provided in the readout area. There is usually an indication of the range in use and the units—volts (V) or millivolts (mV), for example. Such subtleties are quite easy to miss. The number may be the same, but if the unit indication is mV instead of V, the actual reading is very different. Many meters also use the readout to provide an indication of unusual conditions such as "hold," or minimum and maximum values. All this information is in addition to the measured value, its unit, and the polarity. Always refer to the owner's manual if in doubt; there are many variations in meter design.

If the meter is automatic, understand the meaning of any audible or visual indications or warnings. Most digital meters have audible signals. On voltage functions, they usually indicate an automatic range change.

Know what is used to indicate over range. Usually this becomes pretty obvious fairly early on, but not all meter makers use the same identifier. An "OL" is common, but not universal. Some meters flash the number 1.999; others flash the leading 1, but not the other digits.

An over-range indication means the meter is connected to a voltage that is too high for its range selection. When that happens, it should be disconnected from the circuit immediately, and the problem corrected.

Connecting the Cells Into a Group. It is highly unlikely that the measured voltages of all four cells will be identical. Even new cells have small variations in output. This becomes more evident if a digital meter is used, because it is easier to obtain accurate comparisons. Recording the readings is useful, because the data can demonstrate an interesting aspect of electrical theory.

To properly power the 6 V lightbulb, the cells must be interconnected in such a way that their voltages add up to something reasonably close to 6 V. Electrical theory states that the voltages from multiple sources can add or subtract. When two of the cells are connected together at one end only, the voltage across the unconnected ends will be either the sum or the difference in the two cell voltages.

Warning

Interconnecting both ends of the cells incorrectly shorts them out. This is not likely to be dangerous with small alkaline cells, but it will quickly end their useful life. With fresh cells, there is a slight possibility of rupture due to overheating. With nicad (rechargeable) cells or larger batteries, the result can be quite dangerous. Nicad and lead acid cells or batteries can supply very high current, and overheat almost instantly when they are shorted. A solid short circuit on a large battery may cause an explosion.

For the cell voltages to add, which is the desired connection, the positive terminal of one cell must be connected to the negative terminal of the other. This is repeated for the remaining cells, so that each connection is between two poles with opposite polarity. The first and last cells are **not** connected together. This would be a short circuit, as mentioned in the previous warning.

The unconnected terminals of the two end cells are the connection points for the group. If the individual cell voltages have been measured accurately and recorded, the sum of these values should match the output measured across the complete group. These cells are connected in *series.*

The total voltage of other combinations of cells can be found be measuring across only part of the interconnected group. A reading across any two cells will indicate approximately 3 V, three cells will be about 4.5 V, and so on. Once the cells are properly interconnected, the power supply is ready for connection to the control device and the load.

The Conductors. To test the conductors, set the meter to read resistance on the lowest available range. This will be RX1 on an analog meter. It may vary with manually ranged digital meters, though 200 Ω is very common. If the meter is autoranging, it will select the correct range automatically.

Note

When an analog meter is used to measure resistance, it must first be "zeroed" before an accurate reading can be made. This is done by shorting the two probes together and turning the ohms adjustment control provided on the meter. This is typically a round knob or thumbwheel marked "OHMS ADJ" or something similar. The meter is properly set up when—with the leads shorted—the needle rests on the zero, the last scale marker on the right-hand side. This test usually has to be repeated each time a different range is used.

Resistance ranges on analog multimeters are reversed in comparison to the voltage and current scales: The zero end of the ohms scale is on the right. Reading the scale left to right instead of right to left is a common cause of error. (While virtually all portable analog multimeters have reversed ohms scales, some analog bench meters have normal unreversed scales. These meters may also have two ohms adjustments—one for each end of the scale.)

When the probes are separated, the needle should swing back to its normal position at the left-hand mark. The resistance value at this point on the scale is marked ∞, which is the symbol for infinity. An infinite resistance is an open circuit, which is exactly what the meter sees when the leads are separated in the air.

The resistance of the conductors can be checked by connecting the meter leads to opposite ends of each one. As with all meter tests, care should be taken to avoid making contact between your fingers and the metal tip of the meter or other exposed conductor. In some cases this is very difficult to do: There are just too many things to hang on to. Therefore, for resistance tests only, if there is no way to completely avoid making contact, hold only one end. Keeping at least one end free will allow an accurate reading.

Testing the conductors should indicate that their resistance is virtually 0 Ω. A small reading may be obtained, but beware of misinterpreting this value. Most digital meters read a small amount of lead resistance when the probes are connected together. For accurate measurements of very low resistance values, the lead resistance must be deducted from the final reading. Digital meters have no provision for zeroing the meter to compensate for the resistance of the leads, although some have a memory feature that automatically deducts the initial value from all subsequent readings.

In an ideal circuit, the conductor resistance would be 0 Ω. This is virtually impossible, but the resistance should be kept as low as can be practically obtained. The purpose of the conductors is to carry the current around the circuit without interference or power loss. The amount of resistance permissible depends on the amount of current that must be carried and the losses that can be tolerated. Larger conductors have less resistance than smaller ones, and long conductors have proportionally more resistance than short ones. In situations where the conductor runs are very long, the wire size may have to be increased to compensate for the length.

The Control. The control switch can also be checked with an ohmmeter. The resistance across the terminals should be zero when the switch is in the closed position (turned on), and infinite when it is in the open, or off, position. All the switch does is provide a convenient method of breaking the conductive path to shut off the load, and reconnecting it to turn the load on again. It is part of the conductive loop, and, like the conductors, it should not interfere with the power flow when it is in the on position, and should completely block flow when it is off.

Two different sets of terms are used to describe switch contact positions. The terms "on" and "off" refer to the effect the switch has on the load. The term "closed" refers to the position in which the switch contacts are touching to complete or to close the circuit. The term "open" means the contacts are separated to break or open the circuit. On equals closed; off is the same as open. It should be noted that this is backward in comparison to valves. A closed valve blocks flow, and an open one allows flow, reverse to the operation of a switch.

The Load. Assuming the bulb is a normal incandescent type—a wire filament in an inert gas—the lightbulb can also be checked with an ohmmeter. However, this is one of those odd cases where the test results can be misleading. The filament inside the bulb is a resistance wire connected between the screw shell and the contact point on the tip. When a resistance measurement is taken between these two points, an accurate resistance value can be found, but it does not match the real operating resistance of the bulb. The measured value is much lower than the actual operating value. The test proves if the bulb is usable or burned out, but cannot be used to calculate bulb current, wattage, or any other operating value.

The reason for this difference is that the filament resistance increases quite dramatically with an increase in temperature. Resistance increases with temperature in most conductive materials, but in almost all other cases the normal operating temperatures are relatively low. With such a small change in temperature, the change in resistance is too small to have a significant effect. In a lightbulb, however, the operating temperature of the filament is extremely high, and the hot resistance is much greater than the cold resistance. Resistance cannot be measured in a live circuit, and, because the filament cools immediately when the power is shut off, its resistance also cannot be measured directly when it is de-energized. The actual value can only be found by calculation from current and voltage readings.

If a resistance test of the bulb gives a reading of infinity, the filament is an open circuit, and the bulb is burned out. If the bulb is good, the reading will be relatively low, varying with the voltage and wattage ratings. This measured resistance is roughly $^1/_{10}$ of the hot operating resistance. This is a very approximate value, but does provide some idea of what to expect.

FIGURE 4–6 The lower circuit drawing shows the same components and connection as the pictorial. The meter reads the voltage of the power supply.

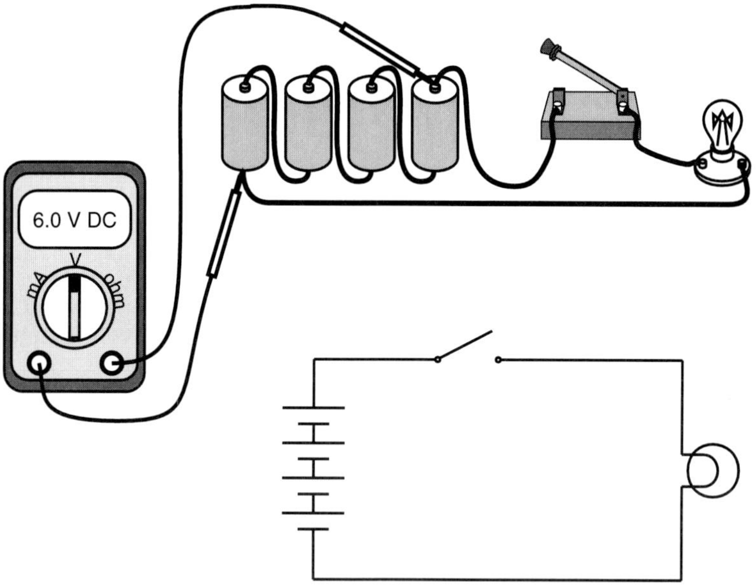

Putting the Circuit Together. Creating a circuit from the component parts is a matter of interconnecting the supply, the control device, and the load with the circuit conductors (see Figure 4–6). There is a basic logic to all circuits that is very much the same whether the circuit is a complex one or a simple one like this example.

To create the circuit, a wire is connected from one end of the cells—traditionally the positive end—to one terminal of the switch. A second wire is connected from the other switch terminal to one side of the lightbulb. A final wire connection is made from the opposite side of the bulb back to the negative side of the supply. This completes the connections for this simple circuit, and, assuming no faulty wires or connections, when the switch is put in the on position, the light will come on.

The order of the circuit layout is somewhat a matter of tradition and consistency. There are several ways in which a simple circuit like this can be connected and still function. However, as circuits become more complex and voltages get to dangerous levels, correct connections become very important. Many of the traditional ways of laying out the circuit and making the connections are not just one of many possible alternatives: They are often the only way the circuit can be connected and be safe. Other connections work, but introduce safety hazards that may not be immediately obvious. This can be difficult to understand, but is an important part of properly installed electrical systems. **The fact that an electrical circuit works does not necessarily prove that it is connected correctly.**

For DC circuits like the battery-powered example in Figure 4–6, it is traditional to start from the positive terminal of the supply, continue through the control device to the positive terminal of the load, and then return to the negative side of the supply. The negative side is treated as a circuit common, and should therefore be directly connected. There should be no switches or other controls that can break the common side of the circuit.

A simple on/off switch has no particular polarity, unless it is an open-blade type or the terminals are identified "line" and "load." If the switch does have open blades, the stationary terminal should be the line side, and the side with the moving blade should be the load side. The incoming supply is the line; the conductor that goes on to the load is, of course, the load side connection. With this connection, the blade is disconnected from the incoming power lead when the switch is open.

Incandescent lightbulbs are not polarity sensitive, so either side could be connected to the positive of the supply. However, this style of bulb has one terminal exposed and one protected. The exposed outer shell should not be connected to the incoming supply lead: It should connect to the circuit common—in this case, the negative of the supply. The protected terminal (the tip of the bulb) connects to the incoming positive lead. One reason for this is that the common side of the circuit may be an exposed conductor, such as the metal frame of an automobile. In a car, for example, all the wiring comes from the positive side and must be protected from accidental contact with the metal car body. Car lamp sockets have the outer shell connected to the frame.

In AC systems, such as the wiring found in a house, business, or industrial building, the common conductor—called the neutral—is connected to ground. It is special because it does not represent a shock hazard. The "hot" side of the circuit must be protected against accidental contact with grounded surfaces. The voltage is also high enough to be a serious hazard, so the system must be installed in such a way as to minimize the risk of human contact. In the lamp socket example, the screw shell of the bulb must connect to the "safe" common, not to the dangerous "hot" side of the circuit. Reversing this connection could result in a serious, possibly fatal, shock to an unsuspecting person doing nothing more than changing a burnt-out bulb.

The examination of this simple circuit example is continued further in the sections that follow.

4.3 Basic Circuit Calculations

Electrical machines and devices are creations of human ingenuity, but electrical behavior follows natural, not human, laws. These natural laws are relatively simple but very rigid. For this reason, electrical behavior is fairly easy to predict.

The apparatus and electrical machines of early scientists were built to test their theories and expand their knowledge, not to transform society, as electricity did later. The discoveries they made are the backbone of modern electrical power systems, but one of their most important contributions was to define units of measure and formulate the rules and equations that are still used to describe and predict electrical behavior.

Ohm's Law

An excellent example of the basic rules of electrical behavior is Ohm's Law. It is named in honor of Georg Simon Ohm, an experimenter from the early 1800s. Ohm studied the nature of electrical resistance. Ohm's Law defines the relationship of the three most fundamental units of electricity. These are potential difference (voltage), the flow of electrical charge (current), and resistance (the degree to which electrical conductors and load devices oppose the current). Obviously a man of science, not a monarch or a politician, the units he chose are both sensible and practical.

Note

Ohm's Law is essentially true for both AC and DC circuits, but AC behavior is more complex. This will be examined in a following section. The logic of Ohm's Law can be summarized as follows:

In a DC circuit the current flow is directly proportional to the voltage and inversely proportional to the resistance.

As a formula, Ohm's Law becomes:

$$I = \frac{E}{R}$$

E = Potential difference, or applied voltage—in volts

I = Current—in amperes

R = Resistance—in ohms

Voltage is the driving force of electrical energy. It is the electrical pressure or potential difference of a circuit. It is usually symbolized by an uppercase "E" or "V". The "E" comes from the term "electromotive force" (EMF), which is a more accurate description of potential difference. The "V," of course, comes from the word "voltage." The use of the letter "V" can be particularly confusing at times. In common usage, the term "voltage" is used interchangeably with "potential difference," "electrical pressure," or "EMF"—all of which represent the same quantity—and the unit for this quantity is also the volt. As a result of this verbal sloppiness, the terms used for the quantity and the unit are the same. Current is in amps, but "voltage" is in volts. When the letter "V" is used to represent the quantity, a lowercase "v" may be used for the unit to differentiate between the two. However, an uppercase "V" is usually considered to be the proper abbreviation for the unit volt.

Current, symbolized by "I," is the intensity of electron flow in a circuit. Electrons flow around the circuit, carrying energy, just as the oil in a hydraulic system carries energy from the pump to the hydraulic motor or ram the system operates. The unit of current measure is the ampere, symbolized by an uppercase "A". One ampere is a flow rate of one coulomb of charge (6.28×10^{18} electrons) per second. As with any measure of flow rate, there must be a unit quantity per unit time. In this case it is a specific number or volume of electrons, called a coulomb, and the time is in seconds.

Resistance is a measure of the electrical "friction" in a circuit. All conductive materials, except superconductors, have resistance. In good conductors the resistance is very low, while in other materials resistance is significant. Voltage will force current to flow through resistance, but the energy used is converted to heat. This is the same as mechanical force overcoming friction. The result is heat generated in the materials where the movement takes place. Electrical resistance is symbolized by the letter "R". Its unit is the ohm, which is symbolized by the Greek letter omega (Ω).

When a material does not conduct at all, it has an infinite (∞) resistance. Such materials are insulators and will block all current flow. Insulators are as important as conductors to the operation of circuits because they are used to confine the current flow to the intended circuit path.

Unit Relationships

Ohm's Law defines the relationship between the circuit voltage, current and resistance. This relationship is a neat 1:1 ratio using the basic units of measure. For example, If a supply with 1 V of electrical pressure, is applied to a circuit with 1 Ω of electrical resistance, the electron flow rate will be one coulomb per second or 1 A.

When calculating with Ohm's Law, remember that the formula can be transposed to solve for any missing value. If any two items are known the third can always be found. The three forms of the Ohm's Law formula are as follows:

$$I = \frac{E}{R}$$

$$R = \frac{E}{I}$$

$$E = I \times R$$

Any of these three formulas can be used, but the appropriate one should be selected to match the situation. The unknown value (the one being calculated) should be on the left of the equals sign. The first formula is in the correct form to find current when voltage and resistance are known. The second is used to find resistance from known values of voltage and current, and the third is used to find the voltage when current and resistance are known.

It should be noted that electrical quantities can vary over an extremely wide range. Electricity is incredibly diverse in application. Some circuits operate at potentials of many thousands of volts, with thousands of amperes of current. In contrast, there are equally useful, though very different circuits that operate at a few volts, and only a few mA or µA current. Resistance varies over a similar range. Values in millions of ohms are nearly as common values so low as to be virtually unmeasurable.

Electrical quantities and calculations have always been in international SI units. The SI system is designed to be as consistent as possible, so all unit divisions are in even multiples of ten. For added convenience, scientific notation and a set of associated prefixes are used to reduce the number of zeros required for very large or very small values. The following list provides the most commonly used prefixes and their numeric equivalents:

Mega (M) = 1,000,000 or 10^6

Kilo (k) = 1000 or 10^3

Milli (m) = .001 ($^1/_{1000}$) or 10^{-3}

Micro (µ) = .000001 ($^1/_{1,000,000}$) or 10^{-6}

The following are examples of how these prefixes are used:

0.001 V = 1 mV (1 millivolt)

0.000025 A = 25 µA (25 microamps)

1000 Ω = 1 kΩ (1 kilohm) (Ω is the symbol for the unit ohm)

Students of electricity must be able to interpret and convert units with prefixes such as those shown in the previous examples.

When calculating values in formulas such as Ohm's Law, it is important to convert the known values to their basic units before completing the calculations. Failure to convert values given in prefix forms, such as kilohms (kΩ), megohms (MΩ), or milliamps (mA), into their basic unit form can cause enormous decimal place errors. When a scientific calculator is used, the conversion can be done automatically by substituting the appropriate scientific notation value for the prefix it represents. These are shown in the list above. Scientific notation uses a base of 10, taken to a specific, even-numbered exponent. Scientific calculators are designed to accept these values directly. The value is entered as two parts: the actual numeric value and then the exponent. The 10 is not entered: It is assumed. The exponent must also be given the correct sign.

Entering values with appropriate exponents automatically converts them into the basic unit. For example, 25 mA (25 milliamps) is 25×10^{-3} amperes. The "$\times 10^{-3}$" replaces the prefix "milli." In a calculator, this would be entered as 25 exponent –3. This value can be entered in one step, and the calculation can be completed without error.

Prefixes are only used to represent the preferred engineering values, which are every third exponent. When only the preferred exponents are used, the values are said to be in engineering notation. Some calculators offer engineering notation as an option. This can be very convenient, because the calculator will give all results using only the preferred prefixes (3, 6, 9, 12, ...). This makes it much easier to convert the final answer to the appropriate written prefix. With the value in engineering notation, no decimal place adjustment is necessary.

When a nonscientific calculator is used, or the calculation is done longhand, it is best to enter the values into the formula in basic unit form before any calculations are done. This avoids the probable decimal place errors and allows for cancellations and simplification in advance of the final calculation. Once the final result is found, it can then be converted into the most appropriate prefix form if required.

Practical Example: A Battery-Powered Light (Part 2)

Voltage Readings in an Operating Circuit. With the lamp circuit connected and operating, additional information can be gathered about the characteristics of such a circuit, and about electrical equipment in general. For example, the multimeter can now be used as a voltmeter to measure voltages in various parts of the circuit. Such tests are usually far more informative than resistance tests, which must be done with no power applied.

With the lamp on, the voltage can be read across the lamp and compared to the voltage across the supply (the cells). In this circuit, there is virtually no resistance in the circuit conductors or in the switch, so there will be no measurable power losses in either of them. Therefore, with the lamp on, the supply voltage and load (lamp) voltage will be identical. With long connecting wires or other resistance in the circuit path, losses would occur and the voltage at the load would be reduced.

With the switch off, the voltage at the lamp will, of course, be zero. There will still be a voltage across the cells, but it may be slightly higher than it was with the lamp turned on. The cells have some internal losses, so their output voltage drops slightly when they supply current to a load. With one small bulb and cells that are in good condition, this voltage difference may be very small. If the current draw is increased by adding a greater load (more bulbs in parallel, for example), this voltage drop will increase, causing the output voltage to be lower than before. This tendency of the supply voltage to drop off slightly under load is true for all types of power sources. The amount of reduction depends on the internal losses and the current demand. Electromechanical generators have voltage regulators to compensate for this, so their voltage output remains relatively constant even when the load varies over a wide range. Direct-conversion devices such as batteries cannot automatically compensate for internal voltage drops.

Another interesting characteristic of circuits and meter behavior can be demonstrated by measuring the voltage across the switch. With the supply connected and the circuit operational but turned off (switch open), a measurement across the switch shows full supply voltage. When the switch is turned on (closed), the light comes on and the voltage across the switch drops to zero. This behavior can be very confusing to the uninitiated troubleshooter. It seems backward, at first, that an open switch has voltage, and a closed one does not.

The reason it works this way can be demonstrated by opening the switch, measuring the voltage as before, and then unscrewing the lightbulb. The meter circuit on the bulb side is through the bulb and back through the circuit common to the supply. When this side of the circuit is broken by removing the bulb, the voltage across the switch drops to zero. This proves the voltage reading across an open switch is due to the connection through the interconnecting circuit conductors and the resistance of the load. Once the load is disconnected, the second connection to the supply is lost, and the voltage disappears.

The other half of this switch voltage experience can be even more confusing. When the switch is closed, current is flowing through the switch to the bulb. This is proven by the fact that the light is on. Yet, although power is applied and current is flowing, there is no voltage across the switch terminals. Again, it is simple to demonstrate why this is so. If the circuit is left on and voltage measurements are made across any of the connecting wires, these readings will also be zero. When the switch it is closed, it acts just like a connecting wire. It has no measurable resistance and carries current with no measurable loss. With no losses, the potential difference across the switch or wire is zero. If the wire had some resistance, a voltage difference would appear. It could be very small, depending on the actual resistance and the current flow, but it would be measurable.

Another interesting test—one that demonstrates a very good troubleshooting technique—is to connect one meter lead to the common side of the supply, and then use the other lead to follow the voltage around the circuit (see Figure 4–7). With the switch open, there will be voltage on the supply side of the switch, but not on the load side, indicating the switch is the point where the circuit is broken. With the switch closed, there will be voltage on both sides of the switch and all the way to the incoming (supply) side of the lamp. In this simple circuit there are not a lot of places to check, but this technique can be applied to any circuit, even a very complex one, with very good results. Measuring from the circuit common, if there is power to the circuit, there will always be voltage up to point where the circuit is broken, but not beyond.

The lessons of these examples, simple though they may be, are some of the most important. Correct interpretation of unexpected readings is critical to efficient troubleshooting. Even minor confusions, like misinterpreting voltage readings across an open switch, can mean a lot of time spent looking for the wrong problem.

Current Readings in an Operating Circuit. Measuring current with a multimeter causes more problems than any type of voltage or resistance test. When a clamp-on current meter can be used, most of these problems disappear, but many times a clamp-on meter either is not available or is unsuited to the task.

FIGURE 4–7 The meter will read supply voltage from common to points 1 and 2, but reads zero at points 3 and 4. Measuring from the common can eliminate a lot of interpretation errors.

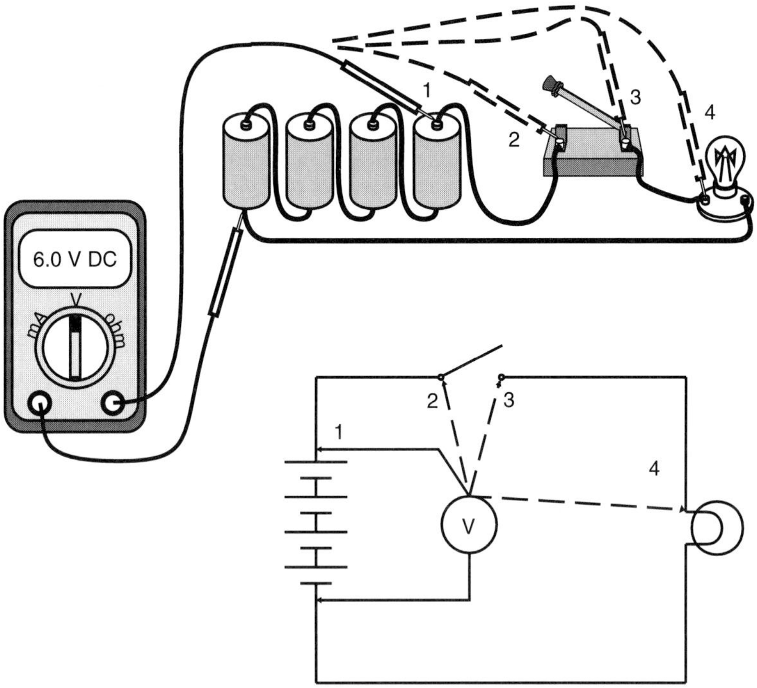

Current is flow. To measure it, the meter must be connected into the circuit path. Therefore, an ammeter must be able to pass current without damage up to the maximum of any particular range selection. It must possible to connect it into the circuit with no significant effect on the circuit operation. The only way this can work is for the meter to have a very low internal resistance.

The low internal resistance of an ammeter can be confirmed by using one multimeter to test the resistance of another one. For comparison purposes, start by measuring the resistance of a voltmeter. An ohmmeter test across its leads will indicate a very high resistance. On an analog voltmeter this will vary with the range selected; on a digital meter it should be constant, typically 10 or 20 MΩ. This is listed as meter impedance, a term that is similar, and closely related to, resistance. This explains why the voltmeter can be connected across any part of the circuit with almost no chance of damage, and no interference to the operation of the circuit. With such a high resistance, the voltmeter draws almost no current from the circuit.

When the multimeter is set up to measure current instead, the resistance will be much lower. Note that the meter leads may have to be moved to different jacks for at least some, if not all, current functions. For milliamp ranges, the actual meter resistance may vary with the range selected, but it will always be relatively low. If the meter is equipped with an ampere input, the resistance of the meter will be lower still, possibly too low to measure with a simple ohmmeter. The high-current (10 A) input of a typical multimeter has a resistance of less than 1 Ω.

It is this low resistance that makes fuses on ammeter inputs such an important multimeter feature. If the leads are plugged into the current jacks, and the meter is connected across a voltage, only a fuse can protect the meter, the circuit, and the user from the effects of the resulting short circuit.

Connecting an Ammeter Into the Circuit. As previously stated, an ammeter must become part of the circuit. It must carry the current it is going to measure. To connect it, the circuit must be opened at a point where the current to be measured normally flows. The meter is connected across the opening. With the meter installed, the circuit is again complete, but this time with the meter as part of the wiring.

In the battery-powered lamp example, a good place for the meter is just after the switch. If the lead from the switch to the lamp is disconnected, the meter can be inserted, with one lead to the switch and one to the load. Clip-on meter leads are a definite asset when using a meter to measure current. For this circuit, the meter would be set to read DC milliamps at a range high enough to accommodate the lamp current. The actual value is unknown, so the best approach is to start at the highest available range, and then change it to the correct range after the first reading.

Although the actual results of this simple test will vary with the lamp used, a typical 6 V flashlightbulb will draw 150 to 300 mA. For the purpose of the example shown in Figure 4–8, lamp current is assumed to be 150 mA.

When the switch is closed, the meter indicates the current flowing through the circuit conductors to the lamp. It also indicates the polarity of the meter connections. With a digital meter, polarity is indicated by a sign in the display. The meter works equally well with either connection. With an analog meter, the meter needle moves down scale if the polarity is incorrect, and the connections have to be reversed. With the correct range selected—the first range above the actual current—an accurate reading can be obtained. The rules for interpreting readout indications and scale markings are the same as for the voltage functions.

Once the current has been found, it can be used with the measured voltage to find the operating resistance of the lamp. As explained previously, the resistance of lamp filament is much higher when it is hot than when it is cold. However, the hot resistance cannot be measured directly, because resistance cannot be measured with power applied to the bulb.

Using the assumed values of 6 V applied, and 150 mA of current through the operating lamp, the Ohm's Law calculation would be as follows:

$$R = \frac{E}{I}$$

$$R = \frac{6\ \text{V}}{0.15\ \text{A}}$$

$$R = 40\ \Omega$$

The current was converted from milliamps into amps for calculation purposes. 150 mA = 0.15 A. The hot resistance of the lamp filament is 40 Ω. This is much higher than the cold resistance that would be found by measuring across the lamp with an ohmmeter.

FIGURE 4–8 The current meter must be part of the circuit path. It is connected in series with the load.

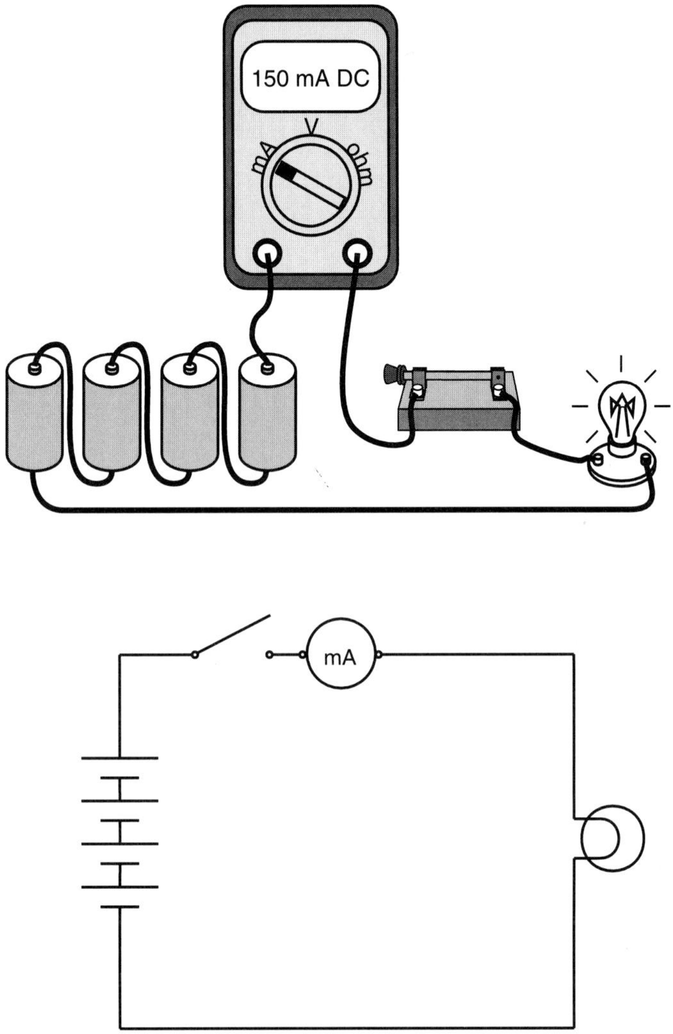

Understanding Short-Circuit Current. There is another important lesson that can be learned from this simple battery and lamp circuit. If the batteries are small—D size alkaline flashlight cells, for example—and the meter has a high current range—10 A is about right—a short-circuit test on the batteries can be used to demonstrate the behavior of the ammeter and the limitations of the supply.

To do this test, the meter must be set up for a high current input. This requires a different jack position for the red meter lead, and a change of meter ranges. The circuit is set up for normal operation, but with a direct connection in place of the milliammeter. With the circuit off, the ammeter is connected across the terminals of the lamp.

Warning

This is an incorrect and potentially dangerous connection. When an ammeter is connected across the circuit the way a voltmeter would be, it will short out the circuit through its very low internal resistance. It is acceptable here because the supply is incapable of producing truly dangerous current. This should not be attempted with rechargeable (nicad) cells or lead acid batteries. They have a lower internal resistance and, as a result, a much greater short-circuit current.

When the switch is closed to apply power to the now-shorted bulb, the ammeter reads the short-circuit current of the group of cells. This test should be kept as brief as possible, just long enough to get a reading. The shorted cells are drained very quickly and may also heat up, increasing the risk of rupture. With D size or smaller alkaline cells, the current is typically less than 10 A—less than 5 A if the cells are nearing the end of their lives. With a larger battery, or nicad cells instead of the alkaline type, the current can easily exceed a safe maximum, and the test should not be attempted.

When a supply is shorted out, it delivers as much current as it is physically able to provide. Even for small batteries this current can be high enough to be a fire hazard. When the supply is a powerful one, the results can be disastrous. This is the reason fuses and circuit breakers are so critical to safe operation of electrical circuits.

During the short-circuit test, the lamp will go out. The short circuit around the bulb holds the voltage very close to 0 V. With no voltage, the bulb draws virtually no current, so it stays off. Almost all the current flows through the short, instead of through the much higher resistance of the bulb.

Repeating the test with a voltmeter across the cells demonstrates one final lesson. When the supply is shorted out, it is severely overloaded. The result is that the output voltage drops off dramatically. This is normal for a "weak" supply. If the load is more than it is designed for, it will be unable to maintain its rated output voltage. In the case of a low-resistance short, the voltage will drop to nearly 0.

Voltage measurements with the circuit under load can be very informative. Whenever there are problems in a circuit that has a weak supply, always check the voltage across the circuit at the load end and, if possible, at the supply end. Check with and without the load connected. If the voltage is much lower with the load connected, there may be problems in the supply, the supply circuit, or at the load end. If the problem is a short, the circuit current will be high, and the voltage virtually zero everywhere in the circuit. When the short is eliminated, the voltage may recover.

If the supply cannot generate enough voltage, the voltage will be low everywhere, even without the load, though it will probably drop even lower when the load is connected. If the supply has a high internal resistance—something has gone wrong inside the supply itself—its output voltage will be low with a load, but may recover almost to normal when the load is removed. This happens sometimes with batteries when only one cell is defective. If the problem is in the circuit, the voltage will be low at the load but normal at the supply end. Different circuit problems are shown in Figure 4–9.

When the load in a circuit is bypassed by a short, the normal result is a tripped circuit breaker or blown fuse. If the short bypasses only a small part of the circuit, it can be difficult to locate. This is not a common occurrence, but it does happen, and with the right circumstance it can be a difficult troubleshooting puzzle to solve.

FIGURE 4–9 A weak, unfused supply can be used to safely demonstrate how different circuit problems cause quite different results.

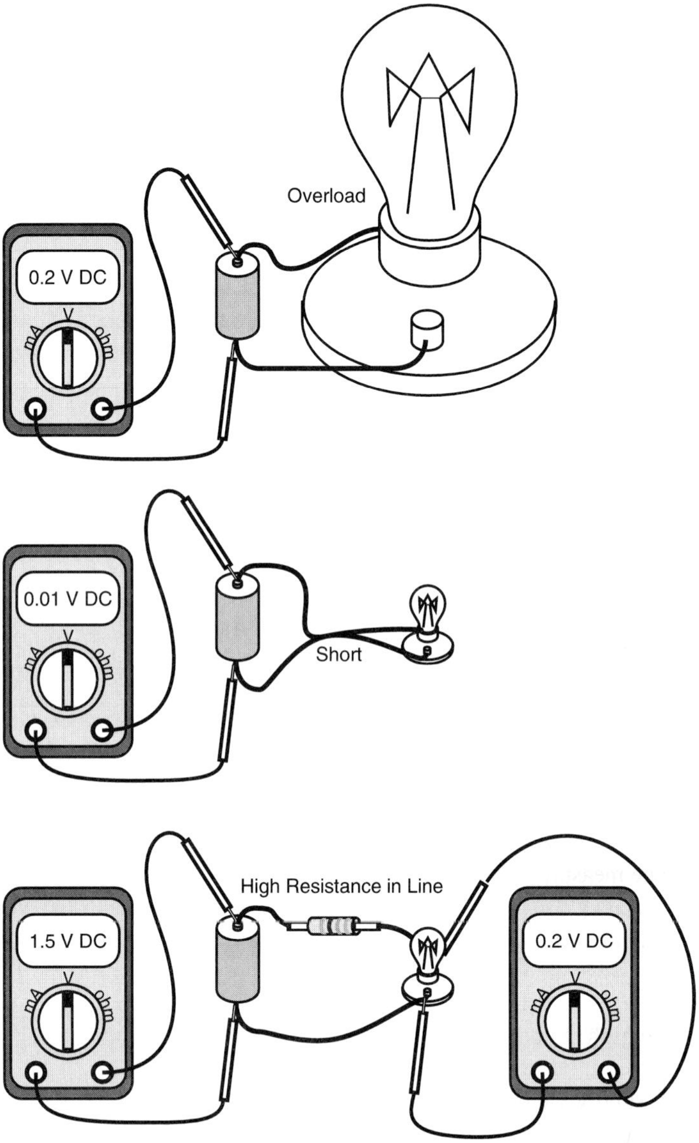

Warning

The final lesson in all of this is that ammeters can be very useful instruments, but must be used with caution. This is particularly true if the meter has a high current range that is not fuse protected. Note that most multimeters can become a short circuit, regardless of the setting of the function switch. All that is required to create this potentially dangerous situation is to insert the red lead into the high-current jack instead of the voltage and resistance jack where it normally belongs.

4.4 Circuits With Multiple Loads

Series Circuits

Although most circuits have a single supply, they often have more than one load. Circuits with multiple loads can be laid out in two basic configurations. These are commonly called series and parallel connections.

In a series connection (shown in Figure 4–10), the source voltage is applied across a group of devices that are connected head-to-tail to form a single conductive circuit path. When the circuit is operating, current flows through all the devices simultaneously. The amount of current is a function of the source voltage and the total circuit resistance. Total resistance in a series circuit is the sum of the individual resistances. Circuit current can be found with Ohm's Law from applied voltage and total resistance.

$$R_T = R_1 + R_2 + R_3 \ldots$$

$$I = \frac{E}{R_T}$$

R_T = Total resistance

$R_{1,2,3}\ldots$ = Individual series resistance values

E = Supply voltage across the group

I = Current flow through the group

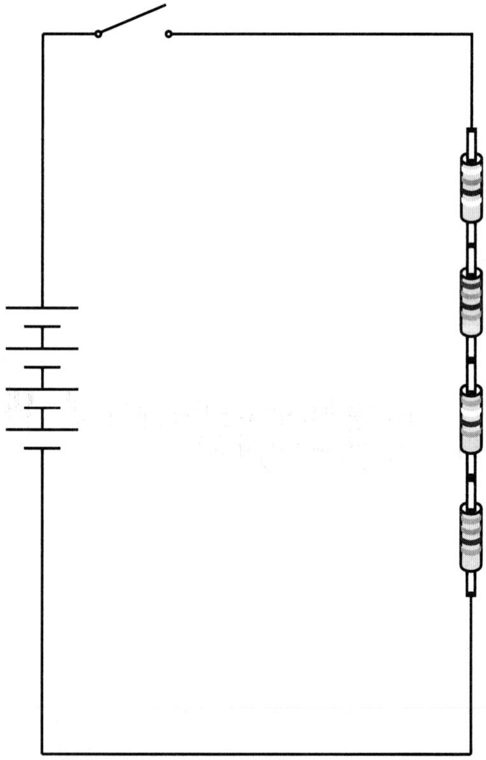

FIGURE 4–10 The resistors of this circuit are connected in series.

In series-connected devices, there is only one conductive path, so current flow is the same in all parts of the circuit. However, the voltage is not necessarily the same for each load in the series group. Unlike the current, the voltage divides among the loads according to their resistance values. These individual voltages appear as a series of potential differences, called voltage drops, across each of the series-connected loads. The voltage drop is a portion of the source voltage. It can be found with Ohm's Law from circuit current and the resistance of the individual loads.

For example, the voltage across one resistor, identified below as R_1, would be its resistance value times the current. The current for the individual loads is the circuit current for the series group. The voltage drop across an individual load is its operating voltage.

$$V_1 = I \times R_1$$

V_1 = Voltage drop on R_1

R_1 = The resistance of load 1

All electrical circuits operate in perfect balance. The voltage applied across the circuit by the supply must all be dropped around the circuit. For this reason, the sum of all the voltage drops around a circuit will always add up to the source voltage.

$$V_T = V_1 + V_2 + V_3 \ldots = E$$

This basic truth is also known as Kirchhoff's Voltage Law, which can be summarized as follows: "Around any conductive loop, the sum of the voltage drops is always equal to the applied voltage." Since each of the series voltage drops is proportional to the resistance that produces it, equal resistances always produce equal voltage drops. For this reason, the voltage drops across a series of equal resistances can be found by dividing the total applied voltage by the number of resistors in series. It should also be noted that the voltage drop on any part of a series circuit is proportional to the resistance of that part divided by the total resistance. This is often referred to as the voltage divider rule and can be expressed as a formula:

$$V_X = \frac{E \times R_X}{R_T}$$

V_X = Voltage drop across resistance R_X

R_X = The resistance across some specific part of a series circuit

E = Voltage applied across the series circuit

R_T = Total resistance of the circuit

For power circuits, such as those containing lights and motors, a series connection is rarely practical. Unless all the loads have the same resistance, they will each have a different voltage. Another serious drawback is that switching the loads individually is almost impossible. Each is a part of the same circuit; if one is shut off, all the rest will also go off.

One of the few common commercial uses of series-connected loads are strings of small Christmas tree lights. The advantage in this application is that the bulbs can be low voltage, even though the supply voltage is relatively high. By connecting a number of equal-resistance bulbs in series across the 120 V supply, the voltage on each bulb can be reduced to a small fraction of the supply value. The total supply voltage is divided equally among the bulbs by the nature of the series connection. There are problems, however. The most annoying characteristic of these lights is that a single burnt-out bulb breaks the circuit path and shuts off the entire string. It is a time-consuming chore to locate and replace the faulty bulb.

Series-connected bulbs are also used in some older incandescent street-lighting systems. Only one conductor runs between the fixtures, instead of the two required for a parallel connection. The conductor is also smaller than required for a parallel connection because it has to carry only the current that is normally drawn by a single bulb. The street-light circuit operates at relatively high voltage from a special constant current transformer. If a bulb burns out, a short-circuiting device closes the circuit at that point, and the remaining lights function normally.

Series connections are also used for the lights in some trains. These trains operate from a relatively high voltage DC supply. There is no simple, efficient way to raise and lower DC voltages, so connecting the bulbs in series is a simple way to reduce the voltage on each bulb.

Practical Example: Resistors in Series

As with the previous example, the following description is intended to provide a practical learning experience, with or without actually doing the experiment that is described. Enough detail is provide that the learning objectives should be realized by reading through the explanations. Practical application tips are also provided where applicable. It may still be very worthwhile to do this experiment, or a similar one, because there is no substitute for experience. Again, it is important to realize that principles can be learned from very simple circuits: Once understood, they can be applied to much more complex circuits.

Resistors and Resistor Color Codes. One of the simplest ways to learn about the series connection is to construct a test circuit, and then measure the performance of the circuit with suitable meters. Lightbulbs are not very appropriate for this type of experiment because of the difficulty in measuring their operating resistance. A better choice is small, fixed-value resistors, such as the color-coded types commonly used in electronics. Other alternatives include fixed or adjustable power resistors. All of these are described below and shown in Figure 4–11.

Color-coded resistors are used in all types of electronic circuits. Manufactured by the millions every year, they are inexpensive and readily available in a wide range of standard values. They are simple devices, but they are a little confusing at first because of the way in which they are marked. Instead of having the rated resistance value stamped on the resistor as a number, it is coded as a set of colored bands. These markings follow a standard color code. Each color represents a specific number, and the position of each colored band determines its meaning. The color-code value is the rated value of the resistor.

The first color band is the first digit, the second is the second digit, and the third is the multiplier (see Figure 4–12). The multiplier is the power of ten that should be applied—or simply the number of zeros that must be added after the second digit. Normal numeric multipliers limit the lowest possible value to 10 Ω. For this reason, gold and silver are used as special decimal multipliers. The fourth band is the rated tolerance of the resistor. Tolerance is a measure of how close the actual resistance must be to the rated value for the resistor to pass the manufacturer's quality control checks. Tolerance is listed as plus or minus some percentage; for example, ±5%. A resistor with this rating should have an actual resistance value that is no greater than 5% above, or any less than 5% below, the rated color-code value.

A chart of the colors and their values is given below. An example follows.

FIGURE 4–11 In the center, a resistor with a marked value, three color-coded resistors 1/2, 1, and 2 W rated, and two 10 W power resistors. The larger power resistors have movable adjustment taps. On the left are four variable resistors: a large one of relatively high power rating, a small one with a long shaft, and two very small ones intended for electronic applications.

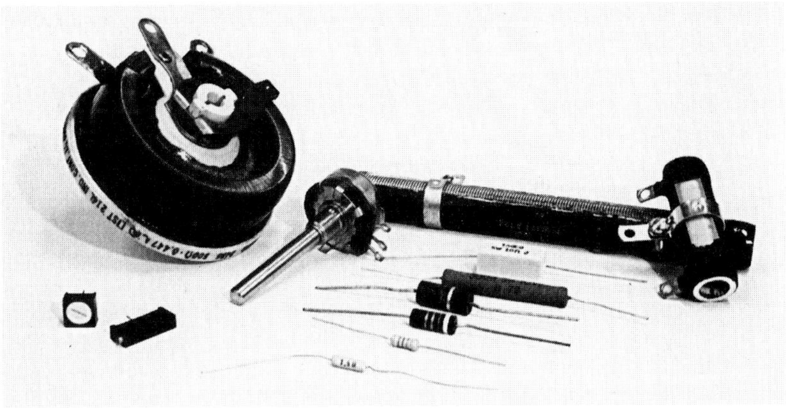

Standard Resistor Color Codes

Black	0	Multipliers for values below 10			
Brown	1				
Red	2	Gold	0.1		
Orange	3	Silver	0.01		
Yellow	4				
Green	5	Tolerance band accuracy ratings			
Blue	6				
Violet	7	Red	±2%	No 4th band	±20%
Gray	8	Gold	±5%		
White	9	Silver	±10%		

FIGURE 4–12 The first two bands represent the first and second digits of the resistor's rated value. The third band is the multiplier—the zeros required to complete the number. Silver and gold complete the colors used. They usually appear in the fourth band as tolerance ratings, but are also used as multipliers (third band) for small resistance values.

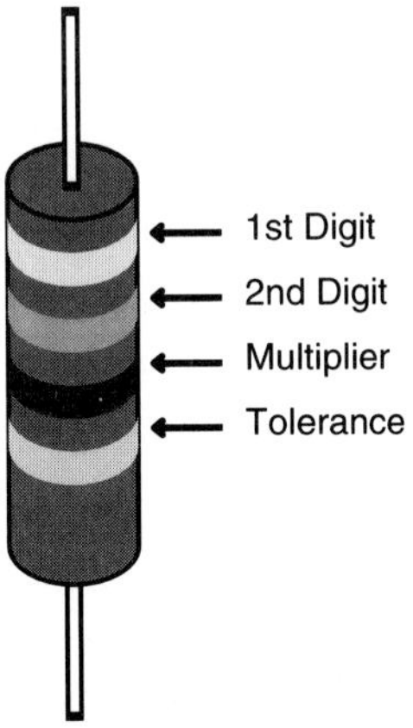

Example: A resistor colored brown/black/green would have a rated value of 1,000,000 Ω or 1 MΩ. Brown is 1, black is 0, and green as a multiplier means five additional zeros (or $\times 10^5$).

Notes About Coded Resistors

The first band is placed closer to one end of the resistor. It is never black, gold, or silver. The smallest color-coded value—disregarding the multiplier—is 10. A 10 Ω resistor is coded brown, black, black, plus the tolerance band. Values below 10 Ω use a multiplier of gold (0.1) or silver (0.01).

Few resistors have an actual value that is exactly matched to the color-coded value. The fourth band indicates by how much the actual value is likely to vary, because it represents the maximum and minimum values accepted during manufacture. For example, a 10 kΩ resistor with a silver fourth band (±10% tolerance) could be any value from 9 kΩ to 11 kΩ and still be within tolerance.

The tolerance rating also affects the variety of rated values available. General-purpose resistors are only made in specific standard values, which are not necessarily even numbered. Each standard value is assumed to cover the full range of values represented by its tolerance rating, and overlaps between standard values are minimized. The minimum acceptable value of one resistor is approximately equal to the maximum allowable value of the next standard value below it, and so on. The closer the manufacturing tolerance, the more standard values there are, because each resistor covers a much narrower range.

The physical size of the resistor has nothing to do with its resistance value. Size is a rough indicator of the power rating, which is a rating of the resistor's ability to withstand the heat generated internally by current flow. The power-handling capacity depends on physical size—which determines the surface area available for heat dissipation—and the materials used in construction—how hot the resistor can operate without damage. There are five common power ratings for color-coded resistors: $^1/_8$, $^1/_4$, $^1/_2$, 1, and 2 W. The power rating is not marked on the resistor and can sometimes be difficult to judge, because a modern $^1/_2$ W resistor is about the size of an older $^1/_4$ W.

Resistors With Higher Power Ratings. Resistors with power ratings above 2 W must withstand higher temperatures and dissipate more heat. They are made from high-temperature materials, usually a wire winding around a ceramic form. They are not color coded. The resistance values are stamped on the resistor as a number. This makes the values easier to read than color codes; however, it is common to use a rather unusual marking method. The letter "R" is sometimes substituted for the decimal point. A 1R0 is 1 Ω, and a 10R is 10 Ω.

Power resistors are available with ratings of 5, 10, 25 W, and more, and each higher wattage rating is physically larger than the previous one. The resistance wire is wound on the form, connected to a terminal or wire lead at each end, and either embedded in a ceramic case or coated with a layer of ceramic material.

Power resistors are manufactured in convenient standard values, usually to a ±10% tolerance. They are also available in both tapped and adjustable styles. Tapped resistors have a series of connection terminals—called taps—that connect to the resistance wire at intervals along its length. These are embedded in the ceramic coating that covers the resistance wire.

Adjustable power resistors are made in much the same way. The resistance wire is wound onto a round ceramic form, usually a hollow tube, and then connected to a terminal at each end of the form. To make the resistor adjustable, a narrow band of the tightly wrapped resistance wire is left exposed along one side of the resistor. A metal collar is fitted over the resistor and secured with a small bolt. The collar has a dimple that makes contact with the exposed surface of the resistance wire.

To adjust the resistor, the collar bolt is loosened and the collar is slid along the length of the resistor until the desired resistance value is obtained between it and one of the end terminals. The bolt is then tightened to hold the collar in place. Adjustable resistors are not intended as a control: They provide a point of adjustment only and should not be changed while they are energized.

For situations where the resistance value must be conveniently variable, with the circuit in operation, a variety of variable resistors are available. Variable resistors are a very common control. In almost every case where a control knob turns or slides, the device inside is a variable resistor. Most variable resistors use either resistance wire or a thin carbon track on an insulating base material. High-wattage types are usually wire wound on a ceramic form; low-wattage types are usually carbon.

For rotary styles, the resistance material is arranged in a circle that is broken at one point. Two end terminals are connected on either side of the gap so the full length of resistance material is between them. A movable wiper makes contact with the resistance material and, through a sliding connection, to the third, center terminal of the resistor. The wiper is connected to the shaft of the adjustment knob. When the shaft is turned, the wiper slides along the circular track of resistance material, varying the resistance between the center terminal and the two end terminals. These resistors are designed to be operated with the circuit energized. The terminals are insulated from the operating knob, and the wiper is designed to provide continuous contact pressure as it is adjusted.

Testing Resistors With a Multimeter. Any experiment with resistors is an exercise in meter use. Each resistor to be used should be measured to determine its actual value, because the manufacturing tolerance allows too much variation for accurate calculations. The meter readings should be made as accurately as possible, taking care to use the meter on its most accurate range. On a digital meter this is the first range above the actual resistance value. If the meter is an analog type, the range should be chosen to keep the meter needle up in the right-hand end of the scale. This will not always be possible, but it is a good goal. The higher scale markings are too close together for accurate readings. The meter should also be carefully zeroed after each range change.

Resistance in a Series Circuit. To build a usable circuit, an assortment of resistors and suitable connecting wires must be available. Batteries may be used, with suitable holders for the individual cells. An adjustable, bench-type DC power supply is much more convenient.

As a first experiment, a string of three or four resistors is connected, head to tail, with the first and last resistor left open on one end. Series-connected resistors form a single circuit path that passes through each resistor in turn. A resistance measurement taken across the group will equal the sum of all the resistance values. This is the total resistance (R_T). It can be confirmed with the following formula for total resistance in a series circuit:

$$R_T = R_1 + R_2 + R_3 \ldots$$

Parts of the circuit can also be measured, with similar results. For example, it there are only two resistors between the meter probes, the measured resistance will be the sum of those two resistors.

Current in a Series Circuit. When a power supply is connected across the series group, it has to force current through all the resistors at the same time. Current does not flow in spurts around the circuit: When electrons move in one part of the circuit, they also move everywhere else at the same time. This is true if the circuit path is one inch long, or 10,000 miles. In the series circuit, if there is current in any one resistor, there is current in all of them, equally and simultaneously.

The amount of current can be measured with an ammeter or calculated using Ohm's Law. The current value depends on the voltage applied to the circuit and the total circuit resistance. It does not depend on any single resistance value. To the supply, the group of resistors looks like one large resistor with a value equal to the total resistance. For accurate results, the voltage used in the calculation should be the actual measured value taken across the group of resistors with the circuit in operation. Ohm's Law would be used in the form

$$I = \frac{E}{R_T}$$

The calculation can be confirmed by installing an ammeter in the circuit, in series with the loads. When all measurements are taken with reasonable care, calculated values will be very close to the measured values, but it is unlikely they will agree exactly. Although electricity behaves with absolute precision, it is very difficult to measure with absolute precision. Fortunately, it is possible to get very close.

An interesting aside to this experiment can be used to clear up a common misunderstanding. If the ammeter is properly connected into the circuit at several different points along the circuit path, the readings will all be the same (see Figure 4–13). Multiple ammeters can also be used, if preferred. They must, of course, be in series with the loads, but can otherwise be placed anywhere in the circuit. This clearly demonstrates that electrical loads, such as the small resistors used in this example, do not consume the electrons that flow around the circuit. The rate of electron flow (the current) is the same everywhere. The electrons that enter the load at one end will exit at the other and continue on around the circuit.

Electrons carry energy in the form of electrical pressure, just like hydraulic oil carries energy as pressure from the pump. The electrons are not fuel: They are not used up by the load, and they only give up some of the energy they have received from the supply. All parts of the circuit are filled with electrons—when one moves, they all move. The supply voltage pushes the electrons out of the negative terminal and around the circuit to the positive terminal. Electrons are not lost when they are forced through a resistance, but the energy they carried is converted to heat.

A series circuit has, by definition, only one circuit path. The rate at which the electrons flow through the circuit is the same at every point. There is no more current flowing into the first resistor than in any of the resistors that follow. This is why a current reading can be taken at any point in the circuit, and the result will always be the same.

Voltage in a Series Circuit. When current flows in a series circuit, voltage measurements across the individual resistors show that each of them receives only a fraction of the total voltage applied across the loads as a group. These individual voltage values are referred to as voltage drops. They represent the portion of the total voltage that is lost or dropped as the circuit current flows through each individual load.

FIGURE 4–12 Current is the same at all points in a series circuit.

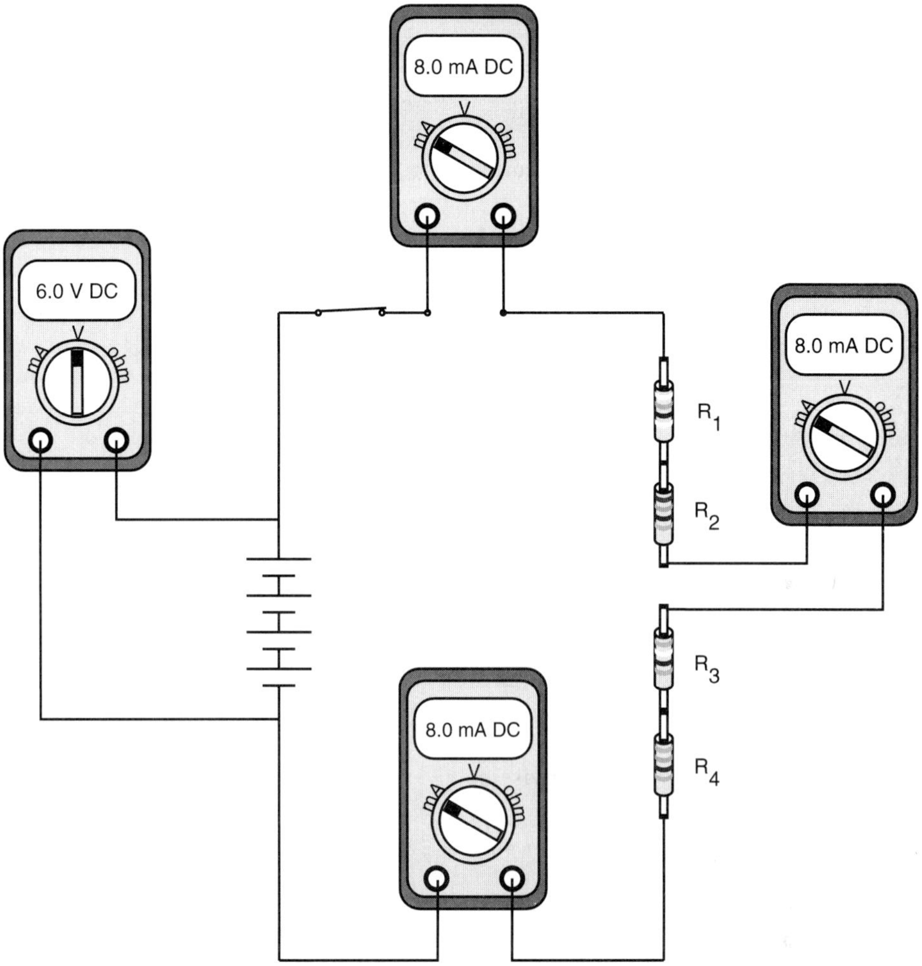

Voltage drop can be measured directly, or it can be calculated. There are several different formulas that can be used. One of the most convenient and consistently useful is Ohm's Law.

$$V = I \times R$$

The E used in all previous versions of this formula has been replaced with the letter "V". The purpose is to differentiate between applied voltage (E) and voltage drop (V). This is not a universal method of symbolizing these voltage values, but it is reasonably common and will be used consistently throughout this book.

The current (I) for a series connection is the same in all parts of the circuit. Therefore, current only has to be found once. It can be measured by inserting a meter at any convenient point in the circuit, or calculated either from the applied voltage and total resistance, or from any of the individual voltage drops and the related resistance value (see Figure 4–14). Once the circuit current has been found, it can be used to calculate voltage drops

FIGURE 4–14 Circuit current can be found from applied voltage and total resistance The voltage drop across one load in a series circuit can be found from the circuit current and the resistance of that load.

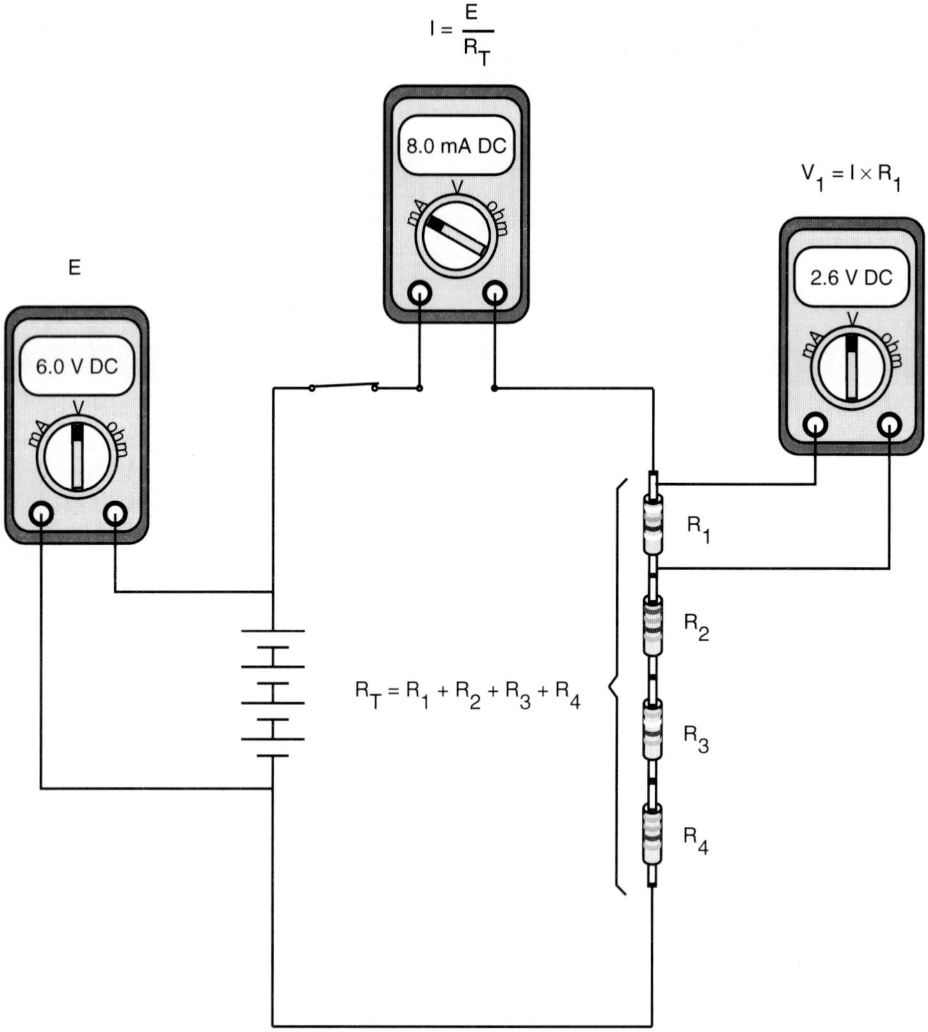

across any of the remaining resistors. When this experiment is done carefully with attention to accurate measurements, the measured and calculated values will be very close.

The sum of the voltage drops around any circuit must add up to the applied voltage. Electrical circuits operate in perfect balance. When a circuit is energized, the current increases until a balance is reached. Balance occurs when the total voltage drop is exactly equal to the applied voltage. This increase from no current to maximum current is virtually instantaneous.

In all circuits, the total voltage applied has to be dropped somewhere. In a series circuit, it is distributed across all of the series-connected loads. These voltage drops are proportional to the various resistance values. The loads with the greatest resistance have the most effect on the current, and therefore have the largest voltage drops.

Current into and out of any point in a circuit is always the same. Electrons are not consumed by the loads through which they pass. The extra energy they carry is lost in the load, but the actual rate at which they flow (the current) is the same entering or leaving. In a series circuit, the loads are all part of the same circuit path. The current is therefore the same in all parts of the circuit. The total energy taken from the power supply is divided up among the loads through the differences in their resistances and the resulting differences in the voltage drops.

Parallel Circuits

In the more versatile parallel circuit, the voltage is constant for all the loads. This makes it easier to design and install equipment and allows for individual control of each load. Just as for the series circuit, the energy into a parallel circuit divides among the various loads through differences in load resistance, but in the parallel circuit this results in different individual load currents instead of different voltage drops.

Parallel connections are made across two points of the circuit that are electrically the same. In a typical parallel connection (shown in Figure 4–15), all of the loads are connected directly across the supply. Each receives the same supply voltage, but draws current independent of all the others. This gives the parallel connection some very important advantages over the series connection. The voltage across a parallel-connected load is independent of the resistance of other loads. The voltage across a series-connected load depends on the combined resistance of all the other loads. Any reasonable change in a parallel circuit—one that does not overload the power supply—has virtually no effect on the operation of any other part of the circuit. Any change in a series circuit can have a dramatic effect on all of the loads in the circuit.

The independence of parallel-connected loads makes it practical to install multiple loads on each circuit. This greatly reduces installation costs, because groups of minor loads, such as room lighting and general-purpose receptacles, can be supplied from a single circuit breaker or fuse, through a shared two-conductor supply cable.

Parallel connections allow the voltage ratings for equipment to be standardized. This makes inexpensive mass-produced electrical goods possible. Once the supply voltage is known, the design of any electrical device can be adjusted accordingly. For example, it is possible to make 100 W lightbulbs for a wide variety of voltages, but each bulb has

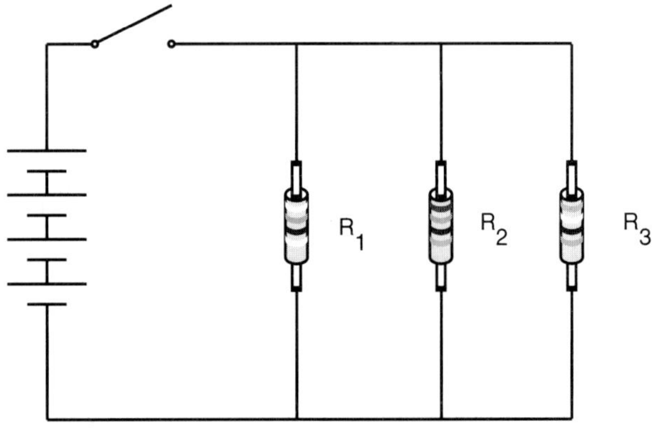

FIGURE 4–15 A typical parallel connection.

to be different. If the voltage is fixed, thousands of identical bulbs can be made at very low cost.

Another major advantage of the parallel connection is that it allows for individual control of the loads. This is true even when multiple loads are connected to the same circuit. Each of them can be turned on and off independently, without affecting operation of the others. For example, there may be several light fixtures on the same circuit, yet they can all be controlled independently.

A Practical Example: Resistors in Parallel

A very informative experiment can be conducted using the same basic materials used in the series circuit example (Figure 4–14): three or four small resistors, a DC power supply, interconnecting wires, two multimeters, and a simple switch. The experiment can be improved considerably by the addition of extra switches and three or four low-voltage lamps with holders.

Each of the resistors should be tested with an ohmmeter, and the actual resistance values recorded for use in later calculations. To put them in parallel, one end of each load is connected to one side of the supply, and the other end to the opposite side of the supply. The connecting wires can loop along from one load to the next, but there must be a complete conductive path—from the supply to each load—that does not pass through any other loads.

With the circuit connected and the supply set to a suitable voltage output, a voltmeter test across the loads will show that each receives the full supply voltage (see Figure 4–16). This is proper for a parallel circuit. The voltage may drop off slightly when the circuit is a

FIGURE 4–16 The voltmeter can be connected across any of the loads, and it will read the full supply voltage. Load voltages, V_1, V_2, and V_3 all equal the supply voltage, E.

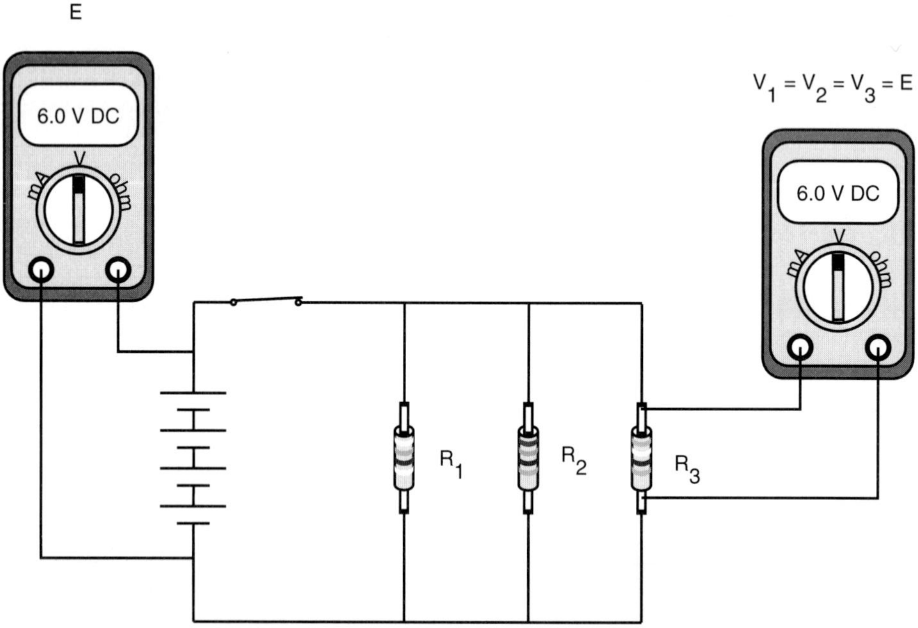

long one, due to the losses in the supply conductors. However, unless the situation is extreme, the voltage drop in the conductors will be only a very small percentage of the voltage applied. In real-world electrical installations, the maximum allowable voltage drop is limited by the applicable electrical codes. On very long circuits, the size of the conductors has to be increased to keep the drop within acceptable limits.

To measure total current in the circuit, an ammeter is inserted in one supply line, ahead of all the loads. The total current is the sum of all the load currents. Each load draws current independent of the others, but the supply must provide current to all of the loads at once.

Individual load currents can also be measured, but it generally requires some ingenuity, particularly in a small circuit. The ammeter must be relocated to read each of the load currents, one at a time. Connecting the meter correctly requires some thought, and possibly some rearrangement of the interconnecting leads. The meter must be inserted at a point where it will see only the desired individual load current, not a combination of two or more loads (see Figure 4–17). Care must also be taken to avoid placing the meter across the supply. This will, of course, short circuit the full output of the power supply through the meter. The usual result of this error is a tripped circuit protector or blown fuse, either in the meter or in the power supply circuit.

The individual load currents should be recorded as they are measured. They can be summed to a total and compared to the initial total current measurement. Any difference should be very small. It is unusual to get a perfect match in these results. The ammeter does have a small amount of internal resistance. This changes the circuit slightly each time the meter is moved, causing small errors in the values. There are also limits to accuracy in all measurements, so small measurement errors are inevitable. Note, however, that large errors are almost certainly due to mistakes in meter placement, misinterpreted readings, or some accidental change in the circuit.

Parallel loads are essentially independent in their behavior, and can therefore be calculated independently. Each of the individual load currents can be found with Ohm's Law. The formula below has been modified slightly by the addition of subscripts to the current and resistance values. For example, I_1 stands for the current in R_1. The calculation is repeated for each load to find I_2, I_3, I_4, … from the respective resistances.

$$I_1 = \frac{E}{R_1}$$

The voltage used in the calculation is the measured load voltage, which is equal to the supply voltage (E). The resistances used are the actual values, measured and recorded before the circuit was built. The calculated individual load currents should be very close to the measured values obtained in the previous step. Small errors are almost impossible to eliminate; large errors usually indicate mistakes in connections or meter settings, or misinterpretation of a meter reading. As before, the sum of the individual currents should be very close to the total.

At this point the only value that has not been calculated or measured is the total circuit resistance. For parallel circuits, total resistance can be confusing, at least initially. Unlike the series circuit, there is no obvious connection between the total resistance and the resistance values of the individual loads. It should be understood that when loads are connected in parallel, each additional load does not add to the resistance of the circuit: Instead, it adds an additional current path. Each time another resistor is

FIGURE 4–17 Although individual meters are shown on each load in the diagram, it would be far more likely that a single meter would have to be relocated as required. To add an ammeter, the circuit must be opened at some point, and the connection completed through the meter. To remove the meter and leave the circuit functional, the circuit must be reconnected.

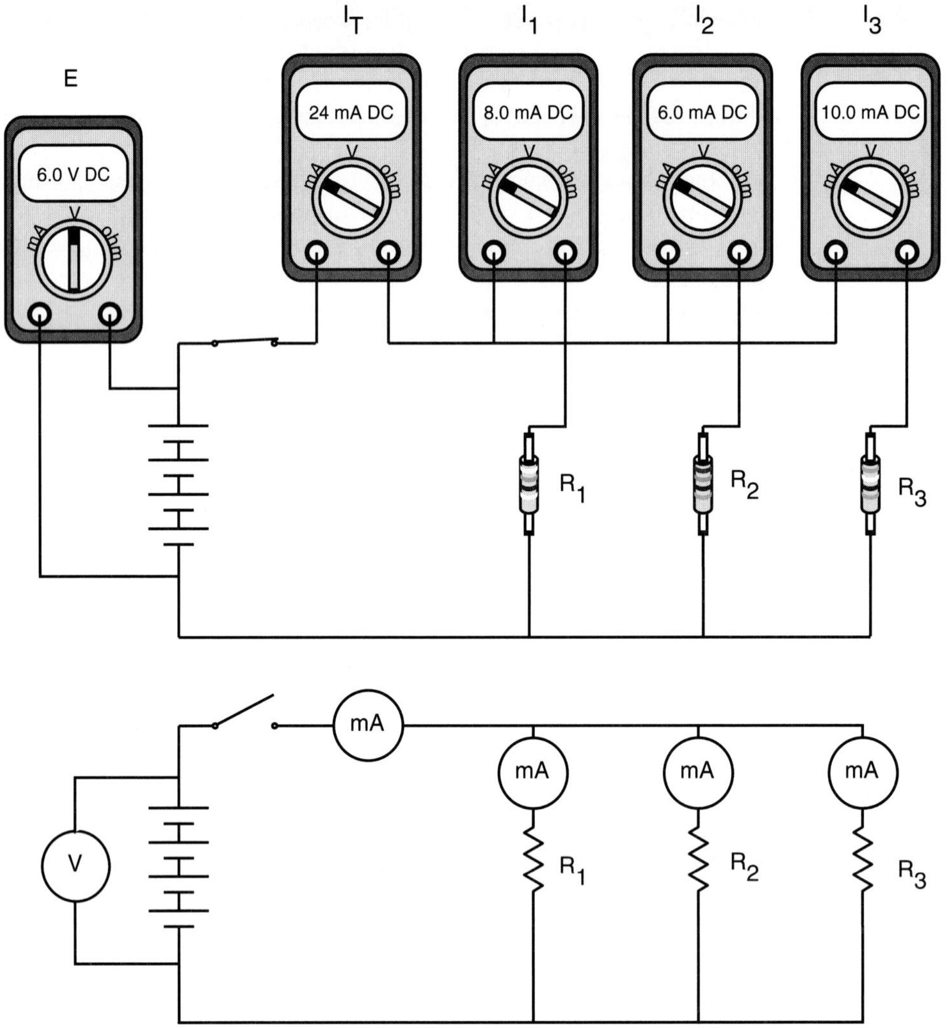

added to the circuit, the total current from the supply must increase. For this to happen, the total resistance must also have been reduced. This is reverse to the effects of adding resistance in series.

Total circuit resistance for parallel-connected loads can be found in three ways, all of which produce the same result (assuming the measured values are accurate). It can be calculated from total current and applied voltage, measured directly with an ohmmeter, or calculated from individual resistance values. To find total resistance from applied voltage and total current, use the Ohm's Law formula with these values. The current value used must be the total for the complete circuit, not for one or more of the individual loads. The resulting resistance value is the circuit total. This formula actually applies to any type of circuit connection. The result is less than the lowest resistance in the parallel group.

$$R_T = \frac{E}{I_T}$$

To find total resistance by measurement, disconnect the power supply and connect an ohmmeter across the circuit in its place. The ohmmeter will then "see" the circuit in the same way as the supply it replaced. The reading obtained is the total circuit resistance, and should be the same as the value calculated from supply voltage and total current. Note that the power supply must be disconnected, not just shut off. Many DC power supplies have components connected across the output terminals that are not disconnected when the switch is in the off position. This makes it impossible to get a proper resistance reading across the circuit.

There are two common formulas for finding total resistance from individual resistance values. The reciprocal formula is efficient and accurate if a calculator is used, and it works for any number of resistors. The product-over-the-sum method may be simpler for circuits with only two resistors, but it must be applied repeatedly for each additional resistance value. When using the reciprocal formula, traditional mathematics might suggest the use of a lowest-common-denominator approach. However, unless the resistor values are small whole numbers, this is not recommended. Using decimal fractions is generally much more satisfactory.

An appropriate calculator—one with either a reciprocal function or an accumulating memory—can be used to solve these problems quickly and accurately. The reciprocal key appears as a [$^1/x$]; an accumulating memory may have up to four keys. The "add to memory" key, usually marked [M+], stores a running total of values; the memory recall key, typically [MR], copies this stored accumulated value back to the calculator display so it can be used in further calculations. There will also be a memory clear function—possibly a separate key such as [MC], or combined with memory recall [MRC]. For accurate totals, the memory must be cleared before any new calculations are started.

An example of each of these methods follows.

$$R_T = \frac{1}{\frac{1}{R_1} + \frac{1}{R_2} + \frac{1}{R_3}}$$

Assume the following values. A 12 V DC supply provides power to four parallel-connected resistors. The resistors are rated at 100 Ω, 150 Ω, 330 Ω, and 1 kΩ. The voltage across each resistor is 12 V. The current for the first resistor is:

$$I = \frac{E}{R_1}$$

$$I_1 = \frac{12\ \text{V}}{100\ \Omega}$$

$$I_1 = 0.12\ \text{A or } 120\ \text{mA}$$

This calculation is repeated for each resistor, resulting in the following values:

$$I_2 = 0.08 \text{ A or } 80 \text{ mA}$$

$$I_3 = 0.0364 \text{ A or } 36.4 \text{ mA}$$

$$I_4 = 0.012 \text{ A or } 12 \text{ mA}$$

The total current is the sum of these:

$$I_T = T_1 + I_2 + I_3 + I_4$$

$$I_T = 0.2484 \text{ A or } 248.4 \text{ mA}$$

Total resistance can be found by measurement, as stated above, by removing the supply and connecting the ohmmeter in its place. This value should be 48.3 Ω. The exact value depends on the actual values of the resistors and the accuracy of the meter used. It may also be affected by other sources of error, such as small resistances in the conductors and connections. Note that the total resistance value is less than the smallest resistor in the parallel group.

Total resistance can also be calculated from total current and supply voltage.

$$R_T = \frac{E}{I_T}$$

$$R_T = \frac{12 \text{ V}}{0.2484 \text{ A}}$$

$$R_T = 48.31 \ \Omega$$

Total resistance, calculated from individual resistances, is:

$$R_T = \frac{1}{\frac{1}{R_1} + \frac{1}{R_2} + \frac{1}{R_3} + \frac{1}{R_4}}$$

$$R_T = \frac{1}{\frac{1}{100} + \frac{1}{150} + \frac{1}{330} + \frac{1}{1000}}$$

$$R_T = \frac{1}{\frac{1}{0.01} + \frac{1}{0.00667} + \frac{1}{0.00303} + \frac{1}{0.001}}$$

$$R_2 = \frac{1}{0.0207}$$

$$R_2 = 48.31 \ \Omega$$

This calculation can be done efficiently in a calculator using one of the following methods. For a typical scientific calculator, enter the first value (100) and press [1/x] to divide this value into 1. Press the [+] key to prepare for the second value. Enter the second value and repeat the same key sequence as before. Repeat this for the remaining resistance values, but on the last one, press the [=] key instead of [+] to bring the calculated decimal values to a total. Press the [1/x] key one last time to divide this total into 1. This completes the calculation. Note that this calculation procedure follows the formula exactly, solving the individual reciprocals, summing them, and then finding the reciprocal of the total.

For a four-function calculator with an accumulating memory, use the sequence 1 [÷] 100 [=], [M+], 1 [÷] 150 [=], [M+], 1 [÷] 330 [=], [M+], 1 [÷] 1000 [=], [M+], 1 [÷] [MR], [=]. This should produce the same results. To avoid the most common errors, make sure the memory is clear before starting, and press the [M+] key only once for each entered value. The [MR] key (memory recall) will bring out the running total that was stored in memory so the final division into 1 can be made.

These techniques work with most calculators. They are not tricks, just good use of calculator features, and they make it much easier to solve problems that would otherwise be very tedious. Both of these methods also maximize accuracy by keeping all intermediate values in the calculator. That way the final answer contains no unavoidable rounding errors.

Total resistance can also be calculated using the product-over-the-sum formula. This technique is restricted to two resistance values at a time, but can be repeated as necessary for multiple resistors.

$$R_T = \frac{R_1 \times R_2}{R_1 + R_2}$$

It should also be noted that the total resistance of a group of equal resistances can be found by dividing the individual value by the number of resistors in parallel. For example, four 100 Ω resistors in parallel have a total resistance of 100 ÷ 4 = 25 Ω.

Combination Circuits

Circuits that combine both series and parallel elements operate with characteristics of both. Calculating all the voltages and current in combination circuits can be very time consuming because there are always multiple steps involved. The circuit must first be simplified to a purely series or parallel form. For circuits that are reasonably well organized, the procedure described below can be used. For circuits that are a confusing mess, refer to the following section about organizing badly laid out drawings. The method described there is a very effective way to accurately redraw a complex, or particularly confusing, circuit, so it can be simplified without error.

Simplifying a Combination Circuit

To simplify a combination circuit, first identify any series or parallel groupings within the larger circuit. Calculate an equivalent resistance value for each of these groups. The equivalent value is the total resistance the group would have in isolation from the surrounding circuit. It can be found with the appropriate formulas used for series and parallel circuits. The original circuit should then be redrawn, with the appropriate equivalent values substituted for the groups they represent. This process is repeated, if necessary, until the simplified circuit contains only parallel- or series-connected loads, but not both.

Once the circuit is no longer a combination of both series- and parallel-connected loads, it is solved completely to find total resistance, total current, and all the possible voltages drops or individual load currents. The circuit is then reconstructed by replacing the equivalent resistance values with the original series or parallel groups. The calculated voltage and current values are retained in their proper places in the reconstructed circuit. Note that the voltage drop calculated for an equivalent resistance is the voltage across the group of loads it represents. Likewise, current through an equivalent resistance will be the current that flows into and out of the group it replaced.

Once the circuit has been reconstructed, the values found for the simplified version can be used to calculate the voltage drops and currents that were previously impossible to find. This process will require multiple steps if there are groups within groups. All calculations should be double-checked by totaling the voltage drops around each closed loop. These must match the voltage applied to that loop. Check current totals into and out of each circuit point: These values must always be the same. (This basic truth about current flow is Kirchhoff's current law.) The voltages across each branch of any parallel grouping must all be the same. The current for any type of series grouping must be the same at each point where the circuit is reduced to a single path (see Figure 4–18).

Bridge Circuits and Similar Problems

When a circuit simplifies to a form where one or more of the loads are part of more than one current path, the circuit cannot be solved using the methods given above: Specialized mathematical approaches are required. One of the more common forms such a circuit can take is a bridge. Because bridge circuits are often used in heating, ventilation and air conditioning electronic control circuits, an example of such a circuit is provided below.

In its simplest form, a bridge contains five loads connected in an H pattern, with the sides of the H connected in parallel across the supply. The load connected across the center of the H is part of two different current paths. The voltages and currents in this type of circuit cannot be calculated by the previous method, because the circuit cannot be reduced to a purely series or parallel form.

Circuits like this are easiest to solve by applying procedures such as those developed from Thévenin's theorem. First, remove the resistor that is of interest and find the voltage across the opening. This "open circuit voltage" is called the Thévenin voltage (V_{TH}). Next, "replace" the power supply with a short circuit, and calculate the resistance "seen" by looking back into the circuit from the same two points. The value obtained by this procedure is called the Thévenin resistance (R_{TH}).

The Thévenin equivalent circuit used to find the voltage and current for the resistor that was originally removed consists of that resistance, in series with the Thévenin resistance, across the Thévenin voltage. The solution to this simple circuit is the voltage and current that resistor experienced in the original circuit (see the example in Figure 4–19).

This example is a very simple application of Thévenin's theorem. For a more detailed examination of this and other methods, consult any good basic electronics text. Analysis of complex resistive circuits is an important part of understanding electronic circuit operation and design. It is far less critical in the study of electrical power, particularly at this basic level. However, a technique such as Thévenin's can be very useful for solving difficult circuits.

FIGURE 4–18 In this simple example, all the resistors are the same value (100 Ω). The first step of simplification reduces the two series groups to their equivalent values (200 Ω each). The second step reduces these two 200 Ω values in parallel (200//200 = 100 Ω), which leaves a series circuit of three 100 Ω resistors on a 30 V supply. The voltage drops will be 10 V each. The 10 V drops can be reinserted into the original circuit, where they can be used to find the remaining values. Note that there are 10 V across the two series groups (2 + 3) and (4 + 5), because they are in parallel.

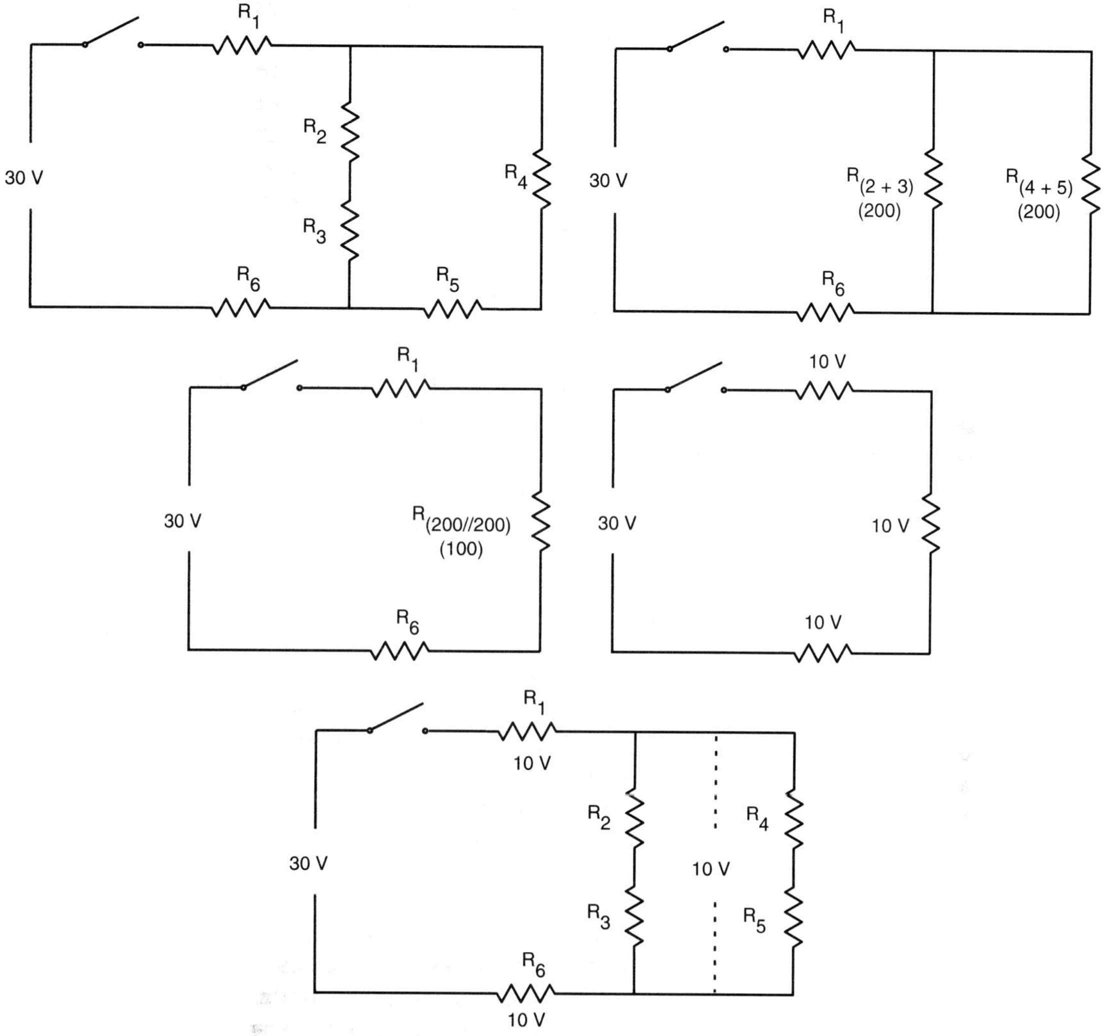

Organizing Badly Laid-Out Drawings

For circuits that are poorly organized, it is best to redraw them into a form that is easier to follow. Unfortunately, it is very easy to make a mistake, with the result that the new circuit is electrically different from the original. There is a method that can be used to avoid this problem.

FIGURE 4–19 The bridge circuit cannot be simplified, using simple deductive approaches, into a form that is only series or parallel. Applying Thévenin, one resistor is removed. The open circuit voltage (V_{TH}) is calculated (the difference between V_2 and V_4), and the circuit is reorganized to find total resistance with the supply shorted. The final circuit is a simple series problem with two resistance values and one voltage.

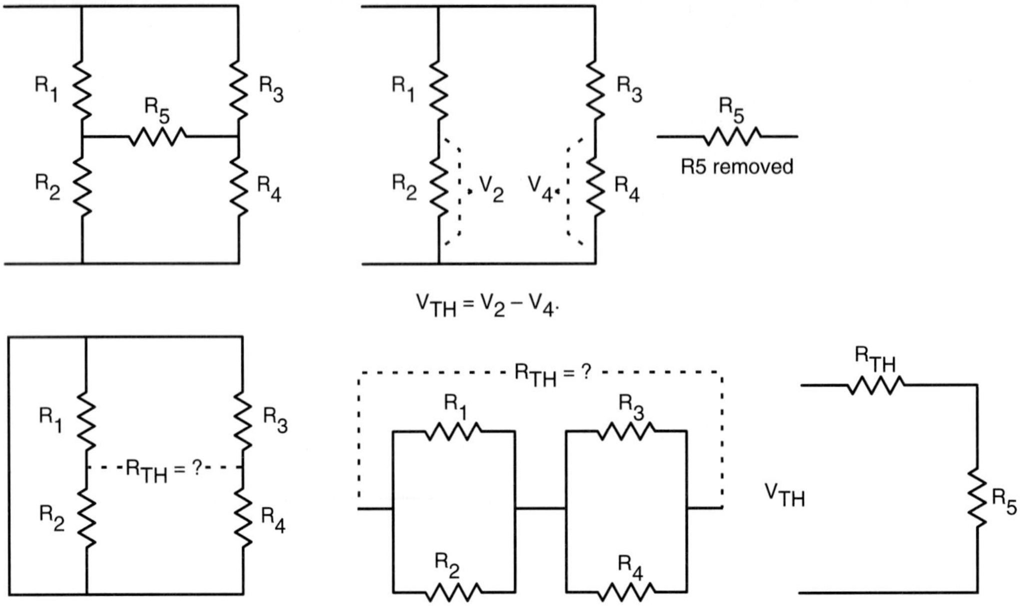

First identify all directly connected points with a unique number. A direct connection means the terminal points are connected by circuit conductors only, not through any other device. The usual procedure is to start at one side of the supply. Mark this point number 1. Locate each device terminal with a direct path back to that point, and assign each of them the same number (in this case, 1). When all of these points are identified—there may be only two, or there may be a dozen or more—choose any convenient device that has a number 1 on one terminal. It is best to pick one that looks like a reasonable start into the circuit, though the choice is not critical. Mark the opposite terminal of this device with a number 2. Locate every terminal with a direct connection back to this second point and assign each of them the number 2 as well. Select another device and begin again, this time with a number 3. Repeat this process as necessary until every terminal is numbered. No terminal can have more than one number; no terminal can be unnumbered. If a device has the same number on each end, it is shorted out and will take no active part in circuit operation.

Once the numbers are assigned, redraw the circuit using the numbers as a guide for layout and as a check against connection errors. Avoid angling conductors or devices. Draw everything in straight lines either vertically or horizontally on the page. Group like-numbered terminals into multiple parallel paths. Organize series-connected devices into straight lines. For a very complex drawing, several rough sketches may be necessary before an easy-to-follow layout becomes clear.

Once the redrawing process is completed, the circuit can be simplified and solved using one or more of the calculation methods described above (see the example in Figure 4–20).

FIGURE 4–20 Applying numbers to common points provides a guide for accurately redrawing disorganized circuits.

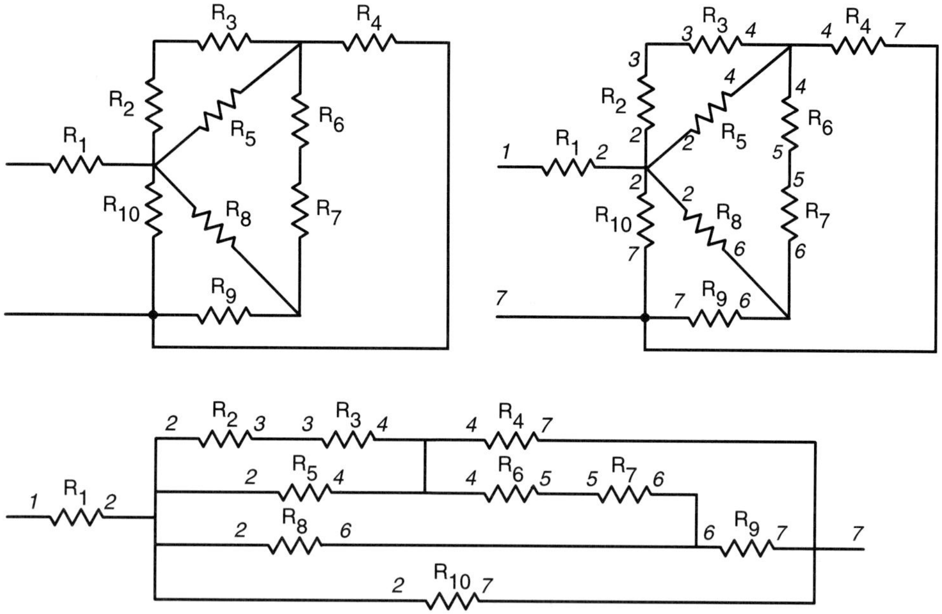

Summary

To find current and voltage in a series circuit, sum the series resistances to a total. Find current from applied voltage and total resistance, and then use this current to find the voltage drops in each load. The sum of the voltage drops is equal to the applied voltage. To calculate current or voltage in a parallel circuit, first apply Ohm's Law to each load individually. The individual load currents can then be summed to find total circuit current. The total circuit current and applied voltage can be used to find total resistance. Resistance can also be found with special formulas using only the individual resistance values, or it can be measured by substituting an ohmmeter in place of the supply. Note that total resistance is always less than the value of the smallest resistance in the circuit.

Combination-circuit calculations are an interesting mathematical exercise, but may be of limited value to personnel involved in electrical service work. However, the underlying principles are worth knowing, because they can be critical to correct interpretation of meter readings taken while troubleshooting. Real equipment often includes combination circuits. Readings taken within these circuits seldom make much sense unless the operation of such a circuit is understood.

Electricity follows the same set of natural laws in a complex circuit as it does in a simple one. Current into any point or circuit section is always equal to current out. Voltage drops around a loop always add up to the voltage applied.

There are important differences between circuits that are entirely series or parallel, and circuits that are a combination of both connection types. In combination circuits, whole sections of the circuit can be simplified to "equivalent" series or parallel values. These individual equivalent values allow the circuit to be solved in a simplified form, and the values taken from this solution can then be applied to the original circuit. This process usually allows the circuit to be solved completely. Circuits that simplify to a bridge or similar pattern that is not totally series or parallel cannot be solved without more specialized approaches.

When simplifying problem circuits or troubleshooting an operating circuit, remember that all solidly connected points of a circuit are electrically common. Electrically common points are always at the same voltage with reference to all other points of the system.

4.5 Conductors and Overcurrent Protection

Conductors and Current Ratings

Even excellent conductors such as copper have some resistance. When current passes through the conductor, energy is lost to this resistance and turned into heat. There are several detrimental results to this energy loss. System efficiency is reduced, because energy lost in the conductors is not delivered to the load. The voltage drop across the conductor resistance also reduces the voltage at the load end of the circuit. Low voltage causes problems for the load, and is not acceptable under code regulations. If the conductor losses will result in too much voltage drop, a larger conductor must be used. This problem occurs in long supply runs, and can greatly increase the cost of an installation.

Conductor heating also causes problems: Even when the run is short and voltage drop is not an issue, the conductor must be large enough—have a low enough resistance—that it will not overheat. Excess heat can damage the conductor insulation and, in more severe cases, can be a fire hazard. Even relatively low temperatures can reduce insulation life. For the conductor to stay at a safe operating temperature, the heat must be dissipated from the conductor to the surrounding material. It must pass through the conductor insulation and any cable jacket, armor, or conduit that may be used to enclose and protect the wiring system. If the run is a typical multiconductor circuit using multiconductor cable or multiple runs of single conductor in conduit, the heating effects of the current in all of the conductors together must be taken into account.

Current ratings for conductors are based on allowable temperature rise and conditions of use. Many factors affect the final value, but the most obvious ones are conductor material and size. These determine the conductor resistance, and therefore the amount of heat generated for each ampere of current. Copper and aluminum are the most commonly used conductor materials. Copper has a lower resistance than aluminum; therefore, an aluminum conductor must be sized larger than a copper conductor to carry an equivalent current. The difference is approximately one trades size—a #4 aluminum to a #6 copper, for example—though this may vary. This does not mean that aluminum conductors are inferior to copper. Both are in common use and both have advantages in certain applications. Copper is more accepted in smaller sizes because it has superior mechanical properties and gives less problems at connections. Aluminum is popular in larger sizes, and is almost universal for aerial applications because of its light weight. These overhead conductors do, however, have some steel in their construction to provide enough strength for the long spans between the supporting poles or towers.

Conductor size affects current capacity because larger conductors have proportionally lower resistance than smaller ones. This simple size relationship holds up quite well with currents up to a few hundred amps. The higher the current, the larger the conductor.

However, there are sensible and practical limits to conductor size. For very high current circuits, it is common to use multiple smaller conductors in place of one very large one. The smaller conductors are more readily available, are much easier to handle and install, and cool better than one larger one because of their greater surface area.

The insulation type and thickness also affect overall current ratings, but there are two separate reasons. First, there is a certain maximum temperature the insulation can handle without gradual breakdown and failure. This limits the maximum operating temperature of the conductor and, therefore, its maximum current rating. Second, the electrical insulation also acts as a thermal insulation, trapping the heat generated inside the conductor. This causes the heat to build up and, of course, also reduces the conductor's current handling ability. These factors are combined into a current rating and a temperature rating for every type and size of conductor.

Conductor temperature ratings are part of the standard specifications marked on the conductor insulation or cable jacket. Common ratings include 60°, 75°, 90°, and 105° Celsius. They represent the maximum allowable sustained temperature for the conductor or cable, and are based on the thermal characteristics of the insulation. The temperature rating affects the overall current rating, independent of conductor size. A high-temperature conductor may have a higher current rating than one with a lower rating, even though they are the same size. It should be understood that this increased current capacity comes from the higher permissible operating temperature, not from a lower resistance. A high-temperature conductor may be less expensive to install, but will have a greater energy loss than the larger conductor that would otherwise have been required.

The conditions into which the conductors are installed also affect the current rating. The current-carrying capacity is reduced when the location adversely affects the natural cooling of the conductor. When a conductor is installed in a location where the ambient temperature is high, such as wiring around a boiler or other heating equipment, the conductor is "de-rated" to a lower allowable maximum current. Conductors buried in thermal insulation are also effected. Conductors run in "free air," such as overhead wiring suspended on utility poles, generally have a higher allowable current rating because of the unrestricted flow of cooling air around them.

Note that some situations require high-temperature conductors because of the conditions into which the conductors are installed, not for the increased current capacity. This is true for conductors in ceiling-mounted lights, particularly for recessed "pot light" type fixtures, where much of the heat from the lamp is trapped inside the fixture. It is also true of heating equipment wiring such as the power and control conductors used in furnaces and similar equipment. Using low-temperature insulations in such situations will always cause problems, because the insulation becomes brittle and breaks away from the conductor.

Overcurrent Protection

All types of conductors must be protected from excessive current. The rating and location of this "overcurrent" protection are an important part of all electrical installations. When the current is kept at or below the conductor's rated maximum, its temperature will stay within an acceptable range. Moderate overcurrent, if it is sustained, will heat the conductor enough to degrade the insulation, reducing its life. Very high current will generate enough heat to cause insulation failure and, possibly, a fire.

The worst situation is usually caused by short circuits in faulty equipment or wiring. A severe fault like a short causes a very high overcurrent. These faults are dangerous and destructive, so the supply must be cut off as quickly as possible when they occur.

Fortunately, it is relatively easy to protect electrical circuits from the dangers of overcurrent. The most common method is to pass the current through a fuse or circuit breaker before it continues on to the external circuit. The best location for the fuse or breaker is right at the supply end of the circuit. With the overcurrent protection mounted as close as possible to the source, the chances of a fault occurring in an unprotected part of the wiring are virtually eliminated. A problem anywhere in the circuit will blow the fuse or trip the circuit breaker, cutting off the power.

The supply end of a circuit is not necessarily the true "source" of electrical energy, it is just the point where that circuit connects to yet another circuit with a higher current rating—typically a larger conductor protected by an overcurrent device with a higher current rating. In most cases this larger circuit will, in turn, connect to yet another circuit that has a still higher rating, and so on, all the way back to the actual supply—the utility generators. (It should be noted here that, in AC systems, transformers are used to change the voltage level. Working back into the system leads to circuits that can supply larger and larger amounts of energy, but along the way there will also be changes in voltage. Each time the voltage is increased, the current required to deliver the power becomes proportionally less.)

In a typical small building, such as a house or small business, the utility supply first comes through either a main breaker or a main switch with a set of service fuses. These supply a set of conductive bars, called busbars, inside the electrical panel. Each circuit used in the building is connected to these busbars through its own circuit breaker or fuse. In this way, each circuit is protected against overcurrent, right at the point of connection to the higher current rated supply.

The need for protection extends to all electrical devices, not just the conductors that supply them. The fuse or breaker used to supply a typical circuit of 120 V outlets is rated at 15 or 20 A. If this rating is too high to adequately protect a piece of equipment, an additional fuse or a breaker with a lower current rating is placed into the circuit ahead of the sensitive circuitry. There are hundreds of examples of this. Internal fuses or breakers, are found in all types of electrical and electronic equipment. They allow the manufacturer to use small conductors and low-current components, without the danger of exposing them to a high-current fault. When a problem occurs inside the equipment, the lower rated internal breaker or fuse will open the circuit long before the circuit breaker in the supply panel can react. This minimizes the damage and disruption, and eliminates a possible fire hazard.

Fuses

A fuse or circuit breaker will automatically disconnect power to the circuit if the current rises above a safe level. Of the two, the fuse is the simplest and, by some accounts, the most reliable. The disadvantage is that the fuse must be replaced if it blows. Where overloads are rare and spare fuses are kept in stock, a fuse may be the cheapest and most effective protection available. In situations where overloads are frequent or spare fuses are a problem, fuse protection can become a serious frustration.

Fuses are categorized according to a number of different characteristics. These include the physical style and dimensions, and the electrical ratings of voltage, maximum current, and current-interrupting capacity, and special features such as current limiting or time

FIGURE 4–21 On the left is a screw-in plug fuse; front to back are a miniature glass fuse, a time-delay cartridge fuse, an industrial HRC II fuse, and a cartridge fuse with knife-blade ends.

delay. The physical styles include plug fuses—which are the screw-in types commonly found in older electrical panels and in most electric ranges—and a variety of cartridge types. Cartridge fuses may have ferrule or knife-blade ends for actual contact with the fuse holder. Some high-capacity fuses must be bolted in place. A variety of small-dimension fuses with relatively low current ratings are commonly found in instruments and electronic equipment. Some common fuses are shown in Figure 4–21.

Codes require that fuses with certain special properties cannot be replaced by other types. For this reason, some fuses are a unique size or shape, or have unusual ends. The holders for these fuses will not accept anything but the correct type. This is not an absolute, however, as there are some classes of fuses that do fit the same holders. There are situations where a blown fuse can be replaced with a different one, with very similar ratings, that is unsuitable and potentially dangerous to use in that application. Incorrect replacements may be made through ignorance, because of availability of one type but not the other, or for reasons of false economy.

Fuse design dates back to the early days of electrical experimentation. The original fuses were short lengths of wire that would "fuse" or melt at a relatively low temperature. Excessive current through the fuse wire would cause it to melt and break the circuit. Different gauges of fuse wire had different current capacities. Fuse wire is still in use in some countries. It sounds inconvenient, but with a well-designed holder it is actually quite simple to use. It is also very inexpensive. Several feet of fuse wire costs only a few cents, and each replacement takes only a few inches.

Replaceable fuses followed fuse wire. These were easier to use and, more importantly, they also contained the arc more effectively. This allowed them to safely interrupt higher fault currents. Some cartridge fuses use renewable fuse elements. The ends of the cartridge can be unscrewed and the fuse link replaced. Most fuses are one-time-use only. Virtually all of the more sophisticated fuses are made this way. This allows the manufacturer to use fuse elements and arc-quenching fillers that would be impractical in a renewable design.

The power circuit fuses discussed here are categorized by voltage, current rating, and interrupting capacity. They are divided into specific classes that are designated by a letter code. The letter codes are based on National Electrical Manufacturers Association (NEMA), National Electrical Code (NEC), Underwriters Laboratories (UL), and Canadian Standards Associations (CSA) specifications. These specifications relate to the physical dimensions and electrical characteristics of the fuse. There are many similarities between U.S. and Canadian ratings, but there are also some differences.

Fuse voltage ratings are generally either 250 or 600 V. These two categories are intended to cover a range of voltages up to and including these ratings. The fuse voltage rating is a maximum value, not an absolute requirement. Circuits supplied from supplies rated at 250 V or less are normally protected by a fuse rated at 250 V, because that is the closest rating. Circuits rated above 250 V (251–600 V) are protected with 600 V rated fuses. A fuse is unsafe in a circuit above its voltage rating, but provides proper overcurrent protection for a circuit operating at a lower voltage.

The current rating of the fuse—the current value printed on it—is generally its maximum pass-through rating. In an environment that is within normal temperature guidelines, the fuse will carry this current indefinitely. How the fuse responds to current values above this rating is somewhat different for different types and classes of fuses, but none of them blow immediately.

Within limits, this delay is desirable, and even essential, for proper circuit operation. Many types of electrical equipment draw high current for a short time when they are first energized. The delay stops the fuse from blowing during this brief start-up period. A simple screw-in plug fuse opens in about three seconds with a 200% overcurrent—twice its current rating. This excessive current draw has no effect on the fuse, as long as it is very short term. Time-delay or dual-element fuses are specifically designed to be slow. They take twelve seconds or more with the same 200% overload. These fuses are classed by UL as time-delay fuses and marked with an identifying letter D. The tripping time for all fuses is inversely proportional to the square of the current: They open much faster on a high current than on a modest overload.

The interrupting capacity of a fuse is quite different from its current rating. Since electricity first became commercially available in the late 1800s electrical systems have grown enormously in size and power. The current that such powerful systems can deliver can easily exceed the 10,000 A interrupting rating of a typical older style fuse. When a low-resistance fault occurs, the current rises faster than the fuse can react, and the arc is so severe the fuse cannot keep it contained. The case may explode or rupture, blasting a fiery arc into the equipment. A fuse failure of this type can be extremely destructive.

Fuses rated for fault currents above 10,000 A are typically rated at 100,000 or 200,000 A. These fuses are classified as high interrupting capacity or high rupture capacity (HRC) fuses. Such fuses are common in large commercial and industrial installations. Most of them have a unique size, shape, or mounting feature to prevent replacement with a different type. This fuse-rejection characteristic is required by codes. One category of fuse, Class K, has the same dimensions as a standard 10,000 A capacity Class H fuse. A second category, Class R, has the same dimensions, but also has a rejection feature in the form of a grooved ferrule or notched blade. A Class R fuse holder will reject standard Class H fuses, but a Class R fuse will fit in a Class H holder.

Never attempt to defeat fuse-rejection features and, if a fuse installation has been upgraded by installing high interrupting capacity Class K fuses in standard holders, do not use the lower rated Class H fuses as replacements.

HRC fuses are made in two categories: HRC Form I and Form II. HRC-I fuses are a standard high interrupting capacity type available in a variety of classes. HRC-II fuses have a much slower response. They are a British design intended to provide short-circuit protection in circuits such as motor feeders, where there is always another form of overload protection. HRC-II fuses do not follow standard dimensions.

Current-limiting fuses obtain their designation by being particularly fast. They must reliably clear a specified overcurrent within one-half cycle of the AC current wave, about 8 milliseconds. This fault current is specified by regulating agencies and is much higher than the normal current rating of the fuse. Current-limiting fuses are used to provide extra protection in circuits where the available fault current is high enough to cause equipment damage. These fuses can be very effective, because not only are there limits to how fast a fuse can react, there are also limits to the rate at which the current can rise. Even with a very low resistance fault, the circuit current cannot reach its maximum instantaneously. Current-limiting fuses reduce the severity of the fault by opening the circuit before the current reaches its peak.

Current-limiting fuses must not be replaced by noncurrent-limiting types. The systems into which they are installed depend on these high-speed fuses for protection during worst-case failures. Serious damage can result if they are replaced with slower acting designs. To protect against this, the NEC requires that fuse holders for current-limiting fuses not accept fuses that are not a current-limiting design.

Current-limiting fuses are often used as backup protection for large circuit breakers. They are also used for current-sensitive equipment such as electronic power supplies and electronically controlled motor drives.

Time-delay fuses, which are typically dual-element designs, are sometimes used on circuits such as motor feeders, with particularly demanding startup current requirements. A typical AC motor has an initial starting current that is not only quite high—six or more times the normal full-load current of the motor—it is also sustained for a relatively long period of time. This is particularly true if the motor has to start a high-inertia load. Dual-element fuses have long delay times when the overload is modest. This allows the motor to start without resorting to a standard fuse with an even higher rating. However, this is not the whole story.

A dual-element fuse is not just slow acting. If that was the case, it might not provide adequate protection during a serious fault. These fuses actually contain two fuse elements connected in series. One has a relatively low current rating with a long delay, the other has a much higher current rating but is much faster acting. The slow-acting element responds to faults up to about eight times the fuse rating. Above that level, the fast-acting element will usually open first. Together they provide good protection—slow response to relatively low level overloads so that difficult loads can get started, and fast response to limit severe high-current faults.

Dual-element fuses are provided in the same classes as normal fuses, therefore a less expensive single element fuse can be used as a replacement. This is not really a hazard, but it is unlikely that such as replacement will last very long. If it is on a motor circuit, it may blow on the first start up attempt.

Dual-element fuses are often used to allow the installation of smaller fused switches. On motor circuits, for example, the alternative to a time-delay type of fuse is a fuse with a much higher current rating. This often results in a larger switch size to accommodate the high-current fuse. Using dual-element fuses can result in a more compact and less expensive installation.

An enormous variety of fuses are available. Changes in system and equipment requirements and fuse technology have produced a vast array of types, shapes, sizes, and ratings, with very little interchangeability. It is highly advisable to have spare fuses on hand, especially for critical equipment or as replacements for unusual types that may be difficult to obtain on short notice. It is also critical they be located where they can be found. Many years can pass before they are needed—at which time they may be needed desperately.

To replace a blown fuse with one of a higher current value, or to bypass the fuse with a wire, coin, or strip of foil, is an invitation to a fire. These are very dangerous practices. However, even the best of intentions can also create a hazard. When selecting a replacement fuse, pay attention to more than just the current rating. The voltage rating must be at least as high as the circuit voltage. Higher is fine, lower is not. The fuse must also have an adequate interrupting capacity, or fault current rating, and if the original fuse was a current-limiting type, the replacement must also have this characteristic. High interrupting capacity and current-limiting fuse designs are mostly applicable to industrial or large commercial applications.

There is some overlap in fuse types because some newer designs do fit in older fuse holders. If the electrical system has been upgraded with a larger supply transformer, for example, the fuses may have been upgraded without replacing the fuse holders and switches. In situations like this it is physically possible, but potentially dangerous, to revert to the older, less expensive fuses.

Circuit Breakers

Circuit breakers acts as switches that open by themselves if the current exceeds their rating. They serve the same purpose as fuses but can be reset when they trip, instead of having to be replaced. They are generally slower and may be slightly less reliable than fuses, and they cost more to install. For most situations, however, they are totally acceptable and are certainly more convenient. Some typical circuit breakers are shown in Figure 4–22.

FIGURE 4–22 From the left front, a two-pole, three-pole, and single-pole breaker. Behind them is a larger three-pole. The small control wires from this large breaker connect to an internal trip coil.

Circuit breakers are particularly well suited to the panelboards of typical homes and businesses. Many of the circuits in these applications supply convenience receptacles—such as the standard 120 V AC outlets mounted in the walls and above counter spaces throughout a typical modern building. There is no way to predict the load on these circuits, because the occupants can connect anything from a clock or radio that requires almost no current to an electric heater that can fully load the circuit all by itself. It is impractical to feed each outlet from a separate breaker or fuse, so an occasional overload is inevitable. With a circuit breaker, the problem can be corrected and the power restored with no cost and a lot less bother than replacing a fuse.

Typical breakers monitor the current in two ways, either of which can cause it to trip. The first method used is based on heat. A small internal thermal element is connected in series with the output terminal. This element heats in proportion to the current supplied to the circuit. The heat causes a bimetallic strip to bend, which in turn acts to "trip" the breaker mechanism. It is calibrated to trip at just above the breaker rating.

The second method is based on electromagnetism. Electrical current always produces a magnetic field, and the strength of this field is always proportional to the amount of current. Magnetic tripping elements are used in most breakers; however, the actual mechanism in the smaller breakers could go unnoticed, even if the breaker was completely open to inspection. The current path inside the breaker is arranged in such a way that the magnetic field around the conductors themselves acts to pull on parts of the trip mechanism.

Neither of these two current-monitoring methods can provide adequate protection on its own. The thermal trip provides excellent protection against low and moderate overcurrents that may be sustained for a long period of time. It also has sufficient time delay to provide moderate, short-term overcurrent to equipment that needs it to get started. It trips faster on higher current faults than low ones, so it tends to protect better when the potential for damage is higher. However, on a severe fault such as a short circuit, the delay is too long to provide adequate protection.

A magnetic trip unit is very fast. While this speed is highly desirable during severe faults, it will not allow many types of equipment to start. For this reason, a magnetic trip is usually calibrated at a much higher current rating than the thermal element. Together the two tripping elements provide good circuit protection. The thermal unit handles all the low-level overloads, and the magnetic trip quickly cuts off high-current faults.

Unlike fuses, almost all breakers can be replaced with one of a higher rating. This is of course a foolish and dangerous thing to do. Although it is much less noticeable than the current ratings, industrial breakers also may have different interrupting capacities. When selecting a replacement breaker or adding circuits to an existing panel, the new breaker must match the panel, plus all the required ratings.

Note

A special tip of interest to all maintenance personnel: Both fuses and circuit breakers are essentially thermal devices. Therefore, the tripping current can be affected by ambient conditions such as an unusually hot location or heat from poor electrical connections. To accommodate a variety of climates and installations, both fuses and breakers are rated for relatively high ambient air temperature. However, high air temperatures can lead to much higher temperatures inside a heavily loaded panelboard. Such conditions are rather obvious, and do cause problems.

Loose connections are a somewhat different matter. They can cause a fuse to blow or a breaker to trip for no particularly obvious reason. The heating is usually quite localized, most often at the fuse clip or the conductor connection terminal. This condition is very easily overlooked until it becomes severe enough to damage the fuse holder, the breaker, or the busbar-to-breaker connection. Whenever a breaker trips or a fuse blows for no obvious reason, the connections should be immediately suspect. Check for signs of heating, and tighten all connections.

Questions

1. List the four elements of a practical circuit and give an example of each.
2. Explain why an electrical circuit cannot function without two conductors between the source and the load.
3. Explain the fundamental difference between a power circuit and a control circuit.
4. Draw a circuit diagram representing a battery-powered light with a control switch to turn the lamp on and off. Use proper symbols.
5. How does the range selection of an analog meter relate to the scales used to make the reading?
6. How can the polarity of a DC circuit be identified using an analog multimeter?
7. What voltage range should be selected when the voltage to be tested is unknown?
8. What is the meaning of a minus sign in front of a DC voltage value in a digital meter display?
9. If a digital meter displays "OL" when reading a DC voltage, what does this mean?
10. Why is it potentially dangerous to short circuit some types of cells or batteries? Identify two types of cells where short circuits are particularly dangerous.
11. When two or more cells are properly connected in series, their voltages add together. What happens if one of the cells is connected in reverse?
12. Describe the process of "zeroing" an analog ohmmeter.
13. What is the resistance across the terminals of an open switch contact? (Assume it is not connected into a circuit.)
14. Is a closed switch in the "off" position or is it "on"?
15. Why is the measured resistance of a lightbulb so much lower than its operating resistance?
16. Which terminal of a lamp base should be connected to the circuit common? Why?
17. List the three basic electrical quantities, the units that are used to measure them, and the correct abbreviations for the quantity and the unit.
18. Convert the following values as indicated:

    15.2 mV = ______ V, 1.5 MΩ = ______ kΩ, 0.085 A = ______ μA
19. What is the voltage across the terminals of a closed switch? Explain your answer. (Assume the circuit is fully operational.)
20. What simple test can be used confirm the internal resistance of an ammeter or voltmeter?
21. Why is it important to have fuses or circuit breakers in circuits with high-current supplies?
22. The voltage output of a weak supply drops off virtually to zero when a circuit is connected, but recovers when the connection is removed. What is a likely cause?

23. Find the actual value of resistors with the following color codes:

    Red, Red, Brown, Silver
    Brown, Black, Orange, Gold
    Brown, Black, Black, Gold
    Orange, Violet, Silver, Gold

    What is the maximum resistance a Brown/Black/Red/Gold-coded resistor can have and still be within the manufacturer's specifications?
24. What is the maximum wattage rating available in color-coded resistors?
25. Resistors convert electrical energy to energy in another form. What is that other form?
26. A circuit has three resistors, 470 Ω, 1.2 kΩ, and 120 Ω, in series across a supply of 24 V DC. Find the total current and the voltage drop across each of the resistors. What would happen to the current and the voltage drops if one of the resistors was removed?
27. List two advantages to connecting loads in parallel instead of in series.
28. State three methods that can be used to find the total resistance of a parallel circuit.
29. Find the total resistance of a group of four resistors in parallel. The individual resistance values are 4.7 kΩ, 560 Ω, 1.2 kΩ, and 1 kΩ.
30. List the basic procedures for applying Thévenin's theorem to a complex circuit.
31. List four factors that affect the current-carrying capacity of a conductor.
32. Explain why overcurrent protection is so important.
33. Explain why fuses are made in only a few voltage ratings, but with hundreds of different current ratings.
34. Explain the term "interrupting capacity" as it applies to fuses.
35. What characteristic makes a current-limiting fuse different from a normal fuse?
36. What are the important points to remember when selecting a replacement fuse?
37. Why are breakers more appropriate than fuses for receptacle circuits, but not necessarily better for protecting a circuit of permanently installed loads?
38. Identify the two types of tripping elements used in circuit breakers and briefly explain the differences in their performance.
39. What is one possible cause when a breaker trips or a fuse blows, but there is no indication of any type of overload?

Problems

Ohm's Law

1. A 240 V water heater element has a resistance of 19.2 Ω. How much current will it draw? What would happen if the voltage was reduced to 120 V?
2. If a toaster is draws 9 A from a 120 V supply, what is the resistance of the element?
3. What is the circuit voltage if R = 10 kΩ and current is .008 A?
4. What is the circuit voltage when R = 15,275 Ω and I = 18 mA?
5. Find the current for a 550 Ω resistor with a supply of 12 V.
6. What is the resistance if 120 V drives a current of 12 amps?
7. Find the size of resistor required to limit current flow to 135 mA from a 24 V supply.

Series Circuits

Find the missing values. Each circuit consists of three resistors in series.

1. Given: $E = 72$ V, $R_1 = 220\ \Omega$, $R_2 = 560\ \Omega$, $R_3 = 270\ \Omega$
   Find: R_T = ______, I =______, V_1 = ______, V_2 = ______, V_3 = ______
2. Given: $E = 24$ V, $R_1 = 100\ \Omega$, $R_2 = 120\ \Omega$, $R_T = 400\ \Omega$
   Find: R_3 = ______, I = ______, V_1 = ______, V_2 = ______, V_3 = ______
3. Given: $I = 114$ mA, $R_1 = 1.5$ kΩ, $R_2 = 270\ \Omega$, $R_3 = 150\ \Omega$
   Find: R_T = ______, E = ______, V_1 = ______, V_2 = ______, V_3 = ______
4. Given: $E = 12$ V, $R_1 = 15\ \Omega$, $R_2 = 18\ \Omega$, $I = 200$ mA
   Find: R_3 = ______, R_T = ______, V_1 = ______, V_2 = ______, V_3 = ______
5. Given: $R_1 = 10\ \Omega$, $R_2 = 12\ \Omega$, $R_3 = 18\ \Omega$, $V_2 = 1.8$ V
   Find: E = ______, R_T = ______, I = ______, V_1 = ______, V_3 = ______

Parallel Circuits

Find the missing values. Each circuit consists of three resistors in parallel.

1. Given: $E = 24$ V, $R_1 = 100\ \Omega$, $R_2 = 120\ \Omega$, $R_3 = 180\ \Omega$
   Find: R_T = ______, I_1 = ______, I_2 =______, I_3 = ______, I_T = ______
2. Given: $E = 120$ V, $R_1 = 1.5$ kΩ, $R_2 = 120\ \Omega$, $I_3 = 308$ mA
   Find: R_3 = ______, R_T = ______, I_1 = ______, I_2 = ______, I_T = ______
3. Given: $R_1 = 270\ \Omega$, $R_3 = 150\ \Omega$, $I_1 = 44$ mA, $I_2 = 21$ mA
   Find: E = ______, R_2 = ______, R_T = ______, I_3 = ______, I_T = ______
4. Given: $E = 72$ V, $R_1 = 2.2$ kΩ, $I_1 = 32.73$ mA, $I_2 = 40$ mA, $I_T = 179$ mA
   Find: R_2 = ______, R_3 = ______, R_T = ______, I_3 = ______
5. Given: $E = 48$ V, $R_1 = 10\ \Omega$, $R_2 = 120\ \Omega$, $R_3 = 470\ \Omega$
   Find: R_T = ______, I_1 = ______, I_2 = ______, I_3 = ______, I_T = ______

Combination Circuits

Find the missing values.

1.

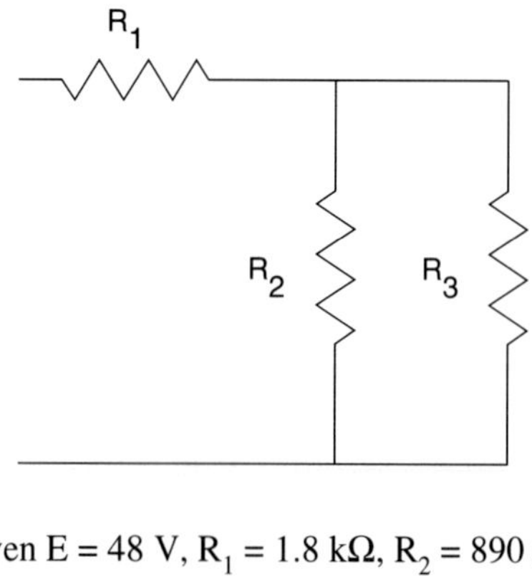

Given $E = 48$ V, $R_1 = 1.8$ kΩ, $R_2 = 890\ \Omega$, $R_3 = 1.25$ kΩ

2

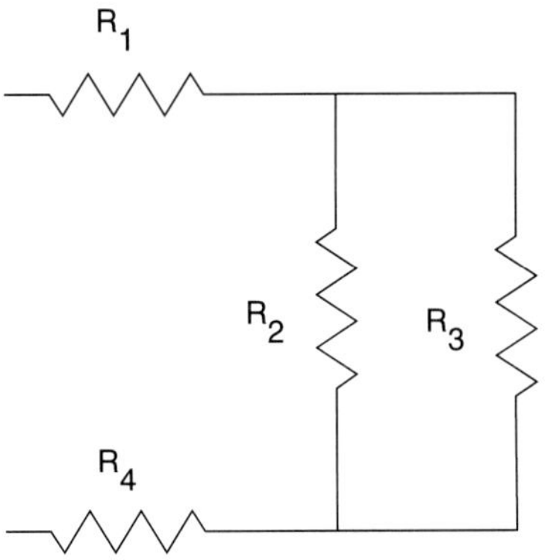

Given E = 12 V, R_1 = 680 Ω, R_2 = 1 kΩ, R_3 = 470 Ω, R_4 = 120 Ω

3.

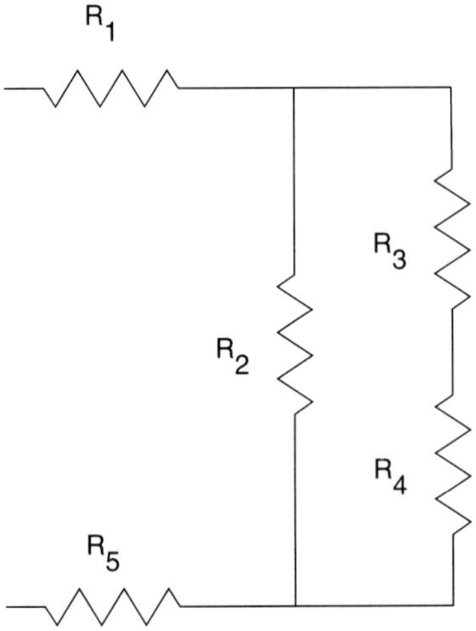

Given E = 72 V, R_1 = 150 Ω, R_2 = 2 kΩ, R_3 = 1.8 kΩ, R_4 = 320 Ω, R_5 = 470 Ω

5 Alternating Current

Objectives

This chapter examines the differences between alternating and direct current (AC and DC), and how these differences affect measuring instruments, circuit operation, and circuit calculations. This is a critical chapter that examines some of the most difficult concepts of electrical theory. A meaningful study of AC equipment is impossible until the principles of AC behavior are understood. After the study of this chapter has been completed, you should be able to:

- Explain the difference between AC and DC
- Describe, in general terms, where AC and DC are found
- State the advantages and disadvantages of AC and DC
- Select appropriate multimeter functions and ranges for use in AC circuits
- Explain the relationship between electricity and magnetism
- Explain what is meant by electrical reactance
- Describe the nature of inductance and inductive reactance
- Describe the nature of capacitance and capacitive reactance
- List the safety hazards associated with capacitance and how to avoid them
- Describe phase shift and explain how it relates to the two types of reactance
- Perform calculations to find unknown electrical quantities in AC-powered circuits

5.1 Introduction

There are two types of electrical power in common usage: direct current and alternating current, commonly known as DC and AC, respectively. Both are widely used, sometimes in unexpected places. It can be very helpful for the service person to know which of the two to expect, based on the equipment and the circumstances. An error in judgment about the type of power can lead to serious problems. Meters react differently to AC and DC, and must be used correctly to avoid mistakes and possible injury or damage. Multimeters set to DC functions do not respond to AC voltage or current. Using AC functions with DC may give a reading, but the resulting values are in error.

Equipment ratings are a mix of AC only, DC only, and AC/DC. Connecting to the wrong type of source can cause serious damage. Most equipment is designed for **either**

AC or DC, **not both**. There are specific exceptions, but in general the two types of supplies cannot be substituted for one another. AC-rated equipment must be supplied with AC only, and DC equipment with DC. They cannot be used interchangeably. That said, there are certain types of electrical components that can operate with either type of supply. Examples include resistive loads, such as incandescent lamps and heating elements, and series universal motors. (This is the type of motor used in noncordless power hand tools such as drills, saws, and routers.)

Unfortunately, even when the load is compatible with AC or DC, there is no guarantee that the equipment can be used safely with either type of supply. Even though a lamp, heater, or tool motor might work just fine on AC or DC, very often the control switches supplied as part or the equipment are rated for AC only. The only way to be sure is to check the nameplate. Tools are marked AC/DC or AC only, both of which are self-explanatory. Lamps and heaters are usually marked with something like 120 V 60 Hz. The 60 Hz means a frequency of 60 cycles per second, and that marking implies that an AC supply is required. The hertz (Hz) unit is relatively new. Older equipment may be marked 60 cycle, 60 cps, or something similar. When the equipment is not marked, but has a standard 120 V AC style of attachment cap on the supply cord, it is safest to assume it is designed for 120 V AC only.

The limitation of the control switch is a problem that extends through all electrical applications. It is relatively easy to make a switch that can handle either AC or DC when it is turned on, but shutting the current off is another story. Direct current is much more difficult to break than alternating current. A switch that is designed for AC use will generally burn up when it is used in a DC circuit at a similar voltage and current. This same limitation applies to other contact devices, such as relays, timers, and motor starters. The contacts are often rated for both AC and DC, but there are usually two very different sets of ratings. The DC ratings are usually much lower than those given for use in AC applications. General-purpose box-mount switches, such as those used in homes and businesses to control lights, are rated AC only.

5.2 Where AC and DC Are Found

The following guidelines can be used to predict which of the two types of power is likely to be present. Although there are exceptions to every rule, understanding the information here can greatly reduce the chance of a nasty surprise.

Electricity from direct-conversion sources such as batteries, solar cells, and thermocouples is always direct current. Equipment powered directly by these sources must therefore be designed for DC. Common examples include equipment on vehicles, cordless tools, and thermocouple-powered safety valves used on gas burners with standing pilot ignition. This type of burner is common in gas-fired water heaters and older style gas furnaces. When the furnace is in a building, other electrical circuits in the furnace may be powered from the AC line; when it is in a recreational vehicle, all the furnace circuits have to be DC. In both cases the thermocouple circuit is DC, and is separate from the other controls.

Power taken directly from a supply utility is always alternating current. Although the utility may use DC for some of its own purposes, all of the electrical energy supplied to their customers is AC. Most portable generators are also designed to produce AC power, although small ones may also have a DC outlet for battery charging. AC generators use some DC internally as part of their operating requirements, but it is normally inaccessible. Automotive alternators are AC internally, but DC externally. The conversion is made by a set of diodes mounted inside the alternator.

Transformers do not operate on DC; therefore, anything powered directly from a transformer is receiving an AC supply. All transformers must have an AC input and will have AC coming out. This output may be converted to DC after it leaves the transformer, but it will be AC at the transformer terminals. For example, a small transformer is used to provide the 24 V AC supply commonly used for heating and air conditioning controls. These transformers are connected to a standard 120 V AC circuit and have a 24 V AC output.

Much of the confusion about transformers and DC comes from the fact that many electronic devices are powered from small DC power supplies that do contain a transformer. These supplies plug directly into the 120 V AC outlet and provide a low-voltage DC, through a lightweight cord, to a small jack that plugs into the equipment. The transformer is not used to convert the AC to DC, it is used to reduce the AC line voltage to a more suitable level before the conversion takes place. This lower AC voltage is converted to DC using a simple electronic circuit that is dwarfed by the transformer. Both the transformer and the associated electronics are mounted inside the power supply case. The output of the device (the voltage and type of current) is described on the nameplate.

Almost all electronic devices must have direct current to operate, but may be powered from an AC supply. The AC is converted to DC, either at the AC outlet with a power supply built into an oversized attachment plug (as described above), or using components built into the equipment. Because many electronic circuits require several different voltages, it is not unusual to find multiple-voltage power supplies and several different DC voltages supplied to the same circuit board.

AC-to-DC conversion is simple, efficient, and relatively cheap. Typically, the AC voltage is reduced with a transformer (often the largest and most expensive part of the power supply), and then converted to DC with diodes. This is called rectifying the AC supply. The rectified DC output that comes from the diodes is full of dips and peaks. These are usually filtered with a capacitor; for a very stable DC output voltage, the current may be passed through a solid-state voltage regulator. For low current supplies this can be done in a very small space. Solid-state control boards used in AC-powered equipment usually have one or more DC power supplies built right onto the board.

Converting DC to AC, while not impossible, is relatively difficult and expensive. The equipment used to do this conversion is called an inverter. It is unusual to find AC equipment powered from a DC source, but there are exceptions. Emergency vehicles are sometimes equipped with specialized equipment that requires 120 V AC. An inverter is installed in the vehicle to supply that equipment from the vehicle's 12 V DC system. Another common application for inverters is in emergency power supplies used for critical equipment. These supplies usually rely on batteries as a source of backup energy. A DC-to-AC inverter is used to provide the necessary AC output.

Equipment nameplates identify the type of supply required (AC, DC, or AC/DC). They also state the necessary voltage or optional voltages, and either the current or wattage the equipment will draw from the supply. This defines what the equipment must receive, but it is not an absolute indicator of what will be encountered on the circuits inside. Beyond the power connection terminals, there may be several different voltages; there may be separate low-current circuits with their own fusing or circuit breakers; and both AC and DC may be present.

A good example of this is a typical residential gas furnace. A basic forced-air furnace without electronic ignition has a 120 V 15 A supply to run the air circulation blower and to power the 120-to-24 V transformer. The transformer, in turn, supplies 24 V AC to the thermostat circuit that operates the main gas valve. This 24 V AC supply is current protected at about 2 A, much less than the 15 A limit of the main supply. The safety shutoff valve inside the main gas valve body is operated from the thermocouple. The thermocouple circuit is DC at about 30 mV and very low current. The basic furnace, with no electronic components at all, has two AC voltages and one DC voltage.

In a more complex furnace, even more variety can be found. The 120 V AC from the supply panel circuit breaker is used for both the main and induced draft blowers. A 120:24 V transformer supplies 24 V AC to the thermostat circuit that controls the electronic ignition and burner control. The thermostat has additional circuits to control the air conditioning system and to provide some options for the operation of the furnace blower. The electronic ignition control manages the burner functions. It opens and closes the two 24 V AC gas valves, ignites the gas with either an electric spark or a hot surface igniter, and monitors for flame during ignition and after flame is established.

These functions involve several different voltages and a mix of AC and DC. Spark ignition systems generate a high AC voltage with a special ignition transformer that is usually built into the control module. The spark voltage is typically 5000 to 10,000 V AC, even though the input to the ignition transformer is only 24 V AC from the thermostat circuit. The high voltage is carried to the igniter electrode through a high-voltage conductor similar to an automotive sparkplug wire. The other side of the high-voltage circuit is to the grounded metal of the spark igniter or the burner.

Silicon carbide hot-surface igniter elements are also used in modern furnaces. These high-temperature-resistance elements require 120 V AC to operate, but their operation is controlled by the 24 V ignition module. Modules of this type have terminals for both the 24 V AC control system and 120 V AC to supply the igniter. A relay in the controller is used to switch the igniter on and off as required. Some care is required when testing at the terminals of the control module, because the 120 V and 24 V terminals are often very close together.

Both spark and hot-surface ignition controllers have a flame-sensing circuit that operates on a mix of AC and DC at a moderately high voltage but extremely low current. This is an unusual situation, because the voltage into the circuit is AC, but the current is DC. This comes about because the system uses the pilot or burner flame as a crude rectifier, converting the AC signal into DC. The current is normally only a few microamperes, too low for many multimeters to read accurately.

When a central air conditioner is incorporated into the furnace, the thermostat has additional 24 V AC circuits to control the compressor motor, the condenser fan, and the furnace circulating air blower. The air conditioning equipment is AC powered, but usually at 240 V instead of 120 because of the relatively high current demand of the compressor. This is, of course, yet another circuit, this time with its own circuit breaker in the supply panelboard. The 24 V AC side of the circuit is usually powered from the same 24 V transformer as the rest of the furnace controls, but air conditioning controls sometimes operate from their own 24 V AC supply. This may be a separate transformer or, more likely, a transformer built into a fan center type of control.

When the furnace is equipped with a motorized humidifier, it also operates at 24 V AC. The power supply for the humidifier is often yet another 120:24 V transformer. Some newer furnaces have a provision for powering the humidifier circuit from the same transformer that operates the rest of the furnace controls. The humidistat that controls the built-in humidifier is usually mounted and wired separately from the main thermostat.

Electronic air cleaners are another common furnace accessory. They may be electrically independent (switched by a sail switch) or integrated into the furnace fan controls. In both cases, they operate from 120 V AC, but this is converted internally into a very high DC voltage. The high-voltage DC is used to charge the electrostatic grid and collection filter. This voltage can be 10,000 V or more—far beyond the range of any normal multimeter—and should be treated with the utmost respect.

When the furnace is controlled by a programmable electronic thermostat, both AC and DC will be found inside the thermostat. The DC supply is used to run the thermostat electronics; the AC brought to the thermostat is part of the normal furnace control system. The 24 V AC circuits are switched by the electronic thermostat, just as they would be by the nonelectronic version it is has replaced. The only real difference, on the 24 V AC side, is that the switching is done with solid-state electronic components instead of a mercury switch. Most of these thermostats have one or more batteries inside. These are used to supply low-voltage DC to the programmable components. Some models operate entirely from battery power. It takes very little energy to run the electronics, so the batteries may last a year or more. In other models, the batteries are there only to preserve the program when the furnace is shut off, or in case of a power failure. In the second case there will likely be at least two different DC voltages, plus the 24 V AC control conductors.

As can be seen from these examples, electricity is very flexible to use, but that flexibility comes with tremendous variation. It's possible to have a mix of both AC and DC at a variety of voltages and currents, all within a single, very common piece of electrically powered equipment. On a modern furnace, with the controls integrated on a single circuit board, all of the voltages, except the 240 V AC circuit to the air conditioning unit, are likely to be present simultaneously on that one relatively small board (see Figure 5–1).

5.3 AC and DC: Differences and Similarities

Much of what is true for DC electricity applies equally well to AC. An AC system uses the same electrons, follows the same rules of nature, and generally performs the same tasks as its DC counterpart. The units of measure used in the study of AC are ohms, volts, and amps, just as they were for DC. However, there are some critical differences in the way these units are used and what they represent. AC meter functions are different than those used for DC. The appropriate function must be selected for the type of current or voltage being measured.

AC and DC power have many similarities, but they are also very different in several important ways. Because of these differences, DC theory does not fully explain AC behavior. Ohm's Law, for example, can be applied to AC circuits, but **only** when it is expanded to include behaviors that are unique to AC operation. DC theory is not wrong, but it is somewhat difficult to apply to an AC circuit without some additional information. Understanding AC behavior requires a good understanding of DC circuit behavior, plus additional knowledge about the ways in which AC is more complex.

A Hydraulic Analogy: AC and DC Compared

In a DC circuit the polarity is constant. In other words, one supply terminal is always positive and the other is always negative. The current—the flow of electrons around the circuit—is always in the same direction. The rate of flow will change with changes in the voltage

FIGURE 5–1 This circuit board provides all the control and burner management functions for a modern gas-fired furnace equipped with air conditioning, humidifier, and electronic air cleaner. All of the various AC and DC voltages required for these functions are found on this board.

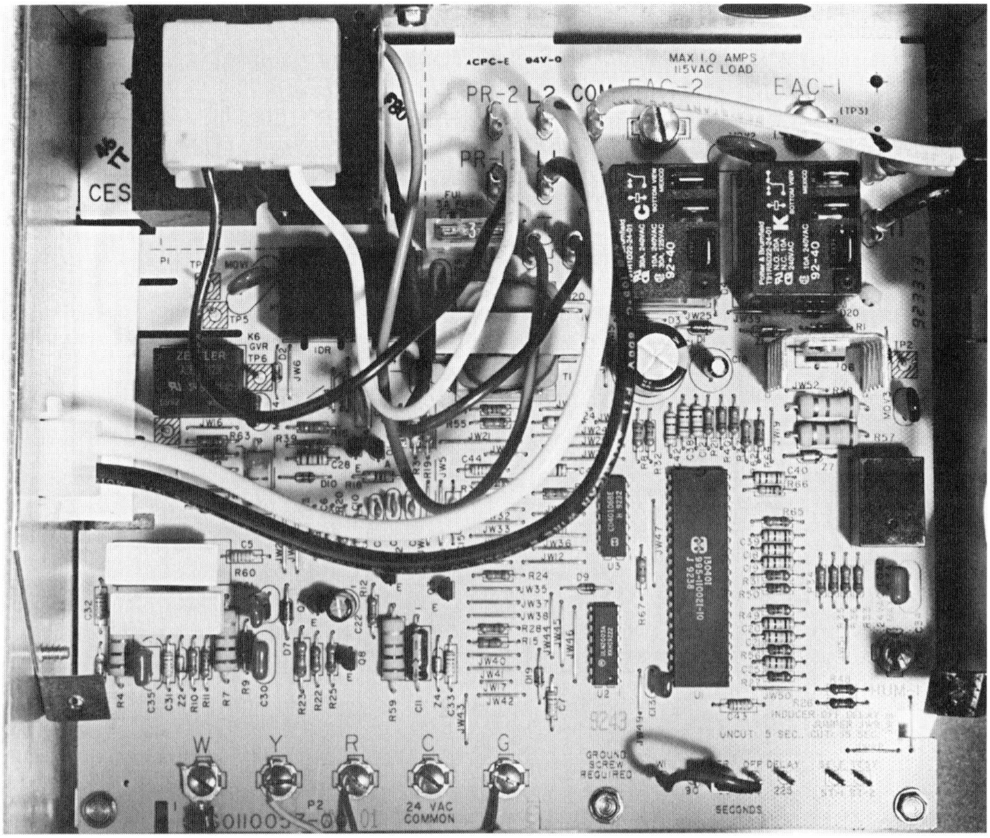

applied or the resistance, but the direction of flow is set by the polarity of the supply, which is usually constant.

In direct contrast to the DC system, both the polarity and the actual voltage of an AC supply are constantly changing. The change is cyclic, meaning it follows a pattern, or cycle, that repeats over and over at regular intervals. The rate at which these cycles occur is kept constant. The following analogy may help explain the differences between the two systems.

In Figure 5–2a, the rotary pump and motor system is comparable to a DC circuit. Pressure and flow are always in the same direction. With a steady mechanical input to the pump and a constant mechanical load on the motor, the pressure of the oil and its flow rate through the pipe will also be constant. Energy is delivered from the pump to the motor in a smooth and continuous manner. Variations in the pump pressure or the resistance of the load will change the flow rate, but the operating principles remain the same.

In a DC electrical circuit, the force or pressure is the source voltage. The working "fluid" of the circuit is the electrons in the conductor, and the flow has a specific direction. With a constant input voltage and a steady load, the current flow and the rate at

FIGURE 5–2 These two hydraulic circuits are analogies for (a) DC and (b) AC systems.

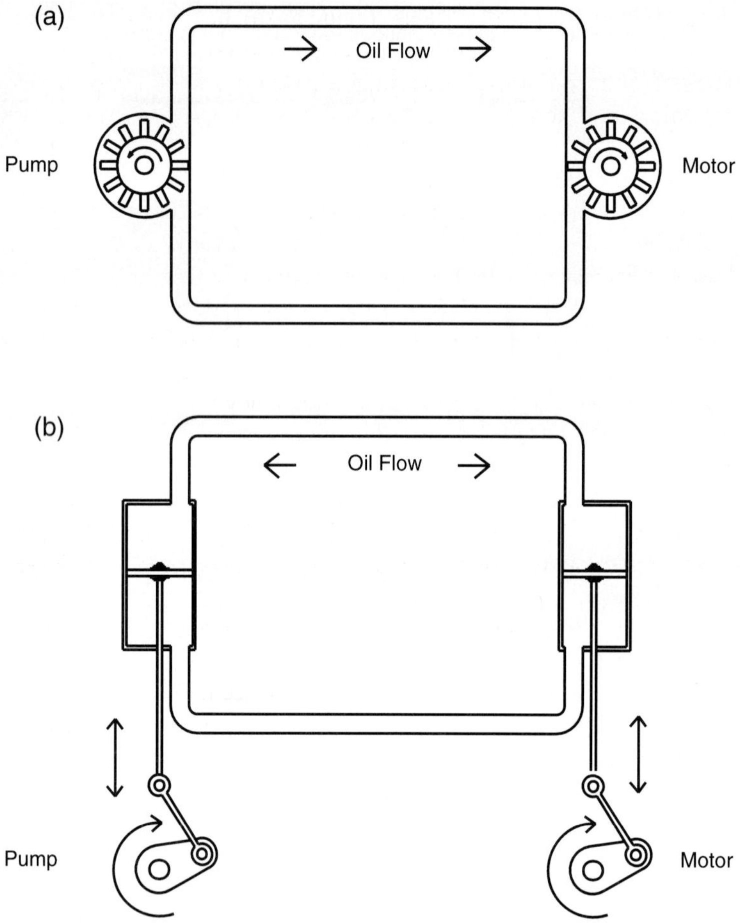

which energy is delivered by the circuit are both constant. As in the hydraulic example, variations in voltage or resistance change the flow rate, but not the fundamental principles of operation.

The piston pump hydraulic circuit shown in Figure 5–2b is an alternating system. Pressure and flow change continuously in both value and direction. Both the speed and direction of the piston vary with the position of the crankshaft. The pressure and flow rate of the oil rise and fall, and change direction, as the movement of the piston follows the rotating crank. Energy is delivered, but it is in pulses that are continuously changing in value and direction.

An alternating (AC) electrical circuit works much the same as the hydraulic system just described. The voltage and current follow virtually the same pattern of changing values, although the reasons for it are different. All AC supplies provide a voltage that is constantly changing in direction and value. The current is driven by the voltage; as the voltage changes, so does the current it pushes around the circuit. The energy delivered by a simple AC circuit is in pulses, just as it is in the alternating hydraulic circuit.

Cycles and Frequency

In the alternating hydraulic example (Figure 5–2b), the changes in the motion and force of the oil follow a sequence of push-pull/speed up-slow down that is repeated in time with the rotation of the pump crank. In this simple system, each revolution of the pump shaft produces one *cycle* of changing pressures and flows. A cycle is defined as one complete segment of a repetitive event. In this case one cycle could start at the point of zero flow—with the piston at the end of its stroke—and would include all the changes that occur from that point on, through a complete rotation back to the starting point. With a constant pump speed, the alternating hydraulic system has a constant number of cycles every second. This cycles-per-second rate is the system's operating *frequency*.

This is also how AC electrical circuits work. Both voltage and current are constantly changing in value and polarity. As in the hydraulic analogy, the change is repetitive, or cyclic, and the rate of repetition is the system frequency. In power systems throughout North America, the standard frequency for AC power is precisely 60 cycles per second, or 60 hertz (Hz). Other standards are used in other countries or in specific applications. Much of the world outside North America uses a 50 Hz standard. There is also an older, 25 Hz standard that still has some appeal in certain applications. A much faster 400 Hz standard is common on aircraft, because 400 Hz motors, generators, and transformers are much lighter than lower frequency equipment with the same power ratings.

The two major power system frequencies are 50 and 60 Hz. There is no clear overall advantage to one over the other. Each has its own trade-offs, and both are very workable systems. Lower frequencies (such as 25 Hz) produce flicker in incandescent lamps and require bulkier transformers. They are sometimes used for very large, low-speed motors. High frequencies have greater transmission line losses, limiting them to short distances between the supply generator and the loads.

AC: The Established System

Judging from the hydraulic examples above, it may appear that an AC system has no particular advantage over a comparable DC system. Actually, it seems to have a distinct disadvantage. An alternating hydraulic system would be very inefficient. The oil has a considerable amount of mass, so a lot of energy would be wasted just changing the speed and direction of flow. Fortunately, electrical systems do not have that problem.

Electrons, unlike oil, have no significant mass; therefore, the electrons can be accelerated, decelerated, and driven in different directions many times every second without significant energy loss. This makes efficient AC systems possible, although it does not explain why they are used.

There are some special characteristics unique to alternating current that make it much more appealing to electrical utilities than direct current. AC generators are simpler to build and require less maintenance than DC generators. They can also be built with much larger power ratings, without encountering overwhelming technical problems. AC systems can use transformers to increase and decrease voltages as required. This is critically important for efficient long-distance power distribution. Transformers are simple, efficient, and very reliable, but they only work on AC. There is no comparable way to change voltages in a DC system.

Alternating current has become the established system throughout the modern world; however, the situation was far from clear when the first electric utilities were established. In the U.S., the AC vs. DC debate became a dramatic competition between Thomas Edison and George Westinghouse. Both submitted contract proposals for a new power plant at Niagara Falls. Edison favored DC. He had many DC systems already in place, giving him the edge in experience. Westinghouse was new to the business but had the young Nikola Tesla—the recognized father of AC systems—in his employ, and so was pushing for the adoption of AC.

Tesla had once worked for Edison, but the two had parted company in disagreement. His ideas about AC were a direct threat to Edison's established market. Both Edison and Tesla were great news items. Edison was already well known, a legendary industrialist with considerable skill at promoting his interests. Tesla was a brilliant engineer and a natural showman. The press had a field day, and in the end Westinghouse won this critical contract. AC, in the U.S. at least, was well on its way to becoming the established system that it is today.

Despite the heated debates and the tall tales told for the benefit of the newspapers, AC was then, and still is today, the better of the two systems for commercial power generation and distribution. It is inevitable that AC would have won out in the end, and probably a good thing it happened almost at the outset. Even though the initial victory was a decisive one for AC, however, the competition between these rival technologies has never really stopped.

Throughout the history of electrical power, AC and DC equipment have competed for dominance in many areas. The competition continues to this day, although advances in power electronics are blurring the distinctions between the two opposing camps. For now it is AC, not DC, that lights our homes and businesses, and turns most of the wheels of industry. Although AC is more difficult to understand, it is also superior for many applications. Many common AC electrical devices are simpler, cheaper, and more reliable than their DC counterparts. DC still has some advantages, especially where the power must come from a portable power supply such as storage batteries.

AC Circuit Basics

All practical electrical circuits, AC or DC, have a source, a conductive path, a control device, and a load. The first of these four basic components is the power *source.* In either system, a practical source must be able to generate an appropriate AC or DC voltage and maintain it under load.

The second of the basic circuit components are the *conductors* that carry the electrons around the circuit. They are usually insulated wire or cable. A wide variety of types and sizes are available and, in general, will carry AC or DC equally well. Voltage and current ratings for the conductor can usually be interpreted as equally applicable to either type of power. Conductor ratings are always maximums, regardless of whether the current is AC or DC. A lower value of either voltage or current is fine; a higher value of voltage will overstress the insulation, and excessive current will cause the conductor to overheat. Either of these lead to failures, and may cause a fire.

Control devices are the third basic circuit component, and switches are the simplest example. Unlike conductors, switches are rarely interchangeable from AC to DC, and may be further restricted to specific types of loads. DC current is more difficult to interrupt than

AC, so the switch must have a fast action and good contacts. Do not use a switch or other control device unless it is specifically marked as suitable for the type of supply—AC, DC, or both. The switch must also have voltage and current ratings that are equal to or greater than the ratings of the load to be controlled. Switches for motors may also have to be horsepower rated.

The final basic component of a practical electrical circuit is the *load.* The load in an electrical system is the device that receives power from the circuit. Loads are usually the reason the circuit was created in the first place. Examples include heating elements, lights, and motors. The design and behavior of various types of loads will be studied in more detail later, but it should be understood that even if all other ratings match, loads designed for AC won't necessarily operate on DC, and vice versa. Most devices have clearly marked nameplates or other markings.

In any operating circuit, AC or DC, the electrons receive energy from the source, carry that energy through the conductors and control devices, and release it to the load. The load device changes the electrical energy into some other form, such as heat, light, or mechanical motion. The fact that the electrons in the AC system are constantly moving back and forth, instead of around and around the entire length of the circuit, does not change the direction of net energy flow (see Figure 5–3).

FIGURE 5–3 The light receives energy from the supply regardless of the direction in which the current is flowing.

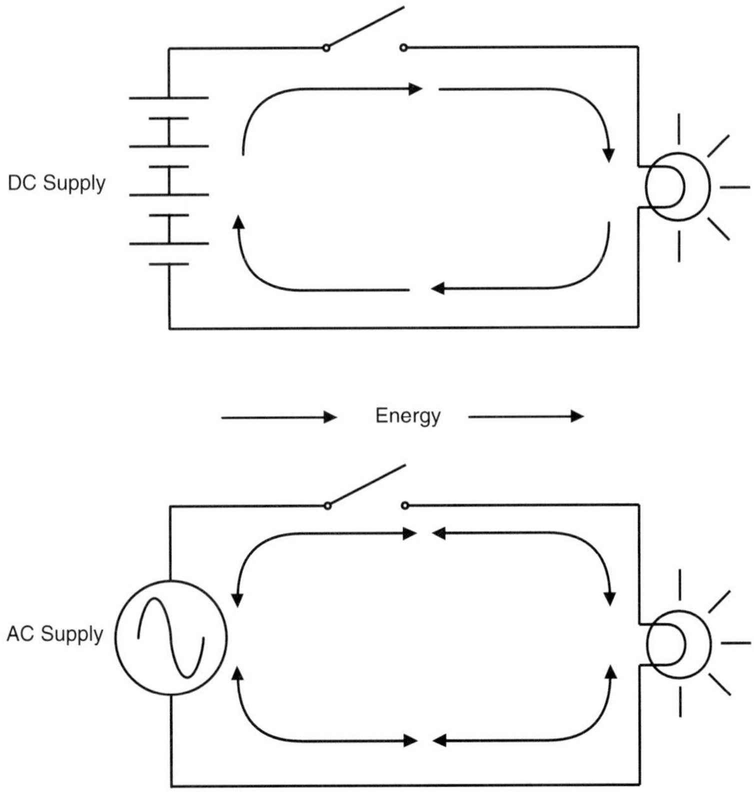

When the circuit voltage forces electrons through a resistance, the energy spent by the supply is given up to the friction effects of the resistance material. This energy is dissipated as heat (and light, in the case of a high-temperature lamp filament). In a DC circuit, the electrons keep moving and are replaced by an endless stream of others that follow. In an AC circuit, the electrons move one way for a time and then reverse. The change in direction does not necessarily change the end result. The electrons are driven by the energy of the source, and give the energy up to the resistor, regardless of the direction in which they move. Energy is not delivered to the resistor when the current flows in one direction, and then recovered when the direction is reversed.

The transmission of energy from a source to a load is what electricity is about. Electrical circuits, AC or DC, exist only to power or control some load.

Sources of DC and AC Electricity

Electricity does not occur naturally in a readily usable form: It has to be generated from some other source of energy. The actual energy source often dictates the conversion process used. Examples include conversion of mechanical, chemical, light, or heat energy. Mechanical energy may be obtained directly from a source such as falling water or the wind, but is also produced from the heat energy resulting from the burning of fossil fuels, or from a nuclear reactor. The heat energy is first converted to mechanical power in an engine or turbine.

Regardless of the source of mechanical power—a water or wind turbine, a diesel engine, a gas or steam turbine—it is used to drive a rotating generator. Wherever possible, the generator is connected directly to the output shaft of the driving turbine or engine. The output can be AC or DC, depending on the generator design.

Direct conversions typically result in a DC output. Examples include chemical energy to electrical energy in an electrochemical cell or battery, conversion of light energy in a photovoltaic cell, and the conversion of heat energy in a thermocouple. (For more details about these DC sources, see "Sources of Electricity" in Chapter 3, p. 68.) Alternating current must be produced in an electromechanical generator, or by rapidly switching a DC supply. The conversion of DC to AC is possible with power electronic devices, but it is complex and expensive, and the output is not as smooth as it is from a rotating AC generator.

Virtually all commercial-scale electrical power comes from rotating electromechanical generators. These can be made to produce either AC or DC, but the most natural output is AC. Even DC generators produce AC internally; the output is converted to DC by an elaborate rotating switch mechanism called a commutator. Commutators are complex and expensive to build and maintain, and they limit the operating voltage and current of the machine. AC generators do not require a commutator, so they are simpler and less expensive to build and maintain. Many generators built specifically for DC output are actually AC generators with electronic AC-to-DC conversion added on. Examples of this include automobile alternators and the generators used in most diesel electric railway locomotives.

Note

The terms "AC generator" and "alternator" are essentially interchangeable. In this text, "AC generator" is used in preference to "alternator," except for specific situations such as the automotive alternator, where the term has become associated with the specific device.

FIGURE 5–4 When the conductor moves through the magnetic field, the electromotive force created drives a current through the meter.

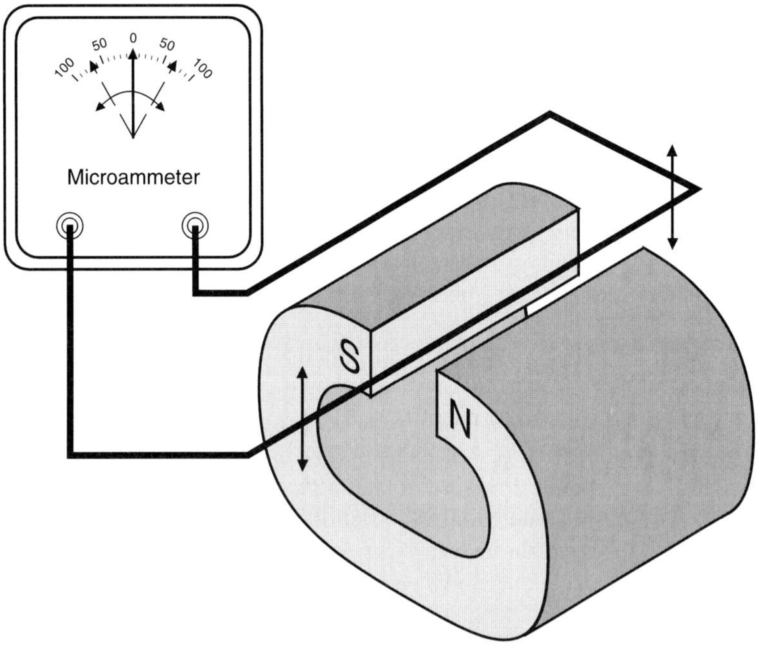

How Generators Work

The fundamental principle on which generators are based is a natural phenomenon that has been exploited by electrical equipment manufacturers for over a century. The behavior is fully understood; the precise reasons for it are not. When a magnetic field cuts through a conductor, the field couples with the electromagnetic character of the electrons. This interaction produces an electromotive force that appears as a potential difference across the conductor (see Figure 5–4). This force has a specific direction that is at right angles to the magnetic field lines themselves and to the direction in which they are moving (see Figure 5–5).

FIGURE 5–5 The magnetic flux, the relative direction in which the field moves, and the resulting electromotive force (EMF) are all at right angles to each other.

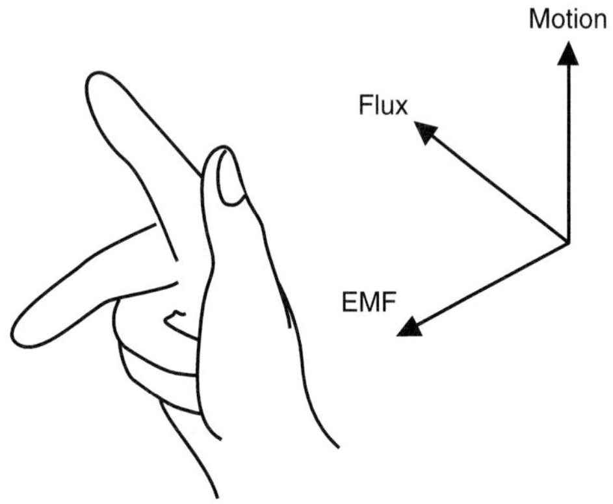

The directional nature of the electromotive force gives the potential difference a specific electrical polarity. The potential difference is the voltage of the system; its polarity depends on the polarity of the magnetic field and the direction in which it is moving.

The process of generating a voltage with a moving, or changing, magnetic field is called electromagnetic induction. The energy transfer from the movement of the field into a potential difference across conductor is efficient, immediate, and unavoidable. It does not come from the magnetic field itself, but from the force that moves it through the conductor. The amount of voltage generated varies with the strength of the field, the speed at which it moves (relative to the conductor), and the length of conductor in the field.

Electrical energy produced by this electromagnetic interaction is the same as electrical energy from any other source, but sometimes the terms used to describe it are slightly different. The magnetic force that creates the field is called magnetomotive force (MMF). The electrical force that creates a potential difference is called electromotive force (EMF), which is essentially just another name for potential difference or voltage. The term "EMF" can also be applied to other sources of electrical energy.

Generators are mechanical machines designed to take full advantage of the relationship between electricity and magnetism. The basic design of an AC generator is relatively simple. The generator driveshaft rotates a number of magnetic field poles past a series of coiled conductors mounted inside a stationary cylinder. As the shaft rotates, mechanical energy is transferred through the magnetic field to the electrons in the conductors. The electromotive force induced in the conductors is the generator's output voltage.

For a DC output, the generator must be built in a way that is somewhat reverse to most AC generators. The magnetic field poles are mounted in the stationary part of the machine, and the assembly that holds the coils of the conductor is rotated inside the field. The voltage induced into the coils is AC, but the connection between these coils and the external circuit is made through a commutator to a set of stationary carbon brushes. The commutator is a series of copper alloy bars arranged to form a symmetrical drum around the center line of the generator driveshaft. The bars run parallel to the shaft and are insulated from each other with thin sheets of mica. Each of the many bars is connected to the generator coils. The entire rotor assembly, the shaft, the magnetic core iron, the conductor, and the commutator is called the armature.

The armature coils are connected to the commutator bars in a repetitive pattern, and the brushes are located at precise points relative to the stationary field poles. The net result is that as the armature turns, the voltage induced in the coils appears at the brushes with a consistent electrical polarity. The polarity of the voltage induced into each set of coils changes each time the coils pass under a different field pole. However, the brushes are positioned so that, just as each coil passes between poles, the commutator bars for that coil pass under the brushes, reversing the connections to the output. The induced voltage is AC, but the commutator and brushes switch connections each time the coil polarity is about to change. The end result is a DC output.

The entire output of a DC generator must pass from the armature to the outside world through the sliding contact between the commutator and the brushes. The brushes ride on the surface of the commutator, held in contact by spring pressure. The surface of the commutator must be accurately machined to avoid brush chatter and unnecessary wear. Understandably, the complex construction of the armature represents a large portion of the cost of a DC generator. The inevitable wear of the brushes and the commutator also greatly increases the maintenance requirements.

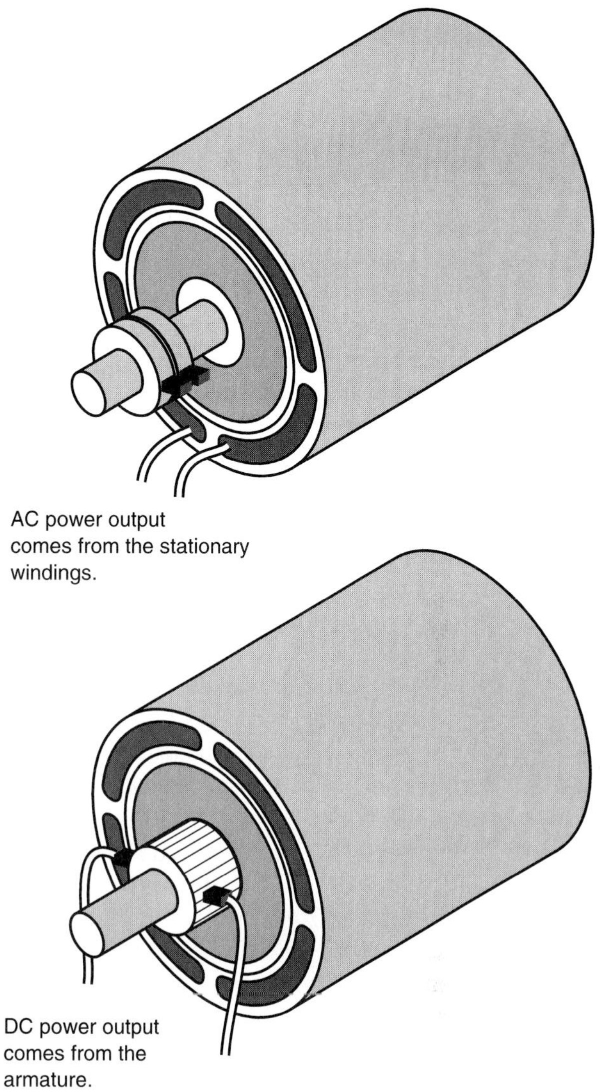

FIGURE 5–6 AC generator output can come directly from the windings. DC output must come through the commutator and brushes.

In contrast to DC types, AC generators are relatively simple in construction, and much more trouble free. The coils in which the voltage is generated are almost always mounted in the stationary part of the machine. This allows a direct connection between the coils and the generator output, permitting a much higher output voltage without the maintenance problems associated with the brushes and commutator (see Figure 5–6).

Some AC generators do have brushes to maintain, but they ride on smooth slip rings, not a segmented commutator. They carry DC current to the rotating electromagnetic field poles that are mounted on the generator shaft. This is a much simpler function than the rotary switch action provided by the commutator. The slip rings are insulated from the shaft, but that is much easier than insulating dozens of individual commutator bars. They also do not carry the output current of the machine, only the much lower voltage and current required by the rotating field.

Another AC generator design eliminates the slip rings altogether. These brushless generators are common in midsize portable generators, such as the diesel-powered units used for remote construction and mining operations. This reduces the maintenance even further, because all sliding electrical connections are eliminated. Brushless designs are not widely used for large utility generators.

Summary. Rotating electromagnetic generators are the only practical way to produce large amounts of electrical power. The operation of the generator is based on the relationship between magnetism and electricity. When a conductor is cut by a magnetic field—resulting from movement of the field, the conductor, or both—a voltage appears across the conductor. Mechanical energy has become electrical energy through the influence of a magnetic field.

The electrical output of a generator can be AC or DC, but all generators used for commercial power are AC. The AC generator is simpler, requires less maintenance, and can be made with a much higher output than a DC machine. Besides, the utility wants an AC output, not DC, so using an AC generator avoids a difficult and expensive conversion.

Practical Exercise: Demonstrating Generator Theory

With a few parts and an appropriate meter, the principles of electromagnetic generation can be clearly demonstrated. The necessary parts include a coil wound on a nonmetallic form, a bar magnet, a soft iron bar, and a suitable meter (preferably a zero-center microammeter, commonly known as a galvanometer). The coil form must be hollow and large enough in diameter to allow the magnet to pass through the center. The coil should have lots of turns, at least a few hundred, although a few thousand is even better. The wire can be very fine: The current produced is in microamperes. The magnet should be reasonably powerful. The iron bar should have a cross section that is at least as large as the end of the bar magnet, and its length should be at least twice the length of the coil.

Variations in the apparatus will affect the results of the experiment, but many combinations are possible. A powerful magnet will compensate for less coil turns or a less sensitive meter, for example. Galvanometers are quite common in educational environments. The needle is at rest in midscale, and the markings read from zero in the center to equal values at both the left and right ends. The meter needle can swing either way, depending on the polarity of the current. All digital meters are inherently center scale and will also work. However, the movement of a meter needle along an analog scale is much easier to interpret than a rapidly changing series of numbers and a periodic change in polarity. A pair of connecting wires, or meter leads with clip ends, completes the materials list.

When the magnet is moved along the outside of the coil to cut the conductors at right angles, a voltage is generated in the coil, current flows through the meter, and the needle swings to one side. When the magnet is stationary, the needle returns to zero. If the coil is moved instead of the magnet, a voltage is generated exactly as before. When the field is held stationary, or the movement is parallel to the conductors, there is no cutting action and no voltage is generated.

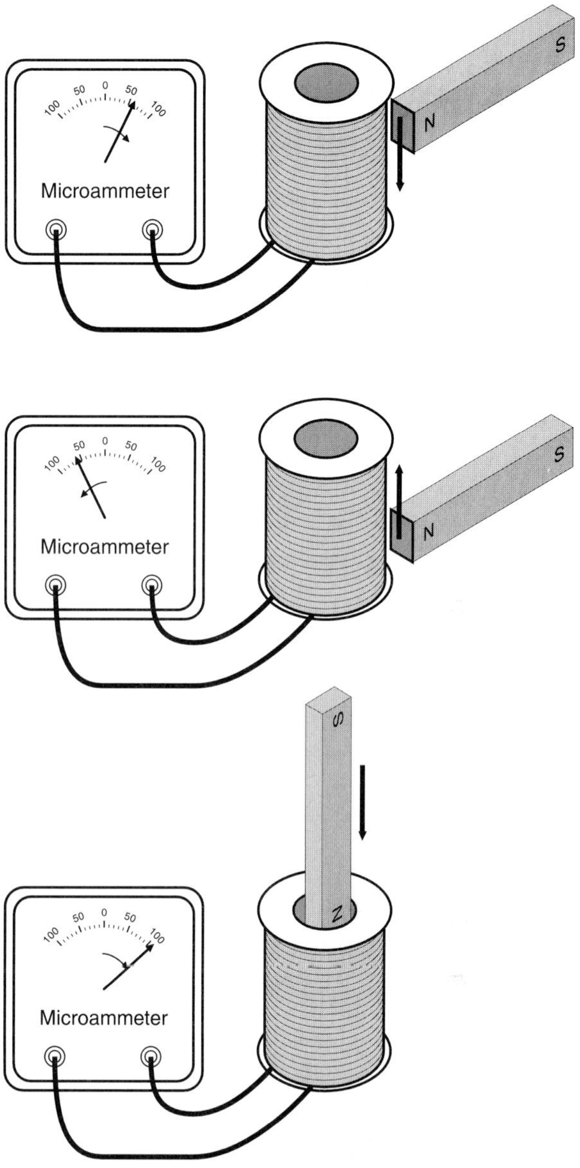

FIGURE 5–7 A demonstration of the electromagnetic induction that occurs when a conductor is cut by a moving field.

Figure 5–7 demonstrates the principles of electromagnetic induction. When there is relative motion between a magnetic field and a conductor in such a manner that the field cuts through the conductor, a voltage will be induced. The field must cut across the conductor, not just pass along its length.

If the experiment is repeated but the magnet is moved in the opposite direction, a voltage is induced as before, but this time the meter needle swings in the opposite direction. The polarity of the voltage is reversed. When the relative direction of motion is kept the same, but the magnet is inverted so that the opposite magnetic pole is against the coil, the polarity is also reversed.

The polarity of the induced voltage is dependent on the direction of the relative motion and the polarity of the magnetic field. When either of these is changed, the polarity of the induced voltage will reverse. There are rules that can be used to predict the polarity based on the direction of motion and the magnetic polarity, but, in general, knowing the correct polarity in advance is not that important. What is important is an understanding that polarity will change whenever the polarity of the magnetic field or the direction of the cutting action is reversed.

Repeating the experiment but changing the speed at which the magnet is moving demonstrates that the voltage induced is directly proportional to the rate at which the field is cut. The faster the magnet is moved across the coil, the greater the voltage.

When the magnet is placed inside the coil, instead of passed across the outside, the field cuts the full length of each coil turn it passes. This results in a greater output than before, because the magnetic lines cut the conductor around its entire circumference, inducing voltage with the same polarity at all points. Before, when the magnet was passed across the side of the coil, the voltages induced by field lines that reached the opposite side of the coil were opposite in polarity and acted to reduce the total output.

Field strength also affects the voltage. A stronger field has a greater flux density; therefore, the same rate of movement causes more flux to be cut by each conductor and induces a proportionally higher voltage. Replacing the magnet with a more powerful one will increase the output. This can also be simulated without a second magnet by holding the magnet at different distances from the coil. Field strength varies as the square of the distance, so it does not take very much change in the magnet's position to produce a large difference in the voltage generated.

Another way to increase the field is to improve the magnetic properties of the path in which the field must form. One way to do this is to place a soft iron bar inside the coil. The iron has a much lower magnetic reluctance than air; therefore, the field will be much stronger and more voltage will be induced.

The number of coil turns also affects the output. When the magnetic field cuts the coil, voltage is induced in each turn. The polarities of these individual voltages are the same, because the same magnetic pole moves in the same direction along the length of the entire coil. Each turn of the coil is a relatively small voltage source, but the turns are all in series, therefore the total coil voltage is the sum of all the individual turn voltages.

As shown by this experiment, there are several variables that influence the amount of voltage produced by a moving magnet-and-coil arrangement. With careful examination, it becomes evident that all these variables effect the same underlying equation: **The voltage induced is proportional to the number of flux lines cut per second.** Each of the factors that changed the voltage also changed either the total flux cut or the speed of the cutting. Any change in the design or operation of an electromagnetic generator that changes the total flux cut per second will also change the output voltage.

An examination of an actual generator reveals a design that incorporates all the aspects demonstrated by these experiments. Each of the generator coils has many turns, and they are arranged so that the maximum practical conductor length is in the field. Both the main coils and the field poles have an iron core, so the field will be as strong as possible. The core iron is continuous except for the narrow gap necessary to allow for rotation of the field. The coils are also arranged so that as a north magnetic field pole passes one side of the coil, the other side is under a south pole. The result is that all parts of the coil generate voltage all the time, and both poles of the rotating field are put to use.

Note

There are appropriate units for all aspects of electromagnetic induction. For example, an electromotive force of 1 V is induced when the conductor is cut at the rate of 10^8 magnetic lines of force per second. Using correct terminology, one "line" of magnetic force is called a Maxwell, and 10^8 lines is called a Weber. The voltage induced is therefore equal to the number of Webers of flux cut per second.

The lessons of this experiment are important ones, central to an understanding of all electromagnetic equipment. The next experiment is often omitted from demonstrations of this type, but is important because it extends the rules of electromagnetic induction in one very critical way. Without this additional information, it is very difficult to understand AC coils and equipment such as transformers.

Returning to the experiment, when the magnet was passed through the coil, the effect on the meter was more pronounced than before (refer to Figure 5–7). With the field cutting each turn of the coil along its entire circumference, more lines of flux are cut per second, and the induced voltage is higher. The following is a variation of this, with slightly more baffling results.

The magnet is removed and the soft iron rod is put in its place. Unless the rod has become magnetized, the meter will not respond to its passage through the coil. However, if one end of the magnet is brought close to either end of the stationary rod, the meter again deflects to one side, indicating that a voltage was induced in the coil. As the magnet approaches the rod, the meter needle swings one way. When the magnet stops moving, the meter needle returns to zero. When the magnet is pulled away from the rod, the meter needle moves again, but in the opposite direction.

There may appear to be little difference between this experiment and the previous one, but there is one very important distinction. In the first instance, voltage was generated by an obvious cutting action: The field cut the conductor. In this example, the voltage was generated by a change in the magnetic field passing through the center of the coil.

It is very important to realize that voltage can be induced without the field cutting the conductor. Any change in the strength of the magnetic field passing through a coil induces a voltage in that coil. This principle is true for an increase or decrease in the field, although the polarity of the voltage will be opposite. The amount of voltage induced is proportional to the rate of change, and, just as before, a change of one weber per second induces one volt.

A crude AC generator (see Figure 5–8) can be demonstrated using the simple apparatus of the preceding experiments. With the iron rod in the coil, the magnet is rotated so that opposite poles pass close to the rod on each half turn. As the magnet rotates, the meter indicates an AC current flow. It swings right and then left, over and over, as alternating magnetic poles create changing fields in the rod and through the coil. With each passing pole, the field through the coil rises, and then falls. The polarity of this field reverses with the arrival of the next pole and, as a result, the polarity of the induced voltage also reverses.

The most natural output of any arrangement of rotating magnets and coils is AC. For a DC output, some type of switching or other conversion process is required. These principles of electromagnetic induction were discovered and described by Michael Faraday, one of the most important of all the pioneers in the field of electricity. His work set the stage for much of the electrical technology we now take for granted.

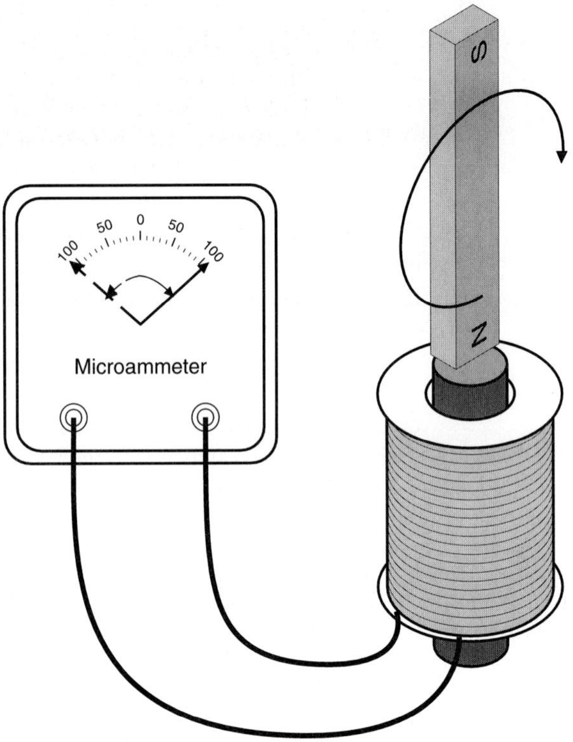

FIGURE 5–8 This crude AC generator demonstrates the basic construction of a real one.

5.4 AC Measurements

The Units

The three primary electrical units, used previously to define DC behavior, are voltage, current, and resistance. In the preceding section on circuit fundamentals, these units were compared to mechanical units used for hydraulic systems. The voltage is, of course, equivalent to hydraulic pressure. It is the driving force of the circuit. Current is flow rate and, like flow rate in a hydraulic system, it is a unit of volume per unit time. In this case the "volume" is coulombs of charge instead of a unit like gallons or liters of oil. The unit for current is the ampere or amp, which is a flow rate of one coulomb per second. Resistance is the electrical equivalent of mechanical friction and, as with mechanical friction, the end result of flow through resistance is energy converted to heat. Electrical resistance has a specific unit called the ohm. One ohm of resistance will restrict current flow to one ampere for each volt of potential difference applied.

The Ohm Is Not Always Resistance

The study of simple DC circuits quite naturally gives the impression that resistance is the only kind of opposition to current flow that a circuit can have. This is not true, particularly in AC circuits. There are other types of opposition that are far more interesting than the simple, but crude, "friction" effects that resistance provides. For this reason it is important to remember that **the ohm (Ω) is not exclusively a measure of resistance in the true, electrical friction, sense of that term. It is used as the unit for all types of opposition to current flow.**

Although it is often used in a broader sense, the term "resistance," as it applies to electricity, is exclusively a measure of the friction effects in the circuit. It is sometimes referred to as "DC resistance," because ohmmeters use a DC voltage to make the measurement. The ohmmeter applies a calibrated value of DC voltage across the circuit and measures the resulting current. The current is proportional to the actual resistance of the circuit, not the total opposition to current flow that the circuit may have. The difference between these two is an important one, which will become clear by the end of this section.

Although there are AC and DC meter functions for the other electrical units, the resistance functions of the meter have no such variation. Resistance measurements can only be done in de-energized circuits and provide a DC resistance value.

Note

There is a fine distinction drawn between true AC and DC resistance. The energy lost due to heat in the conductor is slightly higher for AC than DC, particularly in large conductors. The reasons for this higher AC resistance are complex, and the difference is very small, so—at least for small conductors—it is generally ignored. In most cases, when someone says "the DC resistance of . . ." they mean resistance measured with an ohmmeter and not the total opposition to AC current. This is an important distinction.

Meter Readings of Changing Values

All meters, analog or digital, provide readings that are somewhat different from reality. Analog meters tend to provide a reading that is an average of the actual values. The meter movement is damped to prevent very fast, possibly destructive, needle movements. This damping is provided magnetically or with a small air vane. Digital meters do not read the input continuously; instead, they sample, taking quick periodic checks and displaying the value of that instant. This process is repeated at the meter's sampling rate, which varies widely for different makes and models. High sampling rates provide readings that are closer to reality. Rapidly changing values can be difficult to follow on a digital meter.

Except for the limitations just described, when a meter is set to read DC voltage or current, the reading provided is essentially what the meter sees at the inputs. If the meter, still set to DC, is connected to an AC source, it cannot provide a useful value. For the analog meter, the rapid changes of an AC input are too quick to follow. This, combined with the constant change in polarity, results in a reading of zero volts. This reading, as dangerously misleading as it can be, is not really in error. AC reverses direction every cycle, so the true average is zero. The change is too fast—every $^1/_{120}$ second—for the damped needle to react, so the needle never moves.

A digital meter set for DC but connected to AC may also give a zero reading, but most often attempts to read the values it sees. It samples the changing AC input, producing a series of essentially random numbers that rapidly flicker from one value to the next. Because the polarities of the voltage samples are random, the polarity indication sign will also flicker. Neither the numbers nor the polarity is likely to have any apparent pattern.

Obviously, AC voltage or current cannot be read without some interpretation by the meter. The rapid changes in both value and polarity of a normal AC cycle make it impractical to measure directly. For this reason, all multimeters have separate AC and DC functions for both voltage and current. Analog meters use internal rectifiers to convert the AC to DC, and apply small calibration corrections by using a different scale. Digital meters use different techniques to sample the AC voltage and convert it to a constant. The accuracy of a digital meter is affected by the sample rate and the way the display value is calculated.

AC Meter Readings: Representative Values

AC voltage rises to a peak with one polarity, declines to zero, reverses polarity, and then rises to a peak with this new polarity, declines to zero, reverses polarity again, and so on—over and over again. It would be difficult to construct a meter to display these actual values as they changed, and impractical because the display would be impossible to read. The result would be a vibrating meter needle or a blur of changing numbers on a digital display. It is more convenient, and a lot more useful, to measure AC voltage and current as average values and ignore the cyclic change. To understand the meaning of this average value, it is useful to examine the series of changing values that make up a typical AC voltage or current.

AC voltage and current are almost always displayed as curves that depict the rise and fall in magnitude of a normal cycle of AC voltage or current. These are sometimes called sine curves or sine waves, because the same shape can be produced by graphing the sine values of a rotating radius. The AC sine curve is the line graph that would be produced if a continuous series of actual voltage or current values were plotted on graph paper. The zero line of the graph is drawn horizontally across the center, with one half of the curve above zero and the other half below.

The sine curve plot that represents AC may be referred to as the AC waveform or wave. It is a convenient image, but it should be remembered for what it actually is—just a graphical representation of the changes that occur in a normal cycle of AC voltage or current.

The AC wave has several possible measurement points (see Figure 5–9). The horizontal center line is zero volts. The highest voltage reached during each half cycle, measured from the zero line, is called the peak value (E_{PK}). The full height, from the positive crest to the negative crest, is called the peak-to-peak value. Unfortunately, none of these values are fully representative of the average potential difference or the resulting current. A truly representative average is somewhere between the extremes of zero and peak, but it is not halfway, because the curve is much narrower at the peak.

The true average must be calculated, whether it is taken from graphed values or interpreted by a meter. The mathematical calculation used to generate the average is called root-mean-square (RMS) and the result is called the RMS (E_{RMS} or I_{RMS}) or effective value of the AC voltage or current. Ignoring the mathematical complexity, the RMS values of AC voltage and current are the most convenient to use. An AC RMS value is the functional equivalent of a numerically equal value of DC. This avoids unnecessary corrections when comparing AC and DC circuits. If a resistor is connected to 10 V DC and draws 1 A, it could be connected to 10 V AC and would still draw 1 A—assuming both the AC values are RMS. The heat generated in the resistor would be the same in both cases. Measured

FIGURE 5–9 Measurement points on the AC wave.

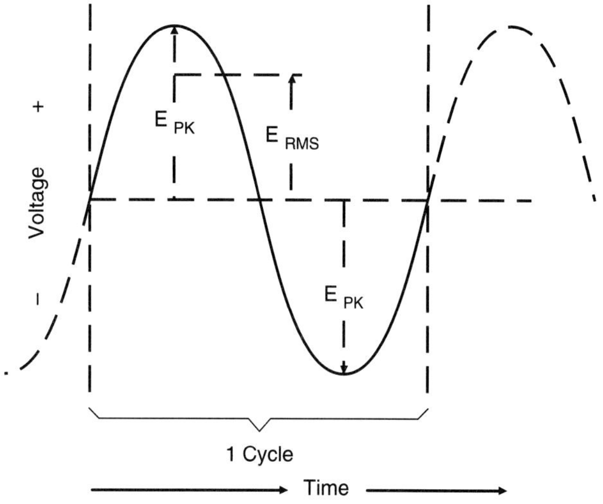

with standard meters—which display RMS values—the AC voltage and current would appear to be identical to the values obtained for the DC circuit. The only noticeable difference would be the lack of any polarity indication for the AC readings.

The RMS value for a true AC sine wave is 70.7% (0.707) of the peak value, measured from the zero line to either crest. This value is a constant for all sine waves. Most of the time it is assumed that the AC source is supplying a sine wave, because that is the output of a well-designed AC generator.

Note

Electrical utilities strive to generate and deliver good, clean sine wave AC to their customers' outlets. Unfortunately, they are not always successful, although it is not usually their fault. Some types of loads, in particular electronically controlled power loads, cause distortions in the AC waveform. When there is too much of this type of load, it can cause problems for the customer, the utility, and sometimes other customers in the area. Remember, however, that unless there are good reasons to believe otherwise, it is generally acceptable to assume everything is normal. The AC supply is usually treated as a source of perfect sine wave voltage, and the load current is assumed to be free of problem-causing distortions.

Assuming a perfect sine wave, the formulas for converting between RMS and peak values are as follows:

$$E_{RMS} = 0.707 \times E_{PK}$$
$$I_{RMS} = 0.707 \times I_{PK}$$

The RMS or effective values accurately represent the AC voltage or current. Calculation is not usually required, because RMS values are the values of normal use. Note that RMS values are a calculated average and have no polarity. **AC voltages and currents are always assumed to be RMS values unless stated otherwise. Equipment ratings are RMS, supply voltages are listed in RMS, and normal meter readings are RMS.**

Warning

Most meters set to read DC will not respond properly when connected to AC. The meter is usually unharmed, but the user may not be so fortunate. Service personnel often have little choice but to stake life and limb on the indications of a meter. A live AC circuit, tested with a meter set to read DC volts, may appear to be dead. It isn't, *but you could be*!

The Oscilloscope

On a meter readout, except for the polarity indications, AC and DC look the same. The meter does not follow the changing AC values, but provides instead a more useful RMS average. In most cases, this is entirely adequate. However, if it is necessary to observe the shape of the wave or note other details, a different instrument is required.

Only a high-speed graphic device like an oscilloscope can follow the AC voltage through its cycle of changes, and display it in a useful way. Oscilloscopes are not generally necessary or practical for electrical service work. However, there are two forces at work that may change that statement. These are the continued increase in the use of electronics in both control and power circuits, and the growing availability of handheld graphical meters.

Note

The last few years have seen rapid advances in both handheld oscilloscopes and the more service-oriented graphical multimeters. As the need for such instruments rises and the cost of owning them falls, they might very well replace the multimeters of today. Although they may be more confusing at first, they are also much more informative than a standard meter.

Oscilloscope Operation

Note

The following information applies primarily to the conventional oscilloscopes found in typical electronics labs and service shops. Handheld oscilloscopes are similar in operation, because they are expected to perform the same types of tasks.

Oscilloscopes vary widely in features and layout, but they all do essentially the same thing. Usually, if one oscilloscope is well understood, it takes a minimum of time and effort to learn to use another. To use an oscilloscope intelligently requires an understanding of what it does, a good grasp of the basic control functions, and some knowledge of the associated terminology.

The following information is based on relatively low-cost two-channel bench scopes typically found in school or college electronics laboratories. It is relatively generic information, intended to introduce oscilloscopes to those with little or no experience in their use.

The Screen

The screen of a typical bench-type oscilloscope is similar to that of a monochrome television tube (see Figure 5–10). As with the television set, the image is drawn on the inside surface of a cathode ray tube by a beam of electrons that originates at the back end of the tube. The inside surface of the screen is coated with a phosphorous material that converts

FIGURE 5–10 A 120 V AC input as seen by an oscilloscope and a multimeter.

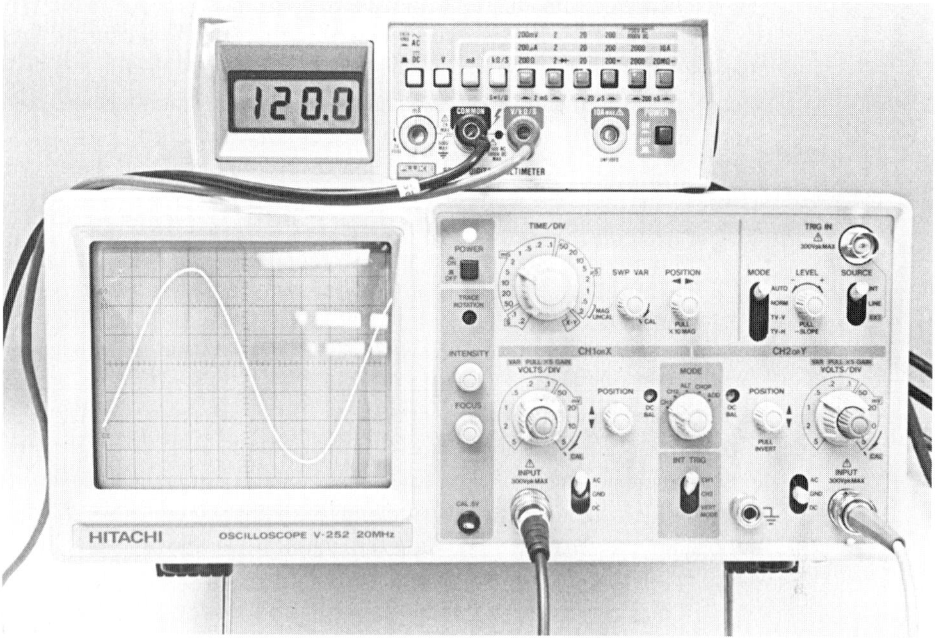

the energy of the electron beam into visible light. When the beam strikes the screen, it excites the phosphorous coating, and a luminous trace, or line, is visible through the glass front of the tube. The electron beam can be deflected, with great precision and at very high speeds, to hit the display screen at any desired location. The vertical deflection of the beam is proportional to the voltage applied to the oscilloscope's probe. The amplification and scaling of the input signal are handled by the internal electronics of the oscilloscope, in association with the settings of the controls.

In a television set, the beam scans the whole screen over and over again, in a repetitive pattern of parallel lines. The scan rate and pattern are fixed, and the image is created by modifying the strength of the beam as it moves. In the oscilloscope, the electron beam draws only the necessary lines: It does not scan any other parts of the screen.

The Controls

The image produced by an oscilloscope is a two-axis line graph. The horizontal (x) axis of the graph is time. The vertical (y) axis is voltage. The screen is marked with horizontal and vertical lines to form a grid, typically 10 divisions wide by 8 divisions high. The majority of the oscilloscope controls can be interpreted in terms of their effect on the legibility, position, and scale of the graph being drawn.

There are several controls that effect only the clarity of the line image. A brightness or intensity control adjusts the intensity of the electron beam, which changes the brightness of the line drawn on the screen. This control often requires some adjustment as other settings are changed. A focus control adjusts the sharpness of the line and, if it cannot be sharply focused across its full length, an astigmatism control is sometimes provided to correct this problem.

A trace rotation adjustment will align the image horizontally with the grid if the image is not perfectly parallel with the grid markings. An ideal trace is a sharp, thin line.

As with any graph, a zero line must be established for both axes. For the vertical axis, many graphs use the bottom of the grid. However, the electrical signals that an oscilloscope is used to display—AC voltages, for example—may change polarity, causing the graph to swing both above and below the zero line. For this reason, the horizontal baseline for the graph can be positioned anywhere, even completely off the screen and out of sight.

Note

This ability to adjust the image right off the screen of the scope can lead to some initial frustration because there may be nothing showing on the screen. A "beam finder" may be provided to help find a lost trace. It "pulls" the trace toward the center of the screen to show where it can be found.

For almost all oscilloscope operations, the x axis represents time. Time cannot be reversed—there is no negative value—so the zero line is usually at the left-hand edge of the displayed image. There is a control for horizontal positioning, and it does provide a limited amount of left/right adjustment of the image on the screen. However, this adjustment does not change the zero line relative to the graph. It just moves the entire graph left to right under the stationary grid markings.

The electron beam draws the line image from the left side (zero on the x axis, which is zero time) toward the right. During normal operation, the line graph is continuously redrawn, but always from left to right. The beam is interrupted to blank the image when it jumps back to the left to start over.

The line graph is scaled by using the oscilloscope's built-in adjustments for voltage sensitivity and time units. Both of these controls have a very wide range of possible settings. The controls for a typical two-channel oscilloscope are shown in Figure 5–11.

The x axis (time scale) of the graph is set with a time base selector switch, usually marked something like "TIME/DIV" or "SEC/DIV" (seconds per division). The typical range of values extends from 1 μs (1 microsecond) or less to 0.2 second or more per division. The trace can also be stopped completely. The time base selected is the same for both channels: They do not have independent time settings. The amount of time displayed on the screen is the TIME/DIV setting times the number of divisions, which is typically 10.

Note

The oscilloscope trace should not be left stopped, as a dot, for an extended period of time. When the electron beam sits for too long on one spot of the screen, it can permanently damage the phosphorous coating.

The y axis (voltage scale) of the scope image always has positive polarity toward the top and negative toward the bottom. The vertical deflection of the trace is determined by the actual voltage of the input signal and the setting of the VOLTS/DIV controls. There are two VOLTS/DIV controls, one for each input channel. They are independent of each other, allowing simultaneous display of two signals with very different voltage levels. The settings of these controls scale the graph vertically. The range of values may start at a few millivolts or less and extend to several volts. This may be extended further by using a probe with a built-in 10 times (10X) multiplier.

FIGURE 5–11 The controls of a typical two-channel oscilloscope.

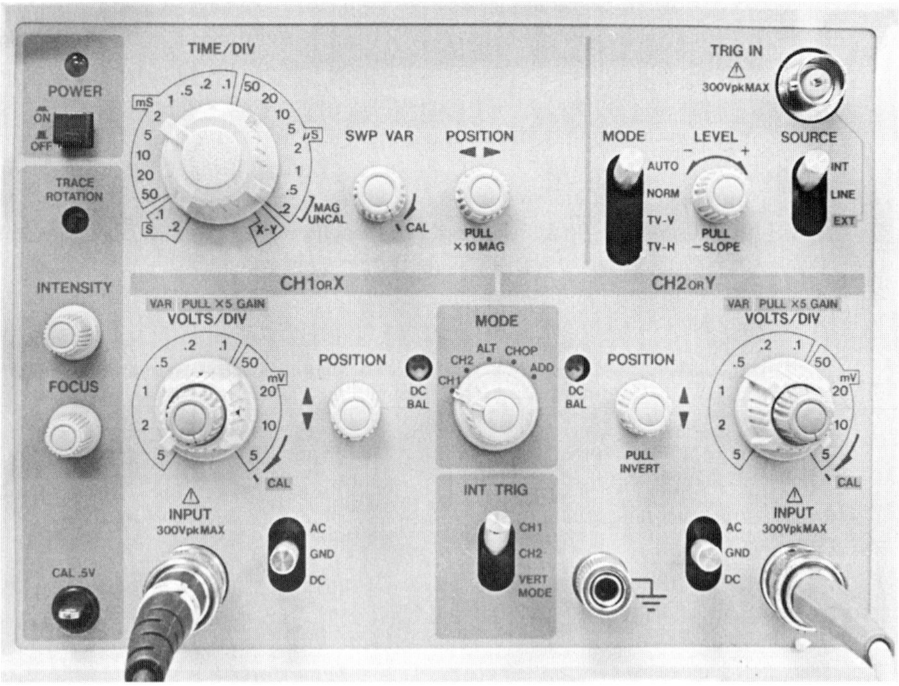

The probe, a more complex version of the meter lead, is often equipped with a switch marked 1X and 10X. In the 1X position, the input signal is passed on unmodified. In the 10X position, the signal into the scope is only one-tenth of the voltage at the probe. This effectively extends the voltage range of the scope by a factor of 10. When this setting is used, voltage readings taken from the scope image must be multiplied by 10 to obtain the actual values.

In normal operation, with a voltage signal applied to the probe, the oscilloscope trace moves from left to right at a rate set by the TIME/DIV control, and is simultaneously deflected vertically by the input voltage as scaled by the VOLTS/DIV control. The screen image is, therefore, a graph of voltage with respect to time.

The events that an oscilloscope is used to view usually occur in very short periods of time. If the display image was only drawn once, it would flash across the screen and disappear, making the instrument virtually useless. However, these signals are typically repetitive, occurring over and over again in regular cycles. The oscilloscope continuously redraws the display to produce a continuous image.

For the image to be stable, the trace must start at the same point on the cycle of the input signal for each successive image that is drawn on the screen. Synchronizing the redrawing of the screen is the role of the trigger circuits and their associated controls. Each time the trace reaches the right-hand side of the screen, it shuts off and jumps back to the left. If it immediately began to redraw, the next image almost certainly would not be aligned with the previous one. The result would be an unstable display that either drifts across the screen or appears as multiple images. To avoid this problem, the triggering circuits hold the blanked-out trace at the left side of the screen until the input signal passes through a specific voltage level. The trace is then released and the image is redrawn. This release point, or trigger level, is adjustable in value and can be set to operate on the rising or falling voltage. Changing the polarities has the apparent effect of inverting the display, but it simply draws the negative half-cycle first. Triggering synchronizes each successive trace of a cyclic input signal, producing a stable display of a rapidly repetitive event.

There are several trigger controls, including trigger level, trigger polarity, an operating mode selector, and a trigger source selector. Trigger level and polarity have already been discussed. Operating modes may include auto, manual, and some specialized settings for television servicing. A typical trigger source switch selects channel 1, 2, or an external source. In most cases the trigger selector is set to the channel where the signal of interest is applied. For example, if the signal is applied to the channel 1 probe, the trigger source selector would also be set to channel 1.

Each channel of the scope also has a separate input selector switch with at least three positions: DC, AC, and ground. In the DC position, the probe is directly connected to the internal input of the scope. In the AC position, DC signals are blocked by a capacitor connected in series with the input. In the ground position (GND), the switch disconnects the probe and grounds the internal input line. Grounding the input in that way ensures that channel has a true zero-input signal. This allows the trace line to be accurately set to the desired zero position on the screen. Because the probe is disconnected internally when the input selector is set to GND, there is no need to disconnect it from the external circuit.

Warning

Unlike multimeters, most bench oscilloscopes have one of the two input leads connected to the chassis, or frame, of the oscilloscope. This is further connected through the ground wire in the instrument's power cord to earth ground. The ungrounded input lead is the tip of the probe; the grounded one is usually a small lead that extends from the probe base or the side of probe itself. The end of this lead is normally equipped with a small alligator clip.

When using an oscilloscope, always remember that the clip lead is a ground wire. If it is accidentally connected to the "hot" side of a grounded circuit, it will cause a short to ground through the clip lead, the probe cable, and the frame of the scope. This can cause considerable damage, usually to the equipment under test, and sometimes to the probe leads.

A similar problem can occur in ungrounded circuits when two channels are used. Both ground clips are connected to the chassis of the scope, which means they are also connected to each other. If they are connected to two separate points in the circuit, the end result is the same as connecting a jumper wire between those points. Again, it is easy to damage circuit components or the scope leads.

The solution is to connect the scope ground clips only to a point in the circuit which is—or can be—connected to the earth ground of the power supply. When more than one probe is used, connect both ground clips to the same point in the circuit.

Some circuits have no suitable grounded point. All the possible connection points are "hot." A quick-thinking service person may realize that if the oscilloscope is not grounded, the ground lead can be connected to a hot point of the circuit without causing a short. A two-wire extension cord eliminates the normal ground connection, and setting the scope on a nonconductive

surface, away from grounded metal, eliminates accidental grounds. The scope lead ground clips can then be connected to a live part of the circuit without shorting things out.

This is a crafty solution to a thorny problem. However, it is also extremely dangerous. When that same individual, who so brilliantly solved the ground problem, reaches back to adjust the oscilloscope, he discovers that the entire instrument is now electrically energized by its ground clip connection to the live part of the circuit.

Later in this chapter, under the heading "Practical Example: Experimenting With Reactance" (p. 188), there is a detailed explanation of how to take and interpret measurements with AC meters and oscilloscopes.

5.5 Reactance

Reactance is an interesting, though difficult, part of all AC electrical systems. Before jumping into this new concept, a brief review of circuit operation might be in order. Up to this point in this text, the loads in example circuits have been resistors. Circuit operation has been described as a supply voltage pushing a flow of electrons (current) through a conductive circuit path that includes a load. The circuit path must be in the form of a continuous loop that includes a power source, conductors to interconnect the various circuit components, the load, and usually a switch to provide convenient on/off control.

A resistive load acts to oppose the current flow and converts the energy drawn from the source into heat. The amount of resistance used must be balanced against the supply voltage to keep the current within a desirable level. The current flow can be predetermined with an Ohm's Law calculation. The heat produced in the resistor is dissipated into the air and surrounding materials. The energy flow is one way only: The resistor cannot reclaim the heat and convert it back into electrical energy.

Fortunately there are many loads where the current is limited by something other than the crude electrical friction effects of resistance, and the electrical energy is turned into something other than heat. These types of loads "react" to the passage of current, or an applied voltage, in much more interesting ways. In reactive loads, the energy conversion is interactive. The load converts the electrical energy into a new form that is not immediately dissipated and lost. Instead, the energy is stored, at least temporarily, and may be returned to the circuit.

Reactance comes in two different forms. One of these is similar to mechanical inertia, which is the tendency of things to sit still when they are stopped, or to keep going in the same direction once they are moving. The other is similar to elasticity, the tendency of most materials to bounce back when pressed or stretched. In mechanical systems, these properties come from the nature of mass—which we often confuse with weight—and from the elastic nature of most molecular structures. Mass acts to oppose change in an object's position, or its speed and direction of motion. Elasticity allows materials to "give" when stressed, and then snap back when released. Both of these properties store and release energy, and interact with any force that affects them.

When a force is applied to a stationary object, the inertia of the object opposes the force applied. As the inertia is overcome, the object begins to accelerate. The energy of the accelerating push is converted into energy stored in the object's mass and motion as inertia.

The inertia acts to keep the object moving at the same speed in the same direction. This inertial energy is not immediately dissipated. It can, of course, be used up and lost as heat generated by friction, but the initial energy conversion was from the force and motion of the push the object received into the stored inertial energy of a moving mass. This energy is recoverable; it can be used to do some useful work.

Just as it takes energy to accelerate a mass, it also takes energy to compress a spring. In this case, the energy is stored in the molecular tension of the spring material and released when the force on the spring is removed. Again, the conversion is interactive, the energy is recoverable, and it is possible to use it to do useful work.

These mechanical properties are quite different. Both affect the physical behavior of virtually everything, separately and in all possible combinations. Such behavior is very dynamic. It can also be difficult to understand, to measure, and to accurately predict.

Electrical reactances can be compared to those of mechanics. Like the mechanical examples, it comes from two specific properties. The first is the relationship between magnetism and electricity. This interaction gives electrical circuits a property similar to inertia. The second is the nature of electrostatic charge, which gives electrical systems a property very much like mechanical elasticity. Like their mechanical counterparts, electrical reactances affect the operation of most electrical systems, singly and in combination. They also can be difficult to understand, to measure, and to accurately predict.

Before electrical reactance can be studied, the underlying forces must be examined. The first of these two is magnetism.

Electricity and Magnetism

Magnetic Forces and Fields. Magnetism is a separate subject, very comparable to electricity, with equal or greater complexity. To examine it in depth is far beyond the intent of this text. However, a look at the basic principles of magnetism should be enough to support a closer examination of its relationship to electricity. It might be helpful to note that, in the study of electricity, it is usually enough to measure the electrical effects of magnetism; direct measurement of the magnetic fields themselves is generally not required.

Almost everyone has played with magnets as a child and in science classes in school. Interesting as these experiences are, they often leave the observer with little more than the knowledge that magnets attract and repel each other, and pick up pieces of iron. A more useful understanding of magnetism begins with viewing the magnet as a source of magnetizing force, instead of a chunk of material surrounded by a magnetic field. A magnetizing force, called magnetomotive force (MMF), always creates a field, but the size, shape, and strength of the field depend on other factors.

The extent and power of the magnetic field is heavily influenced not only by the MMF that creates it, but by the shape, size, and magnetic properties of the surrounding material. This is the "magnetic circuit" in which the field must form. When a magnet is suspended in the air or supported by some nonmagnetic material, the field can be detected and at least indirectly observed by its effects on a compass, loose iron filings, or bits of other magnetic material. Its can be shown to be symmetrical in shape and strength, surrounding the magnet on all sides. Experiments and demonstrations of such fields are common in early science classes, and quite familiar to everyone.

Now, suppose the magnet is imbedded in a block of soft iron, instead of being suspended in the air (see Figure 5–12). The same MMF exists, but the path for the magnetic

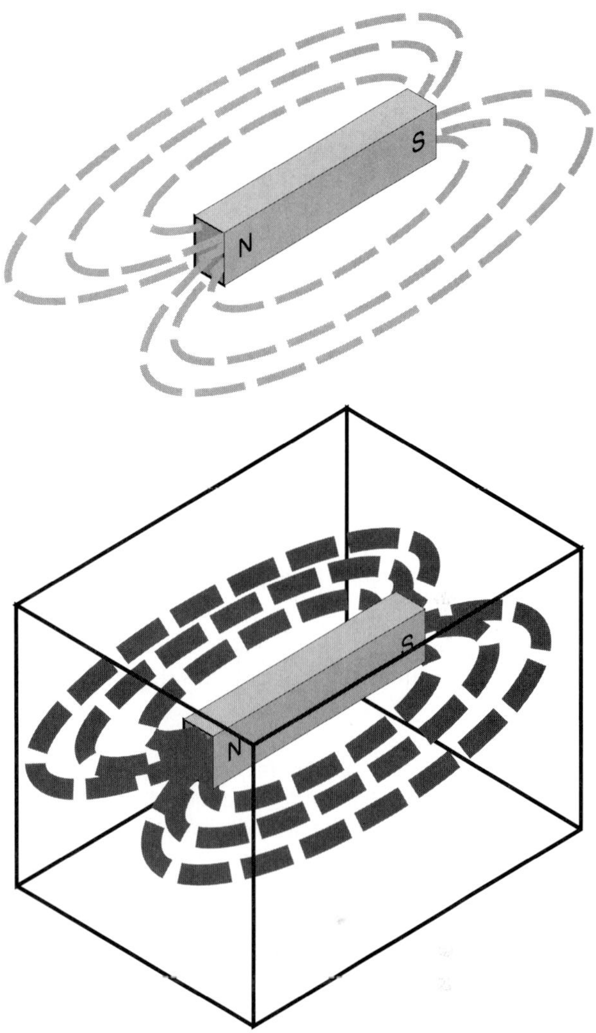

FIGURE 5–12 When a magnet is embedded in iron, the field is much more powerful but is virtually impossible to observe.

field is now composed of solid iron instead of air. The result is a much more powerful field that is impossible to observe. Virtually the entire field is within a "short circuit" path of iron, which, unlike air, is an excellent magnetic "conductor." Good magnetic conductors are said to have a low reluctance to the passage of magnetic lines of force.

The lines of flux that make up the magnetic field have a very strong preference for low-reluctance materials such as iron. Through the use of such materials, magnetic fields can be contained and directed into almost any shape. This allows efficient use of the magnetizing force available and permits the fields to be concentrated at the desired location. It also makes the fields difficult to observe and, for this reason, their power is rarely appreciated. Many electrical devices use this technique. Carefully planned magnetic circuits are built into the equipment, usually from thin sheets of soft iron laminated to the desired thickness. The result is powerful magnetic fields concentrated in the area of greatest benefit, and almost totally contained inside the equipment.

It should be noted that, although the magnetic properties of materials vary greatly, there is no known magnetic insulator. The field can be diverted around an object by surrounding it with shield of soft iron, for example. It can also be short circuited, such as when the metal "keeper" is placed across the ends of a horseshoe magnet, but it cannot be blocked by any known material.

Electromagnetism. A permanent magnet has a natural magnetizing force that is, of course, not adjustable. When an adjustable "magnet" is needed, an electromagnet is used. The simple electromagnet, a coil of wire around an iron core, is a familiar grade-school science device. A demonstration of an electromagnet picking up metal or operating a mechanical bell or buzzer is a popular lesson in the nature of electromagnetism: electrical current creates a magnetic field.

Earlier in this chapter (pp. 146–149), there is a description of an experiment that demonstrates how magnetism can be used to generate electricity. Now the opposite effect, electricity creating a magnetic field, is under examination. Magnetism and electricity are not just related, they are very closely interrelated. When they are in motion, each one creates the other, and the result is a constant interaction between the two. When current flows, a magnetic force is created that encircles the current path. This magnetomotive force (MMF) produces a field around the conductor along its entire length. When a magnetic field moves or changes, an electrical force is generated in any conductive material within the changing field. This is the electrical force we know as potential difference or electromotive force (EMF). If the EMF is across a complete circuit, current will flow.

Note

It is generally believed that the source of all magnetism is in the electrical characteristics of electrons. Electrons carry tiny circular currents all the time. In permanent magnetic materials, these electrons are aligned in such away that the magnetic effects of these currents add together, and a magnetic field is produced. In nonmagnetic materials the electrons are aligned at random, and the magnetic effects of the currents cancel each other.

Current through a conductor generates an MMF that is proportional to the amperage. This force is not particularly large, unless the current is extremely high. The strength of the field around a single straight conductor is too small for most practical uses. However, when the conductor is formed into a coil, the magnetizing force of each coil turn is combined to produce a single, much more powerful, field.

The total magnetizing force of the coil is a product of the current in each turn and the total number of turns in the coil. A coil with 100 turns has 100 times the magnetizing force of a coil with only 1 turn, assuming each carries the same current. When the number of turns is held constant, the MMF is directly proportional to the current. Many different combinations of turns and current can be used to produce the same end result, as Figure 5–13 illustrates. For example, 1 A in a coil with 100 turns of wire will produce the same MMF as a coil of 10 turns carrying 10 A, or one with 1000 turns at 0.1 A.

The field of an electromagnet follows the same rules of behavior as a permanent magnet. Using a bar magnet as an example, the magnetic lines exit from the north pole, travel around the outside of the magnet, and reenter at the south pole. The path is completed, from south to north, inside the source of MMF—the magnet. In an electromagnet, the magnetizing force is inside the hollow center of the coil, running parallel to

FIGURE 5–13 All three of these coils produce the same magnetizing force.

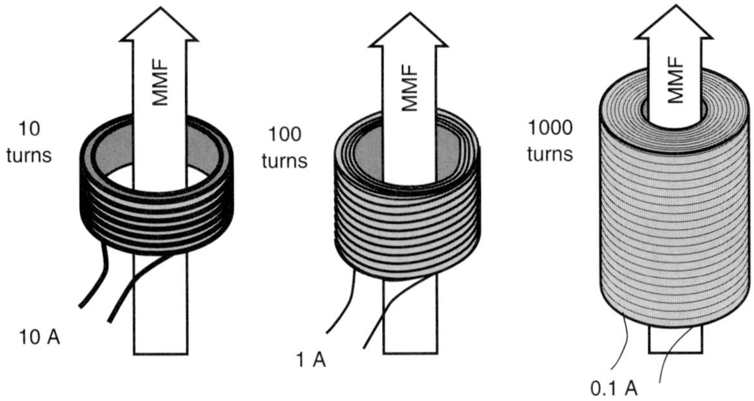

the center line. The magnetic lines of the electromagnet exit at the north end, reenter at the south, and complete the loop up through the center of the coil, exactly as if the coil was a permanent magnet.

A coil is a much more flexible source of MMF than a permanent magnet, because it is electrically adjustable. The north/south polarity of the MMF depends on the direction in which the current flows. The strength of the MMF depends on the level of current flow and the number of turns in the coil. The strength of the resulting field depends not only on the total MMF, but also on the quality of the magnetic circuit in which the field forms. To increase the strength of the field, almost all electromagnets have low-reluctance soft iron cores.

Transformers and motors are excellent examples of electromagnetic devices. To create the powerful fields needed for them to operate efficiently, a magnetic circuit with very low reluctance is essential. This circuit is built from sheets of soft iron, laminated into a heavy magnetic "core." The core is shaped so the field is directed to where it is needed. The core is often very thick, because it must have an adequate cross section to carry the magnetic field efficiently. Much of the size and weight of a transformer or electric motor is the iron core.

Electromagnetic Induction. An electrical current creates a magnetizing force, and magnetism can also generate an electrical force. Magnetomotive force is created only when current flows; likewise, movement or change in the magnetic field is required to generate an electromotive force. The process of generating an EMF from a changing magnetic field is called electromagnetic induction.

Magnetism can generate an EMF in two different ways. First, if lines of flux cut a conductor, an EMF is induced in the conductor. This relationship is often the initial key to an understanding of generators. In all generators, the conductors cut through or are cut by a powerful magnetic field. A single conductor cut by the field of a typical bar magnet would produce only a tiny voltage: Practical generators have hundreds of conductors and very powerful magnetic fields.

Second, if there is a change in the amount of flux passing through a coil, a voltage is generated in the coil (see Figure 5–14). This second relationship is the key to understanding transformer operation and inductive reactance. Transformers link two AC electrical circuits with a shared magnetic field. A changing field is generated by the AC current in one of the coils, and this field is used to induce a voltage in the second coil. This effect

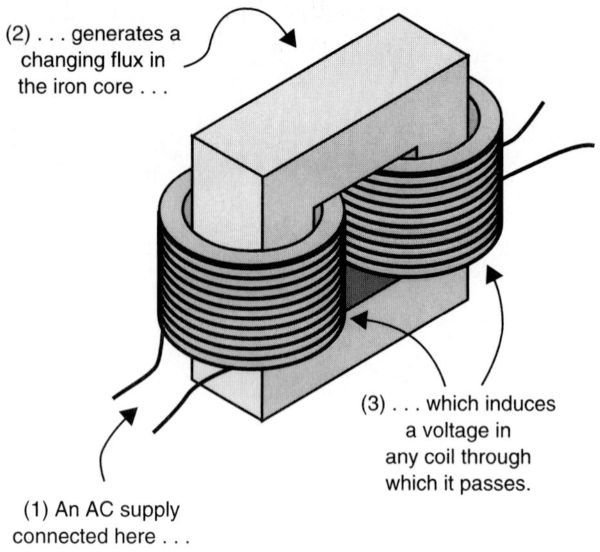

FIGURE 5–14 An elementary transformer demonstrates how electromagnetic induction occurs when a changing field passes through a coil.

is sometimes referred to as mutual inductance. The net result is a transfer of energy from the circuit of the first coil to the circuit of the second. The connection is entirely magnetic, because there is no direct electrical connection between the coils. Although the first coil provides energy to the second coil, they remain electrically isolated from each other. The voltage and current relationships between the two coils can be altered by using different numbers of turns in each coil. (Transformers are examined in a later section of this text.)

Mutual inductance is not the only form of inductance. Even in a two-coil circuit like the transformer circuit just described, the coil that is connected to the supply does not escape the inductive effects of the changing field. The coil that is energized from the AC supply "sees" a changing magnetic field passing through it, in just the same way as the second coil. Even though there is a source voltage across the first coil and current through it, a second voltage is still induced into each of the coil turns. However, this induced voltage is not an exact copy of the supply voltage; it is a mirror image instead.

The induced voltage is exactly opposite in polarity and almost identical in value to the voltage of the supply. Because of the effect it has on circuit operation, a voltage induced into an energized coil is called a counter, or back, EMF. Counter EMF, or back EMF, always acts in direct opposition to any changes in current flow. A declining current will induce a supporting voltage; a rising current will induce an opposing voltage. This opposition is in the form of voltage, but it is produced by changing current and magnetic fields, and always opposes the change from which it is created.

The process of a changing coil current inducing a voltage within that same coil is known as self-inductance. Self-inductance has a powerful effect on the operation of coils in all circuits. In DC circuits, inductance has no effect unless the current changes, so its influence is not continuous. In AC circuits, the current is constantly changing, so the effect is continuous. This opposition to AC current flow is called inductive reactance. It gives all coils a continuous opposition to AC current that is independent of the resistance of the coil wire. Inductive reactance is an interesting form of opposition to current flow because it has none of the energy-wasting, heat-producing effects associated with resistance.

FIGURE 5–15 This hydraulic "inductor" opposes changes in flow.

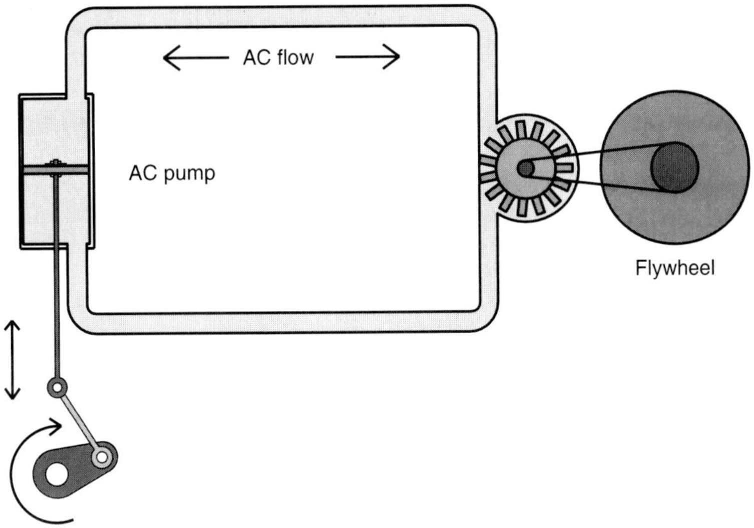

The forces that produce inductance are similar to those of mechanical inertia in a mechanical or hydraulic circuit. In an inductor, electrical energy is alternately stored in, and then received from, a changing magnetic field. The primary byproduct of this process is that the current flow is limited by the effects of the generated back EMF. It is possible to build a mechanical or hydraulic equivalent to an inductor, although it is questionable what application it might have. One possibility is shown in Figure 5–15.

Note that the paddle wheel in the oil line drives the flywheel only; this is not a motor driving a load, only an inductor. It will have little or no effect unless there are changes in the flow rate of the oil. In a unidirectional (DC) system, when the pressure is first applied and the oil starts to flow, the "inductor" will oppose the increase. Then, once the flow reaches its normal operating rate, it will have virtually no effect. When the oil flow is stopped, such as by the sudden closure of a valve, the energy stored in the inertia of the flywheel will be transferred back through the paddle wheel to the oil. This force will try to maintain the flow, this time opposing the decrease rather than the increase. The switching device—the valve—must withstand the pressure generated by the inductor.

If the hydraulic system has an alternating flow (AC), the influence of the paddle-wheel inductor will be continuous. The changing flow must change the speed and direction of the paddle wheel continuously. The effect is very much like that of an inductor in an AC circuit.

Inductance. Inductance is the property of an electrical circuit that produces a voltage whenever there is a change in current through the circuit. Even a straight conductor has inductance, but the effect is very small. The changing field that encircles the current path expands and contracts around the center line of the conductor, inducing small opposing voltages as it cuts through the surrounding conductor material. When the conductor is formed into a coil, the inductance is greatly magnified. The magnetic field that was previously spread along the entire length of the conductor is now concentrated into a single field through the coil. In addition to being much stronger, the field also has a much

FIGURE 5–16 Inductors oppose AC current flow through the effects of a self-induced back EMF.

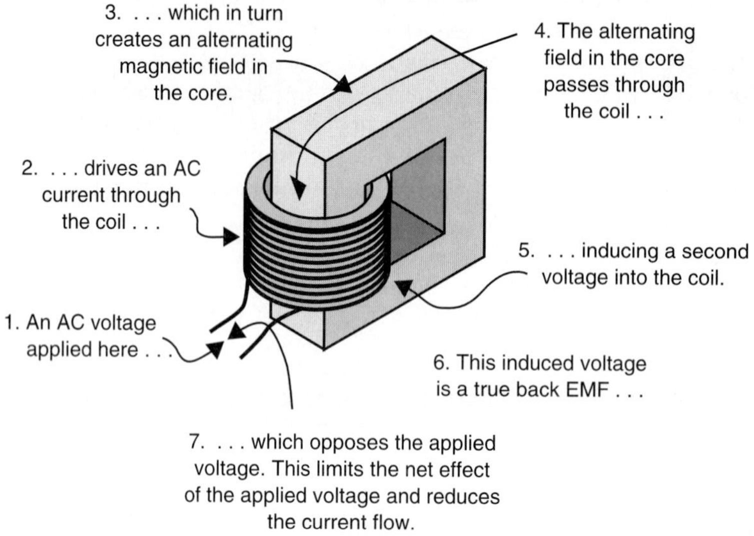

greater influence, because any change acts upon all of the conductor in the coil. Any change in the current creates a change in the coil field, which in turn induces an opposing back EMF in the coil (see Figure 5–16).

The actual value of inductance varies with the electrical and magnetic properties of the coil and the coil core. A coil with a larger number of turns has a greater inductance than a coil with less turns, assuming, of course, that all other factors are equal. A coil with low-reluctance core material also has a greater inductance than one that has a high-reluctance core. If two coils have the same number of turns and the same current flow, the magnetomotive force for the coils is the same. If one of coil has no core material other than air, and the other coil has an iron core, the inductance of the coil with the iron core will be much greater.

The reason for this difference is that, although the MMF is the same in both coils, the magnetic fields created do not have the same strength. With the same driving MMF, the low-reluctance iron circuit has a much more powerful field than a high-reluctance material like air. The stronger the field, the greater the inductance. The effect of an iron core is most pronounced when it is shaped into a complete loop passing through the coil and around the outside. This forms a short-circuited low-reluctance magnetic path, ideal for maximum field strength.

It should be noted that virtually all electrical power devices that use coils do have cores made of magnetic material, and the most commonly used material is iron. The iron chosen is relatively soft, with alloys such as silicon added to improve its magnetic properties. Hard iron and steel are avoided because they tend to become permanently magnetized. This reduces efficiency, because a permanent magnet opposes changes in the strength and polarity of its field.

Most magnetic cores are also laminated. The laminations are relatively thin and may be coated with an insulating varnish to electrically separate the layers. Iron is not only a good conductor of magnetic fields, it is also quite a good conductor of electrical current. Voltages are induced in the iron by the same changing field that induces voltage in the surrounding coil. If the iron is not laminated, this voltage will cause what are commonly known as eddy currents to circulate in the iron. Eddy currents cause a power loss and generate potentially damaging heat in the core. For this reason, the cores of all AC coil equipment are laminated. This includes the cores of simple AC electromagnets, all transformers, and all AC motors. Laminating the core does not totally eliminate the eddy-current losses, but it reduces them to much more acceptable levels. Laminating is unnecessary for simple DC devices, but is required for equipment such as DC motor and generator armatures, because current flow within a DC armature is actually AC.

Note

The unit of inductance is the henry (H). A coil has an inductance of one henry if a change in the current through the coil of one ampere per second induces one volt in the coil. Coils are available in millihenry (mH) and henry values. The physical size of the coil is a very poor guide to its inductance rating. For example, a 1 mH coil designed for 10 A of current would be much larger than a 1 H coil designed for a few milliamps. This is because the 10 A coil must have a much heavier conductor, and must therefore be built larger to accommodate the coil wire.

Inductance only affects circuit operation when there is a change in the current. In a DC circuit these effects are limited to specific conditions, such as when power is first applied or when the circuit is shut off. Inductance is not irrelevant in DC systems, but its effects are always short-lived. For example, when a DC coil is energized, the inductance of the coil will act to oppose the rising current. The net result is a very brief delay in the current reaching its full operating level. This is generally an advantage, because it reduces the impact of adding loads. When the current does reach its normal operating value, the inductive effect disappears.

When the power to the DC coil is shut off, the effect is similar, but the results are much different. Circuit inductance opposes any change in current, rise or fall; therefore, the induced voltage tries to maintain the falling current. Again, the effect is very brief, but this time it has more serious consequences. When the switch opens, the current drops very rapidly, causing a very sudden change in the field. This induces a high voltage that arcs across the switch contacts as they open. This is one of the main reasons that DC coil loads require substantial switches with very good arc-breaking ability.

The high-voltage spike produced by switching off a DC coil not only causes problems for switch contacts, it can also destroy sensitive electronic components, even at great distances from the coil. Fortunately, this spike can be controlled with additional circuit components. Capacitors across the contacts can absorb the energy, greatly increasing contact life. This technique is used in automotive ignition systems. Another common cure is to connect a diode in reverse across the DC coil. Diodes are the electrical equivalent of a one-way valve. When the circuit is operating, the diode does not pass any current. When the connection is broken, the diode acts as a short circuit to the induced voltage, draining the energy away and preventing a potentially damaging spike from traveling back through the supply conductors.

Inductive Reactance. In AC circuits, the current is constantly changing, so any circuit inductance has a continuous influence. The overall effect is called inductive reactance, and it plays an important role in all AC circuits that contain coils. The actual value of the inductive reactance is dependent on the inductance of the coil—described above—and the rate at which the current is changing. In an AC circuit, the rate of change is usually set by the system frequency.

Note

Inductive reactance is a measure of the continuous opposition to current flow that is produced by inductance in an AC circuit. It is measured in ohms, but it is not resistance. The actual value of inductive reactance (X_L) is a function of the inductance (L) and the system frequency (F). The following formula shows how these values interrelate.

$$X_L = 2\pi FL$$

X_L = the inductive reactance in ohms
π = the constant pi
L = the inductance in henrys

An ohm of inductive reactance has the same current-limiting effect as an ohm of resistance, but the overall result is much different. In a resistance, current flow is restricted by "electrical friction," and the energy lost overcoming the resistance is converted to heat. In an inductive circuit, AC current is limited by voltage opposing voltage, force opposing force. When there are two voltages present in the same circuit, the net voltage left to drive current through the circuit is the sum or difference of the two. Inductive reactance reduces the driving voltage of the circuit by countering the supply voltage with an opposing back EMF. The back EMF can be very close in value to the supply voltage. It cannot be measured, because is cannot be separated from the supply voltage, but its effects are very evident.

Inductive Reactance in Action. Inductance and inductive reactance can be very useful or very much a nuisance, but in either case they are unavoidable. Even straight conductors have some inductance, but it is small enough at power line frequencies that it can usually be ignored. When the conductor is formed into a coil, the inductance and the resulting inductive reactance increase dramatically. Desirable or not, coils are very common in electrical machines. Virtually all electromechanical devices operate by converting electrical energy into magnetic fields, and the only practical way to create these fields is with a coil.

The reason for the coil may be the resulting magnetic field, not the inductive effects. However, inductance and inductive reactance will certainly enter into the behavior of the equipment. A wise person works with nature, using natural behaviors and properties to advantage. Successful AC electrical equipment is designed to take advantage of mutual and self-inductance, and all the related behaviors. Good examples include solenoid coils, transformers, and motors.

Solenoid coils for AC use are much different than those used for DC. A DC coil is made with a high resistance—lots of turns of very fine wire—because that is the only way to limit the coil current. An AC coil has fewer turns of heavier wire, because the current will be limited by the inductive reactance. DC coils have the advantage of no vibration and tremendous holding power. AC solenoid coils have the advantage of drawing high current when they are initially energized, but naturally limiting the current to a lower value after the solenoid core has pulled into place. This gives the coil a good pickup force and a lower holding current.

AC coils behave this way because the inductance changes with the position of the moving core. When the core is dropped out, the magnetic circuit of the coil is open, and the inductive reactance of the coil is relatively low. The result is a high initial current, generating a strong field to pull the solenoid core into place. When the core is pulled in, the magnetic circuit is closed, and the inductive reactance of the coil is relatively high. This limits the current to a lower value that is adequate to maintain a strong holding field in the now-closed magnetic circuit. The result is better efficiency and cooler operation.

DC coils, by comparison, draw full current almost immediately, and then stay at that current indefinitely. To prevent overheating, DC solenoid devices often have an external current-limiting resistance switched into the coil circuit just as the solenoid core pulls into place.

Motors and transformers also have inductance, and this is well demonstrated by the fact that the current draw is almost totally unrelated to the measured resistance of the windings. A motor winding typically has a resistance of only a few ohms, measured with an ohmmeter, but the current the motor draws is much less than an Ohm's Law calculation would suggest. (Motors are discussed in detail in a later section of this text.)

In addition to inductance as a natural consequence of coils in equipment, there are occasions where coils are added to circuits specifically for their inductive properties. Many different sizes and styles of coils are manufactured specifically for this purpose. They are rated according to their inductance (in mH or H), their current capacity, and maximum operating voltage. They are sometimes called chokes, choke coils, or reactors. The term "choke" is quite descriptive of their function because they act like a choke to AC current. They are sometimes used for their effect at line frequencies, and sometimes because of their tendency to block fast-changing transients, or high-frequency currents that have been imposed on the system. When they are used in this way they act as selective filters. In some circuits, more complex filters are built up from a combination of coils and other components such as capacitors.

Inductors may have an air core, an iron core, or a core made from some magnetic composite such as ferrite. Air-core inductors are just a coil, usually wound on a nonmagnetic form. No magnetic core material is used. They are preferred where very low inductance values are required, and are more efficient and more accurate at high frequencies. Small air-core inductors are common in electronic equipment; large ones are sometimes found in very high-current feeder circuits, where they are used to reduce the maximum possible fault current. They are also used to trap high frequency communication signals sent on high-voltage utility lines.

Iron-core inductors are most common where larger inductance values are required. They can take several different forms, including toroids—coils wound around a donut shaped core—and conventional core types that are often mistaken for transformers. The giveaway is that they have only two leads connected.

In summary, inductance behaves much like inertia in a mechanical system. Just as inertia opposes changes in movement or flow, inductance opposes changes in current. Inductors store electrical energy in a magnetic field, but they cannot hold it for any length of time. The instant the current is cut off, the field collapses, and the stored energy is returned to the circuit. Although inductance affects DC circuits, it is most widely studied as an AC characteristic. The current-limiting effects of inductive reactance come from voltage opposing voltage. The back EMF induced in the coil by a changing field acts to cancel out most of the applied voltage. This provides a very effective opposition to AC current flow, without the energy losses associated with resistance.

Electricity and Static Charge

The Second Type of Reactance. Capacitance is the second reactive property. It is based on charge storage, and behaves much like a spring or pressure tank in a mechanical or fluid based system. Just as these devices oppose changes in force or pressure, capacitance opposes changes in voltage.

Capacitance comes from the ability of two conductive surfaces, separated from each other by insulation, to accumulate and store an electrical charge. The device most associated with capacitance is the capacitor (also called a condenser), but some value of capacitance is present wherever two conductors are separated by insulation. The reactive effect of capacitance is called capacitive reactance.

Electrostatic Charges. The term "static electricity" means stationary charge. It is usually associated with such things as clothing that clings, the spark generated by walking across a carpet and then touching a metal doorknob, or a flash of lightning. All of these result from an accumulated electrostatic charge. Electrostatic charges occur when electrons are separated from their parent atoms and accumulate in a different location. The area from which the electrons are removed is left with a net positive charge, while the area of accumulation becomes negatively charged. As long as the two charges are insulated from one another, there is a potential difference and stored energy. When the insulation between them breaks down, or some other conductive path is provided, the charges move to equalize the potential difference between them. The result is often a single short-lived spark.

The annoying doorknob type of shock comes from a static charge that builds up on a person's body when shoes pick up electrons from a nylon carpet. Part of the accumulated charge is released to the doorknob or other object that is able to provide a discharge path. This problem is only experienced when the air is very dry. Static charges cannot build up when the humidity is high, because moisture in the carpet and in the air makes them conductive enough to bleed off the charge. Electrostatic charges can have very high voltage levels, but to build up a significant charge, the insulation has to be very good. Sprays that reduce static cling in clothing or protect against static buildup do their job by being electrically conductive.

With the right weather conditions, a static charge develops between the clouds and the Earth. The potential difference may be relative to other clouds, or between the clouds and the Earth below them. When the difference reaches a high enough voltage level, accelerated electrons ionize a path through the air. This provides a conductive path through which the accumulated charges can equalize, and the result is the dramatic discharge we see as lightning. The actual process of lightning is very complex, and is still under scientific study.

Machines do exist that can generate static electricity, some capable of some very impressive voltages. The Van de Graff generators used in schools and science fairs are a good example. Although such devices are excellent for lab experiments and demonstrations, they are not practical sources of electrical energy.

Electrostatic Charge and Current Electricity. All electrostatic charges are electrical in nature, regardless of how they are produced. They have voltage and carry energy. When they are released in a controlled manner, they become more useful current electricity. With a suitable storage device and a voltage source, a charge can be created, stored, and used within a practical electrical circuit. The device commonly used to store charges is called a capacitor. Capacitors are a common part of many AC and DC circuits.

Capacitor Construction. A capacitor consists of two conductive plates separated by a layer of insulation. The capacity or capacitance—the amount of charge it can store—depends on the area of the plates, the distance between them, and the properties of the insulating layer. Capacitance is directly proportional to the area of the plates, and inversely proportional to the distance separating them. The choice of insulation materials is important, because it must be a good insulator that also has strong electrostatic properties. A good insulator is said to have a high dielectric strength. A material that has excellent electrostatic properties has a high dielectric constant.

Note

There is a potential here for some confusion in terms. The word *dielectric* means nonconducting. It is sometimes used interchangeably with the term "insulation" as it applies to electricity. Both words have the same meaning. The *dielectric strength* of the insulation refers to the voltage it can withstand: It is a value in volts or kilovolts per unit of thickness. The *dielectric constant* of an insulation material refers to its electrostatic properties: It is a ratio that compares the electrostatic performance of the material to that of an equal thickness of air.

In capacitor construction, both the dielectric strength and the dielectric constant must be taken into account. As with most things, there are trade-offs. For example, very thin insulation layers lead to compact capacitors with high capacitance values, but the voltage rating is reduced. This is why very large capacitance values are readily available in low-voltage capacitors, but not in high-voltage capacitors.

Small values of capacitance do not require large plate areas and are relatively simple to build. The capacitors may be small disks or beads, and the voltage ratings may be quite high. For larger capacitance values the large plate areas needed dictate a little more ingenuity. A common method of construction uses two thin strips of conductive foil separated by layers of insulation material. Leads are attached to the foil strips, and then the strips are rolled into a tight cylinder and covered by a protective container. This provides a large plate area and a thin dielectric in a very compact size.

For the thinnest possible dielectric, the foils are separated by a sheet of tissue soaked with chemicals instead of a solid material. A DC current is applied, and a dielectric is created by electrochemical action. Capacitors like this are called electrolytic because of the process used in manufacture. They have a higher storage capacity, for their size, than any other capacitor type. Unfortunately, they must also be connected with a specific polarity. Special nonpolarized electrolytics are available for AC motor starting and similar applications, but all others are polarity sensitive. An assortment of capacitors is shown in Figure 5–17.

Warning

Polarized electrolytic capacitors can be identified by the polarity markings on the case. These may be a row of signs, plus (+) or minus (–), around one end, on one side, or on an arrow pointing to one lead. Such capacitors must only be connected to DC, and only with the correct polarity. Incorrect connections or connection to an AC supply will destroy the capacitor unless the voltage is far below the rated voltage. With incorrect polarity, the capacitor conducts some current through the dielectric layer. Heat builds up inside the tightly rolled foils and pressurizes the inside of the case. A sealing plug blows out one end, usually followed by bits of what used to be a capacitor. This can happen very quickly if the voltage is close to the capacitor rating, or may take many minutes if the voltage is lower.

FIGURE 5–17 An assortment of capacitors.

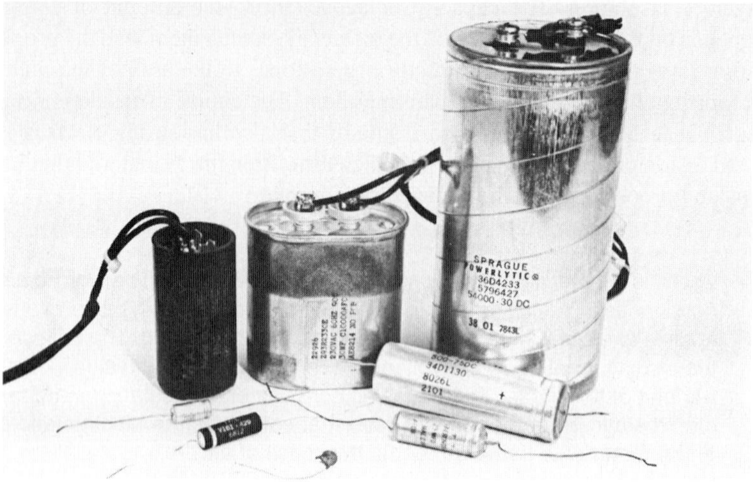

Capacitor Ratings. The unit of capacitance is the farad (F). A capacitor rated at one farad will store one coulomb (C) of charge with one volt applied. A coulomb of charge is a lot—a typical lightning stroke delivers only about 10 to 20 C—so the farad is a very large unit. Capacitors rated in farads are rare. A 1 F capacitor is available for use in very high powered auto stereo systems. The voltage rating is very low, just the 12 to 14 V required for the application. These capacitors are added to these systems to provide high instantaneous power peaks.

Virtually all other capacitors are rated in microfarads, or even smaller units such as nano- or picofarads. A microfarad (μF) is $^1/_{1,000,000}$, or 0.000001, farads. Micro is 10^{-6}, nano is 10^{-9}, and pico is 10^{-12}. (Note that older capacitors are marked MFD instead of μF, or MMF for micro micro—which is picofarads.) It should be noted that physical size is not a very reliable indicator of the capacitance rating. Capacitance is very much affected by the voltage rating of the capacitor and by other physical factors of its construction.

Note

Capacitors are found in many single-phase motor applications, including the motors used in the heating ventilation and air conditioning industry. These capacitors are usually one of three types. For motor starting, the capacitor is an AC electrolytic type designed for short-time duty only. These capacitors are typically housed in a black plastic cylinder with quick-connect terminals on one end. The capacitance ratings are relatively high (100 μf or more) in comparison with their size. Continuous-duty-run capacitors were all oil-filled types a few years ago. These were housed in flat-sided aluminum cans with quick-connect terminals on one end. Oil-filled capacitors are now being replaced by more compact solid-dielectric types.

Capacitor Operation. When a voltage source is connected across a capacitor, electrons are forced onto one plate and drawn off the other. The two plates take on opposite charges; one negative and one positive. This charge difference is a potential difference, or voltage, between the two plates. The capacitor takes on a charge, but it's not like filling a can. It's like filling a pressure tank. The difference in the charges on the plates builds up an electrostatic stress between the plates and the dielectric material. This stress acts as a powerful physical force that opposes any increase in charge. The capacitor does not just accept a charge. The charge must be forced onto the plates by the supply voltage. The electrostatic stress that develops provides the reactive "spring" that makes the capacitor useful..

The capacitor is a true charge-storage device, not just a can full of electrons. It charges until the voltage between the plates is equal to the supply voltage, and then the flow of electrons stops. This process is much like filling an air-storage tank from a pressurized supply line. The tank fills only until the pressure inside the tank is equal to the pressure of the supply line.

The comparison between a pressure tank and a capacitor is a good one, but there are differences. A practical tank and air supply has valves, couplings, and fittings that restrict the rate at which the air can flow into or out of the tank. A practical capacitor has no such restrictions. When it is directly connected to a supply, it can charge or discharge virtually instantaneously. When a capacitor is connected to the supply through a resistance, its charging behavior becomes identical to that of filling an air tank. This technique is often used to build timing circuits. The rate at which the capacitor charges is adjusted with resistance and used as a time reference.

Once charged, the capacitor, like the pressure tank, can be disconnected from the source, and the charge will remain. A voltage check across the terminals will indicate the voltage of the charge, just as a pressure gauge could be used to check the air storage of the tank. Note, however, that the voltmeter will draw a very small amount of current from the capacitor. Therefore, the act of checking the capacitor's charge causes it to discharge. If the capacitor is very small, this can happen very quickly. Large capacitor values can take a long time to discharge through a voltmeter.

This effect can be observed by leaving a voltmeter connected across a capacitor and watching the voltage level decline. The discharge rate depends on the resistance of the voltmeter. A typical digital meter acts as a 10 to 20 MΩ resistor; a good-quality analog may have a 20,000-ohm-per-volt rating (Ω/V), which means its resistance on a DC voltage range is 20 kΩ times the range setting. For example, on a 25 V range, this meter would have a total internal resistance of 500 kΩ.

Warning

A charged capacitor can give a nasty shock. The effect can range from a minor annoyance to possible permanent injury or death, depending on the capacitance and the voltage. Even with nothing connected, the charge on any capacitor will drain down over time, but good-quality capacitors can store a charge for a year or more, so do not make the mistake of assuming they are discharged.

Capacitor Circuit Behavior. Current does not pass through a capacitor, because the plates are insulated from each other by the layer of dielectric material. However, this does not mean current cannot flow in a capacitor circuit. Because it can store a charge, current can flow into and out of the capacitor within the limits of its storage capacity (see Figures 5–18 and 5–19).

With a DC supply, the capacitor will draw current from the supply only as long as it takes to charge up to the supply voltage, and then current flow stops. With no resistance in the circuit to control the flow rate, this can occur almost instantaneously. Once the capacitor

FIGURE 5–18 A DC capacitor circuit and its mechanical equivalent.

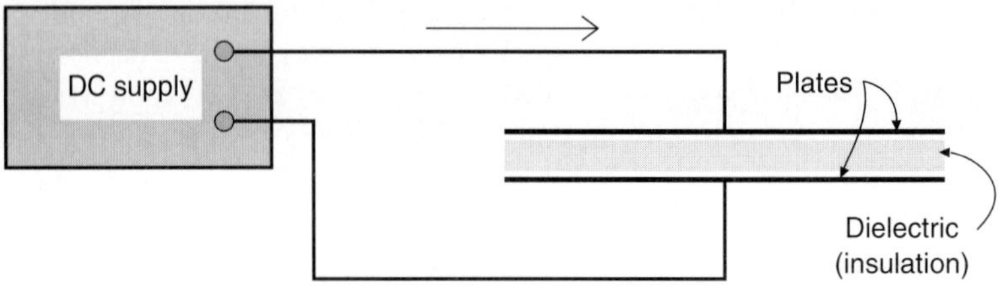

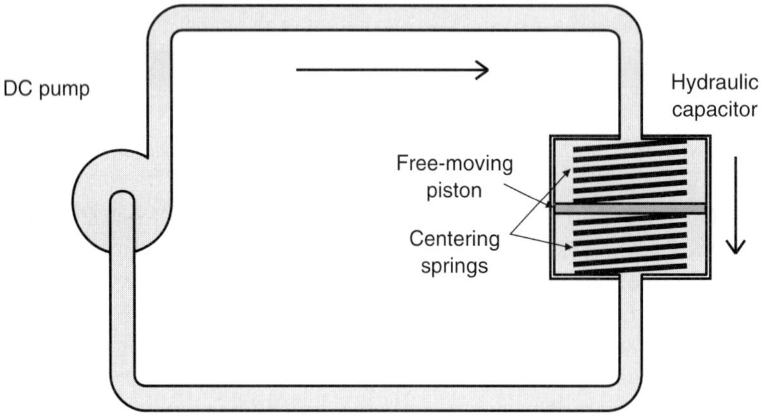

is charged, it has no effect on circuit operation, unless there is a change in the supply voltage. If the supply voltage drops, the capacitor voltage will then be higher than the supply, and current will flow from the capacitor back into the circuit. Should the supply voltage suddenly increase, the capacitor will immediately begin to draw current again until the voltage balance is restored. Capacitors act to resist changes in voltage, and are widely used in DC circuits to help stabilize the voltage level.

With an AC supply, the capacitor will charge and discharge continuously. The result is continuous AC current flow, at least in all parts of the circuit that are external to the capacitor. The current does not pass through the capacitor, but the circuit behaves very much as if it did. As the supply voltage rises toward the peak of the sine wave, the capacitor charges, because its voltage is lower than the supply. When the supply voltage passes the peak of the wave and begins to decline, the capacitor discharges back into the circuit, because its voltage is now higher. The capacitor fully charges, and then fully discharges, each half cycle, first with one polarity, and then the other. A continuous AC current flows in and out of the capacitor, even though no electrons actually pass through the layer of insulation between the capacitor plates.

It should be noted that, although the capacitor circuit has continuous current flow, there is no net energy transfer. The capacitor only stores energy; it does not convert it into any other form. No energy is used and no work is done. A capacitor load has a near-zero reading on a wattmeter.

FIGURE 5–19 An AC capacitor circuit and its mechanical equivalent.

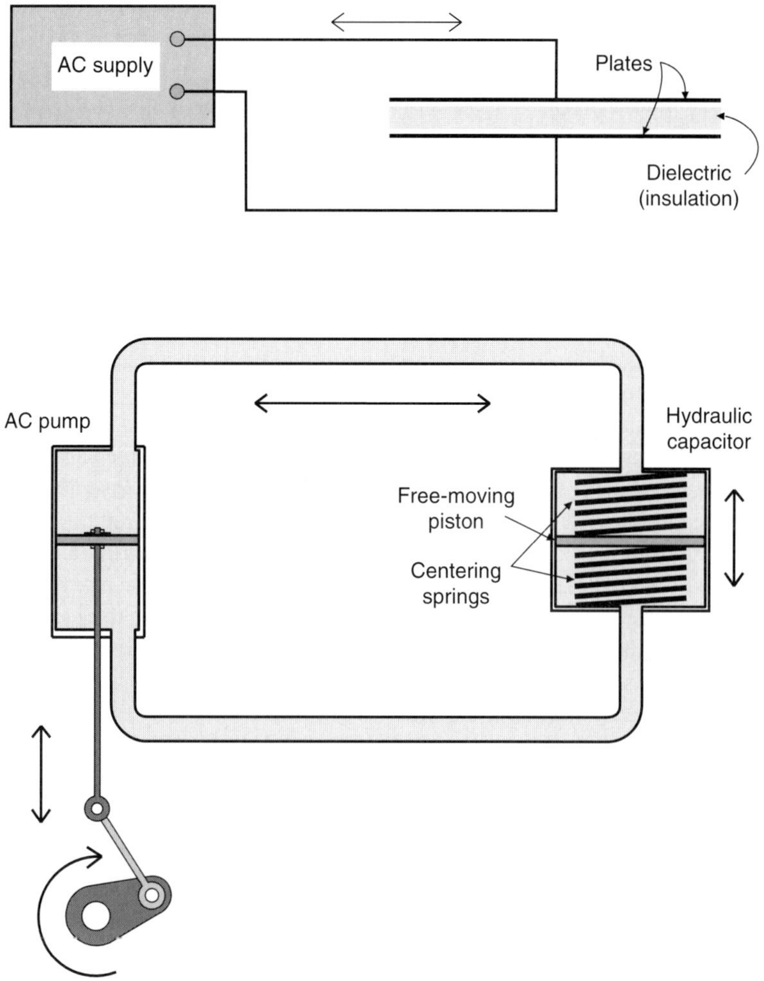

Capacitor Handling Safety. When a capacitor is discharged, the excess electrons on one plate flow through the external circuit to the other plate, where there was a shortage. This process is the equalization of an electrostatic charge and, like a natural static discharge, it can happen very quickly. Unless the discharge rate is controlled, all the energy stored in the capacitor is dissipated in an instant. If the discharge path is through a person's hands or body, the result can be a painful shock. If the capacitor is a large one charged to a dangerous voltage, contact with the capacitor terminals or other circuit wiring can be extremely dangerous.

Before a capacitor is handled, or contact is made with a circuit containing capacitors, the stored charge must be drained off. The traditional method is to short the capacitor terminals with a tool, such as a screwdriver shank or a short jumper. This is quite acceptable when voltages are relatively low and capacitance values are modest. However, at higher voltages, or with larger values of capacitance, it is definitely not advisable to discharge the capacitor by shorting across the terminals. At best there will be minor arcing damage to the capacitor terminals, and possibly to the tool or jumper. More likely there will be major damage to these contact points or to other circuit components, or the capacitor may be damaged internally.

The correct approach—assuming the voltage is within the capabilities of a normal meter—is to first check the level of charge with a voltmeter. **Remember: After the supply is shut off, the voltage across the capacitor is always DC, regardless of the type of supply.** If the supply was AC, the maximum possible charge voltage is the E_{PK} value of the sine wave, about 1.4 times the rated AC value. The minimum value is zero. The actual value cannot be predicted, and may be anywhere between those extremes.

Once the voltage level is known, the level of hazard is much more obvious. If a relatively high charge remains, the capacitor can be discharged through a modest value of resistance. This controls the discharge rate, avoiding any damage. This discharge process can be monitored with a voltmeter, if desired.

Capacitor banks–groups of parallel-connected capacitors—are particularly deserving of respect. Capacitors connected in parallel act as one very large capacitor, because the capacitance values are additive. Most capacitor banks are also associated with high-voltage supplies, so they can store an enormous amount of energy. To reduce the hazard, they are normally equipped with discharge, or "bleeder," resistors connected directly across the terminals of the individual capacitors.

The bleeder resistors have a value high enough that they do not interfere with normal circuit operation, but low enough to slowly discharge the capacitors when the supply is de-energized. There is a time designation for this discharge, typically one to five minutes, after which the voltage should be down to a "safe" level. This level is not zero, nor is it necessarily low enough to totally eliminate the hazard. Even after the time has elapsed, it should never be assumed the capacitor circuit is safe to touch. A voltmeter can be used to check the level of charge and, if necessary, to monitor its continued decline.

The role and behavior of capacitors in practical applications—particularly motor circuits—are examined in the relevant portions of this text.

Applications of Electrostatic Charge. Electrostatic charges have many commercial applications. In most of them, fine particles of material are attracted to some object by giving the particles and the target object an opposite static charge. This technique is used in a number of devices and in certain processes.

Electrostatic air cleaners, the "electronic" air filters used to clean the air in some heating or cooling systems, are a good example of how electrostatic charges can be used in practical applications (see Figure 5–20). A high-voltage source is used to charge a grid that is suspended in the air stream. Any airborne particles passing though the grid also become charged. The same voltage source is used to simultaneously provide the opposite charge to the air cleaner's collector. The charged particles are attracted to a collector, removing them from the air. This process is more efficient than standard filters because the collector can have much larger passages than the size of the particles being removed. Even very fine particles such as pollen can be removed, because they are not being trapped by a physical barrier such as a traditional filter medium. Instead, they are actually pulled onto the collector grid by the attraction created by the static charge.

FIGURE 5–20 The high-voltage DC supply of an "electronic" air cleaner charges the particles in the "dirty" air, so they are attracted to collection plate.

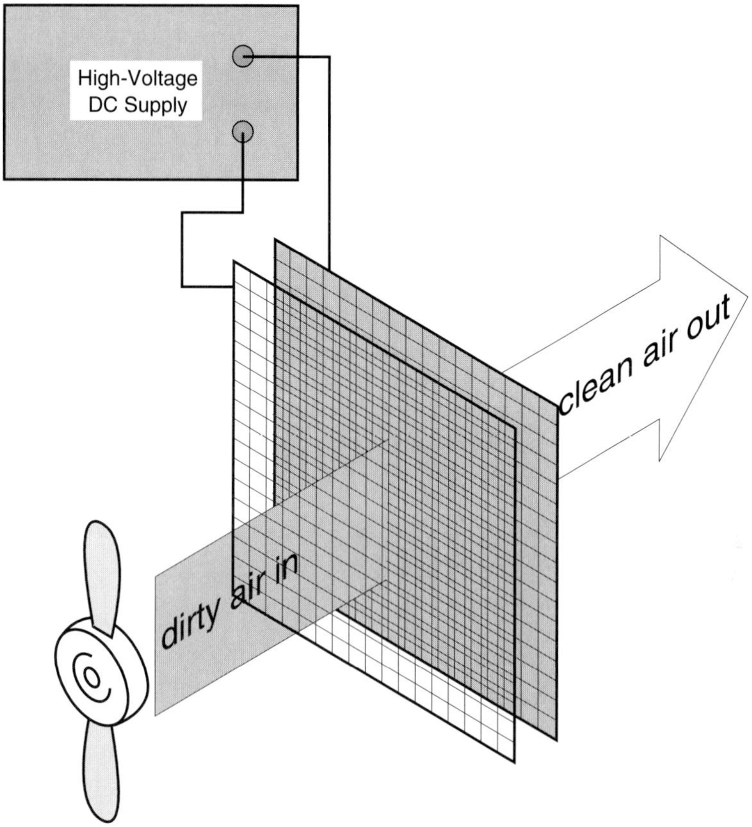

Photocopiers and laser printers also depend on electrostatic charges for their operation. A coated drum is first charged, and then selectively discharged by exposure to light, to form the desired printing pattern. A fine powdered toner is attracted to the areas of the drum that remain charged, and the toner is rolled onto the paper. The image is made permanent with pressure and heat.

Electrostatic charges are also used to reduce waste in spray painting. The paint particles and the object to be painted are charged with the opposite polarities. The airborne paint particles are literally drawn to the surface of the object, resulting in less overspray and better coverage.

Safety Considerations

The primary safety concern with all electrostatic applications is the high voltages involved. To produce a useful static charge, voltages of 10,000 to 50,000 V or more are required. This is far beyond the capacity of normal meters. Even when the specialized instruments required are available, extra caution must be used when working with such high operating voltages. Always read the manufacturer's service literature and heed all warning labels.

5.6 Reactance and Phase Shift

The types of loads covered to this point have been resistive, inductive, and capacitive. Singly and in combination, these three cover the complete spectrum of passive electrical loads. Their construction, characteristics, and general behavior have already been examined. The focus of this section is on the phase relationships that develop between the voltage and current when these loads are connected to an AC supply.

Note

In general, passive loads such as these do not cause serious distortions in the AC voltage or current waveform, but, if they are reactive, they can affect the phase relationship. This means that, when the load receives a sine wave voltage, the current will also be a sine wave, but it may not be exactly synchronized with the voltage wave that drives it. Active loads such as power electronic circuits block parts of each AC cycle, causing distortions in the voltage and current wave. The resulting waveform is no longer a "clean" sine wave. Problems associated with these distortions can have serious impacts on the operation of other equipment, particularly when capacitors are involved. To avoid a very distracting side trip into a much more difficult area of study, these issues will be ignored. In many applications they are irrelevant; in others they are of serious concern, often requiring the attention of specialists in the field.

In any AC circuit, both voltage and current follow a cycle of change that, when graphed, takes on the shape of a sine wave. This characteristic voltage wave is the normal output of a the power utility generators. Current is driven by voltage, so a graph of the current wave has the same general shape and the same cycle of changes.

The term "phase relationship" refers to the relative timing of two cyclic events that operate at the same frequency. If the two cycles pass through zero at the same time and have positive and negative peaks that coincide, they are said to be in phase. If they occur at slightly different times, they are said to be out of phase. The degree to which they are out of phase is referred to as the phase shift. Phase shift can be in either direction, and may be referenced to either of the cycles or to some other related measure of time.

In an electrical circuit, the phase relationship is dependent on the characteristic behavior of the load. If the load is resistive, or behaves like a resistance, the voltage and current will be in phase. If the load is reactive, the normal behavior of the reactance—constantly drawing energy and then returning it to the circuit—will shift the current out of phase with the voltage. The current lags the voltage if the load is inductive, and leads when it is capacitive. The direction and degree of phase shift depend on the nature of the load. It cannot be adjusted or controlled by adjustments to the supply.

The phase shift resulting from reactive loads is a natural occurrence caused by the way in which energy moves through the AC system. As with most natural occurrences, it is usually better to exploit the characteristic in a positive way than to combat its effects. There are many successful devices where phase shift is used to advantage. Good examples are found in the design and construction of single-phase motors. (These are discussed in detail in Part III.)

FIGURE 5–21 One cycle of AC voltage or current is 360 electrical degrees in duration.

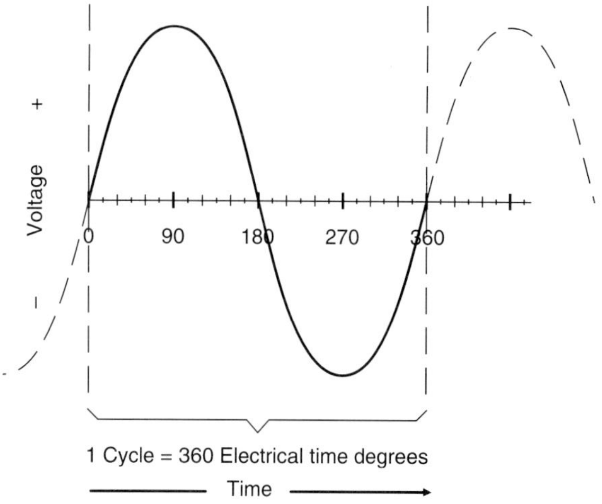

Measuring Phase Shift

The shape of the AC voltage graph is called a sine curve because it can be reproduced mathematically from the trigonometric function sine. Using a fixed starting value and applying the sine function for one complete rotation of angle (360°) will produce a set of values which, plotted on a graph, forms a classic sine curve. Drawing on the mathematical origins of the sine curve, the horizontal baseline of an AC wave is measured in degrees. One full cycle is 360° in length. This can be confusing, because the baseline of the voltage wave is actually a measure of time. In an attempt to clarify this unit, it is sometimes referred to as electrical time degrees (see Figure 5–21).

Although it may seem an inappropriate way to measure time, the use of electrical time degrees to show position on an AC sine wave is consistent throughout the study of electricity. It is relatively simple and quite useful, once the principles are understood. The most common error for novices is confusing mechanical degrees of generator rotation with electrical time degrees on the voltage output graph. This comes about because for the output of a simple two-pole generator the mechanical and electrical degrees are a perfect match. However, if the generator has more than two poles, which is very often the case, each revolution of the generator will produce more than one cycle.

Phase shift is the difference, in electrical time degrees, between one sine wave and another. The "angle" is measured along the baseline of the wave. Although either voltage or current can be used as the reference, for the sake of consistency in this text, phase shift is always referenced to the supply voltage.

For any combination of the three passive load types, the phase shift between the AC supply voltage and the AC current through the load cannot be greater than 90 electrical degrees on either side of the in-phase position. In other words, the maximum possible phase shift between the AC current and voltage is from 90° leading to 90° lagging. This is one-fourth of a cycle either way. The current can be anywhere between 90° lead and 90° lag. It does not have to be right at zero—in phase—or on one of the 90° points. The actual position depends on the mix of load types. An example of lagging AC current is shown in Figure 5–22.

FIGURE 5–22 This AC current is lagging the voltage by 60°.

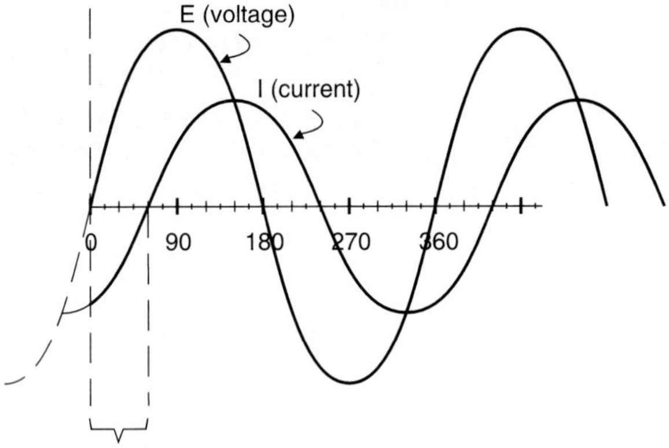

Load Types and Phase Shift

As previously stated, there are three basic types of loads: resistive, inductive, and capacitive. However, these names can be somewhat misleading. Loads are often categorized on the basis of how closely they match the normal behavior of each of these load types, rather than according to the components they actually contain. For example, a load may be described as resistive, inductive, or capacitive on the basis of the characteristic phase shift, regardless of whether it actually contains a resistor, a coil, or a capacitor.

It should also be noted that many practical loads behave as a mix of these characteristics, rather than all of one type. Even for demonstration purposes, it can be difficult to obtain pure loads that match the theoretic models. This is particularly true for inductive loads, because losses in the coil conductor and the magnetic core are very difficult to totally eliminate. Some loads, notably simple resistance circuits or good-quality capacitors, do come surprisingly close to the ideal.

Unless absolute precision is the goal, it is generally not a problem that practical load devices fall a little short of theoretical perfection. However, it is important to understand the concepts and to realize that these limitations do not reflect badly on the theory. Instead they serve as reminders that all but the simplest devices are composites of many different materials and behaviors.

Understanding anything requires a grasp of the basic concepts, plus the ability to combine these concepts in different ways to form a composite whole. In the case of load types on AC circuits, the three basic loads are usually discussed as separate entities. This often extends to typical demonstrations, and even to the mathematical treatments used to study them in greater depth. The three load types are distinctive enough that it is more convenient to examine them individually rather than in combination, but in the real world there is a continual interaction of all three.

Resistive load examples include heating elements and incandescent lightbulbs. Current and voltage are always in phase in resistive loads (see Figure 5–23). Some loads that are not resistive in the traditional electrical friction sense may also draw current that is in phase. These nonreactive loads act like resistance, even though the energy output may not be heat. In general, when all the electrical energy into the load is converted to some other energy form, and none of the energy is returned to the system, the current into that load will be in phase.

FIGURE 5–23 When the current and voltage are in phase, they rise and fall in synchronism. This is the relationship that exists in a resistive AC circuit.

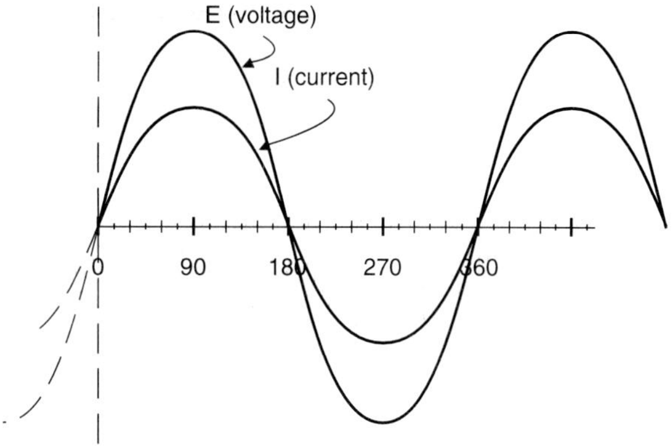

Current and voltage are in phase.

Reactive loads are devices that store and release energy on each half cycle. In coil loads, the energy is stored in the magnetic field; in capacitor, as a charge on the plates. In these loads the current and voltage are never in phase.

Inductive loads are coil loads such as motors and lightly loaded transformers. The reactive nature of these loads causes the current to be shifted out of phase in a lagging direction (see Figure 5–24). A purely inductive load will produce a current shift of 90°, or one-fourth cycle. The amount of phase shift does not depend on the inductance of the coil or the amount of current it draws. All purely inductive loads have the same 90° shift, and the current always lags the voltage.

That said, in a practical inductor the shift is always less than the theoretical 90°. To obtain the complete quarter-cycle lag, the coil would have to be a pure inductance with no losses in the coil wire or in the core. It may be possible to construct a perfect inductor, but it is unlikely to be practical. A coil with no iron core material, cooled to superconductivity, should have a perfect 90° current lag. It has no core losses and no resistance.

In contrast to the ideal, all practical coils have some resistance in the winding and some losses in the core iron. However, they also have a very significant current lag. The lower the losses of the coil, the closer it is to a 90° phase shift.

FIGURE 5–24 A current that lags the voltage by 90° is characteristic of inductive loads.

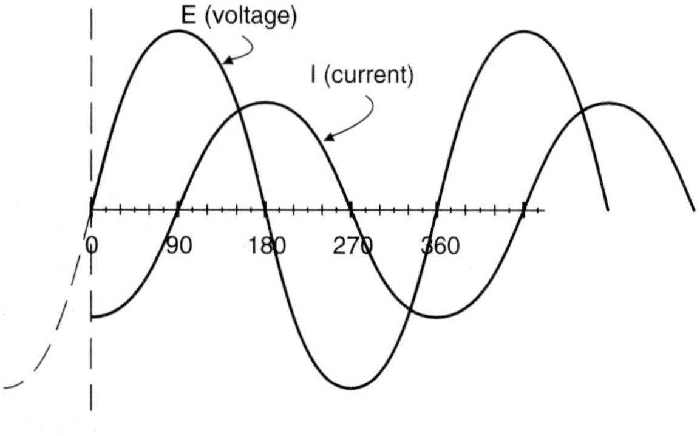

Current is lagging behind voltage by 90°.

FIGURE 5–25 Current that leads the voltage by 90° is characteristic of capacitive circuits.

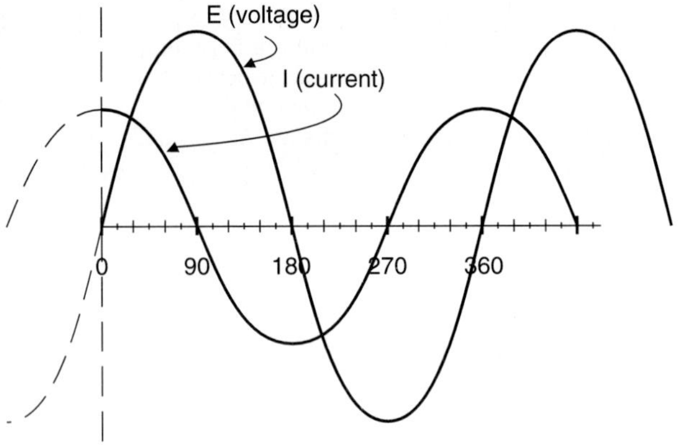

Current is leading voltage by 90°.

Capacitive loads, with a few exceptions, are capacitors. Only a few other situations, such as long power lines and long high-voltage cables, produce significant capacitive load. Capacitors, like coils, are reactive, but the reactance is much different. In this case the load current is shifted out of phase in the opposite direction to that of the coil. In a capacitor circuit the current leads the voltage by one-quarter cycle, or 90 electrical degrees (see Figure 5–25). This means the current is 90° ahead of the voltage that drives it. That seems like a practical impossibility, but with a capacitor load, after the first half cycle, the circuit acts like two sources with current flowing between them. The AC current oscillates between the real supply and the charge on the capacitor. When the supply voltage is increasing, the capacitor acts like a load because it is charging; when the supply voltage starts decreasing, the charged capacitor acts as the supply because the stored charge is returned to the circuit. Capacitors are not perfect: They do have a small amount of loss. The phase shift is not exactly 90°, but it is very close.

Mixed loads produce mixed results. Each type of load, by itself, produces a different phase relationship between the voltage and current. When different load types are combined, each of the loads keeps its own characteristics in its part of the circuit, but the circuit as a whole is a blend. The phase shift is always between the extremes of 90° lead and 90° lag, but can fall anywhere in that range. The actual shift depends on the mix of loads, their characteristics, and their current draw or voltage drop relative to the total.

One of the most commonly encountered combinations is resistance and inductance. This occurs naturally in most coils, in all types of AC motors, and in lightly loaded transformers. When these two characteristics are combined, the current falls between in phase and 90° lag. If the mix is half resistive and half inductive, the phase shift is 45° (see Figure 5–26). More resistance moves the current closer to in phase, and more inductance moves it further toward a 90° lag.

Motors are one of the most interesting examples of this type of load, because the phase shift changes with the mechanical demand on the motor shaft. The inductive portion of the motor current is essentially constant, while the "resistive" behavior is largely dependent on the mechanical power output. The inductive portion of the motor's electrical demand is the energy used to create the magnetic field. This energy is alternately stored in the field and then released back to the system, just as it is for any other inductor.

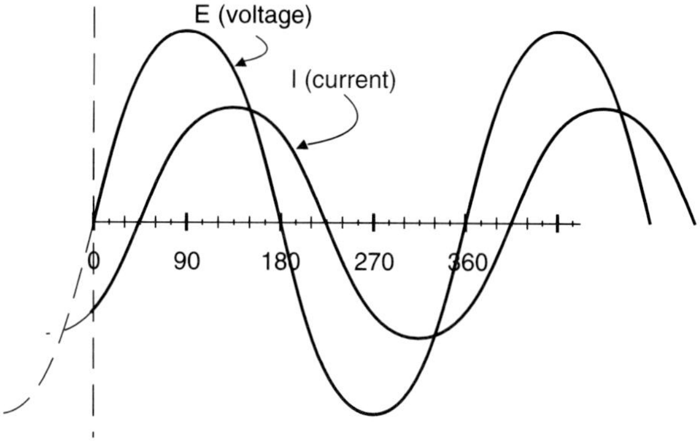

FIGURE 5–26 A 45° shift results when the load is half resistive and half inductive.

Current is lagging behind voltage by 45°.

The in-phase, or "resistive," portion of the motor's electrical demand includes all the electrical energy into the motor that is either converted to mechanical output power or is lost and dissipated as heat. The mechanical output power depends on the mechanical load. The losses include magnetic losses in the core iron, electrical losses in the resistance of the coil conductors, and mechanical losses in the bearings, cooling fan, and so on. The magnetic losses are relatively constant. Losses in the conductors vary with motor load current. Mechanical losses are also nearly constant, because the motor operates at very close to a constant speed. The overall result is that the motor current may lag by 70° or more at no load, but much less at full load. Very large motors may get within less than 10° of in phase; smaller ones are usually around 20° to 30°.

When resistive and capacitive characteristics are combined, the phase angle falls between 90° lead and the zero-degree, in-phase position (see Figure 5–27). The actual degree of shift again depends on how the load is divided between the two load types. This particular combination is not very common in power circuits, because it serves no useful purpose.

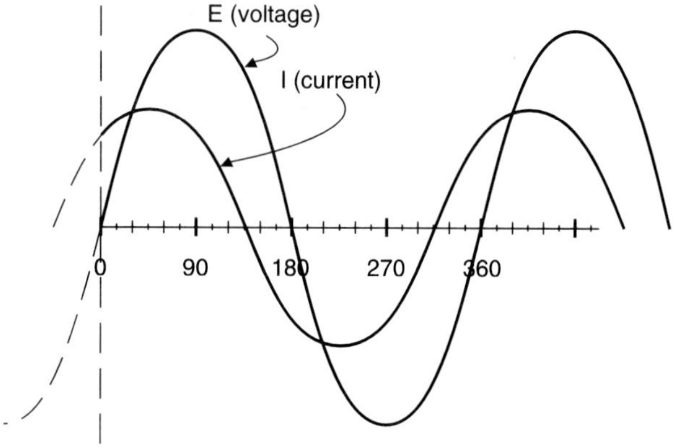

FIGURE 5–27 An equal mix of capacitive and resistive load has a current shift of 45° lead.

Current is leading voltage by 45°.

A far more common mix involves all three types of loads. When resistance, inductance, and capacitance are combined, a curious thing happens. The two reactances appear to cancel each other, leaving only the difference between them, which is then combined with the resistive portion of the load. Because the capacitive and inductive currents are a full 180° (half cycle) apart—90° lead and 90° lag—they supply energy to each other instead of drawing it from the supply.

Within the individual sections of the load, the phase relationships are appropriate for that particular load type. Within a resistive portion, the voltage and current are in phase; in an inductive portion, current lags by 90°; and in the capacitive portions, the current leads by 90°. This is true regardless of the other loads in the circuit. However, in conductors that supply the combined loads, the phase relationship is a composite of the various load characteristics.

For example, assume that a circuit contains all three types of loads, connected in parallel across common supply. The current in the supply conductors cannot lead and lag at the same time. Instead, it is a combination of the total resistive current and the difference between the two reactive currents. The result is a single supply current with the appropriate phase angle. If the reactive portion of the mix is predominantly inductive, the circuit looks like a mix of inductance and resistance. The presence of the capacitor reduces the inductive current portion, but does not move the phase angle into lead. If the capacitor is large enough to dominate the reactive part of the circuit, it supplies all the inductive current, and the circuit behaves as capacitive and resistive combined. If the capacitor and inductor are a perfect match, the circuit acts like a pure resistance (see Figure 5–28).

The behavior of mixed loads can be difficult to accept. It may help to remember that both coils and capacitors alternately store and release energy when they are connected to an AC circuit. These store-and-release cycles occur at different times in the two different loads, making them virtual mirror images of each other. When the capacitor is charging, the coil is discharging its magnetic energy back into the circuit. When the coil is drawing energy, the capacitor is discharging. The two devices complement each other perfectly. Each is capable of supplying current to the other at just the right times. Current flows

FIGURE 5–28 When a circuit contains equal parts of inductance and capacitance, the result is a resistive load.

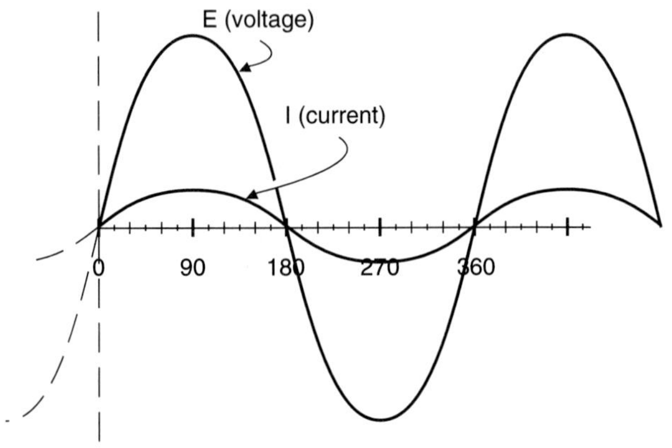

When the inductive and capacitive reactance are identical, the line current is only the resistive portion.

between the capacitive and inductive portions of the load that would otherwise have come from the supply. This reduces the current demand from the supply, but still provides all the necessary reactive current for the circuit.

Note

Resistance is symbolized with an R. Reactance is symbolized with an X, and can be inductive reactance, X_L, or capacitive reactance, X_C. Impedance is the total opposition to current flow in a circuit and is symbolized with a Z. All of these are measured in ohms. The ohm is not just a measure of resistance, it is a measure of opposition to current flow, no matter what the source of that opposition may be.

The Effects of Phase Shift on Circuit Operation. Circuit behavior is far more complex for AC than DC, because of the reactive effects of coils and capacitors. A full mathematical study of reactance is beyond the scope of this text, but a rudimentary understanding can greatly improve troubleshooting skills. Reactance is too important to be ignored, but is often impractical—or impossible—to measure directly with standard instruments. Reactances can have a dramatic effect on the operation of the circuit, and this will be reflected in any voltage and current measurements.

Measured values of voltage and current give a true indication of how a circuit is functioning, but without an understanding of reactance, these measurements may be impossible to interpret correctly. Actual measured values in any circuit with a mix of load types are not consistent with simple DC electrical theory. Once reactive effects are understood, these otherwise puzzling meter readings begin to make sense.

Coming to terms with coil reactance, for example, involves a mixed bag of measured values, calculations, and understanding. For example, in AC coil loads, the resistance of the coil wire is easy to measure, but has almost nothing to do with the current the coil actually draws. The inductance of the coil can be measured with a few newer multimeters, and inductance can be used to calculate inductive reactance, but the inductive reactance cannot be measured directly. The impedance of the coil—its total opposition to current flow—can be calculated from measured values of current and voltage, but it cannot be measured directly either.

In a typical coil, the impedance is primarily a result of the coil's inductive reactance. The coil in an AC-operated solenoid valve, relay, or motor starter can be used to demonstrate this very well. The following test results are quite typical for a small AC coil. It should be noted that larger coils have a much lower resistance. For these larger coils, the difference between the current calculated from resistance only and the actual AC current is even more dramatic.

This test was done on the 120 V AC coil of a small industrial relay. The measured resistance of the coil was 1945 Ω. If the coil was connected to a DC supply, there would be no inductive reactance, and the current could be predicted using Ohm's Law.

$$I_{DC} = \frac{E}{R}$$

I_{DC} = current draw from a 120 V DC supply

$$I_{DC} = \frac{120}{1945}$$

$$I_{DC} = 0.0167$$

The current with a DC supply would be 0.0617 A or 61.7 mA.

When the coil was connected through an AC milliammeter to a 120 V AC supply, the actual AC current measurement was 19 mA, or 0.019 A. The current predicted using Ohm's Law for a DC supply is more than three times higher than the actual AC value.

The AC impedance of the coil can be calculated from the measured values of current and voltage.

$$Z = \frac{E}{I}$$

Z = Impedance, in ohms

$$Z = \frac{120}{0.019}$$

This works out to 6316 W.

Comparing the two results, the impedance is dramatically higher than the measured resistance because of the coil reactance. Because resistance and reactance produce different phase shifts, their effects cannot be added and subtracted in the conventional manner. Using the test results as an example, the reactance value (X_L) in ohms cannot be found by subtracting R from Z. The actual calculation would be:

$$X_L = \sqrt{Z^2 - R^2}$$

X_L = The inductive reactance in ohms

$$X_L = \sqrt{6316^2 - 1945^2}$$

$$X_L = \sqrt{36108831}$$

$$X_L = 6009\ \Omega$$

As this test proves, the primary opposition to current flow in an AC coil circuit is the inductive reactance of the coil. If the coil wire had no resistance at all, it would still draw only 20 mA, because the current would be limited to that value by the 6 kΩ of inductive reactance. Understanding reactance allows coil current measurements to make sense when the measured resistance and DC Ohm's Law calculations don't come close. It should be noted here that DC coils do draw current according to measured resistance. That is why they have so many turns and are wound with such fine wire. The resistance must be kept high to keep the current to an acceptable level.

AC capacitor circuits also defy simple measurement and DC circuit logic. Capacitors can be tested with an ohmmeter, but have no actual resistance value. This test is commonly used to determine if the capacitor is in working order, and is described in detail in the practical example that follows this section. Capacitive reactance can be calculated from the capacitance value, in μF, or from voltage and current readings, but the result may be of limited use. Even when the ohm value of the reactance is known, it can be difficult to apply within the context of a complex circuit. However, if the concepts are understood, field readings are no longer mysterious, which is usually all that matters.

Phase shift is one of the primary effects of circuit reactance; unfortunately, it also cannot be directly measured or observed with standard field instruments. Again, if the effects are understood, there is a much better chance that measurements will be correctly interpreted. The meter readings begin to make sense instead of adding to the confusion.

Reactances often result in voltage and current readings that appear to be impossible. This is particularly true in circuits that contain a combination of both coils and capacitors, but it also occurs in circuits that are a mix of reactance and resistance. The following examples include some of the more confusing situations.

When a coil or capacitor is connected in series with a resistance, the arithmetic sum of the voltage is greater than the supply voltage. This appears to break one of the basic laws of electrical behavior, which states that the voltage drops around a series circuit always add up to the supply voltage. However, the rule is not broken—it only appears to be. To obtain an accurate result, the relative phase angles of each or the voltages must be taken into account. The voltmeter cannot determine the angles; it can only measure each voltage in isolation from all the others. Simple addition of these voltage values is not the correct approach, but the only one that is convenient. Trying to apply simple DC theory to a reactive AC circuit never works out very well.

When capacitors and coils are connected in series, the problem of unexpectedly high voltage drops becomes much more pronounced. Depending on the values of the two reactances and the circuit resistance, it is quite possible for the voltages across the capacitor and coil individually to exceed the supply voltage. Understanding this phenomenon can be critically important when making changes to any circuit with capacitors and coils in series.

Warning

When closely balanced capacitive and inductive reactances are connected in series with little or no additional circuit resistance, the usual result is a catastrophically high voltage, many times greater than the voltage provided by the supply. Smoke and ruin are almost inevitable; personal injury is a distinct possibility.

When capacitors and coils are connected in parallel, the high-voltage dangers of the series connection are gone, but the meter readings can be just as confusing. Instead of voltage drops that are higher than the supply voltage, internal circuit current between the coil and the capacitor may exceed the total current drawn from the supply.

Both of these situations seem to defy common sense. More voltage than is available from the supply across part of a circuit should be impossible (without a transformer), but it happens frequently. More current circulating inside a circuit than is drawn from the supply also seems impossible, but is equally common. What should add up just doesn't, at least not with simple mathematical approaches.

Certainly mathematical methods are available to calculate these values and predict circuit operation. Even the most complex circuits can be reduced to mathematical models. However, within the context of this text, the complexity of these calculations outweighs their usefulness. There are dozens of excellent electrical and electronics texts that cover this area very well.

What is most important is a sufficient grasp of the concepts to allow constructive interpretations of unexpected meter readings. It is also very important to realize some of the safety hazards of capacitors, and of capacitors and coils in series connections. Of secondary importance is some understanding of the roles capacitors play in circuits, given the fact they do nothing useful by themselves.

Note

There is one major issue of reactance and phase shift that, for now, is being ignored. This is the effect it has on power and power measurement. Due to the high cost of wattmeters, the time required to connect them, and a lack of pressing reasons to be bothered, they are rarely used in field service work. Power measurement is rarely a part of routine testing, yet power is a very important issue because it is the basis of all utility billing. When the operating cost of electrical equipment must be predicted or explained, power issues must be dealt with, and serious problems can arise. Electrical power is difficult to measure and to understand. Power is examined in a following chapter, and effects of reactance on circuit power are covered there.

The operation of AC circuits can be difficult to understand. The problems are greatly compounded by the fact that measurements with standard meters are only clues to complex interactions, instead of clear indications of how the circuit is behaving. If the basic natures of these interactions are understood, these clues are much more likely to be correctly interpreted. AC circuit behavior is very dynamic, and occasionally a little puzzling, but that is what makes it so interesting.

Practical Example: Experimenting With Reactance. As with previous practical examples, enough detail is provided that much of the learning that would be gained from actually doing the exercise can be gleaned from the descriptions provided. However, hands-on experience is highly desirable, especially for novices.

The following series of experiments requires an appropriate coil, a few capacitors, an assortment of resistors, two multimeters, and an oscilloscope. It also requires adjustable DC and AC power supplies. A variable autotransformer works nicely for the AC supply, but it should have an isolating transformer. This makes connecting the oscilloscope easier and safer, because it eliminates problems in positioning the oscilloscope ground clip.

The equipment described here is typical of what can be found in many school and college electrical or electronics labs. The component ratings given are only included as a reference point: There are many possible combinations of values that can be used successfully. To conduct this experiment for maximum benefit, it is best if the impedances of the three different load devices are approximately the same. Impedance is the total opposition to current flow in an AC circuit, regardless of whether the opposition comes from reactance or resistance. Impedance is a better term to use than resistance when describing AC circuits or devices, because it helps avoid confusion between true resistance—as an ohmmeter would read it—and the actual opposition to AC flow that is typically due to some combination of resistance and reactance.

In this author's classroom and shops, the materials used include a coil with 1000 turns and a DC resistance of 9.5 Ω. This coil also has a two-piece removable iron core. The coil impedance is around 1000 to 1500 Ω with a 60 Hz supply and the iron core in place. Various capacitors can be used. Something around 2 to 3 μF is a reasonable match for this particular coil. The voltage rating must be appropriate for the supply voltage; a 15 V supply was used for both the AC and DC tests. Several different resistors allow some comparisons to be made. They should include one that matches the coil impedance reasonably well—1 to 1.5 kΩ for this example. A 1/2 W power rating is adequate at 15 V. Higher voltage or lower resistance values may dictate higher power ratings.

A low value of resistance works well for current sensing. Its power rating must also be adequate for the resistance chosen and the maximum circuit current. A 5 Ω sensing resistor has to be rated for 5 W to carry one amp of current. Lower resistance values can have lower wattage ratings, but may not provide a high enough voltage drop for the oscilloscope. A 5 Ω was used in this case.

To check how the available materials match up, they can be tested, or their values can be calculated from their factory ratings. Resistors can be measured directly. Coils and capacitors require a rating, a test circuit, or both. Some calculations are necessary. For example, if the coil to be used has a listed inductance rating, its inductive reactance can be calculated using the formula below. Inductance in the formula is in henrys. Typical coils may be rated in henrys or in millihenrys (mH). When the rating is in millihenrys, this has to be taken into account for the calculation (1 mH = 0.001 H).

$$X_L = 2\pi FL$$

X_L = Inductive reactance in ohms

L = Inductance in henrys

F = Frequency in Hz

In formulas, reactance is symbolized by X with a subscript of L or C as appropriate. L is the symbol for inductance, so inductive reactance is X_L; C is the symbol for capacitance, so the symbol for capacitive reactance is X_C. The line frequency in North America is 60 Hz, and $2 \times \pi \times 60 = 377$. Therefore, for circuits supplied from a 60 Hz line, 377 can be used as a constant. This can be a useful simplification for 60 Hz calculations.

When the inductance is not listed on the coil, or it is an adjustable one with a movable core, the approximate inductance can be found by energizing the coil from an AC supply and measuring the voltage and current. If the coil resistance is ignored, an approximate inductive reactance can be found with the formula

$$X_L = \frac{E}{I_{XL}}$$

E = Supply voltage across the coil

I_{XL} = Current drawn by the coil

There are similar formulas for capacitors. They are marked with a capacitance value in microfarads (μF). To calculate capacitive reactance from this value, use the formula below. Note that the capacitance in the formula is assumed to be in farads, so an adjustment of the decimal position is required. Either use the value in microfarads time 10^{-6} (an exponent of –6 in a scientific calculator), shift the decimal 6 places to the left, or replace the 1 on top with 1,000,000.

$$X_C = \frac{1}{2\pi FC}$$

X_C = Capacitive reactance in ohms

C = Capacitance in farads

Note

There is a good chance that capacitors large enough to match the coil impedance may be electrolytic types. This is a problem, because the supply is AC, and most electrolytic capacitors are polarity sensitive. However, there are solutions. First, if the voltage to be used is much lower than the capacitor's voltage rating, it will probably survive the connection to AC, despite the incorrect polarity. Second, nonpolarized electrolytics are available. For example, starting capacitors used on single-phase motors are nonpolarized, as are the capacitors used in some stereo speaker crossovers. A third alternative is to use normal polarized electrolytic capacitors in pairs. The two capacitors are connected back to back in series, with either the two positive or the two negative ends connected together. The capacitance values have to be doubled when two are used in series, because the total capacitance of the pair is reduced to half the rating of one.

The resistor values can be selected to match the impedances of the capacitor and coil. A perfect match is not required, but the experiment is more convenient if the values are approximately the same. When the calculated power is too great for color-coded resistors—maximum 2 W for the largest color-coded resistor available—then 5 or 10 W power resistors may have to be used instead. If the calculations work out to a rating that is higher than the wattage of the available resistors, a lower supply voltage may be used. To find the wattage rating required, use any of the following formulas:

$$P = \frac{E^2}{R}$$

$$P = I^2R$$

$$P = EI$$

P = the power in watts

Color-coded resistors are power rated according to their physical size. The largest is rated at 2 W, and then they drop in 50% steps to 1, $^1/_2$, $^1/_4$, and $^1/_8$ W, progressing downward by size. Power resistors are not color coded. They are physically larger, and are made from high-temperature materials such as ceramic. They are generally available in 5, 10, or 25 W ratings, and larger.

In addition to the load resistor, one very low value of resistance is used for current sensing. A suggested value would be 1, 5, or 10 Ω, but the actual value depends on the sensitivity of the oscilloscope available and the current through the load. This resistor is connected in series with the load under test. It produces a small voltage drop that is proportional to the current. This voltage drop is used as a "current" signal into the oscilloscope, because the oscilloscope cannot read current directly. The current-sensing resistor carries full-circuit-load current and must drop enough voltage to provide an acceptable signal to the oscilloscope. It should be kept to a relatively small resistance value so that it will not interfere significantly in the circuit operation.

This experiment is in three parts, each one intended to demonstrate the behavior of one of the three types of electrical loads. Both AC and DC voltages are applied. Meters are used to monitor the voltage and current, and the oscilloscope is used to observe the phase relationship between voltage and current for the AC connections. The experiment can also be extended to examine various connections of mixed loads.

For the resistive loads, current for both DC and AC supplies can be accurately predicted using the DC version of Ohm's Law.

$$I = \frac{E}{R}$$

I = current in amps

E = applied voltage (AC or DC)

R = resistance in ohms

For the coil load, with DC applied, the current is limited only by the DC resistance of the coil. This resistance can be measured with an ohmmeter, and Ohm's Law can be applied just as it was for the resistor. If the resistance of the coil is very low, as it often is, the DC voltage may have to be reduced to keep the current to a reasonable limit. If the current-sensing resistor is kept in the circuit for this test, its resistance value should be included in the current calculation.

Checking Out the Oscilloscope. Connect the Channel 1 ground clip to one side of the AC supply and the Channel 1 probe to the opposite side. Note the caution below. Set the supply to 15 V—or the voltage used to match the components available.

Caution

Unless the AC supply is through an isolation transformer, the scope clip must be connected to the grounded side of the circuit. This can be found by testing the voltage from both terminals of the supply to any convenient ground—such as the ground socket of a 120 V outlet or the frame of the scope. If the supply is not isolated, one side should have the full output voltage to ground, the other should be a zero. If the supply is isolated, there may be random voltages to ground from both sides. The oscilloscope ground clip can be connected to either side of an isolated supply without damage or problems. However, regardless of other considerations, the second ground clip must be connected in the same location as the first or eliminated from the circuit. Both oscilloscope ground clips connect to the oscilloscope frame, and are therefore not only grounded, but are also connected to each other. Connecting them to different parts of the circuit will short out that portion of the circuit.

Adjust the oscilloscope to obtain a display of approximately one full AC cycle, as wide and high as possible, without going outside the grid. Make sure the trace is centered vertically on the screen. If the oscilloscope is unfamiliar, this can be a good time to experiment with the controls. Set the triggering for a normal display, not inverted, with the trace starting at the zero line. Use the horizontal position and trigger level controls to ensure the start of the signal is visible on the left-hand side of the screen.

Compare the oscilloscope voltage to the voltage as measured with a multimeter (see Figure 5–29). To measure the voltage displayed on the oscilloscope, the height of one-half the waveform must be multiplied by the VOLTS/DIV setting, and the probe multiplier if applicable. If the oscilloscope has a variable voltage setting, make sure it is set to the calibrate position. The value obtained is the peak voltage (E_P), not the usual effective or RMS value (E_{RMS}). To convert E_P to E_{RMS}, use the formula

$$E_{RMS} = 0.707 \times E_P$$

FIGURE 5–29 With the oscilloscope properly set, a 15 V AC signal displays as a sine wave.

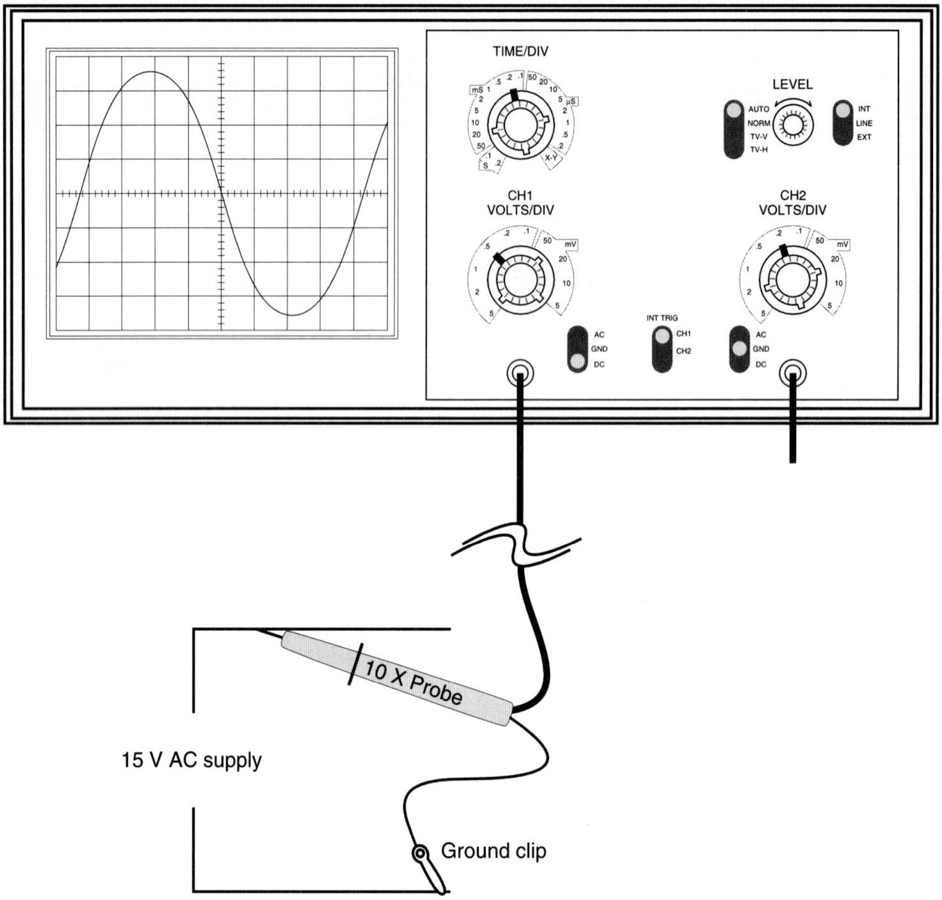

The oscilloscope value after this calculation should be reasonably close to the voltage reading taken with the multimeter. A well-calibrated oscilloscope is an accurate instrument, but it is much more difficult to read than a multimeter. One common cause of error is the parallax that occurs when the user is not looking straight through the grid to the screen trace. The grid is in front of the thick glass that forms the screen end of the tube, while the trace is, of course, drawn on the inside. The trace is further behind the grid than it looks.

The supply frequency is known to be 60 Hz, but it also can be found by measuring the time for one complete cycle on the oscilloscope. The time value is in seconds per cycles, so inverting it—dividing it into one—will give the frequency value in cycles per second (Hz). The procedure is as follows. Convert the physical width of the display into time by multiplying the width of the cycle, in grid divisions, by the time base (TIME/DIV) setting. This time value is called the period (P) of the waveform. The calculated frequency (F) should be close to 60 Hz.

$$F = \frac{1}{P}$$

FIGURE 5–30 With a resistive load, the current is in phase with the voltage.

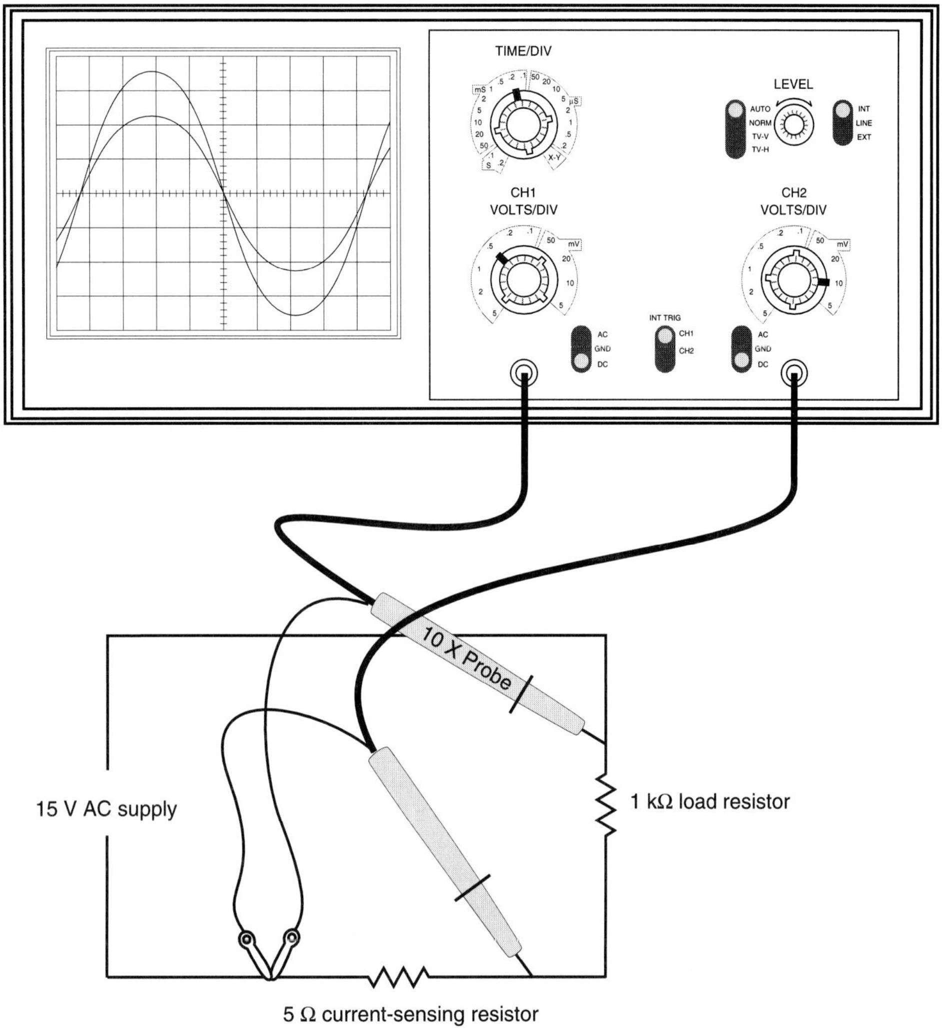

AC and DC Behavior in Resistive Circuits. The test circuit is arranged as shown in Figure 5–30. One multimeter is used as the milliammeter, and the other to measure voltage as required. Both scope ground clips are connected to the same location in the circuit. The circuit is energized with an appropriate AC voltage level.

The current reading of the milliammeter can be compared to the value calculated with Ohm's Law.

$$I = \frac{E}{R}$$

The scope is set up so Channel 1 displays the 15 V of the supply, properly centered at midscreen. Channel 2 is centered and adjusted with the VOLTS/DIV selector to obtain a second sine wave. It is helpful to keep this wave shorter on the screen than the voltage wave, just to avoid confusion about which is which. This second trace may require a very low VOLTS/DIV setting, because the signal is only the small voltage drop across the 5 Ω current-sensing resistor.

The series resistor voltage drop is representative of circuit current. Using the two channels simultaneously demonstrates the phase relationship of the voltage and current waves. The phase relationship is correctly displayed because the oscilloscope graphs both signals on the same time base. With a resistive load and a properly adjusted oscilloscope, the rise and fall of the current wave will be exactly in synchronism with the driving voltage. Both waves cross through the zero center line at exactly the same time, and both the positive and negative peaks coincide. If the scope can be adjusted so the two waves have the same height, they should match perfectly. They are in phase, behaving exactly as electrical theory predicts.

To compare AC and DC behavior, the AC supply can be replaced with a 15 V DC supply. The oscilloscope traces become straight lines—a graph of the unchanging DC values. The current reading with the meter now set for DC current is the same value as it was for the AC supply: For resistive loads, AC and DC behavior are essentially identical.

AC and DC Behavior in Inductive Circuits. The test circuit can be kept the same as it was for the resistive load, except that the resistor is replaced by the coil. The 15 V AC supply is reconnected, and the current and voltage meters reset to read AC.

The coil impedance can be found from the measured voltage and current using Ohm's Law for AC circuits.

$$Z = \frac{E}{I}$$

The impedance is in ohms, and is the composite opposition to current flow that results from the coil's inductive reactance and its resistance. When the resistance of the coil is very small in comparison with the total impedance, it can be ignored, and the impedance can be treated as "pure" inductive reactance (X_L). An accurate value of X_L can be found, if desired, from the formula below. The coil resistance and reactance can never be isolated from each other. They always appear to the circuit as the composite impedance value.

$$Z = \sqrt{R^2 + X_L^{\ 2}}$$

$$X_L = \sqrt{Z^2 - R^2}$$

Z = Impedance in ohms

X_L = Inductive reactance in ohms

R = Coil resistance in ohms

The oscilloscope shown in Figure 5–31 is set up to view both voltage and current, just as before. With the coil connected, the voltage and current traces have essentially the same shape, but the current is displaced to the right. Time is advancing left to right across the screen. The current is out of phase with the voltage, lagging behind by nearly 90 electrical degrees.

If the coil core is removable, this is an excellent opportunity to experiment with the tremendous influence that changes in the core have on the inductive reactance of the coil. Inductive reactance depends on the energy stored in the coil's magnetic field. When the

FIGURE 5–31 When the load is a coil, the current lags the voltage by almost 90°.

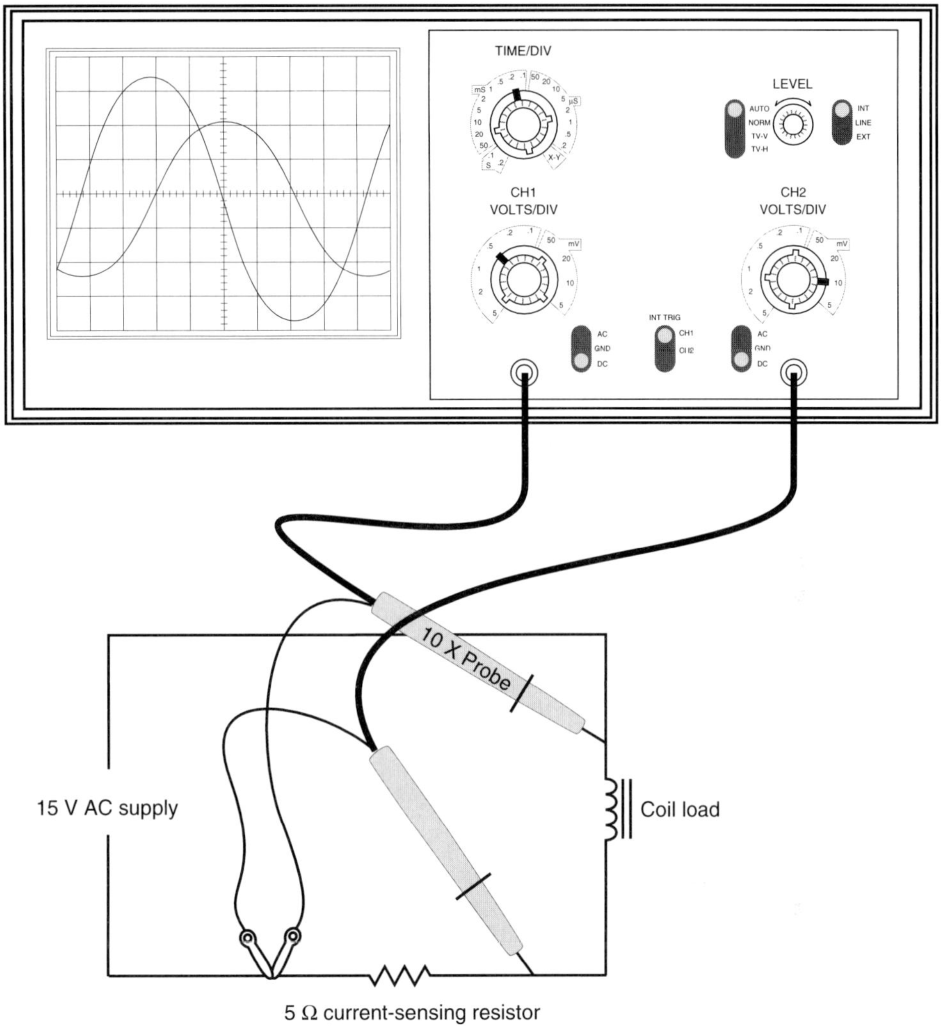

core has a low reluctance, this field is strong, the maximum amount of energy is stored, and the reactive effect is at its greatest. When the reluctance of the core is reduced, the field is weakened, and the inductive reactance is also reduced.

A little experimentation with a movable-core inductor demonstrates the amount of control the core has on coil impedance and current flow in the circuit. With the equipment used in the author's labs, the current can be regulated from 15 mA to over an amp, a control ratio of nearly 100:1. This principle is used to great advantage in stage lighting control and arc welders. Because inductive reactance does not cause an energy loss in the same way a resistance would, it can be used to control a large amount of current very efficiently. Putting an adjustable coil in series with a lamp or group of lamps demonstrates this effect very well.

A final note: Adjustments to the core change the amount of inductive reactance, but they do not change its nature. The phase shift due to reactance remains at 90°. Observed on the oscilloscope, the degree of shift will change slightly, because the coil's resistance is a proportionally larger part of the total impedance when the reactance is reduced.

When the AC supply is replaced with a DC supply, the voltage may have to be reduced below 15 V, depending on the resistance of the coil and the current-sensing resistor. The current draw is predictable from Ohm's Law for DC circuits, because inductive reactance does not have a continuous effect on DC current flow.

$$R_T = \frac{E}{I}$$

The DC current will display much the same as for the resistor; only the values change. Note that the DC magnetic field is very powerful. If the core is a two-part assembly, it is impossible to separate the parts until the power is removed. The current is higher than it was for AC, but DC current also produces a much more powerful magnet because the polarity does not change.

AC and DC Behavior in Capacitive Circuits. Of the three types of loads, capacitance is the least common. Except for the purpose of experimentation, capacitors are never used in isolation from other types of loads. Typical inductive loads, like motors, have an obvious purpose. Capacitors, by themselves, don't do anything useful. In DC circuits, they are most often used for their ability to store a charge. In AC circuits, they are used for their effect on voltage and current phase relationships, and for their ability to provide reactive current to inductive loads.

It is virtually impossible to measure capacitive reactance directly, but it is relatively easy to calculate from current and voltage readings. It can also be found from the capacitance value, but be aware that the μf rating, particularly on electrolytic capacitors, is rarely exact. Other types of capacitors have closer factory tolerances. If a capacitance meter is available, the measured value is likely to be more accurate than the listed value.

Before a capacitor is used in a circuit, it is a good idea to test it first. A capacitance meter is the best method, but this feature is relatively new to multimeters, and a dedicated capacitance tester may not be available. Therefore, anyone involved with electrical equipment should know how to test a capacitor using the resistance functions of a typical multimeter. Analog meters are the easiest to use for this test: A digital will work, but is less convenient. Neither meter can be used effectively without an understanding of how the test works.

An ohmmeter measures resistance by applying a low DC voltage to the device being tested and responding to the current it draws. Because current is inversely proportional to resistance, the resistance scale on an analog meter is in reverse. The meter scale markings are positioned to convert meter current flow into a resistance value in ohms.

When the meter is used to test a capacitor, it actually charges the capacitor with the DC voltage that is normally used to test resistance. The current drawn by the capacitor as it charges is interpreted by the meter as a changing resistance value. On an analog meter the needle swings to the right, and then slowly back to the left. Starting from a fully discharged state, the charging current is high at first, causing the needle to swing upscale, and then, as the capacitor voltage builds up, the current slowly drops to zero. As the current drops, the meter needle swings back to the left.

If the capacitor was connected directly to the voltage source, it would charge almost instantaneously. However, during the ohmmeter test the charging rate is slowed by the internal resistance of the meter. This internal resistance is different for each ohmmeter range, so some experimentation is required. Low-resistance ranges give a quick response, often too quick for a proper test. High-resistance ranges can be so slow that, on large capacitors, the test may drag on for several minutes. This response time is dependent on the meter, the range, and the size of capacitor. The smaller the capacitor, the higher the range required. Good range choices are mostly a matter of experimentation and experience.

The capacitor should be shorted out before the initial test to ensure it is fully discharged. If the capacitor's normal operating voltage is quite high, or the capacitor is very large, discharge it through some value of resistance.

Once the capacitor has been discharged, the meter can be connected for an initial test. Hold the probes on long enough for the needle to get close to the left end of the scale. After the first test, the meter connections can be reversed. As soon as the probes make contact, the meter needle will again swing to the right, but more violently than before. On a moderate-to-high range, a large capacitor may hold the needle there for some time before it slowly returns to the left.

This test can be repeated any number of times, and each time the leads are reversed the same response will occur. When the leads are reversed, the charge on the capacitor has the same polarity as the battery in the ohmmeter. The capacitor discharges through the meter, and the current is high enough to hold the needle well up on the right side of the scale. Once the capacitor has fully discharged, it immediately begins to charge again with the opposite polarity. The needle slowly returns to the left as the charge cycle is completed. Each time the meter leads are reversed, the same sequence is repeated.

When a digital ohmmeter is used for the capacitor test, the needle swings are replaced by a changing sequence of numeric values. This is more difficult to follow at first, but with a little practice it is workable. Reversing the leads does not have the same effect and should not be part of the test. A meter with a capacitor test function that can provide a value in microfarads is far superior to any type of resistance test.

If a capacitor has failed, the test results should indicate a fixed value of resistance. If it has failed "open," the ohmmeter will indicate infinity (∞) Ω; if it has failed (shorted), the reading will be some low value of resistance.

Capacitor tests, as described above, are sufficient to prove the capacitor is in working order. They cannot provide an actual capacitance value but, when the test is conducted properly and the capacitor is large enough to test, there will always be some kind of charging response. If the capacitor has failed, it will appear as a resistance. Tests like this take practice. The best way to gain the necessary experience is to experiment with any meter at hand, testing a variety of good capacitors.

Once the capacitor has been tested, the circuit can be connected, and the experiments with capacitive reactance can commence.

Just as with resistance and inductance, the circuit for the capacitor experiment is the same except for the load device. The coil of the previous test is replaced with the capacitor. A 15 V AC power supply is connected as before, and the circuit current is measured with an in-line meter. The impedance of a capacitor circuit is virtually all capacitive reactance (X_C). Ohm's Law can be modified slightly to find this value. The reactance can also be calculated from the μf rating and the two values compared.

$$X_C = \frac{E}{I_C}$$

$$X_C = \frac{1}{2\pi FC}$$

X_C = Capacitive resistance in ohms

F = Frequency in Hz

C = Capacitance in farads

The oscilloscope is set to display both voltage and current as before. The current wave now leads the applied voltage by 90° (as shown in Figure 5–32). The current crosses through the zero line one-quarter cycle ahead of the voltage. On the oscilloscope, this is to the left of the voltage wave.

Note

It is not unusual for the capacitor current wave to be quite distorted. There may be other contributing factors, but a common reason is that most electrical systems have at least some distortion-producing loads. Voltage distortions typically appear as higher frequency harmonics and voltage spikes superimposed on the normal 60 Hz wave. The capacitor, with its low reactance to high frequencies, acts as a current sink, or low impedance path, for the harmonics and other distortions.

Once the current measurements have been completed, and the wave shape and phase shift observed, the capacitor should also be tested with a DC supply. With the AC supply removed and a 15 V DC supply in its place, the circuit appears to do nothing. There is no current at all. A voltmeter across the capacitor indicates the 15 V of the DC supply. When the power source is disconnected, this voltage remains, and then slowly declines. The supply has charged the capacitor, and when it is disconnected the voltmeter reads the capacitor voltage. The capacitor voltage slowly declines, because it is discharging through the meter. This discharge rate depends on the size of the capacitor and the impedance of the meter used for the test. It can be relatively quick for an inexpensive analog meter, and very slow for a good digital.

It is possible to observe the charging current, but only if the capacitor is relatively large—or there are many of them in parallel—and a sensitive ammeter is used. The capacitor must be fully discharged before the test, and the ammeter must be watched closely when the supply is turned on. The capacitor charges in an instant, and then the current flow stops. An alternative is to slow the charging process by adding resistance to the circuit in series with the capacitor. The charging current and voltage can then be followed very easily with appropriate meters.

FIGURE 5–32 When the load is a capacitor, the current leads the voltage by 90°.

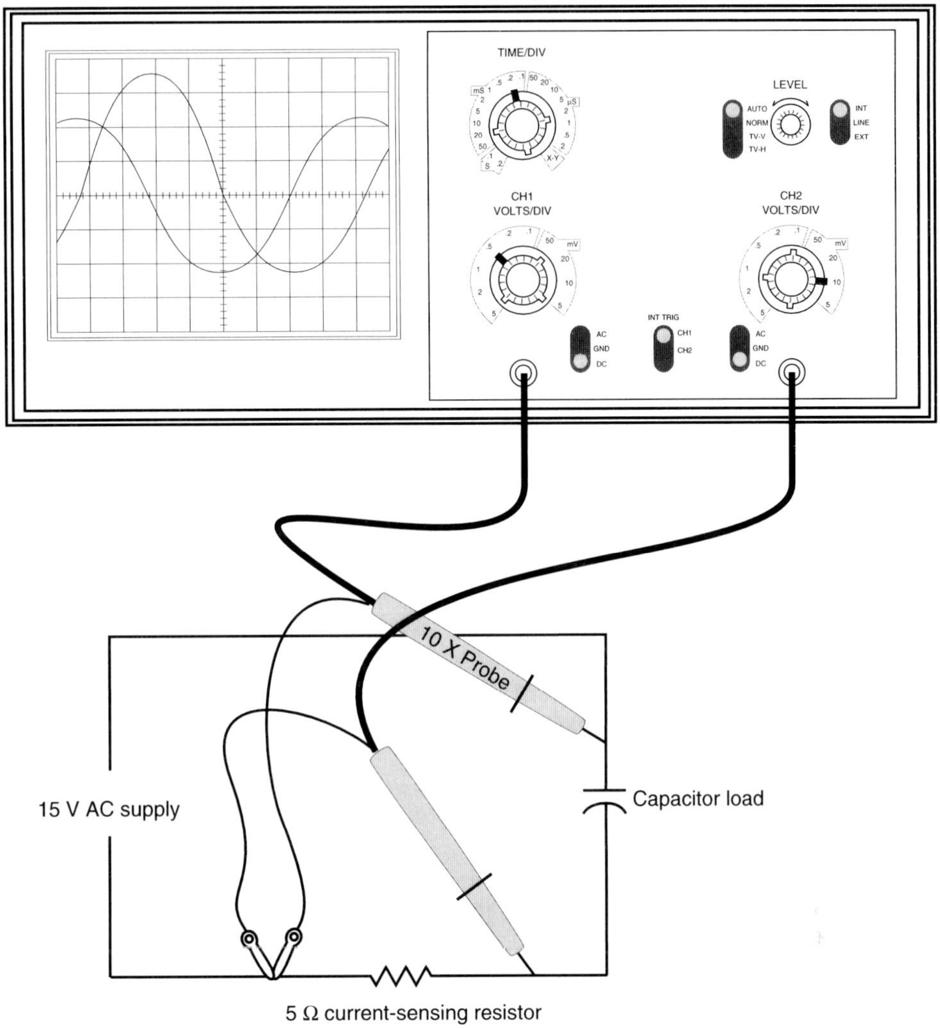

Summary

In resistive loads, electrical energy is used up—converted to heat and dissipated into the air—as it forces the electrons through the electrical resistance of the load. The amount of current flow is dependent on the voltage applied and the resistance value. Resistance is the electrical equivalent of mechanical friction. It is symbolized with the letter R and is measured in ohms (Ω). The unit ohm is often associated with resistance alone, but it is actually used as a standard unit of opposition to current flow.

Resistance is used to convert electricity to heat and, where that is the intended result, the conversion is essentially 100 percent efficient. Electric heat elements, for example, are resistance devices. Virtually all the energy used by the element is released as heat.

Resistance is also added to circuits to limit current or to create a voltage drop. This is an inefficient way to control current because the energy is wasted and the heat generated may cause additional problems. For low-powered circuits, this is usually acceptable; for high-powered circuits, other, more efficient, control methods are needed. Resistors are inexpensive to manufacture and are available in a very wide range of resistance values and wattage ratings.

In inductive loads, current is limited primarily by the self-induced counter EMF generated in the coil. This current-limiting effect is called inductive reactance, which is symbolized as X_L. The unit used for inductive reactance is the ohm. The amount of current flow in a practical inductor depends on the coil inductance, the circuit frequency and, to a much smaller extent, the resistance of the coil winding. Large values of inductance are created by using many turns of wire and very good magnetic cores. The larger the inductance, the greater the inductive reactance.

Inductive reactance has two potentially useful effects: limiting current without power loss, and a lagging phase shift between the voltage and current. The current-limiting characteristic is used to advantage in some high-current power circuits. The lagging phase shift property has a great effect on motor design and operation. (This is discussed in the motor section of this text.)

Capacitors connected in a DC circuit charge up to the voltage across that part of the circuit, and then store the charge until a discharge path is provided. No current can pass through the capacitor, so once it is fully charged, there is no further DC current flow in that part of the circuit. On AC, a capacitor appears to pass current, because there is continuous current in the external circuit. The current does not flow through the capacitor but, because the capacitor is continuously charging and discharging with each reversal of the AC polarity—it acts like a continuous load. The amount of current flow in the capacitor circuit depends on the applied voltage and the total capacitance. The capacitor's opposition to current flow is referred to as capacitive reactance (X_C) and is given a value in ohms. Capacitors with high capacitance ratings take on a greater total charge with each reversal of polarity, so the current flow is higher and the reactance is lower.

Capacitors serve a variety of purposes in DC and AC circuits. On DC, their ability to store a charge is used as a smoothing filter in AC-to-DC rectification circuits. They are also used in timing circuits, because a capacitor with a resistor in series will charge at a predictable rate. Where DC and AC occur together, capacitors are used to separate the two by blocking the DC portion. In single-phase motor circuits, capacitors are used to create a phase shift to produce a rotating field for starting purposes, or to improve performance. In three-phase systems they are often used to reduce power costs by supplying the reactive magnetizing current requirements of the inductive equipment. When the magnetizing current comes from capacitors, instead of from the supply utility, the overall demand on the utility is reduced. This is sometimes called power factor correction. (Power factor is discussed, along with power, in Chapter 8.)

5.7 AC Circuit Calculations

As shown in the examples above, calculations relating to AC circuits can be much more complex than for DC circuits, because the reactive properties of the loads must be taken into account. This is particularly true for circuits with multiple components. The methods and examples below provide the tools required to solve common circuit problems: They will not work in very complex circuits, but can be useful in understanding basic circuit combinations.

When an AC circuit contains only resistance, it can be treated as a DC circuit. Ohm's Law formulas can be applied in the same manner. Although circuits with no reactive loads have some reactance in the wiring and devices, it is generally too small to be important. Common resistive circuits include incandescent lighting and heating elements.

Series circuits:

$R_T = R_1 + R_2 + R_3 \ldots$	Sum of resistances equals the total resistance.
$I = \dfrac{E}{R_T}$	Circuit current equals applied voltage divided by total resistance.
$V_1 = I \times R_1$	Individual voltage drops equal current times individual resistance values.
$E = V_1 + V_2 + V_3 \ldots$	The sum of the voltage drops equals the applied voltage.

Parallel circuits:

$I_1 = \dfrac{E}{R_1}$	Current in any branch equals voltage divided by resistance of the branch.
$I_T = I_1 + I_2 + I_3 \ldots$	Total current equals the sum of the branch currents.
$R_T = \dfrac{E}{I_T}$	Total resistance equals the total voltage divided by total current.
$R_T = \dfrac{1}{\dfrac{1}{R_1} + \dfrac{1}{R_2} + \dfrac{1}{R_3}} \ldots$	Total resistance can be found from individual values.
$R_T = \dfrac{R_1 \times R_2}{R_1 + R_2}$	This formula can also be used to find total resistance.

When a circuit contains only inductive reactance and no significant resistance, the inductances can be converted to their reactive ohm values (X_L) and then treated much as a resistive circuit. Using only X_L values in ohms, all series and parallel formulas apply in the same manner as for resistance. Ohm's Law can be used with X_L in place of R.

$X_L = 2\pi FL$	Converts a value of inductance (L), in henrys, to reactance in ohms at a given frequency (F) in hertz (Hz).

Series circuits:

$X_{LT} = X_{L1} + X_{L2} + X_{L3} \ldots$	Sum of inductive reactances equals the total inductive reactance.
$I = \dfrac{E}{X_{LT}}$	Circuit current equals applied voltage divided by total reactance.
$V_1 = I \times X_{L1}$	Individual voltage drops equal current times individual reactances.
$E = V_1 + V_2 + V_3 \ldots$	The sum of the voltage drops equals the applied voltage.

Parallel circuits:

$I_1 = \frac{E}{X_{L1}}$ Current in one branch equals voltage divided by the branch reactance.

$I_T = I_1 + I_2 + I_3 \ldots$ Total current equals the sum of the branch currents.

$X_{LT} = \frac{E}{I_T}$ Total reactance equals the total voltage divided by total current.

$X_{LT} = \frac{1}{\frac{1}{X_{L1}} + \frac{1}{X_{L2}} + \frac{1}{X_{L3}}} \ldots$ Total reactance can also be found from individual values.

A purely inductive circuit is a practical impossibility, but well-made inductors come close. Most inductive loads are a blend of resistance and inductance. To calculate values for a circuit that combines these two, isolate the resistance and the inductive reactance. If they are in series, the relationship among the resistance, reactance, and combined impedance for the two is expressed by this formula:

$$Z_T = \sqrt{R^2 + X_L^2}$$

This is how practical coils are treated in calculations—as a pure inductance in series, with a resistance equal to the resistance of the coil wire.

The voltage drops across series reactance and resistance do not add in the usual manner. They have phase relationships that are 90° apart, so they are summed as values at right angles.

$$E = \sqrt{V_R^2 + V_{XL}^2}$$

When the resistance and inductance are in parallel—two separate devices—find the individual currents first, and then find the total circuit current as shown below.

$$I_T = \sqrt{I_R^2 + I_{XL}^2}$$

When several inductive reactances and resistances are connected in series, a total ohm value can be found for each type, and then the two totals can be added as shown above. If they are in parallel, the total current for each type is found instead, and then the current totals are summed as above.

When it is impractical or impossible to isolate the two characteristics, but separate reactive and resistance values must be found, the best approach may be to use power readings. A voltmeter, ammeter, and wattmeter are required. This involves power readings and calculations, which are explained in a following chapter on work, energy, and power.

A purely capacitive circuit can be treated much like a resistive or inductive circuit, as long as the capacitance values are all converted to the reactive ohm values (X_C) first. Capacitance (C) is inverse to inductance in that parallel values of C can be summed directly, and series totals must be found from a formula like that used for parallel resistances. However, if the capacitance values are converted to X_C values first, they behave just as other opposition values that are in ohms.

Capacitors in series drop voltage just like a series of resistors or coils. In parallel they draw individual branch currents. It is when they are combined with other types of loads that the calculations become more difficult.

$X_C = \frac{1}{2\pi FC}$ Converts a value of capacitance (C), in farads, to reactance in ohms at a given frequency (F) in hertz (Hz).

Series circuits:

$X_{CT} = X_{C1} + X_{C2} + X_{C3} \ldots$ Sum of capacitive reactances equals the total capacitive reactance.

$I = \frac{E}{X_{CT}}$ Circuit current equals applied voltage divided by total reactance.

$V_1 = I \times X_{C1}$ Individual voltage drops equals current times individual reactances.

$E = V_1 + V_2 + V_3 \ldots$ The sum of the voltage drops equals the applied voltage.

Parallel circuits:

$I_1 = \frac{E}{X_{C1}}$ Current in any branch equals voltage divided by the branch reactance.

$I_T = I_1 + I_2 + I_3 \ldots$ Total current equals the sum of the branch currents.

$X_{CT} = \frac{E}{I_T}$ Total reactance equals the total voltage divided by total current.

$X_{CT} = \frac{1}{\frac{1}{X_{C1}} + \frac{1}{X_{C2}} + \frac{1}{X_{C3}}}$ ... Total reactance can also be found from individual values.

When capacitive reactance is connected in series with resistance, the total impedance can be found using this formula:

$$Z_T = \sqrt{R_T^2 + X_{CT}^2}$$

As with the inductive reactance and resistance combination, if the capacitor and resistor are in series, their voltage drops add as values at 90°.

$$E = \sqrt{V_R^2 + V_{XC}^2}$$

When the capacitance and resistance are in parallel, find the currents for each branch first, and then sum the two currents as follows:

$$I_T = \sqrt{I_R^2 + I_{XC}^2}$$

When the circuit contains both inductive and capacitive reactance, the two different reactances cancel each other's effects. The easiest calculation method varies with the situation, but the following approach is reasonably flexible. When the devices are in series, the total impedance can be found from the difference between the two reactances and the resistance.

$$Z = \sqrt{R^2 + (X_L - X_C)^2}$$

The circuit current and the voltage drops can be found using Ohm's Law:

$$I = \frac{E}{Z}$$

$$V_R = I \times R$$

$$V_{XL} = I \times X_L$$

$$V_{XC} = I \times X_C$$

The voltage drops can be summed as follows:

$$E = \sqrt{V_R^{\,2} + (V_{XL} - V_{XC})^2}$$

When the two different reactances are in parallel, find their individual currents, using Ohm's Law as it applies to these types of circuits. Find the difference between the two currents and treat the result as capacitive or inductive, based on which of the two had the most current. This value is the total reactive current for the circuit.

$I_{XT} = (I_{XL} - I_{XC})$ or $I_{XT} = (I_{XC} - I_{XL})$, depending on which is larger.

Assuming the circuit also contains some parallel-connected resistive load, the total circuit current can be found from the total reactive current in parallel with the total resistive current.

Questions

1. Resistive loads, such as incandescent lightbulbs or heating elements, function just fine from an AC or DC supply. Explain why a commercial lamp or heater may not be suitable for use with DC.
2. List some of the different AC and DC circuits and supplies that exist in a typical modern residential gas furnace and air conditioning system.
3. Define the term "frequency" as it applies to AC electrical systems.
4. Describe the two principal methods by which a magnetic field can be used to induce voltage into a conductor.
5. What variables determine the amount of voltage produced in a conductor by electromagnetic induction?
6. Why do analog DC voltmeters read 0 V when they are connected to an AC supply?
7. RMS values of AC voltage are often referred to as effective values. Suggest a reason why this term is so appropriate.
8. Do typical AC meters functions provide peak, peak-to-peak, or RMS values of voltage and current?
9. What precaution must be followed when connecting an oscilloscope ground clip into a circuit?
10. Compare the direction of energy flow in a reactive circuit to that of a resistive circuit.
11. List the factors that affect the strength of the field produced by an electromagnet.
12. Why do most electromagnetic devices have iron cores?
13. Explain eddy currents as they apply to magnetic cores, and describe how they can be reduced.

14. Explain why AC solenoid coils draw more current when they are first energized than they do after the core pulls into place.
15. Explain the difference between the terms "dielectric strength" and "dielectric constant" as they apply to capacitors.
16. Explain how there can be continuous AC current flow in a capacitor circuit, even though no current actually passes through the capacitor.
17. What safety precautions should be followed when handling capacitors or working in circuits that contain them?
18. Why are special precautions important when working on electrostatic equipment such as electronic air cleaners?
19. Explain how capacitors and inductors can supply current to each other, rather than drawing it from the supply.
20. Why is caution advisable when connecting capacitance and inductance in series?

Problems

1. What current will a 30 mH choke coil draw from a 24 V 60 Hz supply?
2. What current will a 50 μF capacitor draw from a 48 V 60 Hz supply?
3. With a 120 V 60 Hz supply, what is the total current to a 2 H coil in parallel with a 10 μF capacitor? What would the current become if the frequency was doubled?
4. Two capacitors, 20 μF and 50 μF, are connected in series to a 120 V AC supply. What are the voltage drops across each of them?
5. A 4 H coil has a resistance of 250 Ω. Find the impedance it would have in a 120 V 60 Hz circuit.
6. A 25 μF capacitor is connected in parallel with a 25 Ω resistance. What is the total current into the circuit from a 24 V 60 Hz supply? What would the current be if a 250 mH coil was added to the circuit, in parallel with the other two devices?
7. A circuit contains a 500 mH coil, a 10 μF capacitor, and a 100 Ω resistor, all connected in series. What is the total impedance for the circuit when it is operated from a 60 Hz supply? What are the voltage drops on each of the devices if the supply voltage is 50 V at 60 Hz?
8. A 100 μF capacitor, a 250 mH choke coil, and a 50 Ω resistance are connected in parallel to a 120 V 60 Hz supply. Find the current for each device individually, and the total current.
9. A 1 μF capacitor is connected in series with a 1 kΩ resistance. What is the total impedance for the two devices if the supply frequency is 50 Hz?
10. A circuit containing a variety of parallel-connected loads has a total inductive current of 8 A, a total capacitive current of 6.5 A, and a total resistive current of 3.25 A. What is the circuit impedance if the applied voltage is 120 V?

6 Single-Phase and Three-Phase Systems

Objectives

This is a theory chapter that explains the difference between single-, two-, and three-phase systems. This explanation begins at the source, the supply generator, where the three different systems originate. Generators are examined, and some common variations are explained. Common three-phase connections are presented, and the differences explained. After completing the study of this chapter you should be able to:

- Explain the differences in generator construction that result in a single-, two- or three-phase output
- Compare single- and three-phase systems by listing the advantages and disadvantages of each
- List the various voltage combinations available from a three-phase panel supplied with 120/208 V Y and the connections required for each
- Calculate phase and line voltage and current values using the standard mathematical relationships of each three-phase connection
- Describe a test used to check delta transformer connections

6.1 Introduction

Homes, small to medium-size farms, and many small businesses receive only single-phase power. This is not a hardship, however, because most small electrical devices only require single-phase power for proper operation. Single-phase equipment includes power tools, electrical appliances, and all types of electric lighting. Added to that list are the hundreds of different consumer electrical and electronic products that are designed to plug into the standard 120 V AC outlets. Even in industrial environments, single-phase outlets are supplied for portable tools, lights, and other low-powered equipment.

Despite the obvious flexibility and popularity of single-phase AC power, virtually all commercial power is generated as three-phase. It is also distributed primarily as three-phase, and the majority of it is sold and used as three-phase. All large AC electrical equipment uses three-phase power. This includes almost all AC motors above 10 Hp, most large commercial or industrial electric heating elements, and industrial welding equipment.

Lighting fixtures are not three-phase, but in commercial and industrial environments they are split into groups of approximately equal wattage and powered from a three-phase supply. Three equal groups of lighting fixtures appear to the building service as a three-phase load.

Utility power distribution all starts out as three-phase, but may be split off as single-phase feeders into areas where the utility supplies only single-phase services. In North America, homes, most farms, and many small businesses are provided with a dual voltage 120/240 V supply. Three-phase is supplied to larger customers, and to customers who require three-phase for specific loads. Most larger services are three-phase. Unfortunately, there are exceptions.

In modern industrialized nations, only very remote locations have no utility-supplied electrical power. However, while areas that are quite distant from heavily populated cities and towns almost certainly have power, it is usually only single-phase. It costs much more to build a three-phase power line than it does for a single-phase line. Unless there are specific reasons to install such a line, lightly populated rural areas may not have three-phase available. The utility will usually maintain such areas as single-phase only until the load increases or a major installation justifies the cost of the change. Smaller customers who might benefit from three-phase service are unlikely to pay the high cost of such an upgrade. This can be frustrating for farms and businesses that are located in areas where a three-phase line has never been installed.

So what is three-phase, and why is it so widely used? The answer to these questions requires a brief examination of the principles and operation of AC generators.

6.2 AC Generators

Note

The terms "AC generator" and "alternator" are essentially interchangeable. Although they are sometimes used to differentiate between a machine with a rotating field, and one with a stationary field this is not done with any consistency. Automotive alternators are (three-phase) AC generators with an internal rectifier to produce a DC output. They are called alternators to differentiate them from the older DC generators they replaced. In this text, the term "AC generator" is used exclusively, but this does not imply that the term "alternator" is not equally valid.

As discussed in the previous section, to generate an EMF with a magnetic field, there must be a field, a conductor, and relative motion between the two. Any arrangement that meets these requirements will generate a voltage. The most practical way to construct a generator is to arrange the components in a circle so that the moving parts can rotate. Rotary motion is wonderfully smooth and endlessly repeatable. It is a much better option than other possible alternatives, such as the up-and-down motion of a magnetic "piston" moving inside a coil.

All rotary generator designs are based on one of two possible alternatives. One option is to arrange the conductors into which a voltage will be induced around a cylindrical form that rotates inside a stationary field. The second option is to install the conductors in the stationary portion of the machine, and then rotate the field inside them. In both cases, the requirements for generating a voltage are met. In the first case, the moving conductors cut through the stationary field; in the second, the moving field cuts through the stationary conductors.

Both types of generators are built. The first design is generally referred to as a revolving armature machine. It is only used for some small generators, and in special cases where the other type will not work. The second design is generally known as a revolving field machine. This design allows the power output of the generator to come from stationary leads. This is a tremendous advantage compared to a sliding brush and slip ring arrangement, particularly for large generators with high-output voltages and currents.

Single-Phase Generators

In a typical single-phase generator, a winding is installed around the inside of a cylindrical iron core. The iron core is laminated from thin layers of soft iron. The coils that make up the winding are either wound on protruding poles or set into narrow slots cut into the iron. This stationary assembly is called the stator.

The rotating part of the generator, called the rotor, is made to fit inside the stator. The rotor has two or more magnetic poles, also wound on iron cores. When the field coils are energized from a DC supply, the rotor becomes a powerful magnet. The magnetic field of the rotor crosses the narrow "air gap" between the face of the rotor poles and into the stator iron. The magnetic circuit is completed by the stator iron behind the windings.

When the rotor turns, the conductors in the stator are swept alternately by the north and south magnetic poles of the rotor field. Each passing pole induces a voltage in the stationary windings. Because the polarity of the voltage changes with each change in the magnetic polarity of the field, the induced voltage alternates from one polarity to the other. A revolving field AC generator is shown in Figure 6–1. The AC output can be taken directly from the stationary stator windings. The DC field must be supplied through a connection that allows the field assembly to rotate, typically by a pair of brass slip rings mounted on the rotor shaft. The slip rings are insulated from the shaft and from each other. They are matched to a pair of stationary carbon brushes fitted into holders that are supported on the generator frame. The brushes are held in contact with the rings by small springs in the brush holders.

This output voltage of the single-phase generator not only alternates, it also constantly changes in value. It rises to a peak as the field pole passes under the center of a coil, and

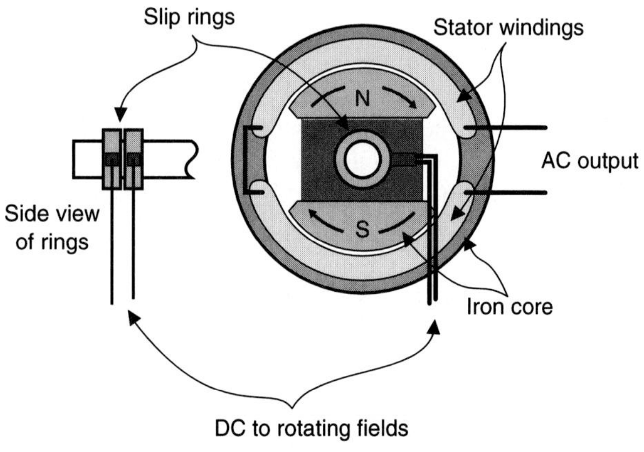

FIGURE 6–1 A revolving field AC generator.

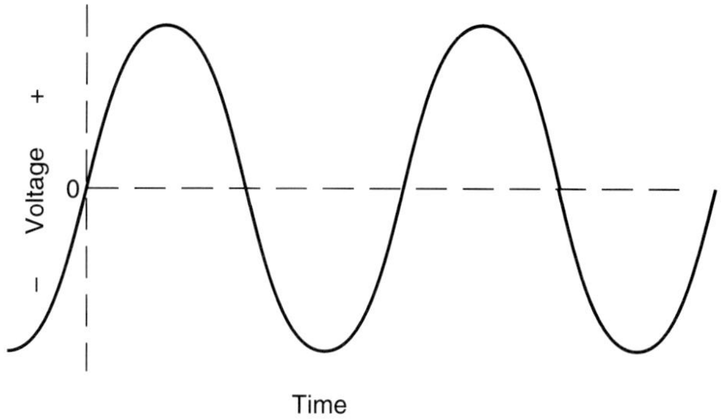

FIGURE 6–2 The output of a single-phase generator plots as a sine wave.

then drops to zero as the pole passes through the gap that separates one part of the winding from the other. The polarity reverses as the opposite magnetic pole moves into position. This process repeats continuously as the generator turns. A graphical plot of the changing voltage is a classic sine wave (see Figure 6–2).

To simplify construction and improve the output wave shape, the stator coils are usually distributed symmetrically across a series of slots, rather than built as protruding poles. To provide control over field strength, which in turn controls generator output voltage, some provision is made to regulate the DC power supplied to the electromagnetic field poles. In a commercial generator, regulating the field strength is the task of the voltage regulator. It monitors the generator's output voltage and adjusts the field as required to correct for any changes.

The output frequency depends on the number of poles in the generator and the speed of rotation. The number of poles matters because each pair of poles generates one complete AC cycle for each revolution of the rotor. A generator designed for high-speed operation usually has only two poles; slow-speed generators may have 60 or more. In most cases, the generator is intended to supply a fixed frequency, such as 60 Hz, and must be driven at a constant speed. The speed is regulated by a governor that controls the power output of the engine or turbine used to drive the generator. This engine or turbine is sometimes referred to as the prime mover.

Two Phases: Better Than One?

A single-phase generator is a capable machine, widely available in small sizes, and perfectly acceptable where relatively small quantities of single-phase power are required. In larger sizes, however, a flaw unique to single-phase becomes intolerable: Single-phase power is pulsating. Just as the voltage and current rise and fall, so does the power output—120 times per second at 60 Hz. These electrical pulsations cause mechanical pulsations on the generator driveshaft that are unacceptable in a large machine.

The best solution to this problem is to add more "phases" to the generator. Each additional phase is an additional set of single-phase windings. For example, a two-phase generator has two identical sets of windings, with the poles of the second set physically located halfway between the poles of the first. Although extra windings are added to the stator, the number of poles on the rotor is unchanged. As the rotor turns, it generates an identical AC output in each of the two sets of "phase" windings. The windings are

FIGURE 6–3 In a two-phase generator, the rotor generates two separate voltages that rise and fall at slightly different times.

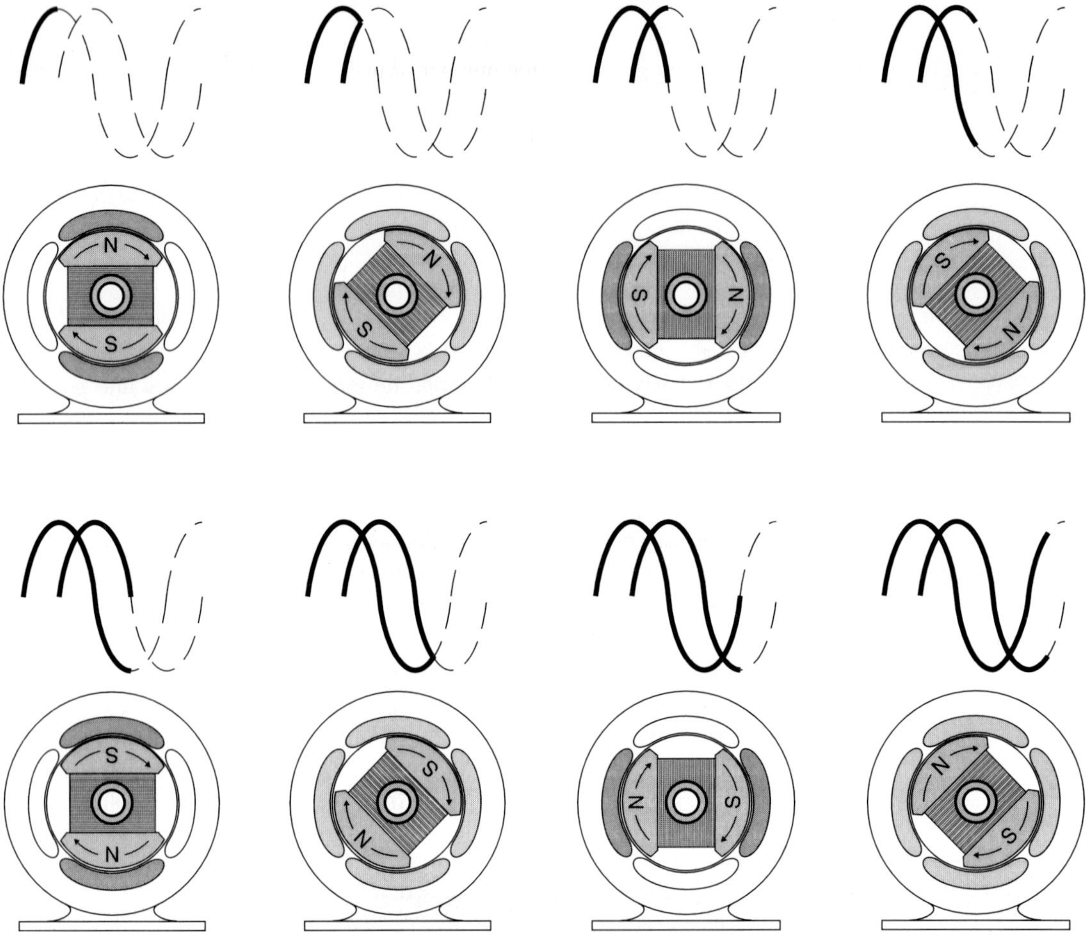

arranged in sequence, the rotor poles sweep past them in sequence, and the voltages generated in the windings also occur in sequence.

The total power output of the generator is a composite of two separate single-phase AC circuits. The power output of each circuit pulsates, but the phase voltages and currents are sequential, so the power pulses do not occur simultaneously (see Figure 6–3). The net result is a much smoother power output.

Two-phase systems have a second distinct advantage over comparable single-phase systems. Not only is the power output smoother, but the sequential voltages can be used to great advantage in certain types of loads. A two-phase motor, for example, is simpler to build and will outperform a comparable single-phase motor. Two-phase generators were once quite popular, although most have been replaced by three-phase machines. In a few areas where large power systems were established very early on, two-phase equipment is still in common use. Two-phase is also used in some instrument applications, such as miniature generators used to detect shaft position.

Three-Phase Generators

The Stator Winding. The basic principles of a two-phase generator can be taken further by adding one more set of windings. The stator of a three-phase generator has three identical sets of windings. Each has the same number of turns of wire of the same gauge, and is arranged to form the same number of electromagnetic poles. In a two-phase machine, the poles of one phase are halfway between those of the other phase. In three-phase machines, the phases are spaced at equal thirds.

In typical single-phase and two-phase machines, the coil groups are wound concentrically—several coils of different diameters laid one inside the other—and can easily be identified and counted. In typical three-phase machines, the poles of the windings are not readily visible. Instead of a series of identifiable poles, there appears to be a single, symmetrical winding composed of many relatively small coils. There are no obvious divisions or breaks in the winding that can be identified as a phase or an individual pole. Figure 6–4 illustrates this difference between single-phase and three-phase windings.

Windings of this type are made up of many identical coils. Unlike the coils of concentric windings, each has the same number of turns and is the same shape and width. The coils are essentially flat-sided ovals, shaped so the long straight sides are parallel and spaced to fit into two stator slots a specific distance apart. The coils are laid into the stator slots in a repetitive, overlapping pattern. The winding starts with a single coil with one side laid into the bottom of a slot. The second coil is placed in the next slot, and so on, until all the slots are filled. When the winding is completed, each stator slot contains two coil sides, one on top of the other. Each coil occupies the bottom of a slot on one side, and the top of a slot on the other.

To create each pole, several consecutive coils are connected. To keep the connections to a minimum, the coils may be wound in groups and installed without cutting the conductor

FIGURE 6–4 The single-phase motor winding on the left (a) has a concentric winding. The four poles are easily identified. A typical three-phase winding (b) is made from identical coils that overlap. There is no visible division between the poles.

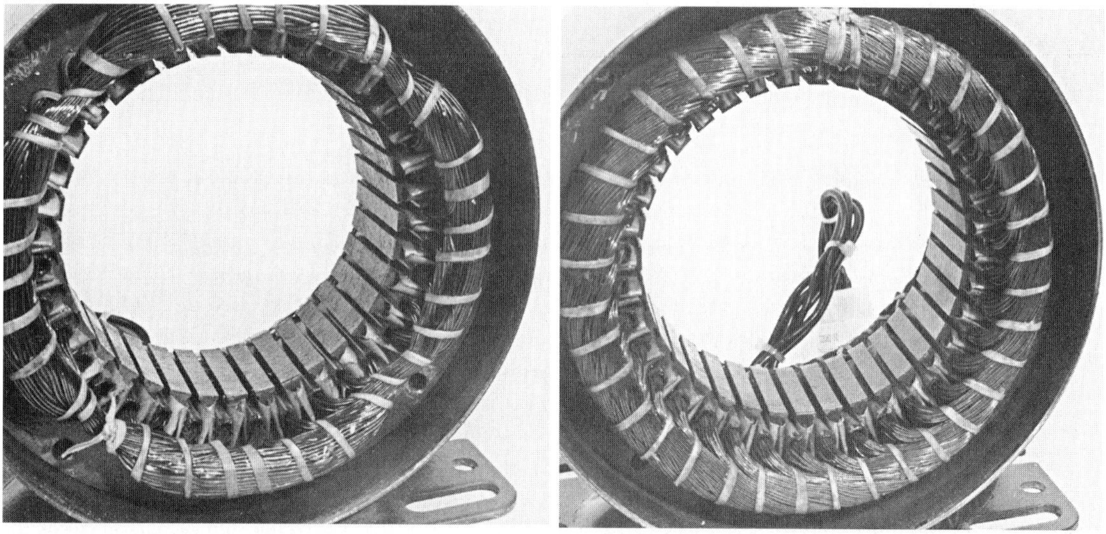

(a) (b)

between them. Each group of coils forms one pole of an individual phase. After installation, the coil groups are interconnected to form the phase windings. Each phase winding consists of every third coil group, or phase pole, around the complete winding. Within a particular phase, the connections are made so that every second pole has the opposite magnetic polarity. The end result is three complete single-phase windings, each with the same number of equal-spaced poles.

Practical Exercise: Exposing the Layout of a Three-Phase Winding. A three-phase winding is entirely logical in construction and quite impossible to appreciate from a visual inspection. However, there is a simple method to identify the location and polarity of each of the pole groups in each phase. All that is required is a three-phase stator, an adjustable DC power supply, and a small compass. The stator can be from a motor or generator—they have exactly the same layout—and must have either 6 or 12 leads. Three- and nine-lead stators have the windings preconnected into a three-phase group, making it impossible to energize only one phase by itself. A piece of chalk or similar temporary marker is a handy addition.

With the rotor removed, arrange the stator so that one open end is facing up. Locate the leads for each phase. If the stator has six leads, they should be marked T1 and T4, T2 and T5, and T3 and T6 for the three-phases, respectively. If it is a 12-lead stator, each phase is divided into two sections. These are numbered T1 and T4, T7 and T10 for the first phase; T2 and T5, T8 and T11 for the second phase; and T3 and T6, T9 and T12 for the third phase. Each of these groups must be connected in series or in parallel to form a complete phase. For example, connecting T4 and T7 creates a series connection in the first phase that would span T1 and T10. Connecting T1 to T7 and T4 to T10 would put the two sections in parallel. The T1/T7 pair would be one end of the phase; T4/T10 would be the other end. The other two-phases can be connected in the same way. Follow numeric order, and do not interconnect one phase group to the other.

Energize one phase of the winding from the DC power supply. Only a small amount of current is required. Locate the stator poles by moving the compass around the inside perimeter of the stator. Identify the magnetic polarity, and mark the center of each pole with an N or S as appropriate. Connect each phase in turn, taking care to use the same electrical polarity each time. For example, if T1 was positive for the first phase, T2 and T3 would have to be the positive side of the connection for the other two-phases. This experiment very graphically demonstrates the placement of the phase poles and the natural symmetry of the three-phase winding (see Figure 6–5).

The Rotor. The rotor of a revolving field AC generator is essentially a powerful electromagnet with two or more poles arranged in a circle around the driveshaft. The core iron may be laminated or, in a generator, it may also be solid. The field windings may be set into machined slots to form poles within a larger mass of core material, or they may be wound on protruding poles. Solid cylindrical rotors with machined slots are preferred for high-speed generators. These rotors are relatively long and narrow. Protruding poles, called salient poles, are generally preferred for lower speed machines. Low-speed generators have more poles. They are much shorter and larger in diameter than similarly rated high-speed machines.

Salient pole rotors may be equipped with a separate outer winding embedded into the faces of the pole pieces. This winding is made of a series of solid copper rods or bars placed a short distance apart around the perimeter of the rotor. The bars are attached at both ends to conductive rings that electrically interconnect all of the bars. This ring of bars is called an

FIGURE 6–5 When the phases of a three-phase stator winding are energized from a DC source, the poles of each phase can be identified with a compass. This is a two-pole stator; there are two poles in each phase.

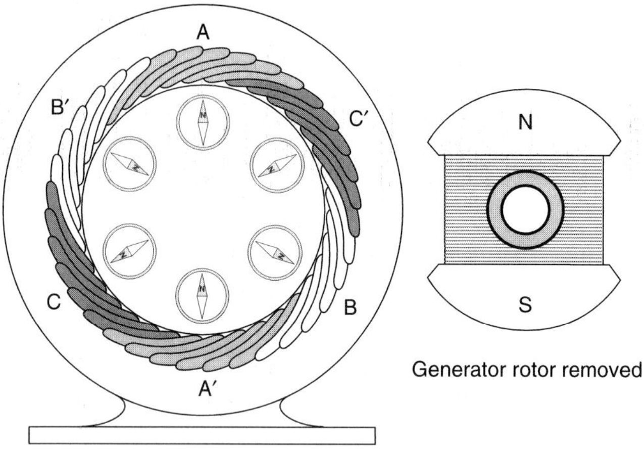

ammortisseur or damper winding. It is similar to the squirrel-cage winding used in most motors. The damper winding reduces oscillations that can occur during sudden changes in generator load.

AC generators and motors are categorized by the number of poles they have. This refers to the number of poles in the rotor, or in any one phase of the winding. It can vary from a minimum of 2 to 60 or more in a very large, very low-speed machine. It is important to understand that, while each phase of a two- or three-phase generator has a full set of "poles," the number of poles on the rotor does not change with the number of phases in the stator winding. A two-pole generator has only two field poles, regardless of whether it is single-phase, two-phase, or three-phase.

Three-Phase Generator Output. In a single-phase generator, the rotor poles pass a single set of windings, inducing a single AC voltage. In a two-phase generator, there are two sets; in a three-phase generator, there are three sets. Each of the phase windings is essentially identical to a single-phase generator winding, but they are laid into the stator in sequence. As the rotor poles pass over one phase winding, they encounter a second winding, and then, a little further on, a third. The phases are equally spaced, but have some overlap. The result is three sequential but overlapping AC voltages induced into three electrically separate windings. Each voltage, viewed by itself, is a normal single-phase AC sine wave. If the three are viewed simultaneously, they are equal in value and frequency, but each successive voltage wave is one third of a cycle behind the preceding one (see Figure 6–6).

The voltage produced in each of the separate phases cannot simply be combined into a single circuit path: Each must have its own circuit. Specific connections are allowed with conductors shared by more than one phase, but there are very definite limits to the possible interconnections. Under no circumstances can any two phases be directly connected on both ends. This would be a short circuit for each phase in the connection.

FIGURE 6–6 The voltage generated in each successive phase is one-third of a cycle, or 120 electrical degrees, behind the preceding phase.

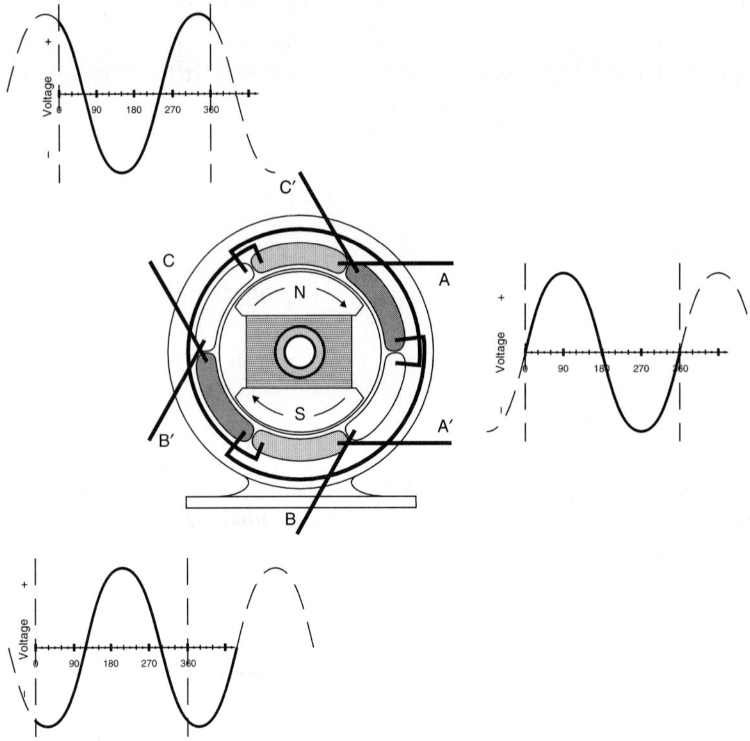

Phase Displacement. When the voltages of a three-phase system are described or compared, they are usually given a phase angle or displacement. One phase voltage is used as a reference—traditionally the one that has been selected as the "A phase" or Line 1—and the timing of the others is then compared to that phase. Phase angles are given in electrical degrees, based on the convention that one cycle is equal to 360°. An electrical degree is really a measure of time. One electrical degree equals $^1/_{360}$ of the time required for one cycle.

For a three-phase system, assuming the A phase is at the start point of 0°, the B phase voltage will start to cross through the zero line at 120°, and phase C at 240°. The separation between the three voltages is 120 electrical degrees. This phase displacement is a result of the physical layout of the generator windings and is always the same, regardless of what is happening in the system.

The phase relationships between the three-phase currents in the system are less dependable than they are for the voltages. Because loads can contain different amounts and types of reactance, the current in any one phase may lead or lag its voltage. If each phase supplies a different load, the phase relationships between the currents can be different than the 120° shift that exists between the voltages. In balanced loads such as three-phase motors, the current are 120° apart, because each phase of the load is identical.

The angle between phases of a two-phase system is 90 electrical degrees; for a three-phase system it is 120 electrical degrees. The two-phase system is not symmetrical. The phase difference between the start of second phase and the reoccurrence of the first is

270°. The three-phase system is symmetrical. The phase difference between any two phases, including from the "last" back to the "first" (C to A), is 120 electrical degrees.

Phase Sequence. The sequence in which the phases occur can also be very important. The phase rotation, or "phase sequence," of the generator output determines the direction of rotation for every motor it supplies. In a remote system, where generators may be subject to reconnection or replacement, it is extremely important that the phase sequence be checked before the system is reenergized. There are instruments made specifically for this test, but a small three-phase motor can also be used. The motor is connected to the system before any changes are made, and the exact connections are marked and recorded. The motor is energized, and the direction of rotation is recorded. After the generator has been replaced, the motor, connected as before, is energized from the new generator. If the motor operates in the same direction, the phase rotation is correct, and the entire system can safely be reenergized. If the motor rotation is reversed, then any two line leads from the generator must be interchanged to correct the error.

Phase sequence becomes even more critical when two or more generators are operated as a group. (This is not at all unusual. Almost all AC systems that have multiple generators operate with all of them connected in parallel.) The phase sequence of every interconnected generator must be the same. Electrically interconnecting two generators with opposite phase sequences is much like trying to couple the shafts of two motors that rotate in opposite directions. It is a disaster.

Interconnecting AC generators is a common but complex task. Special monitoring equipment is required to switch the generators in and out of the circuit safely, and the generators may need special governors to share the total load properly.

Frequency. There is a fixed relationship between the speed of the generator shaft and the frequency of the output. This is true for all generators, regardless of the number of phases. To produce one cycle of AC—from 0 V to maximum positive, back through 0 to maximum negative, and to 0 again—two field poles, one of each magnetic polarity, must pass a stator phase. One rotor pole generates the positive half-cycle; the second generates the negative half-cycle. The number of poles is determined when the machine is designed and built. It is always an even number.

The frequency of the generated AC output can be found as follows:

$$F = \frac{PN}{120}$$

F = frequency in Hz

P = number of poles

N = the speed in rev/min

For any given frequency, the larger the number of poles, the slower the generator has to be driven. Generators powered by water turbines operate at very low speeds. For example, a 60 Hz, 60-pole generator runs at 120 rev/min. Generators that are driven by steam or gas turbines operate at high speeds, and usually have either two or four poles. A 60 Hz two-pole generator runs at 3600 rev/min; a four-pole runs a 1800 rev/min. The generators are made this way to allow them to be directly coupled to the prime mover. Water turbines are naturally low speed; steam- or gas-powered turbines are naturally high speed. Direct coupling eliminates a very expensive gearbox, resulting in a simpler, more efficient generator set.

Single- and Three-Phase Compared

Note

It must be clearly understood that a single-phase system cannot be easily converted to three-phase. Phase converters are available to run three-phase motors from single-phase supplies, but they are fairly expensive, and the results are often imperfect. In contrast, three-phase systems can naturally supply single-phase. Any functional two-wire connection made to a three-phase system is single-phase. Any one phase taken by itself is single-phase. Any combination of two phases that results in a two-wire connection is also single-phase. The single voltage, single-phase supply system (usually 220 to 240 V) used in Europe and many other areas of the world is actually a single-phase connection taken directly from a three-phase distribution line. (This system, and the dual-voltage single-phase system commonly used in North America, are explained further in a the chapter on transformers and services.)

A single-phase circuit requires at least two conductors, not including any safety-related grounds. For a dual-voltage system, such as the 120/240 V supply commonly used in North America, the conductor count, at least for the service, is increased to three. These three conductors are usually the same size, though under certain circumstances the common (neutral) conductor may be smaller.

A two-phase system starts with four conductors—two leads per phase—but this can be reduced to three by using one conductor as a common to both phases. Unfortunately, the common conductor carries more current than either of the other two and must be 41% larger, because the phase displacement is not symmetrical.

A three-phase circuit starts with six conductors—two leads per phase—but this is usually reduced to three. When the system incorporates a neutral conductor, there are four. Again, this does not include any safety-related ground conductors. In a normal three-wire three-phase system, each of the three conductors is shared by two phases. The current in each conductor is normally the same, so all three can be the same size. In a four-wire system, the neutral conductor carries no current unless the loads are unbalanced. The sizing of the three main lines is unaffected by the existence of the neutral. The electrical codes may even allow the neutral to be reduced in size, if the nature of the loads is such that the neutral will have adequate current capacity under all conditions.

A three-phase system makes better use of the conductors than a comparable single-phase system. A balanced three-phase three-wire circuit can supply the same power demand with 25% less total conductor material than would be required with a single-phase. The three-phase circuit has one more conductor, but the total amount of conductive material is lower. This saving is of particular importance when the circuits are very long.

A three-phase system is inherently more complex than a single-phase system. Three-phase loads need three conductors instead of two, so, although the conductor size may be smaller, the cable is more complex and expensive, and may take up more space. A three-phase supply panel must have provisions for three line conductors instead of two, and three-phase circuit breakers and switches must have three poles. This increases the space required by each breaker, and increases the size and cost of the distribution panel. The extra conductor and extra breaker poles only apply to the three-phase loads. Single-phase loads taken from a three-phase panel require the same breaker and conductor provisions as they would from a single-phase panel.

A primary advantage of three-phase is the ability to use three-phase motors. These motors are simpler, more efficient, and require less maintenance than comparable single-phase motors. Due to design problems, single-phase motors are rarely made in sizes above 10 to 15 Hp. By comparison, three-phase motors are available with virtually any horsepower rating for which a supply can be made available. (Standard NEMA ratings go up to 100,000 Hp.) A disadvantage of three-phase motors is the increased complexity in wiring and switching equipment. This is part of the reason that three-phase is rarely used for fractional-horsepower motors.

Summary

Three-phase is an electrical system based on three separate AC voltages generated at equal time intervals in a repetitive sequence. The interconnecting wiring must have at least three conductors, but smooth, continuous power is delivered with 25% less total conductor material than would be required to deliver the same power with single-phase.

Virtually all commercial-scale power comes from three-phase generators. Although three-phase can be produced by electronically switching a DC supply, or by artificially creating a third phase from a single-phase source, a genuine three-phase generator output is superior to either of these methods. All major distribution lines are three-phase. Single-phase distribution almost always originates from a three-phase system.

Three-phase systems can be used to supply single-phase loads. If the voltages match, a two-conductor feed from the three-phase system can be used directly. If the voltages are different, an appropriate transformer can be used, just as it would be to match voltages in a single-phase system.

6.3 Three-Phase Connections

A three-phase generator has three sets of windings—one for each phase—and since each phase has at least two ends, it has a minimum of six leads. However, three-phase is normally delivered with three, or sometimes four, conductors. To reduce the six leads down to three or four, the three separate phases of the generator are interconnected so that each output conductor is shared.

The leads to be connected cannot be chosen at random. Each winding has a "start" and a "finish." The generated voltages occur in a specific sequence of positive and negative half-cycles. If the "start" and "finish" leads of one phase are interchanged, this sequence will be one half-cycle out of phase, because the positive and negative half-cycles have essentially been interchanged. This mistake leads to all kinds of problems, including a very strong possibility of equipment damage.

There are two basic options for interconnecting the three phases. They are called wye and delta, after the characteristic shape of the connection patterns. Each of these two also has a number of variations. The three-phase system is relatively simple in concept, but has enough variables to make it confusing at times. It is a good idea to understand at least the two standard connections and the relationships between the phase and line values of voltage and current. These are examined in detail below.

Note

The connections for three-phase transformers are essentially the same as for three-phase generators. The same basic configurations are used, and the same rules apply. The main difference is the transformer has both primary and secondary windings—which may be connected differently—while the generator has only the equivalent of the secondary windings.

The Star or Wye Connection

Interconnecting one end from each of the phase windings produces what is known as the star or wye connection (see Figure 6–7). The ends chosen for interconnection must be either the "finish" or the "start" ends of each phase. Traditionally the finish ends are used, but in most cases either will produce the same result. A mix of ends is not permissible. Standard lead markings identify the ends of the windings by the numerical order of the lead numbers: The lower number lead is the start of the winding, and the high number is the finish.

This initial connection creates a shared common, called the star point, plus three separate line leads. For a supply, such as a generator or the output of a three-phase transformer, the star point is almost always grounded. A grounded star point may be brought out as a fourth conductor if it is required by the particular application. This grounded conductor becomes the system neutral. The three "lines" are referred to as L1, L2, and L3, or Phases A, B, and C, and represent the primary three-phase connection point. In normal industry shorthand this connection is referred to as a three-phase four-wire wye, which is often abbreviated as a 3∅4W wye (Y).

The wye connection is very popular. Unlike the delta system, it provides a natural ground point around which the three-phases are symmetrically arrayed. This common point, once grounded, becomes a convenient point of connection for a system neutral. With the neutral extended into the distribution system, the 3∅4W Y can supply three-wire three-phase loads, two-wire single-phase loads—at two different voltages—and a three-wire, dual-voltage, single-phase network.

FIGURE 6–7 In a wye connection, one end of each phase is connected together.

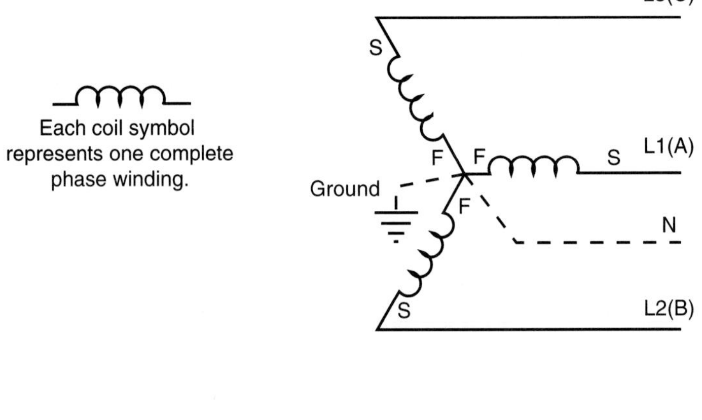

FIGURE 6–8 The 120/208 V wye connection provides many options.

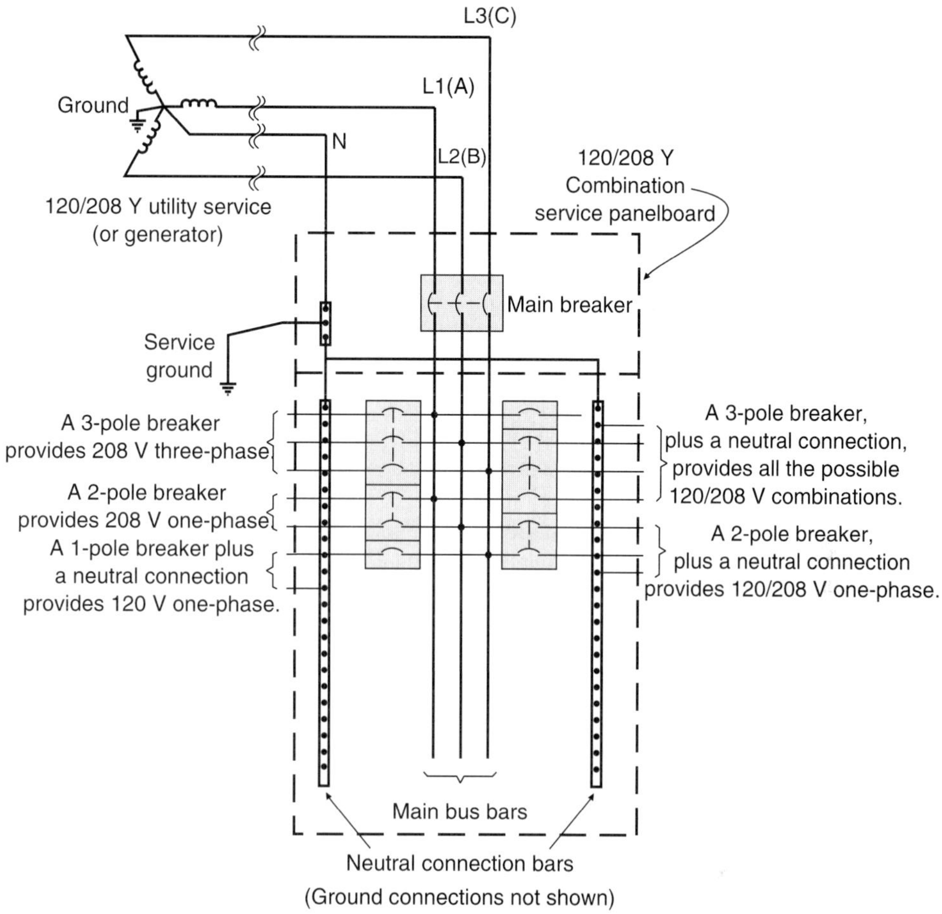

The network connection is a reasonably functional replacement for the extremely popular single-phase three-wire 120/240 V system. The only differences are 208 V instead of 240, and different code restrictions on the size of the neutral conductor. Utility customers that have a 120/208 V network connection in place of the more common 120/240 V system are unlikely to be aware of the difference.

The three-phase four-wire wye is the most flexible of all the possible three-phase connections. All the combinations listed above can be accessed from the same distribution panel using standard circuit breakers. Using the popular 120/208 3∅4W supply as an example, the balanced three-phase loads, such as motors or three-phase heating elements, receive 208 volts through a three-pole circuit breaker. Single-phase 120 V loads connect through a single-pole breaker from one line to neutral; single-phase 208 V loads connect between any two lines from a two-pole breaker. A single-phase three-wire connection includes the neutral and any two line leads (see Figure 6–8).

Phase and Line Values in a Wye Connection

There is enough inconsistency in the use of the terms "phase" and "line" to make their meanings very confusing. The safest assumption is to relate the term "phase" back to the individual phases of the three-phase generator, and the term "line" to the three main conductors used to connect a three-supply to a three-phase load. Following these principles, a "phase" value of voltage is the voltage across one generator phase winding or its nearest equivalent—a transformer winding, for example. In a wye system this is from any line to the star point—or neutral—because each phase is connected to the star point at one end, and one line on the other. Phase current is the current that flows in an individual phase winding.

A "line" value of voltage is measured between any two of the three main output conductors. These are generally referred to as the "line" conductors and must not be confused with the system neutral. A "line" current is the current that flows in any of these lines. Line values can always be measured at a point that is external to the source or load, such as at a supply breaker or switch. Phase values may not be accessible outside the supply or the load.

For a wye connection, the line voltage (E_L) is higher than the phase voltage (E_P), because it is a sum of two phases that are connected in series. Note that the two voltages cannot be added in a conventional manner, because they are 120 electrical degrees out of phase.

When two sine waves are 120° apart, the sum of the values along the two waves is not the same as the sum of their RMS values. One way to calculate the line voltage is to plot two-phase voltages on the same graph, and then sum individual plot points of the two curves. The result is a new sine curve, 30° behind the first of the two-phase voltages, and 1.732 times as high. This is an awkward, though educational, exercise. Thankfully, there are easier methods.

The phase voltages are essentially identical in value, and they have a consistent phase angle of displacement from each other. Therefore, it is a reasonably workable assumption that the line voltage will always be 1.732 times the phase voltages from which it originates. The constant is actually $\sqrt{3}$, which may be rounded to 1.732 if desired.

Phase values of current are those that occur within one of the phase windings; line current is within one of the external three-phase lines. In a wye-connected system, phase and line current are identical (see Figure 6–9). It works out this way because each line is

FIGURE 6–9 Phase and line values in a wye system.

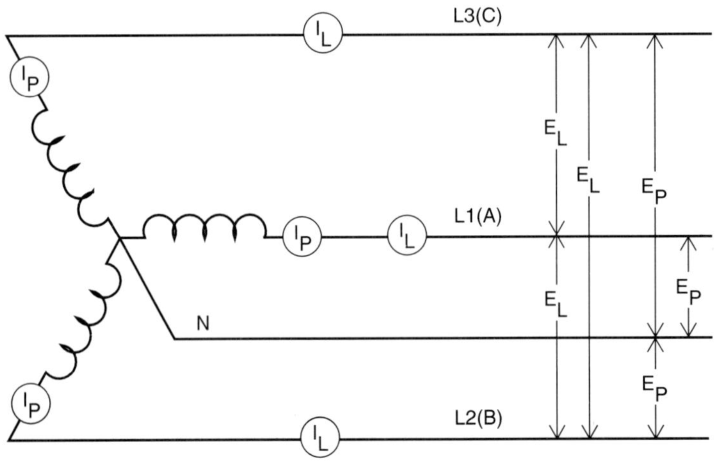

connected directly to one end of an individual generator or transformer phase. There is no other conductor connected to this point, so all the current in any one line must come from the phase to which it is connected.

$$E_L = E_P \times \sqrt{3}$$
$$I_L = I_P$$
$$E_L = \text{Line voltage}$$
$$E_P = \text{Phase voltage}$$
$$I_L = \text{Line current}$$
$$I_P = \text{Phase current}$$
$$\sqrt{3} = 1.732$$

The current relationship of the wye connection can be seen logically from the way in which the line conductors are connected to the individual phases—each phase feeds one line. The voltage relationship is due to the 120° phase angle difference between all of the phases. This explains the odd voltage values common to all three-phase wye systems. The lower voltage is only available as single-phase, but can be obtained from neutral to any of the three lines. This voltage can be very useful for lighting and other single-phase loads. The higher voltage is available for single-phase loads if required. It can be taken from any two lines; the neutral is not used unless the load is a dual-voltage device such as a domestic range or dryer. Three-phase is available at the higher, line-to-line voltage only. The neutral is normally not required by a three-phase load, unless the equipment being supplied also uses the lower, single-phase voltage as part of its normal operation.

The Delta Connection

In a delta connection (illustrated in Figure 6–10), the generator or transformer phase windings are connected in a series sequence that goes something like "finish" of A phase to "start" of B phase, "finish" of B phase to "start" of C phase, and "finish" of C phase to "start" of A. This forms a closed loop resembling a triangle or the Greek letter delta (Δ), from which the connection gets its name.

FIGURE 6–10 In a delta connection, the phases are connected in sequence to form a closed loop.

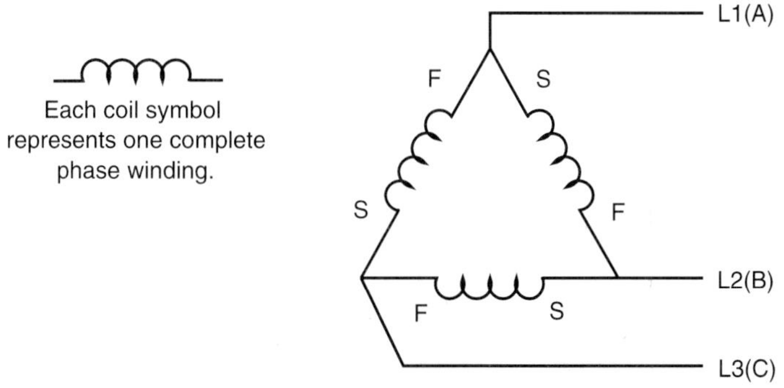

The delta connection appears to be a short circuit but, because of their 120° phase difference, the sum of the three equal phase voltages is 0 V. This can be demonstrated very easily if a three-phase generator or transformer is available. Connect the phase windings in a standard delta configuration, as described above, but leave the last "corner" of the delta unconnected. A voltmeter across the open connection will read the sum of the three interconnected phase voltages.

Warning

For safety reasons, when conducting this test, the voltmeter range should be set for at least double the phase voltage rating of the generator or transformer. The reason for this rather high setting is that an error in the connection of the generator windings, or either the primary or secondary of the transformer, will produce a voltage at the open point that is double the rating of one phase.

In theory, if the delta connection is done correctly, the sum of all three-phase voltages will be exactly 0 V. This is the voltage that should appear across the open point. All of the phases are connected in series, and the meter is across the complete series string. Therefore, the measured voltage has to be the sum of all three phases.

In practice, most deltas will show a relatively small voltage across the open point. This is because for a 0 V sum, all three phases must have absolutely identical outputs, with precisely 120 electrical degrees difference between each phase. There also must be no harmonics or other distortions in any of the wave forms. It is possible to meet these criteria, but quite unlikely. In a generator, even a very small difference in the air gap from one side of the rotor to another can cause a slight difference in the phase voltages. In transformers, even very small differences in the windings or the transformer core can produce enough imbalance to show up on this test.

The test described is a very a good demonstration of the addition of phase voltages. Such a test is often used to field-check delta connections before they are put into service. An incorrect connection—such as the reversal of any phase winding—will cause the sum of the voltages to be twice the phase voltage. If the final connection is made and there is an error, when the system is energized the resulting short circuit can easily cause equipment damage. Unfortunately, there is a complication with the test itself, because the possible voltage may be high enough to present some serious safety concerns.

When the possible test voltage is off the scale of any available voltmeter, one option is to complete the delta loop through a suitable fuse. The fuse should be sized large enough to allow for some circulating current—any voltage imbalance will cause some current to circulate around the completed delta—but not so large that the current can reach a dangerous level if there is a short circuit. If the fuse holds when the system is energized, the circulating current can be safely checked with a clamp-on ammeter. The equipment can then be de-energized, and the final connections completed.

A three-wire delta connection provides only single- or three-phase at the phase voltage of the source. The delta has no natural ground point, and it is sometimes used as an ungrounded system. When grounding is necessary, one corner may be grounded, or the middle of one phase if there is a connection point available. A central ground point can also be created by using a special grounding transformer.

The standard delta connection is a three-phase three-wire system (3∅3W-Δ). Other variations are examined in Chapter 7, "Transformers and Services."

FIGURE 6–11 Phase and line values in a delta system.

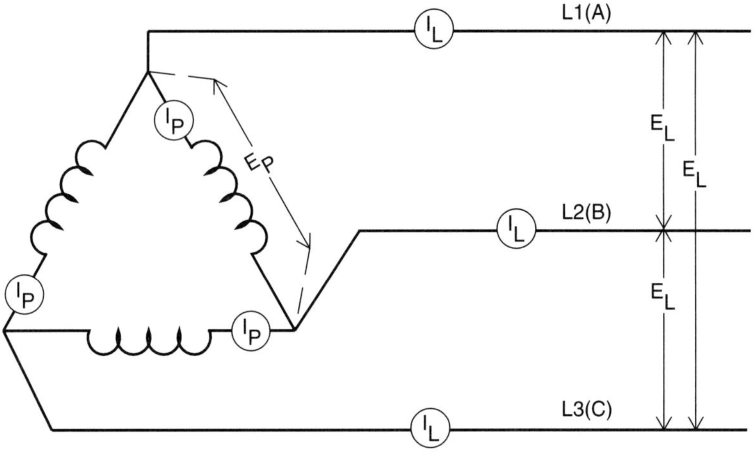

Phase and Line Values in a Delta Connection

In a delta connection, the phase and line voltages are the same (see Figure 6–11). Line voltages are measured between any two external lines, and these are connected directly across one of the phase windings of the supply. The line currents, however, are a summation of currents at the interconnection of two-phase windings. As with the line voltages in a wye connection, these currents occur in phases that are displaced by 120 electrical degrees. As a result, they cannot be summed by simple arithmetic. When they have same value—the load on the delta is balanced—the current in each line is larger than the phase current by a factor of $\sqrt{3}$. The current and voltage relationships for a balanced delta connection can be summarized and calculated as follows:

$$E_l = E_P$$
$$I_L = I_P \times \sqrt{3}$$
$$E_L = \text{Line voltage}$$
$$E_P = \text{Phase voltage}$$
$$I_L = \text{Line current}$$
$$I_P = \text{Phase current}$$
$$\sqrt{3} = 1.732$$

Summary

A three-phase circuit requires a minimum of three lines to carry the power output of the generator or transformer to the various three-phase loads. Each phase of the source by itself is a single-phase supply. The three-phase system combines all three of these phases into a three- or four-wire circuit. Combining the six original leads into three or four conductors simplifies the system and makes better use of equipment and conductors.

There are two basic formats for this interconnection. In a wye or star connection, one end of each of the three phases is tied to a common star point; the other end of each individual phase connects to one of the three-phase lines. A wye connection provides the highest possible three-phase output voltage. Because of the connection used and the way in which the voltages are generated, the wye system line voltages are a combination of two-phase voltages that are 120 electrical degrees out of phase. The result for a balanced system is a line voltage that is $\sqrt{3}$ greater than the voltage across one phase.

There are three line voltages, all of which should be equal. When testing a three-phase system, make a habit of testing all the way around the circuit—A to B, B to C, and C to A. This ensures that the system is properly balanced and that all fuses and conductors are intact.

In almost every case, a wye source is grounded at the star point. This star point connection is a natural common point for the three-phase wye, and may be extended into the system as a neutral. The neutral is not required, except for single-phase loads that are connected across a phase voltage rather than from line to line. However, this is a very common connection in many buildings. If the building has a wye service, the phase voltages can be found from the neutral to each line or, for test purposes only, from any good ground. The ground-to-line voltage will be equal to the neutral-to-line voltage, because the neutral is solidly grounded at the service. Do not connect a load between line and ground, but feel free to check the voltages this way. It is often more convenient, and sometimes more informative, than checking from line to line.

In a wye system, the current in the line is the current in the related phase. When the system contains a neutral, it carries only the unbalanced current of the circuit. Neutral current is equal to the sum of all three line (or phase) currents: In a balanced system this sum will be zero. Note, however, that this sum cannot be found by simple mathematical approaches. There is a phase 120° shift between each current, so adding them is a complex procedure. The important fact to remember is that neutral current is from the unbalanced loads. It is associated with the various single-phase loads connected from line to neutral throughout a typical wye system. The exact value of the neutral current is difficult to calculate, but the worst-case value is equal to the highest line current. This occurs when only one phase is loaded, making the neutral the only return path, or when two-phases are equally loaded and the third has no current.

The delta connection has several variations that will be examined more fully in the following chapter. A standard three-wire delta provides the highest possible line current at the lowest voltage. The line voltage is the same as the phase voltage, and the line current is the sum of two-phase currents—$\sqrt{3}$ times the phase, if they are balanced.

Delta systems are generally less popular than wye systems, because they offer less connection flexibility and do not provide a symmetrical ground point. However, the three-phase delta is a very workable connection and is widely used. It has some distinct advantages that can offset its problems.

When testing a delta system, voltages should be equal from line to line all the way around. Voltages to ground are a little less predictable and sometimes hard to interpret. If the delta is ungrounded, the voltages from a good ground to each of the three lines should theoretically be equal. In practice they may not be. If the system is corner grounded, the grounded line will show no voltage to ground, while the other two will show full line voltage. If the system is a three-phase four-wire delta, the voltages to ground will be the same as from the delta neutral. Two lines will be at one-half the normal line voltage value (normally 120 V), and the third will be at 86.7% (208 V) of the line value. This odd phase is the "high leg to ground" or bastard phase of the four-wire delta and must not be used to supply single-phase loads.

Questions

1. Why does the generator output change polarity?
2. What factors affect the output frequency of a generator?
3. Explain why a four-pole three-phase generator stator has four poles in each of the three-phases, but only four poles in the rotor.
4. What is the phase angle between voltages of any two phases in a three-phase system?
5. What is meant by the term "phase sequence"?
6. Why is phase sequence important?
7. What advantages does three-phase output have over single-phase?
8. What disadvantages does a three-phase system have compared to single-phase?
9. Can single-phase power be taken directly from a three-phase system? What about three-phase from a single-phase line?
10. Sketch a three-phase wye connection. Identify the start and finish of each phase winding, and show where the ground would normally be connected.
11. List all the single- and three-phase connections that are available from a 120/208 V three-phase panel.
12. Explain why the line voltage of a wye connection is less than double the phase voltage, even though it is supplied from two phases in series.
13. How can a delta connection be tested before it is energized to ensure there are no crossed connections? What safety concerns may be associated with this test?
14. Describe the voltage tests that are appropriate for checking a wye-connected system. What are the results if the system is fully functional?
15. Describe the voltage tests that are appropriate for checking a delta-connected system. List some of the variations in delta grounding and the effect they have on such tests.

Problems

1. What is the phase voltage of a 600 V wye connection?
2. What is the phase current in a wye-connected generator supplying 250 A to a balanced load?
3. What is the line voltage of a delta-connected generator with a phase voltage of 240 V?
4. What is the line current of a delta generator if each phase is carrying 150 A?
5. If a 240 V delta-connected generator was reconnected in wye, what would the line voltage be for the new connection? Assume no other changes were made.
6. If the generator of question 5 had a full load current rating of 100 A with the delta connection, what would the full load current rating be for the wye connection?

7 Transformers and Services

Objectives

This section introduces transformers and provides some sense of their operation. It also examines a variety of different types, with a concentration on the transformers used to supply consumer services, and typical power and control circuits. Both single- and three-phase transformers are discussed in some detail, but it should be understood that the subject of transformers is broad enough to justify textbooks devoted to nothing else. After completing the study of this chapter you should be able to:

- Explain how a transformer operates
- Describe, in general terms, the construction of a typical transformer
- Explain transformer turns, voltage, and current ratios, and apply them properly in calculations
- Explain the features that are specific to Class 2 transformers
- Describe the construction of an autotransformer and compare it to conventional two-winding types
- Describe the connection and operation of buck-boost transformers
- Calculate transformer current ratings from nameplate voltage and VA
- State the purposes of transformer grounding and describe any considerations or precautions that should be followed when connecting the grounded conductor into a circuit
- Identify transformer leads from standard markings
- Connect typical multiwinding transformers for different output voltages
- List the various common service configurations and identify the transformer connections associated with each one

7.1 Introduction

An understanding of transformer operation and behavior is important. However, of equal or greater importance is an understanding of how the supply transformer connections define the service. These connections determine the voltages that will be present between each of the supply conductors, and from these conductors to the system's earth ground.

Additional information on control transformers, particularly Class 2 types, is provided in Chapter 16, "Low-Voltage Controls." Small transformers like these are used to provide low voltages for control circuits. This reduces costs by simplifying the control devices and the wiring. Using very low voltages—24 V, for example—can reduce fire hazards and virtually eliminate the risk of electrocution.

7.2 Transformers

Throughout the complex electrical systems that power modern industrial societies, high voltages are used to power major loads and to reduce the current in long supply lines. Lower voltages are used, where possible, for increased safety and to simplify equipment design. Transformers are used to link these circuits together. They can change the AC voltage level, up or down, to match the operating voltage of one circuit to that of another. Transformers also provide a degree of isolation between the two circuits. This helps to protect a lower voltage circuit from the potentially destructive energy of a high-voltage supply. Transformers allow electrical energy to flow between parts of a system that otherwise could not be interconnected.

Transformers range in size from near-microscopic proportions—for use in miniaturized electronic circuits—to house-sized monsters used in high-voltage transmission and distribution. Regardless of their physical size, all transformers operate on the same principles and share many common construction details. However, despite their similarities, standard transformers are usually divided into several distinct categories. The criteria used to differentiate between the various types include the intended application, the type of enclosure provided, and whether the transformer coils are immersed in oil or other insulating coolant, or are cooled only by air circulation.

The transformers described in this section are a reasonable—but not all-inclusive—sampling of the most commonly used types. They include transformers designed for use in power distribution, metering and instrument installations, and control systems. The winding layouts of these transformers include conventional two-winding types, single-winding autotransformers, and the closely related buck-boost transformers used for voltage adjustments.

Theory of Operation

Electromagnetism. Transformers exploit the natural relationship between magnetism and electricity. A flow of electrons—electrical current—creates magnetic forces, and moving magnetic fields induce voltage. These principles are examined in the previous chapters of this section and will be reviewed here briefly because of their role in transformer operation.

There are two closely related methods by which magnetism can be used to produce electricity. The first method involves a magnetic field, a conductor, and relative motion between the two. When the field cuts the conductor, or the conductor cuts through the field, a voltage is induced in the conductor. This is the underlying principle of generators. The second method involves a change in the strength of the magnetic field passing through a coil. In this case there is no need for the field to actually cut through, or make contact with, the conductor. All that is required is a change in the strength of the field that passes through the opening of a conductive loop or coil. When this change occurs, a voltage is induced in the coil. This second method does not require any motion other than an increase or decrease in field strength. This is the underlying principle of transformers.

Both of these methods follow very specific rules. When there is a cutting action, the polarity of the induced voltage depends on the polarity of the field and its relative direction of motion. For a changing field passing through a coil, the polarity of the voltage depends on the polarity of the magnetic field, and whether it is increasing or decreasing in strength. In either case, the amount of voltage induced depends on the rate at which the field is moving or changing. A rate of cutting action, or rate of change, that is equal to 10^8 lines of flux per second will induce an EMF of 1 V. One line of flux is called a maxwell (Mx); 10^8 lines is a weber (Wb).

Note

In reality, a field cutting a conductor and a changing field through a coil are essentially the same actions observed from two different points of view. They are separated here because it is usually easier to understand generators in terms of magnetic fields cutting conductors, and transformers as changing fields passing through coils. The principles of electromagnetic induction are part of the discoveries of Michael Faraday.

Magnetism converted into electricity is only one half the electromagnetic relationship. Electricity also creates magnetic fields. When an electric current flows through a conductor, a magnetizing force—and as a result, a magnetic field—forms around the conductor all the way along its length. When the conductor is formed into a coil, the magnetizing force is concentrated around the center line of the coil. It forms a unified field that passes through the inside of the coil. A straight coil acts much like an adjustable bar magnet. There are magnetic poles on each end of the coil—one north and one south—and the field passes through the coil and returns around the outside. When there is a change in the amount of current, the strength of the field changes proportionally. A change in the direction of the current changes the polarity of the field.

The total magnetizing force of the coil does not depend on the diameter of the coil, but on the current in the coil conductor and the total number of turns. This is an interesting characteristic, because winding more turns on the coil increases the magnetizing force, without any need to increase the current. One amp in a coil with 100 turns has the same magnetizing force as 100 A in a coil with only 1 turn. The total magnetizing force of the coil is a multiple of the current in amps and the number of turns in the coil. The resulting value is in ampere-turns, the standard unit of magnetizing force.

Magnetizing force and magnetic field strength are not the same. These units are much like voltage and current. One creates the other, but the final values are affected by additional conditions. The strength of the field depends not only on the magnetizing force of the current in the coil, but also on the magnetic properties of the materials in which the field will form. This field path is called the magnetic circuit. The reluctance of the magnetic circuit has a powerful influence on the strength of the field. Reluctance is the magnetic equivalent of electrical resistance.

All materials conduct magnetic fields, but some are much better conductors—have lower reluctance—than others. Just as low electrical resistance can result in a higher current, a low-reluctance magnetic circuit can also result in more lines of magnetic flux. More lines of flux equals a stronger field. This is why almost all electromagnetic devices have cores made of low-reluctance material. In power electrical devices, such as transformers, the most common core material is soft iron. The excellent natural properties of iron are further improved by the addition of alloys. The core iron is arranged to form a magnetic "short circuit" for the field by providing a continuous path of iron through the coils and around the outside. The iron has a wide cross section, and is kept as compact as possible. A short magnetic path with a substantial cross section keeps the reluctance low and ensures maximum field strength.

In a transformer, the electrical energy carried by the voltage and current of one coil is converted to a changing magnetic field in the core. The energy of this changing field is in turn converted to electrical energy in a second coil. The two coils are in different circuits, electrically isolated from one another. The connection between the circuits is completely dependent on the shared magnetic field.

Self-Inductance and Mutual Inductance. When a single coil is connected to an AC supply, the changing field generates a back EMF in the coil. This action is known as *self-inductance*—or simply inductance—and the effect it produces is called inductive reactance. (For more information on inductance and inductive reactance, refer to Chapter 5.) When two coils are located and arranged so that the magnetic field created by current in one coil will induce a voltage in the other, the relationship is known as *mutual inductance.*

Mutual inductance involves energy transfer from one coil to another. The energy originates with the electrical supply that feeds the first coil and is transferred to the second coil through the shared magnetic field. The more complete the magnetic link between the coils, the more efficiently they can pass energy from one to the other. In general, the effect of mutual inductance is greatest when the coils are placed in close physical proximity on the same core. A primary goal of transformer design and construction is to make the link between the transformer coils as complete and free of losses as is practical.

Simplified Transformer Model. The basic transformer shown in Figure 7–1 contains two coils on a common iron core. When an AC supply is connected across one of the coils, current flows in that coil, and an alternating magnetic field is set up in the core. The core iron forms a complete loop with a very low reluctance. This ensures that not only

FIGURE 7–1 A simplified model of a transformer.

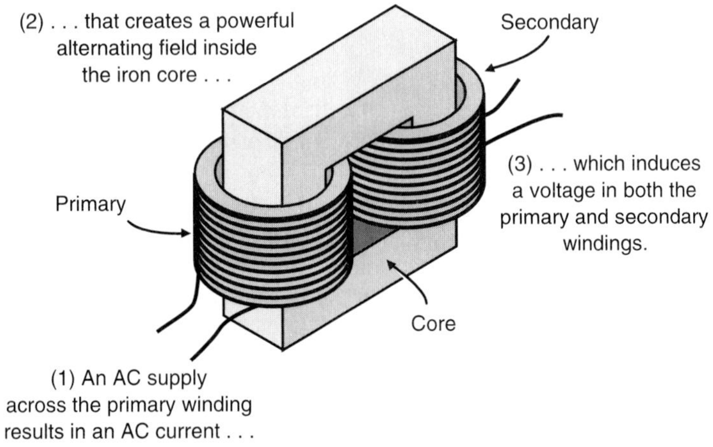

will the field be very strong, but also almost all of the magnetic flux will be contained within the core material. The core passes through both coils; therefore, virtually all of the field also passes through both coils.

Note

In the discussions that follow, the coil that is connected to the electrical supply is referred to as the primary winding, and the second coil, the one that is not connected to the supply, is referred to as the secondary winding. These are standard transformer terms, used in a manner that is consistent with the way they are used in industry. In the basic transformer described here, there are only two coils. In actual transformers, two coils is quite common, but this varies with the transformer design.

The changing field created by the magnetizing force of the first coil—the primary—will induce a voltage in both the primary and the secondary coils. In the primary, the induced voltage is a back EMF that acts to oppose the applied voltage, thereby regulating the current in this coil. This is the same as the inductive reactance effects that occur in a single coil. When the transformer is operating without a load —there is nothing connected to the terminals of the secondary winding—it is essentially an inductor, and it behaves like one. The primary current is almost entirely the *excitation current* needed to create the magnetic field. The excitation current lags the applied voltage by close to 90 electrical degrees.

The voltage induced in the secondary winding (the coil that is not connected to the supply) appears across the terminals of this winding as a generated AC voltage. It does not necessarily have the same voltage value as the supply, but it does have the same frequency and sine wave shape. The actual value of the voltage depends on the relative numbers of turns that make up the two windings. It can be anything over a very wide range, from much less to much more than the voltage of the supply.

It is a fundamental principle of transformers that the volts per turn is essentially the same in all the coils that share the same field. The reasons for this are not immediately obvious and take several steps of logic to explain. In the primary winding, the total applied voltage divides equally across all the turns of the winding. The turns are all in series, and all have the same electrical characteristics, so it is only reasonable that the voltage should divide itself up in this manner.

The voltage induced into the primary as back EMF—in total and turn for turn—is nearly identical to the applied voltage. When the transformer is first energized, the current rises until it is limited by the effects of the back EMF. An operating balance is established with the back EMF only slightly less than the applied voltage. If the two voltages were equal, they would cancel each other and current flow would stop. The difference between them is just enough to supply the losses in the resistance of the primary winding and in the transformer core. The transformer reaches this operating balance within one or two AC cycles.

The voltage induced into each turn of the secondary winding is equal to the voltage induced into each turn of the primary because both windings are influenced by the same changing field. The voltages across each turn of both the primary and secondary windings are therefore almost identical.

Because there are equal volts per turn in both windings, the output voltage of the transformer—the voltage induced in the secondary winding—depends on the relative number of turns in the primary and secondary windings. This is usually expressed as a turns ratio for the two windings.

Assume, for example, the primary coil has 1000 turns, and the secondary coil has 200 turns. If 250 V AC is supplied to primary side, there will be 0.25 V across each turn. The voltage induced into the secondary coil will also be 0.25 V per turn, so the output voltage of the transformer will be 50 V AC (0.25×200). The turns ratio is 1000:200 V, or 5:1. The voltage ratio is the same, 250:50 V, or 5:1.

Note

Actual volts-per-turn values vary from less than one volt in small transformers to several volts per turn in larger ones. In practice it is unusual to know the actual number of turns. This information is of little practical use to the end user and is therefore not part of the normal nameplate data. The transformer ratio is also not supplied on the nameplate, except indirectly as rated voltages for each of the windings. The turns ratio is essentially identical to the no-load voltage ratio.

When a transformer is operated without a load, there is a balance between the excitation current, the resulting back EMF, and the electrical and magnetic losses. At this equilibrium point, the transformer's alternating magnetic field is at full strength, and the induced secondary voltage is at full value. There is no current in the secondary windings, because the secondary circuit has no connected load.

When a load is connected, the secondary voltage drives current through the load and the secondary winding. With current in both windings of the transformer, the magnetic field in the core is affected by two magnetizing forces instead of just one. The magnetic effects of the secondary current oppose those of the primary, and momentarily disturb the operating equilibrium of the transformer. The primary current excites the field, but the secondary current acts to make it weaker. However, any reduction in the field also reduces the primary back EMF, which causes an immediate rise in the primary current and restores the field. The end result of this interaction is that changes in secondary current cause corresponding changes in the primary current. The operating equilibrium of the transformer is automatically maintained.

The transformer's changing magnetic field and the voltages it induces remain nearly constant from no load to rated full load. The primary winding receives energy from the supply and transfers it to the field; the secondary winding accepts energy from the field and transfers it to the load. Only a small percentage is lost in the transformer.

The current relationship between the primary and secondary is not ampere for ampere. A 1 A increase in current demand on the secondary side does not necessarily mean a 1 A increase on the primary side. Magnetizing force is a function of ampere-turns, not current level alone. The relationship between the two transformer winding currents is therefore based on their relative ampere-turns. A change of one ampere-turn in the secondary will result in the same change in the primary.

The no-load primary current is the excitation current that creates the field. As the transformer load is increased, the excitation current and the strength of the field remain virtually constant, while the load-related ampere-turns in the primary track the ampere-turns of the secondary. Disregarding the excitation current (and losses), the ampere-turns are equal in both windings. This gives transformers a current ratio, in much the same way that they have a voltage ratio. The voltage ratio is a result of an equal volts per turn for each winding; the current ratio is a result of equal ampere-turns.

The voltage ratio is directly proportional to the turns ratio.

$$\frac{E_P}{N_P} = \frac{E_S}{N_S}$$

$$\frac{E_P}{E_S} = \frac{N_P}{N_S}$$

The current ratio is inversely proportional to both the voltage and turns ratios.

$$I_P \times N_P = I_S \times N_S$$

$$\frac{I_P}{I_S} = \frac{N_S}{N_P}$$

E_P = Primary voltage

E_S = Secondary voltage

N_S = Secondary turns

N_p = Primary turns

I_S = Secondary current

I_p = Primary current

This relationship works out very well, because the energy the transformer demands from the supply is regulated by the energy that is transferred to the load. With no load, the transformer draws very little energy from the supply. When a load is connected, it automatically draws just enough to provide the demands of the load. Discounting the inevitable losses that occur in any electrical device, the power in the primary supply circuit and secondary load circuit are identical.

Assume a transformer has a 100 V supply and a turns ratio of 2:1 (see Figure 7–2). If the output of the secondary winding is connected to a 10 Ω load, the 50 V induced in the secondary will drive 5 A through the secondary winding and the 10 Ω resistor. Disregarding the no-load current of the transformer, the ampere-turns of the two windings are equal; therefore, the 5 A supplied to the load translates into 2.5 A taken from the supply. The power into the transformer—100 V times 2.5 A—is equal to the power drawn by the load—50 V times 5 A. This is an oversimplification, of course, because it disregards excitation current and losses, but it is an important principle of transformer operation. This relationship is summarized in the following formula.

$$E_P \times I_P = E_S \times I_S$$

FIGURE 7–2 Disregarding losses and excitation current, the power into and out of the transformer is the same

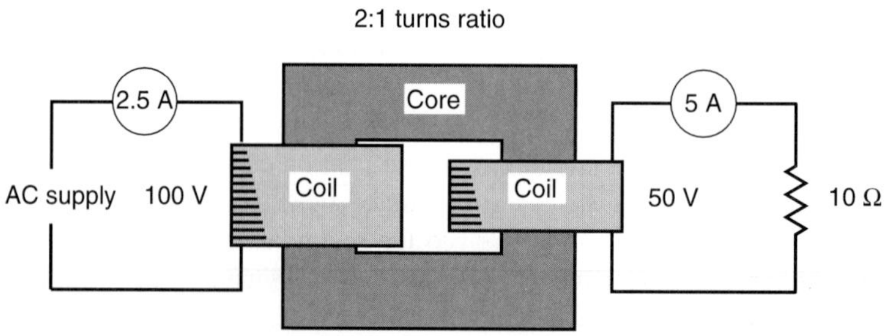

Transformer Behavior

Voltage Ratios. In practice, real transformers have voltage ratios that behave quite closely to the theoretical ideal, as long as the transformer is operated at no more than the rated primary voltage. With rated voltage in, the transformer's no-load output voltage should be slightly higher than the nameplate value. At full load it should equal the rated (nameplate) value; above full load, it should drop below. The no-load voltage is made slightly high to compensate for the voltage drop that occurs as the load is increased. Most of this internal voltage drop is due to the resistance of the primary and secondary windings.

The voltage ratio of the transformer is the same as the turns ratio. Therefore, unless the connection is changed to alter the number of turns, the voltage ratio should be a constant. It should also hold true for voltages other than the rated value. In practice, this works out reasonably well as long as the primary voltage is reduced instead of increased. A drop in the primary voltage will cause a proportional drop in the secondary voltage, but when the primary voltage is increased beyond about 115% of the rated, the ratio begins to change. From this point on, as the primary voltage is increased, the corresponding increases in the secondary voltage become smaller and smaller. At the same time the excitation current begins to rise at an alarming rate. A moderate overvoltage can burn out the primary winding, not from insulation failure due to the high voltage, but from overheating caused by excessive current.

To a voltage that is several times higher than the rated, the transformer will appear almost as a short circuit. This happens sometimes when a transformer is accidentally connected backwards. With the low-voltage winding connected to the high-voltage supply, the transformer may be saved by the overcurrent protection—or it may be instantly destroyed.

The failure of the secondary voltage to keep pace with the primary, and the dramatic increase in the excitation current, are caused by magnetic "saturation" of the transformer core. Any magnetic material, even soft iron, has a limited capacity to carry magnetic flux. As the amount of flux is increased, the core begins to saturate. As it approaches saturation, the magnetic reluctance increases, and each increase in magnetizing force no longer produces the normal increase in field strength. At some point, the core will be fully saturated, and there will be no further increase in flux, regardless of the increases in magnetizing force (see Figure 7–3).

A typical transformer core is designed with only a small safety margin: It is adequate to carry the field strength associated with normal excitation levels, and as long as the primary voltage is reasonably close to the rated value, there will be no problems. However, when the primary voltage is increased above the rated value, there is not enough extra core material to readily accept the necessary increase in field strength, and the core begins to saturate.

Each increase in primary voltage requires a corresponding increase in excitation current, field strength, and back EMF to maintain an operating equilibrium. As the core saturates, the excitation current increases at a greater rate because it takes more current—more magnetizing force—to induce the necessary back EMF. Unless the transformer burns out first, it will reach equilibrium, but the excitation current will be much higher than normal.

In an operating transformer, the excitation current is the portion of the primary current demand that creates the transformer's magnetic field. It is an inductive current that is proportional to the applied voltage, not to the secondary load. Typical transformers are designed for a maximum voltage of about 15% above the nameplate rated value. This percentage is a safety margin and should not be used as a normal operating point.

FIGURE 7–3 A saturation curve is a plot of the secondary voltage against primary input voltage or current. The no-load secondary voltage is proportional to the flux, so the shape of the curve is true for either value. Excitation current rises sharply as the core saturates, because the back EMF is also proportional to flux.

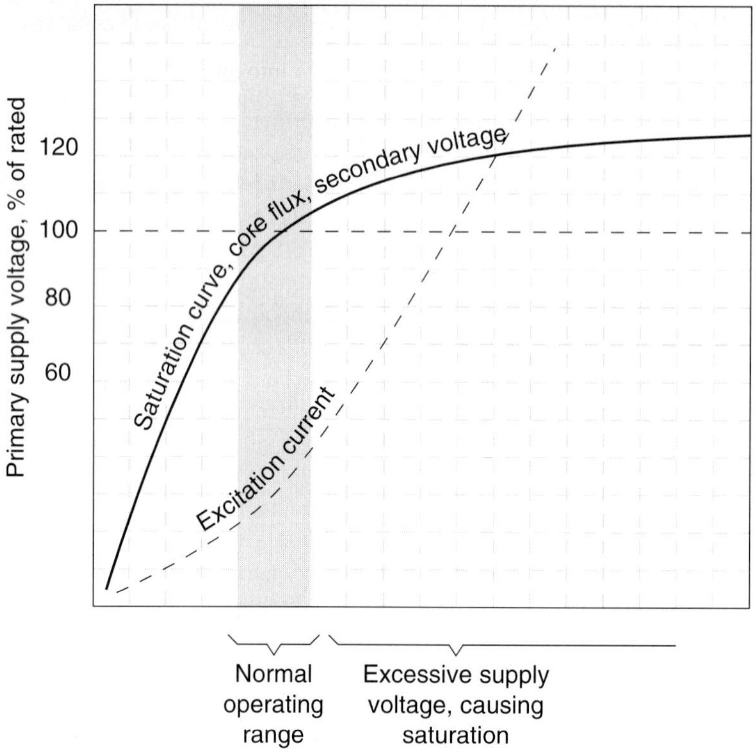

Current Ratios. While real transformer voltage ratios are quite close to the theoretical values, the same cannot be said for the current ratios. Basic transformer theory states that current ratios are inverse to voltage ratios. A transformer with a voltage ratio of 2:1 should have a current ratio of 1:2. At this ratio, for every amp the load receives from the transformer secondary, one-half amp of current must come from the supply into the transformer primary. This relationship is essentially valid, but it is difficult to prove with a practical test. There is a valid reason for this, but the confusion it causes can be very persistent.

The no-load current of a transformer is primarily excitation current and, while it is significant in value, it also lags the voltage by close to 90 electrical degrees. With no load connected to the secondary, the true power load on the supply is only enough to supply the losses in the primary coil and the core. This is a small fraction of the power that is suggested by the amount of current. The transformer draws essentially the same excitation current at all times, regardless of load. Unfortunately, excitation current cannot be measured separately from current that supplies energy to a load.

The transformer secondary current can be in phase, leading, or lagging, depending on the type of load that the transformer supplies. This characteristic is passed on more or less intact to the primary and, with a suitable ratio applied, becomes the load portion of the primary current. This current is added to the excitation current but, because the two currents are out of phase, it is difficult to isolate one from the other.

Attempts to demonstrate transformer current ratios can be frustrating, and there is little point to force-fitting real measurements into an incomplete model. Although the inverse current ratio of the transformer is a valid principle, unless more variables are brought into the equation, current ratio calculations are only approximations.

Despite the current ratio problems described, comparisons of the power in and out of a real transformer can be quite informative. The volt-amp–based calculation of $E_P \times I_P = E_S \times I_S$ is prone to the same problems as the current ratios, but power measurements based on watts, rather than volt-amps, are more revealing. For proper measurement, wattmeters must be placed in the primary supply to the transformer, and between the secondary output and the load. The primary side wattmeter will indicate the true power input, summed with the transformer's internal power losses. The secondary side wattmeter will indicate the true power demand placed on the transformer by the load. Transformer efficiency can be calculated with the following formula:

$$\eta = \frac{P_{IN}}{P_{OUT}} \times 100$$

η = Efficiency in percent

P_{IN} = Power in (true power, in watts, into the transformer primary from the supply)

P_{OUT} = Power out (true power, in watts, from the transformer secondary to the load)

Transformer efficiency varies with load. With no load, it is zero, because there is no power out. It rises steeply up to about 25% of rated load, and then levels off at a relatively high percentage. The efficiency stays high all the way to full rated load and slightly beyond. The efficiency of a typical transformer begins to drop off when the load is increased beyond the full load current rating (see Figure 7–4).

Real transformers are actually quite efficient: Very large ones may reach full load efficiencies of 98% or better. However, all power and efficiency calculations must be based on true power in watts, not on apparent power in volt-amps. This makes measurements more difficult, because at least one wattmeter is required. Attempting to find transformer efficiency without a wattmeter gives very misleading results. When the measured values of current and voltage are used, the excitation current comes into the calculation as if it is in phase. This inflates the input power well above the true value, and gives a calculated efficiency that is much lower than the actual.

Despite the problems, the voltage, current, and power relationships used to describe transformer operation can be very useful. When a change occurs in load current, the resulting change in the supply current is very close to the theoretical ratio. When the load demands energy from the transformer, it is transferred from the supply to the load with very little loss. The resulting changes in the input to the transformer will quite accurately reflect the additional power taken by the load. The transformer will have done its job. The load circuit, operating at one voltage, is coupled to a supply that operates at a different voltage. The power flow is essentially equal on both sides of the transformer, even though the voltages and currents are different.

FIGURE 7–4 Transformers have good operating efficiency across a wide variation in load.

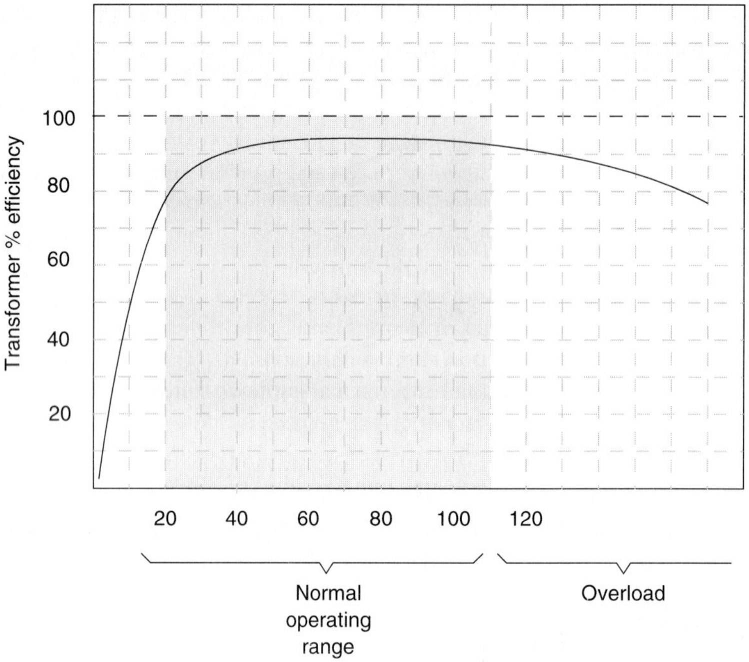

Transformer Terminology

As stated above, the term *primary* refers to the *supply side,* and *secondary* to the *load side.* The primary is always the winding connected to the power supply; the secondary is always the winding that supplies the load. This identification is based on the way the transformer is connected, and holds true regardless of which winding has the highest voltage rating. When there are more than two windings in the transformer, they are usually grouped and identified by their role.

A *step-up* transformer has a primary winding that is operating at a lower voltage than the secondary; a *step-down* transformer is, of course, the opposite. The primary of a step-down transformer is the high-voltage side. Step-down transformers are the most common. For example, distribution, control, and instrument transformers are almost always step-down types. Large step-up transformers are used by utilities to raise the voltage to transmission or distribution levels. Smaller ones are used to supply high voltages to neon tubes in sign lighting, and for spark ignition systems in gas- and oil-fired burners.

The *transformer ratio* is normally the ratio of the primary to secondary voltage. Although it can be mathematically simplified, it is usually listed as the rated values of primary vs. secondary voltage. The primary voltage is listed first, normally separated from the secondary value by a colon (:) or a slash (/). For example, a transformer that is rated for 480 V on the high-voltage side, and 120 V on the low-voltage side, would have a ratio of 480:120 V for a step-down connection, or 120:480 V in a step-up configuration. The voltage ratio is the same as the ratio of primary to secondary turns, but there is no reason to list the turns or the mathematical turns ratio.

Transformers are always reversible: They can be connected as step-down or step-up as long as an appropriate supply is available. The voltage ratio won't always work out quite right, though, because the turns ratio of many transformers is adjusted slightly to compensate for internal voltage drops that occur when the transformer is under load.

There are transformers that have a 1:1 voltage ratio, or equal primary and secondary voltage ratings. The primary and secondary are usually identified and the voltages are listed, even though they are equal. Transformers like this are usually made specifically for use as *isolation transformers,* to separate the load circuits from the main supply. All transformers—except autotransformers, which will be discussed later in this chapter—provide some degree of isolation. The link between the two windings is a magnetic one: There is no electrical connection. This separation can be used to increase safety or to allow different grounding arrangements. However, there is an important difference between true isolation transformers and the isolation provided by a transformer with a more conventional design.

True isolation transformers are equipped with an electrostatic shield between the primary and secondary windings. In a normal transformer, there is no shield: The windings are only separated by normal layers of insulation. Although the insulation does block any current flow from one coil to the other, it does not entirely isolate the secondary winding from the voltage of the primary. A capacitor is two conductive plates separated by insulation and, although they are there for another purpose, the two transformer windings also are conductive "plates" separated by insulation. This creates the equivalent of a small-value capacitor connecting the two windings. It makes a weak electrical link between the windings, which tends to pass the primary side voltage to ground on to the secondary. It will not supply significant current, but it can cause substantial voltages between the secondary circuit and an earth ground.

When the secondary of the transformer is grounded, this problem disappears. When it is not grounded, voltage to ground can be high enough to be a serious hazard. True isolation transformers eliminate this problem, because the grounded electrostatic shield blocks any capacitive coupling between the windings. The shield also helps prevent an accidental connection between the two windings should a fault occur. True isolation transformers allow the secondary circuit to be operated safely without a ground connection.

Transformer taps are connections brought out from intermediate points along the winding. They may be placed in the center of the secondary winding to produce a three-wire output or to make the transformer suitable for certain types of electronic power supplies. Taps are also provided for voltage adjustment. In this case, several are provided, connected at relatively close intervals at one end of a winding or winding section. The taps are usually on the high-voltage winding, where the current is at its lowest. By selecting a suitable tap, the output voltage of the transformer can be adjusted in small-percentage increments. These adjustments are designed to correct for a supply that is slightly above or below the rated value: They are not intended to change voltage from one standard value to another.

Off-line tap changers are manual switches that change tap connections. The taps are connected to switch contacts rather than provided as sets of external terminals or leads. Tap changers are usually supplied with oil-filled transformers because of the problems associated with changing connections under the oil. Because the tap changer must not be adjusted except with the transformer de-energized, the switch is located inside the junction box or, for aerial transformers, under the tank cover. On larger transformers the tap switch may have a provision for a padlock as extra protection against operation by unauthorized personnel. Adjusting off-line taps with the transformer energized will usually destroy the tap changer switch, resulting in downtime and an expensive repair.

On-line tap changers permit voltage adjustment under load. They may be used as voltage regulators and are often provided as optional equipment on large transformers. A typical on-line tap changer is a fairly complex motorized switch assembly. To avoid interrupting the current flow, the tap switch has to momentarily connect to the second tap before breaking the connection to the first one. This connection is a short circuit across the coils between the taps, so it is made through a current-limiting resistance or reactance. The whole tap changer assembly is immersed in insulating oil to control contact arcing and to insulate and cool the tap changer. On-line tap changers must be properly maintained. They are a critical, and usually very active, part of the distribution system.

Transformer Construction

Transformers are constructed to maximize the electromagnetic links between the coils that make up the transformer winding and the iron core material. The two sets of coils must be insulated from each other and from the core iron, but they must also be physically close together. The iron is given a thick cross section, maintained around the complete magnetic path. It is laminated from thin sheets, with the pieces arranged so that successive layers are joined in different places. This makes the magnetic circuit as continuous as possible. The core is also made very compact to keep the magnetic circuit short. Longer paths naturally have more reluctance.

Transformers have a long enough history that the trade-offs of each possible layout are well understood. A few common designs dominate the power transformer industry, because they have proven themselves over many years. Two basic core shapes are the most common. The *shell type* transformer core is in the shape of a squat figure "8" that may be oriented in several different ways, depending on the enclosure and mounting arrangements. The *core type* transformer core is in the shape of a rectangle or "0." In shell type transformers, the windings are coiled around the center "post" of the figure "8." In the core type, the windings are wrapped on two legs of the "0." Both types of transformers are shown in Figure 7–5.

Both core shapes have advantages and disadvantages, and both are in common use. The shell type core encloses the windings more fully and requires no cross connections between two separate coil groups. The core type is a more open design, making is easier to

FIGURE 7–5 Shell type and core type transformers.

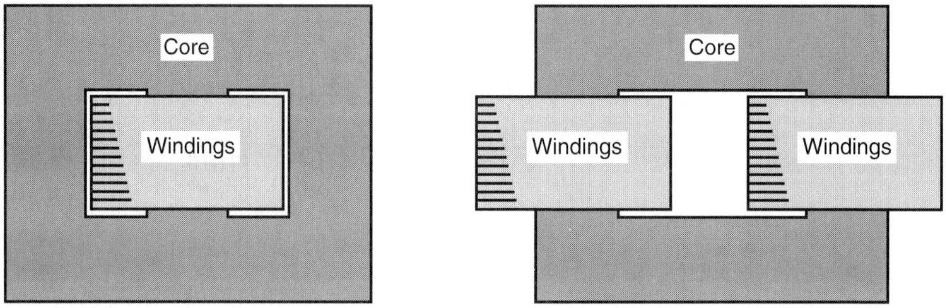

Shell type transformer cores have all the windings around a center post.

Core type transformers have the windings split into two groups, one on each core leg.

create clear paths for circulation of cooling air or oil. In general, shell types are used for smaller transformers, core types for midsize and larger. There are exceptions, of course, and there are other variations besides these two.

The iron used to make the core is relatively soft and is alloyed to improve its magnetic properties. It is laminated to control eddy currents that would otherwise occur inside the core. The laminations may also be lightly coated with insulating varnish. The cores are usually built up one layer at a time from pieces that match to form the final shape. For example, a shell type core may be made from two different shapes, one an "E" and the other an "I." These fit together to make the final figure "8" shape. The coils that form the winding are wound onto a form, and then the core pieces are fitted into the completed winding. The "E" and "I" shapes are set into the winding assembly, alternating sides so that the joints don't line up. The core pieces are aligned to form a block, and then tightly secured to prevent vibration.

Core type transformer cores may be made up of alternating layers of "U" and "I" shaped iron. The performance of the core can be improved by using accurately fitted mitered corners, although this is rarely done in small transformers. The width of the iron sheets may also be varied to create a more circular cross section. This makes better use of the space through a circular coil assembly.

Both high- and low-voltage coils of a core type transformer are wound in sections, with half of each coil on each leg of the core. The coils are also layered to keep the primary and secondary windings close together for maximum efficiency (see Figure 7–6). Placing the high-voltage winding on one leg of the core and the low voltage on the other would be very inefficient. Transformer drawings usually suggest that type of construction, but they are drawn that way for clarity and simplicity.

Instead of using many individual sheets to make up the core, a long strip of iron can be rolled into a circle or flattened "O." Circular cores are wound as toroids, the windings passing through the hole and around the outside. Toroidal transformers have excellent magnetic and electrical properties. They are used for current transformers and small, high-quality power transformers such as those found in some audio equipment. They are expensive to wind, so they are rarely used for general-purpose transformers.

FIGURE 7–6 Two methods of layering transformer windings.

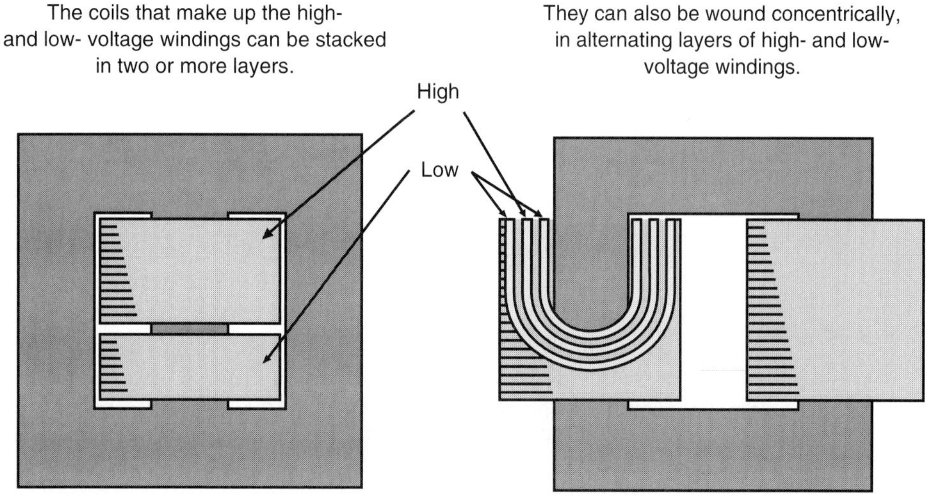

Wound cores are also used in power transformers. A pair of "O"-shaped wound cores are combined to form a variation on the shell type design. The two core sections are placed side by side and clamped into a bolted frame. The winding assembly encircles both core sections, passing through the opening in both of the "O"s. Transformers of this type may be made by winding the iron-strip core through the coil assembly, or the coils may be wound into the core iron using a motor-driven coil form. The coil wire is pulled into the form like fishing line onto a spool.

Most small transformers have shell type cores and a single winding assembly. Oil-filled distribution transformers are often shell type, but may have wound cores in place of the usual stack of laminations. Many larger transformers, including most distribution-size dry-type transformers, are core types. They take advantage of the more open design by creating many vertical channels in the coil assembly, allowing air to circulate freely through the transformer for better cooling.

The cores described above are for single-phase transformers. Three-phase transformers are similar, except that there must be places in the core for three sets of windings, not just one. Core type three-phase transformers may have three or five legs. The three-legged design has a winding assembly on each leg; the five-legged type is essentially identical except the outer two legs have no windings.

In all transformers, the windings must be insulated from the core iron and protected from damage due to absorption of moisture. In small transformers, layers of varnished paper or plastic are used to make a simple form for the windings. This protects them from the sharp edges of the core, while providing the necessary layer of electrical insulation. The coils that form the completed windings are made from varnish-insulated magnet wire that is wound directly onto the coil form. The high-voltage winding has the most turns, but is made from a smaller gauge of wire because it will carry less current. The two windings are applied in layers separated by wraps of heavy insulation paper. A final layer of varnished paper or similar insulation material is used to protect the finished winding.

Larger transformers are much more exacting and complex in their construction. Conductors may be square or rectangular in cross section instead of round. Extra insulation may also be used between coils or coil groups. The coils are usually formed into disks or cylinders, solidly bound or bonded to prevent chaffing from vibrations due to magnetic stress. Disk-shaped coils (sometimes called pancake coils) are stacked on the core leg, with high- and low-voltage sets alternating up the stack. Cylinder-shaped coils slide over the core leg. They have different diameters, so the high- and low-voltage sections fit one inside the other to form several alternating layers. Extra insulation is applied to coils and coil groups, and between the core and coil assembly. Blocking is used to separate the layers and to provide channels for the flow of coiling air or oil. The finished assembly is an impressive and very functional work of art.

Once the transformer is completed, the ends of the windings must be brought out to the connection points provided for external wiring. In small transformers they are simply brought to the termination points and soldered. The external connections are insulated wire leads or terminals and may be the most fragile part of a small transformer. Care must be taken to avoid damage when connections are made.

Most small transformers contain no moving parts or extra components, except for the thermal cutout or fusible link added to the Class 2 types. Larger transformers may have extra leads or terminals for connection options. They also may have thermal sensors, thermostat-controlled cooling fans, cooling radiators, and tap changers to provide small increments of adjustment to the voltage ratio. Except on specialized transformers equipped for on-line voltage regulation duty, the tap changer cannot be adjusted unless the transformer is off line. Failure to follow this requirement can destroy the tap changer.

Types of Transformers

The intended location and application of the transformer will usually dictate the enclosure, the mounting method, and the way the connections are arranged. The transformer may be an open type, or equipped with an enclosure that is suitable for the location (see Figure 7–7). Connections may be leads or terminals, which may be exposed or concealed inside the transformer case or junction box. Open transformers must be mounted inside equipment enclosures. Transformers used in exposed locations indoors must be equipped with a suitable enclosure. The enclosure may be vented, but it must protect the windings and provide a method to enclose the connections. Outdoor transformers must be weatherproof. They are usually oil filled, so the enclosure is essentially a tank. In smaller sizes the tank is sealed. Very large transformers have special provisions for expansion and contraction of the oil. Many have air breathers equipped with desiccant-filled filters to remove moisture that would otherwise contaminate the oil.

Enclosures for small shell type transformers are often no more than covers for the ends. The core naturally surrounds the windings, except at the ends, so the sides of the core are left exposed and only the ends are enclosed. Connecting leads are brought out under one or both of the covers. The covers are often used as connection boxes, with knockouts provided to accommodate cable or conduit connectors. These transformers have built-in mounting provisions, so they can be fastened to any suitable surface.

Distribution Transformers. Distribution transformers, designed for aerial mounting, have bushings with exposed terminals for both the low- and high-voltage connections. The bushing is an insulated sleeve, usually made of porcelain, that provides a passage for the conductor through the grounded metal wall of the transformer tank. A bolted terminal clamp is provided at the end of the bushing to accept the external wiring. The connections are made in the open and left bare. Safety isolation is provided by the location, high in the air on a pole or transformer platform (see Figure 7–8).

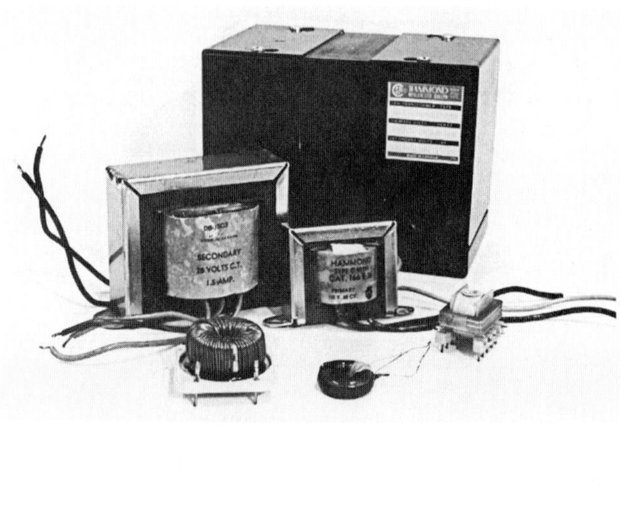

FIGURE 7–7 In front, two small toriods, behind three open transformers with shell type cores; at the back, a shell type transformer with connection boxes enclosing the ends.

FIGURE 7–8 These oil-filled aerial distribution transformers have high-voltage bushings on the top, and smaller low-voltage bushings on the side.

Distribution transformers that are mounted on ground-level pads have covered connection boxes incorporated into the transformer case (see Figure 7–9). The connection box is usually open at the bottom to allow for the entrance of the supply- and load-side cables. There are separate compartments for the high- and low-voltage connections. The high-voltage connections are usually made with special elbow connectors that can be disconnected and reconnected while they are energized. This must be done with a high-voltage "hot stick," but does allow for emergency switching without blacking out a whole neighborhood. The elbow connectors and the special bushings to which they connect also protect and insulate the high-voltage wiring. There are no exposed high-voltage connections, even inside the connection box.

Distribution transformers are available with output voltage ratings to match any standard consumer service voltage, and input voltage ratings that match all standard utility

FIGURE 7–9 The connections for this pad mount transformer are accessed through the two hinged covers on the front. The high- and low-voltage cables enter through the bottom. In this example, the low-voltage cables enter the building service through the two conduits to the right of the transformer.

distribution voltages. They couple the utility to the consumer, isolating the consumer's electrical system from the utility and reducing the voltage to a much safer standard voltage level.

Distribution transformers are designed for very rigorous electrical duty in an outdoor environment. The metal outer case is sealed against contamination, and the core and winding assembly are immersed in a special-purpose mineral oil. The oil keeps the windings clean and dry, and aids in cooling. As the interior of the transformer heats up, the oil naturally circulates inside the tank, carrying heat from the internal parts to the enclosure. Cooling may be through the tank walls alone, or assisted by extra cooling fins or small radiator tubes.

Dry-Type Transformers. Dry-type power transformers are common in large commercial buildings. The term "dry" means the transformer contains no liquids. There is no tank full of oil, only an enclosure with the core and winding assembly mounted inside. Small ones are little more than a core and coil with a junction box built onto one or both ends. Larger ones have ventilated enclosures with a removable panel to provide access to the terminals. Some have thermostatically controlled fans for extra cooling.

Dry-type transformers are preferred for indoor use. Oil-filled transformers are rarely used inside buildings because of the fire hazard. Installation requirements for indoor oil-filled transformers are demanding and expensive. Nonflammable coolants are available, but are banned in most places because they contain PCBs, a well-documented biological hazard.

Instrument Transformers. Instrument transformers are used to power permanently installed metering and monitoring equipment. They are designed for high accuracy at a fixed ratio of the voltage or current being monitored. An instrument transformer for monitoring voltage is called a potential transformer (PT) and looks much like a standard power or control transformer.

An instrument transformer designed for current measurement is called a current transformer (CT) and looks quite different from all other transformer types. One of the more common variations is shaped like a donut and works much like a clamp-on ammeter. The conductor carrying the current to be measured is threaded through the opening, or window, of the CT. There are no electrical connections between the monitored conductor—which acts as the primary of the current transformer—and the secondary, which is a toroid winding on the donut-shaped core. Connecting leads are brought out and wired directly to the appropriate instruments. Current transformers provide a reduced current signal and isolation from the operating voltage of the circuit. A window-style CT, designed for mounting inside an equipment enclosure, may be protected with no more than an outer wrap of varnished insulation paper. For more demanding locations, the CT may be dipped in varnish, or totally encased in plastic or epoxy.

Warning

Current transformers are designed for a current output, not voltage. For proper operation, they must be connected to a very low resistance load, such as an ammeter or the current input of an instrument. They must not be left open-circuited when there is any possibility of current through the primary conductor. If this happens, a very high voltage is induced across the secondary winding. This is a serious safety hazard that also may burn out the CT. It is imperative that the secondary windings of current transformers be shorted out at all times.

Instrument transformers are usually designed to provide standard secondary output values. Potential transformers are most commonly designed for a 120 V output. Their ratios can be 480:120, 600:120, 4160:120, and so on, as required by the circuit. These standards are chosen to work with typical instruments. Current transformers are essentially step-up transformers in terms of voltage, step-down for current. Their ratios are also a part of their specifications. For a current transformer, the most common standard secondary value is 5 A. This produces listed ratios such as 25:5, 50:5, 150:5, or 400:5.

Panel-mounted voltmeters are usually equipped with a 150 V meter movement, with 120 V as the normal operating point (see Figure 7–10). The faceplate is scaled to the voltage on the primary side of the PT, with the 120 V point matching the rated circuit voltage. For example, a voltmeter monitoring a 600 V line would be scaled to 750 V. The 600 V point on the meter face would correspond to an actual meter voltage of 120 V. Connected to a 600:120 V PT, the meter would accurately represent the voltage it was installed to monitor.

FIGURE 7–10 The instrument transformers, a 600:120 PT and a 1000:5 CT, allow low-voltage, low-current meters to monitor a high-voltage, high-current circuit.

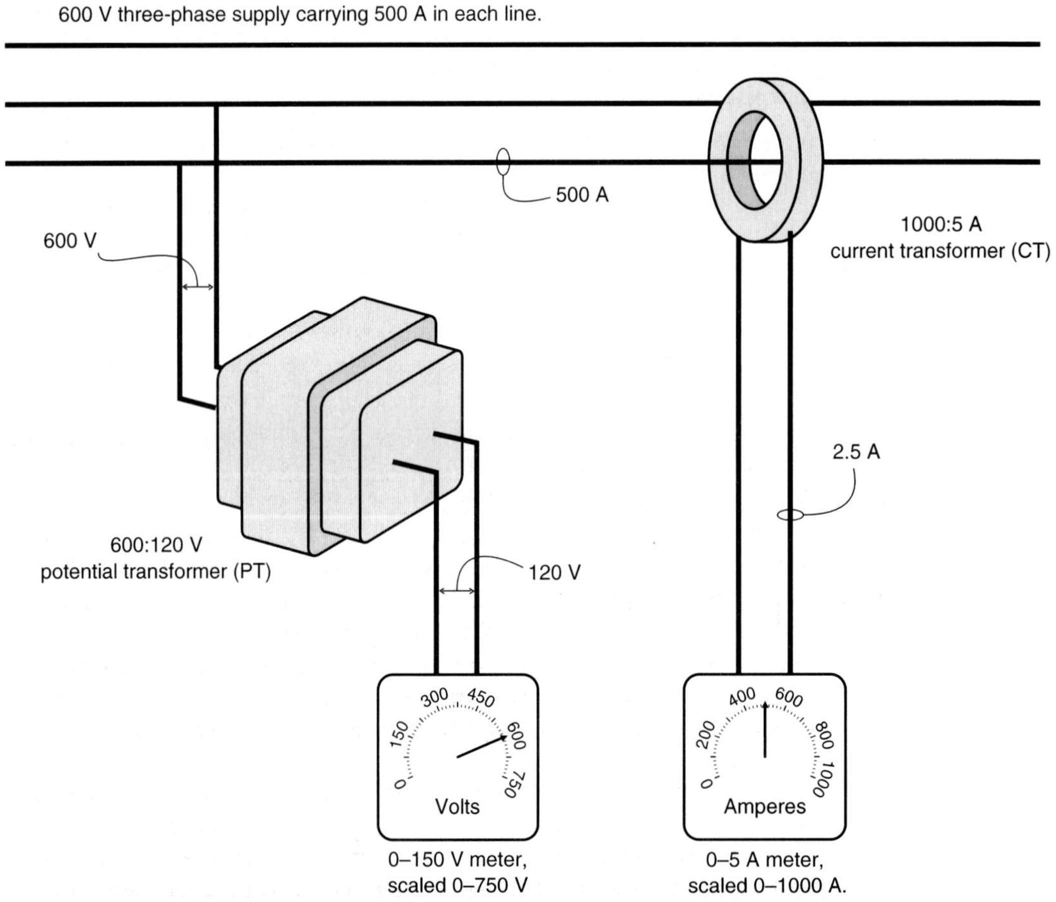

Panel-mounted ammeters are usually designed for a 5 A maximum input. The faceplate can be scaled to any value to match the primary current at the CT, but the meter sees only the secondary current of 5 A or less. For example, a 1000:5 A CT could be mated to a 5 A ammeter with a 1000 A faceplate. With a current of 500 A in the primary circuit, the CT secondary circuit, including the meter, would have a current of 2.5 A, and the meter needle would sit at 500 A on the scale.

Note

The principle of reversibility is true for current transformers as well as voltage types. This can be used to good effect when a high-current source is required for testing equipment such as circuit breakers or motor overloads. A heavy primary conductor is short-circuited through the CT and the breaker or overload. The secondary of the CT is connected to a variable autotransformer. A clamp-on ammeter can be used to monitor the test current as the autotransformer is adjusted. A suitable current transformer can supply several hundred amps from a 5 A variable autotransformer. The test set is compact and easy to use, providing high current under precise control with a minimum of expense.

Control Transformers. Control transformers are designed for a relatively stable output voltage that does not dip excessively under short-term overloads. This is important because typical control circuit loads, such as relay coils and solenoids, tend to draw much higher current when they are first energized than they do in normal operation. This is sometimes called inrush current and is a common problem with almost all AC coil devices. If the voltage drops significantly just because a solenoid is energized, it can cause unreliable performance of the entire control system.

Control transformers are found in every application where the control system is operated at a voltage different from that used to power the main circuit loads. Common examples include industrial motor control, machine tool controls, electrically controlled hydraulic systems, and heating, ventilation, and air-conditioning (HVAC) equipment. There are a number of standard AC control voltages, including 120, 48, and 24 V.

The control power is usually provided by a dedicated transformer connected to the power circuit of the equipment under control. Many industrial motor control circuits are operated at 120 V AC, although other voltages are also used. Control transformers suitable for this type of application are available with primary ratings matching all the standard motor circuit voltages—typically 208, 240, 480, and 600 V. Some transformers have alternative connections that allow them to be used with several different system voltages. This is not a problem, but care must be taken to be sure the transformer is connected for the correct voltage—or it can easily be destroyed.

Typical transformers used for motor control have modest power ratings ranging from 50 to 3000 V amps (VA). The VA rating is a multiple of the transformer's rated secondary voltage and current, so a 50 VA transformer with a 120 V output has a full load current rating of 117 mA. (50 ÷ 120). A 3000 VA (3 kVA) transformer can supply 25 A output at the same voltage.

Control transformers have a shell type core construction, and because they are mounted inside an enclosure with other exposed wiring, they are typically an open style.

The connections on the smaller ones are usually simple screw terminals. A holder may be mounted on the top to accommodate a protective fuse.

Note

Open type transformers often have the terminals mounted on the coil form or a piece of insulating board at the end of the winding assembly. This material is relatively fragile. When the connections are tightened, the terminal must not be twisted, as this can permanently damage the insulation or the winding.

Motor control equipment is usually maintained by personnel with experience and training in the electrical field. The control voltages used are relatively high—typically 120 V or more—because these voltages can drive large control coils without excessive current. Higher voltages also provide more reliable operation than lower voltages, particularly in situations where the control circuits are very long or the field contacts may be exposed to dust and other contaminants. The higher voltages require more protective wiring methods—such as conduit or industrial-rated cables—and the safety hazards are higher. However, in a typical industrial environment with electricians on staff, these concerns are generally not a problem.

The control voltages for systems that are likely to be maintained by "nonelectrical" personnel are usually lower—24 V AC being a very popular choice. Examples include electrically controlled hydraulic systems and HVAC equipment. The control transformers used in these applications usually have power ratings in the 25 to 100 VA range. They have shell type construction and may be open types, for mounting in equipment enclosures, or may be designed for mounting on a standard electrical box.

A popular example of a small shell type transformer is the 120:24 V control transformer used in many HVAC systems. The transformer has a case with a hub or a set screw connector that fits into a standard 7/8" electrical box knockout (see Figure 7–11). This design allows the pair of 120 V leads to be brought out of the transformer directly into the connection box. This complies with the code requirements that the 120 V connections be made inside an approved enclosure. The 24 V terminals are left exposed because 24 V conductors energized from an approved Class 2 supply can be run as open wiring and terminated without an enclosure.

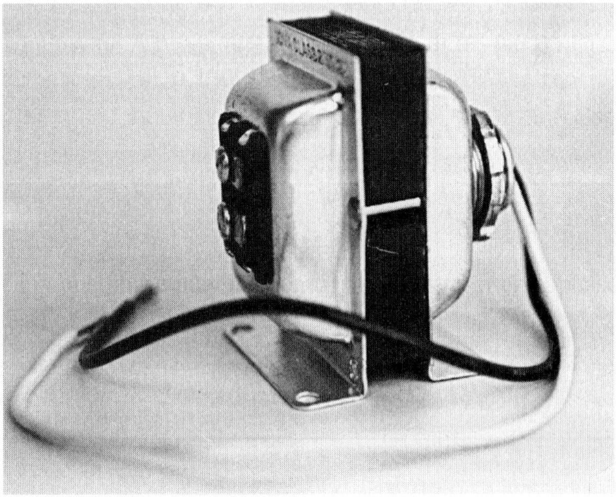

FIGURE 7–11 Enclosed, hub-mount control transformer.

Another common type of 120:24 V control transformer has a permanently attached mounting plate that is designed to fit, as a cover, on a standard 4" electrical box. As with the hub-mounted transformer, the 120 V leads pass through the mounting plate, where they naturally end up inside the box. The 24 V set screw terminals are exposed. This style of transformer can be mounted anywhere a suitable electrical box can be located.

Class 2 Transformers. Very small control transformers with output voltages of less than 60 V may have a Class 2 rating. These transformers are commonly used for signal circuits, such as doorbells or chimes, and for low-voltage control systems such as those used with HVAC thermostats or similar controls. Class 2 designations are defined in both the National Electrical Code and the Canadian Electrical Code. These definitions are quite broad, covering a much wider scope than the circuits discussed here. The definitions provided by the two codes are not identical, but most Class 2 transformers meet the requirements of both installation codes, and are listed by both Underwriters Laboratories Inc. (UL), and Canadian Standards Association (CSA).

The Class 2 designation is all about limiting the current and energy that the circuit power supply can deliver. Both the American and Canadian codes contain comprehensive regulations governing the installation of all types of energy-limited circuits. They define the voltage and current levels allowed, and set standards for the selection and installation of conductors and equipment. These rules are much more relaxed than for regular power circuits. By limiting the energy the circuit power supply can deliver, the wiring and components can be made lighter, simpler, and less expensive. The installation practices can also be less demanding, without exposing the building occupants to increased risk of electric shock or the hazards of fire. A Class 2 transformer is the most popular way of meeting the code requirements for current-limited circuits.

It takes very little energy to operate small coils like those in door chimes, furnace gas valves, or the relays and contactors used in small air conditioning systems. Such circuits can operate at low voltages and currents because of the limited energy demands of the load devices. However, it is not the nature of the loads that makes these circuits safe—it is the nature of the power supply. The power supply is critical because, even though normal operation would never cause a problem, there is always a danger of a wiring fault or equipment failure. A furnace or air conditioning control system, for example, contains a few low-energy loads, a thermostat with light-duty wiring and contacts, and the interconnecting wire and cable. The cable used to connect the thermostat to the other components is typically two or more thinly insulated #18 AWG copper conductors. This light-gauge wire may be stapled directly to the wood framing of the building, and then concealed by the wall covering. If the energy of this circuit is not limited to a low value, the fault current could easily exceed the safe limits of the cable and associated equipment, and the most likely result would be a fire.

Current can be limited by a separate overcurrent device, or by using a supply that is inherently current limiting. A supply that meets Class 2 requirements, with no additional components, is identified as an approved Class 2 source. This is what the Class 2 transformer designation means. Transformers that have a Class 2 rating are self-limiting: They do not require overcurrent protection beyond the fuse or circuit breaker provided by the primary-side power circuit—which is typically rated at 15 A. This breaker or fuse protects the primary supply conductors and reduces the damage in case there is a problem inside the transformer. Protection of the low-voltage equipment connected on the output side of the transformer is inherent in the transformer design. Even with a short circuit across the output terminals, the energy delivered by the transformer is below the maximum set by codes for a Class 2 circuit.

Class 2 transformers may sound like a great idea. They are easy to use and, because they are tamperproof, they simplify certification. They connect directly to a standard 120 V circuit on the supply side, and directly to the low-voltage control circuit on the load side. There is no need for additional fuses or breakers on either the input or output side of the transformer. All the code requirements for a Class 2 power supply are met with a simple, inexpensive transformer. Unfortunately, many Class 2 transformers cannot survive even a brief short circuit across the output. To comply with the Class 2 requirements, the transformer has an internal fuse link, or a winding that is so fine it simply burns out. This means the circuit is fused, but the fuse is built into the transformer.

Internal fuse links, or windings that fail when there is a problem in the circuit, do protect the circuit from damage and prevent a possible fire. However, forcing a transformer replacement where a fuse would have worked is not always the most logical approach. Some manufacturers do use external fuses, even with a Class 2 transformer. There are, no doubt, some service personnel who will spot the fuse and grumble about stocking replacements, but it just might be a better alternative than replacing transformers.

Warning

The Class 2 circuit approval assumes no changes to the power supply. If the transformer is replaced, the replacement must have the same ratings and approvals as the original. If there are two transformers in the circuit, they must not be interconnected. Two Class 2 transformers connected together almost certainly violate the code rules governing the installation. The energy available from the two interconnected transformers is approximately double that of just one. The power supply is no longer limited to the maximum value for which the approval was granted.

Class 2 transformers and example circuits are examined in detail in Chapters 15, 16, and 17.

Autotransformers. While most transformers have at least two windings, the autotransformer has only one. The supply and load share a common winding. There is no electrical separation between them. There is a connection at one end of the winding that is common to both the high- and low-voltage sides of the transformer. The second high-voltage lead connects to the other end, across the full length of the winding. The second lead of the low-voltage side connects at some midpoint, so the low-voltage circuit is connected across only part of the winding. As in any transformer, the volts per turn is equal along the entire length of the winding, so the two circuits operate at different voltages, because each is connected to a different numbers of turns (see Figure 7–12).

The autotransformer provides a voltage change, has a current ratio, and can be used to step the voltage up or down, just like a two-winding transformer. It also follows the same rules as a two-winding transformer. Not only is the voltage the same on each turn of the winding but, if the excitation current is ignored, the ampere turns are also equal in both sections of the winding. Part of the secondary current is conducted, and part of it is transformed. The current in the "secondary" portion of the winding is only the difference between the primary and secondary currents; the rest of the secondary current is

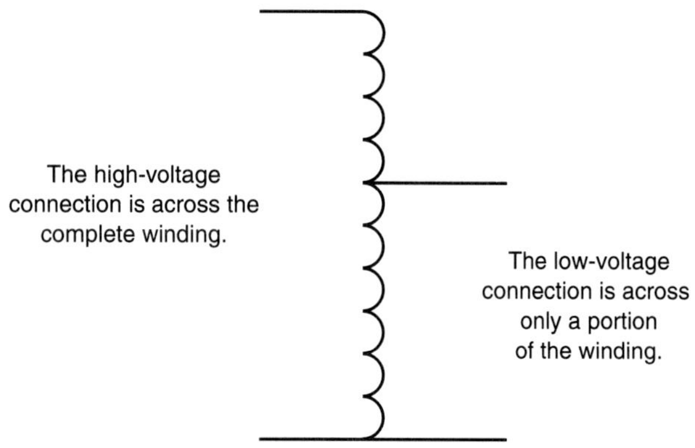

FIGURE 7–12 Autotransformers have one winding that is shared by the high- and low-voltage circuits.

conducted directly from the supply through the "primary" portion of the winding (see Figure 7–13).

Autotransformers are the most compact and efficient of all transformer designs; however, they do not electrically isolate the primary and secondary circuits, and for this reason cannot be used as a general-purpose transformer. Electrical codes are quite specific about where autotransformers are allowed. In general they are restricted to voltage adjustment on a circuit within a system, rather than permitted as a link between two different systems. They may also be used to adapt specific power loads to circuits with different operating voltages, or to reduce the start-up voltage and current to large three-phase AC motors. Autotransformers are also used to provide a limited amount of speed control for certain types of single-phase fan and blower motors.

Specially constructed autotransformers are commonly used as variable-voltage AC power supplies (see Figure 7–14). These devices are so similar in construction to large

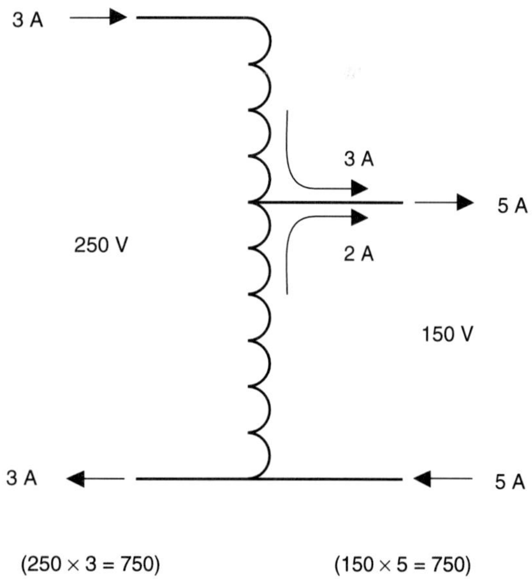

FIGURE 7–13 In this autotransformer diagram, the arrows show how the primary and secondary currents relate. The actual current flow is AC, so it flows in both directions.

FIGURE 7–14 A variable autotransformer is a popular variable AC supply.

rheostats that they are often incorrectly identified. They have a toroid winding, arranged very precisely on a cylindrical core. After the winding is completed, it is pressed down to conform to the special profile used at one end of the core material. This leaves a narrow strip of the winding raised above the rest, which is carefully machined to remove some of the insulating varnish and expose the copper conductor. Each turn of the winding is exposed, but none are shorted together. The exposed edges of the winding form a track for a connecting brush that rotates around a central pivot point inside the circular core.

Turning the adjustment knob rotates the brush so that more or fewer turns of the transformer are included in the secondary circuit. The voltage across each turn of the winding is a small increment of the total. This allows the secondary voltage to be adjusted from zero to the full voltage of the winding. Although the construction and operation of the variable autotransformer is much like that of a large wire-wound variable resistor, the winding is copper magnet wire, not resistance wire. The autotransformer is a much more efficient variable-voltage device than a resistor. It has only a very small amount of power loss, so it also produces very little heat. A disadvantage is that the transformer is an AC device: It cannot be used to adjust a DC input voltage.

Variable autotransformers are used as speed controls for small motors, for dimming commercial lighting circuits, and in automatic voltage regulators. They are very common in laboratories and electrical shops where they are used as a variable AC source. The laboratory-style supplies have a case with a power cord, a power switch, and either binding posts or a 120 V AC-style receptacle. The adjustment knob is often calibrated in percentage (0 to 100) rather than in volts. Many of them are connected to supply an extra 25 to 50% above the supply rating. Variable autotransformers are sold under various trade names, including Powerstat and Variac.

Autotransformers used for speed control of small motors may be continuously adjustable or have several tap connections. Taped windings are simpler and much less expensive than a fully adjustable type of design and, with an appropriate motor (such as a series universal, permanent split capacitor (PSC), or shaded-pole design), can provide multiple speeds. The PSC and shaded-pole motors can only be speed controlled if they are used to drive a variable torque load, such as a fan or blower. (Motors and motor controls are examined in more detail in Part III of this text.)

Autotransformers are also used to adapt power loads to different supply voltages. This application is often found in packaged HVAC systems such as rooftop units. The larger horsepower equipment—the air-conditioning compressors and the main blower motor—are operated at a relatively high three-phase voltage, such as 480 or 600 V, but the low-powered condenser fan motors may be single-phase permanent split capacitor (PSC) units designed for 208 or 240 V operation. These motors may be preferred for the condenser fans over an equivalent three-phase unit. They are cost competitive with a small three-phase motor, can be just as efficient, and use simpler, less expensive controls. They are also used throughout the industry in single- and three-phase units, so they are readily available. PSC motors also have much lower starting torque than a comparable three-phase motor, providing a gentle startup that is much easier on the fan and the motor mounts.

Note

The transformers used for voltage matching may be true autotransformers, or they may be two-winding transformers connected in a buck or boost configuration. Buck-boost connections, a variation of an autotransformer, are described below. These types of transformers are an attractive option for relatively low-ratio voltage adjustments. The resulting autotransformer is a fraction of the size and cost of a two-winding transformer that would serve the same purpose.

Buck-Boost Transformers. Buck-boost transformers are dual-winding transformers connected in a variation of the autotransformer connection. The voltage across the low-voltage winding is used to adjust the total output voltage by bucking (subtracting from) or boosting (adding to) the main supply voltage. They are usually used where the voltage change is relatively small.

In a buck-boost connection, the primary and secondary windings are connected in series (see Figure 7–15). This converts the two-winding transformer into an autotransformer but, unlike a normal autotransformer, one portion of the total winding—the one made up of the low-voltage coils—may be connected so that its voltage is added to the high-voltage portion, or the connections can be reversed so it will subtract.

For any ratio except 2:1, the current in the two sections of the autotransformer winding will be different. For a ratio of less than 2:1, the maximum current is through the portion of the winding that is in series with the load—the low-voltage portion of the winding. The maximum current the transformer can supply is therefore based on the current rating of the low-voltage portion of the winding. This normally works out to a much greater load than would be allowed if the transformer were used in a two-winding mode.

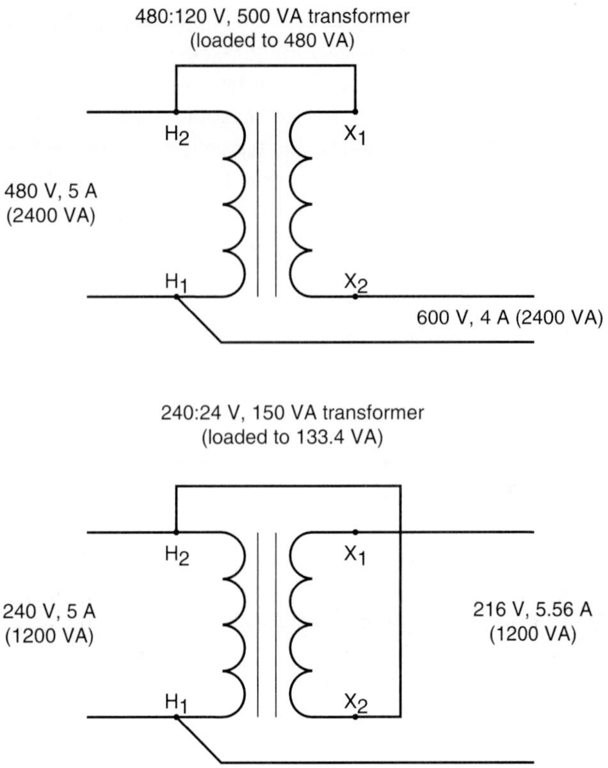

FIGURE 7–15 These examples of a boost connection (top) and a buck connection (bottom) show how a small transformer can be used to adjust voltage for relatively high VA loads.

The load kVA can be much higher than the kVA rating of the transformer, without exceeding its capacity. A surprisingly small transformer can handle the voltage adjustment requirements of a relatively large load.

Transformer Ratings

Transformers provide a ratio change of voltage and current. These ratios are set by the numbers of primary and secondary winding turns and, in theory, the turns ratio and voltage ratio are exactly the same. In practice they are quite close—different only because of internal losses. In theory the voltage ratio is exactly in inverse proportion to the current ratio. This means that the multiple of primary voltage and current should equal the multiple of secondary voltage and current. In practice this does not work out very well, particularly at low loads, because it ignores the excitation current that is required to create the magnetic field. The transformer is a very efficient device, particularly in larger sizes, but trying to prove this from measured current gives very poor results.

In general, all transformers are rated for a maximum voltage, and either full-load current or volt-amps (VA). Volt-amps is the unit of apparent power and is, quite literally, volts times amps. A transformer's VA rating is based on the rated secondary voltage and the rated full-load secondary current. If the nameplate has a VA rating instead of a current rating, the full-load secondary current can be calculated by dividing the VA rating by the rated secondary voltage. The transformer's VA rating is an output value, so secondary values, not primary, are used for the calculation. This holds true regardless of whether the transformer is used in a step-up or step-down application.

$$I_S = \frac{VA}{E_S}$$

I_S = Rated full-load secondary current

E_S = Rated secondary voltage

VA = The transformer nameplate volt-amp rating

When the nameplate voltages are listed specifically as primary and secondary, the transformer manufacturer has probably adjusted the turns ratio to compensate for the inevitable internal voltage drops that occur under load. The nameplate secondary voltage is based on full-load operation, so the actual output voltage will be slightly higher at lighter loads. Most small transformers are built this way, because they have proportionally higher internal losses than larger ones. Manufacturers often list the transformer no-load or open circuit voltage in addition to the full-load rating

When the transformer has listed high-side and low-side voltages, but they are not identified as primary and secondary, the ratio may be as stated. The no-load voltage will be close to the rated value and will drop off slightly as the load increases. A transformer like this can usually be operated in a step-up or step-down configuration with predictable results. A step-up transformer is connected to a low-voltage supply and has a higher voltage output; a step-down transformer reduces a high-voltage supply to a lower one. All transformers are reversible, but the supply voltage used must not exceed the rating of the winding to which it connects. As with other electrical equipment, maximums must be respected. A lower voltage can be used—but not higher.

Note

The VA rating of a transformer assumes full rated voltage. With a lower than rated primary voltage, the VA rating must be reduced proportionally. The limiting factor is current, not volt-amps. The normal maximum full-load current rating—taken from the nameplate or found from nameplate values—is the maximum current rating at any voltage. Current cannot be increased to compensate for reduced voltage.

The full-load current rating of the transformer can be averaged somewhat, because it is based on internal heating. Most transformers operate with a duty cycle that includes times of overload and times of less than full load. Modest short-term overloads are usually not a problem. More extreme currents, such as those caused by a short circuit on the secondary side, will not harm a standard transformer if it is protected by a fuse or breaker with a suitable rating. A Class 2 transformer may be destroyed by a single overload or short, because it may be protected by an internal fusible link.

Transformer Grounding

Grounding at transformers involves more than the normal safety grounds that connect to the metal frames and enclosures of the equipment: It also involves a ground connection to the energized portion of the power circuit. It is this connection that is the least understood, and therefore the most likely to be incorrectly applied.

Almost all AC power supplies have one point of the circuit connected to an earth ground. This is true of both the high-voltage distribution system and the lower voltages supplied for use inside buildings. These ground connections are added to control the level of the voltage to ground, and to prevent a transformer failure from accidentally energizing the low-voltage side directly from the high-voltage source. As a result of this practice, even normal power outlets in buildings have a specific voltage relationship to the ground.

When the system has a neutral conductor, it comes from the point where the transformer secondary is grounded. Because the ground and neutral are connected together, the voltage between ground and the other line conductors of the system is the same as their voltage to neutral.

When the transformer primary is supplied from a circuit that is also grounded—which includes almost all possible AC systems—the primary winding also has a specific voltage to ground. The effect this relationship has on the secondary side depends on the design of the transformer. The primary and secondary windings of a conventional two-winding transformer are not conductively connected to each other. However, unless the transformer is specifically made for isolation purposes, this electrical separation is not quite total. There is a small amount of capacitve coupling between the closely spaced windings that tends to raise the voltage to ground on the secondary winding. This does not affect the voltage between the terminals of the secondary—only the voltage from the secondary terminals to ground.

In isolation transformers, a grounded metal screen or plate is placed between the two windings as an electrostatic shield. This stops any capacitive coupling and makes the isolation complete. This type of transformer can safely provide an ungrounded supply. Ungrounded supplies that are properly isolated are safer than grounded ones, because it is impossible to get a shock from either of the circuit conductors to ground. The voltage from line to line is still as dangerous as before, but the far more likely accidental contact from either line to a source of "ground" is no longer a hazard. Unfortunately, this is not practical on a large scale, but it can greatly reduce shock hazards in specific situations.

Isolation transformers are an attractive option where the likelihood of a shock to ground is high, but the electrical loads are relatively small. They have been widely used to power bathroom razor outlets and in some hospital equipment. They are also common in many electrical labs, where they are used to reduce the shock hazard and to simplify the use of oscilloscopes in AC circuits. Most oscilloscopes have one of the input leads—the ground clip—solidly connected to the oscilloscope frame and through the supply cord to the electrical ground system. This makes is difficult to connect the scope into power circuits with conventional grounded supplies. An isolation transformer in the supply to the circuit under test eliminates the voltage to ground, and allows the oscilloscope ground clip to be connected into one point of the circuit. Isolation transformers usually have 1:1 ratios. A typical laboratory supply, for example, has a 120 V rating on both the primary and secondary windings.

Transformer-isolated razor receptacles are now being replaced with outlets protected by a ground fault circuit interrupter (GFCI). The downside of isolated razor outlets was that the transformer had to be very small to be cost effective, so the outlet could only supply very low powered loads. The newer GFCI-protected outlets provide a very high degree of safety against shocks to ground, and can supply power to high-wattage loads such as handheld hair dryers and curling irons.

Conventional transformers do not have electrostatic shields between the windings. Unless the secondary of the transformer is grounded, the capacitive link between the supply and secondary windings may create a voltage between the secondary winding and ground. In theory this electrostatically induced voltage can rise as high as the primary voltage to ground. In practice it depends on the layout of the transformer windings and is quite unpredictable. Personal experiments conducted by the author using small transformers with 120 V primaries produced voltage-to-ground readings ranging from less than 2 V to over 70 V. These readings were obtained from a random sample of several transformers of the same type, size, and model number.

Connecting the transformer secondary to ground at one point only does not cause a problem. The capacitance coupling the windings is small enough that it produces virtually no current flow. All that happens is the secondary winding and its associated circuits become grounded at that point, making the transformer output into a grounded supply.

The NEC and CEC rules are specific about grounding power supplies such as transformers, and in most normal installations the secondary of a conventional transformer has to be grounded. There are specific allowances for true isolation transformers and specific situations where both the primary and secondary voltage are quite low. For example, transformers with secondary ratings of less than 50 V, connected to a supply that is at less than 150 V to ground, may not have to be grounded.

Note

The 24 V control transformers used in control systems such as those found in HVAC systems do fall outside the code requirements for grounding. Without a specific ruling requiring a ground, the decision to ground or not to ground is left to the discretion of the equipment manufacturer. The result is every possible combination:

- Some systems are not grounded. Ungrounded 24 V control circuits are particularly common in older standing pilot furnaces, with or without air conditioning.
- Some systems are grounded in a conventional manner, with an intentional ground connection at or near the transformer. This ground is usually quite obvious: It is color-coded green, in keeping with the colors assigned by codes, and is marked on the control drawings.
- Other systems become grounded by the nature of the control system to which the transformer connects. Many electronic ignition systems have one of the 24 V terminals connected to the frame of the furnace. This ground connection is required by the control as part of the ignition and flame detection circuits, but has the additional effect of grounding one lead of the 24 V supply transformer. This effectively converts the 24 V system from an ungrounded to a grounded one.

When the ground is through the control module, it is important that no other ground connection is installed. This would ground the circuit at two points, instead of just one. Two different ground points are connected together through the metal of the furnace frame. This allows the secondary current to bypass the interconnecting conductor and pass through the furnace frame instead. This error may or may not cause problems, but it is undesirable.

Warning

When one side of a circuit is grounded, it is very important that no control devices are installed on that side of the circuit (see Figure 7–16). The reason is that an accidental ground can bypass controls that are placed in a grounded line, a potentially hazardous condition. In a normal system, the controls are all connected in the ungrounded line. An accidental ground in that line will short out the circuit and cause the system to fail to a safe condition, rather than a dangerous one.

FIGURE 7–16 Installing controls in a grounded line can be extremely dangerous.

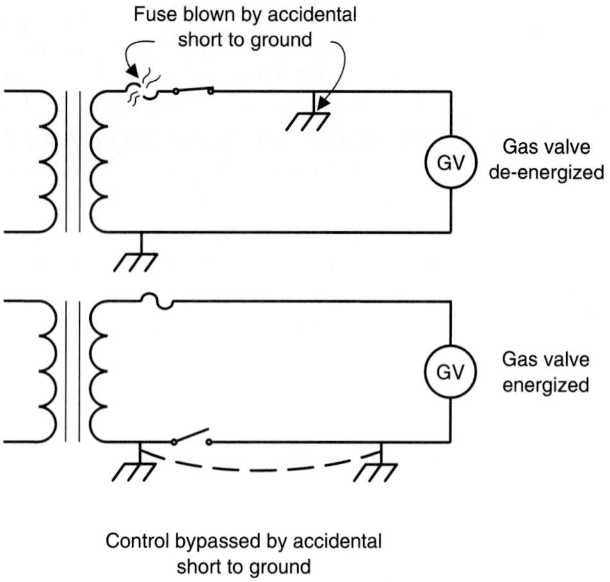

Summary: Grounding Transformer Secondaries. There are several reasons for grounding transformers that have relatively high primary voltages. First, the voltage at which the secondary may float—if it is left ungrounded—can be high enough to damage insulation, and may be a serious shock hazardous. Second, if the transformer should fail, the result can be catastrophic. A fault that connects the primary and secondary windings can apply full primary-side voltage to the secondary-side loads. Without a solid secondary ground, this voltage seeks a path through the secondary-side equipment. An electrical accident like this can seriously jeopardize both equipment and personnel.

In addition to these potential hazards, conductors run outdoors are also exposed to lightning. To protect the building and its occupants, the supply utility grounds both the case and the secondary winding of the transformer, and installs lightning arrestors. The arrestors conduct high voltages to ground before they enter the transformer. This temporarily shorts out the primary supply and trips the closest fuse or other protective device. The power goes out, but no one is hurt. If the arrestors fail, a solid lightning strike destroys the transformer, but the lightning finds a relatively safe path through the secondary ground conductor, instead of a potentially disastrous path through the consumer's building and equipment.

Although the situation is much less dramatic, most indoor transformer secondaries have to be grounded for similar reasons. Once this ground is established, the secondary side circuits have a grounded neutral or common. The grounded line should not contain any control devices because of the danger that they may be bypassed by a simple ground fault.

Grounding is not strictly required in low-voltage supplies that do not have excessive voltages on the primary side. This is very common is 24 V Class 2 circuits. Be aware, however, that these circuits may be grounded or not grounded at the discretion of the manufacturer.

Power Flow in Transformer Circuits

Transformers connect two different electrical circuits that, in most cases, operate at different voltages and currents. In terms of energy flow or power transfer from one circuit to the other, the transformer allows the two circuits to behave much as if they were solidly connected. Although the relative voltages and current levels are different on each side of the transformer and are electrically separate from each other, the energy flows from one circuit to the other with very little interference and only a small amount of loss. The transformer has no preference for the direction in which the energy moves. It can flow equally well from the low-voltage side to the high-voltage side, or from high to low. The actual direction depends on the prevailing conditions of the circuit and can change instantly if those conditions change.

The most common transformer connection is a power supply on one side, and one or more load devices on the other. In-phase power—the true power in watts—flows in one direction only, from the supply through the transformer to the load. Reactive power flows back and forth between the supply and the reactive load, much as it would with a direct connection. In a normal circuit, with no transformer between the loads and the supply, a reactive load absorbs power during one part of the cycle and returns it to the supply on the next. Placing a transformer between the supply and the reactive load does not stop the two-way flow of energy. It has some effect—the coils of the transformer add reactance and resistance of their own—but the fundamental operation of the circuit is the same.

The net result of the transformer's ability to transfer both true and reactive power in either direction is that the supply "sees" the load through the transformer, almost as if the transformer was not there. With no load, or very light loads, the transformer's inductive reactance dominates, but as the load on the transformer is increased, it becomes less and less significant. At full rated load the phase angle between the voltage and current on the supply side of the transformer is essentially the same as it is at the load.

The supply must provide both the reactive and in-phase current, and the transformer must carry them to the load. Although the current is split into these two components for the purposes of calculation and description, there is only one current flowing in the lines and windings of the transformer. This single value of current is a phasor sum of the two out-of-phase components. It is this combined current that the circuits and the transformer windings must carry. This is the reason why transformers are always rated in volt-amps rather than in watts. The conductors must carry all the current, not just the portion that registers on a wattmeter.

While less common, it is also possible for the circuits on both sides of the transformer to have power supplies. The transformer then interconnects two sources. This is generally not done in smaller buildings, but it is common in utility distribution, large commercial and industrial buildings, and other situations where continuity of service is critical. Multiple sources with multiple feeders are used to supply common loads. The transformers in the system may be interconnected on both the primary and secondary sides. As a result, either side can function as the source, and de-energizing the transformer involves more than a single disconnect point. For the safety of workers and continuity of service to the building, special protective measures and equipment isolation procedures are required.

Transformer Connections

Standard terminal markings are not always used for very small transformers, but are quite consistently applied in larger ones. The high-voltage leads or terminals are identified with the letter "H," the low-voltage identification is "X." In addition to the letter, each connection point is also given a number to identify its position in the winding and its polarity relative to the operation of the other windings. In general-purpose transformers, the primary and secondary are not usually identified as such. The transformer can be used for step-up or step-down operation; therefore, the windings are identified only as high- and low-voltage, H and X, respectively. The voltage ratings of the windings are marked on the nameplate.

In a simple two-winding transformer, the leads are marked H_1 and H_2, X_1 and X_2. The polarity of the windings, which is indicated by the numbers, can be described in several different ways. One of the easiest to remember is that, as far as possible, the terminals marked with the number "1" act as if they were connected together. For example, assuming a typical step-down application, X_1 acts like an extension of H_1. When current is flowing out of X_1 to supply the load, it will be flowing into H_1 from the supply. The polarity of the voltage that appears across the secondary, from X_1 to X_2, will therefore be the same as the polarity of the supply voltage from H_1 to H_2 (see Figure 7–17).

Transformer polarity has a number of unexpected effects that become most evident when the primary and secondary windings are interconnected. This can be a direct connection, such as in buck-boost applications, or it can be through a shared ground system. For example, if both the primary and secondary circuits are grounded on one line, the grounded terminals are effectively connected together through the ground system. Which terminals are chosen for grounding will affect the voltage between the ungrounded terminals. When two low-numbered or two high-numbered terminals are grounded, the voltage between the ungrounded pair is the difference between the primary and secondary voltage. When the grounded terminal is the high number on one side and the low number on the other, the voltage between the two ungrounded leads is the sum of the primary and secondary voltages (see Figure 7–18).

FIGURE 7–17 The "polarity" of AC voltages depends on the reference. Comparing the secondary current to the primary, X_1 is always an extension of H_1. Comparing voltages, the secondary may be in phase, or "inverted" to 180° out of phase, by interchanging the leads.

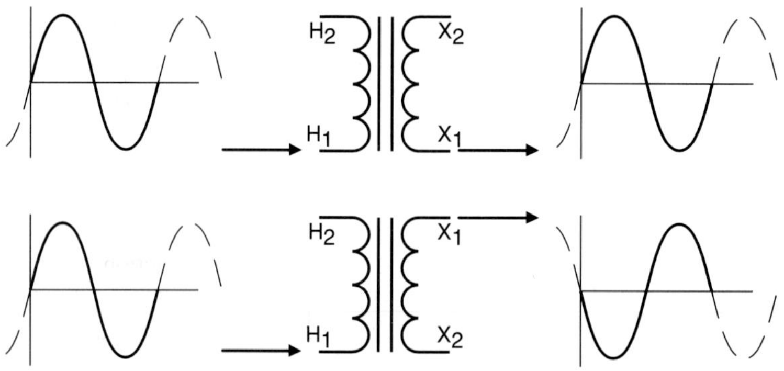

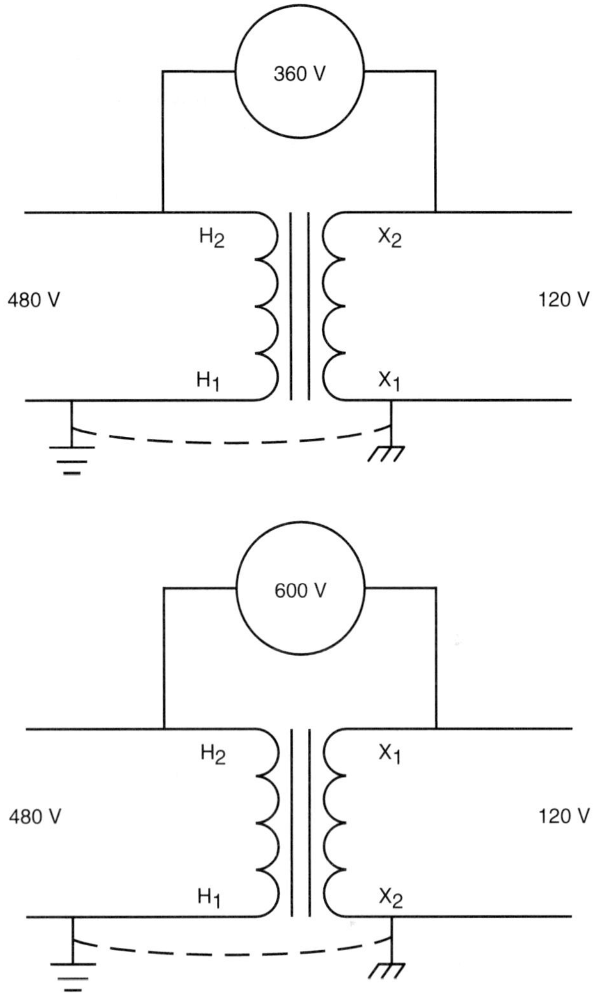

FIGURE 7–18 The voltage between the ungrounded leads of the primary and secondary is either the sum or difference in their voltages. Which voltage occurs depends on the position of the secondary ground.

The voltage between conductors of different circuits may seem unimportant, but it can affect insulation stress in any system where both high- and low-voltage circuits are connected into the same equipment or the same printed circuit board. This happens in many HVAC systems where the 120 V power circuits and the 24 V control circuits share the same circuit boards or control modules. Testing for correct transformer "polarity" is described in the installation and service literature of at least one major HVAC equipment manufacturer. The "correct" connection of a 120:24 V control transformer will produce a voltage of approximately 120 – 24 = 96 volts between the ungrounded 120 and 24 V leads. An incorrect connection will yield 120 + 24 = 144 V. If the transformer leads or terminals are not identified, this can only be determined by trial and test. On some circuit board systems, if this connection is incorrect, the circuit will not function.

Transformer winding polarity also affects the output of buck-boost connections. In these connections the two windings are connected in series. If the connection order is the same as the numeric order—H_1, H_2, X_1, X_2 (H_2 connected to X_1)—the two voltages add, and the connection boosts the output voltage by the secondary voltage value. If X_2 is connected to H_2, the secondary voltage is deducted from the primary, and the transformer is connected to buck, or reduce, the output voltage.

Transformer polarity can be important in two-winding transformers, but when a transformer has more than two windings, it becomes critical. For example, many single-phase power transformers have dual low-voltage windings, and sometimes dual high-voltage windings as well. This makes the transformer more flexible to use, because the number of connection options is increased.

The additional windings are marked H or X as appropriate, with sequential numbers to identify the lead pairs and their polarity. A second high-voltage winding would be marked H_3 and H_4, a low-voltage winding would be marked X_3 and X_4. The polarity of the 3–4 windings is identical to that of the 1–2 windings. When the polarity of X_1 is positive, the polarity of X_3 will also be positive.

The pairs of windings are also identical in other respects. They have exactly the same number of turns made with the same size of wire. To ensure they also experience the same magnetic forces, they are normally "bifilar" wound, meaning they are wound as a pair of conductors laid side by side, instead of two separate windings laid in separately. When the transformer is in operation, the voltage across the winding pairs is identical.

Dual windings permit a variety of connections. For example, a transformer with one 480 V and two 120 V windings can be connected as 480:120, 480:240, or 480:120/240—a dual-voltage three-wire connection (see Figure 7–19). It can also be used to step up voltage from 120:480 or 240:480. Properly done, each of these connections can carry the full rated kVA of the transformer.

When the transformer also has dual high-voltage windings, the number of options is further increased because similar connections are possible on both sides. Extending the example in Figure 7–19 to a transformer rated at 240/480:120/240, the transformer primary can now be connected for 240 or 480 V operation instead of just 480 V.

In every case, the windings are connected in series for high voltage, and parallel for low. The dual-voltage three-wire connection only applies to the secondary side, and consists of a series connection for the windings with outgoing leads from the "center" connection point as well as from the outer ends. Correct polarity must be used when connecting the windings. Series connections are in numeric order with X_2 and X_3 together. Parallel connections have X_1 to X_3 and X_2 to X_4. The same logic applies to dual windings on the high-voltage side (see Figure 7–20).

Many transformers are equipped with taps—extra connections brought out from some point along the winding—to provide optional connections. There are three common configurations for transformer taps, and several variations, including center taps, voltage adjustment taps, and taps, intended to match multiple standard voltages. Center taps are almost always on the low-voltage side. They are supplied on some transformers intended to supply 120/240 V circuits, and at a variety of voltages on small transformers designed for "rectifier" duty as part of AC-to-DC converters.

Voltage adjustment taps are provided on many power transformers to allow small increments of voltage correction. These are usually at some fixed-percentage interval, such as 2.5 or 5%, and are normally on the high-voltage side. On a typical step-down application, selecting a lower tap—less of the total high-voltage winding—raises the output voltage.

FIGURE 7–19 The connection options for a 480:120/240 V transformer with dual secondary windings.

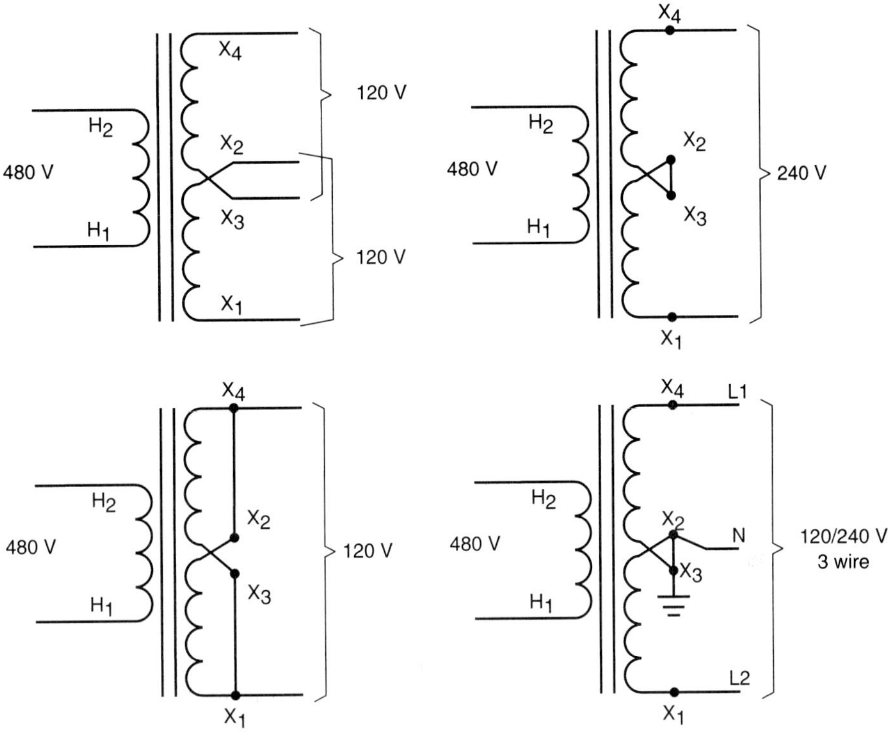

Taps for multiple standard voltages are only used on relatively small transformers, such as the control transformers used in some motor control systems. The transformer can, for example, be wound for use in a 600:120 V step-down mode, with additional tap connections for 480, 240, and 208 V. One transformer can be used with any of these standard industrial voltages. Properly connected, it provides the required 120 V output in each application, but a word of caution is warranted here.

FIGURE 7–20 The connection layout for a transformer with dual primary and secondary windings allows for a variety of different connections.

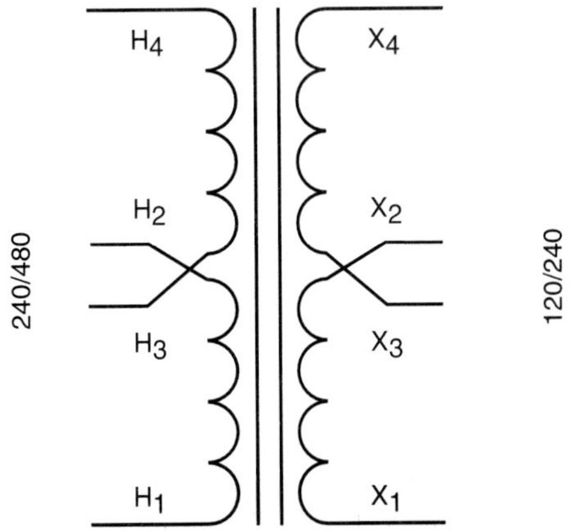

Warning

The unused taps of any transformer winding are energized whenever the transformer is energized, usually at the voltage rating of the tap. For example, if the multivoltage (600, 480, 208:120) transformer described above was connected for 208:120 V operation, each of the other primary side taps would be at its rated voltage, measured from H_1 on the primary side. Therefore, if the transformer has open leads, they must be insulated from each other and from ground. If the transformer has terminals instead of leads, there must be no connections to any terminals except the ones connected for normal use. Accidental contact with these terminals can give a serious shock.

There may be some variations in lead numbering for taps, but there are a number of standard layouts. The transformer nameplate should supply the required information for proper connection. In most cases the numbering is sequential, with the low-numbered terminal—typically H_1—acting as the common. Assume, for example, a 480:120 V transformer with four voltage adjustment taps (see Figure 7–21). The start of the primary winding would be marked H_1, the taps would be numbered H_2, H_3, and H_4, and the final connection would be H_5. One supply lead always connects to H_1; the second connects to the most appropriate of H_2 to H_5. One of the taps, quite likely H_4, will be rated at the nominal voltage of 480 V. If the taps are at 5%, they will be rated for 85, 90, 95, 100, and 105% of 480 V. This may show on the nameplate as voltages of 408, 432, 456, 480, and 504 V, associated with taps H_2, H_3, H_4, and H_5, respectively.

The tap rating should be interpreted as appropriate for an input of that voltage. For example, if the supply voltage is actually 432 instead of the rated 480, connecting the supply between H_1 and H_3 will result in the rated 120 V secondary output, even though the supply voltage is well below the nominal. In a real situation, the available voltage will naturally vary somewhat, and is unlikely to be an exact match for any tap rating. Monitoring the system during a normal load cycle and then selecting the closest tap is usually the most satisfactory approach. Choosing a tap that is slightly below the actual will give an output that is slightly high. This tends to compensate for the natural voltage drop that occurs under load.

FIGURE 7–21 This winding layout is typical for transformers with voltage adjustment taps.

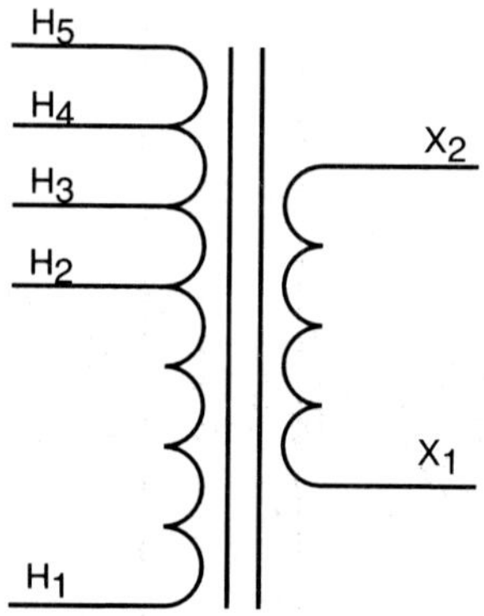

7.3 Services

Note

Some two-phase systems are still in use in areas where extensive generation and distribution systems were established in the early stages of electrification. In these areas, the service voltages and, of course, the number of phases supplied are different than anywhere else. These systems can survive, even with a mix of two- and three-phase equipment, because the two systems can be interconnected with appropriate transformers.

The portion of the electrical system that extends from the utility connection point to the load side of the main service switch or breaker is generally referred to as the consumer's service. The supply voltage and type of supply—single- or three-phase—is based on customer requirements, utility policies, and availability. If no three-phase distribution is available, such as in many remote areas, the utility will not offer three-phase as an option. Residential areas may or may not be serviced with three-phase. Commercial and industrial districts usually have the widest array of choices and the most need for them.

The variations in services include the service voltage, the kVA capacity, the number of phases provided (single- or three-phase), and the connection arrangement. Single-phase services can be two- or three-wire, with three-wire almost universal in North America. Two-wire is used in Europe and in similar systems around the world. Three-phase may be three- or four-wire. Single-phase services may be supplied from a single-phase transformer or from a three-phase supply. These two systems behave somewhat differently, provide slightly different voltages, and have different code requirements and restrictions.

The utility service transformer and its connection arrangement define the consumer's service.

Single-Phase Services

The most common North American single-phase service is supplied from a single-phase transformer with either dual 120 V secondary windings or a single center-tapped secondary rated 120/240 V (see Figure 7–22). In both cases, the midpoint of the secondary is grounded at the supply transformer, and again at the consumer's service disconnect. This grounded conductor is the system neutral. This service provides the flexibility of dual voltages, with the safety of a relatively low voltage to ground. The higher 240 V connection is used for power loads such as heating or larger motors, while the 120 V connections are typically used for lighting and receptacle circuits and loads of 1500 W or less. The 120/240 V three-wire connection is flexible and efficient. The shared neutral is often reduced in size, because it carries only the difference between the currents on the other two lines.

Three-Phase Services

Three-phase services are either delta or wye connected. Both of these systems have advantages and disadvantages. The delta lacks a natural common point and is sometimes not grounded. Wye systems are very symmetrical, making it easier to equalize the single- and three-phase loads and keep the system in balance. They also have a natural common point

FIGURE 7–22 A typical 120/240 single-phase service provides 120, 240 and 120/240 V circuits from the same panelboard.

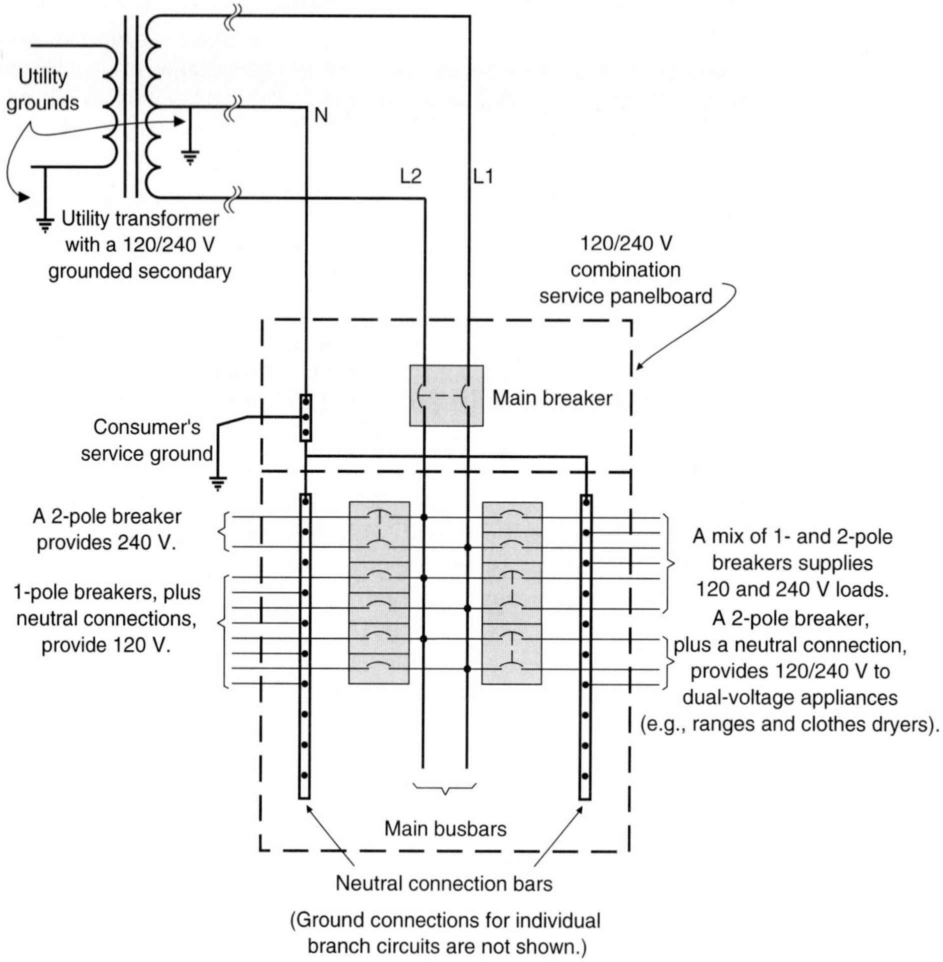

that is very suitable for grounding. Both systems are in wide usage throughout the world. Wye is probably more common, but it is certainly not universal.

Delta Services. A delta service may be supplied from a single three-phase transformer, or more commonly from two or three single-phase transformers connected as a three-phase bank (see Figure 7–23). The primary connection can be delta or wye, because the connection on the primary side does not dictate the connection of the secondary. Three-wire delta services are usually supplied at the standard industrial voltages of 240, 480, 2400, or 4800 V (see Figure 7–24).

Three-wire delta-connected transformer secondaries often do not have a solid ground connection. The AC capacitive coupling from each phase to the metal conduits and enclosures creates an artificial "ground reference." This, plus the effect of the ground fault detectors that are normally mandatory for such systems, may give the impression a solid

FIGURE 7–23 The larger single transformer on the left supplies a single-phase 120/240 V service. The three smaller ones on the right are connected in delta on the secondary side to provide a three-phase three-wire industrial service.

ground does exist. The voltage-to-ground readings may be much the same as for a wye connection with a similar line voltage.

This line-to-ground voltage cannot be used to supply any load. A connection from any line to ground is a ground fault and will appear on the ground detectors as a problem. This can be a cause for serious concern. In ungrounded delta systems, a real ground fault only acts to ground the system; it does not cause a fuse to blow, a circuit breaker to trip, or any

FIGURE 7–24 A three-wire delta secondary has the windings connected head to tail to form a closed loop. Compare this to the photo in Figure 7–23; most of the secondary connections are readily visible.

Various connections are used on the high-voltage side, including a grounded wye.

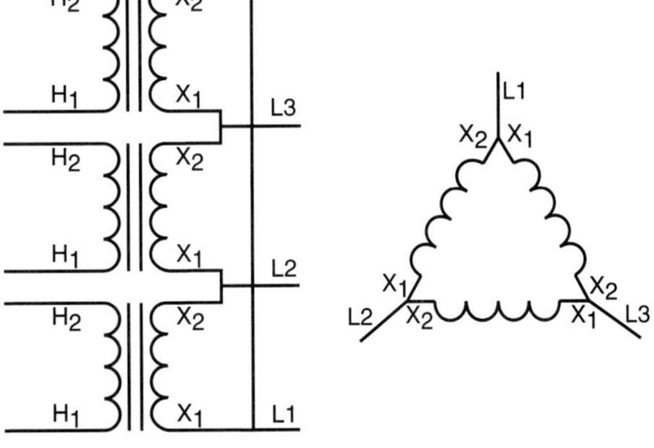

other kind of circuit failure. The first ground fault does virtually nothing, but a second fault can be catastrophic.

A second fault—on a different line—can cause major damage. The two faults are almost certain to be a different points in the building or plant. They connect two different supply lines to widely separated points of the building and its equipment. The fault current will flow between these two points, passing through any possible combination of wiring, metal frames, building structure, or other equipment anywhere in the plant. The end result is completely unpredictable, but it can be dangerous and is potentially destructive.

The four-wire delta connection supplies the most unusual electrical service. A typical installation of this type provides 240 V three-phase three-wire, and 120/240 V single-phase three-wire, from a common three-phase utility connection. The service is usually supplied from two (or sometimes three) single-phase transformers. The primary of each transformer is connected to the three-phase high-voltage line, just as if the third transformer were also installed. The secondary of one transformer is connected to provide 120/240 V single-phase three-wire, the other transformer is connected for an output of 240 V. When there is a third transformer, it is also connected for 240 V, and the three secondaries are connected in a 240 V delta three-wire configuration.

When there are only two transformers, the connection starts out the same, but the third transformer is missing. The side of the delta that would have been completed by the third transformer is left open. This connection is known as an open delta (see Figure 7–25).

The three-phase 240 V output is available from the line outputs of the two or three transformer secondaries. The center tap of the 120/240 V secondary is grounded and

FIGURE 7–25 These two transformers form an open delta connection providing a mix of 120/240 V single-phase and 240 V three-phase through a three-phase four-wire service. The larger transformer provides all the single-phase power, plus its share of the three-phase.

FIGURE 7–26 The three-phase four-wire delta service can be provided with three transformers in a closed delta, or only two connected in open delta.

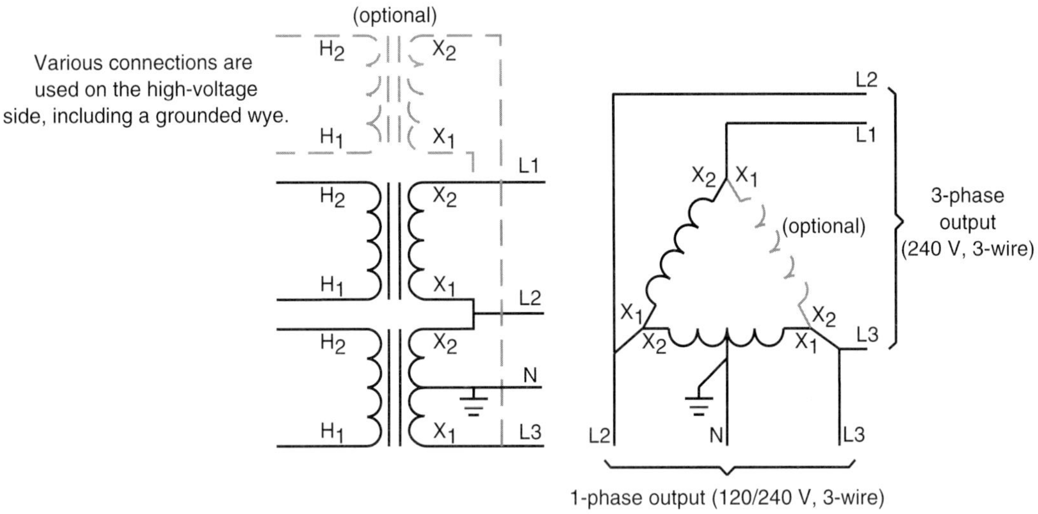

brought out as the neutral of the three-wire single-phase supply. This is the fourth wire of the three-phase four-wire connection. It is not a three-phase neutral, it can only be used with the single-phase loads that the service supplies (see Figure 7–26).

When most of the load is single-phase, the transformer supplying those loads is usually much larger, and only two transformers are installed. When the service supplies a large three-phase load, three transformers may be used to improve performance and increase the total kVA rating.

A four-wire delta service of the type just described can be identified by its separate three-phase and single-phase panels or switches, and by the voltages to ground. Unlike a wye system or an ungrounded delta, there are specific but unequal voltages between the three main lines and the ground or neutral. These readings are 208 V from L1, and 120 V from L2 and L3. Once popular, and still common in existing installations, this service is generally giving way to the 120/208 V three-phase four-wire wye.

Wye Services. The three-phase wye connection finds favor for its naturally symmetrical ground point and a natural tendency to balance multiple single-phase loads. The only downside to the three-phase wye is the "odd" voltages that result from the 120° phase shift between the phase voltages. The popularity of these systems has made these voltages standards of their own, but because they do not match other common voltages, it has complicated equipment manufacturer and selection.

Common wye service voltages include 120/208, 277/480, 347/600, 2400/4160, and 7.2 kV/12.5 kV. Others are used but, at present, these are some of the most popular North American voltages. In all wye connections the lower voltage is only available as single-phase. For services up to 600 V, both single- and three-phase can be taken from the same panelboard or from a common splitter. There is no need to isolate the single-phase from the three-phase, a definite advantage over a four-wire delta.

FIGURE 7–27 This three-phase transformer supplies a four-wire wye service at 120/208 V. A three-phase transformer is cheaper and more compact than multiple single-phase units.

Note

A common standard notation for three-phase wye systems lists the higher voltage first, followed by the letter Y to indicate the connection, and then a forward slash separator and the lower single-phase voltage; for example, 208Y/120. A different approach is to reverse this to 120/208Y. The Y may also be dropped as redundant—only a Y connection can create such voltages—leaving simply 120/208.

A 120/208 V three-phase supply is common for services to apartment blocks and modest commercial and industrial customers (see Figures 7–27 and 7–28). It provides a three-phase voltage usable for motors up to at least 200 Hp, and either 120 or 208 V single-phase for other outlets. Lighting may be 120 V incandescent, and either 120 or 208 V for ballasted lamps such as fluorescent, mercury vapor, metal halide, or sodium vapor. When 240 V are required for resistance heating loads such as apartment ranges and clothes dryers, 208 V is usually substituted, with no change in the equipment. This is lower than the voltage the equipment is designed for but, except for a 25% reduction in the heat output, there are no adverse effects. Single- or three-phase motors must be rated for 200 or 208 V, not just 230 V. Note that many, but not all, 230 V motors are also rated for use at 208 V. Motors with these ratings are readily available.

A 277/480 V supply is common in much of the U.S., while the 347/600 supply is increasingly popular in commercial and industrial installations in many parts of Canada. Motors rated for use on 480 and 600 V three-phase supplies are available from fractional Hp ratings to 400 Hp or more. The relatively high 277 and 347 V single-phase available from these services can efficiently supply the hundreds of ballasted lamp fixtures used in industrial and commercial installations.

Services with these voltages cannot directly supply the 120 V single-phase outlets that are required in all North American electrical installations, so one or more step-down transformers are installed in the building. These transformers are connected much as if they were a utility service. The primary ratings match the available supply voltage—600 or 480 V, for example—and the transformer secondaries are connected to provide a grounded 120/240 or 120/208 V supply. Single-phase types are usually connected to a single-phase 120/240 V three-wire panelboard. Three-phase units are connected to supply a 120/208 V three-phase panelboard. Either system can supply single-phase 120 V outlets in addition to other 240 or 280 V loads.

FIGURE 7–28 Three-phase wye services can also be provided from multiple single-phase transformers. The secondary connections can easily be identified as a four-wire wye.

Single-Phase Services From Three-Phase Supplies

North American power utilities, unlike much of the rest of the world, supply most single-phase services from single-phase transformers. The transformer secondary is center tapped, and the center tap becomes a grounded neutral. This system provides both 120 and 240 V, with the shock hazard limited to 120 V from any line to ground. In rural areas, a single transformer is used for each service, sized to meet the demands of an individual consumer. In urban areas with concentrations of individual homes, transformers usually supply a group of services that are close together. When an individual consumer has sufficient demand or some unique requirements, a dedicated transformer is used for that service.

In European-style distribution, larger three-phase transformers rated 240/415Y or 220/380Y are used to supply both residential and small commercial, institutional, or industrial customers with a mix of 220 or 240 V single-phase and 380 or 415 V three-phase. This system uses fewer transformers with a more extended low-voltage distribution system. It has a higher voltage to ground—220 or 240 V—so the hazards of shock or electrocution are greater.

FIGURE 7–29 A single-phase three-wire network connection can be supplied in place of conventional 120/240 V single-phase. Equipment using the higher voltage receives only 208 V instead of 240. Other differences include changes in the utility meter connections and code restrictions on neutral sizing.

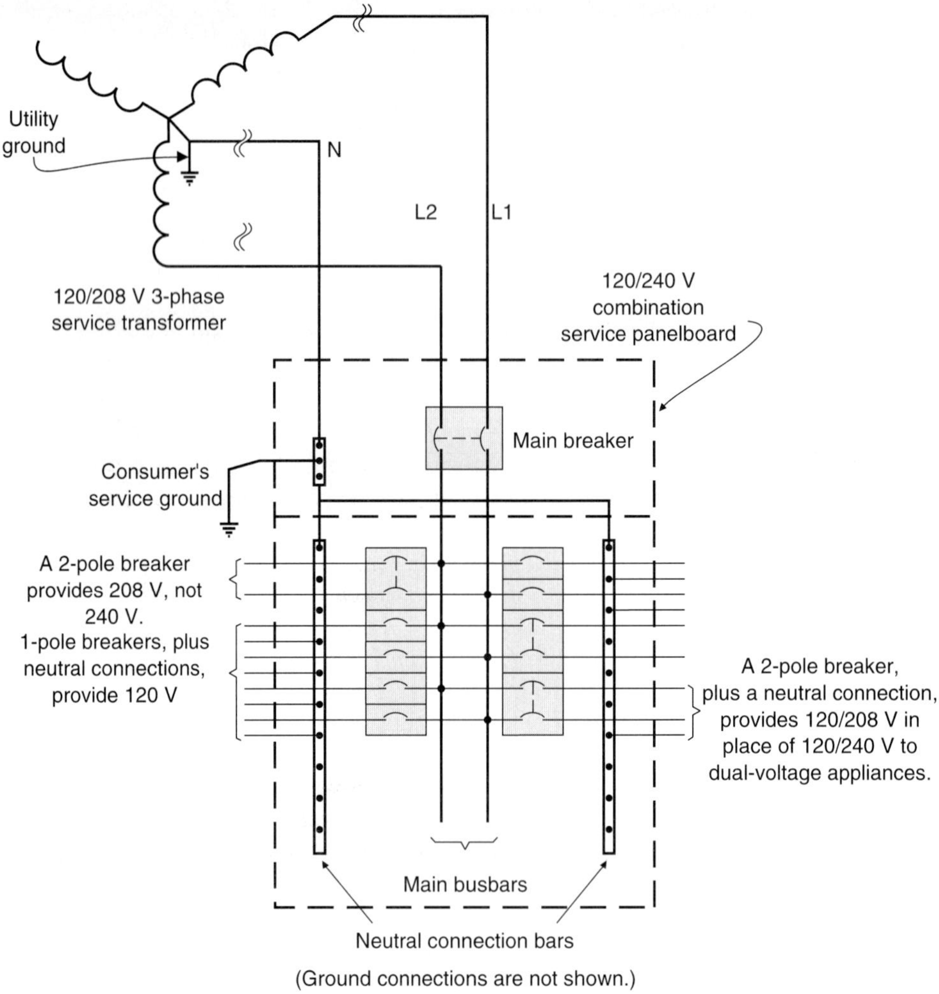

Some North American utilities use a similar system to supply 120/208 V instead of the usual 120/240 V. This is known as a single-phase network and, like the European system, it originates from a three-phase supply transformer instead of the usual single-phase type. The shock hazard of 120/208 system is still only 120 V to ground.

A single-phase network can be identified by the 208 V line-to-line voltage that replaces the expected 240 V (see Figure 7–29). Single-phase motors operating from the higher voltage of this system must be rated for 208 V operation. Although there are suitable motors available, many motors that are designed for 240 V (230 V motor nameplate) operation are not approved for use on 208 V supplies. This can be a problem.

High-Voltage Services

Where many motors are larger than a few hundred horsepower, or any are too large for acceptable 480 or 600 V operation, 2400, 4160 or 7200 V may be used instead. One popular choice is 4160 V, supplied as a 2400/4160Y. The 2400 V single-phase voltage is not used. There are other options, including the higher 4160/7200 or 7200/12,470 V systems. Lower voltage equipment is supplied from transformers that reduce the voltage first to 480 or 600 V, and then to 120/208 V three-phase or 120/240 V single-phase.

Summary

Transformers are dependable, low-maintenance devices that play a key role in all modern AC electrical systems. They are essential to power utilities, because high voltages are a must for cost-effective long-distance power delivery. Using a high voltage permits large quantities of energy to be transferred with relatively low current. This dramatically reduces the size of conductors required and, therefore, power line construction and operating costs. Without transformers to efficiently raise and lower the AC voltage, high-voltage distribution would be virtually impossible.

Most utilities raise the voltage to transmission levels at the power plant, and then reduce it in steps as the system fans out to the customers. The last utility reduction is usually very close to the buildings being served. The connection of this final utility transformer defines the voltage and other characteristics of the electrical service. This information is important to the building owner and to service personnel who maintain the electrical system and any electrically driven equipment it supplies.

There are several different types of services. The two most obvious categories are single-phase and three-phase, but both of these have several different possible combinations. The most popular single-phase service in North America is the 120/240 V three-wire connection. This comes from a single-phase transformer that is grounded in the center of the secondary winding. This grounded conductor is the system neutral. A variation of this is a 240/480 V service for customers that require a higher voltage in an area where three-phase is not available. A 120/208 service is a three-wire network connection off a 120/208 three-phase four-wire wye transformer. It behaves much the same as the traditional 120/240 V service, but can cause problems with motors loads that are rated for 230 V. It reduces the output of 240 V heating elements by about 25%. Elements rated for 208 V are available for some equipment.

Three-phase services are available in much wider variety than single-phase. Wye connections have equal voltages from the ground or neutral to each of the line conductors. They can supply single-phase from any line to neutral, or between any two lines. The voltages between the lines are 1.732 ($\sqrt{3}$) times higher than the voltage from neutral or ground. Many standard voltages are available, including 120/208, 277/480, and 347/600 V. Much higher voltages are also available for customers with high demands. A vast array of motors can be obtained for any of these ratings.

The three-phase delta connection may be available as a three-wire service at voltages such as 240, 480 (or 460), and higher. It may also be available as a four-wire service providing 240 V three-phase and 120/240 V single-phase from the same transformer and service connection. One downside of this system is that one of the three-phase lines is 208 V above ground or neutral, compared to 120 V for the other two. This line has to be isolated from the single-phase system, so a separate single-phase panel must be used. In comparison, a 120/208 V wye service can supply all possible combinations of single- and three-phase loads from the same panel.

Transformers are also widely used within buildings and equipment. They are used with higher voltage services to provide 120 V single-phase loads or specific voltages to special equipment. They are also used to provide low voltages to control systems of all types. In industrial motor control, the control voltage is usually 120 V. In other systems such as HVAC equipment it is usually 24 V.

Transformers are also used to supply instruments that monitor systems that operate at high voltage, high current, or both. A typical potential transformer (PT) provides an accurate ratio of the high system voltage reduced to 120 V. The primary can be any standard voltage up to the highest utility voltages. A small low-voltage PT looks very much like a small power or control transformers. A PT designed for extra high voltage can be an impressive size.

A current transformer (CT) is used to provide an accurate ratio of current reduction from a circuit conductor that is carrying high current or operating at a high voltage. Current transformers are usually designed with a 5 A secondary. They have specific current ratios to match the needs of the system. They are also voltage rated, because they must be insulated from the high-voltage conductor. The secondary circuit of a current transformer must not be opened when the circuit is operating, because a very high voltage will appear at the break. This is a definite safety hazard and often destroys the CT. Strange as it may seem, when they are properly connected, CT circuits are very safe. They have almost no voltage anywhere on the secondary side, to ground or between conductors. The problems occur when changes are made to the system and someone disconnects, or forgets to reconnect, the CT.

Questions

1. Define "self-inductance" and "mutual inductance."
2. Define the terms "primary" and "secondary" as they relate to transformers.
3. What is "excitation current"?
4. Although the changing field in the transformer core induces a voltage into both the primary and secondary coils, these two voltages play very different roles. Explain the difference between them.
5. Explain in step-by-step sequence how an increase in secondary current causes a corresponding increase in primary current.
6. A transformer has a turns ratio or 20:1. If the load receives 12 V and draws 32 A, what is the primary voltage and supply current (ignoring excitation current and losses)?
7. If a transformer is rated to supply 100 A at 240 V, what is its full load kVA rating?
8. An ignition transformer provides 5000 V from a primary of 24 V. If the primary has 100 turns, how many turns are there on the secondary side?
9. A rural distribution line supplies a 14,400:240 V transformer. If the demand on the secondary is 85 A, what is the current in the high-voltage line?
10. If the primary voltage is increased to 150% of the rated value, will the output voltage also be 150% of the rated? Explain.
11. Is the excitation current of a transformer proportional to primary voltage, secondary-load current, or both?

12. What is the maximum overvoltage (in %) permitted on a typical transformer?
13. Why is the actual primary current of a transformer higher than the current ratio suggests?
14. Does a step-up transformer have the high-voltage winding connected as the secondary or primary?
15. What makes a true isolation transformer different from a typical two-winding transformer with similar ratings?
16. Why do ungrounded two-winding transformers sometimes have random voltages from ground on the secondary terminals?
17. Define the term "tap" as it applies to transformers.
18. What precaution must be used when adjusting taps on a typical two-winding transformer?
19. Why do utilities use on-line tap changers on many of their larger transformers?
20. The high- and low-voltage windings in typical core type transformers are split between both of the individual core legs. Why are the windings split up, instead of placing all the high-voltage on one leg and the low voltage on the other?
21. Why is it important to not physically overstress the leads or terminals of small transformers?
22. What is a primary safety concern unique to current transformers?
23. What is the main difference between a Class 2 transformer and a transformer without the Class 2 designation?
24. Why is it incorrect to replace a Class 2 transformer with a transformer that has the same voltage and current ratings but is not marked Class 2?
25. Why is it usually incorrect to interconnect Class 2 transformers?
26. How can the low- and high-voltage sides of an autotransformer share the same winding and still have different voltages?
27. Why do ungrounded transformer secondaries often have a significant voltage to ground?
28. Is it possible to get a shock from the secondary side of an isolation transformer?
29. Explain why it is so important to connect controls only in the ungrounded side of the control circuit.
30. What are the standard letter markings for the high- and low-voltage leads or terminals of a two-winding transformer?
31. What are the two possible voltages between the ungrounded conductors of the high- and low-voltage sides on a 120:24 V transformer that has a grounded secondary?
32. Sketch a connection diagram for a 240/480:120/240 V transformer properly connected to provide 120/240 V output from a 240 V supply.
33. If a step-down transformer has voltage adjustment taps on the primary side, will a lower tap result in a higher or lower output voltage?
34. What cautions should be heeded regarding unused transformer tap leads or terminals?
35. Describe the connections required to obtain 120, 240, and 120/240 V circuits from a typical single-phase service panelboard.
36. A three-phase four-wire delta service can be identified by voltage measurements from the three main lines to the neutral or ground. If the three-phase is provided at 240 V, what are the three voltages to ground?
37. What voltages should be present from line to ground on wye services with the line voltage ratings 208, 480, and 600 V, respectively?

Part III

Electric Motors

The overall intent of this section is to provide the knowledge required to test, service, and select a wide variety of electric motors. References and examples are drawn primarily from the HVAC industry, and only a few, relatively specialized, motor types are not found somewhere in this type of equipment. HVAC manufacturers have adopted, or at least experimented with, almost every type of motor available. This trend is continuing.

This section contains seven chapters. Chapter 8, "Energy, Work, and Power," describes the relationships among various forms of energy and the units of work, energy, and power. This chapter provides a conceptual anchor between the physical world and electricity. Electrical units are summarized and explained using units that are applicable to all other energy forms. Chapter 9, "Motor Fundamentals," examines the general characteristics of electric motors and introduces the related terminology. Chapter 10, "Single-Phase Motors," examines, compares, and explains the array of motors available for single-phase supplies. Common types of single-phase motor controls are also introduced. Chapter 11, "Three-Phase Motors," is devoted primarily to the workhorse of industry, the three-phase squirrel-cage induction motor. The operating theory of these motors is expanded to include the various rotor designs and their effect on motor behavior. Chapter 12, "Three-Phase Motor Control," examines motor protection and control, including multispeed and reduced-voltage starting methods. The principles of electronic starters and controllers are introduced. Chapter 13, "DC Motors," describes conventional, brushless, and stepper motors. It includes an introduction to some of the newest motors to become commercially available. Chapter 14, "Nameplates and Specifications," is a brief summary of the data provided on typical motor nameplates and in manufacturers' specifications.

This section may be used to study theory or as a reference section. Individual motor types can be studied in relative isolation.

8 Energy, Work, and Power

Objectives

This theory chapter provides an overview of energy and energy conversion, and then concentrates on electrical and mechanical forms. It defines the units used, and fundamental mechanical- and electrical energy-related calculations. After the study of this chapter is complete, you should be able to:

- Define electrical units as they relate to electrical energy
- Explain the difference between energy, work, and power
- Explain energy use and relate it to energy costs
- Define the mechanical units as they relate to energy
- Apply units correctly to both linear and rotary mechanical motion
- Define temperature units and perform conversions as required
- Explain what is meant by heat capacity and heating rate
- Apply an understanding of electrical and heat energy units to electric heating
- Define efficiency
- List sources of loss in electrical and magnetic circuits
- Relate energy losses to equipment life span
- List sources of loss in mechanical systems
- Perform efficiency calculations using appropriate data and formulas
- Calculate electrical power in AC and DC circuits
- Explain power factor and perform related calculations
- Explain the principles of power factor correction and list some of the factors involved

8.1 Introduction

One popular definition of energy is "the capacity of a physical system to do work." A more precise definition is "that which is diminished, when work is done, by an amount equal to the work done." Work is defined as "the transfer of energy from one physical system to another." The relationship between energy and work is a circular one. Energy is that property that makes work possible. Work implies a physical change that involves energy transfer from one "system" to another. Energy is measured in terms of work done; work is measured in terms of energy transferred.

Given that energy and work are defined in terms of each other, it seems reasonable that they share common units of measure. The units chosen are based on recognizable physical processes, but can be applied to all activities involving energy. Simple mechanical systems with reasonably large pieces are the easiest systems to manipulate and observe. In such systems, work is described as force acting through a distance. This implies a physical body that is capable of motion; a force applied to that body; and a resulting change in its position, direction of travel, or speed. It is assumed that the body has mass and that the physical change it undergoes is related to the direction of the force applied.

The standard units of energy and work are based on standard values of physical force, applied through standard units of distance. Various units have been proposed and used throughout history. These include the joule, the erg, and the foot-pound. The joule is the unit used in the SI system of units. It is equivalent to one newton of force acting through a distance of one meter. The erg is the unit of the CGS—centimeter, gram, second system. It is a force of one dyne acting through one centimeter. The Imperial unit is the foot-pound, which is one pound of force acting through one foot. All of these units are workable, and, because all are measurements of the same property, they can be converted from one to the other. Electrical systems are defined in SI units, and these are the preferred units of this text.

Energy can take many forms. It can be in a very concrete, observable form, such as energy transferred by a mechanical system to do mechanical work. It can be in a very sensible, but less containable form, such as heat or light. A more transitional form could be electrical energy. Two particularly powerful but mysterious forms are chemical and nuclear energy. In these two forms, vast quantities of energy may be contained in small quantities of material, and the mechanism by which that energy may be released is not at all obvious. All of these energy forms are capable of doing work, which essentially means that they are capable of being converted into some other form.

Energy may also be classed as kinetic or potential. Kinetic energy is the energy of motion; potential energy is the energy of position or condition. A moving object possesses kinetic energy, usually recognized as the inertia of its mass. Most stationary objects have potential energy, due to their physical location and the mutual attraction of gravity. If an object is suspended, it can release energy when it is allowed to descend. The total potential energy is a product of the force of gravity on the object's mass and the distance through which it can fall.

While the object is suspended, there is only a potential for energy release. The energy was stored when the object was lifted into position. That process involved work, and work will be done again when the energy is released during the object's descent. This release may be delayed indefinitely, with no change in the energy available for use. Other examples of potential energy include a compressed spring, a chemical reaction yet to occur, or a stored electrical charge.

One of the most critical properties of energy is whether it is in a form that makes it readily available for use, or in a form that is too diffuse for practical recovery. For example, a quantity of fuel contains chemical energy that can be released by combustion. The energy is in a form that is readily available and relatively easy to release. When the fuel is burned, that energy is converted into heat and light, both of which are difficult to contain. In a very short time, the energy they carry has dissipated into the atmosphere. There is a strong tendency for all energy to become more diffused. Energy consumption is the rate at which energy, in a readily available form, is converted into energy that is beyond practical recovery.

Energy can be released, absorbed, transferred, and stored in endless combinations of action, reaction, and change. Despite the many possible transformations, energy follows a law of conservation: The total amount always remains the same. Unfortunately, readily available or recoverable energy tends to get converted to less recoverable forms and dispersed. Although the total energy is the same as before, no more useful work can be extracted.

In some cases, the release of the energy and its ultimate loss to useful employment occur almost simultaneously. An example would be setting off a charge of explosive. The energy is released in one very brief reaction and is almost instantly dissipated and lost. In other cases, there may be numerous intermediate steps. For example, fuel burned to produce electricity becomes heat energy in the boiler, is converted to mechanical energy when the steam from the boiler goes through the turbine, and then is converted to electrical energy in the generator that the turbine drives. This electrical energy is, in turn, converted to just about every form of energy possible as it is utilized by customers connected to the utility distribution system. At each step, work is done, energy is transferred, and energy is lost to the inevitable inefficiencies of the various system components. The energy is utilized in many different ways—some of it perhaps even stored in potential form. However, once the initial conversion has taken place, the energy released from the fuel is essentially lost to practical recovery.

Power is the rate of energy transfer or conversion, or the rate of doing work. It is a measure of the rate at which energy is put into, or extracted from, some process or system. The conversion or transfer of energy can occur in a transient fashion, or it can be maintained more or less continuously for some significant time period. Power ratings are more applicable to maintained conversions. For example, the output of an electric motor is not a single impulse of force: It is maintained for an extended period of time. The rate at which electrical energy enters the motor is its input power; the rate at which mechanical energy is withdrawn from the motor by the driven load is the motor's output power. The motor continuously converts electrical energy into mechanical energy. Its power ratings are a measure of the rate at which that conversion occurs.

The term "work" is most often related to mechanical systems and is defined in mechanical terms as a force acting through a distance. Power is the rate at which work is done or energy is transferred. It is a multiple of force and the distance per unit time, which is velocity (or speed). An increase in the force or the speed at which a process is taking place translates into an increase in power. With all the losses factored in, power is also a measure of the rate at which the energy source is being depleted.

8.2 Electrical Energy

Electricity is a great connector, a common denominator between sources of energy and distant loads. The energy is transferred by the voltage and current of an electrical circuit. A minimum of two energy conversions is required to accomplish this task. The source energy is converted to voltage by an electromagnetic generator. This voltage is connected to an electrically conductive circuit, and current flows around the circuit through one or more electrical loads. The voltage and current give up their energy to the loads, and it emerges in some other form.

All electrical loads convert electrical energy into some other form. Examples include electromagnetic devices, which produce mechanical force and motion, and chemical processes, where at least some of the energy becomes stored in the chemical properties of the materials involved. Almost all energy conversions also produce some heat. Unless heat is the desired output, it is an energy loss that reduces the efficiency of the system.

The primary attractions of electrical energy are the ease and efficiency with which it can be transported and controlled, and the endless variety of ways in which it can be put to use. These properties have made the generation, distribution, and sale of electrical energy a commercial success. Electrical utilities produce and deliver energy to industrial, commercial, and individual marketplaces that have an insatiable appetite for their product. This appetite seems to be dampened only by the cost.

An electrical utility charges its customers for the delivery and consumption of energy. Rates are based primarily on the quantity of energy that passes from the utility distribution system into the customer's premises. The utility is in the business of producing, delivering, and selling energy. The customer consumes the energy, at least in the sense that it is converted into some other form on the customer's premises. The utility meter monitors consumption by continuously measuring the voltage and current at the point of delivery to the customer. It converts these values into an accumulating measure of energy transfer. The customer is billed according to the amount of energy that has passed through the meter during the billing period.

An understanding of the various units of energy and the concept of power makes electrical bills much more meaningful. This is an essential first step toward understanding how to make the best possible use of this energy source, while keeping costs to a minimum.

Electrical Units

Thankfully, the diversity of measuring systems that plague the physical world is not mimicked in the world of electricity. Electrical measurements and calculations are always done in standard SI units. This text uses SI units and common units that match the data provided on typical electrical equipment nameplates. Some conversions are provided, but not a complete listing.

Electrical systems are based on a working "fluid" of charged particles that can move within a conductive medium. When the medium is a solid, the charged particles are "free" electrons. They are called free electrons because it is the nature of all conductive solids that one or more electrons from each atom moves freely through the material. These electrons slide from atom to atom with very little energy loss. When an electrical potential is applied across a conductive circuit, the electrons move through the circuit much like fluid in a pipe.

When the conductive medium is a liquid or gas, the charged particles are usually ions formed by chemical separations in the solution. These ions move though the fluid carrying energy in much the same way that electrons do in a solid conductor. Solid conductors are the most common materials in a typical electrical system; therefore, the information that follows assumes solid conductors and free electrons, although the behavior of ions in a fluid is essentially the same.

The electrons of the circuit are given energy either by a direct energy conversion or by the influence of a magnetic field. The energy given to the electrons comes from some other energy form, and the equipment or process involved is referred to as the electrical source.

The energy appears as an electromotive force (EMF) across the source terminals and can be measured as a potential difference in volts.

When a complete conductive circuit exists, the EMF causes the electrons to flow through the circuit. This flow is called an electric current. A practical circuit contains an energy "supplier" and an energy "consumer," which are referred to as the source and load, respectively. The interconnecting conductors are selected to keep the transmission losses to a minimum. The electrons flow through the conductors, arriving at the load with most of the energy they received from the supply. The load converts this electrical energy into some other energy form.

Charge: A "Quantity" of Electricity

Before the activities of an operating circuit can be measured and monitored, each of the circuit variables must have defined units, and appropriate measuring instruments must be available. The basis of electrical systems is the working "fluid" of electrons, and the unit associated with this variable is essentially a unit of volume, or quantity, of free electrons. The unit is called the coulomb (Q). It is defined as a specific number of electrons, and is generally referred to as the unit of electrical charge. Electrical charge is measured in coulombs: One coulomb (Q) is 6.24 × 1018 electrons

Circuit current is a measure of the rate at which the charged particles of the circuit are moving through the conductors. This is not a measure of velocity, but instead a measure of volume or quantity per unit of time. It is similar to the flow rate units used with other types of fluids, such as gallons per minute or liters per second. The unit of electrical flow rate is the ampere. The ampere is based on the coulomb, as the unit of quantity, and time measured in seconds. Electrical current (I) is measured in amperes and is a flow rate in coulombs per second. One ampere (A) is a flow rate of 1 coulomb (Q)/second (s), relative to a stationary point.

$$1 \text{ A} = 1 \text{ Q/s}$$

$$\text{I} = \text{Q/s}$$

Electrical Force

The unit of electrical force (EMF) is the volt, which measures the energy level of the electrons in the circuit and can be defined in terms of the energy carried per coulomb of charge. The SI unit of energy is the joule (J). Joules will be examined further in relation to the other forms of energy. Electromotive force is measured as a potential difference between two points of a circuit. A potential difference of one volt (V) of electromotive force (E) exists across the circuit when each coulomb of charge (Q) carries 1 joule (J) of energy.

$$1 \text{ V} = 1 \text{ J/Q}$$

$$\text{E} = \text{J/Q}$$

Power

Power is the rate of energy transfer or conversion, or work done. The unit of power used for electrical calculations is the watt (W). Power units must be based on a unit of energy per unit time. The watt is in joules per seconds. One watt (W) of power (P) equals energy transfer at a rate of 1 joule (J) per second (s).

$$1\ \text{W} = 1\ \text{J/s}$$

$$\text{or } 1\ \text{J} = 1\ \text{W} \bullet \text{s (watt-second)}$$

$$\text{P} = \text{J/s}$$

The watt is a general-purpose power unit, as applicable to mechanical systems as it is to electrical systems. At the present time, most North American motors are rated in horsepower, while European motors are rated in watts or kilowatts. The conversion factor for horsepower and watts is

One horsepower (Hp) = 746 watts (W)

Electrical power can be calculated from volts and amperes. This convenient relationship occurs because of the way electrical units were selected. Electrical energy is supplied at a power rate of one watt when a sustained electromotive force of one volt maintains an electron flow rate of one ampere though the circuit. Power (P) in watts equals the EMF(E) in volts times the current (I) in amperes.

$$P = E \times I$$

While this formula is technically correct in every case, an error is introduced when it is applied to AC circuits, containing reactance, using the values that are obtained by normal measurements. This problem in AC power calculations is explained later in this section. The preceding power formula is derived from the units involved. Potential difference in volts (E) is joules of energy per coulomb of charge (J/Q). Current in amperes (I) is coulombs of charge flowing per second (Q/s). Power in watts is energy transfer in joules per second (J/s).

$$\text{J/s} = \text{J/Q} \times \text{Q/s}$$

$$P = E \times I$$

This equation demonstrates the operation of an electrical circuit and the relationships among the various units. The intent and purpose of all practical circuits is the movement of energy. The joules of energy per unit of charge, times the units of charge flowing through the circuit every second, is equal to the joules of energy per second transferred by the circuit from one point to another.

Energy and Work

The total electrical energy delivered by a circuit is a function of the rate at which the energy is being supplied—the power at which the circuit is operating—and the total time this process is maintained. The energy unit used as the basis for the volt and watt is the joule. In a circuit operating at a power of one watt, energy is delivered at a rate of one joule per second. The total energy transferred—or work done—is measured in joules and can be found by multiplying watts times the total time in seconds. Energy or work done (W) in joules (J) equals the power (P) in watts (J/s) times the time (t) in seconds.

$$\text{W (J)} = \text{P (J/s)} \times \text{t (in seconds)}$$

$$W = P \times t$$

Energy Costs

It is a small step from total energy transfer to energy cost. All that is required is the appropriate cost per unit of energy. This cost is the cornerstone of all utility rates. Although other charges may appear on a typical electrical bill, the key figures are usually the total energy used, the cost per unit of energy, and the accumulated energy consumption for the billing period. The joule is a relatively small unit, and not well understood by consumers that are not accustomed to SI units. The watt and the second are also too small for practical measurements. For these reasons, it is common practice for utilities to measure power in kilowatts (1 kW = 1000 watts) and the time in hours. The basic energy billing unit then becomes the kilowatt-hour (kWHr), which is equal to 3,600,000 joules, or 3.6 megajoules (MJ), of energy. This conversion is derived from the basic units (1000 W/kW × 60 seconds/minute × 60 minutes/hour).

The energy formula can be used, with power in kW and time in hours, to produce the kWHr energy consumption. Adding the utility rate creates a basic formula used for finding the cost of operation for electrical devices and equipment.

$$\text{Energy Cost} = \text{Power (in kW)} \times \text{time (in Hr)} \times \text{the rate (in \$/kWhr)}$$

Figure 8–1 shows a summary of electrical units.

8.3 Mechanical Energy

Electrical energy is almost never used in its original form. Instead, it is converted into something more immediately useful, such as light, heat, sound, or mechanical motion. This makes it impractical to study electrical energy in isolation from other energy forms. Mechanical energy is available energy, related to very tangible forces such as pressure, movement, and mass. It is the stuff of mechanical machines, the most commonsense energy form, and the easiest to observe in action. However, in the larger picture it is just energy in another form, with all the usual ties to all the other forms.

Mechanical Units

Mechanical systems are described with units of mass and distance, physical force, and pressure. As separate entities, these units have little relationship to the nonmechanical world. Electrical systems, for example, would be virtually impossible to describe in such terms. However, as soon as energy enters into the picture, all systems have a common bond and a shared set of units. When calculations are used to quantify work, power, or energy, the mechanical units immediately link with the basic units of electrical systems and all other energy-related processes.

Distance, Speed, and Acceleration

The basic SI unit of linear distance is the meter (m). It can be converted to any other linear distance unit using standard conversion factors.

Linear speed, or velocity, is a measure of distance traveled per unit of time. In the SI system, the basic unit of velocity is meters per second (m/s).

FIGURE 8–1 A summary of electrical units.

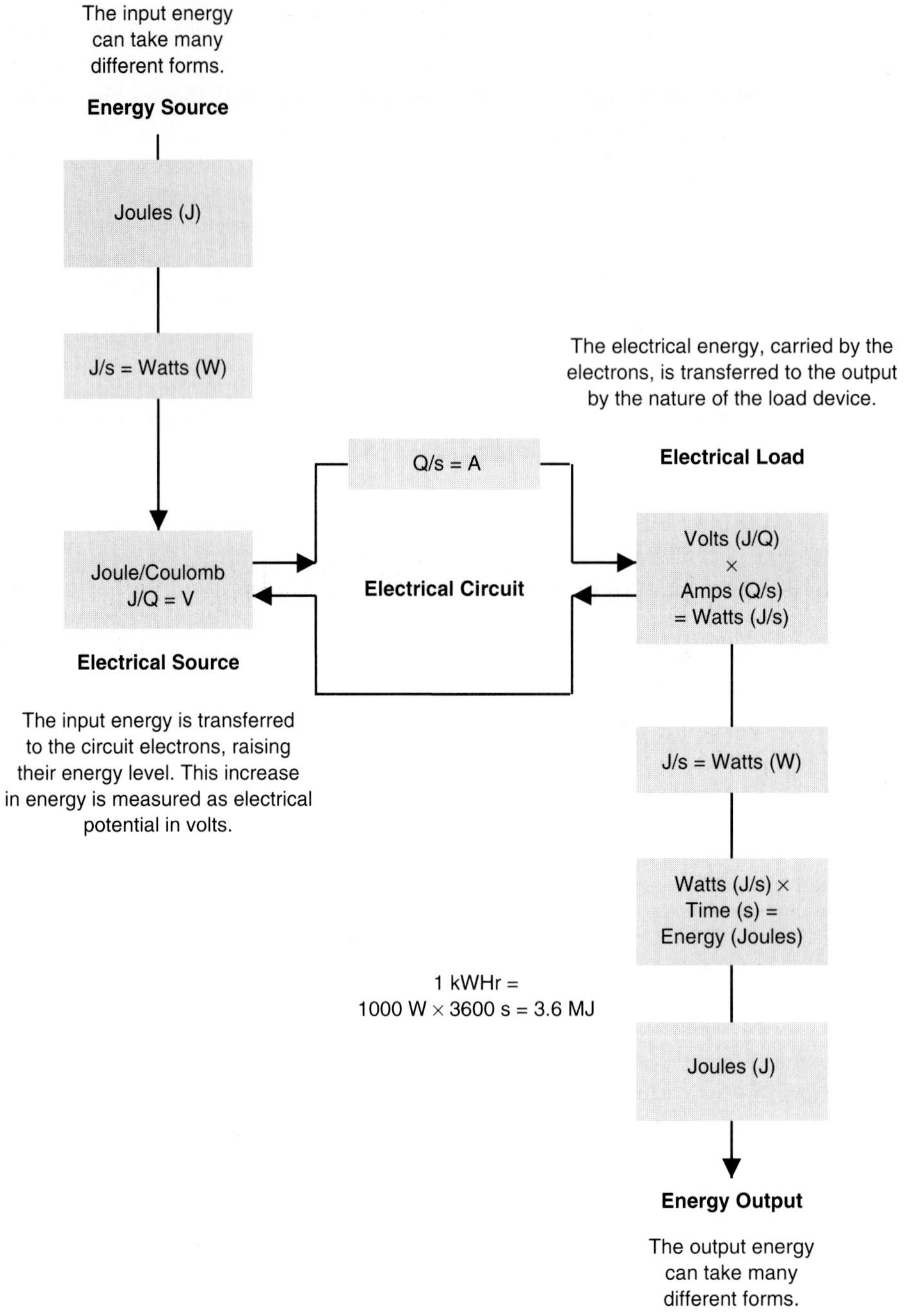

Linear acceleration is the rate of increase in velocity per unit time. It is a confusing unit because the time factor enters into the description twice, once as part of the definition of velocity, and a second time as part of the description of the rate at which velocity is increasing. When an object is accelerating, its velocity (m/s) is increasing at a specific rate per unit time (x/s). The unit becomes (m/s/s), or (m/s^2).

Force and Weight

The basic unit of mechanical force is the newton (N). It is defined in terms of its ability to accelerate a given mass. It is understood in terms of the gravitational force we sense as weight.

One newton is the force required to accelerate a one kilogram (kg) mass at the rate of one meter per second per second (m/s^2). One kilogram (kg) mass, at sea level, presses downward with a force of 9.81 newton (N). This is the accelerating force of gravity and is generally interpreted as a "weight" of 1 kg (or approximately 2.2 pounds). One newton of force is the equivalent of 101.9 grams (approximately 3.6 ounces) of "weight." A free-falling object subjected to the force of Earth's gravity (g) accelerates at a rate of 9.81 meters per second per second ($9.81\ m/s^2$).

Pressure

The SI unit for fluid pressure is the pascal (Pa). One pascal (Pa) of pressure is equal to one newton (N) of force per square meter of area.

$$1\ \text{Pa} = 1\ \text{N/m}^2$$

Because the newton is a relatively small amount of force, and a square meter is a relatively large area, the pascal is a very small unit. The kPa is a much more common form. Kilopascals can be converted to pressure in pounds per square inch if desired. One kilopascal (kPa) of pressure is approximately equal to 0.145 pounds per square inch (psi).

$$1\ \text{kPa} = 0.145\ \text{psi}$$

Work, Power, and Energy

Work is done when force acts through a distance. One newton of force acting through a distance of one meter is called a newton-meter (Nm). The work done in one newton-meter of effort is equal to one joule (J) of work or energy (W). The joule used to measure mechanical work is, of course, the same as the joule used to describe electrical energy—or energy in any other form. Work is a transfer of energy that can be properly measured in these terms. The newton-meter and the joule are interchangeable in calculations. Note that both force and distance are required for work to be done. Force without motion does no work, and no energy is transferred.

One newton-meter (Nm) is equal to one joule (J) of work or energy.

$$1\ \text{Nm} = 1\ \text{J}$$

The Imperial unit of foot-pounds can also be converted to joules.

$$1\ \text{ft-lb} = 1.356\ \text{J}$$

Power, in mechanical terms, is the rate of doing work. For example, imagine one newton of force is applied to an object, maintaining its motion at a velocity of one meter per second. The object is not subjected to a short burst of force, but rather to a force that is

FIGURE 8–2 A summary of linear mechanical units.

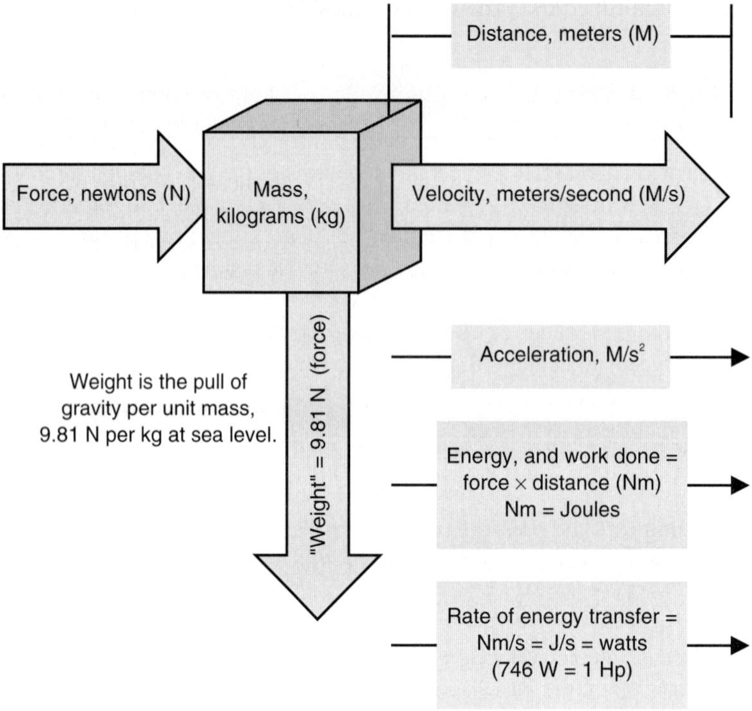

steadily applied. The object does not accelerate, because it presents a continuous resistance to movement that just balances the force applied. One newton is the exact force required to maintain its velocity so that the object does not speed up or slow down. The rate of energy use (and work done) within this system is one newton-meter per second, or one joule per second. This rate is equal to the standard power unit of one watt.

A newton-meter of effort per second (Nm/s) is work and energy at the rate of one joule per second (J/s), or a power rate (P) of one watt (W).

$$P = Nm/s = J/s$$

The common unit horsepower (Hp) is a power rate of 550 ft-lb of energy per second. It has a fixed relationship to the watt.

$$1\ Hp = 746\ W$$

Figure 8–2 shows a summary of linear mechanical units.

Rotary Motion

It is important to note that rotary motion is somewhat different from linear motion, although the two are converted back and forth by all manner of mechanical devices. Rotary force, the twisting power of a driven shaft, is called torque. A description of torque must include a unit of force and a unit of distance. The force is viewed as acting at the end of a lever arm which can rotate around the center axis of the shaft; the distance is measured along the length of the lever. Of course, it is not necessary for the lever to exist, it is only used to define the torque unit.

Note

It is important to realize that although a unit of force and a unit of distance are required to describe torque, it is still only a unit of force, not of work done. No work or energy transfer is automatically implied. The distance is only the length of the lever through which the force acts, not a measure of the distance through which it moves. Torque is a twisting force, and just as linear force can be applied without actually moving anything, torque can also exist without rotation. Force without motion does no work, because no energy is transferred.

A torque equivalent to one newton of force applied to the end of a lever one meter long is defined as a torque of one newton-meter (Nm). As explained above, this is not to be confused with linear motion or work. Torque in newton-meters (Nm) is the functional equivalent of the foot-pound, which is one pound of force acting at the end of a lever one foot in length. These two units can be converted from one to the other as required. Torque in newton-meters may be symbolized by the lowercase Greek letter tau (τ).

Converting the unit of pound force to newtons, and the foot to meters, produces this conversion factor.

$$1 \text{ ft-lb torque} = 1.356 \text{ Nm}$$

When torque acts through a distance, work is done and energy is transferred. The distance through which the torque acts falls on the circumference of an arc. It can be measured as a linear distance along the curve, or in relationship to its rotation as degrees of arc or full revolutions. Both of these approaches can be combined in a single unit by relating the linear distance along the circumference to the radius of the arc. There is a fixed ratio between the radius of a circle and the circumference: Two times the constant pi (π), as described in the formula for finding the circumference of a circle from its radius ($C = 2\pi r$), where C is the circumference and r is the radius.

The SI unit that describes rotary motion is the radian. One radian of rotation has occurred when the radius of any arc has passed through a distance equal to its own length. This may seem to be an awkward unit, but it works very well with the other SI units of force and energy. A newton-meter of torque is equivalent to a force of one newton applied to the end of a lever one meter long. For this force to do one newton-meter of work—which is one joule of energy transfer—it must move through a distance of one meter. This is equivalent to the torque applied through one radian of arc.

Force times distance equals work, or energy. In linear terms this is force in newtons times distance in meters, resulting in newton-meters of work and joules of energy. In rotary terms it is torque times rotary motion, but by defining torque in newton-meters and rotary motion in radians, the relationship between torque and rotation is essentially the same as for force and linear distance. One newton-meter of torque acting through one radian is one newton of force acting through a distance of one meter. The base units of work and energy are the same for rotary and linear motion, all neatly tied to universal units that are applicable to electrical, mechanical, or any other form of energy.

Power in rotary motion is a function of force and velocity, just as it is for power in a linear system. The force is the torque, and velocity is the speed of rotation in radians per second—or revolutions per minute. Power calculation in SI units is simply torque times speed: There are no constants to convert the units. If the torque is in newton-meters and the speed is in radians per second, the result is power in watts.

The SI formula conveniently presents rotary power in the standard power unit of watts. However, most North American motors have their operating speeds listed in revolution per minute (rpm), not in radians per second (rad/sec). Fortunately, the two are relatively simple to convert. One revolution equals 2π radians, and one minute has 60 seconds; therefore, the conversion factor from revolutions per minute to radians per second can be found as follows:

$$1\,\frac{\text{rev}}{\text{min}} = \frac{2\pi\ \text{rad}}{60\ \text{sec}}$$

$$\frac{2\pi}{60} = 0.1047$$

Speed in revolutions per minute (N) times $2\pi/60$—which is approximately 0.10472—is equal to speed in rad/sec. This is often rounded to 0.1047 or 0.105. Unnecessary rounding can be avoided by using $2\pi/60$ instead of 0.1047 or 0.105. Speed in radians per second is usually symbolized by the lowercase Greek letter omega (ω).

Rotary mechanical power in watts can be found directly from speed and torque in SI units.

$$P = \omega\tau$$

P = power in watts

ω = speed in radians per second

τ = torque in newton-meters

This formula can be modified for speeds given in revolutions per minute.

$$P = \frac{2\pi}{60} N\tau \ \ldots \text{ or,}$$

$$P = 0.1047\ N\tau$$

N = rotational speed in revolutions per minute (rpm)

A summary of rotary mechanical units is shown in Figure 8–3.

8.4 Heat Energy

One dictionary definition of heat is "a form of energy, associated with the motion of atoms or molecules, that is capable of being transmitted through solid and fluid media by conduction, through fluid media by convection, and through empty space by radiation." Heat is essentially energy stored in a material as atomic or molecular movement. The amount of energy can change because heat energy is very mobile. It can be transferred by physical contact (conduction), by the natural mixing and circulation of atoms in a liquid or a gas (convection), and through space as radiation. Heat energy can be gained and lost by any of these processes, or all of them acting simultaneously.

There is a natural tendency for heat energy to move from hot objects to cool ones. When an object is to be heated, it is placed in a position to receive heat energy from a

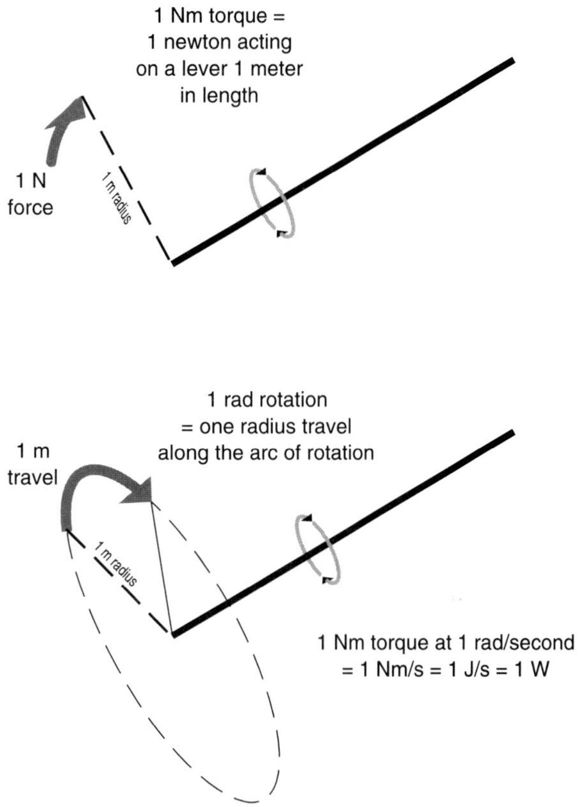

FIGURE 8–3 A summary of rotary mechanical units.

source that is maintained at a higher temperature. When an object is to be cooled, it is placed in a position where it will naturally lose heat energy to something that is at a lower temperature. During the heating process, energy is transferred from the heat source to the object. Cooling is a process of losing heat energy; there is no such thing as gaining cold.

When an object is to be heated or cooled, there are corresponding changes in the activity and energy level of the atoms or molecules from which it is formed. The change in this energy level can be sensed as a change in the temperature of the object. Temperature is an indication of the heat energy level of the solid, liquid, or gas in question. It does not indicate total energy, because it does not take into account the total mass of the material involved or its thermal characteristics.

Temperature Measurement

The SI unit of temperature is the Celsius degree. The Celsius scale is based on a 100° difference between the freezing and boiling points of fresh water at sea level. Water freezes at 0°C and boils at 100°C. One degree on the Celsius scale is the same as on the Kelvin temperature scale. The difference between the two is that the Kelvin scale starts at absolute zero, the temperature at which there is no heat energy remaining. Absolute zero is approximately –273.16° Celsius.

The Imperial temperature scale is based on degrees Fahrenheit (°F). On this scale, water freezes at 32°F and boils at 212°F. Because the Fahrenheit and Celsius scales have different unit values and different starting points, conversion involves more than a single factor. One degree Celsius (°C) equals 1.8 degrees Fahrenheit (°F), but this must be combined with a 32°F difference between the zero points.

To convert Fahrenheit to Celsius and Celsius to Fahrenheit, use the following formulas:

$$°C = (°F - 32) \times {}^{5}/_{9}$$

$$°F = (°C \times {}^{9}/_{5}) + 32$$

Heat Capacity

While temperature is an indicator of the heat energy level, total heat energy also depends on the heat capacity of the material and its mass. Heat capacity is a measure of the energy required to raise the temperature of a specific mass of the material. It can be defined in any of the standard units of temperature, mass, and energy. From the SI system, temperature is in degrees Celsius (or Kelvin), mass is in grams, and energy is measured in joules. The last two are often changed to the larger units of kilograms (kg) and kilojoules (kJ) for convenience.

All materials have a specific heat capacity. In SI units this is expressed in joules per gram per degree (J/g/°C) or kilojoules per kilogram per degree (kJ/kg/°C). In Imperial units it is expressed in Btu per pound per degree F. The specific heat capacity of water is 4.187 J/g/°C or 4.187 kJ/kg/°C. This means that a 1°C rise in the temperature of a kilogram of water requires an energy input of 4.187 kJ. This energy becomes an equivalent value of heat energy stored in the thermal properties of the water.

Heat Rates

Heat is an energy form and can therefore be measured in the standard SI unit, the joule. It can also be measured in the Imperial measure—British thermal unit (Btu)—which is the energy required to raise one pound of water one degree Fahrenheit. The British thermal unit is close to 1 kilojoule (kJ).

One Btu = 1055.06 J or 1.05506 kJ

Heat rates are an energy transfer or power unit. The SI unit is the watt, which is one joule per second. The common non-SI units can be converted directly to watts with the following conversion factors:

One Btu/hr = 0.293072 W (One W = 3.41213 Btu/hr)

One ton of refrigeration = 3517 W (3.517 kW)

Electric Heating

When electrical energy is converted into heat using a resistance element, the conversion is virtually 100% efficient. All of the electrical energy used by the element becomes heat. Therefore, the heat output of electric elements can be calculated directly from the wattage rating, or from measured values of current and voltage. The heat output of an element (in joules) is the wattage of the element times the time in seconds. For example, a 1 kW element will release 1000 joules of heat energy per second, or 3.6 MJ per hour. This is equivalent to 3412.13 Btu/Hr.

8.5 Efficiency

The efficiency of any system is a measure of how well the input energy is converted into the desired output. Virtually all systems have losses. Input energy that does not become useful output is lost in terms of the useful work done. Understanding efficiency requires an understanding of the measurement of energy in various forms, because the type of energy used as an input is very often different than the type of energy at the output.

For example, the input to an electric motor is electrical energy, but its output is mechanical. An electromagnetic generator provides the opposite conversion by changing mechanical energy into electrical energy. Most engines powered by fossil fuels are heat engines of some type. The chemical energy of the fuel is converted by combustion into heat energy, which in turn becomes mechanical energy at the output shaft.

Regardless of the process or the type of energy conversion taking place, efficiency is a rating of the total output energy compared to the total input energy. It is the ratio of output to input expressed as a percentage.

Losses

Energy cannot be lost in the sense of simply disappearing. The total amount of energy is always the same, it only changes forms and is redistributed. However, from the practical viewpoint of a process or machine, energy that enters into the system but does not produce any of the desired output is generally classed as a loss. For example, the heat that escapes from the radiator of an automobile engine is an energy loss. In terms of the original intention of the engine, the portion of the energy (obtained from burning the fuel) that ends up as heat dissipated from the radiator is lost to the process of propelling the automobile.

The conversion of energy from one form to another often involves some type of machinery or complex process, and the conversion is rarely perfect. Not all of the energy entering the process becomes energy in the desired output form. This energy that does not become normal output is usually lost in the sense that it becomes unrecoverable. Total energy losses can also be defined as the difference between the total input energy and the total useful output energy. It is often more convenient to calculate a total in this manner than it is to attempt to measure all the losses individually.

Electrical Losses

Energy losses (illustrated in Figure 8–4) are usually in the form of heat. All practical conductors have some resistance. The passage of current through this resistance produces a voltage drop and a conversion of some of the electrical energy into heat. Because the most common conductor for equipment is copper, these losses are often referred to as the copper losses of the system.

FIGURE 8–4 A summary of losses in a typical electric motor.

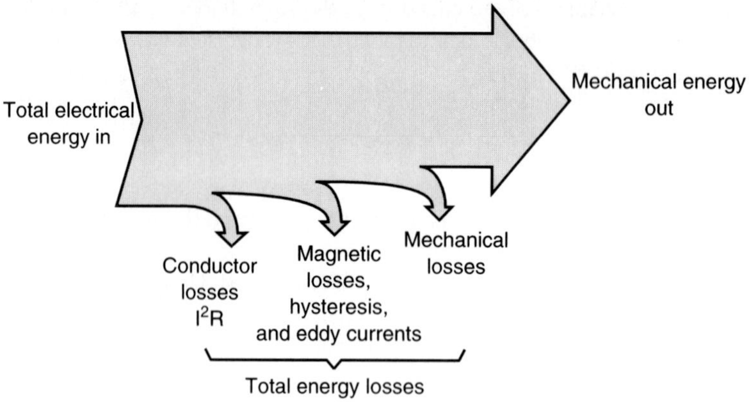

Copper losses are load dependent. They are proportional to the square of the current, as stated in the power formula for current in a resistance.

$$P = I^2 R$$

Copper losses can be kept to a minimum by using conductors of adequate cross section. Larger conductors have a lower resistance and, therefore, result in less energy loss.

Energy can also be lost by dielectric absorption and as static discharge. Dielectric absorption has a small effect in AC circuits with large conductors. Electrical energy lost as a static discharge is called corona and is only significant in systems operating at very high voltages.

Magnetic Losses

When electrical current flows through a conductor, magnetic fields are formed. These fields contain energy. This energy may be returned to the system or lost, depending on the circumstances. In AC circuits the fields are constantly changing. This constant change induces voltages back into the circuit. This is an energy transfer from the electric circuit to the field, and then from the field back to the circuit. One result of this interaction is the current-opposing effects of inductive reactance. Another is a small amount of energy loss that is related to the constantly changing field.

The amount of energy transferred into the field can be very high. In the case of electromagnetic devices such as transformers and electric motors, virtually all of the electrical energy supplied to the device is converted into magnetic energy. To produce such powerful fields, the conductors are formed into coils and placed into a core of magnetic material. Losses that occur are magnetic in nature or related to magnetic effects. One source of loss is stray magnetic fields. When some of the lines of magnetic force "stray" from the desired magnetic circuit, the energy they represent is lost to the process. All magnetic cores "leak" some of the field out into the surrounding environment. These losses are reduced by optimum arrangement of the core, using the best possible core material, and making sure it has an adequate cross section.

There are two other types of magnetic-related losses. One is called eddy currents; the other is called hysteresis. Eddy currents are electrical currents induced in the conductive core material by the changing magnetic field. These currents are similar to the small whirlpools or eddies that can occur in a flowing stream, hence the name. They extract energy from the field, and—because they pass through the resistance of the iron—they convert this energy into heat. The net result is an energy loss and a buildup of heat in the core iron.

Eddy currents are controlled by laminating the core to reduce the cross section of each potential current loop. Because each narrow strip "sees" only a small part of the total field, the voltages induced are very small, and the current path in the strip is much longer than it would be in a thicker core section. The reduced voltage and long, high-resistance path keep eddy currents to a minimum.

Hysteresis losses are related to the effort required to change magnetic polarities in the core material. All magnetic materials have some degree of permanence, which means they tend to retain the field they conduct. Changes in the magnetic field cause changes in the molecular arrangement of the core material. These changes absorb some energy and act to delay changes in field strength or reversals in polarity. The energy lost to this process appears as heat in the core. Hysteresis losses are minimized by using core materials with low-hysteresis characteristics.

Losses and Life Span

Energy losses almost always end up as heat. This is not only wasted energy, it also can have a tremendous effect on the maximum power output and the life span of the equipment. In electric motors, for example, the heat buildup in the windings and in the iron core must be drawn out of the motor to prevent it from overheating.

The life span of the motor is directly related to the operating temperature. There are several classes of insulation used on electrical conductors, but all of them have a maximum temperature rating. As the motor temperature rises to that point and beyond, the life of the motor is shortened. This reduction in life span is not linear: A small sustained increase in temperature can result in a dramatic reduction in motor life. For this reason, motors are rated for a maximum ambient temperature, a maximum temperature rise, and a maximum current. When the motor is operated in such a way that it is unable to cool itself properly, its life span is greatly reduced. This can be a result of overcurrent, high ambient temperature, blocked air flow, or any combination of these factors.

Mechanical Losses

In mechanical machines, energy losses are primarily due to friction. The lost energy appears as heat, which is usually absorbed by the structure of the machine and eventually dissipated into the surroundings. Heat is a common cause of mechanical failure. When the heat builds up rapidly enough to cause damage, it has to be forcibly drawn off by some type of cooling system.

Mechanical losses in electrical motors are typically a combination of bearing friction and windage. Bearing friction can be reduced by good bearing or bushing design and proper lubrication. Windage losses include surface friction of the moving parts in contact with the air, and energy put into driving built-in cooling fans. These losses can be reduced by streamlining parts and using good design practices for the fans. Extremely large machines are sometimes encased in a gastight shell filled with hydrogen instead of air. This is for improved cooling, but also reduces windage losses. Hermetic refrigeration compressors run in an atmosphere of refrigerant, and the rotor usually has no cooling fins or fans because the motor is adequately cooled without them.

Efficiency Calculation

In an electromechanical device, such as an electric motor, the total energy lost to heat is due to the combined mechanical and electrical losses. The amount of loss due to one or the other varies with operating conditions. Copper losses are current dependent, and therefore vary with load. Magnetic losses depend on the strength of the magnetic field, which is generally dependent on voltage rather than current. In most cases the voltage is held constant, so these losses tend to be much the same at no load as they are at full load. Mechanical windage losses are affected by speed, which is constant for some motors and variable for others. Bearing losses tend to be relatively small. They can be affected by speed and by loads that increase the lateral forces applied to the bearing.

The mechanical losses of the driven equipment are usually much greater than for the electric motor that provides the power. When the motor is coupled to the load through a belt, chain, or gear train, the losses in that part of the drive can easily exceed the mechanical losses that occur in the motor itself.

When machines and systems are compared, overall efficiency is very important. Because the most efficient machine has the lowest losses, it has to be better designed, better constructed, or both. It almost always costs more to purchase, and almost always costs less to run.

Efficiency is normally expressed as a percentage and can be found from power input and power output.

$$\eta = \frac{P_{OUT}}{P_{IN}} \times 100$$

η = Percent efficiency

P_{OUT} = Power output

P_{IN} = Power output

Because power input is the sum of the total power out plus any losses in the system, it is sometimes convenient to substitute this into the formula.

$$\eta = \frac{P_{OUT}}{P_{OUT} + P_{LOSSES}} \times 100$$

η = Percent efficiency

P_{OUT} = Power output

P_{LOSSES} = Power losses (total)

When calculating machine efficiency, the power must be in watts or an equivalent unit. The electrical "power" unit volt-amp (VA) is often referred to as apparent power, and that is exactly what it is. Apparent power is not necessarily equal to the true power in watts, and has no place in efficiency calculations.

8.6 Electrical Power

Electrical power is a function of the voltage and current of the circuit, but only in very specific circumstances is it a simple multiple of the two. In AC circuits it is affected by circuit reactance, and further complicated when the energy is supplied from a polyphase system, such as a three-phase supply.

In reactive circuits power flows both ways, and the net power delivered is the difference between power flow from the source and power flow back to the source from the reactive portions of the load. This is a relatively simple concept, but it makes measurements difficult. The direction of the power flow affects the phase relationship between the current and voltage, but the phase shift change cannot be detected by measuring these two values separately. Instead, the current and voltage must be read simultaneously by an instrument that either measures the phase shift or takes it into account. The usual instrument is a wattmeter, something that is often encountered in schools but is not particularly common in the field. Unfortunately, without appropriate instruments, it may be impossible to measure energy usage.

Three-phase systems deliver power with three sequential voltages and simultaneous current flow in three or four conductors. The conductors are shared by three related, but largely independent, circuits. Total power flow is the sum of the power delivered by each of the three circuits; measuring it demands a very good understanding of three-phase connections, specialized meters, and a fair amount of time. It involves multiple readings, multiple single-phase wattmeters, or a complex and costly three-phase power meter. If the meters are standard laboratory types, instrument transformers will almost certainly be required. The connections for the wattmeters are very specific. It is extremely easy when connecting the meters to make mistakes that result in measurements that only appear to be correct.

The implications of not being able to read power properly and conveniently may have a minimal impact on service work. However, it is an issue that cannot be avoided in dealing with energy management, or in any circumstances when an accurate cost analysis is required. The three-phase measurement problem can be solved with appropriate instruments. The impact of reactive loads can be unimportant in some situations and critical in others. Dealing effectively with the operating cost of reactive loads requires an understanding of the equipment, the utility metering system used, and utility billing policies.

DC Power

In a DC circuit, current flow does not reverse, so the energy normally flows only from the supply to the load. Power can be measured accurately by monitoring the voltage and current, and multiplying these two values. Assuming both are in basic units, volts and amps, the answer will be in watts.

Power in all DC circuits can be found with

$$P = E \times I$$

(The development of this formula was discussed on p. 282)

The power formula, sometimes called Watt's Law, shares two variables with the Ohm's Law formula. By substituting values into Watt's Law, two power formulas can be generated. For example, since $E = I \times R$, $(I \times R)$ can be substituted for E with the following result:

$$P = I^2R$$

Following the same logic, $I = \frac{E}{R}$. Substituting $\frac{E}{R}$ for I produces

$$P = \frac{E^2}{R}$$

All of these formulas generate the same result. Which is correct depends on the variables provided. For example, if current and voltage are known, $P = E \times I$ is the logical choice. $P = I^2R$ is used when current and resistance are known. These formulas can be transposed like any other formula, but the square root function must be taken into account. For example, if P and R are known and the voltage must be calculated, the formula can be transposed as follows:

$$P = \frac{E^2}{R}$$

$$E^2 = PR$$

$$E = \sqrt{PR}$$

Such formulas can be convenient to use but are easily forgotten. When the development of the formula is understood, it can almost always be reproduced when required. It is also possible to solve most problems more than one way. Using just the basic Ohm's Law and Watt's Law formulas may involve more steps, but the results are the same.

AC Power

In an AC system, the current constantly reverses, spending an equal amount of time moving in each direction. This by itself does not cause the flow of energy to reverse. For example, current flow through a resistor produces heat, without regard to which way the current happens to be flowing.

However, reactive loads, such as coils or capacitors, store energy by converting it into a form that makes it recoverable. A coil accepts energy until the current through the coil is at its peak. At this point the magnetic field of the coil is at its maximum. Then, as the AC current declines, the field begins to collapse, inducing a voltage back into the coil. The result is that the energy stored in the coil's magnetic field is released back into the circuit. This store-and-release pattern is repeated each half-cycle.

In a capacitor, the charge on the plates and the total energy stored increase until the AC supply voltage reaches its peak. At that point, the capacitor voltage is equal to the peak voltage of the supply. As the supply voltage starts to drop, the capacitor voltage is higher, so the stored charge flows back into the circuit. This charge-and-discharge pattern occurs on each rise and fall of the AC voltage, once each half-cycle.

Energy is exchanged between the circuit and the coil or capacitor on each half-cycle of the AC wave. The energy is stored differently in each of the two types of loads, and, although there are many similarities to the results, there are also some important differences. Both types of reactance draw current that is out of phase with the supply voltage, but in coils the current lags the voltage, and in capacitors it leads. This gives the two types of reactance complementary characteristics. Each is interesting enough on its own; when they are used together, the results can be amazing.

When the voltage and current are in phase, such as in resistive loads, the energy flow does not reverse. It flows from the source to the load at all times. When the voltage and current are out of phase, the direction of the power flow reverses twice each cycle, each time the reactive load releases energy back to the circuit. The reverse flow cannot exceed the forward flow—the reactive load only stores energy, it cannot create it—but the two can, at least theoretically, be equal. In a perfectly reactive load—one that produces a full 90° phase shift between voltage and current—the forward and reverse flows are equal, and the net energy transfer is zero.

Power is the rate of energy flow, and it is always a product of voltage and current of the circuit. However, when calculating power in an AC circuit, it must be remembered that the usual RMS values of AC voltage and current are averages created for convenience. The real values are constantly changing. When AC power is calculated properly from all the instantaneous values along the voltage and current waves, it becomes obvious that AC power flow is also a changing value. When the power values are plotted on a graph, the result is a sine-squared wave, similar in shape to the voltage and current from which it is created (see Figure 8–5).

The total power transfer from the source to the load is the average value of the AC power wave. Unlike the voltage and current, the power "wave" is not on the center line of the graph. It may be totally positive—all above the line—or partially negative—up to half the wave falls below the line. Positive values represent power flow from the source to the load; negative values are power flow back to the circuit from reactive loads.

The graph depicts the actual values as they would occur in a circuit. The percentage of positive power is directly related to the phase angle between the current and the voltage of the circuit. This is not an angle in the usual sense. It is measured in electrical degrees, which are actually a unit of time referenced to the length of one cycle. One complete cycle is assumed to be 360° in duration. A phase angle of 90° is therefore one quarter of a cycle.

The phase angle is referenced to the first zero crossing of either the voltage or current. In most texts, including this one, the reference is the voltage. From this reference point the current may be in phase, leading, or lagging. When current leads the voltage, it occurs earlier in time, shifted ahead by the reactive nature of a capacitive load. When the current lags the voltage, it has been delayed, shifted behind the voltage by the reactance of an inductive load. The theoretical maximum phase difference in any reactive circuit is 90° lead or lag. This is rarely, if ever, attained, but it is not unusual to get close.

The portion of the power flow that is positive or negative is set by the phase angle. At a phase angle of zero—voltage and current in phase—the power flow is entirely positive. This occurs in a purely resistive circuit, or one where the two reactances are equal. When the phase angle reaches 90° in either lead or lag, the positive and negative power flows are equal, and net power flow is zero. This would occur in a purely capacitive or inductive circuit. At in-between values of phase angle, the net power is also at some value between maximum and zero.

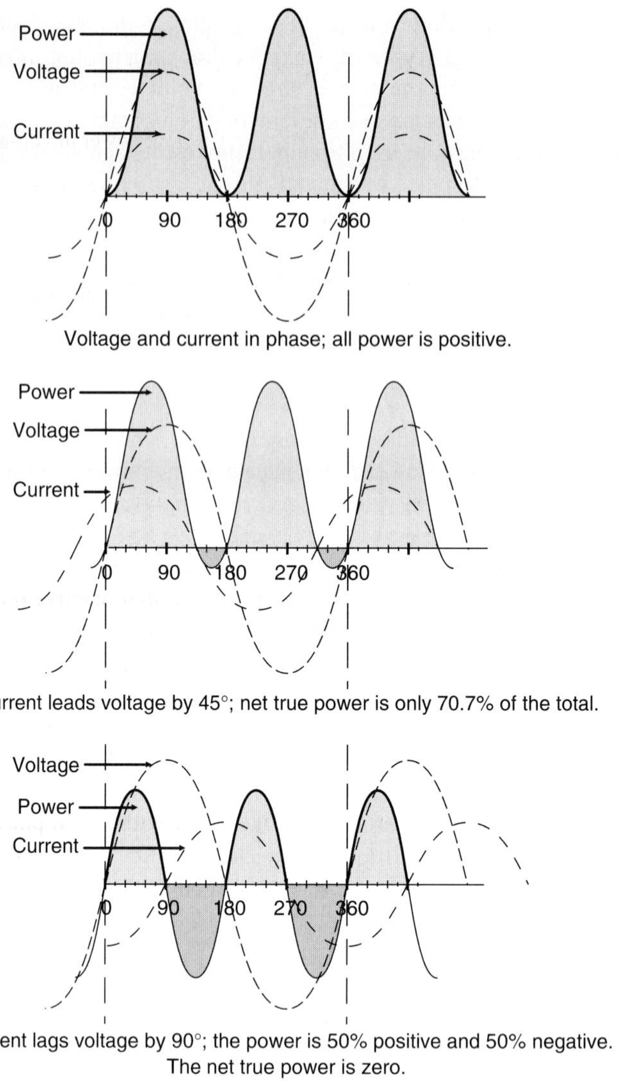

FIGURE 8–5 When the AC power curve is developed from the instantaneous values of voltage and current, the result shows the effect of reactance. In the resistive example (top), power is totally positive because none of it is returned to the system. In the first reactive example, a 45° leading current (center) produced by a mix of capacitive reactance and resistive load shows the "negative" power returned by the reactive portion of the load. The pure inductance depicted in the final example has a net energy transfer of zero because the "perfect" reactance returns all the energy it receives.

Power Factor

The net power of an AC circuit is referred to as the true power. This value represents the actual energy transfer from the supply, and is therefore the wattage of the circuit. The true power (P) is a multiple of the RMS current and voltage times the cosine of the phase angle. The angle is referred to as theta (θ) in most electrical texts. This cosine value is also the circuit power factor (Pf).

$$P = E \times I \times \cos \angle\theta \ldots \text{ or}$$

$$P = E \times I \times Pf$$

$$\angle\theta = \text{the phase angle}$$

$$Pf = \text{Power factor}$$

The power rate that the circuit appears to have, based on current and voltage readings, is called apparent power and is measured in volt-amps.

$$S = E \times I$$

S = Apparent power in volt-amps (VA)

Power factor is also a ratio of true power to apparent power, a relationship that expresses very well what power factor represents.

$$Pf = \frac{\text{Watts}}{\text{Volt-Amps}}$$

$$Pf = \frac{P}{S}$$

Pf = Power factor

Note: With a phase angle of 0°, cosine and power factor equal 1. When the angle is 90° lead or lag, they equal 0. Power factor can be expressed as a numeric value or in percent and should include a lead or lag designation as appropriate. Capacitive loads have a leading power factor; inductive loads are lagging.

The third form of AC power is called reactive power and is measured in volt-amperes reactive (VAR). Reactive loads draw current and have voltage drops, but the energy they receive is returned to the supply. This circulating flow of energy is sometimes called *watt-less power* because it does no work.

Apparent power is the sum of the total true and reactive power for the circuit, but these values are 90° out of phase, so they cannot simply be added. A convenient calculation technique is to place all the power values on a "power" triangle as shown in Figure 8–6.

A power triangle is constructed with the in phase (true) power on the base, reactive power on the vertical side, and apparent power on the hypotenuse. The phase angle in electrical degrees is equal to the physical angle between the base and hypotenuse. Any

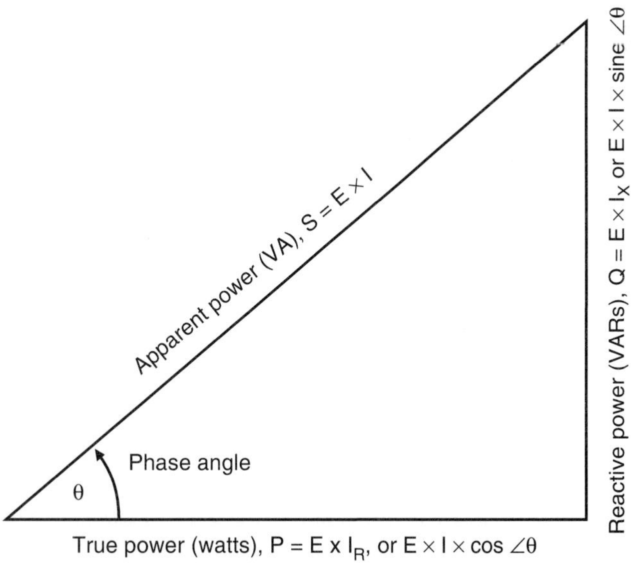

FIGURE 8–6 A power triangle shows the relationships between true, apparent and reactive power.

geometric or trigonometric relationship that can be applied to a right angle triangle, can be used to solve for missing values.

Apparent power is the product of circuit voltage and current, with phase shift ignored. It is the most meaningless of the three AC power values, and the easiest to find. No special instruments are necessary, because only RMS voltage and current values are required.

True power can be measured directly with a wattmeter. Power factor can be found from the ratio of true to apparent power, and phase angle can be found from power factor.

$$\angle\theta = \cos^{-1} \times \text{Pf}$$

In theory, power factor can range from a minimum of 0 to a maximum of 1, in lead or lag. In practice lagging loads are much more common because there are so many devices that use coils, and so few that have a large amount of capacitance.

Reactive power (Q) can be calculated from true power and apparent power using the right angle triangle relationship. Lead or lag should be indicated as appropriate.

$$Q = \sqrt{P^2 + S^2}$$

$$Q = S \times \text{Sin} \angle\theta$$

Q = reactive power in reactive volt amps (VAR)

$\angle\theta$ = the phase angle in electrical degrees

Power factor can have a major impact on the utility bills paid by large customers. The supply utility must provide all the current demanded by the customer, regardless of whether it is supplying true power to a load that does work, or wattless reactive power that circulates in the system without doing anything useful. Customers with a lot of reactive load draw relatively high current in comparison to their true power load. The utility equipment must carry the full line current, even though much of it is reactive current that does not register on a normal kilowatt hour meter. The circulating reactive current increases losses in the system, lowering efficiency and reducing the effective capacity. To obtain proper compensation, most utilities use special meters and levy additional charges. They may also insist that customers maintain their power factor within certain limits.

Revenue meters designed for monitoring customers with reactive loads have two measuring elements. One responds to the apparent power in kVA, the other to true power in kW. The kVA is usually recorded as a peak. The kilowatts are recorded as an accumulating energy consumption in kilowatt hours. Utility policies vary widely, but a typical bill includes kWHr consumption plus additional charges for peak kVA.

Note: The triangle calculation method can also be used to solve for current and voltage in circuits that contain a mix of in phase and reactive loads. In parallel circuit calculations the adjacent and opposite sides are the in phase and reactive currents, and the hypotenuse is the total current. For series circuit calculations the adjacent and opposite sides are the voltage drops across the in phase and reactive portions of the load, and the hypotenuse is the line voltage.

This is a form of phasor addition. Phasors are straight lines with a length scaled to represent a quantity, at an angle representing the phase relationship to a reference. For typical parallel problems the reference is the line voltage. For series circuits it is the line current. Values that are in phase fall on the reference line; reactive values are drawn up or down at 90°. The individual phasors are summed using geometry or trigonometry.

Single-Phase AC Power Calculations

The formulas given below are applicable to any two-wire single-phase AC circuit. They can also be used for dual-voltage circuits, such as the common 120/240 V three-wire connection. To find power in a three-wire circuit, calculate each half separately, using the two line currents and the voltages from line to neutral. Ignore the neutral current. If the power factor is different in each part, find the totals for true power and reactive power, and then use these totals to find the total apparent power.

Any number of true power values can be added to find total watts. The same is true for reactive values, which can be added to find total reactive power (VAR). Different apparent power values cannot be added unless the power factor for each of them is the same. Total apparent power can be found from separately calculated totals for true power and reactive power. This is a phasor sum, not simple addition.

$$S = E_L \times I_L$$

$$P = E_L \times I_L \times Pf$$

$$Pf = \cos \angle\theta$$

$$Pf = \frac{P}{S}$$

$$Q = E_L \times I_L \times \sin \angle\theta$$

$$S = \sqrt{P^2 + Q^2}$$

Where:

E_L = Line voltage
I_L = Line current in amperes
$\angle\theta$ = Phase angle in electrical degrees
Pf = Power factor
S = Apparent power in volt-amps (VA)
P = True power in watts (W)
Q = Reactive power in volt-amps reactive (VAR)

The last formula above describes the power triangle. Any other right angle triangle relationship can also be used when calculating power, phase angle, and power factor values.

Power factor is always either leading or lagging, based on current referenced to the voltage. Lead or lag should be stated whenever power factor is given or calculated. Whether the power factor leads or lags depends on the dominant reactance in the circuit. When the load is mostly inductive, it will lag. It will only lead if the load is capacitive, or in a special load such as an overexcited synchronous motor. When power factor is stated for an obvious load, such as a motor, the lagging designation is often dropped.

Three-Phase AC Power Calculation

Three-phase power is included here to complete the section on power and to provide the required tools for three-phase motor calculations in the next section. The formulas given below can be successfully applied with only a rudimentary knowledge of how three-phase systems work, but they are limited to balanced loads.

Most three-phase loads, including motors, are designed to draw equal power from each of the three supply lines. Such loads are inherently balanced. However, three-phase systems often supply a mix of three-phase and single-phase loads, leading to differences in the power supplied by each of the three phases. To calculate power in an unbalanced system, the total load usually has to be broken down into its component parts. The true and reactive powers for each of the individual single- and three-phase loads are calculated and then summed to totals. Total apparent power can then be calculated from the total true and reactive powers. Individual apparent power values cannot be added unless they have the same phase angle. Unbalanced load calculations, while not necessarily difficult, are time-consuming and generally outside the intent of this book.

Three-phase calculations are normally based on line values, not phase values. Line values are measurements taken external to the load, using only the three main line conductors. Line current values are measured in each of the three supply lines. Line voltages are measured from line to line, not from line to neutral or ground. Loads that are supposed to be balanced—such as motors and three-phase heating elements—are often treated as balanced loads even though small variations in the equipment and the three supply voltages may introduce minor differences. Relatively small variations can be accommodated by averaging the three line readings. Large variations are usually an indication of problems in the system.

Three-phase systems use three (and sometimes four) lines to supply power from three sequential voltage sources. The lines are shared, so power calculations based on line values must be corrected to compensate for the overlaps. A constant has to be added to all three-phase formulas that use line values. This constant is the square root of 3 ($\sqrt{3}$), which is equal to approximately 1.732. The resulting general-purpose three-phase formulas are very similar to their single-phase equivalents.

$$S = E_L \times I_L \times \sqrt{3}$$

$$P = E_L \times I_L \times \sqrt{3} \times Pf$$

$$Q = E_L \times I_L \times \sqrt{3} \times \text{Sin}\angle\theta$$

Power factor for the three-phase circuit can be found with the same formulas used for single-phase calculations.

$$Pf = \frac{P}{S}$$

$$Pf = \text{Cos}\ \angle\theta$$

Overall power factor is the ratio of the total true power to total apparent power. Lead or lag is determined by the dominant characteristic of the reactive load. As with multiple single-phase circuits, any number of true power and reactive power values can be added to produce total watts and VARs. Apparent power values cannot be added unless they have the same power factor. The three-phase power factor cannot be found from phase angle unless it is the same in all phases.

8.7 Motors, Efficiency, and Power Factor

As with all things, AC motors, three-phase or single-phase, operate with an efficiency of less than 100%. Most motors also operate at a power factor that is less than 1 and lagging. There are two notable exceptions to this. The first is the wound-field synchronous motor, which is essentially an AC generator operated as a motor. The power factor of this motor can be adjusted across a wide range from lead to lag. The adjustment is reasonably independent of motor load, and can used to correct and control the power factor of a complete plant.

The second exception is the single-phase permanent split capacitor (PSC) motor. This motor will be studied in detail in Chapter 10. The PSC motor has a permanently connected run capacitor that acts to supply the motor's reactive magnetizing current. With an appropriate design, PSC motors can operate at unity power factor. It should be noted, though, that not all PSC motors are built for optimum power factor or efficiency. These motors are a popular choice for direct-drive fans and blowers in the $^1/_4$ to 15 Hp range.

Efficiency and power factor are separate values, and both vary with motor load, but although power factor affects the operating efficiency of the system, the two must not be confused.

Motor Efficiency

A motor converts electrical energy into mechanical energy. Its efficiency is a function of the total input energy compared to the total mechanical output. The difference between the input and output energy is the sum of all the electrical and mechanical energy losses.

Efficiency can be described and calculated with the following formula:

$$\eta = \frac{P_{OUT}}{P_{IN}} \times 100$$

η = Percent efficiency

P_{OUT} = Power output

P_{IN} = Power input

Input power must be in watts. For a three-phase motor, this usually requires two or three single-phase wattmeters, or one designed to read three-phase. Output power is mechanical and presents its own measurement difficulties. Mechanical power is a function of shaft speed and torque, both of which require specialized test equipment. For general testing, the easiest out is to assume that the motor develops its rated horsepower and simply convert this nameplate rating into watts. This provides a reasonably accurate full-load value if the motor is receiving its full nameplate rated voltage.

When mechanical output has to be measured—this is the only way to find efficiencies at partial load—a tachometer and an appropriate dynamometer or prony brake are required. The tachometer is used to measure the shaft speed; the dynamometer or prony brake loads the motor mechanically while measuring the output torque. Torque may be measured in pounds force feet (lb_f ft), ounce force inches (oz_f in), or in the SI unit of newton-meters. In keeping with the SI nature of electrical units, calculations here are based on SI units. However, because North American motor nameplates provide data in horsepower and revolutions per minute, and dynamometers are likely to be calibrated in pounds foot or ounce inches, appropriate conversions are provided.

$$P_{OUT} = Hp \times 746$$

$$P_{OUT} = \omega \times \tau$$

$$P_{OUT} = 0.1047 \times N \times \tau$$

P_{OUT} = Output power in watts

ω = Speed in radians/sec

τ = Torque in newton-meters (Nm)

N = Speed in rev/min

If torque is in pound force feet or ounce force inches, these can be converted to newton-meters:

$$1\,lb_f\,ft = 1.35582\text{ Nm}$$
$$1\,oz_f\,in = 0.00706156\text{ Nm}$$
$$1\text{ Nm} = 0.7375621\text{ lb-ft}$$
$$1\text{ Nm} = 141.6119\text{ oz-in}$$

Mechanical losses within the motor include the friction losses of the bearings, and windage losses of the moving parts such as the shaft, rotor, and cooling fans. Electrical losses include any power losses due to winding resistance. These are often referred to as copper losses, or I^2R losses, because that is the usual formula for finding them. The total copper loss includes losses in both the stator and rotor conductors. There are also losses in the core iron, including eddy current and hysteresis losses in both the stator and rotor. Iron losses are sometimes lumped in with the electrical losses.

Because most AC electric motors run at a nearly constant speed, the mechanical losses tend to be relatively constant from no load to full load. In calculation problems, mechanical losses are either given or can be found by deduction from other values. In real-world tests, losses are calculated from experimental data.

The copper losses—I^2R in the stator and rotor conductors—vary with load, because currents in both the stator and rotor are dependent on the amount of mechanical load on the motor. Losses are proportional to the square of the current, which helps explain why motors overheat so quickly when they are overloaded. The iron losses—eddy current and hysteresis losses in stator and rotor iron—vary with field strength. The field strength is primarily dependent on the voltage, not the current, so it is also constant. This is difficult to explain, but essentially true. In calculation problems, iron losses, like mechanical losses, are usually found by deduction from other data.

There are many high-efficiency motors in the marketplace. These motors are usually built to somewhat better specifications than standard motors. Improvements may include larger gauge wire in the stator windings, optimized rotors, more iron, or iron with better magnetic properties, more attention to winding details, better bearings, improved fan design, and more precise air gaps (the space between the rotor and the stator). These premium motors demand premium prices, but they almost always pay for themselves several times over during their operating life.

Motor Power Factor

When a motor is operating with no mechanical load connected to the shaft, the power supply must provide only the magnetizing or excitation current, plus enough in-phase current to carry any losses. The power demand used to magnetize or excite the motor field is entirely reactive. The power demand consumed by the losses is all in-phase, true power in watts.

The reactive load is essentially constant whether the motor is operating at no load or full load. It is a reasonably acceptable simplification to find the reactive load from no-load measurements of true and apparent power, and then use this value as a constant for further calculations.

The true power wasted in the motor losses increases somewhat as the motor load increases, but it is only a small fraction of the total power demand created by the effects of a mechanical load on the shaft. Only the copper losses change significantly as the motor load is increased.

FIGURE 8–7 This simplified motor model shows how motors appear to the supply.

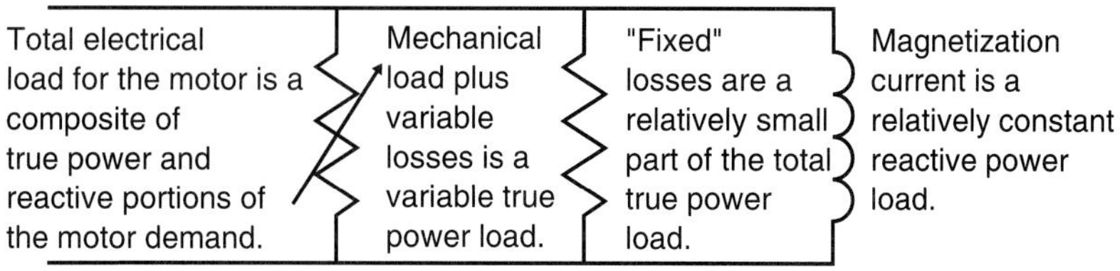

Approximate circuit model for a typical motor.

A workable, though oversimplified, model of motor behavior is a circuit with three individual parallel loads, an inductor, and two resistors (see Figure 8–7). One of the resistors has a fixed value; the other is variable. The inductor represents the nearly constant reactive load of the motor's magnetizing current. The fixed resistor is the constant true power load of the iron and mechanical losses. The variable resistor is the effect of mechanical load on the motor, plus the copper losses. The inductor (motor field excitation current) has a greater current demand than the fixed resistor (iron and friction losses). The variable resistance (motor load current) always demands a little, but when it is adjusted to represent full motor load, it draws much more than either of the other loads.

Like the motor it represents, this circuit has some interesting characteristics. First, with no load the motor is predominantly inductive. The power factor is very low and lagging. This is the normal demand of a typical motor with little or no mechanical load. When load is added to the motor, it has the effect of demanding a lot more true power from the power supply. This is the energy that is used to supply torque to the motor shaft to overcome the mechanical resistance of the driven load. At full load the motor demand is predominantly true power. This makes the power factor much higher, although it still lags.

The net result is that the motor power factor is very low at no load, but rises as the load is increased toward the full rated output. At no load, a typical three-phase induction motor may have a power factor of less than 0.3 (30%) lagging. At full rated load, a large motor may have a power factor in the 80%+ range, or even into the 90s. Small motors usually have slightly lower power factors than larger ones.

Power Factor and Efficiency

Power factor is a function of the phase angle ($\angle\theta$) between the input current and voltage. The phase angle is, in turn, a function of the balance between reactive and resistive elements of the load. Power transferred to the load is often viewed as component parts—the in-phase or resistive power in watts, the out-of-phase reactive portion in volt-amps reactive, and their phasor sum—the apparent power in volt amps. Power factor can be found from the phase angle ($\cos \angle\theta$) or the ratio of true power to apparent power.

Efficiency is based entirely on true power in watts and is a ratio of power input to power output. It is affected by power factor, but not at all the same. Power factor affects efficiency because, when it is less than 1, more current flows in the system than is required to supply the true power load. This additional current results in higher losses, with no additional power delivered as output. Increased losses without an increase in output translate into lower efficiency.

Power Factor Correction

Low power factor impacts on the efficient use of power production and distribution equipment such as generators, transformers, conductors, and switchgear. The system must be sized to carry the total current demand, including any extra caused by poor power factor. For this reason, most utilities charge customers with poor power factor for more than just kilowatts of energy usage. This usually affects only commercial and industrial customers, but can have a dramatic effect on their operating costs.

Many such customers are forced by the economics of punishing utility bills, or by enforced utility policy, to correct their power factor to some minimum level. Power factor correction is a process of reducing the reactive power demand relative to the true power demand. This usually involves making adjustments to reduce the reactive load, and finding alternative sources of reactive power so that it does not have to come from the utility supply.

There are a number of approaches to power factor correction. In some cases, changes in operating procedures or equipment use can provide some improvement. For example, reducing the number of lightly loaded motors and transformers will help. If such adjustments can be made without loss of production, gains in both power factor and overall efficiency can be made. This is usually worth investigating, but more extreme measures are often necessary.

By far the most common way to improve power factor is to use capacitors as a source of reactive power. When a motor is coupled with suitably sized capacitors, the reactive power that provides the motor's magnetic field comes from the capacitors rather than from the supply. This reduces the current demand and improves overall power factor. The capacitors may be installed at each motor, at the motor controller, or in banks connected at the building service or subservices.

The optimum capacitor location, in terms of energy savings, is at the motor. With the capacitors next to the motor terminals, the current is reduced in the entire system. This method is not the most popular because it is the most difficult to design and install, particularly in a retrofit. The second most efficient solution is to install the capacitors at the motor control center. The motor feeders do not benefit from the reduction in current, but the control center current demand is reduced. Again, the installation may be complicated by the number of smaller installations. It should also be noted that, in both of these cases, the existing motor control equipment must be capable of switching a capacitive load.

The most popular method, especially for retrofitting, is to install banks of capacitors with an automatic controller. Installation is simpler, and the banks can be engineered to match the circumstances. This type of installation is not without its problems, however, and any major investment like this should be approached with caution.

Power factor correction capacitors are rated in kilovars (kVAR), a rating of the reactive power they can supply. When they are used in individual sets, they are sized to match the motor kVAR requirements; when they are used in banks, they are usually switched in groups by a controller that senses system power factor. The total capacitance is automatically adjusted to match the total kVAR demand.

Note

Older capacitor installations often run into unexpected power quality problems when new electronically controlled equipment is installed. This situation is now well understood and can be accommodated by any reputable company with experience in the field. Unfortunately, when problems such as these do arise, the end results can be exasperating and expensive to correct.

There is another, old technology solution to poor power factor: One or more synchronous motors can be installed. They are usually connected to a load that is always in demand, such as an air compressor or an important pump or fan. Properly sized to the needs of the plant, the synchronous motor can provide mechanical power while stabilizing the plant power system and correcting the power factor at the same time.

Synchronous motors have been called the "prima donnas" of electric motors. They tend to be large and expensive, but are very reliable performers. In some cases they are operated without load, used only for their power factor correction ability. Operated in this way, they are sometimes referred to as synchronous condensers. There is some renewed interest in these motors because, unlike capacitor banks, they do not contribute to harmonic and other power quality problems.

8.8 Practical Considerations for Power Delivery

All mechanical and electrical systems involve trade-offs in design. In mechanical systems, a common trade-off is a reduction in size or weight at the cost of strength and durability. In a hydraulic system, it could be operating pressure vs. oil flow rates. Small high-pressure systems may be preferable to large ones operating at lower pressures, but the high pressure is more difficult to contain. In electrical systems, the central trade-off is voltage vs. current. High voltages require better insulation and more sophisticated control equipment; low voltages allow for simpler wiring and components. Low-voltage systems are also much safer, but they need a proportionally greater current flow to provide the same power. Extremes always cause problems, regardless of the system.

The appropriate balance of mechanical force and movement, or electrical voltage and current, depends on practical rather than theoretical considerations. Extremely high mechanical forces or electrical voltages are usually avoided. They are very difficult and expensive to contain and control, and can be very dangerous. Systems are taken to extreme levels when there are no other acceptable options, but the costs can be staggering. High-voltage distribution is a good example. The power utilities have no other practical choice, but the physical size of power line towers, the length of the insulator strings, and the sophistication and cost of high-voltage breakers and switches attest to the difficulties.

Over the years, standards have developed. The many different levels of voltage and current are reasonably well matched to their applications. Voltages are kept low for safety reasons, and only raised to higher levels when it is necessary. A 24 V control system, for example, is simple, inexpensive, and adequate for many applications. However, it is very limited in the power it can deliver, and quickly runs into voltage drop problems if wiring runs become very long. A 120 V control system has a much greater flexibility than one operated at 24 V, but it is also more dangerous. The hazards of electrocution and fire are increased, and making the system safe increases installation and component costs.

When motors are required, 24 V is impractical beyond very low-powered drives such as those used for damper and valve actuators. A 120 V system does not fare much better. One or two horsepower is about the limit. Doubling the voltage to 240 extends the motor range to 10 to 15 horsepower, the upper limit for single-phase. Three-phase motors extend the range to almost any usable horsepower rating, but as the power rises, so does the voltage. It is usually impractical to operate continuous-duty motors at extreme current levels, and high-voltage is the only alternative.

Questions

1. Briefly explain the relationship between energy and work.
2. Give an example of potential energy.
3. Define energy "consumption" in general terms.
4. What is meant by the law of energy conservation?
5. Define "power."
6. Describe how electricity "connects" energy sources to distance loads.
7. List several properties of electricity that have contributed to its commercial success.
8. Does a utility meter measure power or energy?
9. Briefly describe the movement of energy through a simple electrical circuit, starting at the input energy of the source and ending at the output energy of the load.
10. Define the coulomb.
11. Define the ampere in terms of time and coulombs.
12. Show the relationship between volts, amps, and watts in terms of joules, coulombs, and time.
13. What SI units are shared by both mechanical and electrical systems?
14. Define the unit torque in general terms and on the basis of SI units.
15. List the losses that occur in a typical electric motor.
16. The electrical unit of apparent power cannot be used in efficiency calculations. Why?
17. Only in very specific circumstances is electrical power (watts) a simple multiple of volts and amps. What are these circumstances, and in what situations is the calculation more complex?
18. Explain why it is usually impossible to measure AC power using only voltmeters and ammeters.
19. List the three forms of the Watt's Law formula.
20. What is meant by the term "electrical degree"?
21. What are the three units of "power" used for AC circuit calculations? Provide a formula for each of these units, and explain what measurements would usually be required to find them.
22. What is power factor in terms of apparent and true power, and in terms of the phase relationship between the voltage and current?
23. What is the difference between a load with a power factor of 75%, leading, and one that is 75% but is in lag?
24. Why do power utilities charge customers extra when they have a low power factor?
25. Why do larger control systems use 120 V instead of the safer 24 V supplies used in many smaller systems?

9 Motor Fundamentals

Note

A truly meaningful understanding of motors can only be built on a foundation of several different, but related, areas of knowledge. These include an understanding of electrical basics, electromagnetic relationships, inductive and capacitive reactance, and mechanical and electrical power. All of these are described in detail in the preceding chapters of this book.

Objectives

After the study of this chapter is complete, you should be able to:

- Describe the physical characteristic of typical electric motors using appropriate terminology
- Describe the differences between various types of AC and DC motors with regard to their basic construction, the terminology used to describe them, and their speed control characteristics

9.1 Introduction

Electric motors are a major capital investment, a primary energy user, and a major contributor to maintenance and service costs. The forces that drive motor research efforts today include demands for increased efficiency and better control of speed and power output. There is also increased pressure to reduce motor maintenance. There are several emerging technologies and intensive efforts into refinements of existing designs.

High-efficiency motors are in demand, not only for large industrial applications, but also for the motors used in commercial and residential situations, such as those used in HVAC equipment. Energy costs are high and steadily increasing. The price premium for a high-efficiency motor can often be recovered in less than one year.

Better speed control has been an issue with electric motors since their invention. An adjustable-speed electric motor that is reasonably priced, efficient at all speeds, highly reliable, and virtually maintenance free, is—in the minds of many—the holy grail of the electric motor industry. Many of the latest motor and motor control developments are attempts at attaining this goal.

Electric motors are complex. In this chapter and those that follow, there is no attempt to provide mathematical models of motor behavior, or the minute details of motor winding layouts. The intent is to provide good general knowledge, covering the majority of commercially available motor types. Calculations are limited to those required for comparison tests that may be conducted in a reasonably well-equipped college laboratory. These calculations also provide a basis for comparisons of motor operating costs, an issue that is central to energy management.

9.2 A Brief Overview of Electric Motors

An electric motor accepts electrical energy as voltage and current, and converts it into mechanical rotation and torque. The techniques employed to accomplish this conversion are deceptively simple and quite ingenious. They have also been well understood for a very long time. Like many successful devices, electrical motors are a product of inventive genius, craftsmanship, and decades of engineering experience and refinement. The operating principles of the majority of modern motors are virtually identical to those of much earlier models. In most cases, even the physical construction is much the same. There have been important improvements in insulation materials and many small details, but the basic designs have changed very little since the days of Thomas Edison and Nikola Tesla, the men who pioneered and built the first successful commercial electric motors.

All motors must produce rotation. This may seem like a silly statement, because it is obvious, but the initial challenge in building any electric motor is to produce continuous directional torque and rotation. The conventional DC approach, pioneered by Edison, uses a stationary magnetic field mounted in the frame of the motor. The windings in the rotating armature are energized through a sliding connection between insulated copper bars and carbon brushes. The current in the armature windings creates a field that is out of alignment with the stationary field, causing the armature to turn. To keep it going, the bars and brushes act as a switching system that changes the winding connections. The two fields are kept out of alignment, even as the armature rotates.

The assembly of copper bars is called the commutator. Individual commutator bars, or segments, are wired to different sections of the armature winding. As the armature turns, the stationary brushes slide across the surface of the commutator, constantly changing the connections to the windings. This technique is used in all conventional DC motors, and has been adapted for use on AC in the series universal motor. The universal motor is a modified DC design that tolerates AC, but does not take advantage of AC behavior.

All other AC motors are based on Tesla's work: They use the changing nature of AC to advantage, creating rotation without a commutator. Instead of switching connections to a wound armature, a rotating field is created in the stationary stator windings. The rotor either locks onto the moving field or is dragged along close behind. None of these motors can operate directly from a DC supply. AC motors are simpler in construction than DC types, but the design gives them a constant operating speed. This is fine for many applications, but a serious limitation in others.

Electric motors can be divided into three basic categories: single-phase AC, three-phase AC, and DC. Each of these can be further divided into various types or subcategories. These are the traditional divisions, and they are a very serviceable way of cataloging older designs. However, it is questionable if they are as universally applicable as they once were. All motors, including the latest types, can be placed under one of these headings, but some of them are very radical departures from the other motors in the group.

For example, a brushless DC motor is powered from a single-phase or three-phase AC supply. In construction, it is more closely related to a three-phase synchronous motor than a traditional DC motor. In operation it is vaguely similar to older DC designs, but is entirely dependent on a sophisticated electronic controller. Such motors are virtual categories unto themselves, but may be best understood in comparison with older designs.

Traditional DC motors come in three basic types with a few variations. Only one, the permanent magnet DC motor, is common in the HVAC industry. It is most often found in small sizes as the drive motor in some actuators. In addition to the traditional DC motors, there are several types of motors—including the brushless DC design mentioned above—that use DC internally but require a dedicated electronically controlled power circuit to provide the necessary switching for motor rotation. These electronically commutated motors are also finding their way into this and other industrial fields. They are technologically advanced motors that may have a very bright future. The present drawbacks are high initial cost and a yet to be established history of performance and reliability. It also takes time for manufacturers, installers, and service personnel to become comfortable with new technology.

Three-phase motors are the largest motor category based on installed horsepower, and one of the smallest in terms of the number of different types. There are only three common three-phase motor designs, but one of them overshadows the others by a very wide margin. There are some variations in the rotors used with these motors, but very little variation in the design of the stator—the stationary portion containing most of the windings. Three-phase motors dominate most industries because of their rugged reliability and good operating efficiency. The main drawbacks to three-phase motors are the need for a three-phase supply and the lack of a simple, efficient speed control. Speed control is possible, with a variable-frequency electronic controller. This is a well-established technology, but it is also very complex. Like most high-powered electronic controllers, variable-frequency drives can also cause serious power quality problems.

The greatest variety of motors is in the small ones, particularly those designed for operation from single-phase AC supplies. The HVAC industry uses almost all of the different types available, including some variations that are unlikely to be found anywhere else. The advantages of one type over another are not always obvious, and there are broad overlaps where any one of two or three different motor types can provide acceptable service in a given application. However, each of the motor types has fairly distinct advantages and disadvantages over the others. In the material that follows, each type of motor is described in general terms for comparison purposes, and in greater detail for a full understanding of how they work.

Although single-phase motors are the most common small AC motor, they have inherent disadvantages when compared to three-phase designs. Due to these limitations, almost all AC motors larger than 10 Hp are three-phase. Larger single-phase motors are available, but are not commonly used. When a three-phase supply is available, it is usually used for all but the smallest motor tasks.

The fact that motors move makes them interesting, at least initially. Unfortunately, a visual examination, even with the motor completely disassembled, provides very few clues as to how it works. Electric motors are deceptively simple in construction. Many are little more than a lot of wire wound onto an iron frame, a metal cylinder on a driveshaft, and a couple of bearings installed in removable end covers.

For curious individuals, a first encounter with an electric motor may leave them puzzled. Those who love to tinker with mechanical devices are usually frustrated. A careful examination answers very few questions, and there is very little, if anything, to change or adjust. There are ways to expose the mysteries of motor operation, but a visual inspection by itself is not very satisfying. However, visual inspections are still important. An understanding of the physical layout, combined with some appropriate tests, will eventually lead to a clear understanding of how these motors work.

The AC Stator

The stationary part of an AC motor is called the stator (see Figure 9–1). Much of the stator's weight is the iron core used to carry the motor's magnetic field. The core iron is laminated, as it is in all AC current devices, to minimize the eddy currents that would otherwise occur. The outer frame of the motor is usually cast iron or formed from sheet metal. It encloses the core iron and winding assembly, and provides support for the ends of the motor and for the base.

All stator windings have electromagnetic poles, but they can be created in different ways. They may be visibly distinct pole pieces—called salient poles—protruding inward from the frame, or they may be created by placing coils of wire into a series of narrow slots cut into the stator iron. Salient poles are used for most series universal and shaded-pole motors, and for most DC motor types. Slotted stators are used for most other motors.

FIGURE 9–1 At the back are three four-pole AC stators: from left to right, a resistance or capacitor start, permanent split capacitor (PSC), and a three-phase. The squirrel-cage rotor in front is used with all of them. These are experimental lab motors, but they are very similar to the commercial versions.

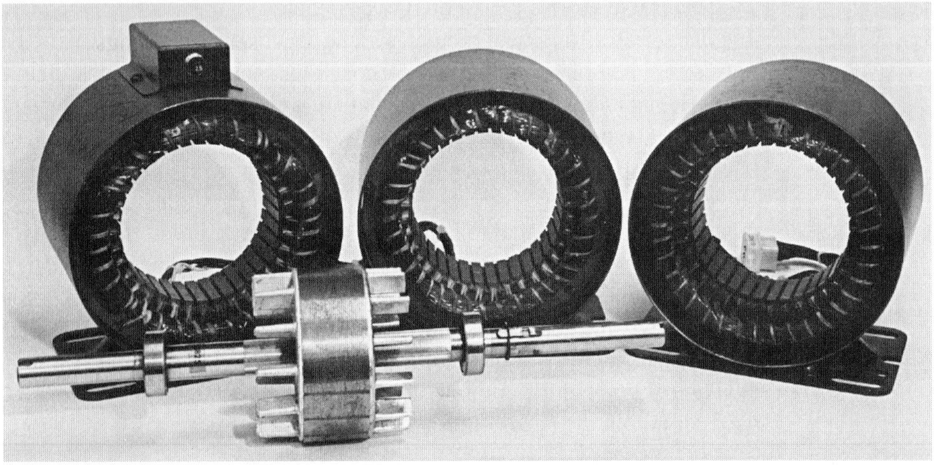

The individual core laminations are stamped from sheets of soft iron alloy made for this purpose. For a slotted core, each lamination has a series of evenly spaced notches cut around the inside of the rotor opening. The notches have a slight lip on the top inside edge. Once the laminations are stacked and secured, the notches form the winding slots, and the lip holds a retaining wedge that is inserted across the top of the coils to hold them firmly in place.

Before the coils are installed, each slot is lined with heavy insulation paper. Once all the coils are secured in place and connected to make the completed winding, they are coated or vacuum impregnated with a heat-activated insulating varnish. The entire assembly is then baked to harden the varnish. The varnish helps exclude moisture and contaminants, and keeps the windings from moving due to the magnetic forces they experience during normal operation. Any physical movement can wear away the insulation and cause the motor to fail.

The stator windings form a number of distinct magnetic poles in the stator iron. The physical width of each pole varies with the total number of poles in the motor, but even the narrowest would span several core slots. Each pole may consist of a single concentrically wound coil distributed across several slots, or several identical coils installed in an overlapping pattern. The number of slots spanned by the coil group gives each pole the desired width. The actual pole locations may not always be visible. For motors with overlapping coils, each coil is identical, and the winding appears as a symmetrical assembly with no obvious divisions anywhere. The individual poles are created by interconnecting the coils into groups. Three different types of windings are shown in Figure 9–2.

All magnets have two poles, so the stator winding poles are also in pairs. The most common conventional motors have two, four, six, or eight poles. There can be more, but very large numbers of poles are usually only used for very large motors directly coupled to low-speed equipment. The extra poles increase the motor diameter and add significantly to the cost. Four poles is the most popular choice, but two- and six-pole motors are also common.

When motors are built for two- or three-phase supplies, they have a complete set of winding poles for each of the phases. Each of the motor poles is duplicated in each of the phase windings. For example, a four-pole two-phase motor would have eight distinct poles in the stator—four for each phase winding. A three-phase four-pole motor has twelve distinct poles, four for each phase (see Figure 9–3).

Many single-phase motors have a second set of windings referred to as the start or auxiliary windings. These provide an artificial second phase, and are placed just as they would be in a two-phase motor. In some cases the second winding is rated for intermittent duty and is only used for starting. In other cases it is rated for continuous duty and is energized at all times.

FIGURE 9–2 Three different types of windings.

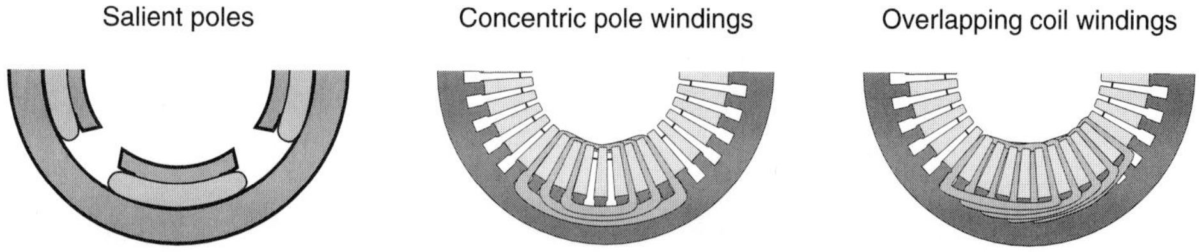

FIGURE 9–3 These motors have different winding arrangements, but all are four-pole. The salient pole is the most obvious. The single-phase stator is identical in layout to a two-phase. The three-phase on the right has each phase winding shaded for identification. There are four poles in each phase.

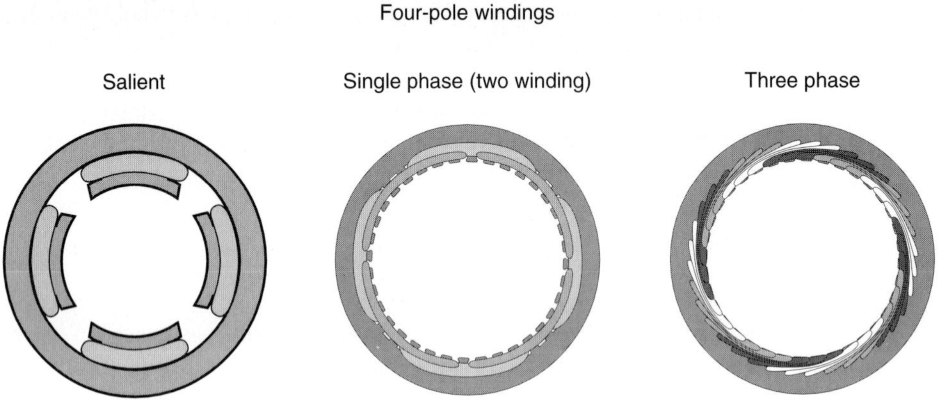

Most three-phase motors use identical coils overlapped to form a symmetrical winding assembly. Each pole of the motor consists of several series-connected coils. To reduce the number of connections, the coils that form each pole are often wound in sequence without cutting the conductor, and then installed as a group.

Note

A DC supply and a compass can be used to expose the elegant simplicity of a three-phase motor winding. With the rotor removed, one phase is energized from a low-voltage DC supply. Moving a compass around inside the stator quickly identifies the polarity and location of each magnetic pole. Energizing each of the phases in turn, and marking the resulting poles, provides an accurate picture of the winding layout for the motor. The test works on any six-lead motor; other lead arrangements are more difficult to demonstrate because the phases are internally interconnected. (This test is described in Chapter 6.)

Single-phase motors usually have concentric windings made with several different-size coils wound from a continuous conductor. The widest coil sections fit into slots that are a full pole width apart; the smaller ones fit inside the larger ones, filling in the empty slots. The motor poles can usually be identified and counted from a visual inspection of the stator. Single-speed single-phase motors with slotted stators and distributed windings have either one or two complete sets of windings. When there are two sets, the motor winding appears to have double the actual number of poles. These motors need a second set of windings to provide a rotating field, either for starting or as part of normal operation.

The poles of this second winding are always centered halfway between the poles of the first. Each motor pole is duplicated in the second winding, so the correct count is the number of poles in either of the windings, not the total number in both. In some motors the second winding is easily identified because it is wound with a lighter gauge of wire. However, this is not a completely reliable indicator because in other motor designs the two windings are virtually identical. Two-speed and multispeed motors are more complex because they contain extra windings related to the optional speeds.

Most shaded-pole and series universal motors have protruding pole pieces—salient poles—instead of a slotted stator core and distributed windings. Identifying the number of poles in these motors is simply a matter of counting them. The poles are not duplicated for a second or third phase, as they are in other motors.

The AC Rotor

Note

The rotors described here are primarily for "induction" motors. These motors are characterized by having no direct power connection to the rotor. Rotor current is driven by voltage that is induced into the rotor conductors from the stator field. Induction motors include most single-phase and three-phase types.

The rotating part of all AC motors, except the series universal, is referred to as the rotor. The rotor is part of the magnetic circuit of the motor and is therefore made of laminated iron, just as in the stator. Some rotors have a conventional winding made with coils of magnet wire set into slots, much like the winding in an AC stator. Examples of motors with wound rotors include single-phase repulsion motors and three-phase wound-rotor motors. Neither of these is very common.

The most common AC rotor has a very simple "squirrel-cage winding," usually made of aluminum alloy bars cast into holes in the rotor laminations. The bars may be embedded deep enough in the iron that they are invisible. The ends of the bars are interconnected by short-circuiting rings at each end of the rotor. These rings are cast at the same time as the rotor bars, and may include cast fan blades. Large motors may have an assembled squirrel-cage winding made from copper or brass bars installed individually in the rotor iron and secured to the end rings by welding or brazing. Squirrel-cage rotors are the most widely used rotor type: They are relatively inexpensive to build, simple, rugged, and very effective.

Armatures, Commutators, and Brushes

Series universal motors are a modified DC motor design, so the rotating portion is called the armature. (Armature is the standard term used for conventional DC motor rotors.) The armature is an assembly consisting of the motor shaft, a laminated iron core slotted lengthwise to receive the windings, and a commutator to which the windings connect. The commutator may be flat or drum-shaped, and consists of a series of copper segments separated by insulating mica. Each segment is electrically insulated from the other segments and from the shaft. Each of the many coils of the armature winding connects to two of the commutator segments (see Figure 9–4).

Motors with commutators also have brushes. These are usually carbon based, although blends of harder materials are sometimes used. The brushes contact the moving commutator to provide supply current to the armature windings. As the armature turns, different commutator segments pass under the brushes. The brushes and the commutator work together as a rotating switch, changing connections to the armature so the motor continues to develop torque in the desired direction of rotation.

FIGURE 9–4 On the left, a repulsion motor stator and rotor. On the right, a DC motor and armature. Compare the complexity of these two wound rotors with the squirrel-cage rotor in Figure 9–1.

Other Motor Details

Many motors have shaft-mounted fans to circulate cooling air through the rotor and the motor windings. The fans may be internal, external, or both, depending on the motor enclosure. The fans may be separate metal or plastic assemblies or, for squirrel-cage rotors, may be a series of blades added as part of the casting process.

The ends of the motor are referred to by various names, including end plates, end bells, and end shields. They enclose the windings and support the bearings on which the motor shaft rotates. The bearings may be ball, roller, or sleeve types. The end bells may also support accessory switches, terminal blocks, or other equipment. End bells are usually cast, and then machined to fit the stator frame and the bearing assembly. On most small motors they are secured by a set of four bolts that extend through the motor from end to end.

Note

When a motor is disassembled, each end should be marked in such a way that it can be returned to the same position it was in originally. This helps to ensure proper bearing alignment and a uniform air gap. Shallow file cuts or matched punch marks—one on one end, two on the other—are common marking methods. The file cuts are made across the separation line to simultaneously mark the end bell and the frame: This allows accurate realignment. This method should not be used if it will cut through protective coatings and expose the motor frame to corrosion.

Standard motor enclosures include open, drip-proof, totally enclosed, submersible, and explosion-proof types. There are also many specialized situations where the motor is an integral part of a factory-built assembly. A prime example is hermetically sealed refrigeration and air conditioning compressors. The motor and compressor are assembled as a unit and then sealed inside a common housing. The motor has no separate enclosure; the rotor and windings are exposed to the refrigerant and lubricating oil. Cooling is provided by the refrigerant gases and by conduction through the housing.

9.3 Motor Speed Control

Most AC motors run at a nearly constant speed that is set by the number of poles and the supply frequency. A two-pole motor runs the fastest; for each additional pair of poles, the speed is proportionally lower. Two-speed operation can be provided using two windings or special winding reconnections. Adjustable speed is generally only possible if the supply frequency can be varied.

Note

Creating a variable-frequency output from a fixed-frequency utility supply is a complex task, but advances in power electronics have made it relatively commonplace. Variable-frequency motor controllers are commonly known as a variable-frequency drives. They are not normally used for single-phase motors, only for three-phase.

Two single-phase motor designs—the permanent split capacitor motor and the shaded pole—are used as adjustable-speed fan motors. They are not naturally adjustable-speed motors, but can be used with good effect on a variable torque load, such as a fan. The speed is varied by reducing the voltage to the motor windings, and thereby limiting the motor's maximum output torque. The motor runs at a lower speed because it cannot develop enough torque to bring the load up to full speed. It does not stall, because of the torque characteristic of a fan load: The lower the speed, the less torque it takes to keep the fan turning. The motor output torque and the torque demand of the fan are in balance at a lower than normal operating speed. This type of application is successful because the motor torque characteristics at reduced voltage are a reasonable fit for the fan load. It is a special case, though. This type of speed control is not possible with other types of loads.

Universal motors that run on AC or DC, conventional DC motors, and motors that have dedicated electronic controllers are usually adjustable speed. Their maximum speed is affected by the number of poles, but there are other variables involved. Virtually all conventional DC motors can be speed controlled through voltage adjustment, but the series universal is the only motor that can provide truly adjustable-speed operation from an AC supply, without the need for an elaborate controller. This motor is a modified DC design that behaves much like a typical DC motor: A change in voltage produces a corresponding change in speed. The series universal motor can also operate successfully over a very wide range of speeds. The maximum speed for small-diameter universal motors is far beyond the limits of other AC motors.

Questions

1. Define the following terms as they relate to electric motors: stator, rotor, armature, end plate or end bell, commutator, salient, laminated, eddy current, and induction motor.
2. What method is used to seal the motor windings against moisture and contaminants?
3. Why is it important to mark the position of the motor ends before disassembly?
4. Are typical AC motors adjustable or constant speed?
5. What type of AC motor has the highest operating speed?

10 Single-Phase AC Motors

Objectives

After completing the study of this chapter you should be able to:

- Identify the various types of single-phase motors
- Compare the different single-phase motor types by stating their operating characteristics
- Explain the purpose and operating principles of start switches
- Explain the purpose and operating principles of start relays
- State the operation of thermal disk motor protectors
- Compare the operation of bimetallic and melting alloy overloads
- Explain the difference between a manual motor starter and a motor rated switch
- Explain the difference between a relay and a contactor
- Demonstrate an understanding of overload heater selection and installation
- Draw diagrams of basic motor control circuits and explain their operation
- Demonstrate how to reverse a typical single-phase induction motor
- Explain why single-phase motors may be difficult to reverse while they are running
- Draw connection diagrams for various motor types and demonstrate an understanding of the connections they represent

10.1 Introduction

Single-phase motors are made in greater variety than any other motor type. Single-phase power is also the most popular electrical service in North America. It is usually supplied as a dual-voltage system offering 120 and 240, or 120 and 208 V, depending on the supply configurations. Higher voltages are also available from typical commercial or industrial three-phase systems. With such universal availability, it is only natural that motors have been developed to operate from single-phase supplies. Unfortunately, single-phase AC power is not the ideal source for a motor. To this point in history, no one has produced a single-phase motor that is ideal for all applications. For this reason, there are many, quite different designs, each with its own strengths and weaknesses.

Single-phase motors can be grouped by type and, in some cases, into variations within these groups. The three basic designs are the series universal, squirrel-cage induction, and repulsion motors. The series universal motor is usually operated on single-phase AC, but it is not exclusively an AC motor. It is a modified DC motor design and has the unique ability to operate directly from either an AC or DC supply. Repulsion motors share many of the features of induction motors, but produce torque in a radically different way. Most repulsion motors are actually a hybrid design, sometimes called repulsion-start induction-run. They start as a repulsion motor, but run as an induction type. They are still manufactured, but see limited use due to their high cost. Repulsion motors are prized for their high starting torque and tolerance to voltage dips on weak electrical supplies. Many old repulsion motors are still in service.

The most popular single-phase motors are squirrel-cage induction motors. These motors depend on the changing nature of the AC supply to produce a rotating field in the stator windings. Producing a rotating field is relatively simple if the supply is three-phase. Each phase is used to energize one of three sets of windings. The voltages are sequential, so the windings are energized in sequence and the field naturally rotates. (This is examined more fully in Chapter 11 "Three-Phase Motors.") Producing a rotating stator field with a single-phase supply is more difficult, but a number of methods have been developed that are still in common use. Single-phase motors in this category include the split phase, capacitor start, capacitor run, and shaded-pole designs.

10.2 The Series Universal Motor

The term "universal" refers to this motor's ability to run from either an AC or DC supply; the term "series" refers to the connection of the field windings and the armature windings of the rotor. The windings are connected in series across the supply. The name is often shortened to "universal motor" (see Figure 10–1).

FIGURE 10–1 The field poles and armature assembly of a small, two-pole series universal motor. The brushes, bearings, and outer casing have been removed.

The series universal motor is commonly found in portable AC power tools, vacuum cleaners, and small appliances. It is the only electric motor that can operate from either a DC or AC supply without an elaborate electronic controller. This motor can also provide true adjustable-speed operation through changes in voltage alone. This is a feat that very few AC motors can perform well, if at all. To a series universal motor, adjusting the voltage has essentially the same effect as adjusting the throttle on a gasoline engine. The voltage is proportional to the available torque output, so if the voltage is reduced with a load on the shaft, the motor slows down. Reducing the voltage to other types of AC motors while they are under load will usually cause them to stall. The series universal motor is also the only conventional 60 Hz AC motor capable of speeds above 3600 rev/min. Series universal motors with small-diameter rotors can reach shaft speeds of 30,000 rev/min or more. Good examples of these small high-speed motors are found in routers and laminate trimmers.

Although the motor operates on AC or DC, variable-speed tools are rated AC only. This is because the speed controller in the switch only works with an AC source. Attempting to run variable-speed AC tools from a DC source destroys the control switch, not the motor. Although 120 V DC supplies are uncommon, they are provided on some portable, engine-driven welders. A good number of fine power tools have been damaged or destroyed by users who failed to take heed of the DC markings on the welder outlet. Note that cordless tools are DC powered and use low-voltage DC motors. These motors are similar in appearance to universal motors, but are not designed for AC use.

The series universal motor is not an induction motor. Power is supplied directly to stationary field poles installed in the motor frame and, through the brushes and commutator, to the armature windings. The energized field poles set up a magnetic field around the armature. This field is established in a circuit of iron that includes the motor frame, the field poles, and the armature core. The only break in the circuit is the narrow air gap between the surface of the armature and the inside of the pole faces.

The commutator and brushes provide the connection between the power supply and the coils that make up the armature winding. The armature coils are interconnected so that when current flows through the circuit, a magnetizing force is created in the armature iron that has the same number of poles as the stationary field. The location of these poles depends on the relative position of the brushes, which are normally located so the magnetic poles of the armature are halfway between the poles of the stationary field. The normal forces of attraction and repulsion act to align the fields, causing the armature to turn.

If the armature and main fields were allowed to become magnetically aligned, the motor would stop. This does not happen because, as the armature turns, different commutator segments pass under the brushes, changing the connections between the supply and the armature coils. The armature field is kept in position—constantly out of line with the stationary field—by the switching action of the commutator and brushes. Because the armature field is always out of alignment with the stationary field, there are always magnetic forces producing torque and making the motor run (see Figure 10–2).

When the universal motor is powered from a DC source, the field polarities are fixed, when it is powered from an AC source, the polarities are constantly reversing. The motor runs much the same with either supply, because both the armature and the field polarities change at the same time. The torque is always in the same direction, even though the direction of the current and the polarity of the magnetic fields are constantly changing.

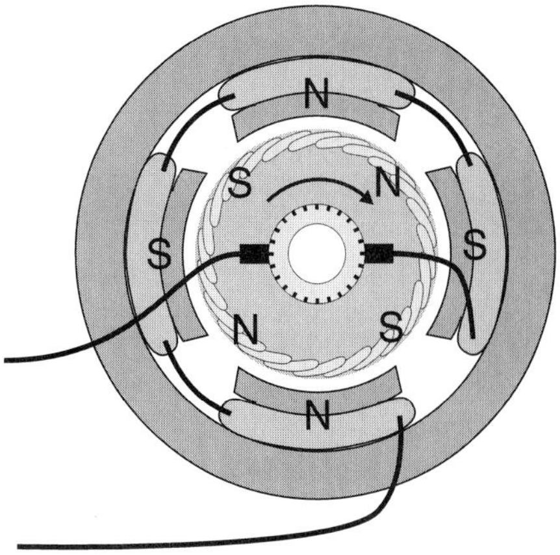

FIGURE 10–2 In a series universal motor, the armature field is always out of alignment with the stationary field poles. The interaction of the two magnetic fields creates torque. Note the series connection of the field poles and armature circuit.

A universal motor is classed as an adjustable-speed motor because it can be speed controlled with a variable voltage. It is also a variable-speed motor because it does not maintain a constant speed under conditions of changing load. When this motor is put under load, it slows down dramatically. When the load is removed, it speeds up again. Most other AC motors have a relatively constant speed behavior, meaning the speed changes only a small amount from no load to full load.

Universal motors can produce high torque, but only at low speeds. Anyone who has used a single-speed electric drill understands the behavior of a series universal motor very well. The bit speed is high until drilling starts, and then as the speed drops, high torque is produced. If the bit jams, the torque output of a large drill can be high enough to be dangerous to the operator.

The high no-load speed of the series universal motor can lead to motor failures. In theory the motor will accelerate until it self-destructs; in practice, there are physical and electromagnetic limitations that keep the speed down, but they are not always capable of protecting it from damage. For this reason most universal motors, particularly the larger ones, are permanently connected to some type of load. If the load is removed with full voltage applied to the motor, the armature may literally fly apart. This is generally less of a problem for small motors, because the small-diameter armature can withstand a much higher speed, and the natural drag of the bearings, brushes, and the cooling fan is usually enough to keep the speed within limits. With larger motors these loads are less significant, and the maximum allowable speed is much lower.

There are a variety of controls used to take advantage of the adjustable-speed characteristic of the universal motor. Multiposition switches and tapped windings can provide several increments of change. Likewise for a similar switch and either a tapped resistor or several different resistance values. Rheostats can also be used if the motor is quite small. Electronic controls are common because of their small size, wide control range, and low cost. Only resistance-type controls can operate on either AC or DC, and then only if the switch is rated for DC use. Because AC supplies are universally available and DC is relatively rare, most variable-speed controllers are designed for AC use only.

Single-speed control is usually managed with a simple manual switch. Reversing requires reversal of the field or armature connections and usually involves a double-pole double-throw (DPDT) switch wired into the armature circuit. The switch is cross-connected for polarity reversal, and is usually separate from the on/off control. When a single control switch is used, it

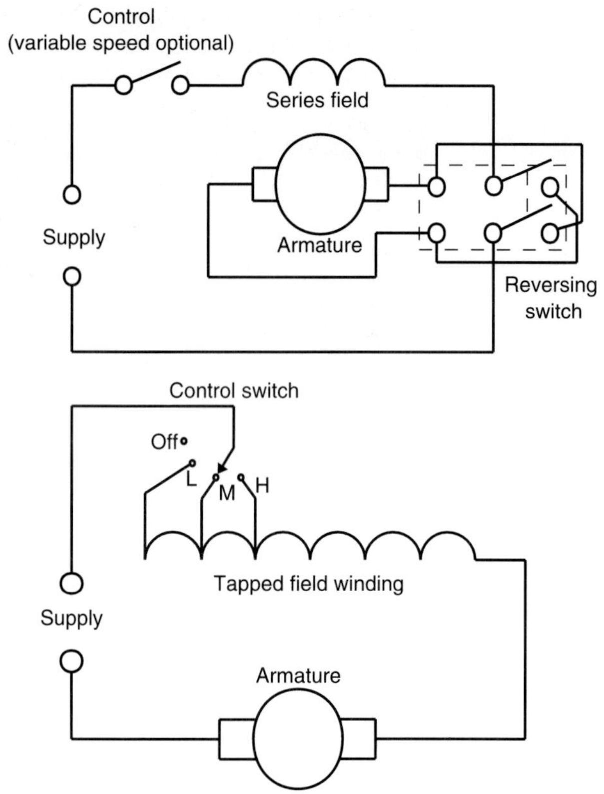

FIGURE 10–3 The universal motor can be reversed with a single-pole double throw (SPDT) switch (top). One way to provide multiple speeds is to use a tapped winding and a

is typically a DPDT with a center off position. Because of the ease with which the control can be shifted from forward to reverse, even unintentionally, this may be less satisfactory than separate switches (see Figure 10–3).

A mechanical governor is also used with some universal motors to provide speed control and speed regulation. This type of control has been used for many years in one popular brand of home-kitchen mixer. It demonstrates the durability of a technologically crude but well-executed design. A simple flyweight governor is connected to a set of contacts that are in series with the motor. A capacitor and a resistor are connected across the contacts to reduce the arcing and provide a reduced current flow when the contacts are open. The spacing of the contacts is adjusted by a spiral ramp inside the control knob. The closer the contacts are to each other, the higher the speed must be before the governor can open them. When the motor slows due to increased load, the governor responds by applying additional power to the motor, maintaining a relatively constant speed under changing load conditions. This control is a simple but remarkably successful device providing a full range of regulated speed adjustments. It is one of the few common motor controls that provides an automatic governor function.

The series universal motor is available in a wide range of sizes up to a maximum of about 2 Hp. It has good low-speed torque, is capable of very high speeds, and has high power output for its size and weight. The motor's weak points are the brushes and commutator: Commutator damage usually results in the motor, and consequently the tool or appliance, being discarded. For this reason, the series universal motor is unsuitable to situations demanding long periods of continuous operation.

Series universal motors require routine periodic maintenance. For long motor life, particular attention must be paid to the bearings, brush wear, and accumulation of carbon dust. Bad bearings cause the commutator to vibrate, resulting in destructive arcing and possible brush damage. Badly worn or sticking brushes cause poor contact between the brush and the commutator and, again, serious arcing.

The build up of carbon worn off the brushes is also a problem, because carbon dust is conductive. It can short out commutator bars and may even affect the insulation of the armature and field windings. Replacement brushes should match the originals. The bearings or bushings should be checked for looseness and for adequate lubrication, and should be replaced if necessary. Do not overlubricate: Excessive oil or grease is deadly to electric motors. If the windings become coated with sticky lubricant, the accumulation of dust soon blocks air passages and may cause premature failure from overheating.

10.3 Induction Motors

When a continuous loop of conductive material is placed in a changing magnetic field, a voltage is induced in the conductor, and current circulates around the loop. This is a natural occurrence that is part of the way that a magnetic field and electric current interact. The voltage induced depends on the rate of change in the field strength: It does not matter how the change is produced. The field itself can be changing in strength, location, or polarity, or the change may result from movement of the conductive loop. There may also be several different factors occurring simultaneously.

Regardless of the means, if there is a change in the strength of the field passing through the coil, the end result is a voltage induced into the coil. When the conductive loop is an open circuit, the induced voltage appears across the opening, but there is no current. When the loop is a closed circuit, current flows, and a second magnetic force is created. The polarity of this new magnetic force is opposite to that of the original field. This behavior is described by Lenz's Law, which can be summarized as follows: "The polarity of the induced voltage across a coil is such that it tends to circulate a current that will oppose the change of flux inside the coil."

This principle of electromagnetic induction can be used in several interesting ways. For example, if a magnetic field cuts through a sheet of conductive material, and the field is moved, a force develops that tends to pull the material along. This can be demonstrated with a suspended aluminum or copper plate and a permanent magnet. Although the magnet does not attract the plate, when it is moved in a way that causes the field to cut through the plate, it will tend to drag it along (see Figure 10–4). This technique can also be used to produce the opposite effect. If the plate is moving and the field is held stationary, the resulting force acts as a brake.

Induction disks are used in several applications. For example, in many automobiles the speedometer cable spins a magnet inside an aluminum drum, turning it against the force of a coiled spring. The indicating needle is attached to the drum. The faster the magnet spins, the further the needle advances along the scale. Another example is found in standard kilowatt hour meters used by power utilities to measure energy consumption. The meter contains a flat aluminum disk that turns when energy passes through the meter. The coil and core assembly inside the meter produces a moving field that is proportional to the flow of electrical energy. This field turns the disk against the braking action of a stationary permanent magnet. The meter is calibrated by adjustments to the permanent

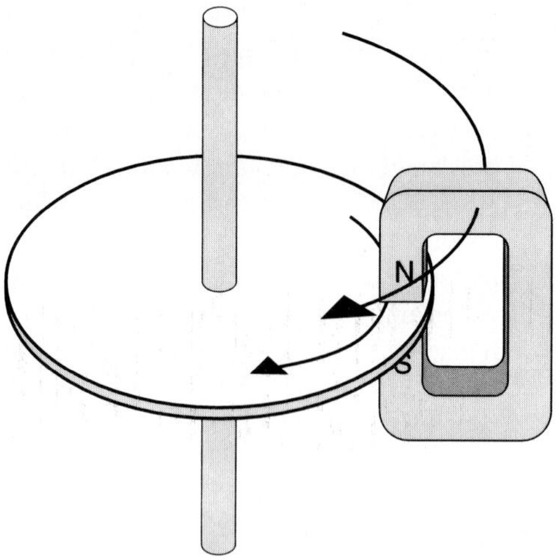

FIGURE 10–4 The moving magnetic field induces circulating currents in the plate. The interaction or the moving field and the magnetic forces created by these currents will drag the disk along.

magnet so that each rotation of the disk represents a specific amount of energy. The disk rotations are recorded as an accumulating count on the meter dial. A similar arrangement is used in other electrical metering and monitoring devices.

The reason a conductive, but nonmagnetic, disk is affected by a magnetic field is that any movement of the field induces voltages and produces circulating currents in the conductive material. The disk is a continuous sheet, not a defined shape, so the loops of current define themselves. Any number of voltage and current loops can develop as the field moves across the plate, but the result is essentially the same as if the conductor were formed into defined coils. Voltage is induced, current flows, and magnetic forces develop that interact with the original field.

When the movement of the field is linear, this force is along the line in which the field moves. When the field moves in a circle and a suitable "rotor" is placed in the field, the result is an electric motor. Given a revolving field, there are numerous rotor designs that will create torque and rotation. An aluminum or copper cylinder will work, as will an iron drum with heavy copper plating. The iron would improve performance by providing a low-reluctance path for the magnetic fields. However, the most successful rotor design is a laminated iron cylinder with a ring of conductive bars embedded around the outer edge. The bars are shorted together at both ends to create a large number of conductive loops. This design is known as the squirrel-cage rotor.

The Squirrel-Cage Rotor

The squirrel-cage rotor has a simple, very rugged construction (see Figure 10–5). A series of copper or aluminum bars are embedded in a cylindrical, laminated iron core. The conductive bars are interconnected on each end by shorting rings. If the iron core could somehow be removed, the bars and rings would look much like the exercise wheels placed in hamster and squirrel-cages. Typical rotor construction involves a stack of punched metal disks with cutouts around the edge for the bars, and a hole in the middle for the motor shaft. The stack is usually given a small amount of twist, or "skew," so the rotor bars will be at a slight angle to the stator slots. A mold is used to contain the molten aluminum

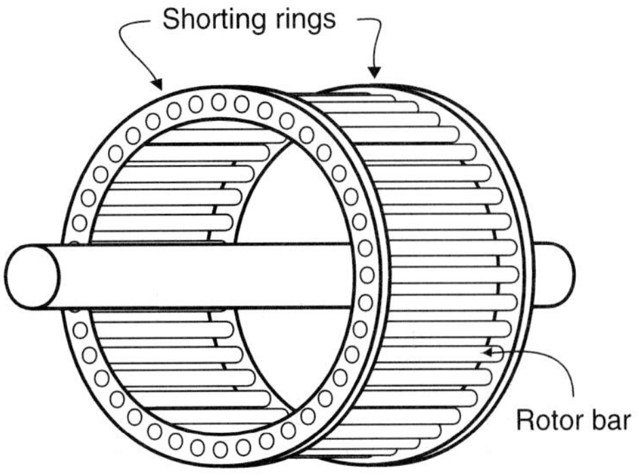

FIGURE 10–5 A squirrel-cage rotor with the iron core material removed.

alloy, and the bars and end rings are cast in place as a single unit. If the motor has internal air-circulating fans for cooling, these may be cast as part of the end rings. The completed assembly is pressed onto the motor shaft.

Large squirrel-cage rotors may also be fabricated with heavy copper or alloy bars, with the shorting rings welded in place. This type of construction allows for the creation of gaps in the iron to form cooling passages through the rotor at various positions along its length.

The squirrel-cage is the electrical part of the rotor; the iron is included to provide a better magnetic path for the stator field. The iron is laminated to reduce circulating eddy currents, which would be a waste of energy and would cause the iron to heat. Skewing the bars at a slight angle improves the motor's performance by reducing torque pulsations that cause vibration and noise. These pulsations are caused by magnetic field concentrations around the rotor bars and along the slots in the stator. The interaction between these two is smoother when the bars are skewed, rather than exactly parallel to the slots.

The operating dynamics of a squirrel-cage rotor are more complex than the simple structure suggests. (They are examined in more detail in Chapter 11.) Several "standard" rotors are used in three-phase motors; single-phase rotor characteristics are less standardized.

A few key points about these rotors should be fully understood, particularly in regard to single-phase motor operation. When a squirrel-cage rotor is placed in a *rotating field*, voltages are induced in the conductive loops formed by the rotor bars and the short circuiting rings that interconnect the bar ends. The result is much the same as for a solid sheet of conductive material. Current circulates in the rotor, creating magnetizing forces that interact with the stator field. The squirrel-cage rotor "attaches" itself magnetically to the stator and is pulled along. It can never quite catch up, so it always runs just a little slower than the field.

The difference between the speed at which the stator field rotates and the speed of the rotor is called *slip*. Slip is normally expressed as a percent of the speed at which the field rotates. For a typical induction motor, the slip is less than 5% at full load, but it may be higher for certain motor designs.

When the rotor is placed in a *changing field*—note the subtle but critical difference between a changing field and a rotating field—the result is a similar magnetic interaction. There are rotor voltages and currents, and considerable magnetic and mechanical stress and vibration, but no specific direction of rotation. The rotor tends to oscillate and may or may not start to turn. If there is any mechanical load on the shaft, the rotor stays in one spot, and the motor hums from the vibration. The motor will burn out if it is left energized for very long.

However, even though the field is not rotating, if the rotor is started in either direction, it begins to develop significant directional torque. It will speed up and run as if it had a true rotating field. **The ability of the squirrel-cage rotor to run in a changing field is the basis of several very popular single-phase motors.** Some provision must be made to start the motor, but once it is up to speed it performs quite well. This behavior is as simple to demonstrate as it is difficult to explain. It is primarily a result of rotor inductance, combined with the fact that the rotor is turning. Although the conductive loops in the rotor are very short, they still have both resistance and inductance. The inductance delays the circulating rotor currents, causing them to lag behind the induced voltages that produce them. Once the rotor is turning, this delay causes the magnetizing force to be offset from the main field. Because it is always offset to the same side—opposite the direction of rotation—the force produced by the magnetic interaction is always in the same direction, and the motor runs.

To demonstrate this behavior, temporarily disconnect the start winding of a split-phase or capacitor start motor. Briefly energize the motor to see if it starts. A clamp-on ammeter is useful because it will show the high current drawn by the motor. With absolutely no load, the motor may start to turn, but the amount of torque produced is almost nothing. If the shaft is given a spin in either direction, it comes up to speed and runs normally.

The speed at which the stator field rotates is set by the number of poles and the supply frequency. This is called the *synchronous speed*, and it is as true for a changing field as it is for a field that actually rotates. Even at no load, a squirrel-cage rotor can never exceed the synchronous speed. If it were turning at the same speed as the field, it would "see" the field as stationary. There would be no voltage induced in the rotor, and therefore no rotor current, no magnetic force, and no torque.

Although the physical characteristics of the rotor are fixed, its electrical characteristics vary with the speed of rotation. The stator field is supplied at 60 Hz, but the rotor "sees" a lower frequency because it is moving. The frequency of the field—as it appears to the rotor—changes from the 60 Hz line frequency with the motor at rest to 5 Hz or less when the motor is upto full operating speed. This change in frequency has a tremendous effect on the starting current and torque, and is part of the reason why most AC motors draw such high currents when they are started.

Once the motor is running normally, a very small change in speed produces a large change in the stator and rotor currents, and in the motor's output torque. This is the reason typical squirrel-cage motors have a nearly constant operating speed. The no-load speed is limited to something slightly less than synchronous speed, and only a small reduction is required to bring the motor current and torque up to the rated full-load value.

The squirrel-cage rotor is used in almost all AC motors, both single-phase and three-phase. Exceptions include series universal, repulsion, synchronous, and wound rotor motors.

A Summary of Squirrel-Cage Rotor Characteristics

All squirrel-cage motors are classed as induction motors because all the electrical energy used by the rotor is obtained from the stator field by electromagnetic induction. There are other types of induction motors, but they operate on the same principles as the squirrel-cage rotor motor. They are made differently to accommodate other operational considerations.

Placed in a revolving field, a squirrel-cage rotor runs just slightly below the speed of the main field. The difference in speed is called slip, and it varies with the motor load. In most motors, the variation between the no-load speed and the full-load speed is less than 5%. Some motors, such as the shaded pole and permanent split capacitor types, have a greater slip, so their speed variation is greater.

The squirrel-cage rotor operates in a changing field, in much the same manner as it does in a rotating field. The efficiency is slightly lower, and, because it will not develop significant starting torque, the motor has to have some special provision for starting.

The squirrel-cage rotor has no directional preference; it runs equally well in either direction. The direction of rotation is set by the direction of the rotating field or, in the case of a changing field, by the direction in which the motor starts.

The ability of the squirrel-cage rotor to keep running, once started, in a field that is not actually rotating, is particularly difficult to understand and explain, but is very easily demonstrated. This is the normal operating mode of a split-phase (resistance start) or capacitor start motor.

The squirrel-cage rotor is only used in AC motors; in a DC field it act as a brake. AC motors are sometimes stopped by temporarily connecting the windings to a DC supply.

The Shaded-Pole Motor

The stator of a shaded-pole motor is similar in construction to the stationary portion of a series universal motor. Both have distinctly separate field poles that protrude inward to face the rotor or armature. The inner faces of the poles are curved to match the shape of the rotor and, with the rotor in place, there is only a narrow air gap separating the two. As with most magnetic devices, the poles are in pairs, so there is always an even number. Shaded pole motors are made with two, four, six, or more poles, depending on the desired operating speed.

Shaded pole motors with only two poles are sometimes built with a rectangular iron core assembly and only a single winding. These are often called skeleton motors, because of their very open construction and the lack of an external housing to enclose the motor parts (see Figure 10–6). The entire motor is built onto the iron core, which does double duty as the motor frame. The winding is located at one end of the core and serves to magnetize both poles, with opposite polarities. A typical example has sleeve bearings mounted in shallow end caps that contain a small reservoir of oil-soaked packing. The oil wicks onto the motor shaft through a small hole in the bearing. These motors are usually mounted using bolt holes provided in the frame.

Skeleton motors are also supplied with attached gear reducers. The inside motor bearing is eliminated, and the motor shaft is extended through a bearing in the gear box. The motor frame is bolted to the gear case, which in turn is mounted to the equipment. The motor and gear reducer are a single assembly that cannot be taken apart. The motor shaft is pressed into the first gear inside the lubrication-filled case, making it very difficult to remove without damage.

Skeleton motors are quite popular in HVAC applications. They are used to power small fans, such as combustion air fans in residential furnaces, and, with gear reducers, as the drive motors for electrical damper and valve actuators. They are also found in small appliances and in some consumer electronics equipment.

Shaded-pole motors are also made in a more traditional motor style, quite different in appearance from the two-pole skeleton types. They have a cylindrical shape like other motors and have full end covers to protect the windings and support the bearings. This

FIGURE 10–6 This skeleton-frame shaded-pole motor has an attached gear reducer to provide a reduced output speed. The copper shading coils are embedded in slots on opposite sides of the squirrel-cage rotor. One is visible just above and to the right of the shaft end.

style of motor may have two or more poles, depending on the rated speed. Low-speed versions tend to be shorter, but larger in diameter to accommodate the large number of poles required.

All shaded-pole motors, regardless of size or configuration, use the same method to create a rotating magnetic field. A short-circuited "shading coil" is embedded in the face of each field pole. As AC current flows through the main winding, it creates an alternating magnetic field in the core. The portion of this field that passes through the shading coil induces a voltage into that coil. The voltage is very small, but the resistance of the coil—one short turn of heavy-gauge copper—is so low that a relatively high current circulates in the loop. The magnetic forces created by this current oppose changes in the shaded portion of the main field, delaying both its formation and decay. As a result, the motor field rotates, sweeping across the face of each pole, toward the shaded end.

The location of the shading coils determines the direction of rotation, which is always toward the shaded ends (see Figure 10–7). To reverse rotation, the shading coils must be relocated to the opposite side of each pole. The usual way to do this is to disassemble the motor and turn the stator portion over. This "moves" the shading coils to the opposite side of the pole faces, effectively reversing the rotation of the magnetic field. Most shaded-pole motors are made so the stator can be reversed and the motor reassembled. This is a workable method of reversal if there is no need to change the rotation after the motor is installed.

Many small shaded-pole motors are classed as "impedance protected." This means they will not draw excessive current, even when the motor is overloaded or stalled. On small fans, for example, if the bearings become dry or something jambs in the fan blades, the motor will stall, but it will not overheat to the point of damage or fire hazard. In most cases it can be salvaged with a little cleanup and a few drops of lubricating oil.

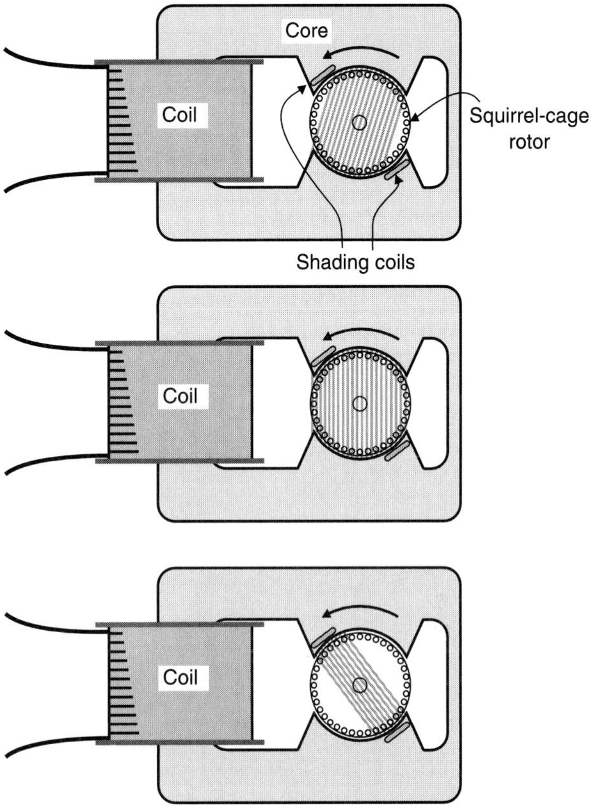

FIGURE 10–7 The delaying effect of the shading coils causes the field to rotate. Direction of rotation is always toward the shaded end of the poles.

The synchronous clock motor is one of several interesting variations on the shaded-pole design. Synchronous motors run with no rotor slip. The squirrel-cage rotor is replaced with a rotor that locks onto the revolving field instead of slipping to a slightly lower speed. The synchronous rotor may be permanently magnetized or made from a material that tends to stay magnetized. The poles of the rotor match the position and number of poles in the rotating stator field. The opposite poles attract, and the rotor becomes "synchronized" to the stator field, turning at exactly the same speed. The speed of rotation for the stator field is fixed to the frequency of the AC supply. Small synchronous motors are very well suited to clock drives, motorized timers, audio turntables, and other applications requiring a precise operating speed.

Another variation on the shaded-pole motor is sometimes used to operate small valves or dampers. These motors have a different layout than other shaded-pole motors, but operate on the same principles. They often have aluminum disk rotors in place of a squirrel-cage or permanent magnet type, and are "impedance protected" against burnout. This characteristic simplifies the design, because the motor can be allowed to stall at the end of the valve or damper travel. The motor is energized to open or close the valve, and left energized to hold it in position. When the motor is de-energized, the valve or damper automatically returns to its normal position, driven by a built-in return spring. This type of arrangement is common in zone control valves for hydronic heating systems, and for small dampers in larger HVAC installations.

The reversible shaded-pole motor is an exception to the general rule that shaded-pole motors are not electrically reversible. Instead of moving the shading coils, the motor is equipped with two sets, one on each side of the pole piece. The best known example of these unusual motors is used in the HVAC industry to drive electrical actuators used for dampers and valves. The motor is quite similar to other two-pole skeleton-style motors, but the shading coils are made of many turns of fine wire, instead of the usual single turn of heavy wire or strap. There are also two shading coils on each pole instead of the normal one.

The shading coils are wired into opposing pairs, and the leads are connected to the contacts of an external controller. When the contacts short one pair of coils, the motor runs like any normal shaded-pole motor. When the other pair is shorted, the motor reverses; when they are both open, the nondirectional field acts as a brake. This holds the damper or valve in position when the control system is not calling for any adjustment. The motor is not very powerful, but several stages of gear reduction give the actuator arm a considerable amount of force. These motors cannot be examined without destroying the actuator, because they are immersed in oil inside the actuator gear case.

Under specific circumstances, the speed of a shaded-pole motor can be varied using voltage-control methods similar to those used with the universal motor. However, despite the similarity in the control methods, the effects on motor operation are very different. Unlike the universal type, the shaded-pole design is not a true adjustable-speed motor. Like other AC induction motors, the speed of rotation for the motor's magnetic field is determined by the supply frequency and the number of poles in the motor. This is very different from the series universal motor, where the motor speed is set by a balance between the motor output torque and the demands of the load.

Changing the voltage to a shaded-pole motor does not change the speed of the revolving field. It only reduces the maximum output torque. When the motor is driving a suitable variable-torque load, such as a fan, reducing the voltage will reduce the motor speed. As the voltage is reduced, less current flows, the rotor torque drops off, and the fan slows down. Because the fan requires less torque to operate at a lower speed, the motor will not stall. The motor continues to run, but there is a lot of "slip"—the rotor turns more slowly than the rotating field. For this to work successfully, the load must have a suitable torque characteristic—typical of a fan or some impeller pumps—and the motor and load must be carefully matched. The controller must be restricted so that it cannot reduce the voltage too far, or the motor will stall. Motor torque varies roughly as the square of the applied voltage, so the minimum operating voltage may be relatively high.

One inexpensive control method consists of extra windings installed in the motor stator. Extra tap leads are brought out from this winding to a multiposition selector switch. The switch provides three or more different connections, each with a different amount of the extra winding. Changing the amount of extra winding effectively adjusts the voltage and current to the motor. This type of control is common on portable fans.

The shaded-pole motor is most commonly used for small fans, such as those used for bathroom ventilation and circulating fans in domestic refrigerators. Larger shaded-pole motors, typically rated at 1/4 or 1/3 Hp, are sometimes used on direct-drive blowers such as those used in residential furnaces. They may also equipped with extra tapped windings to provide several operating speeds. This motor is somewhat less expensive than a permanent split capacitor motor, but is also less efficient. The controls for these two motors are essentially the same.

Note

This shading technique used for motors is also used to reduce vibration in AC electromagnetic coil applications. Examining the pole face in a relay or motor starter with an AC coil will reveal one or two shading coils embedded in each iron pole. The delay they produce prevents the magnetic field from dropping to zero at the end of each half-cycle. Without the shading coils, AC electromagnets have a severe chatter.

The Split-Phase (Resistance Start) Motor

The motor described here as a split-phase or resistance start motor is also known by other names. Most authoritative texts and manufacturer's catalogs refer to it as a split-phase or standard split-phase motor. It is also called a resistance-start induction-run motor, which may be reduced to the acronym RSIR.

Examining the appropriateness of these names just adds to the confusion. This motor uses additional resistance in one winding to "split" the single-phase voltage into two currents that have a different phase shift. It is one of several motors that create a "split phase," but it is the only one to accomplish the split with resistance alone. As for the "induction run," not only does this motor run as an induction motor, it also starts as one. (Actually, all but two of the common AC motors start and run as induction motors. Repulsion-start/induction-run motors use a repulsion armature for startup, and then become induction motors by shorting out the commutator. Three-phase synchronous motors usually start as induction motors, and then make a transition to synchronous operation when the rotor is energized with DC.)

In any case, both "split-phase" and "resistance start" are used in this text, depending on the context. There is no intention to cause more confusion or promote one title over another. "Resistance start" is used where it helps differentiate this motor from similar designs.

The resistance start motor is one of two common motor designs that takes full advantage of a characteristic that is unique to an induction rotor like the squirrel-cage: This is the ability of the rotor to continue to develop torque in the direction in which it is started, without the benefit of a true rotating magnetic field. With a squirrel-cage rotor, a rotating field is necessary to start the motor, but not to keep it running. Even three-phase motors will operate on single-phase, once they are up to a reasonable speed. Their output torque is reduced, but they run.

Although three-phases are better, a genuine rotating field can be created with just two. Therefore, if the single-phase source can be "split" into two phases, and these two phases are used to energize appropriately placed motor windings, a rotating field can be produced from single-phase power. The split-phase motor gets its name from this process.

Note

It should be clearly understood what is meant by a second phase. In a genuine two-phase system, a second voltage that lags behind the first voltage by $^1/_4$ cycle (90 electrical degrees) is created in the supply generator. The two voltages are delivered through a system with two different circuits, and are used to power two separate windings in a two-phase motor. The result is a rotating field (see Figure 10–8). The two different voltages are said to be separate phases because, although they have the same frequency, wave shape, and voltage value, they are displaced from one another by a consistent amount of time—or phase shift.

FIGURE 10–8 For a two-pole, two-phase field, a full rotation occurs in one complete cycle of the supply. Each phase powers one pair of poles. As the voltage rises and falls, the magnetic fields created, change in strength and polarity. The result is a smoothly rotating magnetic field.

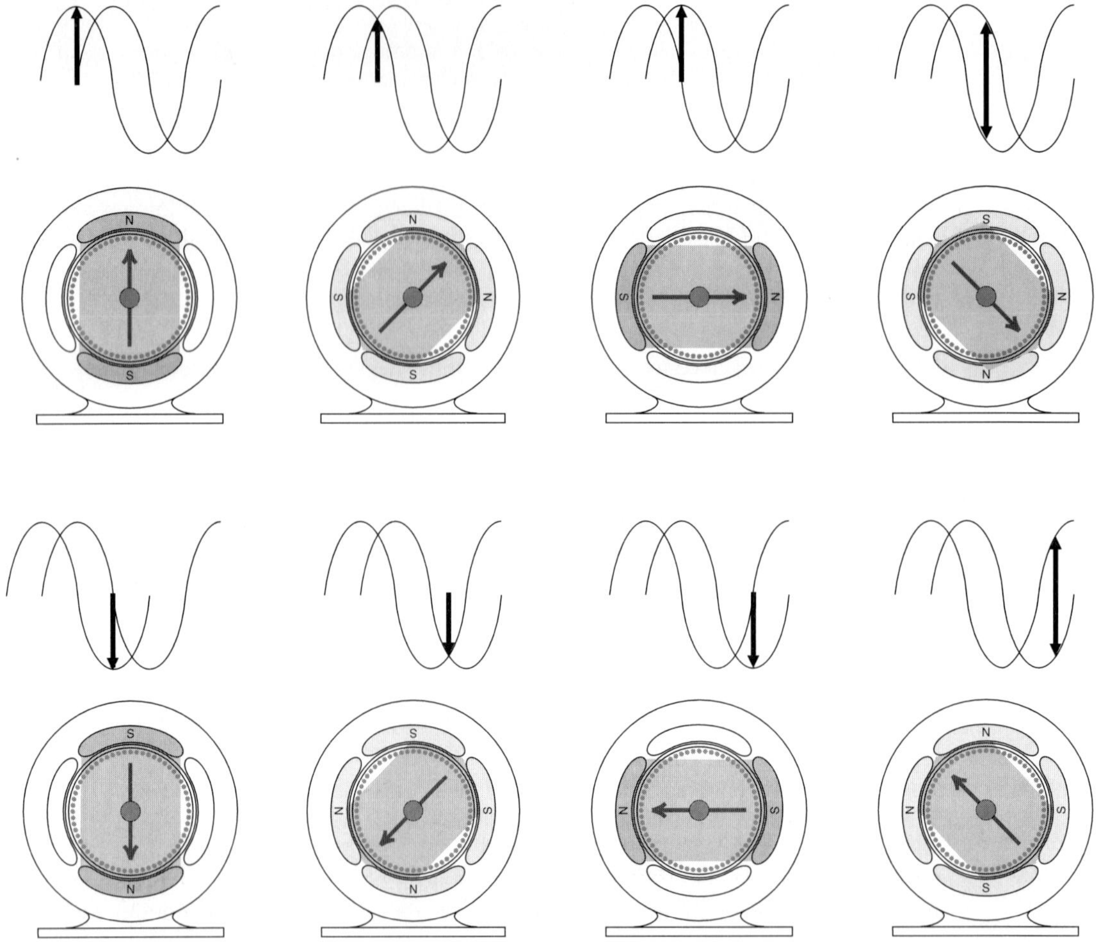

A genuine two-phase system has two separate circuits. However, it is current, not voltage, that creates the motor's magnetic fields. Therefore, if the single-phase voltage can be used to create two currents with different phase relationships, a two-phase rotating field can be created from a single-phase supply.

In an AC circuit. the voltage and current are not necessarily in phase with each other. The current may lead or lag the voltage, depending on the amount and type of reactance the load contains. When loads with different characteristics are connected in series, the phase relationship depends on the blend of the reactive characteristics involved. For example, if a resistor and coil are combined, the current lags at an angle that is less than 90° and greater than 0° (in phase). If the resistor is predominant, the phase angle is closer to being in phase, and if the coil is predominant, the shift is more lagging. When a capacitor is used instead of the coil, the current moves into lead. With all three components used together, the phase shift is set by the difference in the reactances, combined with the total resistance. Capacitive reactance cancels an equal amount of inductive reactance, and vice versa.

When loads with different characteristics are connected in parallel, each of the parallel branches can have a different phase shift. In a circuit branch that is predominantly inductive, the current is lagging. In a branch that is more resistive or capacitive, the current is more in phase, or in lead. The result is different currents, all out of phase with one another, produced by the same supply voltage. This is the technique used in split-phase motors to create two current "phases" from a single-phase supply.

The resistance start motor has two sets of windings, both with complete sets of poles. The *run* winding is designed to operate continuously. In a typical split-phase motor it is formed from concentrically wound coils containing a large number of turns of relatively heavy-gauge wire. There is one set of coils for each motor pole, spaced at equal intervals around the stator. The coils are connected as a group, with opposite polarity on every other pole. The run winding is laid into the stator slots first, placing it deep into the core iron. The heavy conductor gives it a low resistance; the large number of turns gives it a high inductance reactance.

The poles of the *start* winding are located halfway between the poles of the run winding. It is also wound with concentric coils connected for alternating polarities; however, there are fewer turns than the run winding, and the conductor is several gauge sizes smaller. The start winding is laid into the stator slots after the run winding, overlapping it as necessary. The much thinner conductor gives it a more resistive characteristic, and because it also has less turns, its inductive reactance is less than for the run winding. (For a picture of a split-phase stator winding, refer to Figure 6–4.)

In typical resistance start motor, the current in the primarily inductive run winding lags the supply voltage by 60° or more, while the current in the more resistive start winding lags by about 30°. The difference between the two currents is only about 30°, far less than the ideal 90° of a true two-phase system, but adequate to create a rotating field (see Figure 10–9). Although there is only one supply voltage, the two winding currents are truly out of phase with each other. The single-phase supply has been "split" into two, and the end result is a relatively simple motor with significantly better performance than a shaded-pole type.

As explained previously, the motor will run without the start winding, which is only required to provide starting torque and to help bring it up to speed. Once the motor is

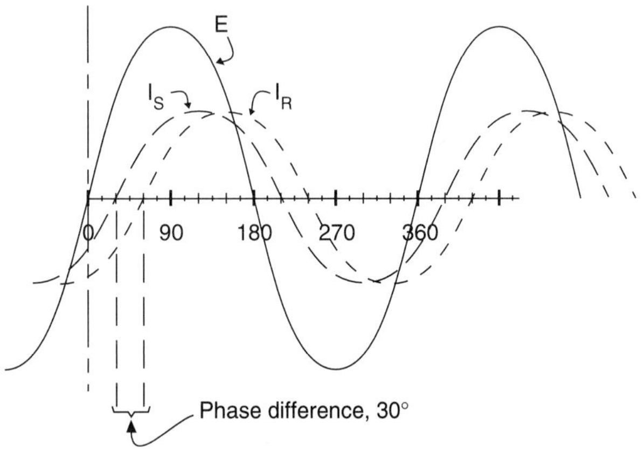

FIGURE 10–9 The applied voltage (E) produces current in the run winding (I_R) and in the start winding (I_S). Because the two windings have different characteristics, the two currents are approximately 30° out of phase.

turning fast enough to operate on the run winding alone, the start winding is de-energized. The relatively high resistance of this winding causes a high power loss and generates a lot of heat. If it is kept energized too long, it will burn out.

The method used to disconnect the start winding varies with the motor design and the application. General-purpose motors usually have an internal switch operated by shaft speed. Welded-frame appliance motors also use speed switches, but they are mounted externally, to the motor frame, to allow for replacement. Sealed motors, such as refrigeration compressors and some submersible pumps, have a starting relay mounted external to the motor.

For a resistance start motor, the starting relay is usually operated by motor current, which is much higher when the motor starts than when it is up to speed. The high start current activates the relay, closing the contact in the start circuit. As the motor nears full speed, the current drops off, and the relay drops out, disconnecting the start winding from the supply. (Relays and start switches are examined in more detail later in this chapter.)

Resistance start motors are common general-purpose induction motors with modest starting torque, fairly simple construction, and modest power output for their size and weight. They cannot be speed controlled, but two-speed models are readily available at extra cost. They are widely available in fractional-horsepower sizes, but are unlikely to be found in sizes larger than one horsepower. They can be electrically reversed by interchanging the connections to either the run or start winding, if the connections have been brought out to the motor connection terminals. (It is traditional to change the start winding leads, not the run. Following the instructions on the motor will result in reversal of the start winding connections.) The resistance start motor cannot be reversed while it is running, unless extra controls are added or the motor is specifically designed for "instant reverse."

Two-speed resistance start motors may have a consequent-pole arrangement or dual-run windings. All consequent pole motors have a 2:1 speed ratio, such as 1725/865 rev/min. Dual-winding motors may have a 2:1 ratio, but are typically something like 1725/1140 rev/min. Most dual-winding motors have two run windings and a single start winding. The start winding is designed for the higher of the two speeds, but is used for starting the motor with either speed selected. The start switch is more complex than in a single-speed motor because of the extra switching required.

A consequent-pole motor has only one run winding, but it is arranged to allow for two different connections. On high speed, the winding current flows in the opposite direction around every second pole, forming a normal, N-S-N-S... type of pole sequence. On low speed, the winding is reconnected so that the current flows the same way around every pole. This produces four identical poles but, as a consequence of these poles, there are four additional poles of the opposite polarities located in between the others. This effectively doubles the total number of poles and halves the motor speed.

A special-purpose two-speed resistance start motor is manufactured specifically for the HVAC industry. It is used in two-speed belt-driven blower applications. The difference between this motor and a general-purpose two-speed design is that a special provision has been made inside the motor to allow both the high- and low-speed terminals to be energized simultaneously. On a general purpose motor, this could burn out one or both of the windings.

This special motor runs at high speed with either both speeds or only the high speed energized. This characteristic allows the furnace manufacturer to provide an optional continuous low-speed blower operation, plus the higher normal operating speed, using typical single-speed blower controls. The continuous low speed is controlled by a manual switch, and the motor is automatically switched to high during a heating or cooling cycle. For the heating mode, a single-throw fan control selects high speed; for air conditioning a relay is used. (These controls are described in detail in Chapter 17.)

Resistance start motors are most commonly used to drive small centrifugal pumps and belt-driven fans and blowers. They are also common in large appliances such as washers and dryers, and for some small, hermetically sealed refrigeration compressors. They are very common general-purpose motors that are readily available from 1/4 to 3/4 Hp, and occasionally up to 1 Hp.

In manufactured items, these motors may be somewhat different than in their general-purpose form. Typical appliance motors, for example, have welded frames and cannot be taken apart. It should also be noted that although a typical resistance start motor has a low to moderate starting torque, many of the larger appliance motors are high-start-torque versions. This unusually high torque is obtained by allowing the start current to exceed NEMA guidelines for motor of that size.

In an appliance application the start current is usually not a problem, because the horsepower rating is typically 1/2 Hp or less, and the motor is not required to start frequently. The choice of this motor over the capacitor start motor described below is one of economics. The actual cost of building high torque resistance start appliance motors is among the lowest of any single-phase motors type.

The Capacitor Start Motor

Although the resistance start method works fine for many applications, the phase difference between the two winding currents is typically 30° or less, far from the 90° ideal. Fortunately, a phase difference of close to 90° is not that difficult to obtain. If a relatively large value capacitor is added in series with the start winding, that branch of the circuit becomes predominantly capacitive. The current in the very inductive run winding lags the supply voltage, while the current in the now capacitive start circuit is slightly in lead. With a correctly sized capacitor, the phase difference between the run and start winding currents can be very near the ideal (see Figure 10–10). For a capacitor start motor, the start winding and its capacitor are still used only for start-up duty. Once the motor is up to speed, the start winding is de-energized, and the motor operates on the run winding alone.

FIGURE 10–10 In a capacitor-start motor, the phase angle difference between the start and run winding currents approaches the ideal 90°.

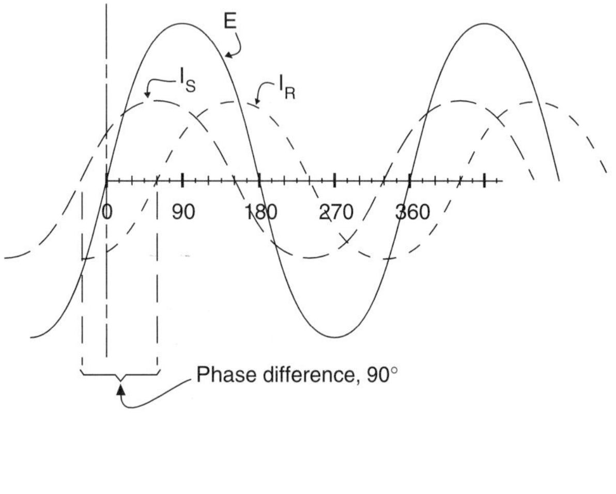

The improvement in the phase shift of the start winding current may seem like a relatively small change, but the results are often well worth the additional cost of the capacitor. The near-90° phase difference between two winding currents gives the capacitor start motor a dramatically higher start torque, with a start current that is actually lower than that of a comparable resistance start design. This combination makes the capacitor start motor less prone to burnout when starting heavy loads.

Except for its superior start torque, the capacitor start motor is essentially the same as a resistance start motor. The start winding and capacitor are disconnected during normal running, so the motor runs as a single-winding induction motor. Its running characteristics are therefore identical to a comparable resistance start motor. Its efficiency is limited by the single-winding operation.

Like the resistance start type, the capacitor start motor is relatively simple and inexpensive to build. The two motors are almost identical in construction. There are two sets of windings—one continuous-duty run winding and a second, intermittent-duty start winding. The windings are made from concentric coils distributed in slots in the stator iron. The run winding is laid in first at the bottom of the slots. The start winding is set in last, halfway between the run winding poles. The start winding is usually wound with a smaller gauge wire, giving it a higher resistance, but the difference between the two windings may be less than for a typical resistance start type. The main difference between the two motors is the capacitor added to the start circuit. General-purpose capacitor start motors can often be identified by the humped metal capacitor housing on top of the motor, but this is not an entirely reliable indicator. The capacitor may also be mounted in a separate location or in an oversized connection box.

Neither the start capacitor nor the start winding is rated for continuous duty. The start circuit must be disconnected from the supply when the motor is close to full-speed operation. This can be done with a speed-operated switch or a current relay, just as for the resistance start motor. A potential relay can also be used in preference to a current relay. This relay has some advantages over the current type, and it is the more popular of the two for motors with a starting capacitor. (The operation of the potential relay is explained later in Section 10.4 of this chapter.)

Like the resistance start motor, the capacitor start motor cannot be speed controlled, but is available in a two-speed version. Although it is quite possible such a motor exists, this author knows of no capacitor start motor designed for simultaneous energization of both speed leads. Always assume that the controls to the motor must de-energize the speed in use before the second speed can be energized.

A capacitor start motor can be reversed by interchanging the start winding leads. Starting the motor in either direction is a relatively simple control problem that can be solved with standard components. However, changing the motor's direction while it is running is much more difficult. Direction of rotation is determined at startup by the start winding connection relative to the run winding. When the motor is running, the start switch or relay contact keeps the start winding isolated from the line. Reversal of the connections to either the start or run windings will not reverse the direction of rotation until the motor speed is low enough to reengage the start winding.

A specially built "instant-reversing" motor and the appropriate controls can be purchased from many motor suppliers. These motors incorporate relays, or special speed switches, that permit "instant reversal" with suitable external control equipment. There are other ways to solve the reversing problem, but they involve special controls and can be quite expensive. For example, a standard motor can be combined with a "zero-speed" switch that senses when the motor has stopped. When the motor is reversed, the controls de-energize the motor and reverse the start winding, but the speed switch automatically delays restarting until after the motor has stopped.

Capacitor start motors have a wide range of applications, including larger blowers, compressors, pumps, and conveyors. They are readily available in fractional and integer sizes to about 10 Hp. Larger ones are also available, but see limited use because three-phase motors are less expensive, more reliable, and more efficient.

The Capacitor Run Motor

Capacitor run motors are also called permanent split capacitor (PSC) motors. They usually have slotted stators and two sets of distributed windings, one placed halfway between the poles of the other. The winding layout is very similar to the start and run windings in a resistance or capacitor start motor; however, the gauge size and the number of turns in the second winding are not always reduced. The capacitor is connected in series with one winding. There is no start switch or relay.

Capacitor run motors can usually be identified by their continuous-duty capacitor and the lack of a start switch or relay. The winding connected to the capacitor must be designed for continuous operation, because it is not disconnected from the circuit after starting. For this reason this winding may be called an auxiliary winding instead of a start winding. Auxiliary more accurately describes the service this winding provides, but both terms are in common use. The gauge of wire and the number of turns in the auxiliary winding can vary a lot from one motor to another. The manufacturer may select anything from a relatively light-duty auxiliary winding and a small-value capacitor to a full-blown second run winding that is electrically and magnetically equal to the first.

The running capacitor, connected in series with the auxiliary winding, is usually an oil-filled type, though non-oil-filled styles are becoming increasingly common. Oil-filled run capacitors of this rating are almost always flat-sided gray metal cans with quick-connect wire terminals on one end. Newer types of run capacitors are quite similar to other non-oil-filled capacitor designs, complete with nonmetallic cylindrical cases in the manufacturer's choice of colors. The physical size of the capacitor varies with the capacitance and with the voltage rating. High voltage ratings typically require physically larger capacitors because the dielectric between the plates must be thicker. The capacitance value is usually only a crude indicator of the motor horsepower. There are many variables in the design of a motor; adjusting one affects all the others. The run capacitor used for one make or model of motor may be quite different in size to that of another, even though the motors are rated for the same horsepower and operating speed.

The capacitor is sized for optimum running conditions. Ideally, it produces a 90° phase difference between the currents in the two windings, when the motor is operating at full rated load. With this degree of shift, the motor runs as if it had a two-phase supply. Unfortunately, the capacitance required for best running performance is a much lower value than would be used for optimum starting torque. For this reason, the PSC motor produces much less start torque than it would with a larger capacitor. This is a necessary trade-off: A capacitor large enough for good start torque is far too large for acceptable running performance.

For applications such as fans and blowers, the PSC is a proven performer. The low starting torque is actually an advantage. A fan driven by a PSC motor comes up to speed gently, with no disturbing rush of air or light-dimming current draw. The motor has a better power factor, lower start and running current, and, in most cases, slightly higher operating efficiency than any other single-phase motor. It also runs more quietly and, because it has no start switch or relay, is simple and reliable. The continuous-duty capacitor and second winding add to the cost, making it more expensive than the shaded-pole, resistance start, or capacitor start motors. The PSC motor is not well suited to belt drives, so it is normally connected directly to the fan or other load.

With a variable torque load that is properly matched to the motor performance, PSC motors can be built to provide multiple speeds or can be speed controlled, within reasonable limits. Multispeed versions have extra windings, tapped at intervals to provide three to six different speed settings. As in the shaded-pole version, the extra windings function as a series impedance that reduces the voltage across the main windings. The result is a reduction in motor current and maximum output torque.

Multispeed PSC motors are generally available from $^{1}/_{4}$ to $^{1}/_{2}$ Hp or more (see Figure 10–11). They typically have six poles, giving them a maximum full-load operating speed between 1000 and 1100 rev/min, appropriate for modestly sized blowers such as those found in residential furnaces. They are compact, mounting directly inside the blower wheel, with little exposure beyond the wheel end. The direct-drive design eliminates the inefficiencies of the older belt drive, and reduces the size and cost of the blower assembly.

When blowers are belt-driven, an adjustable pulley is used to change blower speeds. For direct-drive blowers, adjustments are provide through the multispeed capability of the motor. In applications such as furnaces, this is very important. The blower speed has to be adjusted to maintain proper air flow through the heat exchanger. Because of variations in duct resistance, this has to be done in the field, using a measured temperature differential

FIGURE 10–11 This $^{1}/_{4}$ Hp PSC fan motor has extra windings to allow multispeed operation. The motor has six poles and a nameplate speed rating of 1050 rev/min. This speed is at full load with the high speed selected. No other rev/min ratings are provided because they are load dependent.

FIGURE 10–12 Two- and three-speed control can be provided with appropriate relays and a multispeed fan motor. The contacts of the controls must be arranged so that only one motor speed lead is energized at a time.

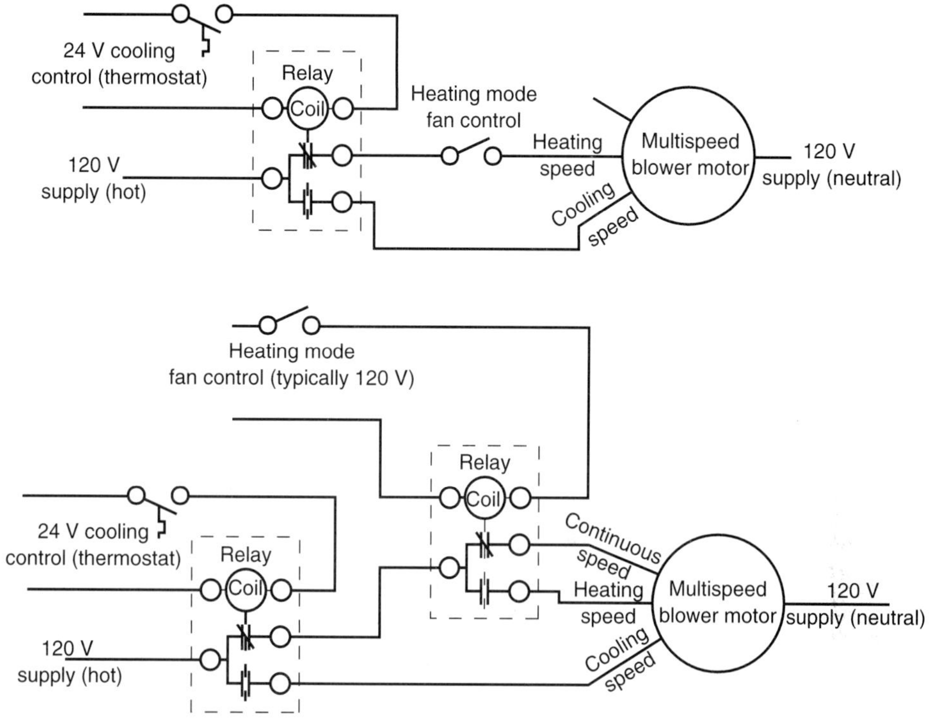

across the heat exchanger. A speed is selected to provide a temperature rise that is within the manufacturer's specifications. This ensures maximum heat exchanger life and operating efficiency.

Typical furnace controls for multispeed shaded-pole or PSC motors include one SPDT relay, for two-speed control, or two SPST relays for three-speed operation. Two-speed operation provides separate speeds for heating and cooling. The third is a low speed used for continuous air circulation. The control contacts are connected so that it is virtually impossible to simultaneous energize two or more motor speeds (see Figure 10–12). This would burn out the motor. There are numerous variations to these controls. (Other arrangements are shown in Chapter 17.)

Warning

Multispeed motors must have the unused speed leads disconnected from any other leads or equipment, and insulated to prevent accidental contact with other circuit components or grounded metal. When the motor is running, all the leads are energized, even the unused ones. The voltage from the circuit common—typically the neutral—to each of these leads will be different. The actual voltage may be significantly greater than the line value: It varies with the speed that is in current use and the speed represented by the lead under test. Connecting any leads to other equipment, to ground, or to each other may damage the equipment or the motor.

In newer furnace designs this problem is handled by providing terminals for all the motor leads, including the ones that are not used. This may be in the form of a two-sided connection block, or a set of active terminals plus unconnected "parking" terminals for all the unused leads. If the motor has open leads, the unused ones have to be capped with wire connectors or taped with electrical tape.

This type of multispeed motor will not change speeds without a suitable load. The change in speed is caused by reduced motor torque supplied to a variable torque load. At less than full speed, the motor is operating at reduced voltage with a lot of rotor slip. For successful operation, the load must have a reduced torque demand at lower speeds. Most fans and blowers, and some impeller pumps, have this kind of behavior, making them suitable candidates for variable-speed or multispeed PSC drives. If the load has a constant torque demand, such as a conveyor, hoist, or positive displacement pump, a motor like this is unsuitable; any attempt to reduce the speed would cause it to stall.

A single-speed PSC also can be speed controlled with an external controller that varies the motor voltage. The controller is usually an autotransformer with several taps, or an electronic circuit with a variable output voltage. The controller must have a preset minimum voltage high enough to keep the motor running. If the voltage is reduced too far, the motor may stall, overheat, and burn out. Motor output torque varies roughly as the square of the applied voltage, so the acceptable voltage reduction is usually no more than about 50% of the full rated value. Single-speed PSC motors are available from very small, subfractional sizes to sizes up to 10 Hp or more.

PSC motors are normally reversed by interchanging the connections to the auxiliary winding. With a few exceptions, they cannot be reversed while they are operating, but the reasons are different than for the resistance and capacitor start motors. Those two motors cannot be reversed until they have slowed down enough to reset the start switch, or relay, and reconnect the start winding. In a PSC, the auxiliary winding is always energized through the run capacitor. Because this winding is always in the circuit, it would seem reasonable to assume that reversing the leads should make it change direction, even if the motor is running. However, this does not work very well.

When the auxiliary winding leads are suddenly reversed with the motor running, most PSC motors continue to run in the same direction. But they do not run very well and usually trip off on overload. The same thing can happen if the motor is spinning backwards when it is started. This can happen with some scroll compressors, for example. These compressors have relatively low start torque requirements so they can be driven successfully with a PSC motor. Under certain conditions, if the pressure in the system is not allowed to equalize between starts, the compressor can spin the motor backward, causing it to start in reverse. The resulting overload trip can be difficult to explain.

Not all PSC motors refuse to reverse properly while they are running. Small single-speed PSC motors with identical run and auxiliary windings are often equipped with a SPDT reversing control that changes which of the windings is connected to the capacitor. The two windings are connected to the supply common at one end, and the capacitor between them on the other. When the control connects the supply to one of the windings, it effectively places the capacitor in series with the other. When the other winding is energized, the capacitor is in series with the opposite winding, and the motor is reversed. When the torque demands of the load are not excessive, this motor can provide instant reversing (see Figure 10–13). A motor of this type is used in some electric actuators used for controlling

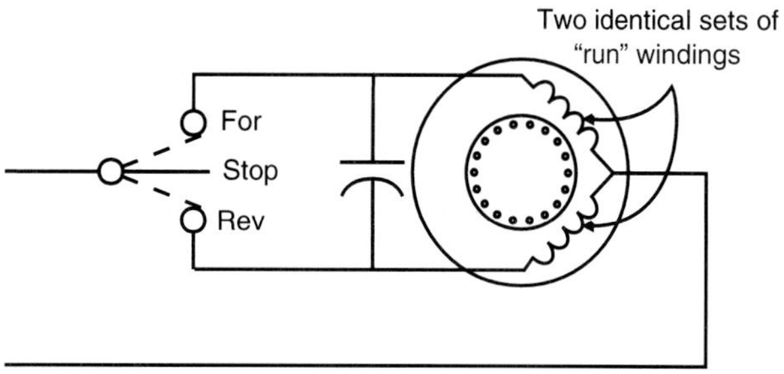

FIGURE 10–13 A small PSC motor may be designed for "instant" reversal. The control selects direction by changing which of the windings is energized directly, and which is in series with the capacitor.

dampers and valves. The controller has a special SPDT balancing relay with a center off position that allows the actuator to stop anywhere in midstroke.

Like most motor types, there are many possible variations in PSC design. In some of these motors, the capacitor is large enough to supply all the reactive current needed by both windings, giving the motor a full-load power factor of one. This is the only motor, other than the three-phase synchronous, capable of a perfect power factor at full load, with no extra components. Other PSC motors have a lighter second winding and operate much like a resistance or capacitor start motor. Their efficiency is similar to these other motors, but the start torque is lower.

PSC motors compete with the shaded-pole motor for fan drives and other small-motor applications. In theory the PSC should be a superior motor, but as with many things, either one can provide excellent service if it has been well designed and constructed. The PSC generally has a lower operating current because the capacitor improves the power factor. The shaded-pole has a little simpler winding and no capacitor to fail. Both motors have very low starting torque, so they are both sensitive to bearing problems or anything that may require significant start torque.

Note

When testing any motor with a run capacitor, remember that the capacitor is in series with one inductance (the auxiliary winding) and in parallel with another (the run winding). Because of the way these two types of reactive loads interact, the voltage drops across the auxiliary winding and the run capacitor can exceed the supply voltage—sometimes by quite a lot. The capacitor—with the auxiliary winding in series—is also connected in parallel with the run winding. Instead of an unexpectedly high voltage, there can be significant more current flow than expected. At no load, the internal motor current may exceed the current from the line.

The Capacitor-Start/Capacitor-Run Motor

The capacitor-start/capacitor-run motor is also called the two-value capacitor motor. It combines the superior starting torque of the capacitor-start motor with the higher performance and improved operating efficiency of the capacitor-run design. The most common type uses an intermittent-duty start capacitor connected in parallel with the continuous-duty run capacitor. The start capacitor is switched out of the circuit by a start switch or relay when the motor approaches its normal operating speed. As an alternative to a second capacitor, an autotransformer can be added to the run capacitor circuit. The run capacitor's effective value is changed by altering its applied voltage with the autotransformer. The voltage is higher at the start, increasing the amount of charge the capacitor can store and, in effect, changing the capacitance of the circuit. The connection is changed by a start switch or relay to return the capacitor voltage to normal as the motor approaches full speed.

General-purpose capacitor-start/capacitor-run motors usually use a speed switch to disconnect the start capacitor. Fully enclosed motors, or motors for special applications such as submersible pumps and refrigeration compressors, usually use potential relays. The relay coil is wired across the start (auxiliary) winding, and operates from the winding voltage drop. This voltage changes as the motor comes up to speed because of the series-connected capacitor. The pickup voltage of the relay—the voltage at which the coil opens the contacts—can be much higher than the rated line voltage. Once the relay operates, it is kept energized by the normal running voltage of the windings. This is a continuous-duty winding, so it is never disconnected from the line. The run capacitor is always in series, so the voltage across both the capacitor and the winding can be much higher than the supply voltage. When testing voltage across the auxiliary winding or run capacitor, it is a good idea to be prepared for a higher voltage than the rated line value.

Capacitor-start/capacitor-run motors are commonly used in relatively demanding applications such as positive displacement compressors and pumps. They are quite popular for large single-phase refrigeration or air conditioning compressors, and for submersible water pumps for water wells. These two applications use separately mounted capacitors and a potential relay for switching the start capacitor. Submersible pumps have the capacitors and relay in a control box mounted on the surface, often a considerable distance from the pump. Because of the separate auxiliary winding circuit, the pump must have an extra supply conductor between the motor and the control.

The controls for a capacitor-start/capacitor-run motor are essentially the same as for a capacitor-start motor. It is usually reversed by interchanging the start (auxiliary) winding leads. Like the capacitor-start motor, it cannot be reversed while it is running, and the consequences of trying to reverse it can be more serious. When this is attempted with a capacitor-start motor, nothing happens until the motor is shut off, and no damage results. With a capacitor-start/capacitor-run motor there may be problems. The auxiliary winding is still in the circuit, but with only the run capacitor connected. If this winding is suddenly reversed, the motor continues to operate very poorly in the same direction as before. It will usually draw excessive current and may trip out on overload.

The capacitor-start/capacitor-run motor is available in fractional and integer sizes to 10 Hp or more. Many larger horsepower single-phase motors use this design instead of just a start capacitor, because the run capacitor and continuous second winding improve overall performance.

The Repulsion Induction Motor

The repulsion motor has the highest start torque of any single-phase motor; it also has the lowest start current relative to the torque it produces. This makes it most suitable for very demanding loads, such as pump jacks and positive-displacement compressors. The repulsion motor has been used for belt-drive refrigeration compressors, but is unsuitable for hermetically sealed units because it has brushes and a commutator. Repulsion motors are also used where voltage is difficult to maintain, such as in silo unloaders and barn cleaners. They are somewhat less affected by voltage drop while starting than other motor designs.

FIGURE 10–14 The repulsion motor stator is similar to that of a split-phase motor, but there is no extra start or auxiliary winding. The rotor is similar to a series universal motor, but the commutator assembly is more complex. (See Figure 10–15.)

Although there are motors that start and run as repulsion motors, they have all but disappeared from the scene. Even the more popular repulsion-start/induction-run motors discussed here are now quite rare.

The stator windings of a typical repulsion motor look much like those used in induction motors, but there is usually only one set (see Figure 10–14). The motor can usually be identified by its wound rotor, brushes, and commutator. The rotor looks similar to the armature of a series universal or DC motor, but there are several critical differences in the way it is used. First, the brushes are interconnected by short jumpers, but do not connect to anything else. This is a type of induction motor: There is no direct connection between the rotor and the supply. Second, if the motor is a true repulsion-start/induction-run motor, not a straight repulsion motor, a close examination of the commutator reveals a very complex construction (see Figure 10–15).

Typical repulsion induction motors are equipped with an internal "necklace" of copper segments located underneath the commutator bars. The necklace is usually visible at the open end of the commutator. A spring holds the necklace away from the commutator bars until the motor is nearly at full speed, and then it snaps open to short circuit all the commutator segments. With the commutator shorted, the motor runs as an induction motor, very much as if it had a squirrel-cage rotor. Most repulsion-start/induction-run motors also have a mechanism to lift the brushes off the commutator when the necklace is in the running position. This greatly extends the brush and commutator life, because they are only in contact during startup.

Typical repulsion motors are well-built, high-performance motors that are expensive and quite rare in new installations. The motor is mentioned here because of the unique procedure used for reversing one of the more common variations. Electrically reversible repulsion motors are reversed in much the same way as resistance start, capacitor-start, or

FIGURE 10–15 This close-up of the repulsion motor rotor shows the edge of the necklace inside the commutator. This spring-loaded ring of copper segments short-circuits the commutator to convert the motor from repulsion start to induction run.

similar motors. However, many repulsion motors can only be reversed by changing the position of the brushes. On these motors, the brush assembly is held in position by a retaining screw that is accessible through a removable cover in the motor end bell. With the retaining screw removed, the brush rigging can be shifted through a small arc until the appropriate holes line up and the retaining screw can be reinstalled. This simple operation reverses motor rotation; no other changes are necessary.

10.4 Start Switches and Relays

Several single-phase motors have some type of start-to-run transition. It usually involves a change in the motor connections, but it may also be more complicated. The necessary changes are almost always done automatically, just before the motor reaches its normal operating speed. The most common starting devices are speed-operated start switches and current- or potential-operated starting relays. Speed-operated switches are usually mounted inside the motor; starting relays are mounted external to the motor.

There are several variations in the start-to-run transition and in the way it is handled. Resistance start motors have an intermittent-duty start winding that must be disconnected from supply. Capacitor start motors are the same, except for the addition of an intermittenduty start capacitor in the start winding circuit. The commutator of a repulsion-start/induction-run motor must be shorted and, in most cases, the brushes have to be lifted off the commutator at the same time.

Some PSC motors have a resistance device that temporarily bypasses the run capacitor. This allows them to start like a resistance start motor, giving them a better start torque. Two-value capacitor motors use a much higher value of capacitance during the startup period. This usually comes from a start capacitor that is switched out when the motor is nearly up to speed. PSC motors may be converted to two-value capacitor types by adding an aftermarket "hard-start kit." These kits are also available to convert resistance start motors to capacitor-start.

In general, the transition from start to run mode occurs at about 75 to 85% of the motor's rated full-load speed. The transition should be positive and consistent, and the starting device should not reset itself back to the start mode until the motor has been shut down or has slowed to a much lower speed. This helps to ensure that the motor does not suddenly drop back to the start mode during a mechanical overload. If that happens, an overloaded motor can cycle between start and run modes, destroying the start circuit.

To accomplish the transition at the right speed, the start switch or relay must sense a motor variable that changes predictably as the motor accelerates. Obviously, one of these is shaft speed, and speed-operated switches are very common in general-purpose motors. A second variable is motor current, which is much higher at startup, but drops as speed increases. A third variable is the change in voltage across a start or auxiliary winding that is connected in series with a capacitor. All of these are used. Some are only applicable to specific motors. All of them have advantages and disadvantages.

Speed Switches

Speed switches are typically operated by centrifugal flyweight devices mounted on the motor shaft (see Figure 10–16). The weight or weights are arranged so they stay in place

FIGURE 10–16 A typical speed-operated start switch has a centrifugal operator mounted on the shaft, and a switch mounted on the back of the motor terminal block. When the motor is at rest, the flat disk of the centrifugal operator presses against the extended finger of the switch contact, holding the contacts closed. When the motor reaches 75 to 85% of full speed, the flyweights snap out, pulling the disk clear of the contact fingers and allowing the contact to open.

until the critical speed is reached, and then snap out quite suddenly to provide a positive opening of the start switch contacts they control. The usual arrangement consists of the shaft-mounted centrifugal operator and a separate start switch mounted inside the motor end plate. The switch assembly for a single-speed motor usually has a single contact, held open by spring action. The contact is not enclosed, except for the fact that it is mounted inside the motor. An arm from the switch extends toward the center of the motor, where it makes contact with a flat disk that is part of the centrifugal device.

When the motor is at rest, the disk presses against the switch arm, holding the start switch contact closed. At 75 to 85% of the motor's rated full-load speed, the weights of the centrifugal operator swing out, snapping the flat disk inward, clear of the contact arm. The start switch contact springs opens to disconnect the start circuit. The mechanical noises made by the contact arm riding on the surface of the disk, followed by the weights moving into their running position, give these motors a very characteristic startup sound. At normal operating speeds the centrifugal operator is completely clear of the contact arm, and the start switch is held open by a small spring. This almost totally eliminates wear on the switch fingers and reduces motor noise.

The centrifugal operator is designed so that it does not reset until the motor shaft slows to a much lower speed. Once in the run mode, it does not drop back to its startup position until the motor speed is well below the normal operating range. This prevents the start winding from reengaging under moderate overload conditions. A severe overload will bring the speed down enough to reenergize the start winding, but the load has to be so high, and the reset speed so low, that it is unlikely the extra torque will be enough to reaccelerate the motor. The motor will simply stall and, assuming it has proper overload protection, will be automatically disconnected from the line. Figure 10–17 shows the location of the start switch contacts as they might appear in a drawing for a single-speed motor.

Some motors have more complex start switches. There may be two or more contacts, and they may be a mix of normally open and normally closed types. The extra contacts can be part of a two-speed motor circuit or a special-purpose motor control. In some cases an extra start switch contact is used in a control circuit that is not part of the motor

FIGURE 10–17 Two different motor diagrams showing the location of the speed switch contacts.

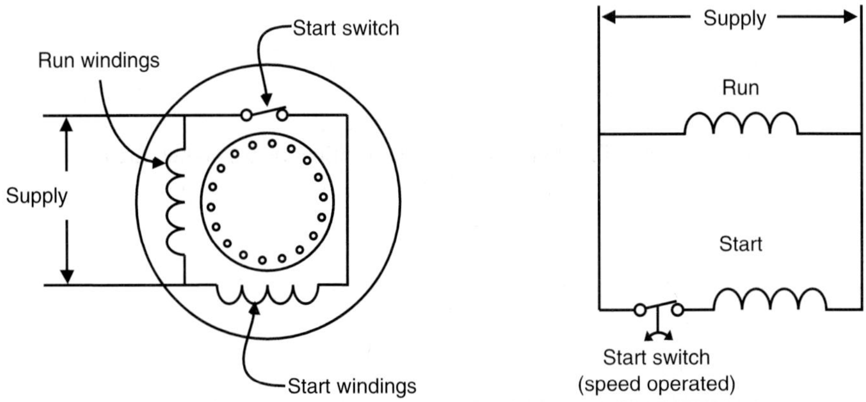

operation. In domestic clothes dryers, for example, a normally open contact is used as an interlock in the heating element circuit. It prevents the heating element from becoming energized before the blower and the drum are up to speed.

There are other applications where a speed switch is used to prove the motor shaft is turning, sometimes even with motors that do not require a start switch as part of their normal operation. A good example is the combustion air blower motor used in some packaged heating systems. Combustion air blowers are used with burners that depend on mechanical draft rather than the natural draft provided by a chimney. The combustion air blower draws in fresh air for burner combustion, and forces the combustion gases out through the vent. Because the burner has no natural draft, the blower is crucial to safe burner operation, so the burner control system must monitor either the air flow or the operation of the combustion blower.

There are several ways this is done, including the use of a speed-operated switch, much like a start switch, on the shaft of the blower motor. The switch contact is a normally open interlock in series with the burner control. It prevents the burner control from starting the ignition sequence until the combustion air blower is up to speed. If it opens during normal burner operation, it will de-energize the gas valves to shut the burner down. Because these motors are often quite small, they are likely to be either shaded-pole or PSC types that do not normal have a start switch. The switch is added as part of the burner control, not part of the motor circuit.

A variation on this idea uses an electronic detector and circuit board in place of the mechanical speed switch. The sensor is a Hall effect device that detects magnetic fields. A small magnet is bonded to the motor shaft and the sensor is mounted close enough to the shaft to can detect the passage of the magnet when the shaft is turning. The sensor is wired back to a separately mounted circuit board. The system operates just as a mechanical speed switch would, but uses a noncontact sensor and an electronic circuit instead of a centrifugal operator and a standard contact.

In most fractional-horsepower motors, the speed switch is mounted on the motor terminal board. This is a fiber plate that is visible inside the connection box. It supports the connection terminals, the start switch contacts, and the solder connections for the leads from the motor windings. The start switch, centrifugal operator mechanism, and internal motor connections can be inspected by carefully removing the end plate. Before removing either of the end plates, they should first be marked for easy realignment. The bolts are removed, followed by the opposite end plate and the rotor. The connection end plate cannot always be pulled straight off, but instead must be worked free, and then carefully tipped to one side to avoid damage to the internal wiring. There are several connecting wires from the stator winding to the switch and terminal board. When the motor has an internal overload, it is usually mounted separately on the same end plate.

Note

Whenever a motor is taken apart, any spacers on the shaft should also be accounted for and replaced on the same end. They often come off inside the bearing housing and can easily be lost. Improper alignment can affect the start switch, because it changes the spacing between the switch fingers and the centrifugal operator.

Centrifugally operated switches are the most common start switch for general-purpose motors. Replacing a start switch is one of the few nonmechanical motor repairs that is reasonably practical for the average skilled worker. Notes about the connections, and a bit of soldering, are the main requirements. Have the replacement switch on hand first if possible. Mark the end plates for alignment, and remove the opposite end bell and the rotor. Do not damage the winding with the sharp edges of the rotor fan. Mark all leads as to their locations on both the new switch and the old one. Desolder the leads on the old switch and resolder them to the new one.

Repulsion-start/induction-run motors also use centrifugally operated flyweight mechanisms, but they have a different job to do and may be quite different in design. Repair of a repulsion motor is best left to an experienced motor technician. They are mechanically and electrically complex.

Current Relays

Many induction motors naturally draw a higher current at startup than they do during normal running. Start current is particularly high for resistance start motors, somewhat less for motors with start capacitors. Low-start torque motors, such as PSC and shaded-pole fan motors, have comparatively low start current. Small "impedance-protected" shaded-pole motors draw much the same current regardless of shaft speed or motor load. For each motor type, the peak value of the starting current is reasonably consistent on each start, regardless of the motor load.

Current-operated start relays are most often used on resistance start motors, in applications where a speed-operated switch is unacceptable. Common examples are hermetically sealed refrigeration compressors that are small enough to use resistance start rather than capacitor-start motors. The relay is usually a simple solenoid arrangement with a few turns of relatively heavy wire and a single contact in series with the start winding of the motor. When the motor is at rest, the weight of the solenoid core holds the contact open. On startup, the high current draws the solenoid plunger upward and closes the contact. This energizes the start winding to get the motor started. When the motor is up to speed, the current drops to normal running values. The running current is too low to hold in the solenoid, so it drops out again, opening the contact and de-energizing the start winding.

Current relays have the disadvantage of having normally open contacts. Each time the motor starts, the contacts must close to supply current to the start winding, and then reopen to break the start circuit. This is much harder on the contacts than if they were closed when the motor started, and only had to break the circuit after the current had dropped to a relatively low level.

Many of these relays are gravity operated. They must be mounted with the coil vertical and right side up. They often mount directly onto the compressor terminals, and may be combined with an overload protection device built into the same housing. The relay contact can usually be tested by turning the relay upside down to close it, and right side up to open it. If the contact opens and closes as it should, and there is continuity through the coil, the relay is probably in working order (see Figure 10–18).

FIGURE 10–18 Two drawings of a resistance start motor with a current-operated start relay. The high start current closes the relay contacts each time the motor starts. They reopen as soon as the current drops to normal.

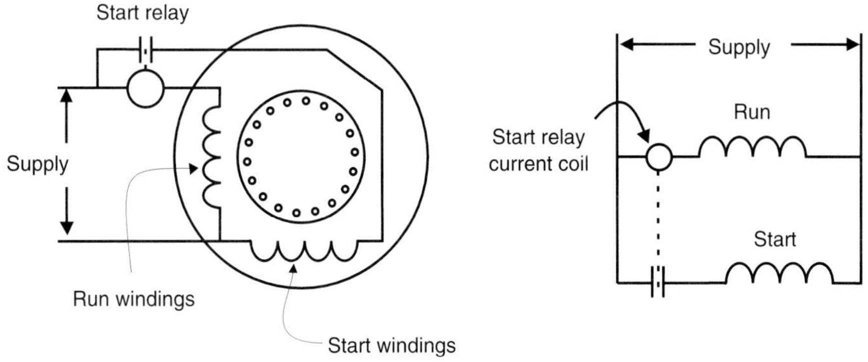

Hot Wire Relay

The hot wire relay is a relatively unique current-operated device that operates as a combination starting relay and overload device. Unlike the electromagnetic current relay, its contacts are closed when the motor starts. They open when the motor current passing through the "hot wire" element heats it enough to make it expand. The expansion of the element allows the start contact to spring open. The motor current continues to pass through the wire element, keeping it hot enough that the contacts don't reset.

If the current is higher than normal for an extended period of time, the element heats even more, expanding further and tripping a second contact to shut off the motor. The hot wire relay must have time to cool down before the motor can restart.

Hot wire relays may be found on some older equipment, but are now obsolete.

Solid-State Start "Relays"

There are also solid-state devices that can behave as current-operated "relays." The simplest of these is part of a family of solid-state components called thermistors (see Figure 10–19). All thermistors have one thing in common: Their resistance changes with changes in temperature. If the thermistor has a positive temperature coefficient (PTC), the resistance rises with an increase in temperature. If it has a negative temperature coefficient (NTC), the resistance of the material is highest at low temperatures, decreasing as the temperature rises.

Thermistors are used for temperature sensing and for certain current control and "switching" applications. Those that are used for temperature sensing are very small. They cannot carry significant current, but their small size allows them to respond quickly to changes in the temperature around them. Thermistors used for power applications such as current control or motor starting must conduct substantial amounts of current. They are physically larger and may have quite different resistance values than thermistor temperature sensors.

FIGURE 10–19 These solid-state start "relays" are positive temperature coefficient thermistors. Their cold resistance is low enough for them to act much like closed contacts. When they heat up from current flow, the resistance rises dramatically, making them act much like open contacts.

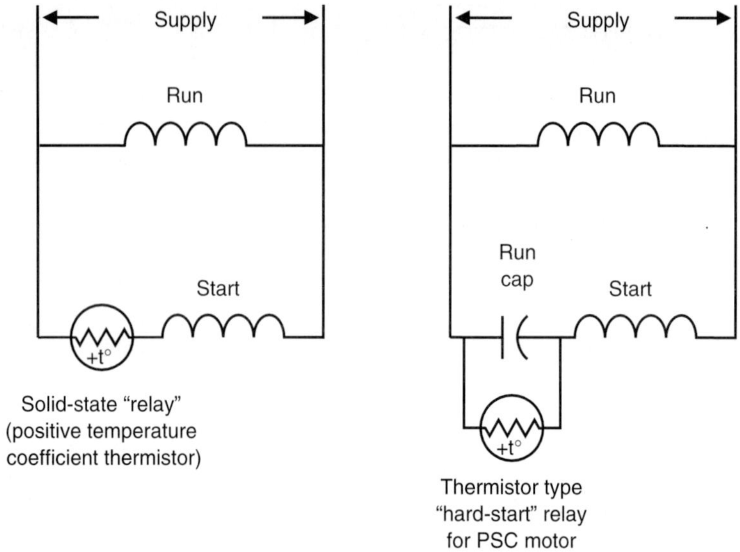

A PTC thermistor can be used to provide an effective start switch with no moving parts. It is used in place of a current relay, connected in series with the start winding of a resistance start motor. The "cold" resistance is only a few ohms, so the thermistor conducts the start winding current with very little interference. However, this current quickly heats the thermistor material, causing the resistance to increases to a "hot" value of several thousand ohms.

The cold resistance is low enough to allow essentially normal operation of the start winding, and the hot resistance is high enough that the winding is virtually disconnected. Once the motor has started, there is enough current to keep the thermistor in a high-resistance state, but not enough to interfere with normal motor operation. The resistance of the thermistor is so much higher than the winding impedance that almost all of the voltage is dropped across the thermistor. This leaves very little voltage across the winding, and therefore very little energy to cause it to heat.

Solid-state start devices are sometimes used with capacitor-run (PSC) motors as well. A typical PSC motor has a better running performance than a resistance start motor, but a lower start torque. To increase the initial torque, an appropriate PTC thermistor may be connected in parallel with the terminals of the run capacitor. When it is cold, the thermistor temporarily bypasses the capacitor, effectively converting the PSC motor into a resistance start type. The motor-starting current heats the thermistor, raising its resistance to a high enough value that it is virtually out of the circuit. Enough current continues to flow through the thermistor to keep it in the high-resistance state, and the motor operates as a normal PSC design.

Thermistor "start relays" do not reset immediately when the motor is shut off. The thermistor material must cool down before it goes back to the low-resistance state required for startup.

There are other types of solid-state motor-starting relays that are more sophisticated than the simple thermistor. Totally enclosed devices with no moving parts, they connect to the motor windings in place of the normal start switch, and perform a similar function.

Potential Relays

Voltage-operated "potential relays" are used for capacitor-start or capacitor-start/capacitor-run motors. The relay coil is connected across the start or auxiliary winding, which is in series with one or more capacitors. The voltage drop across the winding—and the relay coil—is lowest at startup, rising as the motor comes up to speed. This rising voltage operates the relay and opens a normally closed contact wired in series with the start capacitor.

The operation of the potential relay is relatively simple; the reason it works is more difficult to understand. A motor winding is essentially an inductor. However, it is magnetically linked to the rotor winding, much like the primary winding of a transformer is linked to the secondary. When the rotor is stationary, it "sees" a 60 Hz magnetic field passing through the shorted loops of the squirrel-cage winding. The induced voltage is relatively high, creating a high rotor current. With a large energy flow from the stator winding to the rotor, the effective impedance of the stator winding is relatively low.

As the motor comes up to speed, the rotor begins to catch up with the rotating stator field. The field is still changing at 60 Hz, but the rotor "sees" a much lower frequency. This changes the magnetic link between the rotor and stator windings. Less voltage is induced, so the rotor current is lower. There is less energy transferred from the stator windings, so the effective impedance of the stator winding is much higher than at startup.

Impedance is the total opposition to current flow in an AC circuit. It is the net effect of all the resistive and reactive characteristics of that circuit. A circuit with a low impedance draws a relatively high current because there is very little opposition to current flow. In a motor, the effective impedance—the opposition the winding has from the point of view of the supply—is partly dependent on the electrical behavior of the rotor, which is in turn affected by motor speed.

One result of the low startup impedance is that the motor current is at its highest when the motor is first energized. A second result is that in the auxiliary or start winding circuit—because it is in series with the capacitor—the voltage drop across the winding is also at its lowest. The impedance of the capacitor—its capacitive reactance—is not effected by the speed of the motor. Therefore, when the impedance of the winding is low, more of the supply voltage is dropped across the capacitor, leaving less for the winding (see Figure 10–20). As the motor speeds up, the impedance of the winding rises, causing the voltage drop across the winding to increase also.

Because of the way series-connected capacitors and inductors interact, the voltage drops across both the winding and the capacitor can be very high. Under the conditions found in a typical PSC motor auxiliary winding circuit, it is quite normal for both of these voltages to be well in excess of the supply voltage. This is the reason that run capacitor voltage ratings are so high, and potential relay pickup voltages are usually higher than the rated supply voltage.

The start winding and the capacitor are in series; therefore, the logic of electrical theory states they must add up to the supply voltage. These high voltage drops seem to disobey this law, but that is not the case. The current is the same through both devices—they are in series—but their voltages are out of phase. The capacitor voltage lags the current;

FIGURE 10–20 As the speed of a motor changes, so does the effective impedance of the windings. When a capacitor is in series with the winding, the voltage drops across the capacitor and the winding will also change. This characteristic is used to advantage in potential-operated starting relays.

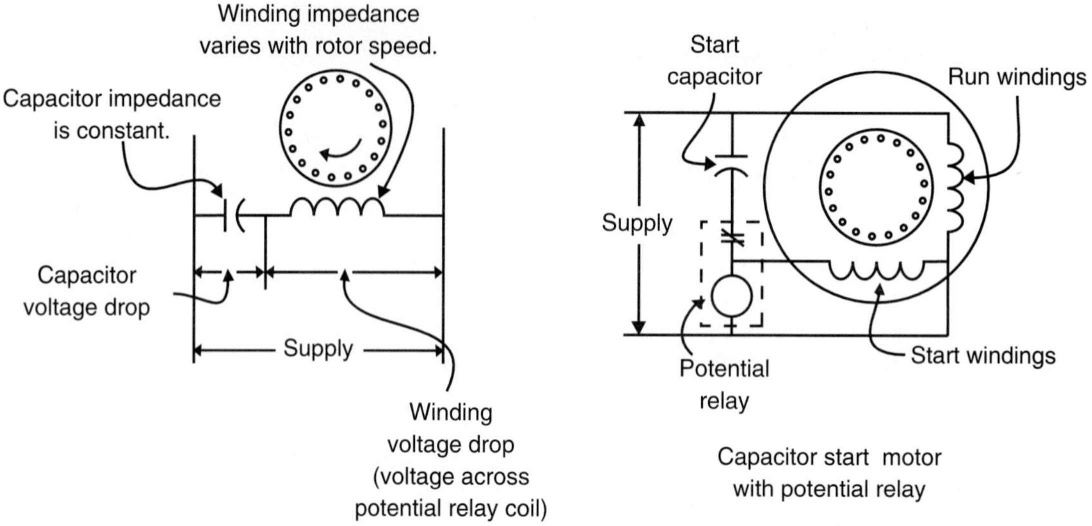

the coil voltage leads the current. Out-of-phase voltages cannot be added with simple mathematics, because the peaks and zero crossings do not coincide.

For capacitor start and capacitor-start/capacitor-run motors, the changes in impedance and voltage drop are consistent and predictable. A potential relay, set to operate at a specific winding voltage, will be triggered at essentially the same motor speed each time the motor is started. Potential relays have calibrated "pickup" voltages that allow the motor manufacturer to match them to the motor characteristics. Once the relay is energized, the voltage must be reduced to a much lower level before it will release. The low "dropout" voltage is important because it prevents the relay from reenergizing the start circuit if the motor is overloaded. The pickup and dropout voltages are listed in the relay specifications. When a relay is replaced, the ratings of the new one should match those of the original.

For a capacitor-start motor, the relay contacts completely disconnect the start winding from the supply (see Figure 10–21). The relay coil, still across the now-disconnected winding, is kept energized by the voltage induced into the winding by the motor's alternating field. This voltage is less than the supply voltage, but higher than the relay dropout voltage, so the relay stays picked up. In a capacitor-start/capacitor-run motor, the relay only disconnects the start capacitor; the auxiliary winding stays in the circuit, still in series with the run capacitor. The voltage drop across the winding keeps the relay coil energized.

Potential relays serve essentially the same purpose as current relays, but are very different in construction. The potential relay coil is wound with many turns of fine wire, and is connected in parallel with the start winding, not in series with the run winding. The potential relay coil operates at a relatively high voltage, but draws very little current. The cur-

FIGURE 10–21 Two different potential relay connections, shown in ladder diagram format.

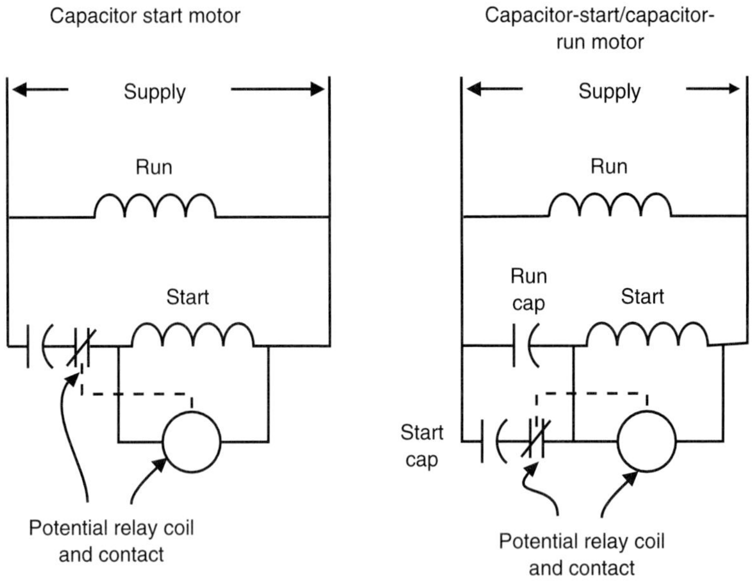

rent relay has to carry the motor current continuously, not just during startup. Unlike many current relays, which must be mounted right side up, potential relays are less likely to be sensitive to the way they are mounted. The contacts are held in their normal position by a spring, not by gravity.

Thermal-Delay Relays

Some motor-starting relays operate from heat generated in the "relay" by the motor current. These relays are essentially normally closed time-delay relays. The contacts open after a specific time period, instead of sensing some motor characteristic that is related to its speed. The contacts may operate more slowly than they would with other types of relays, but this is usually not a problem.

Summary

For resistance start, capacitor-start, and capacitor-start/capacitor-run motors, internal speed-operated switches are the most common type of start switch. Some motor manufacturers are also using solid-state electronic relays, or offering them as an option in place of the traditional speed switches. Where a speed switch is impractical, such as for sealed refrigeration compressors and submersible pumps, some type of external starting relay is used instead.

Current relays, or some type of thermal-delay relay, can be used for resistance start motors. Potential relays are preferred for motors with start capacitors, partly because their normally closed contacts take less punishment.

When resistance start refrigeration or air conditioning compressors have problems starting, a "hard-start kit" can be added as an option to replacing the compressor. Typical kits add a start capacitor and a relay of some type to disconnect it when the motor is up to speed. When PSC motors are used for these compressors, the "hard-start kit" is a little different. It may add additional capacitance in parallel with the run capacitor, or bypass the run capacitor with a resistance. A PTC thermistor across the capacitor makes the motor start as a resistance start type. This improves the start torque, but not as effectively as additional capacitance.

Certain types of start relays have become associated with specific motor applications. Sometimes the type of start relay used has more to do with previous practice than with superior performance. As with most things, a start switch or relay that is well designed and well made provides many years of trouble-free service.

10.5 Single-Phase Motor Protection and Control

The term "motor protection" refers to equipment and connection techniques used to prevent damage to the motor under abnormal conditions, such as excessive current due to mechanical overload. "Motor control" refers to manual or automatic equipment and methods used to start, stop, reverse, or make other intentional changes in the way the motor operates. Motor protection may involve built-in or external devices, or some combination of the two. Some elements of motor protection are normally integrated into the motor control devices and circuits.

Motor Protection Requirements and Methods

In some cases, single-phase motors are not provided with any protective device other than the fuse or breaker that supplies the circuit. This may be permissible when motor overload is either very unlikely or will be immediately obvious to the equipment operator and is not a fire hazard. Common examples include some small direct-drive fans, small appliances, and power tools. In most cases, however, some form of overload or overheat protection is desirable, if not essential, to reduce the risk of equipment damage and eliminate potential fire hazards.

Motor overload protection can sometimes be provided by using a current-limiting motor design, such as the "impedance-protected" motors used in small fans. These small shaded-pole motors do not draw excessive currents, even if the motor is stalled. However, impedance protection is impractical or impossible in all but the smallest of motors. Protective devices are then added to the motor or the motor circuit. The most common of these are built-in thermal protectors and external motor overloads. Both are widely used, and although better systems also exist, one of these two options is adequate for most installations.

Built-In Thermal Protectors

Many types of motor protection devices are thermally operated, typically combining a low-resistance heating element with heat-operated contacts. The heating elements are usually connected directly into the motor circuit and carry the full motor current. The heat-operated

FIGURE 10–22 A typical thermal-disk motor protector has the operating contact mounted right on the edges of the disk. When the disk is hot enough to "pop" to the opposite curvature, it opens the contacts. Note the symbol and the terminal numbers.

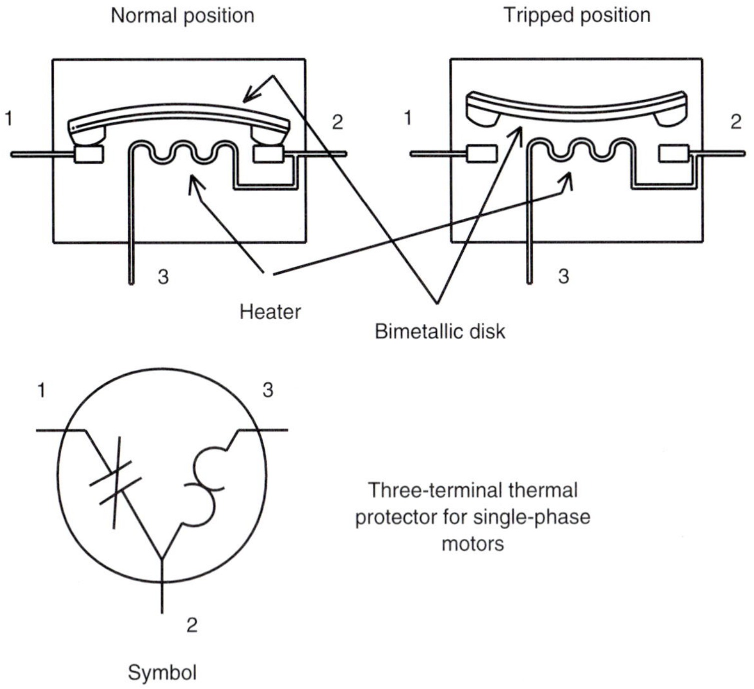

contacts are normally closed, opening only when the heat buildup becomes excessive. The amount of heat varies with the current to the motor; therefore, the operating point of the contacts can be calibrated to match a specific amount of motor current.

Thermal devices are popular for overload protection because they are simple and reliable, and have a natural time delay. This delay is important because it allows the contacts to remain closed through short periods of high current—such as a normal motor startup—yet trip properly on a sustained overload. A thermal overload device that is properly matched to the motor approximates the motor's thermal behavior. Any condition that causes the motor to overheat also causes the overload to overheat and to trip the motor off the line.

Built-in thermal protectors are usually some type of "thermal disk"—a round piece of bimetallic material, ranging in size from 1/2" to 1" or more. Bimetallic material is a laminate of two different metals that have different rates of thermal expansion. A bimetallic strip bends when it is heated. A typical bimetallic disk is slightly curved, like a shallow dish. When it is heated, it "pops" to the opposite curvature, because the material on the inner layer expands at a higher rate than the outer layer.

In one common type of thermal protector, a pair of contacts mounted on opposite edges of the bimetallic disk makes contact with a second pair of contacts that are mounted in the insulating material that forms the case (see Figure 10–22). The disk is anchored at the center and, when it is cool, both sets of contacts are held firmly together to complete the circuit. When the disk is heated enough to "pop," the two contacts mounted on edges of the disk snap away from the stationary contacts. They separate like a fast-acting, double-break disconnect switch, opening the circuit and shutting off the motor.

The heating element or "heater" that initiates the contact operation is a short piece of relatively heavy resistance wire mounted underneath the disk. The resistance is not very high, but it carries full motor current and is located very close to the disk. This type of thermal protector is compact, inexpensive, and very reliable. The trip point is not adjustable, but protectors are made in a wide range of current ratings.

Built-in thermal overload protectors may be either self-resetting or manual-reset types with a pushbutton that resets the contacts. Which of these two is best depends on the motor application. The self-resetting type requires no intervention when the motor trips. Once it has cooled off, it resets itself, and the motor automatically restarts. If the condition that caused the initial overload has not cleared, the self-resetting overload continues to cycle the motor off and on until the problem is corrected—or the motor is destroyed. If this is a likely occurrence, a self-resetting overload should not be used. However, in applications where the cause of the overload is almost certain to clear itself after a brief delay, self-resetting overloads may be very suitable. Self-resetting overloads are often used on small refrigeration and air-conditioning compressors.

Manually reset overloads force the operator to attend to the problem. The motor cannot be restarted until the overload has cooled and the reset button has been pushed to re-close the contacts. This not only protects the motor from multiple restarts under impossible conditions, it also protects the equipment operator from a sudden, unexpected startup. For safety reasons, manually reset overloads are the only acceptable type on virtually all operator-attended equipment. Examples include power saws, drill presses, lathes, and all types of materials-handling equipment. Any motorized equipment that can pose a safety hazard if it is started without warning must not be equipped with self-resetting overload protection.

Built-in thermal overloads have advantages and disadvantages when compared to the separate motor protection devices to be examined next. Their advantages include simple construction, low cost, and the ability to respond directly to motor overheating as well as to overcurrent. Because the overload is mounted inside the motor case, it is affected by the heat of the motor, as well as the heat produced by current through the internal heater. If the motor is hot for any reason, the overload device trips much more easily than if it is cold. It also takes longer to reset. This provides protection that is closely matched to actual motor conditions. An external overload device is mounted away from the motor, so unless some type of thermal sensor is installed inside the motor as part of the motor protection and control circuit, no overheating protection is provided.

Most built-in motor overloads are "direct-acting line break" controls. This means they are wired directly into the motor supply circuit, and when they open, the contacts break the supply line voltage (see Figure 10–23). As good as they are at their task, motor-mounted thermal overloads of this type have two limitations: There are practical maximums to the current and voltage ratings, and they cannot be integrated into a complex multimotor control system. However, there is a variation of this basic design that provides some of the advantages of internal overload without these disadvantages.

Motor-mounted thermal protectors can be equipped with "pilot duty" contacts instead of the usual line break power contacts (see Figure 10–24). The heaters are in the motor circuit and carry motor current as before, but they are electrically separate from the contacts. The contacts are normally closed, just as with the line-break type, but they are rated for pilot duty only. They are not used to disconnect the motor directly. Instead, they are wired into an

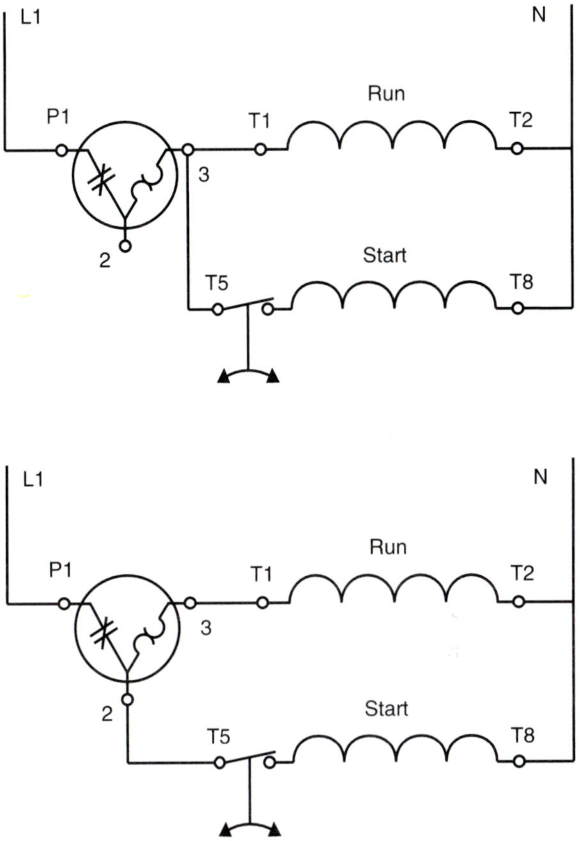

FIGURE 10–23 Two different connections for a three terminal protector in a single-voltage motor. In the first connection, the heat protector carries run and start winding current. In the second it carries run winding current only.

external control system, where their opening trips a motor contactor. Overloads of this type are found on some larger single-phase and three-phase refrigeration compressors. They are mounted in the connection box, external to the sealed compressor unit.

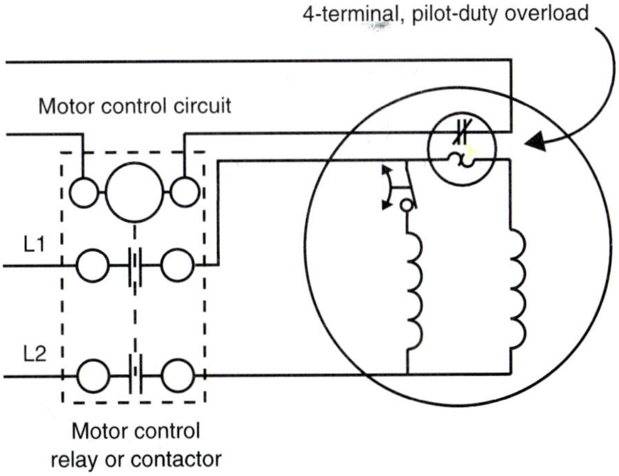

FIGURE 10–24 A four-terminal pilot-duty overload does not break the line. Its pilot-duty contact must be wired into an external control circuit, where it breaks power to the coil of the motor control device.

External Overload Protection

Motor overload protection can also be provided by an external overload device. The simplest and most common external overloads are thermally operated, much like the motor-mounted protectors. The overload device must not trip during short-term overcurrent periods, such as a normal startup, but must trip for even a modest overload that is maintained long enough to endanger the motor. Thermal-based motor overloads, with their natural time delay, have proven to be a reliable and cost-effective form of motor protection. The external type of protector may be mounted some distance from the motor, and still provide a good approximation of the motor temperature as long as the ambient conditions at the motor are not radically different from the conditions at the overload.

The heater and thermal element combination used for external overloads is somewhat more refined than in the direct-acting motor-mounted version. Thermal disks are rarely used because external overloads have only pilot-duty contacts, so snap action is not necessary. Thermal-disks are also difficult to adjust, and adjustable trip points are a feature of all external overloads. This way, one overload device can be matched to a variety of different motor current ratings.

Most external overloads use either a melting-alloy or bimetallic-strip element (see Figure 10–25). A "heater" is connected in series with the motor winding, just as in the

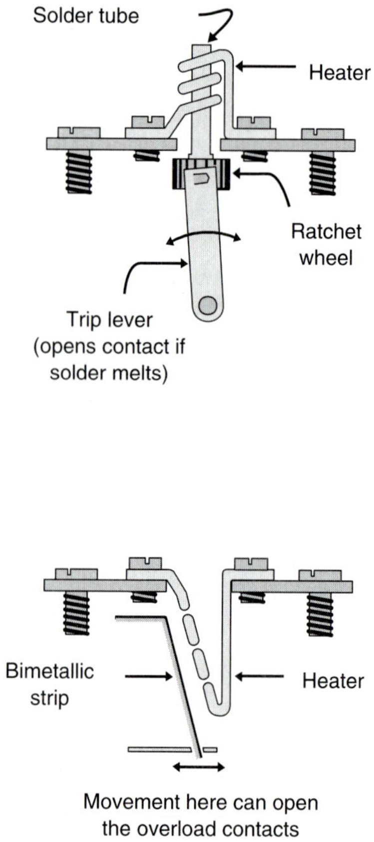

FIGURE 10–25 The latch mechanism of a melting-alloy overload is held by a small ratchet wheel that is soldered in place. Heat melts the solder, allowing the wheel to turn and releasing the overload contact. A bimetallic-type overload trips a latch if the bimetallic strip bends too far.

built-in protectors, and these elements respond to the heat it generates. Melting-alloy overloads of this design are sometimes called "solder pots" because the heart of the device is a small ratchet wheel with its axle "soldered" inside a tube. Under normal conditions, the wheel is held firmly in place by the solidified solder; under overload conditions, the heat from the heater element melts the solder, allowing the ratchet wheel to turn. The wheel is part of the latch mechanism, and when it turns, the latch is released, the starter contacts open, and the motor is tripped off the line. The device is constructed so the solder stays inside the tube, even when it melts. The solder only has to cool before the starter can be reset and the motor restarted.

Bimetallic elements bend to trip a latch mechanism that holds the contact closed. They are built in greater variety than melting-alloy overloads, because of the many ways that the bimetallic element and the overload heater can be integrated into a control device. Like the melting-alloy overload, the bimetallic type can be reset once it has had time to cool.

Built-in motor protectors are selected and installed by the motor manufacturer to match the motor's current demand and thermal characteristics. External overloads are normally part of the main motor control device. They are usually built and marketed by a different company than the motor, so they have adjustments that allow them to be matched to the specific situation. Different overload settings are provided through a series of interchangeable heater elements, a mechanical adjustment to the overload device, or some combination of these methods.

10.6 Manual Starters

A manual starter is essentially a switch with contacts designed for motor duty and a built-in overload device. Typical examples are operated by a toggle handle or double pushbuttons, with one button for ON and the second for OFF. Whether it is operated by a toggle or pushbuttons, the contacts of the manual starter latch closed through a mechanical linkage that can be released by the overload device. The contacts are spring loaded, so they open if the latch is released. The contacts are fast-acting, whether they are released by an overload trip or by normal operation of the switch.

Many manual starters use interchangeable heaters and melting-alloy tripping units. A wide range of heaters is available, allowing an accurate match to the motor ratings. It is quite common for the heater and solder element to be integrated into a single replaceable part. Bimetallic elements may also be used.

10.7 Electromagnetic Controls

Relays and Contactors

Electromagnetic controls are built around electromagnetic relays or their larger cousins, the contactors. Although the actual physical layout varies a lot, all relays and contactors operate on the same principle. They have a coil with a two-piece iron core. One part of the core is fixed; the other is allowed to move. The moving "armature" of the core may be a solenoid type that moves in a straight line, or a clapper type that swings on a hinge point. The armature is mechanically linked to one or more sets of contacts. The contacts may be identified as normally open (NO) or normally closed (NC), according to their status when the coil is de-energized. When the coil is energized, the mechanical movement causes the contacts to change state. The normally open contacts close, and the normally closed contacts open. The relay or contactor is essentially an electrically operated switch.

It is important to realize there is no electrical connection between the relay or contactor coil and the contacts. This allows the coil to be part of a different circuit than the load that it controls. These two circuits are usually referred to as the power circuit—containing the motor or other power load and the control circuit—which contains the control devices and the contactor coil. The two circuits are electrically separate, but linked logically and operationally by the electromagnetic action of the coil. When the control circuit energizes the coil, it magnetically operates contacts that control the motor or other power load.

The control circuit can be at a different voltage than the power circuit. It can also be a different type of supply (such as DC instead of AC). However, the relay coil voltage and supply type must match the ratings of the control circuit. Coils are specifically designed for AC or DC operation at a specified voltage, and must operated within a few percent of these values. AC and DC coils are not interchangeable. Most general-purpose relays can be purchased with a variety of standard coil voltage ratings. Motor contactors usually have field-replaceable coils that can be matched to any standard industrial control voltage. DC coils may not be available for all contactors, because the physical construction of a DC electromagnet is somewhat different than for an AC device.

Connecting a coil to a supply that does not match its ratings can result in improper operation or damage to equipment. If the voltage is too low, the coil will not develop a strong enough magnetic field to operate the contactor. If the voltage is too high, the coil draws far too much current and, unless some other protective device such as a control circuit fuse opens quickly enough to protect it, the coil is destroyed. It is important to check the coil voltage of any relay or contactor that is being added to a circuit or installed as a replacement. The old and new relay can look identical and still have different coil ratings. These ratings are usually marked on the coils themselves, or on the case if the relay is totally enclosed. The coil rating is almost always visible without taking the relay or contactor apart.

The contact ratings of a relay or contactor are a separate issue, quite unrelated to the coil ratings (see Figure 10–26). Contact ratings are essentially the same as switch ratings, representing the maximum allowable voltage and current. Lower values of either voltage or current are perfectly acceptable. Note, however, that some types of loads are more difficult to switch than others. Resistive loads, such as heaters and incandescent lights, are generally the easiest. Inductive loads, which include anything with a coil in it, are much more difficult. Motors are similar to other coil loads. They need a very capable set of contacts for reliable control.

Contacts ratings are stated in different ways. General-purpose relays are usually marked AC, DC, or AC/DC. The maximum voltage and a maximum "noninductive" current are also listed. There may be several different ratings, based on different voltages and on AC or DC use. The DC ratings are typically much lower, because a DC current is generally much more difficult to break. AC contacts that are specifically designed for motor use are usually rated in horsepower, again with different ratings provided for each applicable operating voltage.

Relays designed for motor control are limited to small motor sizes, but can be very useful and cost-effective when the control needs are more complex than can be provided by a simple switch or manual starter. Relays are widely used to control the fractional- and subfractional-horsepower motors found in HVAC equipment. Relays are usually used with motors that have internal overload protection, because the relay has no built-in motor overload device.

FIGURE 10–26 The contacts of this relay are rated for 1 Hp 120 V AC, 2 Hp at 240 V AC. The coil is rated for 120 V AC, but other coil voltages are available.

Contactors are essentially large relays specifically designed for controlling power loads. When they are combined with an overload device, they are called magnetic starters. Contactors and starters are made for use with single-phase motors, but are available in greatest variety for three-phase equipment. Three-phase starters can be adapted to single-phase control if desired, although one contact set may go unused. (The construction and operation of three-phase motor starters and thermal overloads are examined in more depth in Chapter 12.)

Magnetic Starters

A magnetic motor starter consists of an electromagnetic contactor (described above) and one or more overloads equipped with pilot-duty contacts. The overload assembly and the contactor are usually factory-interconnected to form a single unit. The contacts of the overload are wired into the coil circuit of the contactor. They are normally closed, opening on overload to cut power to the coil and electrically "tripping" the motor off the line. Although many overload devices provide the option of automatic or manual reset operation, automatic reset should never be used where a sudden restart of the motor could be a safety or fire hazard, or could cause equipment damage.

Both melting-alloy and bimetallic overloads are used with magnetic starters. As with manual starters, melting-alloy overloads depend on a wide range of heaters for trip-current settings. Bimetallic types often have a calibrated adjustment, and some have both interchangeable heaters and an adjustment dial. On bimetallic overloads with fixed heaters, the adjustment knob is normally calibrated in amperes. For overloads with both replaceable heaters and an adjustment dial, the actual current rating of the overload depends on the heater chosen and the dial setting. Therefore, the dial is usually calibrated in percent, rather than in amperes. This design allows a full range of current adjustments with only a few heaters. In either case, the overload can be accurately set to match the motor current.

Selecting Overload Heaters

Replaceable overload heaters are not normally marked in amperes. They carry a manufacturer's part number that must be cross-referenced to a table of full-load motor current ratings. Overload charts or tables are usually supplied with the starter so that the appropriate heater can be selected. When the starter is supplied with an enclosure, the chart may be permanently attached inside with other manufacturer's data. At the time of purchase, the supplier usually asks for the motor nameplate data so the correct heaters can be provided with the starter.

In many cases, the overload can be selected right off the charts with no more data than the full-load current rating. However, there are situations where a proper selection is more involved. Other nameplate data such as the motor service factor and the specific electrical code requirements may enter into the final decision.

10.8 Control Circuits

Single-phase motors are made in greater variety than other motor types. The methods used to control them, beyond simple start/stop functions, depend greatly on the type of motor and the application. General-purpose motors are used for equipment that is custom built or designed with standard motor mounts of some type. The motor may be selected to power a new piece of machinery or as a replacement unit. These motors are almost always one of the five induction types—shaded-pole, resistance start, capacitor-start, PSC, or capacitor-start/capacitor-run.

A mixed bag of control equipment is available for these motors, including controls designed for specific applications and a variety of general-purpose control components that may be adapted to different applications. The more specific types of controls are relatively easy to select and apply. The more general-purpose ones can be much more flexible to use, but a good understanding of electrical equipment is needed to apply them successfully.

Some types of motors—such as series universal motors—are rarely sold off the shelf, but are commonly incorporated into tools, appliances, or other manufactured devices. The controls are provided as part of the equipment and, like the motors, are not readily available as general-purpose items. The controls are often very ingenious, but simple enough to make them cost-competitive in the mass market.

Manual Control

The simplest and least expensive off-the-shelf control is a general-purpose manual switch. These may be suitable for small motors where overload protection is either built-in or unnecessary. However, their use is restricted by electrical codes, because they do not have a specific horsepower rating. Good-quality general-purpose switches have a reasonable life expectancy if the motor is small and is not started and stopped very often. Switches designed specifically for motor duty are essential for larger motors, or where frequent starts and stops occur. They are designed for this type of service and are rated in horsepower.

An important step up from a manual switch, the manual motor starter provides horsepower-rated switch contacts and overload protection in one package. They are well suited to controlling a single motor from one location, where no additional control features are required. They are the least expensive way of providing basic on/off motor control, plus properly sized overload protection, for motors that do not have built-in overloads. Manual starters are simpler and less expensive than magnetic motor starters, but cannot be used to provide more than on/off control from a single switch location. They also cannot be integrated into a larger control system.

Magnetic starters allow for very flexible control options. Virtually any type of motor control function that can be provided by contacts can be provided with a magnetic starter. (These starters, and typical control circuits used with them, are examined in detail in the Chapter 12.)

Warning

For motors that operate at more than 120 V—typically 208, 240, or higher—it is absolutely imperative that the disconnect switch be a two-pole device. Using a single-pole control switch or single-pole manual starter on equipment that is powered from a 208 or 240 V supply is illegal and extremely dangerous. Shutting off the motor leaves it energized from one single-phase line, at 120 V to ground. This is a deadly trap for service personnel because the motor appears to be de-energized, but is still a shock hazard. Two-pole switches break both "hot" supply lines, totally isolating the motor from the supply.

Automatic Controls

Automatic control for many common applications can be provided with direct-acting, horsepower-rated switches built for the purpose. Examples include float switches for liquid-level control, pressure switches to cycle compressors and pumps, and line voltage thermostats suitable for direct control of small motors. A manual disconnect should be provided ahead of the automatic control. For motors that do not have built-in overloads, the disconnect may be a manual starter incorporating overload protection.

For larger motors, a magnetic starter or appropriate horsepower-rated relay is used, and the pressure switch or other control then serves as a "pilot device." The contacts of the pilot devices are used to switch the coil circuit of the motor starter instead of operating the motor directly. For example, refrigeration compressors are usually controlled by a contactor that is, in turn, controlled by a thermostat or pressure switch.

More complex control systems can be built from standard components. For fractional-horsepower motors, generic relays with suitable contact ratings are usually the least expensive approach. Such relays are often found inside electrical enclosures and on circuit boards that are part of motorized equipment. Relays do not provide overload protection, but this can be provided by other means, such as selecting motors with built-in thermal units. Larger motors require the higher current ratings of contactors or motor starters.

Reversing

Reversing can be accomplished with a manual reversing switch, such as a drum controller, or a reversing starter. However, unless a suitable control is devised or a special motor is used, reversal of single-phase motors can be a problem. A typical reversing controller changes the connection to the start winding. For resistance start and capacitor-start motors, moving the control directly from forward to reverse, without allowing the motor to stop first, has no effect. The motor continues to run in the same direction because the start winding is disconnected until the motor slows almost to a stop. PSC and capacitor-start/capacitor-run motors can be damaged by reversal without stopping the motor first.

Reversing Drum Controllers. There are large, very complex drum controllers used for special applications, but the one discussed here is a simple manual switch that can be used to reverse single-phase, three-phase, and DC motors by using various connections. These controllers are quite standardized components available from several companies. Multiposition rotary switches are also available for this purpose. They may be available in wider variety, with more convenient contact arrangements, than the drum controller shown in Figure 10–27.

FIGURE 10–27 The contact arrangement for a typical reversing drum controller are shown at the top. This control can be used for a number of different motors. The connections shown here are for a single-phase, single-voltage split-phase motor.

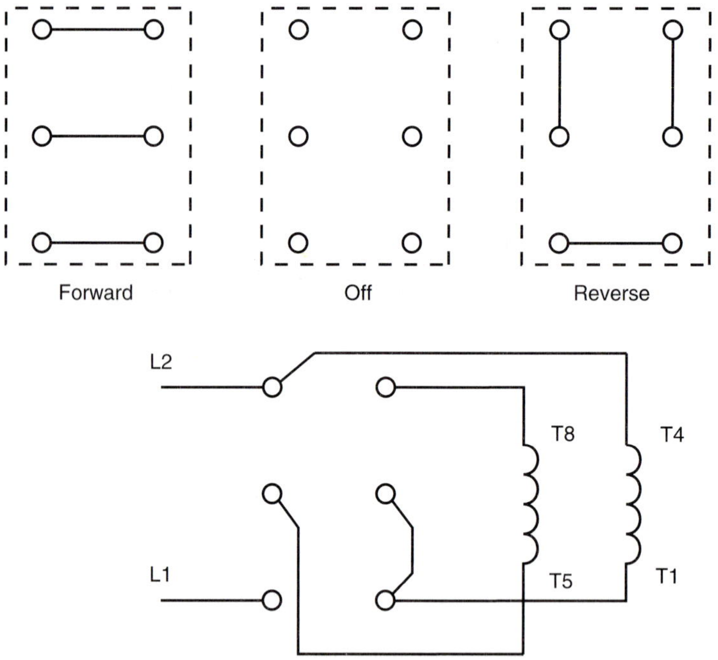

Connections for reversing a single-phase motor:
The switch reverses connections to the start winding, T5–T8.

10.9 Motor Drawings

There are many variations to the symbols and layouts used to show motors and their connections on control drawings. Even the winding symbols are not consistent, although they are usually rather obvious. In more formal electrical drawings, four "bumps" on the coil symbol is standard for a "normal," parallel-connected field winding, such as the stator winding of a single-phase or three-phase motor, or the parallel-connected "shunt" field of a DC motor. These "bumps" may be simple half-circles or more elaborate interlocked arcs. A winding symbol with three bumps indicates a series-connected field winding, and two bumps symbolizes the commutating or compensating windings found in some series universal and wound-field DC motors.

Motors are often drawn in a semipictorial format that attempts to indicate the type of motor, or the arrangement of the windings and, possibly, the connection of the start switch contacts. In drawings of this type, it is common to have the start or auxiliary windings shown at right angles to the main winding to symbolize their placement inside the stator. When the motor is a dual-winding two-speed type, there will usually be two run windings shown connected to separate external connection terminals.

When motor windings are drawn for clarification of alternate connections, the drawings are usually less pictorial, taking on more of a ladder schematic format. Three-phase windings may be drawn in a wye or delta shape (more on this in the section on three-phase motors) but single-phase motors are usually easier to understand if the windings are shown in a relatively formal ladder style. This is, of course, far from a universal standard. Motor nameplates show only a listing of the required connections The drawings used in manufacturer's literature are often in a semipictorial format.

For examples of single-phase motor connections, refer to Figures 10–28, 10–29, and 10–30, and the explanations below. In a typical single-phase motor, the run winding is directly across the supply. The start winding is in parallel with the run, but is switched out after startup in resistance start and capacitor-start types. Only the start capacitor is switched out in the capacitor-start/capacitor-run motor. PSC motors have no start switch: The run capacitor is in the circuit at all times.

Dual-voltage motors have dual run windings, but usually only a single start winding. The run windings are numbered in sequence, T1-T2 and T3-T4. They are connected in series for the highest voltage, and in parallel for a low voltage. A single start winding carries the markings T5-T8; a double one is marked T5-T6 and T7-T8 for the two halves respectively. With only one winding, it is rated for the lower voltage and connected across only one of the run windings during startup. This way it gets approximately the rated voltage.

Thermal protectors usually have three terminals. The "extra" terminal is brought out between the heater and the contact. This way, windings that are connected to the second terminal are cut off when the overload opens, but do not draw current through the heater. For a dual-voltage motor, only one run winding is actually monitored for overload current when the motor is connected for low-voltage operation. Both are in series for the high-voltage connection, so they are monitored as a pair. This connection allows the same overload device to be used for both voltages, even though the current to the motor is twice as much on the low voltage connection as it is on high.

Current relays have a normally open contact in series with the run winding, or in the motor line supplying both windings (see Figure 10–31). The current relay contact closes on the high start current, opening as the current drops off with the motor's rise in speed.

Potential relays are voltage operated and have normally closed contacts. The coil is across the auxiliary (start) winding, and operates from the voltage that appears across that winding (again, see Figure 10–31). This winding is in series with the motor capacitor, so its voltage drop varies as the effective inductance of the motor winding changes with the motor speed. This winding voltage is low when the motor starts up, and rises as it approaches the normal operating speed. The relay picks up at a preset voltage that occurs when the motor speed is about 75 to 85% of the rated speed. It holds in on the voltage that remains across the winding after the start capacitor is out of the circuit.

FIGURE 10–28 Typical winding connections for series universal and shaded-pole motors.

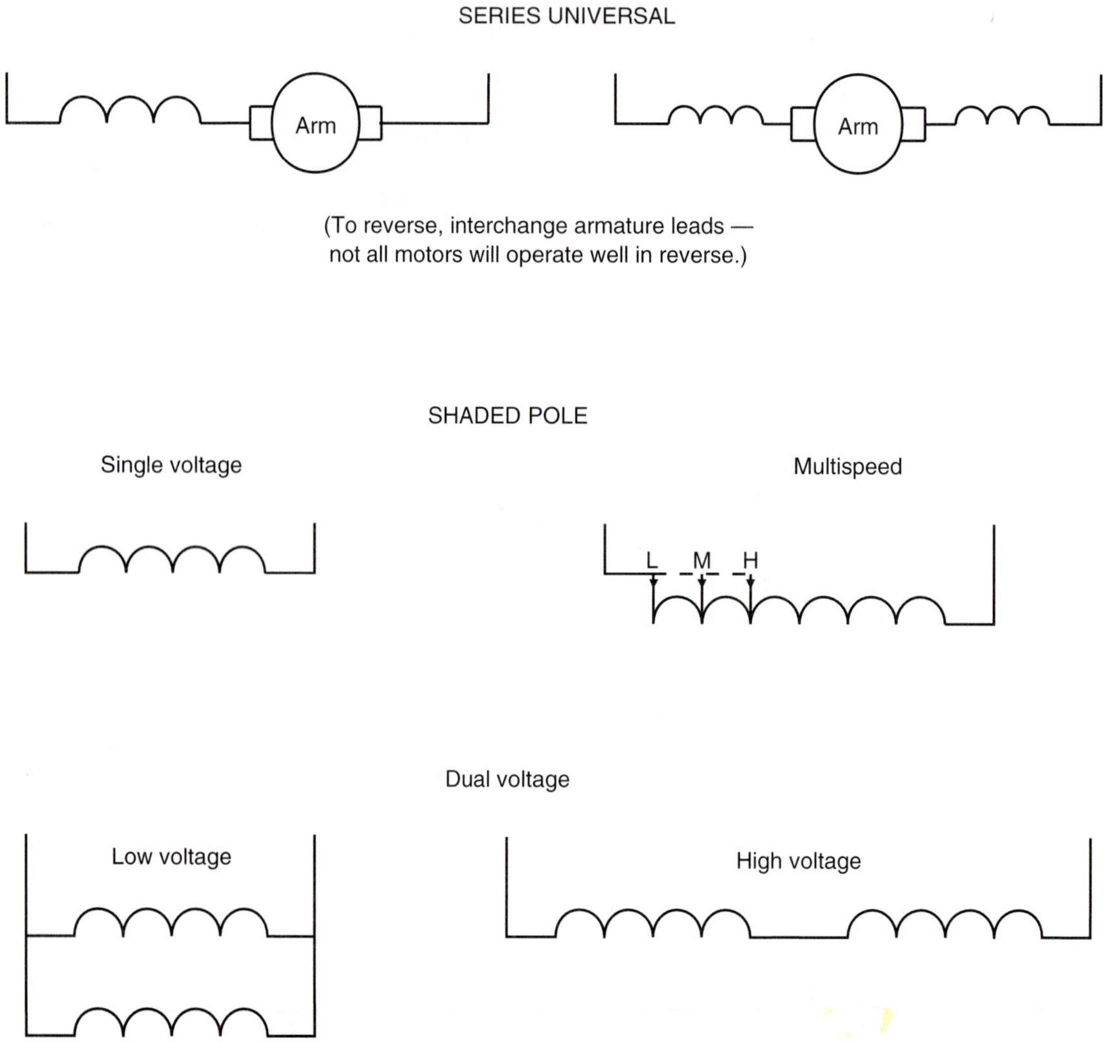

FIGURE 10–29 Typical winding connections for PSC motors.

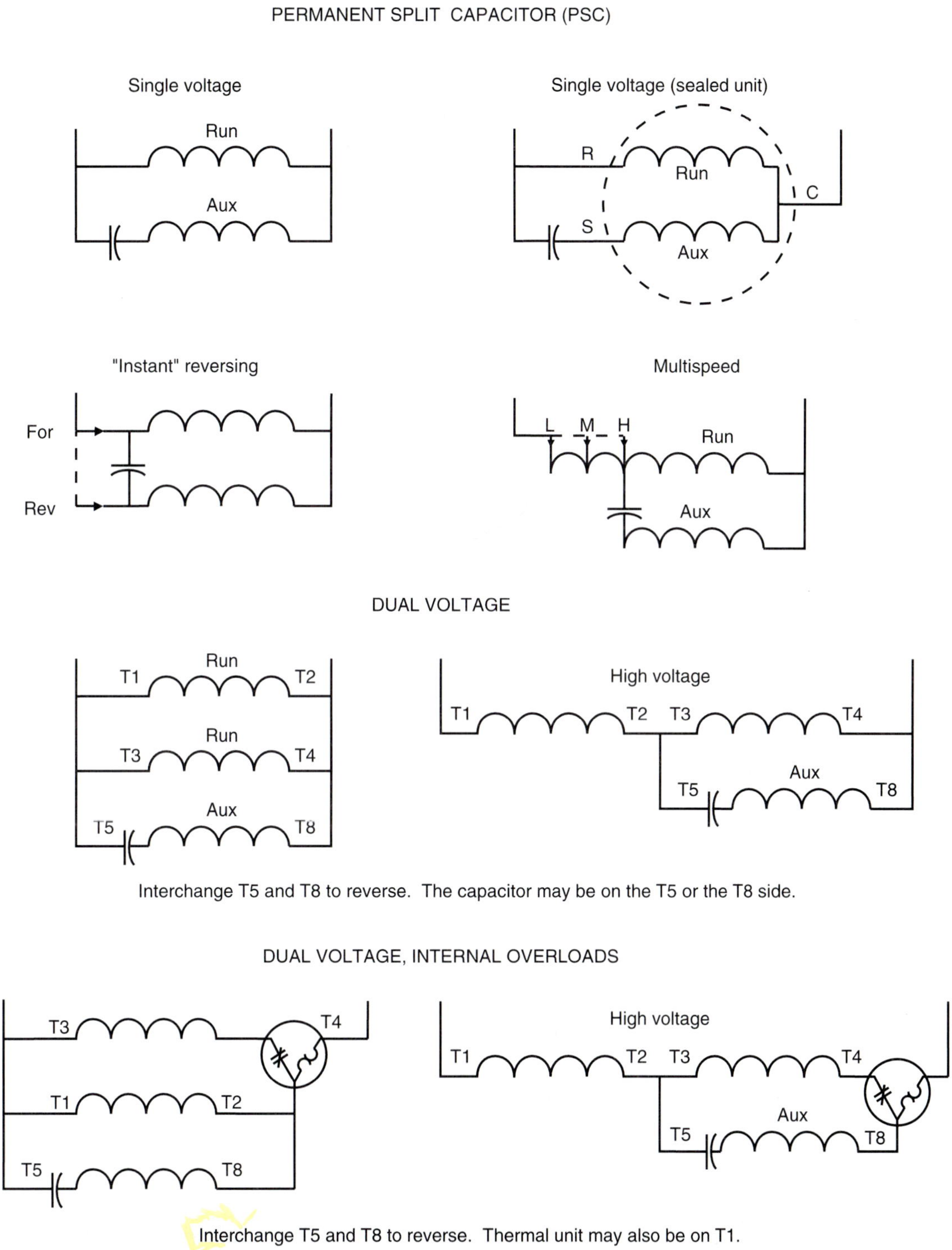

FIGURE 10–30 Typical winding connections for split-phase and capacitor-start motors.

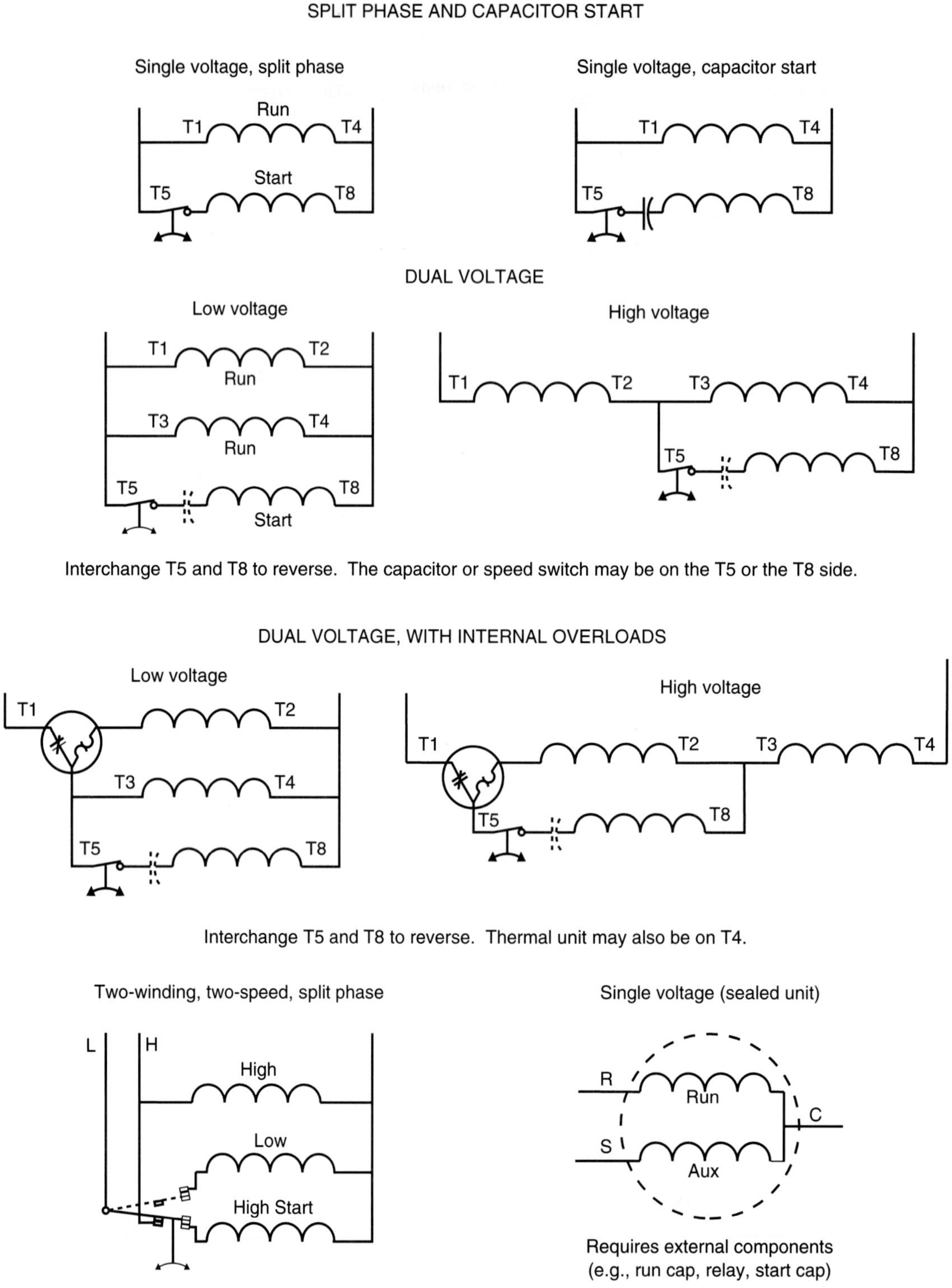

FIGURE 10–31 Typical connections for starting relays. The motors on the right are sealed units such as refrigeration compressors.

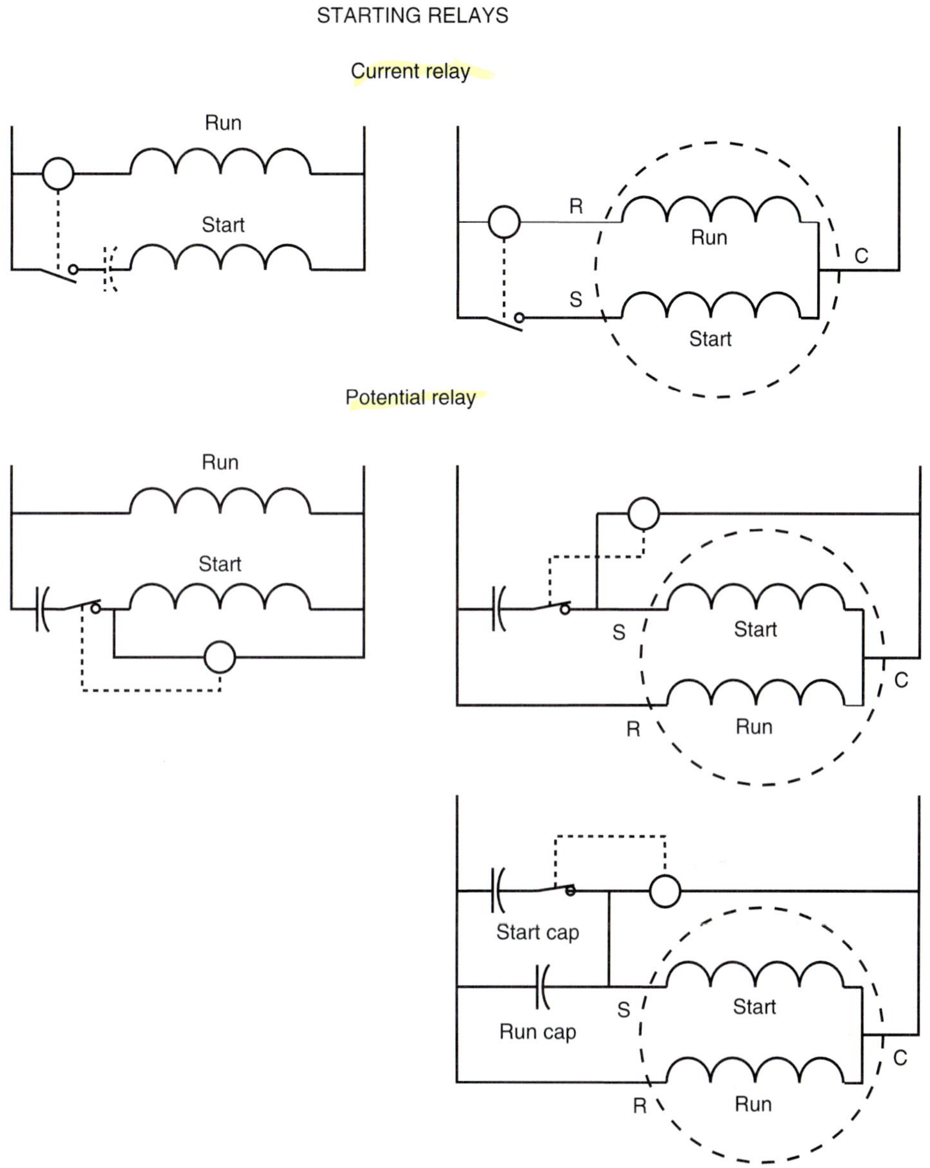

Questions

1. List the operating characteristics of a series universal motor.
2. What are the main items to check when performing routine maintenance on a series universal motor?
3. How does a squirrel-cage rotor behave when it is placed in a rotating field? What if the field is only changing, not rotating?

4. Why does a squirrel-cage rotor always run slightly below the synchronous speed of the stator field?
5. How can the rotor frequency be less than the frequency of the stator supply?
6. List the characteristics of shaded-pole motors.
7. Explain how a rotating field is created in a shaded-pole motor.
8. What procedure is used to reverse the rotation of a typical shaded-pole motor?
9. What is meant by the term "impedance protected" on the nameplate of a motor?
10. How can two currents with different phase angles be created by the same voltage?
11. What technique is used in capacitor-start motors to increase the phase shift between the start and run winding currents?
12. List the main operating characteristics of the PSC motor.
13. Why is there only one speed listed on the nameplate of a typical multispeed PSC fan motor?
14. What precautions are necessary when connecting a multispeed PSC or shaded-pole motor?
15. What is meant by the term "variable torque load"?
16. Is it possible for the run capacitor of a motor to have an operating voltage drop that exceeds the supply voltage? Explain.
17. What is the procedure for changing the direction of rotation for a repulsion motor?
18. At what speed does the start-to-run transition take place in a typical split-phase motor?
19. On single-phase motors with a start-to-run transition, it is important to have a wide differential between the start to run changeover speed and the speed at which the circuit resets itself to the start condition. Explain why.
20. Explain the operation of a coil-type current relay.
21. Why is it important how a typical current relay is mounted? What could happen if the relay was mounted upside down?
22. Explain the difference between a PTC and an NTC thermistor.
23. How does a PTC thermistor, in parallel with the capacitor of a PSC motor, improve the motor's start torque?
24. Draw a motor connection diagram, using correct symbols, to show the connections for typical examples of the following: speed-operated switch, current-operated relay, potential relay, and PTC thermistor used for resistance start and PSC motors.
25. Under what conditions is it possible that a single-phase motor may have no special thermal or overload protection, and still be installed to code?
26. Why are most motor overload protection devices based on heat-operated contacts?
27. Define what is meant by a "normally open" contact.
28. What is the role of the control circuit in an electromagnetic motor control system? What is the role of the power circuit? How are the two different in terms of their connection to the control relay or contactor?
29. Why is it so important to match the coil voltage of a motor contactor to the motor control circuit voltage?
30. Some adjustable overloads are marked in percentage, some are marked in amperes. Explain the reason for this difference.

11 Three-Phase Motors

Objectives

This chapter describes three-phase motor operation and explains the differences among various motor types. It also identifies the winding arrangements used for typical motors and the connections required for normal operation. After completing the study of this chapter you should be able to:

- Explain how a three-phase supply creates a rotating field
- Describe the winding layout for a two-pole motor and explain what the differences would be if the motor had four or more poles
- Explain why increasing the number of motor poles reduces the motor's operating speed
- Define, in general terms, the meaning of locked rotor torque, pull-up torque, breakdown torque, and slip
- Describe briefly the physical differences between Design B, C, and D rotors
- Compare the operating characteristics of Design B, C, and D rotors with regard to their torque, speed regulation, and efficiency
- Draw representative connection diagrams for three-, six-, nine-, and twelve-lead motors, showing the most common variations
- Explain how the number of motor poles is changed in a consequent-pole two-speed motor
- Explain the difference between the terms "constant torque," "variable torque," and "constant horsepower" as they apply to multispeed motors

11.1 Introduction

To the power utility, smooth generator output and higher power transmission efficiency make a three-phase system vastly superior to single-phase. To the consumer, the greatest advantage of a three-phase supply is the ability to use three-phase motors. A three-phase system delivers three equal voltages in a symmetrical, continuously repeating sequence. Applied to an appropriately constructed motor, these voltages provide a natural rotating field, with none of the special windings, capacitors, start switches, or relays associated with single-phase motors.

The three-phase induction motor is the workhorse of industry. Three-phase motors are compact, rugged, and efficient. They are also very reliable, requiring only a minimum of maintenance over many years of service. They are available in a full range of voltage and power ratings, from fractional horsepower outputs to ratings in the thousands of horsepower. NEMA standard ratings for three-phase motors go up to 100,000 Hp.

11.2 Three-Phase Motor Stators

All three-phase motors use the same basic layout for the iron core and the stator winding. The only major variation in the electrical parts of the motor is the choice of rotor. Other differences are minor refinements to better match motor operation to specific load characteristics, or to improve overall efficiency. Mechanical features vary with environmental considerations, mounting method, and vertical or horizontal orientation. Special bearings and oversize shafts are available for special situations such as severe shaft loading, high vibration, excessive torque shock, or similar problems.

Environmental considerations such as dust, explosive atmospheres, exposure to weather, and washdown requirements dictate the type of enclosure used. Mounting can be to standard dimensions for bases or flanges, or the motor may have custom mounts for special-purpose situations. Vertical or horizontal orientation affects bearing choice and mounting provisions. Ambient temperature can affect the enclosure used, but is more likely to determine the insulation class. The class of insulation used sets the maximum operating temperature of the motor. Available insulation classes are A, B, F, and H, each of which is rated for progressively higher operating temperatures. Note, however, that the temperature rating of the insulation is based on internal winding temperature found by an accurate measurement of the winding resistance. It is not based on a temperature reading taken with a thermometer on the motor case.

The magnetic and electrical portions of a three-phase motor stator are essentially identical in construction to the stator of a three-phase generator. The laminated iron core is slotted around the inside surface to receive the windings. The windings are made up of groups of identical coils, laid into the slots and interconnected to form three separate phase windings. The poles of each successive phase are in a physical 1-2-3-1... sequence, repeated over and over all the way around the stator. The ends of the windings are usually connected to flexible leads that are brought into a terminal box mounted on one side of the motor. Larger motors or those with sealed enclosures may have the windings connected to rigidly mounted terminals. The actual number of leads or terminals, and their relationships to the coil groups and windings, will vary with the type of motor connection and the voltage ratings, but the markings are highly standardized.

The stator has three sets of windings, one for each phase, and each phase winding is made up of two or more coil groups. These are usually connected in series, but parallel or series-parallel connections may also be used. The number of coil groups depends on the number of poles in the motor. If, for example, the motor is classed as having two poles, there will be two coil groups or "poles" for each phase, six groups in all. These are arranged in a sequence that repeats only once in the full distance around the inside of the stator. Each of the phases has two poles of opposite magnetic polarity, located directly across the stator from each other. Assume the stator is viewed from one

FIGURE 11–1 A two-pole, three-phase winding consists of six groups of coils, connected to form three equally spaced phases with two poles each. A standard winding symbol may be used to represent the complete winding of each phase.

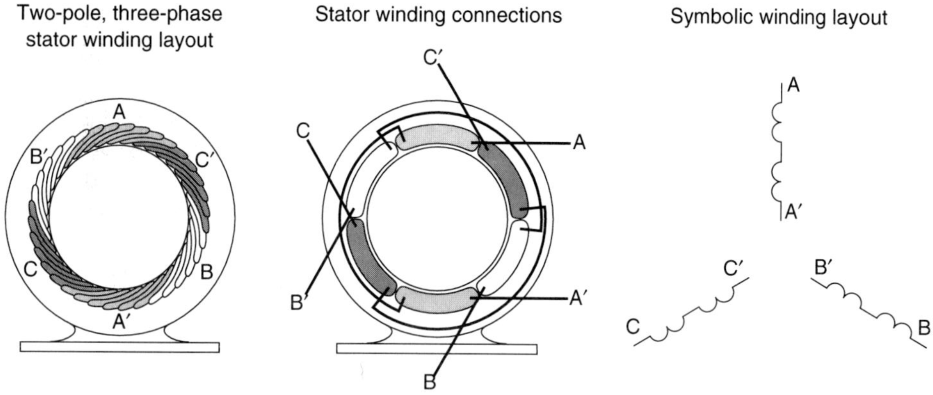

end, and the poles of each phase are identified as A-A', B-B', and C-C'. The A, B, and C poles are the "start" end poles; the matching A', B', and C' poles are the "finish" end poles. The sequence around a two-pole stator would then be A, C', B, A', C, B', and back to A. This places the pole pairs windings at equal thirds, or 120° (mechanical), intervals around the stator. The windings are evenly spaced and perfectly symmetrical (see Figure 11–1).

The three winding phases are connected to the three incoming voltages of the three-phase supply. The three separate AC voltages of the supply have the same sine wave shapes, voltage values, and frequency, but they overlap in time. Each waveform is identical, but displaced from the other two by a third of the total time for a cycle, or 120° (electrical). This overlap creates a situation where there is always current in at least two phases, and very often in all three. The current that enters on one phase line must exit from some combination of the other two. The actual division of current at any instant in time depends on the phase voltage relationships at that time. The sum of the three line currents will always be zero, because the total current in and out of the motor must be equal.

As the phase windings are energized, a unified magnetic field moves around the stator from one phase pole to the next. The rate at which it moves is set by the frequency of the incoming three-phase supply and the number of motor poles. It should be noted that the stator field does not change in strength or polarity as it moves. As the current in one phase rises, the preceding one declines. The overlapping sequence of rise and fall, and polarity reversal, makes a smooth transition from one set of motor phases to the next. The result is a field—with the same number of magnetic poles as any one phase—that rotates smoothly around the center axis of the stator (see Figure 11–2).

Frequency is the rate at which the voltage and current cycles occur, so it stands to reason that doubling the frequency would also double the speed of the field. The effect of the number of poles is not as obvious. Suppose the motor has four poles instead of the two used in the previous example. The number of winding groups or poles in each phase is now four, and the total number of "poles" in the stator is 12—four for each of three phases. The rotating magnetic field that results has four poles instead of only two.

FIGURE 11–2 The current flow in each phase varies continuously as the three-phase sine wave goes through its cycle. The result is a complex but repetitive pattern of changes that produces a rotating magnetic field. In the diagram above, one-quarter on an electrical cycle translates into a one-quarter revolution of the field. If the motor had more poles, the mechanical rotation will be proportionally less.

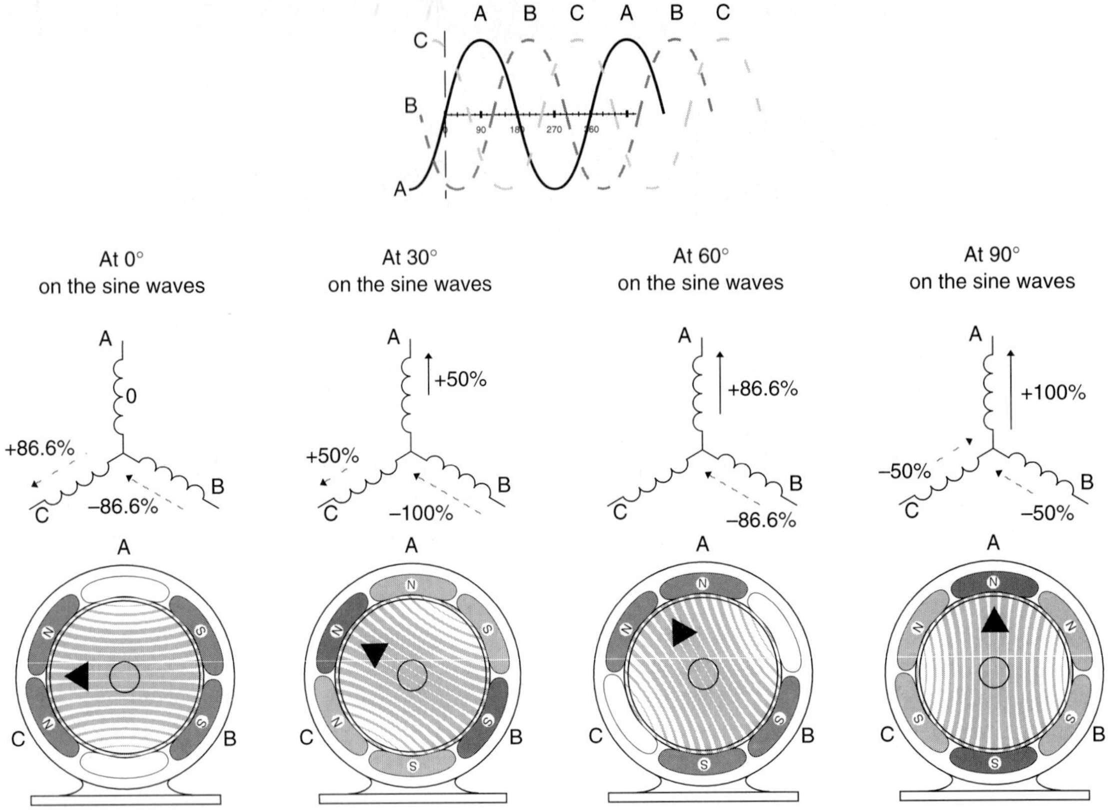

In a two-pole motor, the rotating field has only two poles (one north and one south), and a complete AC cycle moves the field from phase to phase through one complete mechanical rotation. In a four-pole motor, the sequence of the winding polarities and phases repeats four times around the stator instead of twice. The magnetic field of the stator now has four poles, N-S-N-S, instead of a single N-S pair. A complete AC cycle moves the field past a set of six winding poles, just as in the two-pole motor. However, in a four-pole motor, this moves the field only half of a mechanical revolution. The field rotates as before but, with twice as many poles, it moves at half the speed (see Figure 11–3).

The number of poles in a motor can be any even number, but there must be enough physical space in the stator core to hold all the coils of the completed winding. Each pole must be reproduced for each of the three phases. This can take up a lot of space, which explains why very low-speed motors tend to be shorter in length, but larger in diameter. There are practical limits to the maximum number of poles. Low-speed motors are more expensive and far less common than higher speed versions.

FIGURE 11–3 When the stator has four poles, a complete cycle of the AC supply only moves the field one-half turn.

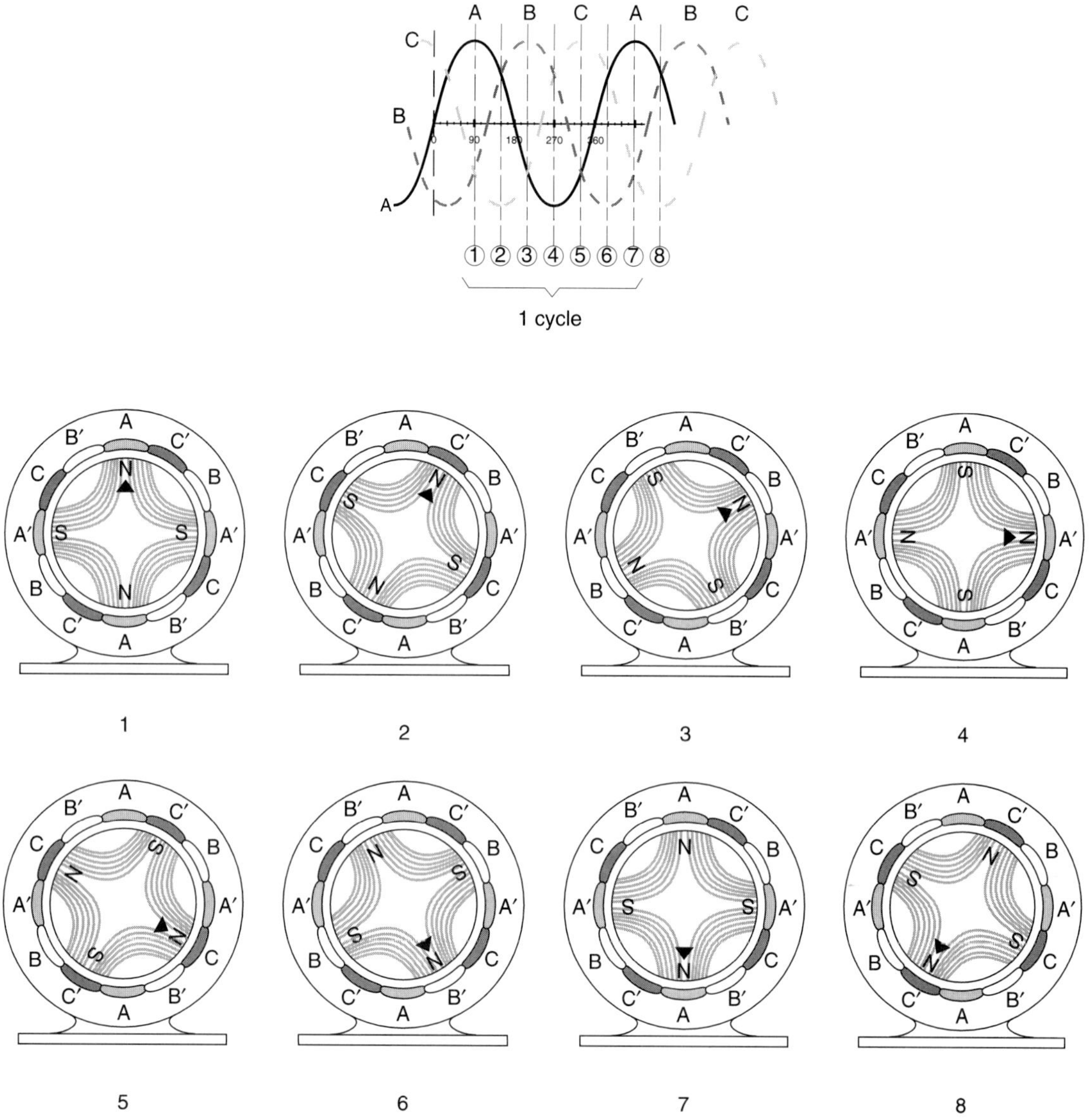

11.3 Types of Three-Phase Motors

There is not as much variation in three-phase motor design as there is in single-phase types. The primary reason for such variety in single-phase motors is the difficulty in producing rotation from a single AC supply voltage. What is difficult to do with one phase is easy with three. The relatively simple stator winding layout just described converts a three-phase supply into a natural rotating field: No special windings or other auxiliary equipment are required.

The vast majority of three-phase motors use a squirrel-cage rotor. Motors based on this rotor are simpler, less expensive, and require less maintenance than any other three-phase type. The alternatives include the wound-rotor motor and the three-phase synchronous motor. These two motors are used for special situations in which their unique characteristics justify the additional cost and maintenance. The squirrel-cage motor will be examined in depth after a brief explanation of the two other motor types.

Wound-Rotor Motors

The wound-rotor motor uses the same stator as the squirrel-cage design, but the rotor is made up of interconnected coils, similar in layout to the three-phase winding used for the stator. When the motor is operating, the rotating field of the stator induces voltages into the rotor windings. In a squirrel-cage motor, the rotor conductors are short-circuited by the end rings. Rotor currents flow very freely, opposed by only a small amount of rotor bar resistance and inductive reactance. In the wound-rotor motor, the rotor windings are connected to an external circuit through shaft-mounted slip rings and a set of stationary brushes.

The three slip rings are mounted just past the end of the rotor windings, on one end of the rotor shaft. To accommodate the extra space needed for the slip rings and their brushes, typical wound-rotor motors have an extra deep end bell on one end. The brushes are mounted inside the end bell, in alignment with the shaft-mounted rings. One or more access openings with removable covers allow for inspection and routine maintenance such as replacement of the brushes.

The rotor leads originate at the brushes and are usually terminated in the motor connection box. The stator is connected to a suitable three-phase supply, just like any other three-phase motor. The rotor is connected to a separate external controller.

Although there are more sophisticated controllers available, most wound-rotor motor controllers are just a variable-resistance three-phase load. Old wound-rotor drives had large, manually operated drum controllers. A single operating handle on the controller provided a full range of motor speeds and control of direction. There were several handle positions for both forward and reverse, and a central stop position. Inside the controller there were dozens of cam-operated contacts wired directly into the motor circuits. The stator circuit was switched directly by contacts in the controller or by external contactors. Selecting forward or reverse provided the necessary line connections for either rotation.

The rotor connections went through the controller to a three-phase bank of resistors. As the handle of the drum controller was moved from notch to notch, the cam-operated contacts would shunt out sections of the resistor bank, changing the resistance of the rotor circuit. These changes altered the motor's output torque and speed.

The wound-rotor motor controls are interlocked, so the motor is started with the rotor resistance at maximum. This provides the lowest possible starting current and the highest starting torque. When the motor is through the starting phase, the rotor circuit resistance can be reduced to increase the operating speed.

Newer wound-rotor motor controllers also use resistor banks, but the switching is done with magnetic contactors, and the drum controller is replaced with a much more compact multiposition switch or a similar device. There are other control alternatives, including a sophisticated electronic system that recovers the energy that would normally be lost as heat in the resistor banks.

The wound-rotor motor is not particularly common in new installations, but it may still be found on hoists, electrically powered cranes, and in similar applications. It is an interesting motor, one of the few AC types that can provide some speed control without the use of electronics. The ability to control motor starting current and torque is also a very attractive feature. No other three-phase motor can provide the same degree of control, even with a complex controller.

Three-Phase Synchronous Motors

Synchronous Reluctance Motor. The three-phase synchronous motor is made in two basic styles. The simplest is the synchronous reluctance motor, which uses a modified squirrel-cage-type rotor. The rotor has sections machined away to create very distinct raised lobes that are parallel to the rotor shaft. These lobes form "low-reluctance" paths through the rotor iron (see Figure 11–4).

The term "reluctance" is the magnetic equivalent of resistance. When parts of the rotor are cut away, the magnetic reluctance through these areas is much higher than through the lobes that are left. The number of lobes on the synchronous reluctance rotor is equal to the number of poles in the motor's stator winding. The low-reluctance lobes "lock on" to the rotating stator field, so the rotor runs with no slip, maintaining exactly the same speed as the stator field. The only advantage this motor has is its absolutely constant operating speed. It is only used in relatively low-power applications where this speed characteristic is a necessity.

The synchronous reluctance motor has some serious drawbacks. The lobed rotor gives it a very rough starting characteristic. Unlike other motors, which tend to start quite smoothly, this motor has a strong tendency to "cog" until it reaches full speed. Cogging is severe torque pulses that occur when the poles of the rotating field catch up to, and then pass, the rotor lobes. This naturally occurs when the motor starts, because it is impossible for the rotor to accelerate instantly from zero to full speed. It can reoccur if the load torque rises above the pull-out torque of the motor, causing the rotor to slip poles. If the load is not reduced immediately, the motor will stall.

Synchronous Motors. The second type of synchronous motor is much more common than the reluctance type. It is usually referred to simply as a synchronous motor, although it could also be called a wound-rotor synchronous or DC-field synchronous motor to differentiate it from the reluctance type.

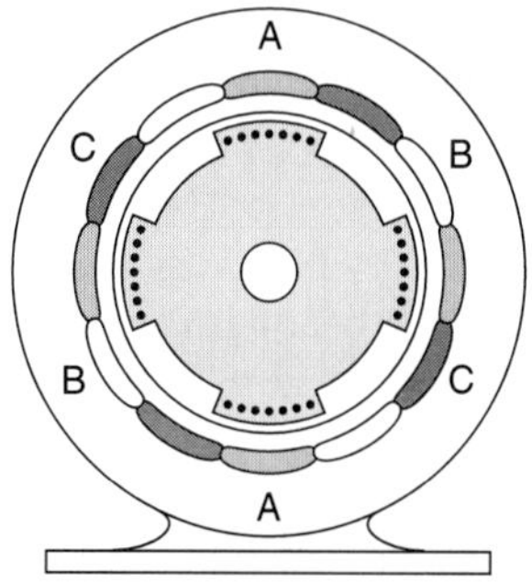

FIGURE 11–4 The synchronous reluctance motor has a squirrel-cage rotor that has been modified to create specific low-reluctance paths through the rotor iron. These paths match the number of poles in the stator field. The rotor does not start very smoothly, but it quickly locks to the rotating field. It runs in synchronism, rotating at exactly the same speed as the stator field.

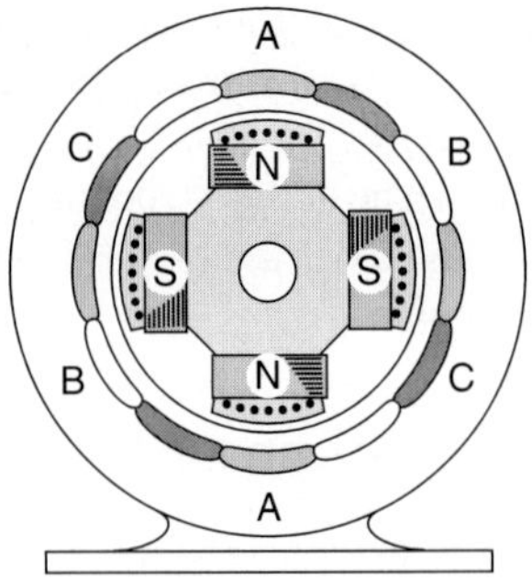

FIGURE 11–5 The synchronous motor rotor has wound magnetic poles, like those used in an AC generator. When the motor is almost up to speed, these are energized and the rotor locks onto the stator field. Like all synchronous motors, it runs at exactly the same speed as the rotating magnetic field.

In some respects, the synchronous motor is the ultimate in three-phase motors. It is essentially identical in construction to a three-phase generator, and can operate as a generator if required. (Actually, most motors will act as generators under the right operating conditions.) It has distinct rotor poles, usually salient (protruding from the central supporting core), with a single-coil winding wrapped around each pole. The number of poles in the rotor matches the number of poles in the stator field. The coils on each pole are interconnected, usually in series, to form the complete rotor winding. The electrical polarity of the pole windings alternates, so the magnetic polarity of the poles also alternates—N-S-N-S—all the way around the rotor (see Figure 11–5).

Once the motor is up to speed and running normally, the rotor windings are energized from a DC supply, creating a powerful field in the rotor that locks onto the stator field. Like any synchronous motor, it runs in synchronism with the stator field, turning at exactly the same speed.

The three-phase synchronous motor has a high operating efficiency, plus the unique ability to run with an adjustable power factor. Synchronous motor power factor is regulated by changing the DC "excitation current" to the rotor field. When the excitation current is high, the motor is said to be overexcited and the power factor is leading. When this current is low, the motor is underexcited and operates in lag. The power factor can be regulated over a wide range. This is the only three-phase motor that can run with a leading power factor.

The ability to control power factor is the most attractive feature of the motor. It is also the reason behind most large synchronous motor installations. When these motors are used, they are often the largest motors in the plant. When they have been installed to stabilize the plant power system and to regulate the overall plant power factor, they are usually coupled to relatively constant loads, such as large compressors or pumps that are always needed. They run almost continuously, and are shut down only for periodic maintenance.

Synchronous motors are available in speed ratings up to 3600 rev/min, but they are also available with very low operating speeds. Low-speed versions of these motors are rather odd-looking. The stator is a relatively narrow ring with an unusually large diameter, typically constructed with a very open design. There are no true end bells to enclose the motor or support the bearings. The rotor has obvious salient-type electromagnetic poles, mounted on a circular frame called a spider. The rotor is suspended inside the tall, flat stator on a pair of pedestal bearings that may even be separately mounted—not part of the stator at all. Slip rings mounted on the motor shaft connect the rotor field circuit to a set of brushes mounted on a stationary brush rig. To supply the DC excitation, some units have a small DC generator mounted on one end of the motor shaft.

Higher speed versions of the synchronous motor are more conventional in appearance. When they are relatively open, they can be identified by their distinct wound-rotor poles. When they are fully enclosed, it may not be possible to differentiate them from a more conventional motor without a close examination of the motor nameplate or the associated motor controls.

Synchronous motors require a special startup procedure and a more elaborate control system than a conventional squirrel-cage motor. They cannot be started with the rotor fields energized. Instead, the rotor circuit has to be connected across a "field discharge" resistor to protect it from high voltages that would be induced by the moving stator field. The rotor field is never left with an open circuit: The resistor is kept in the circuit until the motor is nearly up to full speed, and then the DC excitation is connected and the resistor is switched out. With DC applied to the rotor, it pulls into step with the stator field. The motor is said to be "synchronized." This is its normal running mode, and it can run that way indefinitely.

The most common method used to bring the rotor up to speed is to start it as a squirrel-cage induction motor. All synchronous motors have a squirrel-cage winding embedded in the face of the rotor poles to dampen rotor oscillations that can occur from normal load fluctuations. This winding can also serve double duty as a starting winding, allowing the motor to be started as a conventional squirrel-cage motor.

Large synchronous motors with shaft-mounted DC exciter generators are sometimes brought up to speed by temporarily operating the DC generator as a motor. Neither the stator nor the rotor is energized until the rotor is up to speed. A few even have dedicated "pony" motors for starting purposes.

Obviously, there are several steps to starting a synchronous motor, and these can involve controls and control functions that are not encountered on other motors. Older synchronous installations used manual controls that had to be operated in specific sequence for a proper startup. On newer motors, the sequence is fully automatic.

Many of the newer machines are brushless types, so maintenance requirements are much the same as for a similarly enclosed squirrel-cage motor. The brushes are eliminated by using a small stationary field AC generator mounted on the motor shaft. A voltage regulator controls the field strength of this exciter generator, and it in turn supplies the main motor field. Because the exciter generator is an AC machine, rectifier diodes are mounted on the rotor between the exciter generator armature and the main DC field poles to convert the excitation to DC.

Squirrel-Cage Motors

Squirrel-Cage Rotors. The squirrel-cage rotor is a made of circular iron laminations with conductors embedded around the circumference. These conductors, or rotor bars, are usually made from copper or aluminum alloy bar stock, or cast in place out of aluminum alloy; in some special-purpose motors, the bars may also be made of brass. The rotor bars are all connected together by shorting rings around each end of the rotor. In a cast rotor, the rings and bars are cast together as one solid piece (see Figure 11–6). Fabricated squirrel-cages must have the bars bonded to the rings by welding or brazing. As unlikely as it may seem for such a rugged structure, rotor bars do sometimes break, usually at the junction with the end ring. Broken rotor bars affect motor performance. They can be detected with relatively simple equipment, if the motor shaft is accessible and can be slowly rotated during the test.

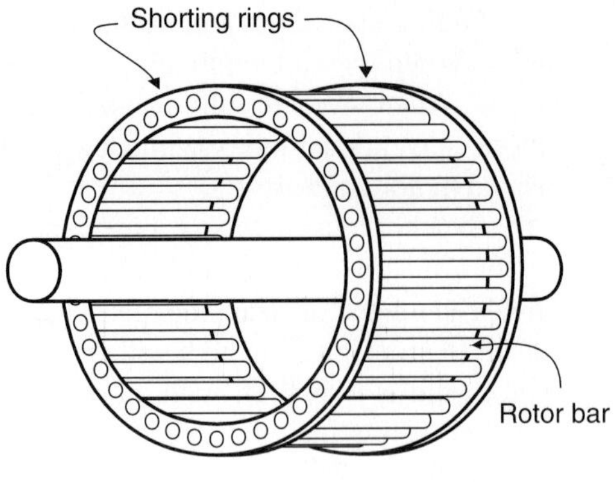

FIGURE 11–6 The squirrel-cage rotor with the iron removed to show the rotor bars and end rings.

The rotor bars may be visible or completely out of sight below the surface of the rotor iron. They are often "skewed" at a slight angle to give the motor smoother operation. When the bars are left parallel to the shaft, they are also exactly parallel to the windings and slots in the stator core. There are small variations in the stator field associated with these slots. The rotor bars will pass through these areas more smoothly if they are at a slight angle.

Variations in Squirrel-Cage Rotor Design. Although there are a few exceptions, almost all squirrel-cage motors have one of four standard rotor designs. These are classed by NEMA as Designs A, B, C, and D. Design D is further broken down into three categories based on the amount of slip the rotor has at rated full load: The divisions are 5 to 8%, 8 to 13%, and 13% or more. Each of these rotor designs has a different rotor bar resistance and a different characteristic behavior. The differences between the rotor types are examined below. However, there are some terms that must be understood first, before further explanations can make any sense.

Locked rotor current is a term that appears on some nameplates as LRA. It refers to the current the motor will draw with full rated voltage applied and the rotor held motionless. This is the current the motor draws if it is stalled, and is also the peak current that occurs briefly during a normal startup. Very high locked rotor and starting currents can be difficult to manage and may require special control equipment.

Locked rotor torque, or start torque, is the amount of torque the motor produces from a standing start. It is typically greater than rated running torque, and it varies greatly with the rotor design. Most three-phase motors have very respectable start torque. For some applications, the torque is too high for the load, and steps have to be taken to avoid mechanical damage or annoyances like spilled product when the motor starts. In other cases, where the initial torque demands of the load are exceptionally high, a high torque rotor design may have to be selected to get the load up to speed in an acceptable length of time.

Pull-up torque is a low point in the motor's speed–torque curve. The torque outputs of Designs A, B, C, and low-slip D rotors have a dip that occurs at around 20 to 30% of the motor's rated speed. The maximum torque the motor can produce at this speed is actually less than it produces at the very start, before the rotor begins to turn. If the torque demand of the load exceeds the pull-up value, the motor will not be able to accelerate the load through this portion of the startup power curve. This can happen with certain types of loads. For example, a motor driving a heavily loaded conveyor belt may start the load moving, but be unable to bring it up to speed. Conveyors have a relatively constant torque demand at all speeds. If the motor has enough start torque to get it moving, but the pull-up torque is too low, the belt will start but will not accelerate past about 20 to 30% of its normal speed, and the motor will trip on overload.

Breakdown torque is the peak torque the motor develops. For Designs A, B, C ,and low-slip D rotors this peak torque occurs at 70 to 80% of the motor's rated synchronous speed. The breakdown torque is typically much higher than normal running torque. This gives these motors a considerable torque reserve. Brief periods of excessive load will not cause serious problems as long as they are below the breakdown rating. The breakdown value is much higher than the motor torque at any other point; therefore, when the motor stalls, it does so quite suddenly.

Full-load torque is the torque the motor is rated to supply continuously, assuming the normal criteria of rated voltage and an adequate supply of cooling air. The motor's normal operation, from no load to full load, is on a very small part of its torque–speed curve. The full-load speed is typically less than 5% below the synchronous speed, and the full-load torque is normally lower than startup, pull-up, or breakdown torque ratings.

Speed regulation is a rating of the difference between no-load and full-load speed. A motor with good speed regulation has a nearly constant output speed from no-load to full-load operation. Most AC induction motors have good speed regulation, although some are better than others. The synchronous motor described above has perfect speed regulation, because it has an absolutely constant speed for all normal loads.

The terms defined above—locked rotor current and torque; pull-up, breakdown, and full-load torque; and speed regulation—are used in the rotor descriptions that follow below. They must be clearly understood for the differences between the various rotors to be appreciated (see Figures 11–7 and 11–8).

Design A rotors are relatively rare. They are very similar in behavior to Design B, but have a higher locked rotor current and higher pull-up and breakdown torques. They are best suited to loads with infrequent starts and stops. The high locked rotor current rating means high starting currents that may have to be controlled with a reduced-voltage starter.

Design B is by far the most popular of all rotor designs. This rotor provides a good balance of start torque, breakdown torque, speed regulation, and operating efficiency, and is suitable for a wide variety of applications. It is usually started with full voltage, but reduced voltage may also be used. Design B rotors are not suitable for loads with cyclical torque pulsations. They are best suited to applications with infrequent starts.

Design C provides higher start torque than a design B rotor. The rotor is usually constructed with a double squirrel cage—two sets of rotor bars, one inside the other. The outer set of rotor bars has a relatively high resistance to provide improved starting torque without excessive starting current. The inner set of bars provides good full-speed characteristics such as low-slip, good breakdown torque, and good efficiency. Design C rotors are sometimes used for loads such as compressors that are difficult to start, but run for extended periods of time at full load. The Design C rotor meets these needs better than any other rotor type. Design A or B rotors would not have the necessary start torque, and a Design D is unsuitable for continuous loads.

Design D rotors have a single cage with high-resistance bars—such as brass instead of the usual copper or aluminum—producing a motor with very high start torque. Unfortunately, the motor also has a high slip, resulting in poor speed regulation, and is less efficient due to higher power losses in the rotor. It may not be rated for continuous duty at full-load because the

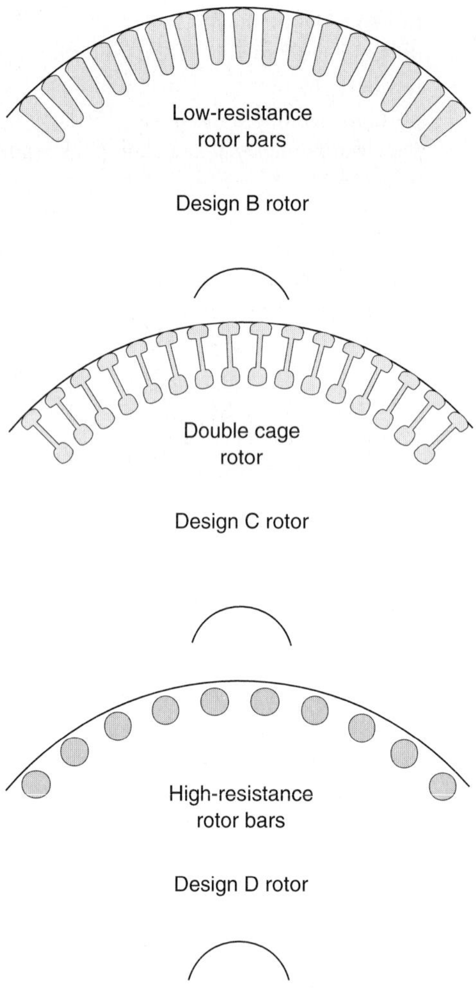

FIGURE 11–7 Relatively minor changes in the shape and resistance of the rotor bars produce significant changes in motor performance.

high-resistance rotor bars may overheat. Design D motors are used on loads that are difficult to start, but less demanding once up to speed. A common application is a cyclical load, such as a punch press, where a massive flywheel delivers the power, and the drive motor only starts it up and keeps it going. Design D rotors are subcategorized by their percent slip.

The characteristic torque–speed curves for the three most common rotor designs can be graphed on a diagram such as Figure 11–8 for comparison purposes. Initial torque is the motor locked rotor or start torque, the low point of the curve is the pull-up torque, and the peak of the curve is the breakdown torque. High-slip Design D rotors have a smooth torque curve without pull-up or breakdown values.

For a motor to start the load, its start torque must exceed the initial torque requirements of the load by a reasonable margin. To accelerate the load to full-speed successfully, the pull-up torque must also be greater than the demand. Once the load is up to speed, the motor will not stall unless the torque demand exceeds the breakdown peak. Normal motor operation is in the very narrow band between the synchronous speed and the point where the torque curve intersects the full-load line.

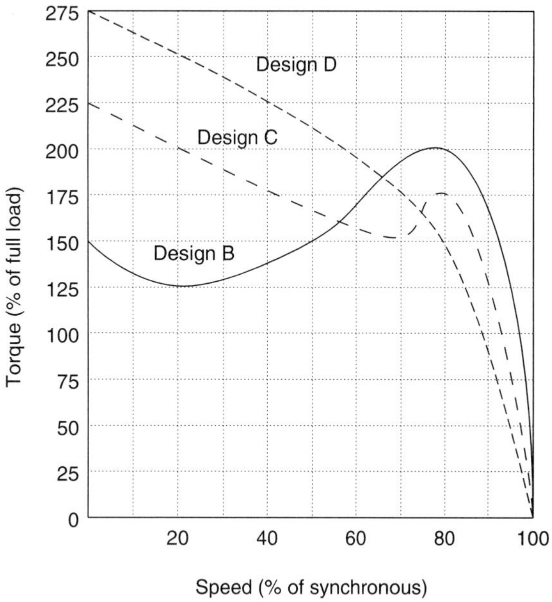

FIGURE 11–8 Approximate speed–torque curves for Design B, C, and D rotors. The motor's normal operating range is a narrow wedge from the intersection of the 100% rated torque line to just below 100% of synchronous speed. Start torque is at 0 rev/min. Pull-up torque is the low point of the curve before it reaches normal speeds. Breakdown torque is a peak at some speed below the normal operating point.

11.4 Motor Speed and Torque

Three-phase motors are generally available with several different speed ratings. These are determined by the number of motor poles and the supply frequency. In North America the supply frequency is normally 60 Hz, so the motor nameplate lists that frequency and the related operating speed. Motors that are designed for both 50 and 60 Hz operation have two different speeds listed. The slower of the two is for the 50 Hz supply.

For each cycle of three-phase AC current, the stator field moves past two poles in each phase. For a two-pole motor, this means one complete revolution of the field for each AC cycle. A four-pole field moves half that distance, and for each additional set of poles it is proportionally less. There is a set mathematical relationship between the number of poles and the frequency and speed at which the stator field rotates.

The speed of the rotating field is called the synchronous speed. Poles are listed in terms of magnetic poles in the rotating field, not the total number of poles wound into the stator to create the field. (In a three-phase motor, this is three complete sets, one for each phase). Frequency is based on a single voltage cycle, not on the combined voltage peaks and polarity changes of all three supply phases.

The synchronous speed of a motor can be calculated as follows:

$$N_S = \frac{120F}{P}$$

N_S = Synchronous speed in rev/min

F = Frequency in Hz

P = The number of motor poles

This formula is developed as shown below. Two poles per cycle is the rate of advance of the field, 60 sec/min converts the time unit of the frequency to match the time unit of the speed, and the poles per revolution is the number of poles that make up one full mechanical rotation inside the stator. Canceling the units shown below will produce the formula given above.

$$N_S\left(\frac{\text{rev}}{\text{min}}\right) = \frac{2\left(\frac{\text{poles}}{\text{cycle}}\right) \times 60\left(\frac{\text{sec}}{\text{min}}\right) \times F\left(\frac{\text{cycles}}{\text{sec}}\right)}{P\left(\frac{\text{poles}}{\text{rev}}\right)}$$

Poles must be in pairs, so a motor must have at least two. The field of a two-pole motor connected to a standard 60 Hz supply rotates at a synchronous speed of 3600 rev/min, which is one revolution per cycle. Note, however, that unless the motor is a synchronous type, this is not the shaft speed. The shaft speed is the speed of the rotor, which for a squirrel-cage motor is always somewhat less than the synchronous speed of the field. Squirrel-cage rotors require some degree of slip for the rotor to develop any torque.

With no-load, the motor speed is close to the synchronous speed. On large motors, particularly low-speed models, the no-load speed may be so close to the synchronous speed that a typical tachometer may not be accurate enough to detect the difference. With smaller motors, the no-load slip is more noticeable because the mechanical drag of the bearings and cooling fans is a greater proportion of the motor's output power. The motor slows down a little to produce the torque needed to drive these mechanical losses.

Once the motor is mechanically loaded up, things change. At full load, with rated voltage and frequency, the motor should run at its rated nameplate speed. This may vary slightly with the make, model, size, and motor type, but for typical general-purpose motors it is less than 5% below the synchronous value. Certain types of motors have different slip percentages, and therefore different nameplate speed ratings.

Rated motor speed drops in inverse proportion to the number of poles. Four-pole 60 Hz motors have a synchronous speed of 1800 rev/min. Increasing the number of poles to six drops this to 1200 rev/min, and for eight poles the synchronous speed falls to 900 rev/min. In general, only special-purpose motors are made with more than eight poles. As the number of poles increases, the relative diameter of the stator must be increased to accommodate the additional windings. The windings cannot just be reduced in size to fit the space available, because the magnetic strength of the fields must be maintained to avoid reducing the motor output horsepower.

For the motor to have the same output horsepower, each reduction in speed must be matched by a proportional increase the full-load torque. The motor must be designed for this higher torque output, through some of the increase is relatively automatic. There is a complete set of magnetic poles, all of equal strength, rotating inside the motor at any time that it is running. Increasing the number of poles while maintaining the magnetic strength of each pole naturally results in proportionally higher torque. For example, a four-pole motor has four sets of poles acting to pull the rotor around, as compared to only two in a two-pole motor.

The relationships among motor speed, torque, and horsepower are often misunderstood. For example, assuming the same horsepower in each case, a motor rated at 3450 rev/min has half the output torque of a motor rated at 1725 rev/min motor, and only a third of the torque of a motor rated at 1150 rev/min. However, since output power is a function of torque and speed, the output of all three of these motors—the amount of work they can do per unit of time—is identical.

One way of understanding this is to envision each motor connected to drive an identical low-speed load such as a conveyor belt. Each motor is coupled to the conveyor through a suitable speed reducer. The reduction ratios are selected to give the same final conveyor speed for each motor. Assuming the same efficiency for each of the speed reducers, all three motor drives will produce identical results. The conveyor will operate at the same speed, carrying the same load and, assuming motors are of equal efficiency, the cost of operation will also be identical.

Two- and four-pole motors are generally the least expensive and the most common. Six- and eight-pole motors are popular for certain types of loads such as direct-drive blowers and fans. The extra cost is easily justified by the gains in mechanical simplicity and reduced size of a direct-drive vs. that of a belt drive.

The choice of motors for any application is dictated by a balance of several factors. A suitable motor must be available at a reasonable purchase price, and the motor characteristics must match the requirements of the load. For a direct drive, the motor torque must be at least adequate for all operating conditions, and the speed must be a perfect match. For a belt, chain, or gear drive, the ratio of the speed change must produce the desired results. Higher motor speeds mean lower ratios for high-speed drives and higher ratios for low-speed drives. The initial and long-term costs of a mechanical speed change must be factored into the decision if no standard motor speed is suitable.

Many loads can be driven directly by typical two-, four-, or six-pole motors. For small motors, six or eight poles cost extra; more than eight poles may be virtually unavailable. When the desired load speed is above 3450 rev/min, or below 1150 or so, a direct drive may not be practical. If the horsepower is relatively high, the cost of a low-speed motor may be justified, compared to the cost of an appropriate gear reducer. However, it should be noted that low-speed motors have a lower starting torque than motors designed for higher speeds. If the load is quite difficult to start, a higher motor speed and a gear box may be the only choice, even if a low-speed motor is available.

11.5 Motor Voltage Ratings

The motor windings must be designed for a specific voltage and, naturally, they are made to match the preferred standard service voltages. However, for reasons that may have more to do with tradition than logic, they are normally given a rating that is somewhat below the nominal voltage of the system to which they will be connected. For example, motors designed for 120 V supplies are nameplate rated at 115 V. Motor nameplates typically display voltages such as 115, 200, 230, 460, and 575 instead of the standard nominal voltages of 120, 208, 240, 480, and 600. This practice leads to a lot of confusion about the correct voltage value to use in describing a system or motor.

Note

There is no clear-cut answer for every situation but, unless it is a specific reference to a motor nameplate rating, it is probably best to state voltages as the nominal utility voltages of 120, 208, 240, 480, and 600. Older standard values should be avoided except where they are still in use.

It should be noted there are three specific standard voltage ratings—120, 240, and 480—with a 1:2 or 2:1 ratio to each other. The lowest of these, the 120 V, is not normally supplied as a three-phase voltage, but is universally available in single-phase. The 240 V rating is available in both single- and three-phase, though 240 V three-phase services are becoming less common. The 480 V rating is not commonly supplied in single-phase, but is widely used in three-phase.

Note

One exception to this generalization about 480 V single-phase is found in some packaged HVAC systems. It is fairly common to find a 460 or even 575 V rated single-phase motor—usually a PSC type on a direct-drive condenser fan—in a system that uses three-phase motors for the refrigerant compressors and the main air blower. However, this is a special case. The manufacturer has decided to use a specific single-phase motor in a three-phase system. Possible reasons for this decision include the much more gentle start torque this motor provides, and the availability of six- and eight-pole PSC motors in relatively low horsepower ratings.

Motor windings are in sections that are interconnected inside the stator. This makes it relatively easy to provide a motor with two different voltage ratings, as long as the voltages have a 1:2 ratio. This practice simplifies motor stocking and provides for more flexibility in application. For this reason, many motors are built with extra leads that can be connected for operation from two different voltages. For single-phase motors, the standard is to provide an option of operation from either a 120 or 240 V supply, with a motor nameplate rating of 115/230 V. For three-phase motors, the usual combination is for either 240 or 480 V supplies, with a nameplate rating of 230/460 V.

Note that 240 V is getting pretty close to 208 V, which is now a very popular three-phase supply voltage. To extend the range of applications for their 230 and 230/460 V rated motors, many manufacturers design them so they also provide acceptable performance from a 208 V supply. It is quite common for 115/230 V single-phase and 230/460 V three-phase motors to have 208 V included in their nameplate ratings. Labeling practices vary, but examples include, "Approved for use at 208," "115/208–230," and "208–230/460." Note, however, that motors designed specifically for 208 V operation are rated and labeled at 200 V, not 208.

The 208 V rating for 230 V motors should be viewed as a compromise solution. There is nothing wrong with this practice but, to be fair, expectations should be lowered slightly when the motor is operated at this lower voltage. Motor performance is always affected by a change in voltage, even if that voltage falls within the allowable nameplate ratings. When a motor is under load, a reduction in voltage will cause a rise in current and operating temperature and a decline in motor efficiency. Driving a 208–230 V motor to the same power output on 208 V, as would be expected from a 240 V supply, will result in a higher current and a higher operating temperature.

All other things being equal, motor life is directly related to sustained insulation temperature. High temperatures shorten motor life dramatically. The operating temperature should be kept in mind in all motor applications—even more so when the voltage level is near the bottom of the nameplate ratings. Note also that the margin allowed for voltage drop in the supply circuit—which is 10 V (240–230) for the higher nominal rating—is virtually nonexistent when the motor is operated at the lower 208 V rating. If the voltage drops to less than 208 V, the motor will almost certainly be in trouble unless there is a compensating reduction in load.

At least one manufacturer also permits an unusual reconnection of their smaller 208–230/460 V motors for an additional rating of 575 V. This is an impressive range of optional voltages. A single motor can be used on all of the most common standard three-phase service voltages used in North America. Operation of the motor at the 575 V rating does involve a slight reduction in the maximum output horsepower. For example, the 10 Hp version is derated to 9.6 Hp.

Motors with a 575 V rating—designed for operation from a 600 V supply—are usually made specifically for that voltage and no other. There is no standard voltage with a 2:1 ratio to 600 V, so the manufacturer has no reason to provide optional connections. Motors with this voltage rating are readily available up to 400 Hp or more, and can be obtained with ratings far beyond that level.

In general, there is very little electrical equipment that falls between 600 and 2400 V. Most motors made for voltages between 600 and 2400 V are part of specialized equipment made to meet the needs of specific situations. Normally, if a large enough 600 V motor is not available, the next step up is a rating of 2400, 4160, or (particularly in the U.S.) 6900 V. Although 7200 V distribution is not used in Canada, it is common in much of the U.S. This is the normal supply voltage to a 6900 V rated motor. Such high motor voltages may seem extreme to personnel that have not worked on high-horsepower electrical equipment, but motors with these ratings are commonplace in large commercial buildings and in horsepower-hungry industries.

11.6 Motor Connections

There are standard lead markings and connections for virtually all three-phase motors, and enough variations to cause a lot of confusion. The information that follows examines the most common examples. It is not a comprehensive listing, but should be adequate for most situations. It is standard manufacturing practice to provide a description of permissible connections and the related voltages on the motor nameplate. This information is usually adequate for proper connection of the motor. In unusual circumstances, such as lost motor lead numbers, it can be very helpful to understand the logic behind the numbering scheme that is used.

Three- and Six-Lead Motors

Motors designed for operation at only one voltage may have three leads, labeled T1, T2, T3, or six leads, marked T1 to T6 (see Figure 11–9). Three-lead motors have the necessary winding interconnections premade internally. There is no way to tell how the windings are actually connected, but that information is essentially irrelevant. The only requirements are that the motor be connected to the supply voltage for which it is designed, and for the

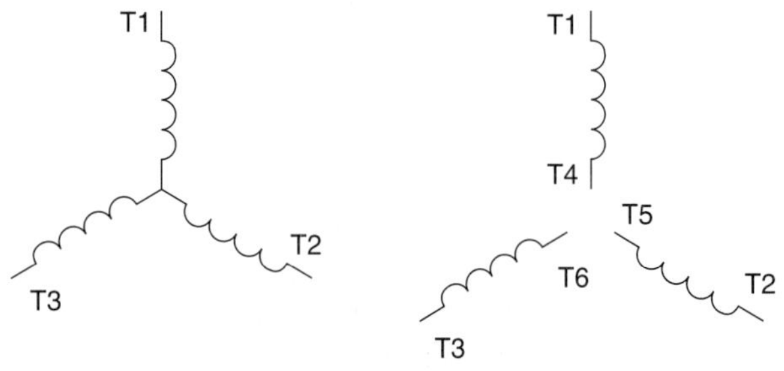

FIGURE 11–9 Lead numbering for three- and six-lead motors.

correct direction of rotation if that is applicable. Beyond that, there are no connection options. The incoming lines connect L1 to T1, L2 to T2, and L3 to T3, or any two can be interchanged to reverse rotation.

Note

The connection of the three incoming lines (L1, L2, and L3, also called phases A, B, and C) to the motor terminals T1, T2, and T3 is a recurring theme in all three-phase motor connections. The instructions for motor reversal—interchange any two supply leads—are also the same in every case. The motor connections may be reversed in the motor connection box or at the controller. The controller is often the most convenient point because it usually has open screw terminals or lugs. The leads at the motor are usually less accessible and may be insulated with tape.

For six-lead motors, the winding interconnections must be made in the field. The six leads represent the two ends of the three sets of windings that the three-phase motor contains. Opposite ends of each phase have numbers with a difference of three, resulting in leads marked T1 and T4, T2 and T5, and T3 and T6, respectively. Testing the leads with an ohmmeter will show continuity between the lead pairs, with equal resistance values for each of them. On a normal six-lead motor, there will be no continuity from the leads of one winding to those of any other winding.

The order of the two numbers used to identify each phase winding indicates the winding polarity. This is important, because polarity must be respected when the motor is connected. For example, the numbers T1 and T4 cannot be arbitrarily interchanged without causing serious problems. Reversing the leads will reverse the current flow and the magnetic polarity of each pole in that phase. Needless to say, the motor will not operate properly under such conditions.

Note

The numerical order of the lead markings is used to indicate correct winding polarity in DC and AC, single- and three-phase motors, and in other types of equipment such as transformers. Always respect number order and, if lead numbers are lost, do not assign the numbers at random based on identification of lead pairs. The results can be very different from what was intended.

North American six-lead motors are usually intended for delta operation, and typically have only one listed operating voltage. The actual connection is listed on the nameplate, so it is a simple task to identify whether it is a wye or delta. European-style 50 Hz motors may show nameplate voltages for both delta and wye connections. Typical dual-voltage ratings displayed on North American motors have 1:2 ratios. Delta- and wye-connection voltage ratings have a 1:1.73 ratio. An example would be a rating of 220/380 V or 240/415 V. When listed that way, the lower of the two voltages is the rating for the individual windings. The motor can be connected in delta at the lower voltage, and in wye for the higher voltage. The difference between these two connections is explained below.

Note that a motor approved for both 50 and 60 Hz operation will show two sets of voltages on the nameplate: one for 50 Hz, and the other for use on 60 Hz supplies. For example, a 50/60 Hz motor could have voltage ratings of 380 and 460 V. These are both standard voltages in their own areas of use. The motor will provide essentially the same output on 380 at 50 Hz as it does with 460 V at 60 Hz. It will, of course, run more slowly on the lower frequency—approximately 1425 rev/min, as compared to 1725 for a four-pole motor.

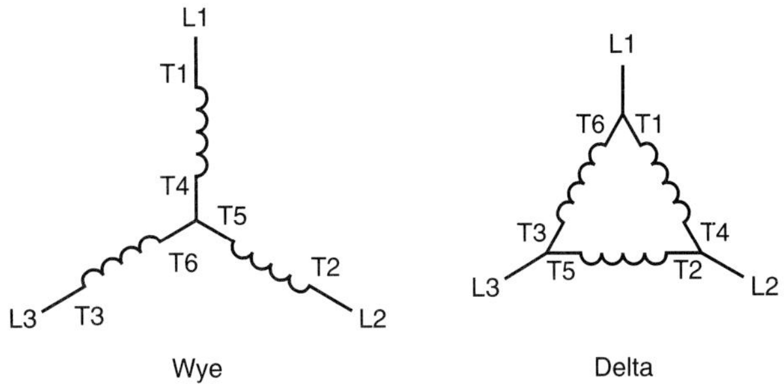

FIGURE 11–10 Six-lead motors may be designed for connection in delta or wye. In wye, three leads are interconnected to form the star point. The star point of a motor has no other connections. In delta, the windings are connected in a specific numerical sequence that forms a triangle.

In general, operating a motor at 60 Hz requires a higher applied voltage than is required to run the same motor at a 50 Hz rating. Winding impedance is proportional to frequency, so the higher the frequency, the higher the voltage needed to produce the same current level. Motors built for 50 Hz operation are common in imported equipment or equipment that is designed for export.

The delta connection is named for the Greek letter delta (Δ), which is essentially an equilateral triangle. This is representative of the actual connection, because in a completed delta all three phases are linked in a continuous series loop. As with any connection, winding polarity cannot be ignored. For all delta connections, the leads must be in a start-finish, start-finish, start-finish order all the way around the three-phase circuit. The final lead connects back to the open end of the first winding, closing the loop. The order of the motor terminal numbers indicates the start-finish polarity. The lower number of each lead pair can be regarded as the winding start, the higher number as the finish. For a six-lead motor designed for delta operation, the connection would be T1-T6, T2-T4, and T3-T5. T1, T2, and T3 would also be connected to the supply, as usual (see Figure 11–10).

For a six-lead motor intended for wye operation, the connections would be T4-T5-T6 tied together and insulated, and T1, T2, and T3 connected to the supply lines as before. The three highest numbered leads form the star point; the lower numbered leads go to the supply. On low-voltage motors, the windings can usually be reversed without harm, if they are all reversed. This would make T1-T2-T3 the star point; and T4, T5, and T6 the supply connections. There would be no advantage to this connection, however, so it is best to go with the traditional method.

There are many situations where the star point of a wye is connected back to the supply, but this is not done in motors. Instead, the star point connection is insulated from all other connections and from ground and is left free to "float" at whatever voltage may occur. In normal operation, the star point will be within a few volts of ground potential.

North American six-lead motors may be encountered in systems that use wye-delta starters (also called start-delta starters). This motor starting system is examined in Section 12.5 of Chapter 12. Essentially it involves starting the motor with the windings connected in wye, and then changing the connection over to a delta configuration after the motor is up to speed. This type of controller only works with a six-lead motor (or a twelve-lead motor made for dual-voltage operation) that is designed to operate with the windings connected in delta. Such systems are sometimes used in areas where the supply utility restricts the maximum allowable startup current. The wye-delta starter is only one of several reduced-current starting systems.

Nine- and Twelve-Lead Motors

North American motors designed for dual-voltage operation usually have nine leads or, less commonly, twelve leads. A nine-lead motor has the last three winding ends (T10, T-11 and T12) preconnected internally, while a standard twelve-lead motor has T10, T11, and T12 connected in the motor terminal box. There are only a few instances where having access to all twelve-leads makes any difference.

Note

AC generators use the same winding layouts and numbers as AC motors. However, it is usual to bring out all twelve leads instead of only nine. Unlike typical motor connections, bringing out the last three connection points provides the end user with considerably more flexibility in the ways the generator can be used.

The numbering system used for three-phase nine- and twelve-lead motor windings is standardized and consistent. It is based on the assumption of a full set of twelve leads, even though the last three ends may not be available in the connection box. As with six-lead motors, the lead pairs that connect to opposite ends of each winding always have a numeric difference of three, and winding polarity always follows the number order. Nine- and twelve-lead windings follow the same general rules, but are split into two equal parts to accommodate two possible operating voltages.

The first half of each phase winding is numbered just like a six-lead motor. The second half is numbered in a sequence that is three digits higher than its twin. The A phase is made up of two parts marked T1-T4 and T7-T10. The B phase is marked T2-T5 and T8-T11, and the C phase is marked T3-T6 and T9-T12.

When this numbering scheme is shown in diagram form, it is usually drawn as a sequential inward spiral of lead numbers on a wye-shaped connection diagram (see Figure 11–11). This may give the impression that the system is specifically designed for motors with wye connections, and that the windings in a motor designed for a delta connection might be numbered differently. This is not the case. The numbering system is as consistent as possible for all three-phase motors, regardless of the final connection configuration.

FIGURE 11–11 The standard terminal markings for three-phase nine- and motor windings.

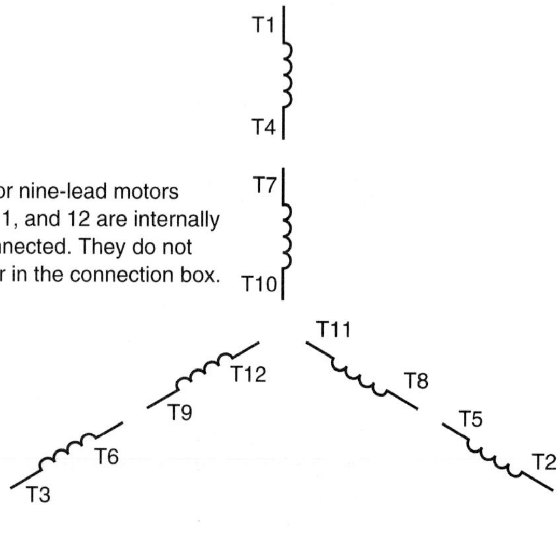

Splitting up the motor winding provides the option of two different operating voltages. If each of the phase windings pairs are connected in series, the final motor connection is for the highest of the two nameplate voltages; if they are connected in parallel, the lower of the two voltages must be used. In each case, the voltage across the individual winding sections (T1-T4, for example) is the same. For this reason, assuming the voltage at the terminals is maintained at the appropriate value, a dual-voltage motor runs equally well with either of the two connections. The motor draws twice as much current at the lower voltage as it does from a higher-voltage supply, but this does not imply a higher operating cost or a less efficient motor. If the voltage supplied to the motor terminals matches the nameplate ratings for the connection used, there will be no difference in power output, or the operating cost, between the low- and high-voltage options. The differences are in supply availability, and in the current and voltage ratings of the supply conductors and control equipment.

When installing a new or replacement motor designed for dual voltages, always make sure that the connection is correct for the voltage in use. Never assume previous connections are correct. If the motor is connected for low voltage, and is inadvertently supplied from the higher voltage, damage to the motor or motor control equipment is a very likely result. If the motor is connected for the high voltage but supplied with the low, it will stall at about a quarter of its rated load. With a load connected, the motor may not come up to speed, and can quickly overheat. With the overloads set for the much higher current rating needed for low-voltage operation, the incorrectly connected motor may burn out without tripping the overloads.

Wye Connections. In a typical 230/460 V motor, either all twelve or, more commonly, only nine leads are brought into the terminal box. Motors wound for wye operation require the winding groups to be in series wye for 460 (480) V, and in parallel wye for 230 (240) V operation (see Figure 11–12). In the parallel wye, the two star points are usually not tied together. The result is two sets of wye-connected windings in parallel, rather than three sets of parallel windings connected in wye.

When the motor is provided with twelve leads, the final lead ends, T10, T11, and T12, must be connected to form a star point for both high- and low-voltage connections. For motors

FIGURE 11–12 Twelve-lead wye connections.

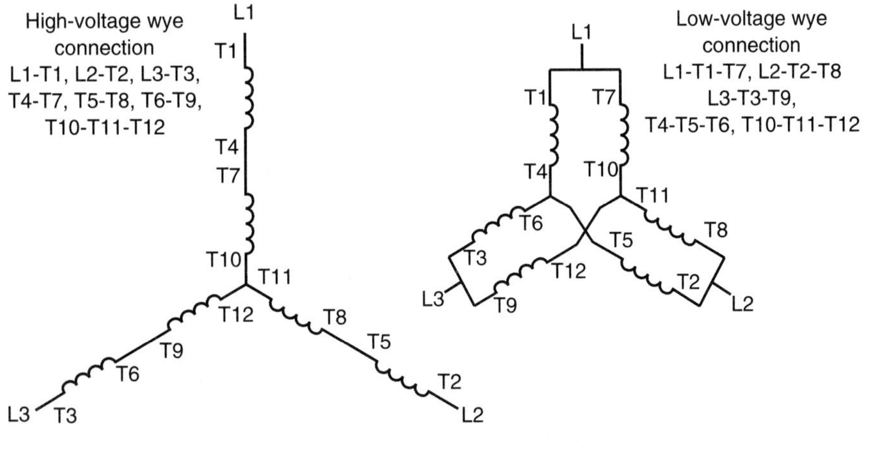

FIGURE 11–13 Nine-lead wye connections.

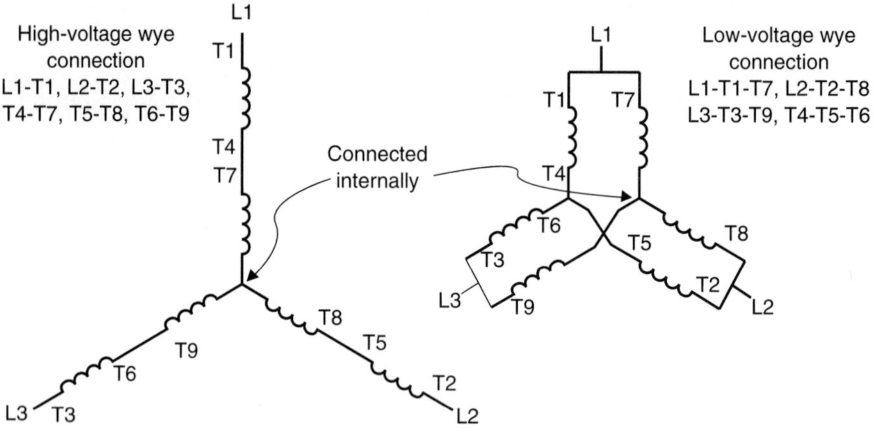

with only nine leads, T10, T11, and T12 do not appear inside the motor connection box; they are preconnected inside the motor (see Figure 11–13).

Delta Connections. It is physically possible to connect a twelve-lead motor in either wye or delta, but the appropriate connection must be used for the motor windings to receive their rated voltage. The manufacturer's nameplate connection diagram will indicate which is correct. Figure 11–14 shows high- and low-voltage twelve-lead delta connections. The series delta, which would be the high-voltage connection, will have the individual phases connected exactly as for the wye connection—T1-T4 series connected to T7-T10, T2-T5 series connected to T8-T11, and T3-T6 series connected to T9-T12. This requires only three connections—T4 to T7, T5 to T8, and T6 to T9—to form the winding groups. These three groups are then connected in a standard start to finish, start to finish… delta configuration. These connections are T10 to T2, T11 to T3, and T12 to T1. As usual, T1, T2, and T3 would also be connected to the incoming supply lines.

For the low-voltage delta, the phase groups would be in parallel instead of series, with the connections T1 to T7, T4 to T10, T2 to T8, T5 to T11, T3 to T9, and T6 to T12. The parallel groups are then connected in delta and to the supply. In the final connection, there are four motor leads with each of the supply conductors. This can require some creative

FIGURE 11–14 Twelve-lead delta connections.

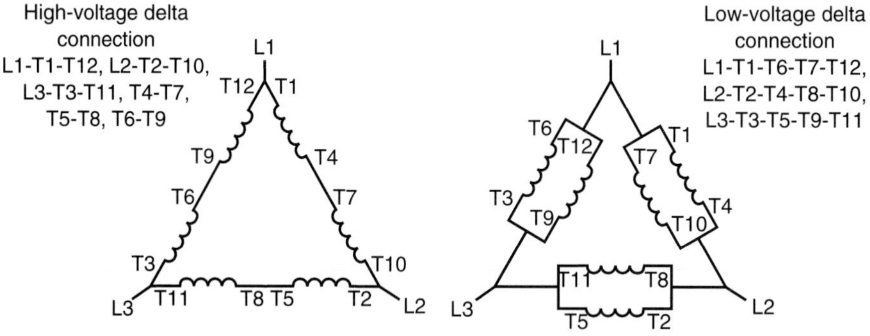

FIGURE 11–15 Nine-lead delta connections.

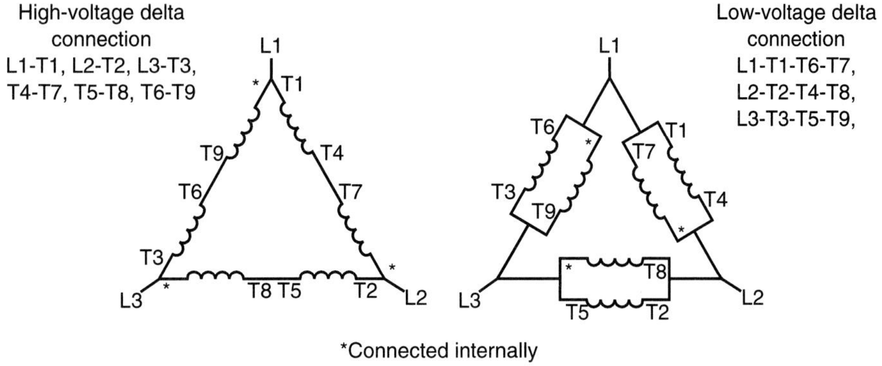

connection and taping skills if the motor is equipped with open leads instead of some other provisions for termination.

Given a large motor with a compact junction box, it can be a real challenge to make good, properly insulated connections, and still get the cover back on the box. It is important in situations like these to pay attention to the physical arrangement of the leads before making any connections. A little advance planning can reduce the frustration level and produce a more professional result. As a general rule, the motor leads should never be shortened, and the connectors should be removable. It should never be necessary to cut the connectors off to disconnect the motor. If this has to be done, the leads may soon be too short for a proper connections, and the motor may have to be sent to a repair shop to have new leads installed.

A nine-lead motor can only be connected according to the original design. The final connection must be appropriate for the internal connections of T10, T11, and T12. When the motor is designed for a wye connection, these three leads are preconnected into a star point. When it is intended for a delta connection, T10, T11, and T12 are preconnected to T2, T3, and T1, respectively, just as they are for the twelve-lead delta connection described above (see Figure 11–15). The only difference is that the leads are connected internally and are not accessible in the connection box.

For a nine-lead delta motor, the high-voltage connection is T4 to T7, T5 to T8, and T6 to T9, to series connect each of the phase windings. T1, T2, and T3 are connected to the supply (and, internally, to T10, T11, and T12). For a low-voltage supply, the two parts of the phase windings must be connected in parallel. The end result is T1, T6, and T7 connected to one supply line; T2, T4, and T8 to the second; and T3, T5, and T9 to the third. The supply lines are, as always, connected to T1, T2, and T3. This connection is slightly less bulky than for a twelve-lead motor, but can still be a challenge to connect on a high-current motor.

Although lead markings are standardized and most motors will be one of the types shown in Figures 11–12, 11–13, 11–14, and 11–15, there are other variations. Always check the motor nameplates and connection diagrams carefully.

Larger motors sometimes have rigidly mounted terminals instead of flexible leads. When the motor is designed for dual voltage, the terminals are arranged in a grid pattern so that a set of solid metal straps can be used to make the winding interconnections (see Figure 11–16). To accommodate the necessary variations, the terminals are placed in what

FIGURE 11–16 Dual-voltage motors with fixed terminals often have jumper straps supplied to simplify connections. These are examples of how the terminals and straps are arranged.

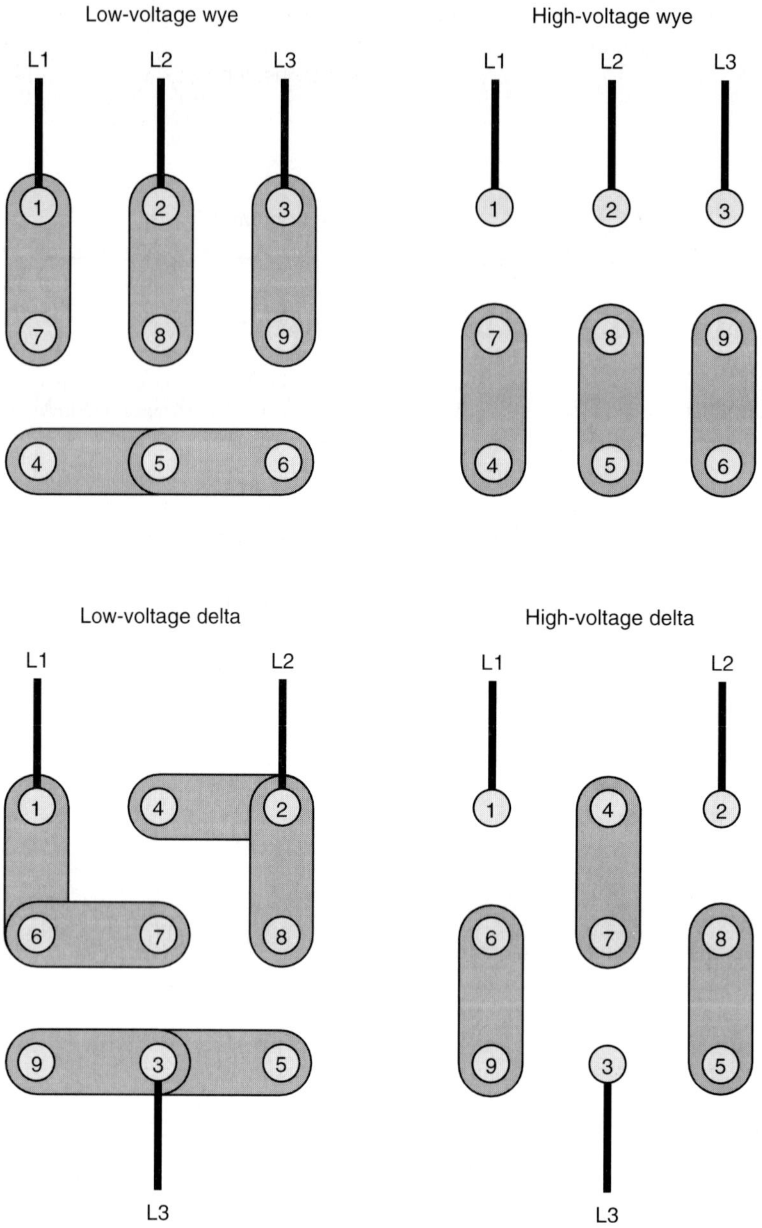

appears to be a random order instead of a numeric sequence. The order chosen allows the equal-length jumper straps to be placed directly over the correct terminals for either low- or high-voltage operation. These straps are supplied with the motor. Different connections require a different number of straps. Any extras should be doubled up on the terminals so they are always with the motor.

Note

Be very careful when tightening these connections. It is alarmingly easy to twist off the terminals on some motors. The studs are made from a relatively soft copper or alloy that is much easier to break off than a typical steel stud. Check the manufacturer's literature for torque recommendations, if possible.

11.7 Multispeed Three-Phase Motors

Although single-speed versions are far more popular, three-phase motors are also available with two or more operating speeds. The maximum number of speeds is four, but two is the most common. These extra speeds are made available in two different ways. The first is to reconnect a single winding to produce double the original number of poles. The speed combinations available this way always have a ratio of 2:1. For example, a four-pole to eight-pole scheme gives approximately 1725 and 850 rev/min, and a two-pole to four-pole transition gives 3450 and 1725 rev/min.

The number of motor poles is doubled by creating what are called consequent-poles. In a normal connection, every second pole in each phase has the opposite polarity. For consequent-pole operation the windings are reconnected to reverse the current in every second pole. This makes the magnetic polarity of every pole the same, instead of alternating north and south around the stator. The result is not a single magnetic polarity magnet, which is impossible. Instead, "consequent-poles" of the opposite polarity form between each of the normal poles. This effectively doubles the number of motor poles, reducing the operating speed to half of what it was originally (see Figure 11–17).

FIGURE 11–17 Consequent-pole motor fields are formed by reconnecting the windings to produce identical magnetic poles in every "normal" motor pole. The result is a complete second set of poles with the opposite polarity, halfway between the originals.

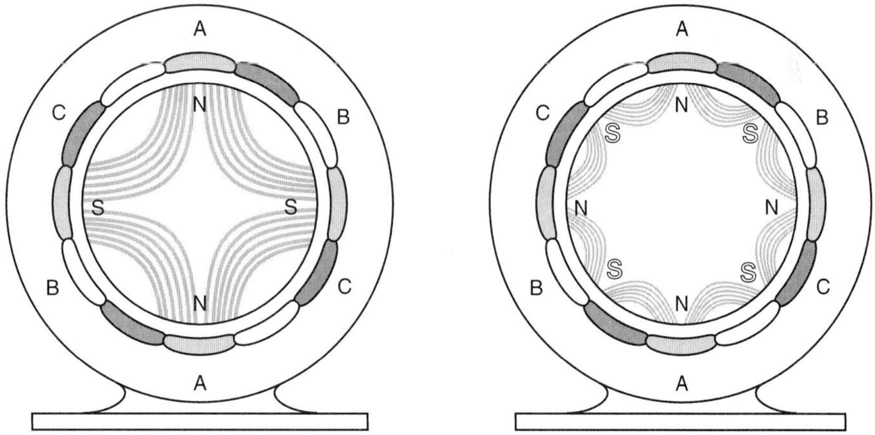

Normal four-pole field

Reversing connections to every second pole winding creates four identical poles and, as a result, four additional consequent poles of the opposite polarity.

The second method of providing extra speeds is to use different windings for each speed. Many ratios are possible with this method, but the most common is less than 2:1. For example, a four-pole/six-pole combination would give speeds of approximately 1725 and 1140 rev/min. For more than two speeds, these techniques are used in combination. A two-winding motor that can be reconnected for consequent-pole operation on both windings has a total of four possible speeds.

A two-winding motor is generally more expensive than a single-winding model, but the controller is also less complex and less expensive. The total cost of the motor and controller combination should be considered before one motor is chosen over the other.

There is another, somewhat specialized multispeed three-phase motor called the pole amplitude modulated (PAM) motor. The advantage of this motor is that it can offer speed combinations with ratios that are impossible or impractical to provide with more traditional methods.

Note

In many cases, an electronic variable-frequency controller connected to a single-speed three-phase motor is the best choice for motor applications that require changes in operating speeds. These controllers are complex, but the technology is quite mature, and reliability is a matter of record. Such systems provide fully variable speed with a standard motor, in comparison with two- to four-speed choices with a special multispeed motor and controller. Many variable frequency controllers also incorporate advanced motor protection features that are a costly option with conventional controls. Variable-speed controller and motor combinations are generally referred to as variable-frequency drives. They are cost competitive with traditional multispeed motor and controller combinations.

When a motor has a provision for changing the speed, some thought must be also be given to the torque required at each of the various speeds. The maximum output torque of the motor has to match the needs of the mechanical load. Different types of loads respond differently to a change in speed: A motor that is a perfect match for one load may be completely unsuited to another.

There are three variations on the torque–speed relationship, each involving a different internal motor connection. A *constant torque motor* is one that can supply approximately the same torque at each of its operating speeds. The maximum output horsepower of this motor is lower when it is operating at the lower of the two speeds. For example, if the high-speed to low-speed ratio is 2:1, then the horsepower available at the low speed will be only half the rating at the high speed. This motor is best suited to applications such as hoists, positive displacement pumps, conveyors, material mixers, or other loads where the torque demanded by the load is essentially the same at any speed.

The second variation is called a *variable torque motor*. With this design, the maximum motor torque is directly proportional to its speed. The maximum horsepower output of this motor is approximately proportional to the square of the speed. For example, if the motor has a 2:1 speed ratio, the available horsepower output at the lower speed is only one-fourth of the rating at the higher speed. This type of motor is best suited to applications such as fans, blowers, and centrifugal pumps. These types of loads have a torque demand that is proportional to the operating speed.

The third variation is called a *constant horsepower motor*. This motor is less common than the other two, because there are fewer applications. The motor can provide proportionally more torque at the low speed compared to the high speed. Assuming a 2:1 speed ratio, the maximum low-speed torque would be double that of the high-speed torque. Such motors are sometimes used on machine tools and metalworking equipment. An example of the this type of performance curve would be a metal lathe. Low cutting speeds are more appropriate for large-diameter pieces or for heavier cuts. Both of these operations increase the torque demand on the motor. Higher speeds are associated with smaller material diameters and shallower cuts, which, of course, require less torque from the drive.

When you are selecting a multispeed motor both the speed ratio and the speed–torque relationship must be appropriate for the driven load. Constant torque and variable torque two-speed motors are widely available in both single-winding and two-winding styles.

Questions

1. What factors determine the speed at which the stator field rotates?
2. Why do motors usually have no more than six or eight poles?
3. What is the main attraction of a three-phase synchronous motor (with a wound-rotor, not a reluctance type)?
4. Define the following terms as they apply to squirrel-cage rotors: "Locked rotor torque," "pull-up torque," "breakdown torque," "synchronous speed," and "slip."
5. Compare the operating characteristics of Design B, C, and D rotors. List one application for each.
6. Calculate the synchronous speed for a six-pole 50 Hz motor.
7. If a squirrel-cage motor has a nameplate speed of 825 rev/min, how many poles does it have?
8. A 60 Hz motor runs at exactly 1200 rev/min. What type of motor is it? How many poles does it have?
9. A 50 Hz motor is rated for 1450 rev/min. How many poles does it have, and what would its approximate speed be if it was connected to a 60 Hz supply?
10. Two 10 Hp motors are rated for 3450 and 1725 rev/min respectively. Which of these motors has the highest rated output torque?
11. Explain why a 208–230/460 V motor may run hotter on 208 V than from a 230 or 460 V supply.
12. List the common nominal utility voltages below 2400 V, and the motor nameplate voltage ratings that would be appropriate for operation from each of them.
13. Why aren't standard three-phase motors available with full-load speed ratings of more than 3600 rev/min?
14. What change is required to reverse the rotation of a three-phase motor?
15. The leads of a nine-lead wye motor are disconnected and separated from one another. A continuity test is used to determine the internal interconnections of the leads. Which of the nine leads will show continuity to each other?

16. Since the AC winding current is constantly reversing, why does it matter if the numbers on a winding are reversed? (Why not switch T2 and T5, for example?)
17. Draw connection diagrams for wye and delta six-lead motors.
18. A six-lead motor has a nameplate voltage rating of 575 V and a connection diagram that indicates it is intended for delta operation. Would there be any problem if the motor was connected to a 600 V supply with the windings connected in wye? Explain.
19. Assume two identical motors, rated 15 Hp, 230/460 V, are properly connected and supplied. One is running from a 240 V supply, and the other from a 480 V supply. What differences should there be in the fuse ratings, control equipment, and supply conductors for the two motors? Should there be any difference in the operating efficiency or the operating cost, assuming identical loads are connected?
20. Sketch drawings for wye and delta nine-lead motors on both high and low voltage.

12 Three-Phase Motor Control

Objectives

Aspects of motor control that are essentially unique to single-phase motors are covered in Chapter 10. Additional motor control information of a general nature or specific to three-phase motors is included in this chapter. After completing the study of this chapter you should be able to:

- Describe the control and protective functions provided by typical motor control circuits
- Describe the main features of a typical motor contactor
- Explain the difference between "overcurrent" and "overload" protection as these terms are commonly applied to motor circuits
- Explain the operation of thermal overloads of both the bimetallic and melting-alloy types
- List the considerations and cautions relevant to the selection and installation of overload heaters
- Draw a diagram for a basic motor control system using maintained-contact pilot devices
- Draw a diagram for a basic motor control system using momentary contact controls
- Identify common symbols used in schematic diagrams
- Describe the layout of a schematic wiring diagram and the referencing system used to relate coils and contacts
- State the reasons why reduced-voltage starters are used
- Describe the basic principles employed in primary resistor, wye-delta, part winding, and autotransformer starters
- Describe, in general terms, the operation of an electronic "soft starter"

12.1 Introduction

Motor control equipment includes both control and motor-protective functions. The complexity of control components and the total control system depend on the needs of the driven machinery. The protective devices should be selected after considering the total cost of a motor replacement, including issues such equipment downtime. These considerations should be balanced against initial costs and maintenance problems associated with the protective equipment.

Please note that electric motor installations, like all electrical installations, must comply with the regulations stipulated by local inspection authorities. This generally means compliance with the National Electric Code (NEC) in the U.S., or the Canadian Electrical Code (CEC) in Canada. There may also be additional code supplements or amendments at the local and state or provincial levels. Both the NEC and CEC are quite detailed in the motor area, providing very specific rulings on virtually all possible installation scenarios.

12.2 Typical Control Functions

A typical three-phase motor control system provides some or all of the following functions and features. The control system must be able to start and stop the motor. It may also have to provide reversing, jogging, and braking. It must protect the motor and the motor circuit against damage due to overcurrent or overload, and protect the motor from overheating and undervoltage. The control system must also provide logical functions such as ensuring correct operating sequences or appropriate time delays. The controls for individual motors may also be required to function as an integrated part of a larger control system.

Start/Stop Control

Conventional motor controls are based on mechanically operated contacts. Very small motors may use relays, but for a three-phase motor the main control is usually either a manually operated switch or starter or an electromagnetic contactor. Start/stop control involves connecting or disconnecting the power supply to the motor. For three-phase motors, the main control device must simultaneously switch all three supply lines. Three-phase motors do not use a neutral, so it is not connected to the motor even if there is a neutral in the system.

A suitable control must be able to perform this switching action repeatedly under typical motor load conditions without undue wear or degradation of the contacts. It must also be able to successfully make and break the circuit under worst-case conditions, such as a solidly blocked rotor. There are limits to the number of operations that should be expected of any control device, and motor control equipment is no exception. Unfortunately, there is an added confusion caused by different rating systems competing in the same marketplace.

Note

NEMA rates motor contactors and starters as general-purpose devices suitable for a certain maximum horsepower. In contrast, IEC rated contactors and starters are intended to be selected by type of duty and expected number of operations. This can make comparisons of components difficult. An IEC starter, selected on the basis of horsepower only, may be less expensive and also less durable than a competing NEMA model. This is not because one is necessarily better than the other, but because the rating systems are different. The NEMA design allows for selection on the basis of horsepower only for all but the most demanding service. More factors must be considered if an IEC device is to be properly matched to the control requirements.

Many manufacturers make both NEMA and IEC rated equipment. In addition to these two major rating systems, other equipment manufacturers—such as those supplying the HVAC industry—make contactors of their own. These may be rated by other criteria, such as the motor's locked-rotor ampere draw. With no consistent common rating system, it can be very difficult to fairly compare one type of contactor with another.

Reversing, Jogging, and Braking

Reversing a three-phase motor requires that two of the line leads be interchanged. This is normally done with a second contactor that has the necessary cross connections. Reversing starters are manufactured as a unit. The contactors are mounted on a common base and mechanically interlocked, so that it is not possible for both to close at the same time. This eliminates the possibility of a disastrous short circuit.

Jogging is a momentary energization of a motor to allow small movements. A jog button, in comparison with a start normal button, only energizes the motor while the button is pressed. The circuit is de-energized as soon as the jog control is released. Jog controls are usually provided where some manually controlled positioning may be necessary. For example, up and down jog buttons are sometimes provided on freight elevators to allow the operator to level the car with the floor.

Braking is another option that is essential in certain motor-driven applications. Braking can be provided in several ways. A convenient and compact approach is to use a motor that is factory equipped with an electrically controlled brake. The brake assembly is usually integrated into the motor case by extending the end bell opposite the drive end of the motor shaft. Because the brake is an integral part of the motor, it does not require additional mounting or alignment. These brakes are usually spring applied and electrically released.

Separate brakes are also available. They are usually mounted on the same base as the motor and must be aligned with the motor shaft. This type of brake requires a motor with the shaft extended out both ends. Most of these brakes are drum types, with springs to apply the brake shoes. They are released electromagnetically by a solenoid coil.

Specialized electric eddy current brakes are sometimes used on hoists to control descent of the load. These brakes have no mechanical wear—the drag comes from eddy currents induced into a conductive drum or disk. A degree of braking may also be provided by injecting DC current into the AC motor winding. This is sometimes offered as an option on electronic motor controllers, but can also be added to conventional control systems. With a DC field, the squirrel-cage rotor acts as a brake.

Motors can also be "plugged" to a stop by electrically reversing the motor while it is running. Normally this would result in the motor stopping and then immediately reversing, but for this type of control, a special speed switch is connected to the motor shaft. When the stop button is pressed, the controller reverses the motor, but the speed-operated plugging switch cuts the power to the motor just before it starts to run backwards. Plugging is a very effective braking method, but is makes severe demands on the motor, the controller, and the mechanical parts of the drive.

Overcurrent and Overload Protection

In circuits that are not installed specifically to power motors, the fuses or circuit breakers are sized to match the maximum current rating of the circuit. In this way, the conductors and the associated equipment of the circuit are properly protected from a sustained current in excess of their maximum ratings.

The protective device—the fuse or circuit breaker—has an inverse time/current characteristic. The time it takes to open the circuit is inversely proportional to the square of the current. Very high current faults are cleared quickly; with lower current values, there is a significant time delay.

Under normal conditions of temperature and heat dissipation, the fuse or breaker will carry its rated current indefinitely. When the current is only slightly higher, they should eventually trip, but the delay may be many minutes or even hours. At progressively higher currents, the delay becomes less and less until the limits of the protective device are reached. At that point, neither the melting of the fuse element nor the opening of the breaker contacts can happen any faster.

It might seem that even moderate overloads should be cleared as quickly as possible, but this is neither practical nor desirable. It is entirely normal for many types of electrical loads to draw excessive current for short periods of time, usually when they are first energized. This is true for many common loads, including standard incandescent lightbulbs, many heating elements, coil devices such as transformers, solenoids and motors, and almost all electronic equipment.

Lightbulbs draw high initial current because the cold resistance of the filament is only about one-tenth of the resistance at the normal operating temperature. This makes the initial current about 10 times the rated value, but it last for only a very brief time. The current drops to normal as soon as the filament gets up to temperature. Transformers draw a high current until their magnetic field is fully established. Transformer inrush current lasts for only one or two cycles, but may be six or more times the normal operating levels. The actual inrush current is a random value that is impossible to predict, because it varies with the polarity and voltage of the AC sine wave at the instant the switch contacts actually close.

Solenoid and relay coils draw high initial currents because, when they are first energized, their magnetic circuits are open. The armature—the moving part of the electromagnetic core—is separated from the rest of the core iron. This open magnetic circuit has a much lower inductive reactance than when the two parts are drawn together. The coil current drops to normal as soon as the armature has pulled into place. This behavior gives AC coils a powerful initial field for fast pickup, and then a lower holding current that reduces coil heating and power loss.

Most electronic equipment draws a high initial current because of the large filter capacitors used as part of the AC-to-DC conversion in the power supply. (Virtually all electronic equipment is DC powered. The AC is converted to DC internally, or in a separate power supply that is part of the power cord.) The power supply capacitors fully discharge each time the equipment is shut off, and must be fully charged when it is turned on again. The initial charging current can be quite high but, again, it drops to normal very quickly.

As these examples demonstrate, many common electrical devices could not operate from a circuit protected by a very fast-acting circuit breaker or fuse. Their excessive current draw lasts for a very short time, but certainly long enough to trip a breaker or blow a fuse that has no built-in delay. The short intentional delay in the response of the protective device solves this problem with no ill effects. A brief overcurrent does no damage unless the current is extremely high, much higher than the six to ten times rated that is typical of these loads.

Of course there are limits. The time delay must not be so long that the circuit conductors or other equipment can overheat. Sustained overload, even when the current is only slightly higher than the circuit's safe rating, will cause problems, because the heat buildup will damage the insulation.

Motors raise protection problems similar to those of other types of equipment, but are a more extreme case. A typical three-phase induction motor may draw six times the nameplate current rating (full-load running current) of the motor. This is not necessarily higher than for other types of loads, but it is sustained for a much longer period of time. While other types of loads drop to normal operating currents within a few milliseconds, motor current stays high until the motor is up to full speed—and that can take many seconds.

In addition to their startup demands, motor circuits are expected to absorb periodic short-term overloads without tripping. In many motor applications, brief mechanical overloads are part of normal equipment operation. The overcurrent device must not trip, even though the circuit current is well above the normal operating level.

In summary, for a motor to provide acceptable service, the circuit overcurrent protection must stay closed through a relatively long period of high current during startup. It must carry frequent periods of modest overload and supply the motor's full rated current indefinitely. Yet, to protect the circuit, it must open when the current is high enough to cause damage. Not surprisingly, it is virtually impossible to provide the necessary protection with a single device. A breaker or fuse rated to protect the motor circuit at normal full-load operating levels would never hold through the startup period; rated high enough to carry the startup current, it will fail to protect the circuit from overheating due to a more modest, but sustained, overload.

Although there are several ways to solve this dilemma, the traditional solution is to split the necessary protection into two separate but related functions. Overcurrent protection is provided by a fuse or breaker; overload protection is provided by a separate overload device provided as part of the motor or the motor controller. The primary purpose of the overcurrent protection is to open the circuit on high-current faults such as shorts and grounds. The overload protection is intended to disconnect the motor when the current is excessive but not extreme. It protects against problems such as a sustained mechanical overload.

Overcurrent protection must be very fast, because of the potential for damage from extremely high currents. This function is provided by breakers or fuses, as it would be in any other circuit; however, the current ratings used are much higher than they would be for a circuit supplying other types of loads.

Overloads must also be cleared promptly, but some delay is necessary to permit normal starting currents and short-duration overloads. Overload protection is usually provided by thermally operated overload devices mounted in or on the motor, or at the motor controller. Like other overcurrent devices, typical overload protection has an inverse time/current characteristic. A severe overload, such as a stalled motor, will be cleared much faster than a moderate overload.

The choice of fuse or breaker for overcurrent protection depends on the features desired and acceptable cost. Breakers have an advantage for circuits with more than one ungrounded line—single-phase supplies with no neutral, or three-phase—because a trip on any line automatically breaks all the supply lines at the same time. When the circuit is protected by fuses, one fuse may blow, leaving the equipment energized from the remaining lines. With a three-phase motor this can lead to "single-phasing," which is operation of a three-phase motor on single-phase power. This can be a serious problem because, with a modest mechanical load, the motor can overheat and burn out without tripping the overload or overcurrent protection. Fuse protection—normally in the form of a fused disconnect switch—has the advantage of a lower initial cost. Additionally, because they have no moving parts, fuses can usually clear a severe fault faster than a breaker.

In many cases, each motor has its own overcurrent and overload protection. This provides the best protection for the motor and the motor circuit, because the ratings of the overcurrent can be closely matched to the demands of the load. However, this is not always practical. Under certain conditions, such as when a single piece of equipment contains several motors, a single overcurrent device may supply power to the whole group. This is permitted by electrical codes, as long as the required conditions are met. The codes dictate the minimum size, the maximum length, and the acceptable installation methods for each of the individual motor supply conductors. Separate overload protection for each of the motors may also be required.

Selecting the correct size and type of breaker or fuse to protect a motor circuit is a relatively complex procedure governed by a whole set of electrical codes. These rules detail many variables, including the characteristics of the fuse or breaker, the type of motor, and the motor duty cycle. Duty cycles relate to the starting frequency, reversing, or plugging duty, and whether the mechanical load is continuous, intermittent, or cyclic. Although the codes must cover every possible situation, typical motor applications tend to fall under the same generalizations.

Most motors are started infrequently and operated for extended periods of time, usually at or near their full rated load. For these situations, the overcurrent rating for a typical general-purpose circuit breaker is approximately 250% of the full-load ampere rating (FLA) of the motor. When fuses are used, this value is approximately 300% of the FLA for a conventional, nondelay fuse. For time-delay fuses, such as those with dual elements or HRC Form II types, the current rating is lower. These fuses are designed primarily for motor circuits. They are less sensitive to short-term starting currents, so the current rating can be lower than would be selected for more conventional types.

Note

Manufactured equipment, including units with multiple motors, usually has nameplate data that defines the circuit requirements for the equipment as a whole. In this situation, the supply conductors and the overcurrent protection are sized to the nameplate requirements of the unit, not the individual components it contains. Repairs or replacements within the equipment are normally based on matching the ratings of the item being replaced, regardless of whether it conforms to the rules that apply to other installations.

In motor circuits, overload protection is usually not provided by the fuse or breaker. The overloads are separate from the overcurrent device. They may be mounted on the motor housing, inside the motor, or at the motor controller. Conventional overloads are thermal units that monitor motor current by passing it through a low-resistance "heater element" that is integrated into the overload device. The heat produced by excessive current bends a bimetallic element or melts a bit of trapped solder, opening a contact and breaking the circuit to the motor. The overloads are typically designed to open at 115 to 140% of the motor's FLA rating, depending on other motor characteristics.

The respective roles of the overcurrent and overload in motor circuits are often misunderstood. Typically, the overcurrent device is rated much higher than the current-carrying capacity of the conductor it supplies. The only reason this is permitted is that the circuit also contains an overload device set to a much lower current value. Although the overload device is incorporated into the circuit to protect the motor, it also limits the maximum load current that can occur on the circuit conductors. For this reason it is the rating of the overload device, not the overcurrent protection, that is used to select the motor supply conductors. For most circuits that supply other types of loads, the current capacity of the conductor and the rating of the overcurrent device have to match.

Overheating Protection

The primary purpose of all overcurrent and overload protection is to protect the circuit conductors and the motor from overheating. For this reason, protection against overheating is, in many cases, assumed to be covered by overload protection. However, there are situations where this is inadequate. Electrical codes may require additional overheating protection, particularly in situations where the ambient temperature at the motor is normally higher than at the overloads, or where the ambient conditions at the motor can leave it without adequate cooling. Examples include areas where heavy dust accumulations can block the motor's air passages, and sealed refrigeration compressors where a loss of refrigerant circulation can leave the motor without its normal cooling medium.

There are several methods of providing overheating protection. When a motor is equipped with built-in thermal overload protection, a degree of overheating protection is also provided. Overloads are thermal devices; when they are mounted at the motor, they are directly affected by motor temperature, as well as by the current. This type of protection is common on smaller motors, such as fractional-horsepower single-phase types, and is also used in some three-phase units. Motor overloads that are mounted externally, but in contact with the motor case (such as those used with some hermetically sealed refrigeration and air conditioning compressors), operate much the same as the internal types.

Overheating protection may also be provided separately. One of the simplest methods consists of one or more small thermostats inserted into the motor windings at the time of manufacture. These thermostats have normally closed contacts that open at a suitable preset temperature. If the motor has two or more thermostats, they are connected in series. The thermostat leads are brought into the motor connection box, and are connected into the external control system in such a way that an open thermostat contact will stop the motor.

As an alternative to internal thermostats, the motor manufacturer may provide imbedded temperature sensors such as thermistors or thermocouples. These require an external transducer and monitoring equipment, but have the advantage of providing an actual temperature measurement, as opposed to only a preset trip point. With appropriate auxiliary equipment, motor temperature can be monitored, and alarm points can be set to warn the operator prior to actually tripping the motor off the line. This is a more sophisticated system than simple overheating protection, and it only common on larger motors. It should be noted, however, that the size and cost of sophisticated electronic monitoring and control equipment is dropping continuously. A variety of integrated electronic motor monitoring and protection devices are readily available now, and may eventually replace conventional overloads and overcurrent protection for all sizes of motors.

Control Logic

Most motor control systems can be divided into two distinct circuit functions. The power circuit contains the supply breaker or fused switch, the supply conductors, a controller such as a motor starter with the main power contacts and the overloads, and, of course, the motor. The control circuit usually contains a power supply, such as a control transformer, and all of the control devices. The transformer may be eliminated if the control voltage is the same as the operating voltage of the motor. Control components include manual controls such as start/stop switches; automatic sensing controls such as thermostats or pressure switches; sequence controls such as timers, counters, and auxiliary relays; and final control elements such as motor starter coils and solenoid valves.

The way in which the controls are interconnected determines the logic of the control system. Control logic is based on what each control device does in the system. It determines the sequence in which the various devices operate, and how the equipment responds to failures. Basic systems include start and stop controls, and essential protective functions such as disconnecting the motor on overload. More complex control logic includes such functions as interlocks, time-delays, sequence control, and safety-related items such as emergency stops and limit switches.

Interlocks are used to ensure specific conditions are met before a piece of equipment is energized. This can include switches on covers or doors that must be closed, and contacts from other circuits to ensure related equipment is operating first. Interlocking contacts are also used for such things as a pumps or blowers that must be started in sequence.

Time-delays are used in a variety of ways. Examples include allowing a system to stabilize prior to initiating the next step in the process, and ensuring a minimum time between motor starts. Timers can also be used to schedule daily and weekly operating modes.

An example of a sequence control would be the ignition and shutdown procedures used for larger burners. The sequence is automatic, but it must always occur in correct order, with each step completed before the next one starts. A typical burner might go through the following steps: combustion blower startup, proof of combustion air flow, prepurge time-delay to clear the fire box of any combustible gases, burner monitoring system prechecks, pilot gas and igniter on, proof of pilot, igniter off, main gas on, proof of main flame, pilot off, and continuous burner monitoring. The shutdown sequence might include all gas shut off, a post-purge time period, and finally a shutdown of the combustion air blower. Sequences such as these are repeated for every burner cycle and must be absolutely fail-safe. They are usually provided by a custom controller based on a clock driven-timer, or by an electronic logic board.

Emergency stops are added to system where there must be some way for a human operator to cut power to the system. They are installed in addition to normal stop controls, and are usually located in one or more convenient locations, often very remote from the normal controls. In extreme cases, emergency trips can extend for miles. For example, on long conveyor belts there must be some way to shut off the belt at any point along its length. The emergency stop controls are connected to a continuous pull line on one or both sides of the belt. Pulling the line opens the stop circuit and shuts off the belt. The switch that actually trips the circuit usually has to be reset locally before the system can be restarted.

Limit switches are used to detect conditions outside of normal operation. For example, a temperature limit switch can be used to shut off a burner on high heat exchanger temperature. A mechanical limit can be used for such things as detecting a material pileup on a conveyor system or in a loading chute.

Undervoltage Release or Protection

It is generally a good idea to disconnect motors from their supply if the voltage drops significantly or is suddenly cut off. reduced-voltage to an unloaded motor is not necessarily a problem. However, if the motor is carrying some mechanical load, low voltage greatly reduces its maximum output torque and may cause it to stall. When a motor stalls, it quickly overheats due to excessive current and loss of air circulation.

A no-voltage condition, typically a tripped circuit breaker, blown fuses, or a utility power failure, should also automatically disconnect the motor. The loss of power is not the problem. The trouble starts when the power is restored. First, the breaker or switch that is used to restore power may be expected to start the motor, even though it is not designed or rated for such rigorous duty. Second, the motor may be started without warning from a remote location. Depending on the application, this may be completely safe or extremely hazardous. A motor driving a woodworking machine should not be started by anyone except the operator. A motor driving an enclosed pump or compressor can normally be started without warning, with no danger to anyone.

Two slightly different protective functions are commonly used to protect the motor and the power system. One also protects people and property in situations where an unexpected restart could be dangerous. When a control system disconnects the motor on an undervoltage condition, but immediately restarts it when the power is restored, it is called *undervoltage release*. When the control system disconnects the motor and does not restart it without operator intervention, it is called *undervoltage protection*. Undervoltage release is commonly used on systems where an unexpected restart is not a potential hazard to the equipment or personnel. Undervoltage protection is used in situations where a restart is a potential hazard.

Undervoltage protection is also used to prevent simultaneous restart of a number of motors, regardless of the safety hazard. Unless the supply is designed specifically to allow for it—a highly unlikely situation—starting a whole group of motors will very likely trip off the main supply. It can be difficult enough to supply the surge of current required by one motor: Attempting to start many motors at once can easily exceed the capacity of the main fuses or circuit breaker.

Undervoltage tripping is essentially inherent in magnetic starters, because a typical coil will release at about 85% of the rated voltage. This provides either undervoltage protection or release, depending on the layout of the control system. Note, however, there are exceptions. The power to the coil circuit of the starter contactor must be from the same supply as the motor. This is usually the case, because control power is typically supplied by a transformer that connects to the motor supply. However, the control circuit is only single-phase, so the transformer is connected to only two of the three supply lines. The third phase is not monitored for undervoltage unless an additional control, such as a small relay, is added for that purpose. This additional protection is particularly valuable on systems that use fused switches instead of circuit breakers, because the possibility of losing only one line is much higher when fuses are used.

Manual motor starters do not provide undervoltage tripping unless an undervoltage trip coil is provided in the starter or in the supply breaker. Internal tripping coils of this type are available as an option on many industrial-grade manual starters and circuit breakers. Either can provide undervoltage protection, but not undervoltage release.

Note

Breakers equipped with undervoltage coils should have a visible explanatory label. The breaker cannot be closed unless there is voltage on the undervoltage coil. When there is a power failure or a problem in the control circuit, the breaker trips, but cannot be reset until the undervoltage coil is reenergized. Anyone unfamiliar with this type of protection may, quite naturally, get the impression the breaker is defective. This is a common protective feature in many electronic motor control systems, because it provides a convenient disconnect point for a system that has no normal contactors.

Starters

The simplest type of motor controller is a manual starter. These are common on single-phase circuits, less so on three-phase. They have the advantage of lower cost and simplicity. They can provide overload protection and, on some models, can also provide undervoltage protection. They cannot be remotely controlled and, therefore, cannot be integrated into a more complex control system. They are ideally suited to a single-motor application where the control requirements are limited to on/off switching and basic motor protection.

The main difference between a manual starter and a switch is that the starter is equipped with overload protection that can trip the contacts to the open position. The mechanical linkage that controls the contacts is connected to a set of thermal overloads in such a way that an overload releases a latch and the contacts open (see Figure 12–1). Optional undervoltage coils act through the same linkage to release the contacts, unless the coil is energized. The overloads use interchangeable heaters that can be selected to match the motor current.

Most manual starters are compact devices, complete with their own enclosures. They are wired directly into the motor circuit, and mounted just like a control switch. They may be equipped with start/stop pushbuttons or some type of toggle. Unlike magnetic starters, they usually do not have a separate button to reset the starter after an overload trip. On most manual starters, the overloads are reset when the stop button is pressed or the toggle operator is moved into the off position.

Although manual starters are readily available and entirely suitable for certain applications, most three-phase motors are controlled by some type of magnetic starter. A magnetic

FIGURE 12–1 This three-phase manual motor starter has start and stop pushbuttons. The overload heater has been removed to expose the shaft of the melting-alloy element (lower left). A latch arm contacts the toothed wheel on the end of the overload element. The wheel is held stationary by a small amount of trapped solder, which melts on overload to release the latch arm and trip the main contacts.

FIGURE 12–2 This magnetic starter has the cover removed to expose the three sets of power contacts (top right) and one auxiliary contact (top left). One overload heater has been removed (bottom right) from this bimetallic-style overload block. The reset button is in the center bottom of the starter.

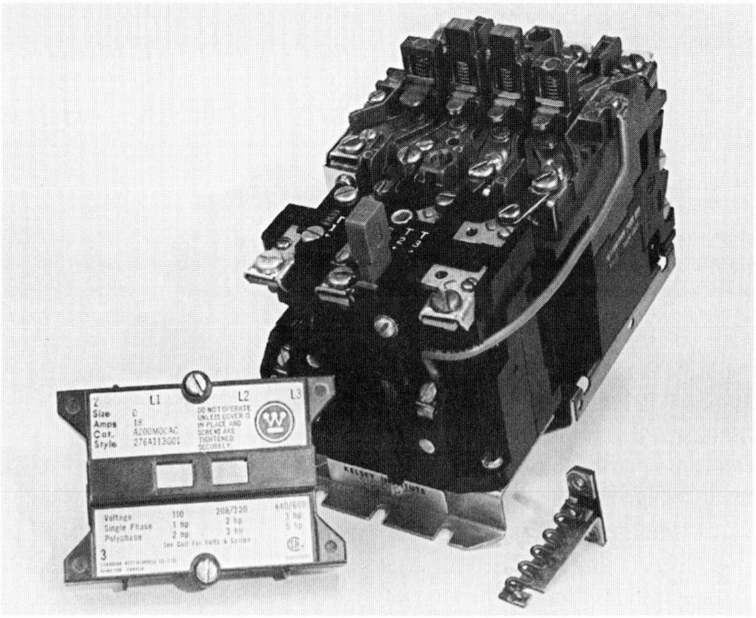

starter is a motor duty contactor combined with a set of overloads (see Figure 12–2). Typical three-phase motor contactors have three sets of normally open power contacts, but they are also available with four or five. One normally open auxiliary contact is standard; more can be added as required. Auxiliary contacts are designed for pilot duty in the control circuit. They are not intended to switch a motor load.

The contactor and overloads may be physically connected together to form a single unit, or mounted separately and wired together in the enclosure. The only advantage of one method over the other is convenience. The overloads trip the contactor electrically, through a normally closed overload contact wired into the contactor coil circuit. Unlike the manual starter, no mechanical connection between the two is required, so they can be mounted separately if desired.

A combination motor starter includes a fused disconnect or breaker, a starter, and, usually, a step-down control transformer, all built into a suitable enclosure. An external switch handle operates the disconnect switch or breaker. It is mechanically interlocked to the enclosure door, so it cannot be opened with the switch in the on position. There is a provision to bypass this interlock, but unauthorized access is restricted because a tool (usually a screwdriver) is required to release the latch. The switch handle is also equipped with provision for a padlock. With a lock in place, the switch cannot be energized, and the cabinet cannot be opened. Combination starters provide all the necessary motor control and protection functions in a single unit. They also meet the regulations for restricted access and provision for safety lockout. Many options are available to tailor the starter to specific requirements.

Overloads

Manual starters have a latch mechanism that holds the contacts closed against a spring. Internal thermal overloads trip the latch mechanically, opening the contacts if an overload occurs. On magnetic starters, tripping is electrical. The overloads have one or more contacts wired into the coil circuit of the contactor. There may be a single overload block with three thermal elements, or three separate overload units, one for each phase. Each overload device has at least one normally closed contact wired in series with the contactor coil circuit. When an overload trips, its contact opens, cutting power to the coil and opening the main controls to shut off the motor.

There are two types of overload mechanisms in common use. A bimetallic type depends on the bending action of a bimetallic strip that is heated: The bimetallic elements are arranged so that at a certain point they release the latch on the overload contact—or the main contacts, in the case of a manual starter. The second type of overload element uses heat to melt a small amount of an alloy, such as solder, that is trapped in part of the latch. The usual layout is a small-toothed wheel that turns freely when the alloy is hot, but is held stationary when the alloy is cool enough to be solid. These two mechanisms are used in a wide variety of ways, but in each case the principles are the same.

Three-phase starters are usually equipped with one overload element in each line. When the elements are combined into a single overload unit, they usually share a single set of overload contacts. When the elements are mounted separately, there must be a contact in each one. These are wired in series so that any one of them can break the circuit.

The heat that acts on the bimetallic or melting-alloy elements comes from small resistance elements—commonly called heaters—that are connected in series with the motor supply conductors. The heat produced is proportional to the motor current. The tripping point is set by the resistance of the heater and the characteristics of the bimetallic or melting-alloy device.

Typical North American overload units have a single normally closed overload contact. European-style overloads usually provide an additional, normally open, contact for alarm purposes. Most bimetallic overloads have adjustable trip current settings. Some have fixed heaters built into the overload unit; others accept a range of interchangeable heater elements. Melting-alloy overload units have interchangeable heaters.

Most bimetallic overloads can be set for either manual or automatic reset. Manual reset is by far the most common choice. When an overload occurs, the overload contact stays open until it is manually reset. In most installations, the overload reset buttons are accessible without opening the motor control cabinet. Extension buttons are mounted in the cabinet door, in line with the reset buttons on the overload units inside. When the automatic reset option is used, the overload contacts automatically reclose once they have cooled down. This setting should be used with caution. In many control situations, resetting the overloads is potentially dangerous. It is almost always best to force the operator to be aware of the trip. Automatic reset is not an option on melting-alloy-style overloads.

Overloads are simple, reliable devices that provide adequate protection for most motors if they are properly matched to the characteristics of the motor. Resist the temptation to adjust overloads or change overload heaters to compensate for nuisance trips. This invariably ends up as a premature motor rewind or motor replacement.

Selecting and Installing Overloads

Overload settings, obtained through replaceable heaters or by adjustment, are usually based on the motor's nameplate full-load current rating and the service factor. As a general rule, the heater rating is 125% of full-load rating for motors with service factors of 1.15 or more, 115% for motors with a service factor below 1.15. However, these are only common values used for general-purpose motors. There are many other variations for specific types of motors and special applications.

As a matter of interest, these overload values are based on two rules of thumb: Overload tolerance, the accuracy of a typical overload device, is approximately ±10%. Typical motors operated within their ratings can tolerate a continuous overload of rated current times the service factor. A setting of 10% above the service factor is 125%. This corresponds with the overload tolerance plus the service factor.

In many cases, motor overload settings are determined not by reference to the appropriate code books, but from charts supplied with the starter. For overloads that use interchangeable heaters, these charts cross-reference the motor full-load current ratings to a specific heater by part number. When a starter is purchased, the equipment supplier will generally ask for the motor ratings so the correct heaters can be supplied at the time of purchase. This is an acceptable practice as long as code requirements are met.

Adjustable overloads have a specific range of values. When the overload device has fixed heaters, the scale of the adjustment knob is in amperes. When the overload device has both interchangeable heaters and an adjustment, the scale is in percentage. The actual setting is the overload heater rating times the percentage setting.

When installing new heaters, first safely isolate the equipment. Make sure that all of the heaters have the same ratings, and then check for correct orientation. Most heaters will not fit properly unless they are correctly oriented in the overload block. However, at least one manufacturer uses heaters that fit four different ways, each position providing a slightly different overload rating from the same heater. Extra care must be taken with heaters that seem to fit more than one way in the holder. As usual, reading the manufacturer's information may not only be interesting and informative, it can also be a great time saver.

Make sure the heater mounting screws are properly tightened. Watch for any signs of previous heating. If the connection has overheated in the past, scrape the contact surface clean before installing the replacements.

Note

As with all things electrical, overload connections must be clean and properly tightened. Poor connections at the heaters or other overload terminals will generate extra heat, causing nuisance trips at less than rated overload currents. This is an important consideration when doing an initial installation, changing heaters, or trying to determine the cause of a trip that occurred for no apparent reason. When the problem is loose connections, they can often be spotted by a visual inspection or by feel—with the usual cautions against contacting energized conductors, of course!

Heat from bad connections can also cause similar problems with fuses and breakers. Poor contact in a fuse holder will heat the end of the fuse, lowering the current at which it will blow. Once they have been overheated, the fuse clips can lose spring tension, leading to further problems. Fuse contact surfaces should be kept clean, and all connections tight. Clip clamps are available that fit over the end of weak fuse clips to clamp them tight.

Almost all breakers also incorporate thermal tripping elements, and are therefore affected by heat in the same way as overloads. Poor contact on either side of the breaker lowers the trip setting, because heat is conducted from the connection into the breaker. This problem should be corrected promptly. If it is left too long, the breaker connection terminals or the contact surfaces between the breaker and the panelboard bus bar may be permanently damaged.

12.3 Motor Control Drawings

There are several types of electrical control drawings—and many variations of each. Three general types are pictorial, wiring, and schematic. Each of these has its own strengths and weaknesses. In many cases, two or more different types of drawings are required to fully explain a piece of equipment.

To be effective, a drawing must convey accurate information in a format that is as easy to follow as the complexity of the equipment permits. When available drawings fall short of this ideal, it is often worth the time and effort to redraw critical sections of the original, or to make a completely new drawing in a more usable format. This is particularly the case when the same equipment is encountered frequently, or when the existing drawings are worn out or unreliable. Much-used drawings often become difficult to read. Drawings that are not updated may be inaccurate due to unmarked field revisions.

Redrawing control system diagrams can also be a powerful learning tool. Reorganizing an old drawing into a revised format usually leads to a very good understanding of how the system works. A computer assisted design (CAD) program—even a relatively simple one—can greatly facilitate this process. It should have a library of standard symbols, or allow the user to make their own. This avoids the tedium of redrawing the same symbols over and over. Common drawing symbols are shown in Figure 12–3.

Types of Drawings

Pictorial drawings are used for very simple circuits or to detail specific components. They are often used in explanatory literature and training manuals. The devices are drawn much as they appear in real life. Interconnecting wiring is usually shown in a layout that is similar to the actual routing used, but organized to avoid unnecessary crossings. The wires are often colored or marked with identifying patterns to simplify tracing the circuit. A legend may be provided to identify components and wiring.

Pictorial drawings are excellent when they are used to show specific details. They can also be invaluable when unusual devices are used, or where a good understanding of mechanical operation is essential. In simple systems they make acceptable—but not outstanding—general-purpose drawings. They are of limited use with complex control systems, because the wiring shown on the diagram quickly becomes a maze. They can be time-consuming to create, and equally difficult to reproduce or revise.

Wiring diagrams show much less cosmetic detail than a pictorial. The components and devices can usually be identified by their general shape and electrical connections, but there is no attempt to show unnecessary detail. The location of the various components on the drawing is at least roughly the same as the actual physical placement of the components within the equipment. The way in which the wiring interconnects in the equipment normally corresponds to the layout implied on the drawing, but the routing may be different.

FIGURE 12–3 Common drawing symbols.

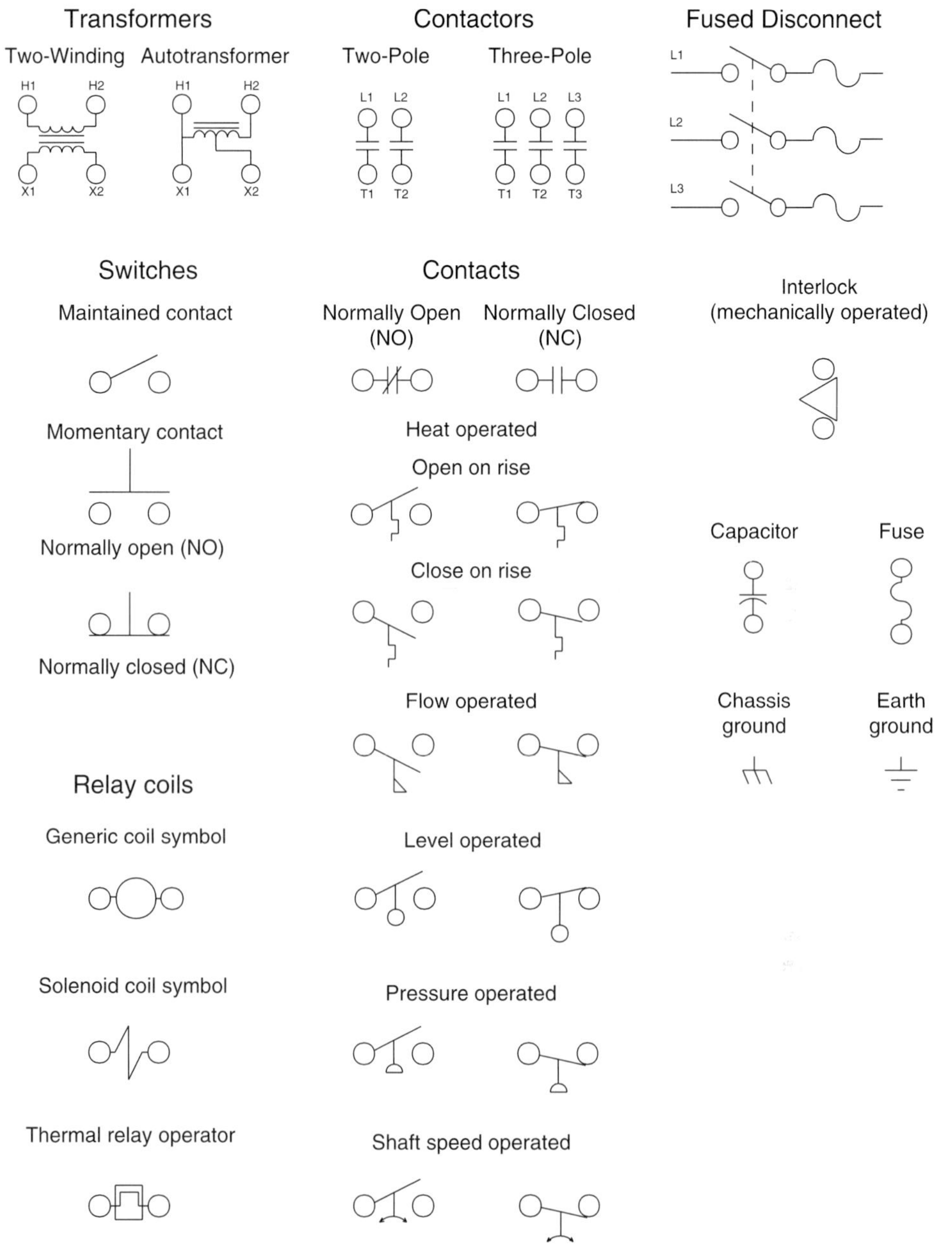

For example, each wire should connect to the actual terminals in the same order that they are shown in the drawing, but the routing of the wires may be reorganized to avoid confusing overlaps and crossings.

Wiring diagrams are primarily assembly drawings in which electrical operation is not usually shown in any detail. An assembler following a good wiring diagram can correctly wire the equipment with no knowledge of how it actually works. The internal workings of the various devices and components—such as contacts, relay coils, and so on—may not be shown. Instead, they are usually drawn as simple boxes, with only the terminals identified to show connection points.

Wiring diagrams are particularly helpful for locating components and specific terminals. For this reason they can be a valuable part of the manufacturer's service literature. However, they make poor general-purpose service drawings. It is time-consuming to follow a complex control sequence through the mazelike layout of a typical wiring diagram. More importantly, it is virtually impossible to take in complete sections of the circuit as a single visual image. This makes a quick and accurate analysis of the circuit very difficult. The fundamentals of the circuit can be seen with effort, but the subtleties remain virtually invisible. Service personnel with extensive experience can follow a wiring diagram with relative ease, but only when the equipment is familiar.

Thankfully, most manufacturers also supply a schematic diagram, typically in some variation of the ladder diagram format. A schematic diagram emphasizes clarity of operation by showing devices in correct electrical order. All relevant electrical components are shown. This includes manual switches; automatic control contacts; indication devices such as pilot lights and audible alarms; and the coils of relays, solenoids, and starters. Motors are usually shown with enough detail to understand any special connections, such as multiple speeds, built-in overloads, or internal thermostats. Symbols are used extensively in place of realistic-looking representations of the various components and devices. This results in a less cluttered drawing and more useful information in a compact format. For a comparison of the three drawing types, see Figure 12–4.

In a schematic, the devices are arranged on the drawing with little regard for their physical location in the equipment. This is a necessary price for the operation to be shown clearly. The disregard for physical location extends to individual electrical components within circuit devices. Relays, for example, often have their coils and contacts widely separated on the same drawing. An identification system of letters and numbers is used to show the relationships between the scattered components. Coils, for example, are marked with one or two letters and a number, such as 1K, 2K, or CR1, CR2, etc. The contacts are given the same identification, sometimes with an additional number that is unique to that contact. For example, 1K1 and 1K2 would be two different contacts operated by the same coil (1K).

With the coils and contacts appearing at different locations throughout the drawing, it can be difficult to locate all the contacts that relate to a particular coil. To alleviate this problem, larger ladder drawings have each line of controls numbered sequentially (see Figure 12–5). These numbers are part of a cross-reference scheme for the various relay coils and their contacts. Beside each coil there is a list of all the line numbers where the associated contacts can be found. Once this system is understood, all the contacts of a particular relay can be quickly located, even though they appear in many different locations on the drawing.

On a ladder diagram, the power supply is drawn as parallel lines on opposite sides of the drawing, like ladder rails, and the switches, contacts, coils, etc., are arranged across the drawing like ladder rungs. Ladder schematics may be oriented horizontally or vertically. Both formats are common, and both are equally functional. Sometimes both orientations are used in the same drawing set. In three-phase motor control drawings, for example, the schematic is usually split into power and control functions. If both are shown on the same sheet, the power circuits are often drawn with the opposite orientation to the control functions.

FIGURE 12–4 These three drawings—a pictorial, a wiring diagram, and a schematic—show the same motor control circuit in three different formats.

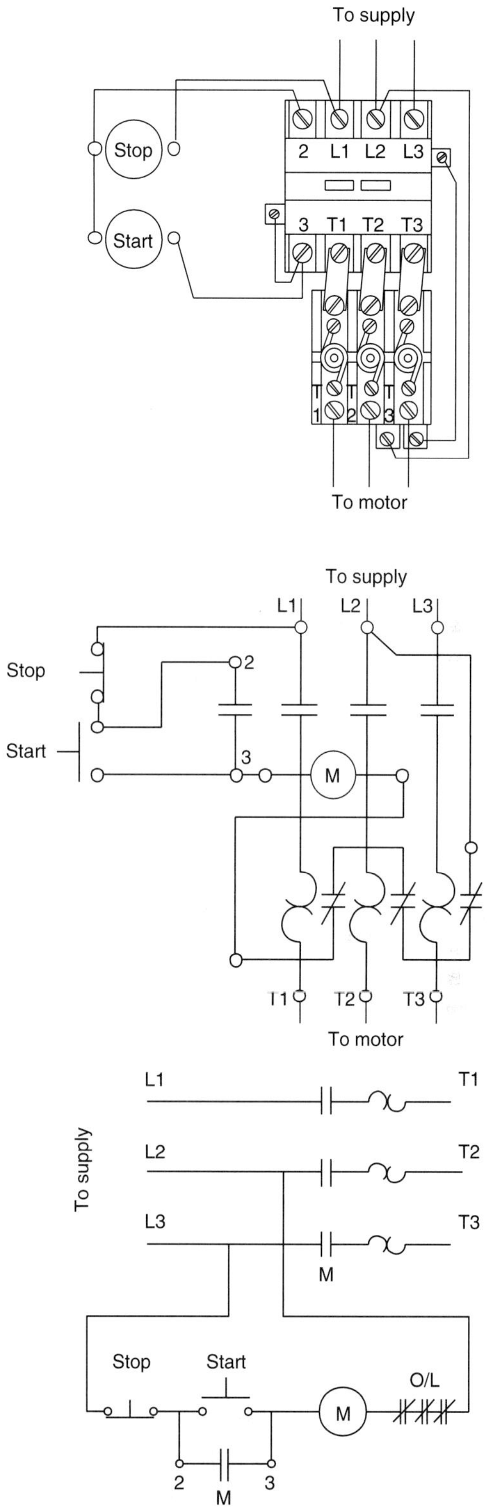

FIGURE 12–5 The numbers down the left of this ladder drawing are line numbers. The operating coils are all on the right, and references beside them indicate which line(s) of the drawing contain the contact(s) associated with that coil. Each contact is identified by a coil number and a contact number. The "power circuit drawing" reference for the bottom four coils indicates the contacts appear on a separate drawing sheet.

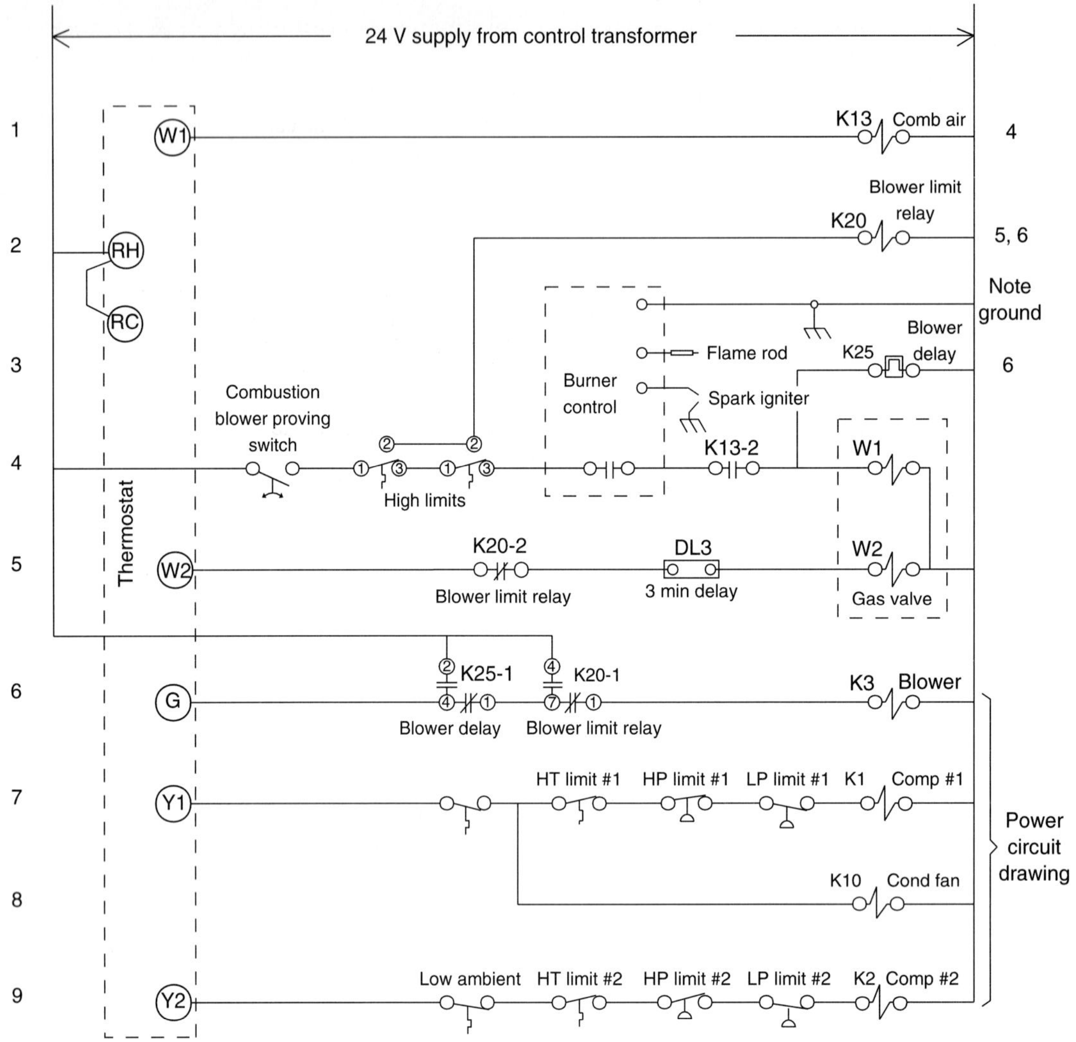

In vertical, two-wire ladder diagrams, the power supply is at the top of the diagram. The circuit common is normally on the right-hand side of the page, and the line or "hot" conductor is on the left. The common is the side of the supply that goes more or less directly to all the loads' coils, indication light, and other output devices. Control devices—switches and contacts—are normally between the line or hot side and the load. The common is the grounded side of the circuit, if the circuit has a ground.

The logical flow of the diagram is from the top left corner, reading each line from left to right. It should be possible to read the drawing line by line, much like one would read a book. The sequence may not be quite that simple, but good drawings are arranged with this in mind.

When a horizontal format is used, the common is normally at the bottom, the power supply is on the left, and the natural flow of the drawing is from top left to bottom right. Each line of the drawing is read from top to bottom.

Circuits with more than two wires, such as 120/240 V single-phase or any type of three-phase, are awkward to show on a ladder diagram because three or more lines must be shown, and multiple crossings become unavoidable. A more popular method for three-phase circuits is to group the power loads and show them in a nonladder format. The three power lines are drawn close together, and taps are added for each load, as required. This provides a neat and orderly presentation of the power circuits; the ladder format is retained for the controls (see Figure 12–6).

FIGURE 12–6 When power and control circuits are shown separately, they often have the opposite orientation. The horizontal power circuit shows the supply and a common circuit breaker (top left) feeding the control transformer, two motor starters, and their motors. The control circuit shows the coils for the starters, their overloads, and the associated start/stop pushbutton controls. Contacts are referenced to their coils by common markings. There are no line numbers because the drawing is essentially self-explanatory. The M1 contact ahead of the M2 coil is an interlock that makes it impossible to run M2 unless M1 is also running.

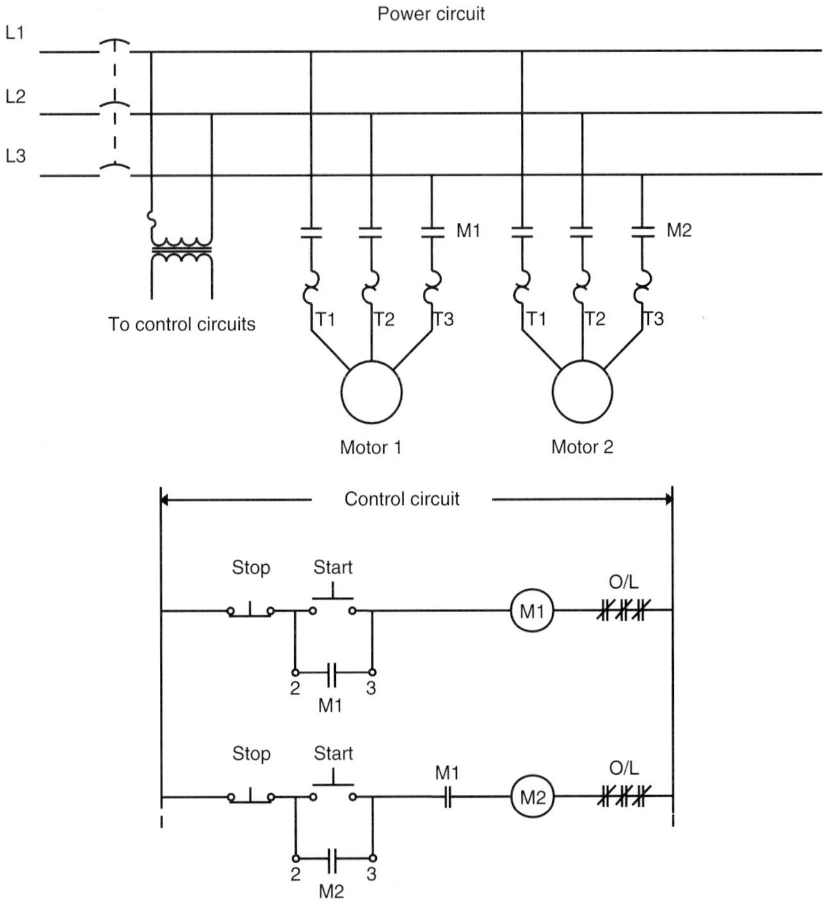

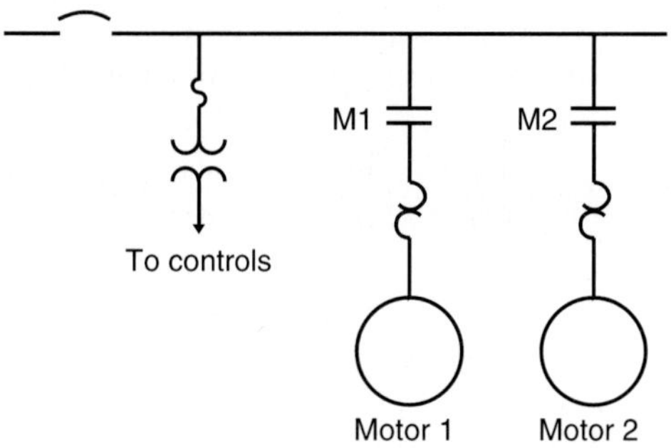

FIGURE 12–7 This one-line diagram is a simplified form of the power circuit shown in the previous drawing.

All three-phase systems require at least three power lines, so drawing the power conductors as individual lines is often unnecessary. The power circuit may be reduced to a single-line format, with all three supply lines represented by a single line (see Figure 12–7). Interconnecting equipment is shown with similar symbols, but only one pole is shown. Where required—such as on the primary and secondary sides of a transformer—delta or wye connections are noted. Single-line diagrams are most popular for showing the layout of large systems on a single drawing sheet. For example, the essential components and interconnections of the distribution system for a large building or an industrial plant may be shown with considerable clarity on a single drawing of modest size.

Summary

The are at least three basic types of electrical drawings, described here as pictorial, wiring, and schematic. Each has its strengths and weaknesses. Pictorials show equipment in an easily recognizable fashion, and can provide small details that other drawings miss. Wiring diagrams do not show electrical operation clearly, but they relate well to the physical locations of components and show the actual wiring layout. Schematics show electrical operation and very little else. They can be difficult to relate to the physical equipment, but are unsurpassed as a tool to understand the electrical operation of the system.

Manufacturers' drawings vary a great deal. Simple equipment often comes with a single general-purpose drawing that combines characteristics of two or more of the types discussed. More complex equipment usually comes with at least a wiring diagram and a ladder-type schematic. This is a good combination, because the two complement one another. The wiring diagram aids in component location and identification, and shows actual wiring layout, while the schematic shows the electrical operation.

In wiring diagrams, some attempt is usually made to distinguish wiring at the power circuit voltage from wiring that is at the control voltage. Wiring that must be installed in the field is also marked separately from internal equipment wiring. When different colors or wire numbers are used in the equipment, these markings are also carried over to the wiring diagram. The different voltages and the functions of the various circuits are generally not grouped on the drawing, because it is impractical to do so.

Schematics may show power or control components only, or both control and power circuits combined. The components in the drawing are usually grouped by purpose and voltage, and then arranged with the operating sequence as the guiding principle. Actual situations vary enough that many different layouts are possible. One result of the "operation before location" philosophy of schematics is that the coils and contacts of relays very often end up in different sections of the drawing. Small schematics simply identify the parts with a consistent system of letters and numbers; larger drawings include line numbers and cross-references for coils and contacts.

Industrial schematics for large electrical systems that are built on-site often include more detail than equipment drawings. Wire numbers and terminal identification are usually carried over from the wiring diagram. In large industrial settings, experienced service technicians rarely look at the wiring diagrams. All the information they normally require is on the schematics.

Reading schematics requires more skill and knowledge than other drawings because of the greater use of symbols and the code-like component identification. However, once these hurdles are cleared, schematics are far superior to all other drawing types as a tool to understanding control logic and the sequence of operation. Unfortunately, not all equipment comes with good schematic drawings. A knowledgeable service person should be able to produce a well-organized schematic from a wiring diagram or pictorial.

Although several attempts have been made to standardize drawing layouts and symbols, none are universally accepted. Some manufacturers follow national or international standards; others have adopted internal "standards" unique to their equipment. Standards also change over time, so layouts and symbols often vary with the age of the equipment. Service personnel who work on only one make of equipment quickly become familiar with the drawings and drawing practices of that manufacturer. When they are confronted by equipment from other manufacturers, some adjustments are necessary. Imported equipment may be accompanied by drawings that are crudely translated from some other language and drawing standard. Even the basic symbols can vary.

No matter what type of equipment, or the drawings supplied, **understanding what the equipment is supposed to do is critical to understanding how the control system works.** Unfamiliar systems always take time.

12.4 Motor Control Examples

Control Voltages

Motors usually operate at the highest practical voltage available in the facility. For motors operating at 600 V or less, the control systems can use the same voltage, but there are many advantages to using a lower one. A low-voltage control system provides isolation from the power circuit it controls. This protects the control system components and reduces the risks to operators and maintenance personnel. The most common approach is to provide control power from a relatively small step-down transformer that is connected to the load side of the motor overcurrent device. With this connection, both the control circuit and the power circuit are de-energized when the motor disconnect is opened.

Two of the most common control voltages are 120 and 24 V AC. A 120 V control system is generally more reliable in an industrial setting. Relay and starter coils are often relatively large, demanding significant power to operate, and control circuits also tend to be very long, increasing problems with voltage drop in the conductors. Many industrial environments also contain dust and chemical contaminates that can coat and corrode the electrical contacts inside the field devices. A higher control voltage system is more likely to operate reliably despite minor contact contamination. A 120 V control system is also much more expensive to install than a 24 V system, because the installation must meet more restrictive code requirements.

Twenty-four volt controls are preferred for systems—such as HVAC equipment—that are serviced by "nonelectrical" service personnel. The safer 24 V system is simpler and cheaper to install. With a current-limited supply—typical of these systems—the wiring codes are much less restrictive. The control components are generally simpler and do not generally require box mountings to enclose the connections. A 24 V control is reasonably reliable, particularly where the operating coils are relatively small, the distances between the devices are not extreme, and the contacts are not subject to excessive contamination.

Control-Related Tasks

Motors are useful only when they drive something, and in most cases some type of automation is involved. The controls must respond to a variety of physical conditions such as flow, temperature, pressure level, and the operating status of other motors. In addition to providing automation, the control system must also protect the motor from damage—and personnel from injury. Safety-related functions are a critical part of almost all control systems. Safety considerations include controls for unusual occurrences—emergency stops, automatic limits to temperature, travel, pressure, etc.—as well as preventative functions, such as equipment interlocks to prevent unsafe operation, and provisions for safety lockout of equipment that is shut off for maintenance.

There are many variations to motor control, particularly when multiple motors are used to drive equipment that is either mechanically interconnected or depends on the control system for correct sequence of operation. However, complex controls are built from simple circuits that perform specific tasks. Large control systems are usually collections of basic circuits that provide similar functions to many different pieces of equipment.

Control Circuit Components

One of the main building blocks in any motor control system is the motor starter and its associated components. Starter circuits provide the means to connect and disconnect the motor from the power supply. The starter coil drops out—on low voltage—protecting the motor from low-voltage conditions, and the integrated overload device cuts power to the starter coil to protect the motor and its supply conductors from damage due to overload.

The control circuit examples presented below are typical of industrial installations. Larger HVAC systems often contain similar circuits for the main motors. Smaller systems such as packaged rooftop units may also have similar circuits, although the overloads are more likely to be mounted at the motor.

Standard motor starters consist of a contactor and an overload block. Three-phase types are three pole, with three power contacts and at least one normally open auxiliary contact. Some small to midsize contactors look like four-pole devices because the three main contacts and the one auxiliary contact are laid out side by side, as if all four were designed to be part of a power circuit (see Figure 12–8). Larger contactors have separate, much smaller auxiliary contact assemblies. Extra auxiliary contacts may be mounted on almost any contactor. Refer to Figure 12–2 for a photograph of a starter similar to the one in Figure 12–8. A wiring diagram for the same starter is shown in Figure 12–9.

The main contact terminals are labeled L1, L2, and L3 on the line side, and T1, T2, and T3 on the load (motor) side. The auxiliary contact is usually labeled 2 and 3 for

FIGURE 12–8 The top view of a "typical" small motor starter with standard terminal identification for the line and motor connections, and the auxiliary contact.

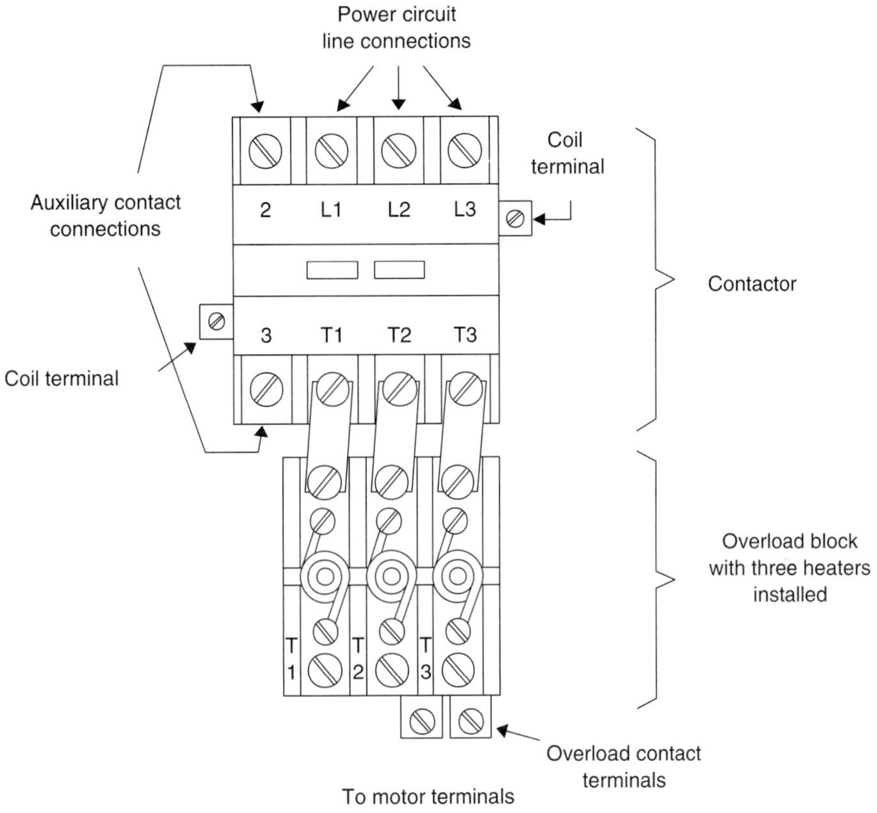

NEMA equipment; 13 and 14 is common on IEC equipment. The coil terminals on NEMA starters may or may not be labeled. A1 and A2 are standard IEC coil markings. When the overload block is connected directly to the contactor, the T1, T2, and T3 terminals are on the load side of the overload. When they are mounted separately, short cables are used to connect the contactor to the overloads. In both cases the motor leads are wired to T1, T2, and T3 on the load side of the overload device.

Contactor coils of NEMA-style starters are often factory connected, in series, with the overload contacts between terminal 3 of the auxiliary contact and the line side power terminal marked L2. This is the normal connection when the coil is rated for line voltage. A line voltage control system is powered directly from L1 and L2. When the control voltage is different than the line, the L2 connection must be removed, and that end of the coil connected to the neutral (grounded side) of the control transformer. This places the overload contacts in series on the grounded side of the control coil. This is against conventional wisdom about connecting controls on the grounded side of the circuit; however, the connection is extremely short, so the chance of a ground fault is virtually zero. It should be noted that IEC control circuits place the overload contact ahead of all the other controls, instead of after the coil as is done in NEMA control circuits.

FIGURE 12–9 This is a wiring diagram for the motor starter shown in Figure 12–8. Note the symbols used for the contacts, coil, and overload heater. The overload contact has a slash to indicate it is normally closed.

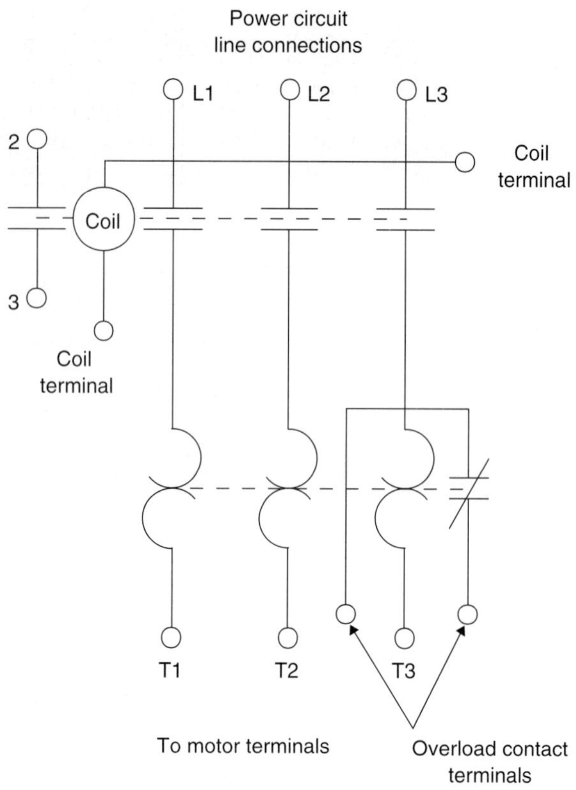

In every case, the voltage rating of the coil must match the control voltage that is going to be used. If the coil rating is incorrect, the coil will either burn out (the voltage applied is too high) or the contactor won't pick up (the voltage applied is too low). Coil voltage is marked on the coil itself, not on the contactor nameplate, and is usually visible without taking the contactor apart. Coils are available for all standard control voltages, including common line voltages. Although it is somewhat simpler to use line voltage controls, it is also more dangerous, and can complicate the installation of certain components such as indication lights.

Two-Wire Control

Two-wire control is commonly used with maintained-contact pilot devices that sense physical changes such as temperature, pressure, or liquid level. When the pilot contact closes, the contactor picks up to start the motor; when it opens, the motor stops. The pilot device is used as a simple SPST switch that automatically turns the coil circuit on and off (see Figure 12–10).

Two-wire controls provide low-voltage release. The contactor coil drops out if the voltage goes too low, but picks up again as soon as the voltage returns to normal. Extended periods of low voltage can burn out the coil if it isn't taken out of the circuit. The control circuit is supplied from only two lines; therefore, the third wire of a three-phase system is not monitored for low voltage.

FIGURE 12–10 In this control system, a maintained-contact thermostat (or similar pilot device) controls the coil of the motor starter. When the control circuit is closed, the coil pulls in the main contacts to start the motor. The overload contacts act to open the control circuit when an overload occurs. The auxiliary contacts (2–3 in the schematic) are not used.

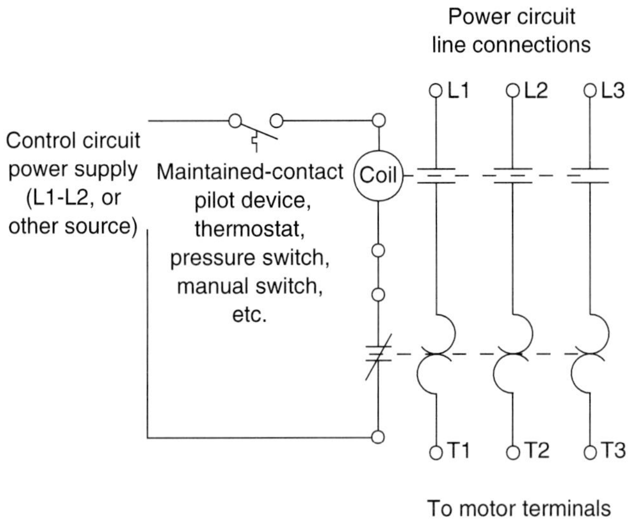

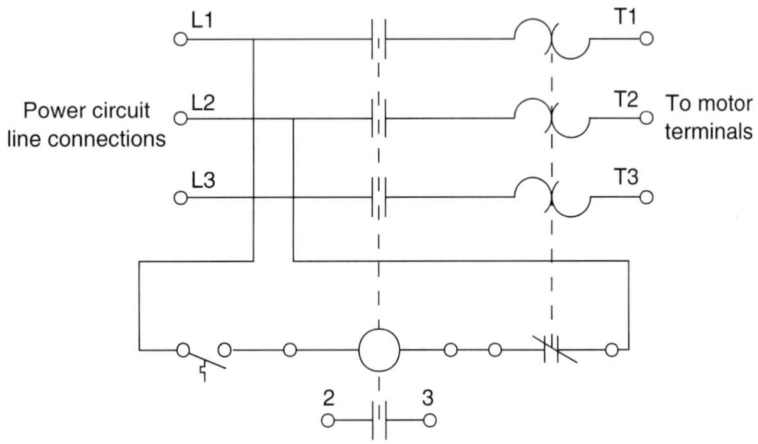

On overload, the normally closed overload contacts open to remove coil power and shut off the motor. An overload reset typically triggers an immediate restart, because the pilot contacts are usually still closed. Auto-reset is an option on many bimetallic overloads. Using it can be convenient, potentially destructive, or extremely dangerous, depending on the situation. With auto-reset overloads, the motor restarts as soon as the bimetallic overload element cools down. This feature should be used with extreme caution.

FIGURE 12–11 A typical hand-off-auto control provides local manual control, plus automatic operation through a two-wire maintained-contact pilot device. This type of control is common on float-controlled pump circuits, thermostat-controlled fans, etc.

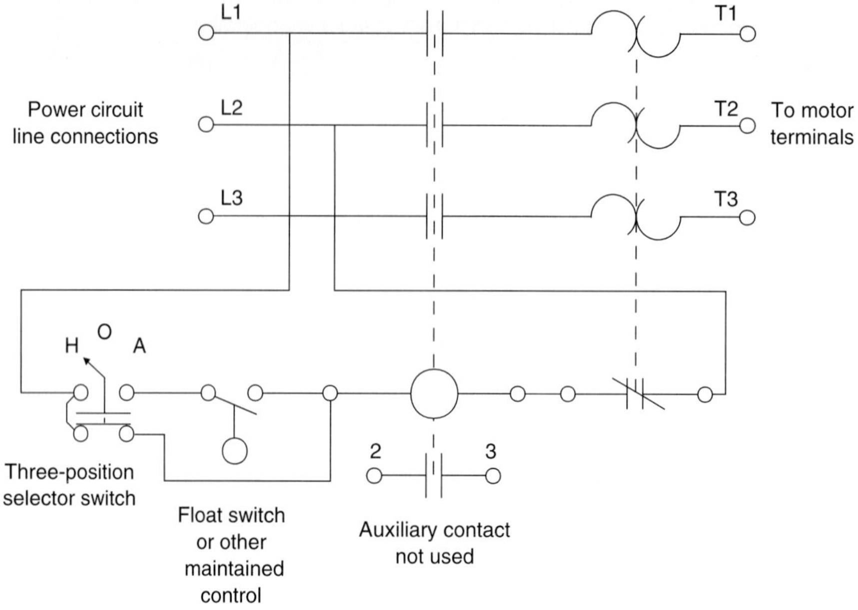

Hand-Off-Auto Control

A common application of two-wire control is the hand-off-auto system used with sump pumps, air compressors, and similar equipment (see Figure 12–11). Like other two-wire controls, it is normally used with a SPST maintained-contact pilot device. A three-position selector switch provides a manual or "hand" mode and an "off" position, in addition to the "auto" selection used for normal automatic operation.

In the "off" position the selector disconnects all control power, shutting the system down. In the "auto" position the control circuit is routed through the pilot device to provide normal, automatic operation. In the "hand" position the pilot device is bypassed, turning the system on manually. In this position the motor is energized continuously. Hand-off-auto circuits provide a simple but effective control for many applications.

Three-Wire Control

Typical three-wire systems have separate start and stop buttons, which are usually mounted together in a start/stop station—an enclosure containing the start and stop pushbuttons and, possibly, additional controls or indication lamps. A green button is traditionally used for start, and a red one for stop. Both switches are momentary-contact types, and each may be equipped with multiple contacts. The contact that performs the actual start function must be normally open; for the stop function, the contact must be normally closed.

FIGURE 12–12 A typical three-wire control system uses momentary-contact manual pushbuttons to start and stop the motor. Once a start is initiated with the start button, the control circuit is sealed in by the auxiliary contact.

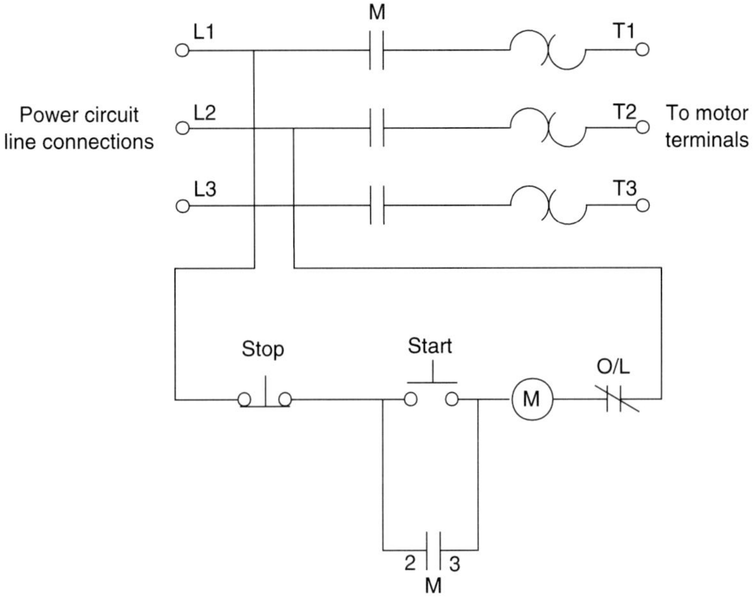

The normally open auxiliary contact on the starter is used as a seal-in or holding contact and is wired in parallel with the start switch. Pressing the start button completes the coil circuit initially, but this contact opens as soon as the button is released. The coil does not drop out, because the circuit is maintained through the seal-in contact. Because the coil circuit is maintained by one of its own contacts, this system provides low-voltage protection. If the circuit is broken or power is lost, even very briefly, the coil drops out. The seal-in contact is then open and the motor will not restart until the start button is pressed again. This circuit is illustrated in Figure 12–12.

Low-voltage protection can be very important. It protects against sudden restarts, which could be dangerous to equipment and personnel, and prevents simultaneous restart of many motors. In an operating plant, a momentary loss of power stops everything. If the control circuits provide only low-voltage release, when power is restored all the motors may attempt to restart at the same time. In most cases this would be a disastrous overload, probably tripping the main supply.

When a motor must be controlled from several locations, the three-wire control system can be extended as required. The control switches are all momentary contact, so additional controls can be added with very little complication to the circuitry. All the stop switches have normally closed contacts and are wired in series. All the start switch contacts have normally open contacts and are wired in parallel. Any one of the stop switches can break the circuit to shut off the motor; any start switch contact can momentarily energize the coil to start the motor (see Figure 12–13).

FIGURE 12–13 This system provides stop/start control from two locations. Pushbuttons Start 2 and Stop 2 would be mounted in a separate enclosure, some distance from the other control station. Any reasonable number of control locations can be added in the same way.

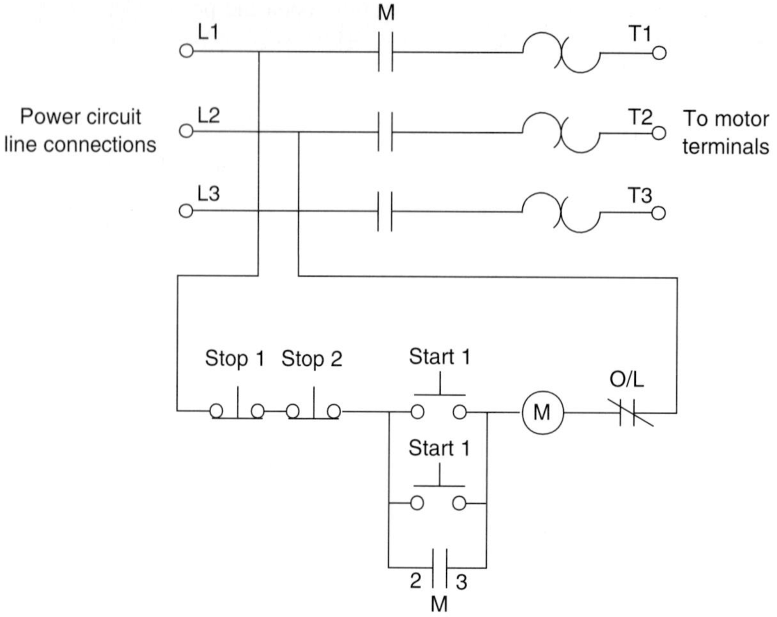

12.5 Reduced-voltage Starting

Virtually all three-phase motors are designed for full-voltage "across the line" starting, and this method is the most widely used. However, with larger motors, some type of reduced-voltage starter may be required. Reducing the motor voltage on startup may be necessary to meet utility requirements or to protect the driven load from the mechanical shock of a full-torque start.

A large motor may draw enough current during startup to cause a noticeable dip in the voltage in the building and the surrounding area. For this reason, the utility may require some type of current-reducing starter to lessen the disturbance to their system. Reducing the voltage also reduces the motor starting torque. While this can be a problem in some cases, in many situations the reduced start torque is desirable. It lessens startup problems with loads that are sensitive to jerking (such as materials handling equipment) or have a tendency to surge (such as loads suspended on long cables). It can also increase the service life of other types of loads by reducing the mechanical stress on the equipment.

Note

The term "reduced-voltage starting" is commonly used to cover various methods of reducing motor starting currents and/or the torque. Not all of these methods actually reduce motor voltage, which can lead to some confusion in terminology.

Open- and Closed-Transition Starters

All of the reduced-voltage-type starters have two or more stages of operation—in most cases a start mode followed by a normal run mode. The transition from one mode to the other may be open or closed. Open-transition starters open the power circuit temporarily when they change from start to run. Closed-transition starters keep the motor connected to the line in some way at all times, even during this transition. A closed-transition starter causes much less electrical disturbance than an open-transition type, because opening and closing the motor circuit during the start period can produce significant dips and spikes in the line voltage.

Some types of starters are inherently closed-transition, and some can be built both ways. Closed-transition versions of starters that can also be made for an open transition are invariably more complex and expensive than the open-transition models. Inherently, closed-transition starters tend to be the simplest and the least expensive, but may also provide the lowest levels of performance.

Auxiliary Timers

There are many variations and control arrangements possible for each type of starter, but a common requirement is a timer to signal the transition from start to run. The timer may be a separate device, or it may be some type of auxiliary on the main contactor. Auxiliary timers are available as an add-on for many makes of contactors. They are usually pneumatic types, using an internal bellows or "dash pot" (piston and cylinder) arrangement to create the delay. The bellows or dash pot is spring loaded. When the contactor picks up, the spring forces the air from the bellows through a small port. The timed contacts operate just at the end of the spring travel, opening or closing as required. The time-delay is set with a screw or knob that adjusts the flow rate through the port.

Special Contacts

In addition to the usual normally open and normally closed contacts, some starters also require contacts with special operating characteristics. For example, autotransformer starters often use a "late break" contact. On pickup the late-break contact operates much like a standard normally open contact, but on dropout it stays closed longer, overlapping the transition of standard contacts.

Other examples of special contacts include "early make" and "fleeting." Early-make contacts are self-explanatory. Fleeting contacts close only briefly, while the contactor or relay armature is still in motion. They are used to supply a single pulse when the contactor opens or closes.

Reduced-voltage Starter Examples

There are several methods used to reduce motor starting current and torque during the startup period. Primary-resistor starters connect the motor through a set of series resistors. A reactor starter uses series-connected inductors—called reactors in this type of application—in place of the resistors. (Reactor starters are rarely used on anything except high-voltage or very large horsepower motors, so no examples are provided here.) Wye-delta (or star-delta) starters only work with six- or twelve-lead delta-connected motors. The motor windings are connected in wye for starting purposes, and then shifted to delta for normal running. Autotransformer starters use a tapped autotransformer to reduce the motor voltage.

The part-winding starter is generally lumped in with these other starters because it accomplishes much the same thing, although it does not actually reduce the voltage to the motor. It can only be used with dual-winding motors—such as dual-voltage (230/460 V) models—operated from the lower of the two voltages. Note, however, that the motor has to be approved for use with a part-winding starter: Not all dual-voltage motors are suitable. Part-winding starters have the advantage of simplicity and low cost. The disadvantages include low starting torque and a danger of overheating the winding that is energized first.

Autotransformer starters provide the highest torque per ampere drawn from the line. They also have the advantage of being adjustable through several selectable taps. Primary-resistor starters provide a smoother transition from start to run, because the amount of voltage reduction varies with current through the series element, and this drops as the motor accelerates. Primary-resistor starters may also be adjustable. Wye-delta starters have the appeal of no extra voltage reducing equipment. They have a low starting torque, are not adjustable, and also require a special motor, so they are usually supplied as part of a system.

Part-winding starters are usually the least expensive. Primary-resistor starters are also low in cost, but only in small sizes. For larger sizes, the costs of the various types of starters are usually quite similar.

Primary-Resistor Starters. A primary-resistor starter has high-current resistors connected directly between the motor and the supply. The main contactor starts the motor through the resistors; a second contactor shunts the resistors out of the circuit after a preset time-delay. There are numerous variations including single-stage, multistage, and carbon-pile types, which can be varied by mechanical compression of the carbon. Liquid resistor designs use containers filled with a conductive solution that acts as the resistance element.

The series resistance reduces the voltage applied to the motor and, therefore, the motor current and starting torque. The resistors are thermally protected, because if the timer circuit failed to shunt them out, they would quickly overheat.

Primary-resistor starters use two full-sized contactors, one set of overloads, a set of resistors, and a timer (see Figure 12–14). Thermal cutouts near the resistors provide the overheating protection. The starter cabinet must be large enough to accommodate the equipment and provide sufficient circulation for cooling the resistors. Primary-resistor starters are inherently closed-transition.

Wye-Delta Starters. A wye-delta starter requires a 6- or 12-lead motor designed to run with the windings connected in delta. The controller connects the motor in wye for start-up. This limits the applied voltage to about 58% of the line value, and drops the starting torque to about 33% of the normal full voltage value. After a preset time-delay, the motor connection is changed to delta for normal operation. Both closed- and open-transition types are available.

The combinations of contactors used varies with the manufacturer and the starter design. An open-transition starter must have a minimum of five poles in the start mode, and six in the run. Common examples use two three-pole contactors for the run mode, plus

FIGURE 12–14 This primary-resistor starter uses a timed auxiliary contact on the main contactor (M), which energizes the coil of the second contactor (S) after an appropriate delay.

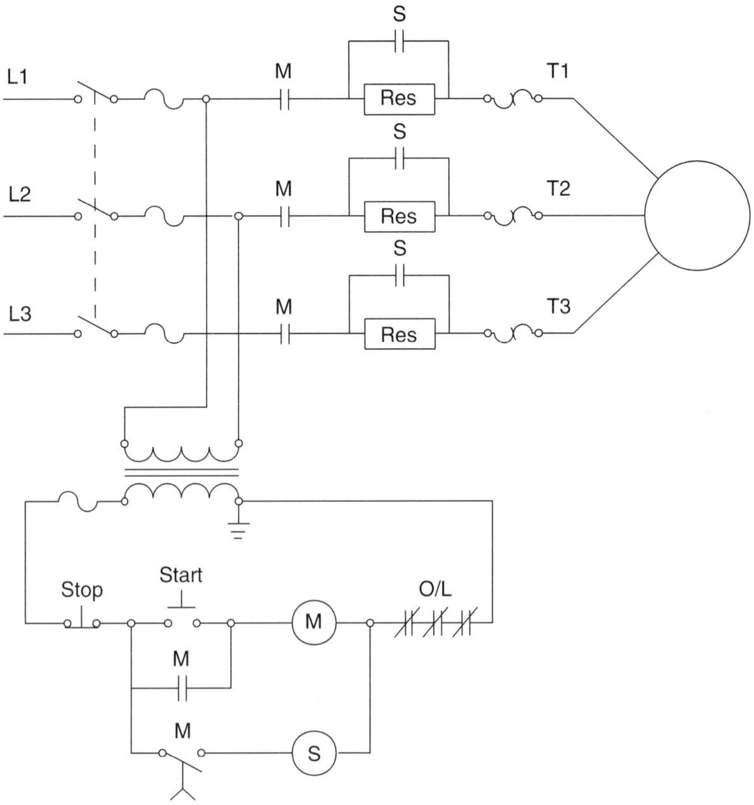

either a two-pole or another three-pole to connect the star point during starting. The star-point contactor is mechanically interlocked with the second run mode contactor to prevent the possibility of a line-to-line short. A timer is used to control the transition from start to run.

The closed-transition version of this starter has current-limiting resistors and an additional contactor. The wye-delta starter usually has only one set of overloads, although separate start and run overloads may also be used. Figure 12–15 shows an open-transition wye-delta starter.

Autotransformer Starters. The autotransformer starter is the only reduced-voltage starter that can provide more current to the motor than is drawn from the line. A transformer that steps down the voltage also steps up the current. There are typically three taps on the transformer (80%, 65%, and 50%) to allow some choice in the degree of voltage and torque reduction. Both open- and closed-transition designs are available. The transformer used is typically a core type that is surprisingly small for the current it delivers. It is an autotransformer—there is no separate primary and secondary, so there is one winding on each core leg.

FIGURE 12–15 This open-transition wye-delta starter has three contactors. Two of them (S and 2M) are mechanically interlocked to prevent a line-to-line short. The start sequence is S, followed almost immediately by 1M to connect the motor in wye; after the delay time, S drops out and 2M picks up to connect the motor in delta.

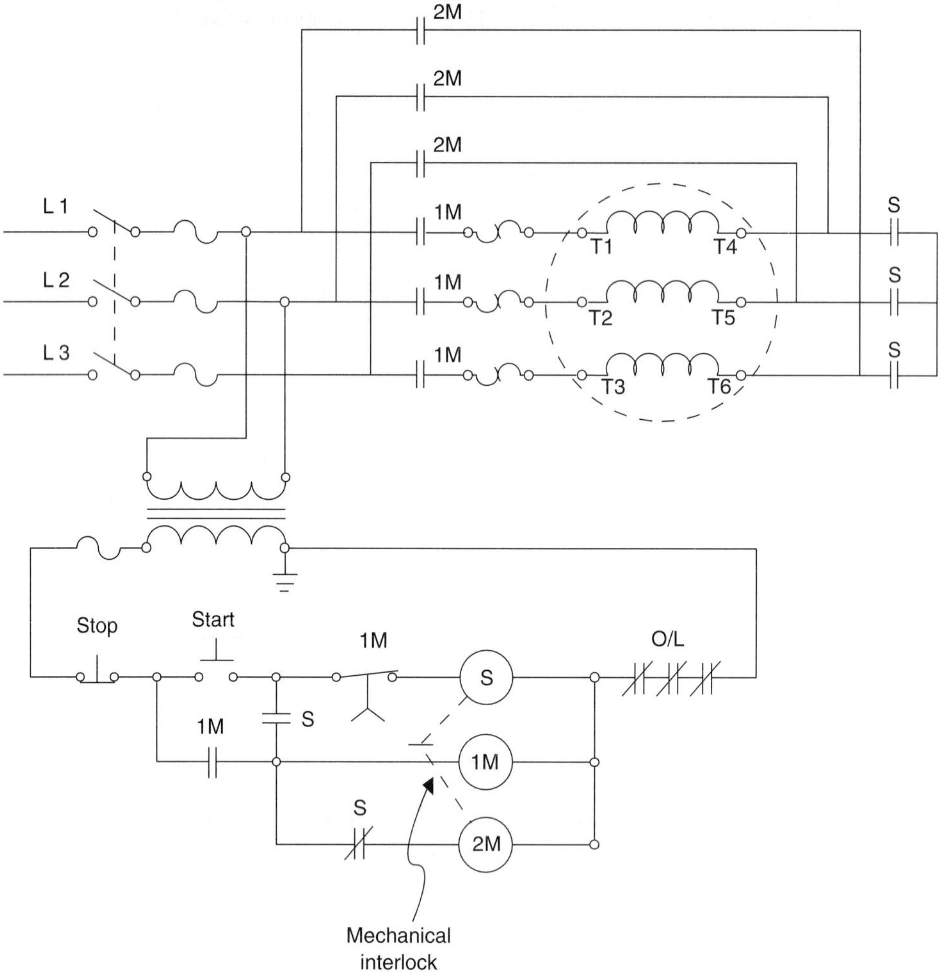

An open-transition version of this starter needs a minimum of five poles for start mode, and three for run. These may be provided with only two contactors, if one has five poles. Some manufacturers use three contactors, using standard two- and three-pole contactors in place of the less common five-pole type. Closed-transition starters require three contactors. They can all be three-pole, or one—the star-point contactor—can get by with only two. The star-point contactor must be mechanically interlocked with the second-stage run contactor to prevent the possibility of a line-to-line short.

This starter also uses a timer to control the transition. Closed-transition connections require a special late-break auxiliary contact to ensure a proper overlap of the run

FIGURE 12–16 This autotransformer starter uses a late-break contact to provide a slight overlap of 2S and R during transition. The operating sequence is 1S then 2S, to put the motor across the transformer secondary. R is kept off by the normally closed 1S contact. After the timer times out, 1S drops to open the star point. Then R picks up to shunt out the remaining transformer winding and put the motor across the line. The late-break contact drops 2S just an instant after R closes to take the transformer out of the circuit. During the transition, the motor is briefly connected in series with part of the transformer winding, until R closes. This is not a problem because the star-point is already open.

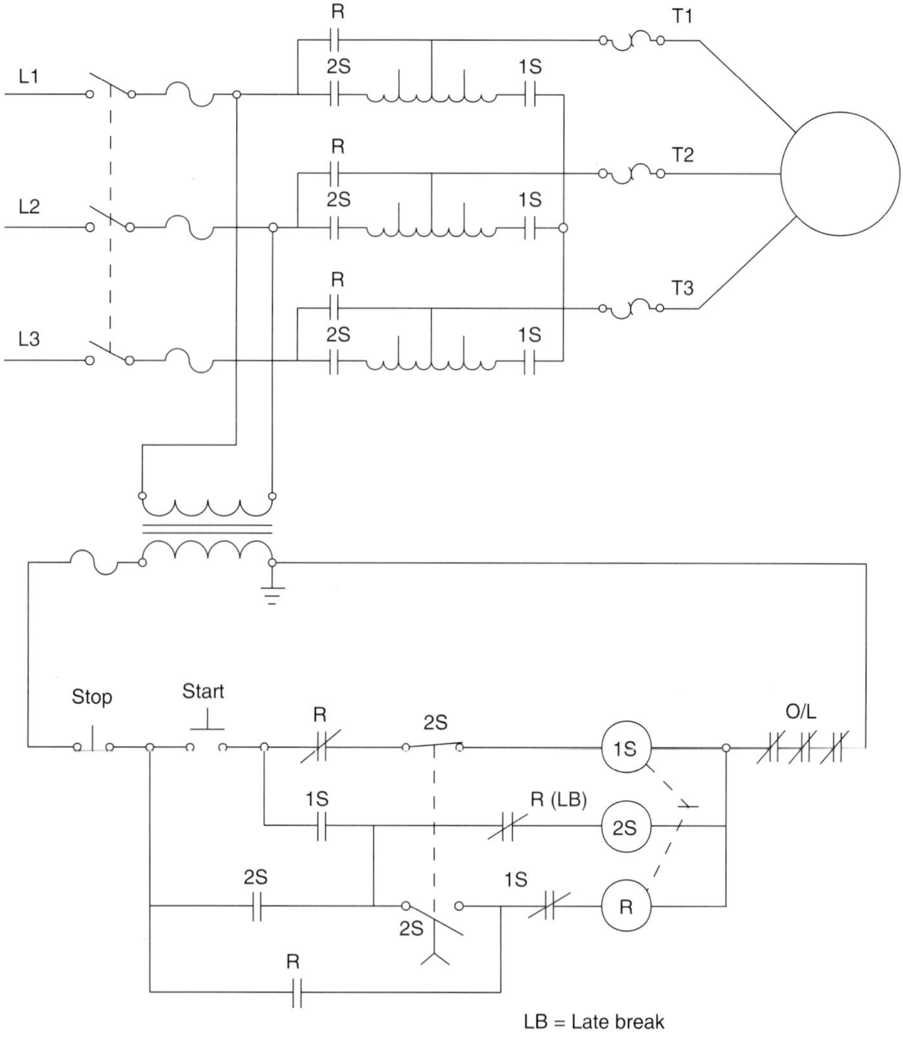

and start contactors. The drawing in Figure 12–16 shows a closed-transition autotransformer starter.

The drawing in Figure 12–17 is an example of an open-transition starter. It has one three-pole and one five-pole contactor. The transformer may be simpler and cheaper, using only two coils instead of three.

FIGURE 12–17 This open transition autotransformer starter is somewhat simpler and cheaper than the closed transition version shown in Figure 12–16.

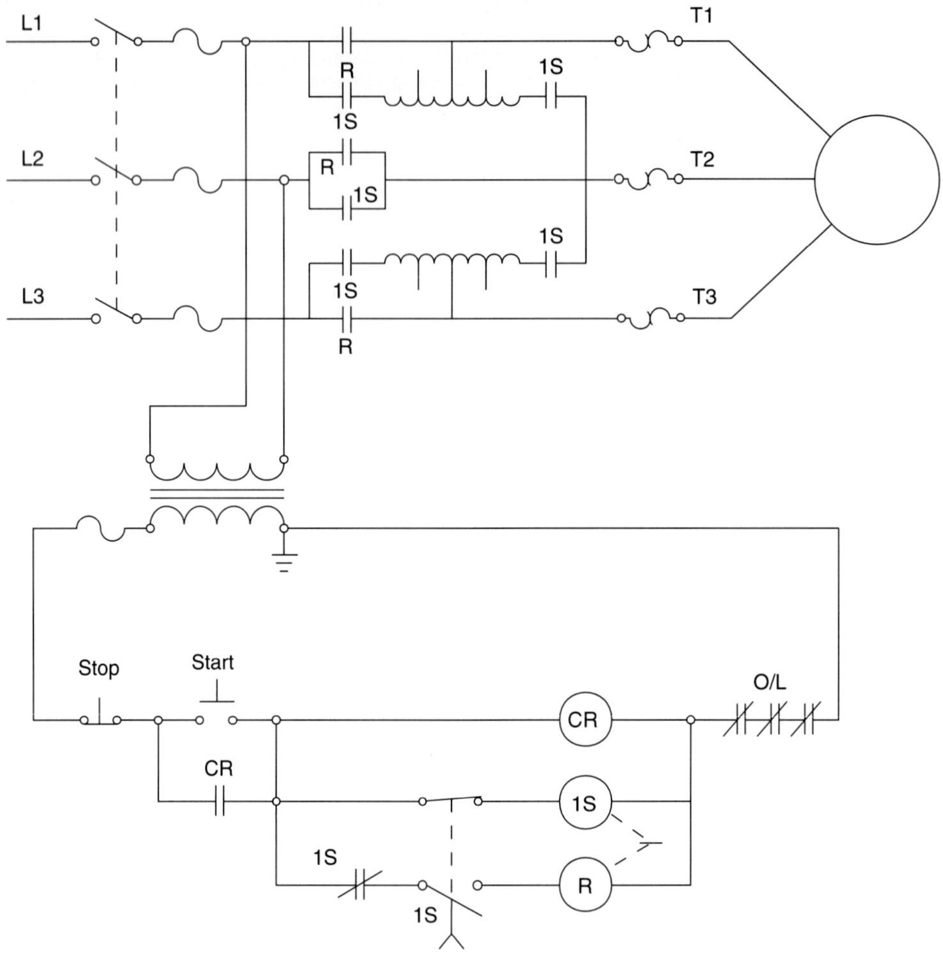

Part-Winding Starters. Part-winding starters are not really reduced-voltage starters, but are used to solve the same problem. The motor used is usually a dual-voltage type, and must be approved for part-winding start. The supply line voltage must match the motor's lowest voltage rating, which is the voltage rating of one winding set—not two in series. Part-winding starters use two half-size contactors, two sets of half-rated overloads, and a timer (see Figure 12–18). This starter is inherently closed-transition.

Note

It is not uncommon for the timer in a part-winding starter to be bypassed by frustrated field service personnel. This essentially converts the starter to a full-voltage, across the line start. This action is usually prompted by starting problems. The motor torque with only one winding may be too low to start the load. This can result in nuisance trips from motor overload or overheating of the motor. Bypassing

FIGURE 12–18 The part-winding starter is a relatively simple system that energizes one set of motor windings at a time.

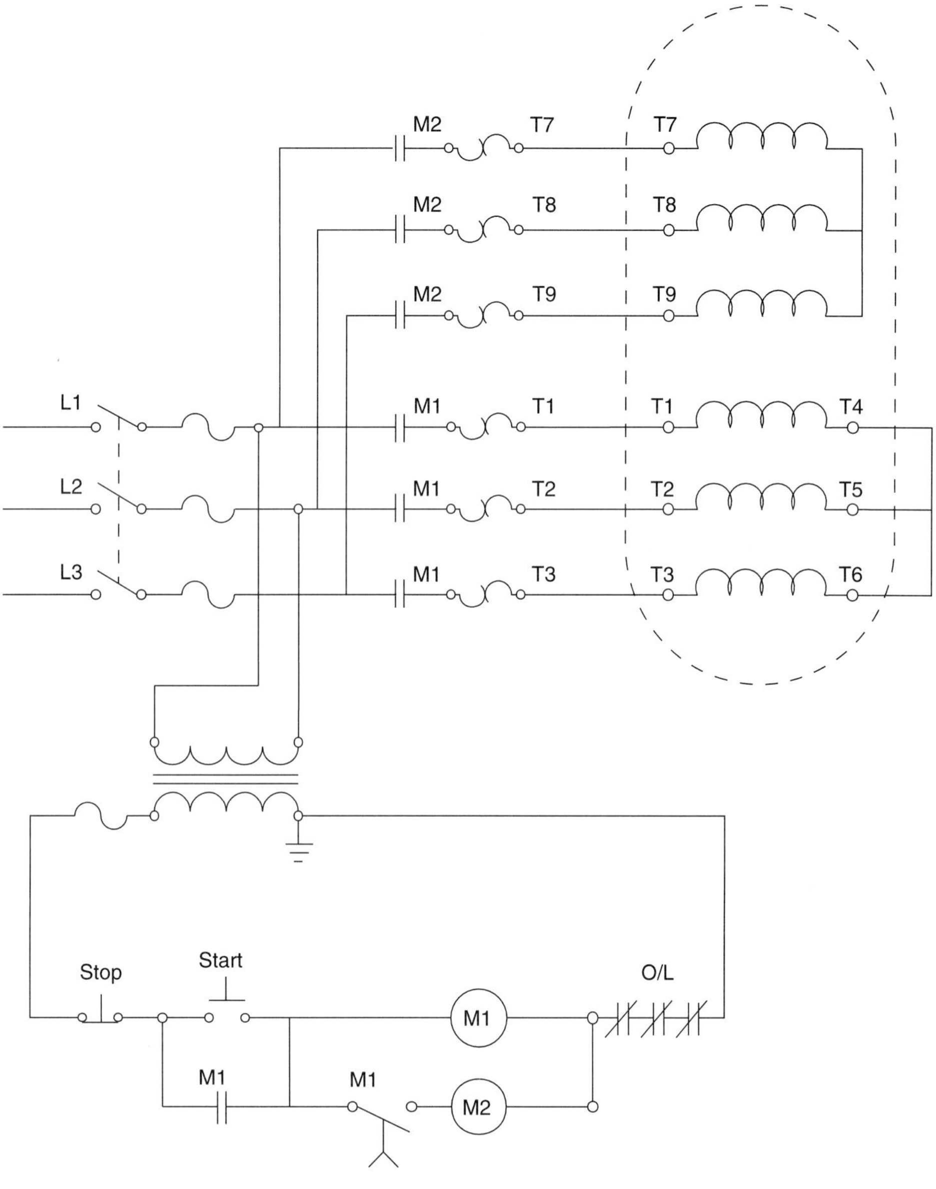

the timer defeats the original purpose of the starter and may be in violation of agreements with the supply utility.

Solid-State Starters. Solid-state starters use solid-state switching devices and, in most cases, a technique called "phase control" to apply an adjustable voltage to the motor. The switching devices are typically silicon-controlled rectifiers (SCRs) installed in series in each motor supply line (see Figure 12–19).

FIGURE 12–19 An SCR soft starter, shown here in block diagram form, uses a technique called phase control to send only a portion of the AC sine wave to the motor. The voltage is "ramped up" by adjusting the firing point of the SCRs. This provides a smooth startup that can be adjusted to suit the application.

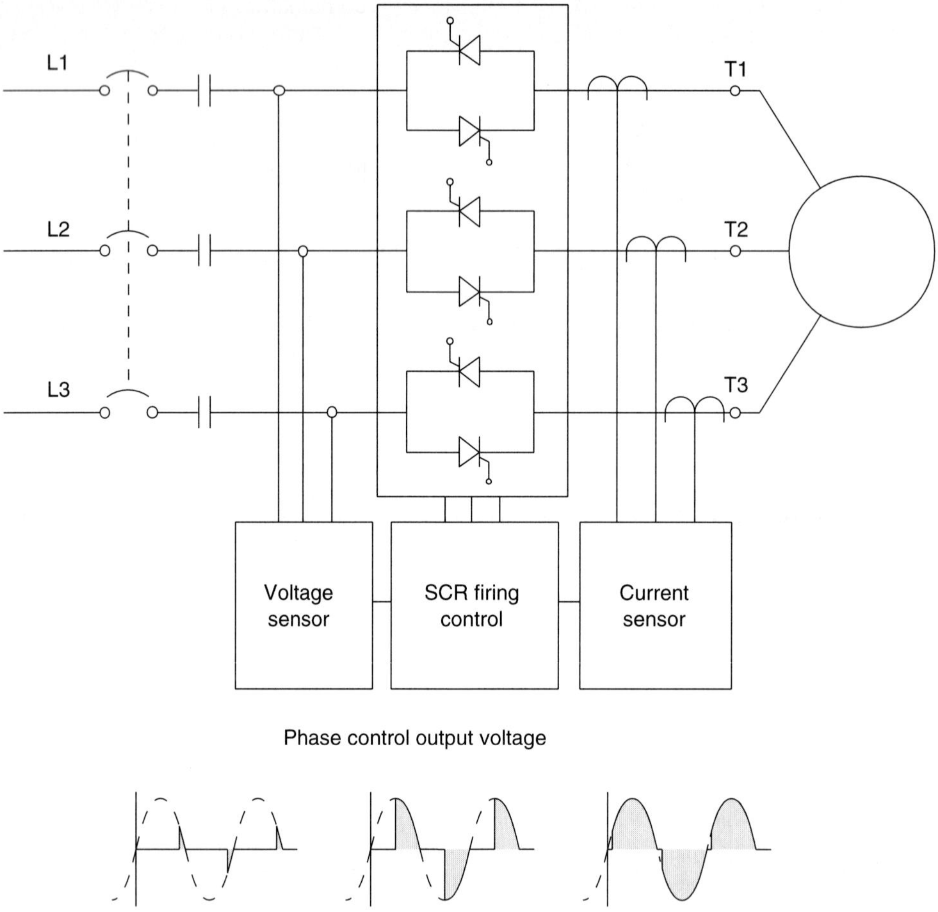

An SCR turns off—acts as an open switch—each time the current passes through zero. This occurs naturally at the end of each half of the AC cycle. The SCR then stays off until a pulse is applied to its gate lead to turn it on—make it conductive again. Every half-cycle, the SCRs have to be triggered back on again for the current to flow to the motor. If they are turned on at the very beginning of the next half-cycle, the motor will receive full voltage. If the turn on is delayed, the voltage applied to the motor is reduced. The time-delay is not long—a half-cycle takes only about 8.3 milliseconds—but accurate control of such short time-delays is neither difficult nor expensive. For a gradual start, the turn-on delay starts at some preset minimum—say 6 ms—and then it is gradually shortened to zero. This gives a smoothly increasing voltage from the preset minimum up to full line value.

While conventional starters depend on simple timers to control the transition from starting to running, the electronic starter usually offers a number of more sophisticated options. The rate at which the voltage to the motor is increased may be controlled by time, current, motor shaft speed, or some combination of all three. A time-based system increases the voltage at a specific rate that is field adjustable to match the application. This is often combined with current limiting to prevent the motor current from becoming too high when load conditions are abnormal. For example, a conveyor drive set for starting an empty conveyor would probably go into current limit if it were started with the conveyor loaded. The current-limit feature would act to increase the motor's startup time, compensating for the additional load.

The rate at which the voltage is increased can also be based on actual motor speed, if the motor is equipped with a feedback tachometer. Such tachometers are available as options on some motors, or may be field mounted on the motor or the drive. The tachometer is usually a very small generator with a voltage output that is directly proportional to the shaft speed. It is wired back to the controller, where its voltage signal is used as a speed reference. The rate of acceleration may also be combined with the current-limit.

Reduced-voltage electronic starters are often called soft starters. They are very compact, cost competitive, and offer features, such as the current-limit control described above, that are not available for any other type of starter.

Some solid-state starters have a bypass contactor to connect the motor directly to the line at the end of the startup period. Undervoltage tripping may be provided by a shunt trip coil in the breaker or through a separate contactor.

Questions

1. Compare, in general terms, NEMA and IEC rating systems for contactors.
2. Explain why the two contactors used in reversing starters are mechanically interlocked.
3. Describe three methods of providing brakes on electric motors.
4. Explain why an instantaneous fuse or circuit breaker could not be used to power many common electrical loads.
5. Explain the difference between "overcurrent protection" and "overload protection" as these terms are applied to motors.
6. List the advantages and disadvantages of using circuit breakers in place of fuses as motor overcurrent devices.
7. Motor circuit conductors are selected according to the maximum current rating of the (overcurrent or overload) device. (Choose one.)
8. List two methods of providing overheating protection for a motor.
9. Give an example of an interlock.
10. Give an example of a system that uses a specific control sequence.
11. Explain the differences between interlocks and limits.
12. Give two reasons why the motor control system should disconnect the motor when the supply voltage is low or shut off.
13. Explain the difference between undervoltage protection and undervoltage release.

14. Describe a situation where it may be impossible to reset a perfectly good circuit breaker.
15. Explain how a melting-alloy overload trips the motor off the line when the control device is a magnetic starter.
16. What precautions should be taken when installing overload heaters?
17. On a typical ladder-style drawing, what are the number references on the right-hand side?
18. In a vertical ladder diagram, the circuit common should be on the (left or right) side of the drawing. (Choose one.)
19. Explain how to use the line number references on a large ladder schematic to locate the contacts of a specific coil.
20. What are the advantages of a low-voltage control system?
21. Why are 120 V controls preferred to 24 V in most industrial environments?
22. What are some of the advantages of 24 V control systems over systems operating at higher voltages?
23. Describe the operation of a two-wire control circuit. Explain why this system does not provide low-voltage release.
24. Describe the operation of a three-wire control system. Explain how this system provides low-voltage protection.
25. Why are three-wire control circuits unsuited to maintained-contact controls such as thermostats, pressure switches, and floats?
26. Why are reduced-voltage starters used on some motor installations?
27. Describe the operating principle of a pneumatic timer.
28. Explain the difference between "open-transition" and "closed-transition" as these terms apply to reduced-voltage starters.
29. Explain the operation of a late-break contact in comparison with a normal contact.
30. List the operating steps of a closed-transition autotransformer starter from the instant the start button is pressed until the motor is running normally with full voltage applied. The steps should include each contact operation in both the control and power circuits, listed in order of occurrence.

13 DC Motors

Objectives

After completing the study of this chapter you should be able to:

- Explain the operation of a conventional brush-type DC motor
- List the advantages and disadvantages of conventional DC motors compared to their AC counterparts
- Explain how AC supplies are used to supply power to DC motors
- Describe the basic operation of electronically commutated motors, including brushless DC, switched reluctance, and stepper motor types

13.1 Introduction

DC motors are particularly useful in specific situations. These include applications that are based on DC—such as battery-powered tools and other portable equipment, automobiles, trucks, etc.—where batteries are an essential and central part of the electrical system. They are also popular for driving very heavy equipment—such as locomotives and giant mine haulage trucks—needing high torque, precise control, and smooth acceleration over a wide speed range. Only in very recent times has it become possible to build AC motor drives that are capable of powering such demanding loads. These drives combine specially constructed AC motors with complex, power electronic frequency controllers.

In general, a DC motor with a suitable controller is capable of full-load torque from zero to full-rated speed. This torque can be smoothly applied and fully controlled across the complete range of operating speeds. DC motors that have wound fields are also capable of full-rated horsepower output at well above their rated base speed. Small-diameter DC motors often have a top speed that is much higher than is possible with a comparable AC motor.

DC motors are available in an extremely wide range of sizes, either as bare motors suitable for direct drive, or with factory-fitted gear reducers (see Figure 13–1). Small ones are found in instruments, robots, electronic devices, tools, appliances, and toys. Monsters with up tp 10,000 Hp or more provide variable speed power to equipment such as steel rolling mills, mining hoists, locomotives, and mining shovels.

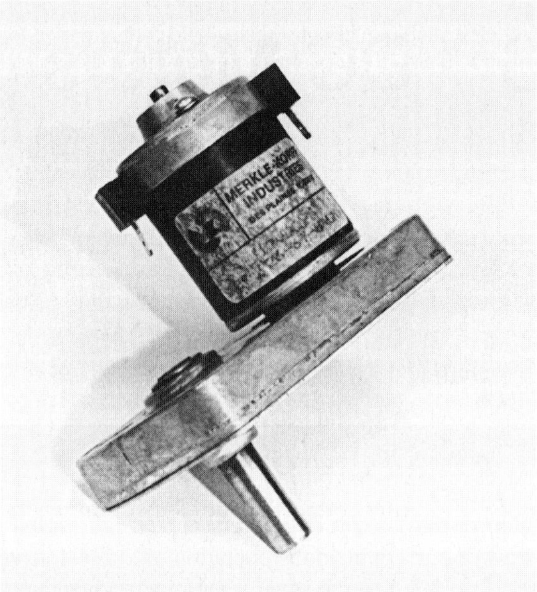

FIGURE 13–1 A small DC motor with gear reducer attached.

DC motors are generally more expensive than AC motors, and more demanding to maintain. However they also have some distinct advantages. It is difficult and often impractical to control the speed of a typical single- or three-phase AC motor. In comparison, the speed of a DC motor can be adjusted by regulating the supply voltage to the motor armature. With an appropriate controller, the DC motor can be speed controlled and reversed very smoothly, even under full load.

With a suitable electronic controller and a feedback device, a conventional DC motor can provide excellent servo performance. A servo motor must be able to accelerate, stop, and reverse very smoothly. It may have to operate for extended periods of time at any speed in its operating range, or constantly change speed and direction. This level of control is generally more difficult, and often impossible, to provide with any type of AC motor.

Such high performance is rarely needed in HVAC applications. However, small DC motors are often used to position dampers and valves. They tend to be more compact and more powerful than comparably sized AC motors. Very small, economy-grade DC motors are also cost competitive with comparable AC designs, despite their greater complexity. They are only rated for a few thousand hours, but for many applications, such as damper or valve drives, this can represent many years of service before they need replacement (see Figure 13–2).

13.2 DC Motor Construction

All DC motors must incorporate some method of switching the direction of current flow in parts of the motor winding. The motor will not keep turning unless appropriate changes are made to the magnetic polarity of either the rotating or the stationary portion of the motor. These changes must be synchronized to the position of the shaft, and

FIGURE 13–2 This damper positioner uses a small DC motor (cylinder in center left). A 9 V battery (just above the motor) supplies closing power on power failures.

occur repeatedly as it turns. Conventional DC motors use electromechanical commutators and brushes that are built into the motor. Electronically commutated motors use external electronic components to do the switching, and control them with a position sensor referenced to rotation of the motor shaft. (Electronically commutated motors are examined later in this chapter.)

In conventional DC designs, the main magnetic field is built into the stationary part of the motor. This is usually referred to as the motor *frame* rather than as the stator—the AC motor term. The frame holds the *field poles* and enough iron to provide a good magnetic path. The field poles may be permanent magnets or electromagnetic poles wound on iron pole pieces (see Figure 13–3). In either case, the field poles are positioned around the inside of the motor frame with alternating magnetic polarity. There must be at least two poles, with the number of poles increasing in multiples of two. Two-, four-, and six-pole motors are the most common, although much larger numbers are possible.

The rotating portion of a conventional DC motor is called the *armature.* It is the most complex part of a DC motor. A laminated iron core is pressed onto the motor shaft, and a commutator is mounted at one end. The armature windings are installed in the rotor slots, bound in place, and terminated on the commutator.

The *commutator* is usually drum-shaped and consists of a series of narrow, wedge-shaped copper bars mounted around the shaft. The commutator bars, or segments, must be insulated from each other and from the motor shaft. The insulation is usually thin sheets of mica sandwiched between the bars and in the dovetails used for securing them to the anchoring rings. Each of the many coils of the armature winding must connect to at least two commutator segments. On small motors, this connection is usually made by crimping or soldering the conductor into a small slot cut into the end of the bar.

The ends of the DC motor are usually called end bells or end shields, just as for an AC motor. They support the bearings on which the shaft rotates, as well as the brush rigging. On smaller motors, the brushes can often be accessed through removable insulating caps on the outside of the motor. Very small motors are not always designed for servicing; they have extra long brushes in holders that are not designed to allow brush replacements. On larger motors, the brushes must be accessed through inspection openings around the end of the frame or end shield. These may be left open or covered with solid plates or vented covers.

FIGURE 13–3 This DC motor frame has four wound field poles. The iron is laminated, but many DC motors use solid iron frames.

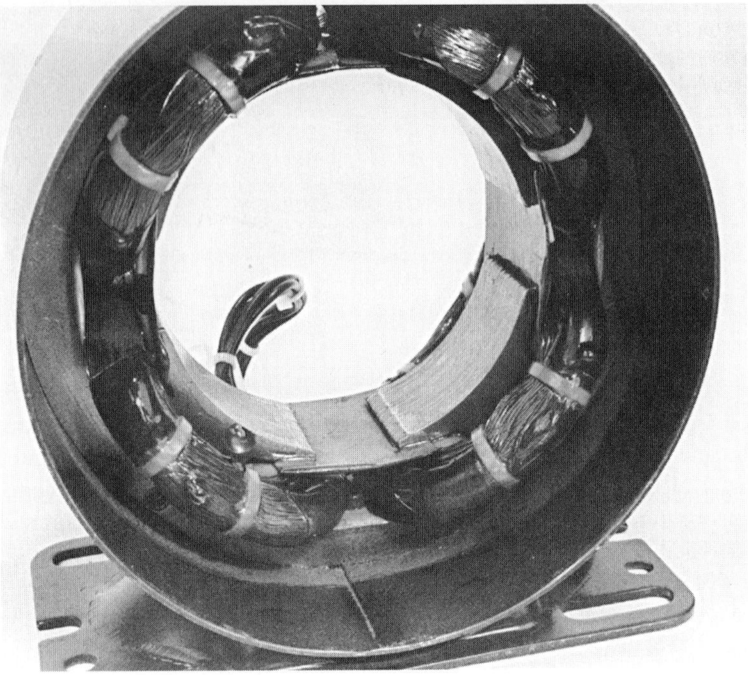

13.3 DC Motor Operation

DC motors produce torque by the interaction of two magnetic fields. One field is produced by the stationary field coils mounted in the motor frame, and the second is created in the armature by current directed into the armature windings via the brushes and commutator. The magnetic interaction of the two fields produces a force that causes the armature to turn. With no commutator, the armature would simply turn until the fields pulled into alignment and then stop.

With a commutator and brushes, rotation is constant. The moving commutator and stationary brushes act as a rotating switch that constantly changes the connections to the armature windings. These changes maintain the position of the armature field, regardless of its rotation. As a result, the stationary field and the field in the armature stay out of alignment, even as the armature turns (see Figure 13–4).

Reversing the connections to either the field or the armature will reverse the polarity of one of the two magnetic fields, also reversing the direction in which the motor rotates. In most cases, the armature connections are reversed, instead of the fields. If the motor has a permanent magnet field, there is no choice, because the field polarity is fixed. All larger motors have wound field poles. The majority of smaller motors have permanent magnet fields, although there are exceptions.

The smaller conventional DC motors found in typical HVAC applications have permanent magnet fields. Permanent magnet motors are more compact, slightly more efficient, and simpler and less expensive to produce. The magnets are usually molded

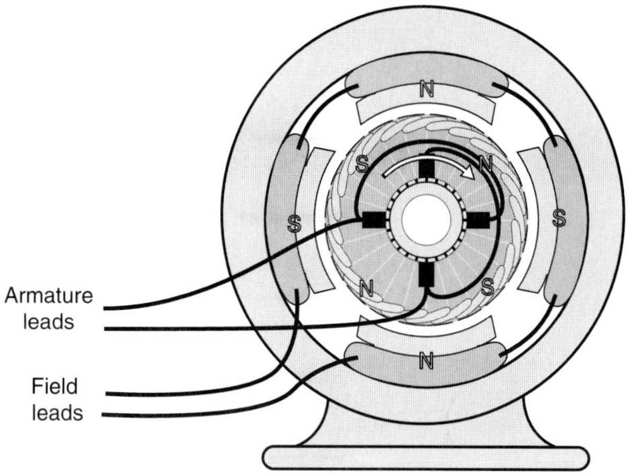

FIGURE 13–4 DC motors produce torque and rotation from the interaction of the main field and the field produced by the armature current. The two fields are kept out of alignment by the switching action of the brushes and commutator.

ceramic types. Sometimes they are made in one piece and magnetized after they are molded. This makes it impossible to see the separation between the fields. In spite of appearances, a one-piece magnet can have almost any number of poles. They can be counted with a small piece of iron, such as a screwdriver, or identified with a small magnet or compass.

DC motors can have starting-current problems that are even more severe than for AC motors. Small DC motors can often be started directly from the supply, but larger ones must have the starting current limited in some way. Special starters are made for this purpose, but they are rarely used. DC motors are almost always chosen because of their superior speed control. To utilize this feature, an electronic controller is used, and one of its features is control over starting current. Therefore, while starting a DC motor can be a serious problem with conventional controls, a modern electronic controller reduces the voltage to the motor during the start period, and gradually increases it as required to bring the motor and the load up to speed.

Except for battery-operated equipment, almost all DC motors are powered from an AC supply that is converted to DC within the motor controller. Up until quite recently, it was actually easier, more efficient, and less expensive to provide a regulated DC output from an AC supply than from a DC supply. This is changing now, but many modern DC motor controls depend on an AC-to-DC conversion as part of their normal operation. They cannot operate from a DC supply.

13.4 AC Supplies for DC Motors

AC-to-DC conversion is simple and inexpensive. For relatively small motors, a few diodes or a one-piece bridge rectifier are connected into the AC line between the supply and the motor. They draw AC current, but pass DC on to the motor. If the motor is matched to the voltage from the rectifier, it can be connected directly to the output. DC motors are quite tolerant of the pulsating output of the rectifier diodes, so no filtering is required (see Figure 13–5). Almost all small motors have permanent magnet fields, so direction can be controlled by changing the polarity of the supply. A simple relay control system can turn the motor on and off and change directions.

FIGURE 13–5 A diode is the electrical equivalent of a check valve: It only conducts when the anode (the arrowhead side) is positive. The result is that half of the AC wave is blocked by a single diode (top) or redirected by the diodes in a bridge. The input to the rectifier is AC but the output is DC. The "freewheeling" diode provides a path for reactive energy generated by the motor.

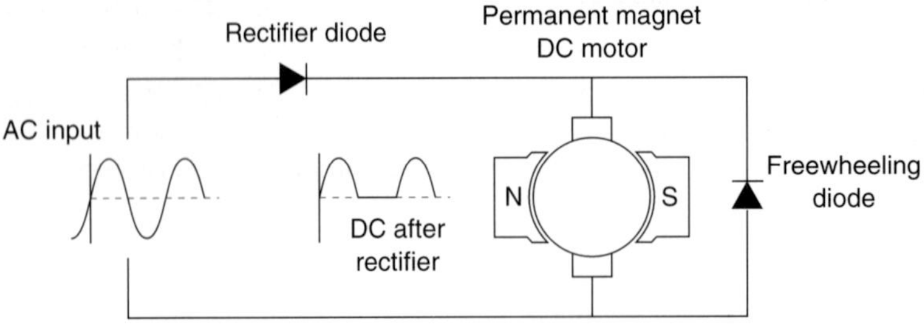

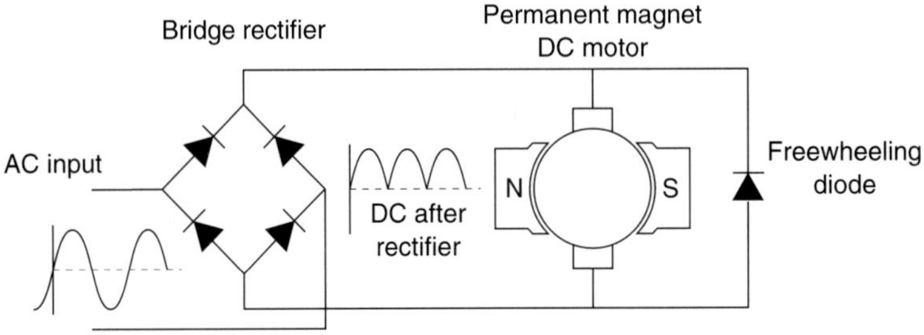

Simple drives like the one shown in Figure 13–6 may be found on some actuators used for damper or valve positioning. The DC motor, equipped with a small gear reducer, is compact, powerful, and cost competitive with a suitable AC replacement. For two-position control, limits can be built into the actuator, or mounted on the equipment, to shut the motor off at the end of the stroke. For more sophisticated control, the motor can be equipped with a shaft-mounted sensor that provides a signal to an electronic control circuit. Typical sensors send a string of pulses, an exact number for each turn of the

FIGURE 13–6 A simple relay system can provide forward reverse control of a PM DC motor

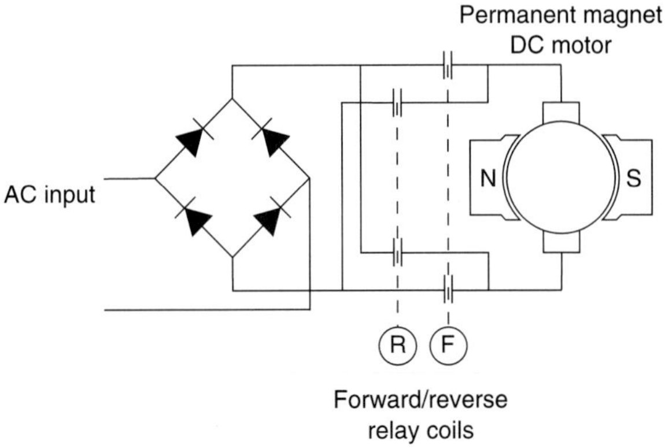

motor shaft. The control circuit counts the pulses and calculates the position of the actuator from the number of turns. To prevent accumulated errors or problems from unexpected power failures, the counter is reset to zero when the damper actuator hits a limit switch at the end of the stroke.

For larger motors, or more precise control of smaller ones, a phase control system can be used to vary the voltage output of the AC-to-DC conversion. Silicon-controlled rectifiers (SCRs) can be used in place of two or more of the rectifier diodes, or in series with the pulsating DC output (see Figure 13–7). By controlling the triggering of the SCR gate circuit, the average DC output voltage can be varied over a wide range. This technique is extended to the conversion of three-phase supplies for more power output. SCRs are available with very high voltage and current ratings, so this method can be applied to very large motors.

FIGURE 13–7 There are several variations of SCR-based phase control. All provide a variable voltage to the DC motor from a suitable AC supply. The SCR firing circuits are not shown.

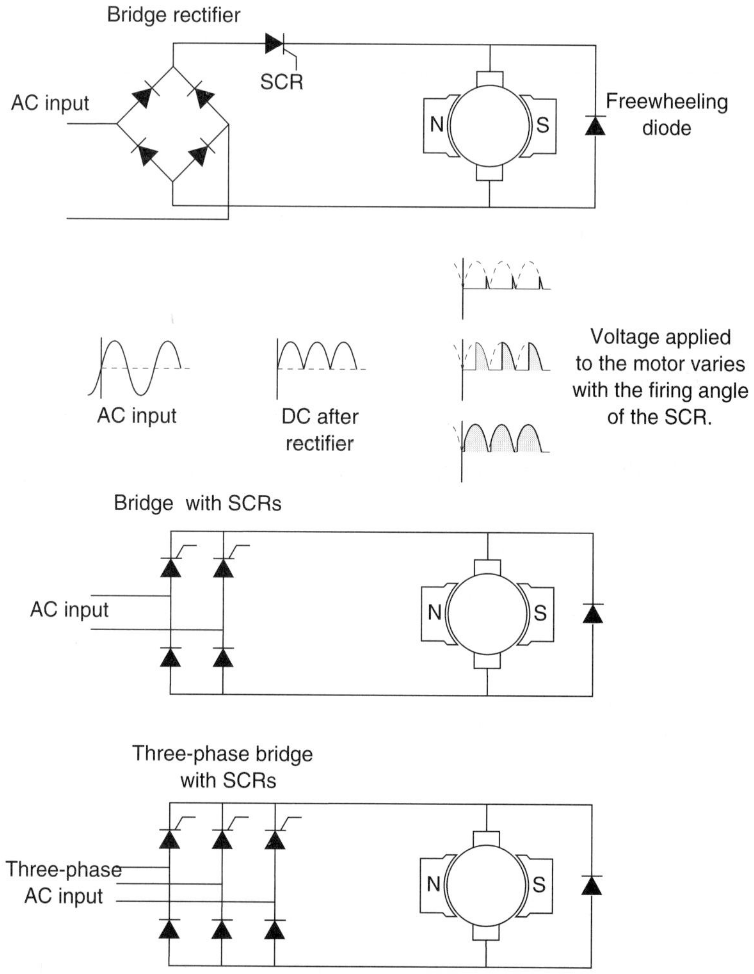

Large DC motors are complex and expensive machines requiring the careful attention of an knowledgeable and experienced maintenance staff. Small motors are often treated as disposable, but periodic attention to bearings and brushes can greatly extend their working life. Very small ones are not designed to be serviced or repaired.

13.5 Electronically Commutated Motors

In addition to the conventional DC motor described above, there are several types of electronically controlled motors. These motors are often categorized with DC motors because, although they usually require an AC supply to the electronics that drive them, they do not operate on AC in the traditional sense. These motors are found in a wide array of applications, including HVAC equipment. They are being introduced into this field in response to a need for higher operating efficiencies and better control of fan and blower outputs.

All conventional DC motors have the same drawbacks. The commutator and brushes are high-maintenance items subject to normal wear and sensitive to load, humidity, and contamination. The electrical switching action that must constantly take place as the motor turns limits the voltage and current which the motor can tolerate. Brush position is critical. The brushes must be located on the neutral plane of the magnetic field to control arching on the commutator. This neutral plane tends to shift with changes in load and speed. To compensate for this tendency, extra field poles called commutating poles may be required between the main field poles. On very large motors, special compensating windings embedded into the pole faces are also used. These extra poles and windings add to the cost, complexity, and maintenance requirements of the motor.

Conventional DC motors are also difficult to cool. The highest current, and therefore the greatest heat, occurs in the armature windings. This is the rotating part of the machine, so all the heat must be carried out of the armature by the cooling air; it cannot be conducted directly through the stationary frame to the outside surface of the motor. Larger DC motors, particularly those operated as variable-speed drives, must be equipped with separate cooling blowers to remove the heat buildup in the armature.

For all its faults, however, the DC motor can be a remarkable performer. Combined with a relatively simple electronic controller, it can provide excellent torque and speed characteristics, including smooth reversal under full load. With a high-performance controller and appropriate feedback, it can attain a level of control that is almost impossible to duplicate with any type of AC motor, regardless of the sophistication of the controller used. Because of the many positive traits of DC motors, different motor technologies have been developed to provide similar performance, without the maintenance problems. These designs generally fall into three major groups: brushless DC, switched reluctance, and stepper motors.

Brushless DC Motors

Most motors classed as brushless DC designs have permanent magnet rotors and three sets of windings in the stationary part of the motor. The windings are laid out much like the winding in a three-phase motor stator. There are individual salient (protruding) poles, instead of the distributed windings used in three-phase stators, but the physical arrangement and sequence of the poles are essentially the same. The permanent magnet rotor is typically a molded ceramic magnet with as many poles as required. As with other ceramic magnets, the individual poles cannot be seen by visual inspection. The rotor magnet is a single molded piece, magnetized into individual poles after the molding process. Typical brushless motors have more poles than comparable small three-phase motors.

The brushless DC motor has an external electronic controller that energizes the windings in sequence to create a rotating magnetic field. The permanent magnetic field of the rotor locks onto the rotating stator field and turns the shaft at exactly the same speed. The frequency of the switched DC supplied to the motor stator varies with the motor speed, just like the frequency of the current inside a conventional DC armature. The controller takes the place of the commutator, receiving its switching signals from a shaft-mounted optical or magnetic sensor. Motor speed is regulated by changing the voltage to the motor, which changes the current and output torque. The motor operates in balance between output torque and mechanical demand, much like a conventional DC motor.

Like all electronically commutated motors, the brushless DC motor cannot operate independently from its controller.

Brushless DC motors can be very simple, inexpensive motors—computer power-supply fans are a good example—but they can also be very sophisticated high-performance devices. For higher performance the motor and its controller are combined with additional controls. For example, for a brushless DC servo application on a computer-controlled machine tool, the motor controller takes its orders from the computer that controls the operation of the machine. The computer determines the speed and direction of rotation required, and the motor controller holds the motor to those parameters. The motors and controllers used for this type of application can respond very rapidly to the demands of the computer program. The brushless design provides very high performance in a compact, low-maintenance package.

The brushless DC motor shown in Figure 13–8 is an example of a different but specific application of this technology. These motors are relatively small—$^1/_3$ to 1 Hp—with an integrated electronic control package coupled to the motor. They are used to drive blowers in high-efficiency furnaces and similar equipment. The motor and controller mount as a unit inside the blower wheel, which is mounted directly to the motor shaft. This motor is much more expensive than a conventional AC motor, but less expensive than any other motor-and-controller combination capable of providing the same features.

The integrated controller monitors the motor torque and speed, which allows it to maintain a constant airflow through the heat exchanger of the unit with no external sensors or auxiliary controls. Multiple speed/torque/airflow parameters can be programmed into the motor controller to match the requirements of the equipment. For example, a modern high-efficiency furnace can have several operating modes. These might include continuous air circulation, two stages of heating, and one or two stages of cooling. The optimum airflow is different for each mode, and can be difficult to maintain with a conventional motor. With a programmable motor of this type, conventional furnace controls can call up the correct program, and the blower will supply the appropriate airflow.

FIGURE 13–8 The controller for this brushless DC motor is built into an extended end bell of the motor. This motor is designed specifically for the HVAC industry for use in variable-speed fan drives.

The ability of the motor to adjust to changing conditions is particularly valuable in systems with multiple zones. As the zone dampers open and close, the static pressure of the supply ducts will change, but the blower will continue to maintain proper airflow. The necessary controls are provided as part of the motor package, with no extra equipment to be field installed. Other types of fan motors cannot adjust to such changes without one or more sensors in the airstream, combined with a variable damper or variable-speed motor drive.

In general, brushless DC motors are efficient, quiet, low-maintenance, and very reliable. They are superb performers when variable-speed control is required. They are initially expensive, however, and cannot operate without a complex electronic controller.

Switched-Reluctance Motors

Switched-reluctance motors are a relatively new technology that is very similar to the brushless DC design (see Figure 13–9). The main difference is in the rotor, which is a laminated iron assembly with protruding lobes in place of permanent magnet poles. The lobes lock onto the rotating stator field, because the lower magnetic reluctance paths that they provide naturally align themselves with the field. The rotor is somewhat simpler and lighter than that of a permanent magnet motor. Switched-reluctance motors, like all electronically commutated motors, cannot operate without a dedicated controller. The controller is complex; the motor is relatively simple.

How well switched-reluctance motors will be accepted is yet to be determined. They are very similar to brushless DC motors in appearance and behavior, but the engineering is different and, so far, less proven. Unlike the permanent magnet motor, the switched-reluctance motors turns freely when it is not powered.

FIGURE 13–9 Brushless DC and switched-reluctance motors use very similar technology, but have different performance characteristics.

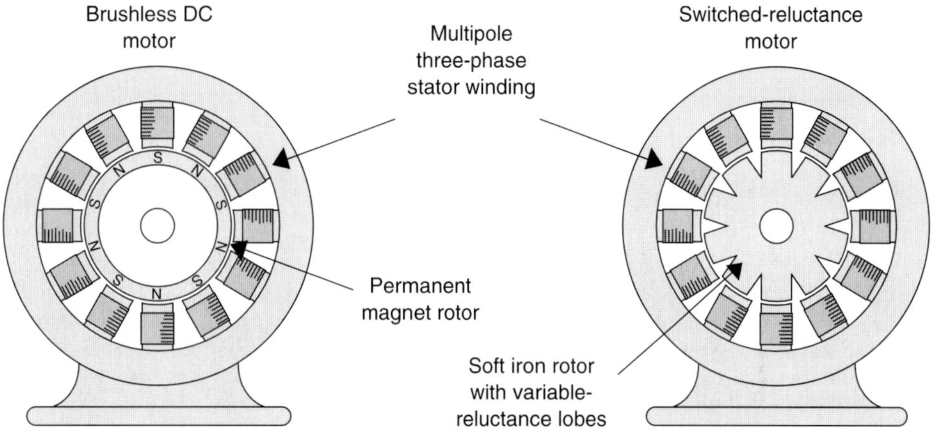

Stepper Motors

Stepper motors (see Figure 13–10) are similar to other electronically commutated motors, but they are not primarily designed for continuous rotation. They are capable of continuous rotation, but actually move in small, incremental steps. Unlike most other motors, they can be left energized but motionless indefinitely. While stationary, they are not really stalled, just holding that position until the controller sets them in motion again. Other types of motors begin to rotate when the motor or controller/motor combination is energized. A typical stepper motor control energizes the motor in a stationary position until it receives a direction signal and series of pulses that tell the controller how fast and how far to advance the motor. The controller then steps the motor at the appropriate rate until the pulses stop.

As long as the motor is not forced out of synchronism with the control signal, it will rotate exactly the correct number of steps in either direction. This is the primary appeal of the stepper motor. It is a relatively simple and reasonably priced actuator that requires no feedback signal. As long as the mechanical demands of the mechanism are within the capabilities of the motor, it can provide reliable positioning with a high degree of accuracy, without the complications of a position sensor.

Stepper motors are made with individual step angles as large as 15° or more, or as small as 1.8° or less. These two step angles represent motor ratings of 24 and 200 step per revolution, respectively. The number of operating steps can be increased with microstepping controllers that regulate the current to the motor coils. By regulating coil current, instead of simply switching the coils on and off, a microstepping controller can provide multiple positions between the natural electromagnetic steps of the motor. A 200-step motor with a microstepping controller may have 25,000 individual steps per revolution.

Stepper motors are made in three different physical layouts. The two most common types are the "canstack" permanent magnet type and the hybrid motor, which is a combination of permanent magnet and variable-reluctance designs. The variable-reluctance motor is similar in principle to the permanent magnet stepper, but has a toothed rotor with no permanent magnetic properties. Variable-reluctance motors are rarely used in industrial applications.

FIGURE 13–10 This hybrid stepper motor is about the size of a coffee mug. The two rotor sections have 50 teeth (poles) each. The stator has only eight poles, but they are divided into smaller segments by teeth formed in the iron core laminations. The stator teeth are spaced so that if there were a complete set there would be 48, two less than either rotor.

The canstack motor is the most popular in nonindustrial applications because it provides reasonable performance in a simple, low-cost design. The stator in the canstack motor is made up of two sets of multitoothed "cans" set one above the other (see Figure 13–11). Each can contains a simple flat field coil wound on a form that fits inside the donut-shaped stator can. When the coil is energized, the soft iron can is magnetized with one polarity on the "top" and the opposite polarity on the "bottom." Since the teeth originate alternately at the top and bottom of the can, they have alternating magnetic polarity. This type of construction provides relatively large number of magnetic poles with a very simple construction. The two cans that make up the complete stator are misaligned so that a moving field can be created by energizing the coils in the sequence.

The stator windings are usually classed as two-phase—which means there is one coil in each can—and have four motor leads. Motors with more than four leads are sometimes referred to as four-phase motors, but they are essentially two-phase motors with either dual- or center-tapped stator coils. The number of leads brought out varies from five to eight, depending on how the connections are made to the two stator windings.

The rotor of the canstack motor is usually a molded cylinder that has been magnetized with many poles. Each of the rotor poles is a narrow strip running parallel to the shaft. There

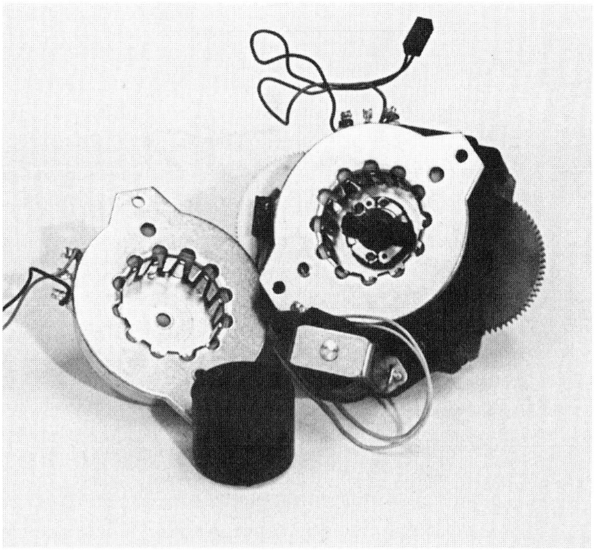

FIGURE 13–11 This canstack stepper motor was removed from a popular damper actuator. The two stator "cans" have been separated to expose the internal teeth that form the stator poles. The details of the rotor (front) cannot be seen in this photo. Its outer surface is a cylinder of ceramic material that is magnetized to form many narrow poles.

is no visual indication of the presence of these individual poles. The rotor looks and acts like a simple ceramic magnet until it is checked very closely. One simple way to demonstrate the true nature of the rotor is to slide a small permanent magnet around its surface. The magnet will be alternately attracted and repelled by each of the poles in sequence.

Hybrid steppers (see Figure 13–10) use a single high-powered permanent magnet in the shape of a ring. The magnet is mounted around the motor shaft, between two multi-toothed rotor sections that look much like fine-toothed gears. On a typical motor there are 50 teeth on each rotor section, and the two rotor sections are one tooth out of alignment. This creates a total of 100 poles with alternating north-south polarities. The stator of the hybrid motor has a relatively low number of salient poles, typically eight, but each pole face is grooved to form teeth similar to the teeth on the rotor sections. The pitches of the teeth in the rotor and stator are slightly different. A typical stator would have two teeth fewer than one rotor section if a complete set was installed. The result is that the rotor teeth are in perfect alignment with the stator teeth in only four places, two on each rotor section.

These alignments occur at the quarter points; for example, 0° and 180° at the front, 90° and 270° at the back. Even though only four points are fully aligned, the permanent magnet rotor locks onto the teeth of the stator. By changing the polarity of the stator windings in a sequence, the rotor can be advanced one quarter of a tooth pitch at a time in either direction. A typical hybrid motor with 50 teeth on each rotor will have 200 discrete steps per revolution—a 1.8° step. This is a very common motor.

Note

Hybrid steppers have stainless steel shafts because they must be nonmagnetic. This makes them relatively easy to bend, so be careful not to drop them. **Do not disassemble a hybrid stepper motor to satisfy your curiosity.** Removing the rotor can end the motor's useful life. The rotor is often magnetized in place, and once it has been removed it is permanently weakened.

There is another class of steppers that use permanent magnets and wound coils to move along either a linear or circular path of metal teeth. This system produces a relatively high force for its size, so it can be used to drive significant loads without any gearing. Rotary versions are used for direct-drive positioners; linear versions are used for applications such as two- or three-axis automated drilling machines. The linear "motors" make up the tracks of the drill carriage. The motors are controlled by a computer to position the drill for each hole. Linear motors are fast and accurate; they can provide a relatively long travel without the problems associated with lead screws or other linear positioners.

Stepper motors are very common in all types of computerized equipment because they are naturally a digital type of control. They are widely used in computer peripherals such as printers and disk drives, and are also common in some of the newer damper and valve positioners used in HVAC applications. They are almost always operated in "open loop" mode with no feedback. To avoid positioning errors associated with loss of power or equipment malfunction, the control is designed to rezero itself when the motor is driven to one end of the stroke.

Industrial stepper motor drives often use more sophisticated controls, including feedback where it is warranted. These motors are found in many types of equipment, including some automated machine tools.

Questions

1. What is the purpose of the commutator and brushes in a conventional DC motor?
2. Electronically commutated motors do not have a commutator or brushes. What is used in their place?
3. Why do many modern DC motor drives require an AC supply for operation?
4. Why are stepper motors preferred over conventional motors for many types of positioning?
5. What precautions are warranted when working with hybrid stepper motors?

14 Nameplate Data and Operating Conditions

Objectives

This is a reference chapter. After completing the study of this chapter you should be able to:

- Interpret the data provided on a typical motor nameplate
- State, in general terms, what motor manufacturers regard as acceptable operating conditions

14.1 Nameplate Data

Not all of the information listed below is found on all nameplates. The data provided varies with the size and type of motor, as well as its intended applications. For an example of a typical motor nameplate, refer to Figure 14–1.

Model numbers are an internal manufacturer's model or catalog number. They can be used to cross-reference the motor to other specification sheets, connection diagrams, recommended controllers, etc.

Horsepower (HP) is the rated mechanical-output power of the motor. It may also be given in kilowatts (kW). One horsepower is 550 ft-lbs per second, or 746 W. Rotary mechanical power is a multiple of motor torque and shaft speed, so motor torque can be calculated from rated speed and horsepower. Low-speed motors must develop proportionally higher torque to produce the same output power. Remember that the rated torque is maximum continuous torque at rated full-load speed. The starting torque of single-phase motors depends on the starting method used. The starting torque curve of a three-phase AC motor depends on the rotor design (Design A, B, C, or D). In general, starting torque and breakdown torque (the point at which the motor is so overloaded that it will stall) are both higher than full-load running torque. (Motor torque curves are discussed in Chapter 11.)

The motor must be matched to the load that it will drive. Certain loads, such as hoists and conveyors, have a relatively constant torque characteristic. The amount of force required to keep the load moving is more or less independent of the speed. Of course, extra torque is needed to accelerate the load, but at any steady speed, the physical load is much the same.

FIGURE 14–1 The type of data shown on this nameplate is typical, but a rating of 208-230/460 volts at 60 Hz is much more common than the 50/60 Hz at 208-220/440 volts of this example.

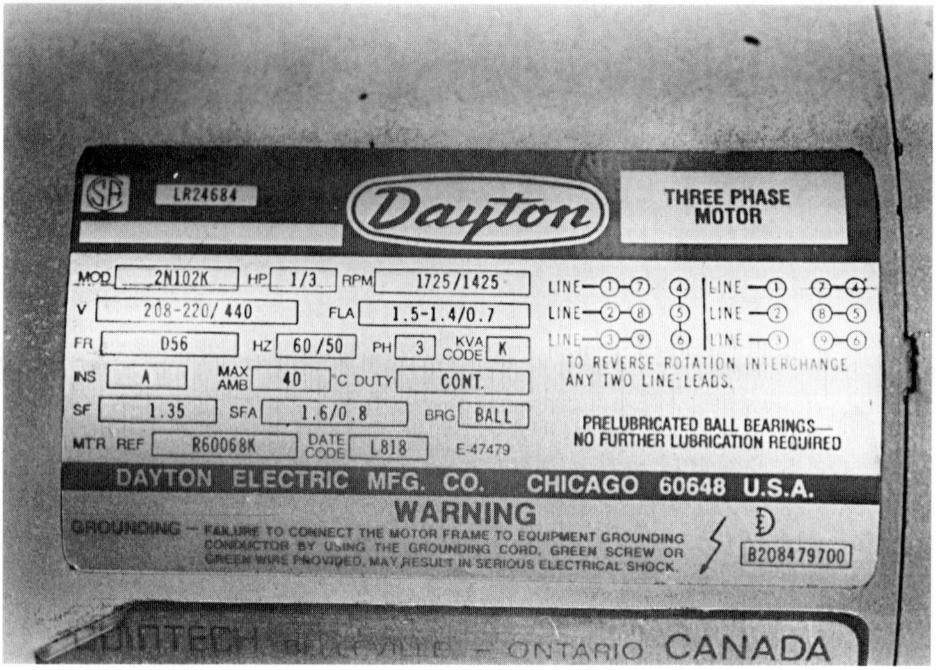

Other loads have a variable-torque characteristic—the amount of torque the motor must supply increases with the speed. For these loads the power demand is proportional to the square, or cube, of the speed, so the motor output must increase exponentially as the speed of the load is increased. Common examples of variable-torque loads are fans and blowers and centrifugal pumps or compressors. If they are equipped with variable-speed motors or an adjustable ratio drive (such as an adjustable belt pulley), the motors can easily be overloaded. The motor current must be checked when the speed is adjusted.

Rated speed (RPM) is the full-load speed of the motor. The no-load speed of a typical AC motor will be slightly higher. Some types of motors, specifically PSC, shaded-pole, and, to a lesser extent, three-phase induction motors, can be speed controlled by varying the voltage **if they are connected to a variable-torque load.** This is not a particularly good method of speed control and is rarely used on anything except fractional horsepower motors. Variable-frequency controllers can provide excellent speed control, but are much more complex than variable-voltage supplies.

The *number of phases (PH)* can be 1, 2, or 3. Single-phase motors above 10 Hp are rare. Three-phase motors are common from fractional to very high output ratings. Two-phase is usually restricted to very specific areas where two-phase distribution is still in use.

Frequency (Hz) is the design power supply frequency in hertz (Hz). It can be 50, 60, or 50/60, depending on the intended market for the motor. Fifty Hz motors are used in Europe and some other parts of the world; North American supplies are all 60 Hz. The supply frequency affects the motor speed, so motors rated 50/60 Hz have two RPM values. The lower one is for the 50 Hz supply.

Variable-frequency motor controllers are readily available for a wide variety of three-phase motors. They change the motor speed by changing the frequency. The controller converts the incoming AC to DC, and then back to variable-frequency AC. Some small variable-frequency controllers accept a single-phase supply; larger ones are all three-phase.

Voltage (V) is the design supply voltage. In most cases the nameplate will read something like 115, 208, 230, 460, 575, or some combination of these. Ratings like 220 and 440 are also used, but in 60 Hz motors they are an older standard.

Frame numbers for the motor follow very specific standard designations. They define the mounting provisions, including bolt sizes and spacing, and the placement, length, and size of the shaft. A replacement motor with the same frame designation as the original will physically fit the original mounting bolts and have the same shaft size and location.

Note that motors also come with several different enclosure types. These include open, drip-proof (DP), totally enclosed fan cooled (TEFC), explosion-proof, and submersible. These examples are defined industrial standards.

Bearing numbers may be included on the nameplate. This can be convenient because replacements can be selected prior to disassembly.

Full-load amps (FLA) is the motor's rated current demand when it is providing full rated output to the mechanical load. If the motor has multiple voltage listings, it will also have multiple current listings. Dual-voltage motors operated at the lower voltage draw twice as much current as at the higher voltage. The output and input power are the same, as is the cost of operation. The disadvantages of the lower voltage include larger motor supply conductors and higher current fuses or circuit breakers and motor control components.

The full-load ampere rating of the motor is used to determine the size of the supply conductors, the overload settings, and the overcurrent protection. Electrical codes contain the necessary rules and tables for these selections.

Small changes in supply voltage or imbalances in voltage between the phases can have significant effects on the current demand and output power.

Locked-rotor amps (LRA) is the motor's rated current draw with the rotor stalled. This rating is common nameplate data on refrigeration and air conditioning equipment, but may not be included on the nameplates of general-purpose motors. Hermetically sealed refrigeration and air conditioning compressors are not rated in horsepower because there is no shaft available to supply an external load. Their nameplates include the full-load and locked-rotor amps. The locked-rotor rating is used, in conjunction with the full-load ampere and voltage ratings, to select motor control contactors and other circuit components.

KVA Code letters can be used to find the locked-rotor current of the motor. The codes follow a standard NEC (National Electric Code) chart that gives the locked-rotor rating in kVA per Hp.

A	0–3.14	F	5.0–5.59	L	9.0–9.99	S	16.0–17.99
B	3.15–3.54	G	5.6–6.29	M	10.0–11.19	T	18.0–19.99
C	3.55–3.99	H	6.3–7.09	N	11.2–12.49	U	20.0–22.39
D	4.0–4.49	J	7.1–7.99	P	12.5–13.99	V	24.0 or more
E	4.5–4.99	K	8.0–8.99	R	14.0–15.99		

Insulation class letters define the type and temperature rating of the insulation. There are actual temperatures associated with each class, but there are variations in the way these values are unitized. The temperature ratings are approximately A = 105°C, B = 130°C, F = 155°C, and H = 180°C for the four classes. Realistically these ratings are almost impossible to use in the field, because they represent the maximum hot spot temperature ratings for the winding insulation. These values are the basis for motor temperature ratings.

Rated temperature rise is no longer provided on motor nameplates. It is the maximum allowable winding temperature above ambient. The design ambient is usually 40°C, with 50°C a common rated rise. This translates to a 90°C maximum operating temperature, but the rating is based on internal conductor temperature obtained indirectly from winding resistance measurements. It does not mean that the motor can be operated with a surface temperature of 90°C on the outside of the frame.

Maximum ambient is the maximum allowable temperature of the air around the motor. It is typically 40°C, but other values are used with motors that have higher temperature insulation.

Operating temperatures are directly related to motor life. For a "normal" life expectancy, the temperature of the winding conductors should not exceed the rated value for that insulation at any point in the winding. Each 10°C rise above that value will reduce the motor life by approximately 50%. This reduction is exponential: At 20°C above the rated, the expected motor life drops to 25%, and at 30°C it is 12.5%.

Duty is the motor's rated duty cycle. Most motors are rated for continuous duty. However, when the motor is connected to a load that does not operate continuously, it may be designed to match the application. For example, the motor used to open and close an elevator door is not continuous-duty rated. It has to supply high torque for a few seconds, and then may do nothing for a minute or more. Duty cycles can be described using terms such as "continuous," "intermittent," etc., but these terms have specific meanings regarding the number of minutes per hour. Duty cycles affect conductor and overcurrent ratings and the settings of the overloads.

Service factor (SF) is the rated acceptable continuous overload. Some motor nameplates may include a service factor amp rating, which is the full-load rating times the service factor. The motors maximum continuous-0output power is its rated Hp times the service factor. This rating assumes all normal operating requirements are met (see below). Operation at the maximum allowed by the service factor rating may result in reduced efficiency and lower-than-nameplate speeds. The motor also runs hotter, with a reduced margin of safety for temporary overloads or other problems. The highest service factors are normally available on open and drip-proof motor enclosures because they are much easier to cool than totally enclosed types.

14.2 Acceptable Operating Conditions

The following are typical of the assumptions upon which motor nameplate ratings are based.

- Voltage within ±10% of rated
- Line-to-line voltages within ±3% of each other
- Frequency within ±5% of rated

- Current within rated (short-term overloads excepted)
- Adequate cooling airflow unrestricted by physical barriers, accumulations of dirt or other factors
- Cooling air at or below the design maximum ambient temperature rating
- Operating altitude not more than 1000 m (3300 ft) above sea level

More conservative ratings would not be out of line. A 10% variation in line voltage translates into a 20% variation in maximum torque. A 3.5% variation from one line to the other can increase internal heating by as much as 25%.

Motor temperature rise is based on internal conductor temperature found by calculation from accurate resistance measurements, not on an indirect temperature measurement. An adequate flow of cooling air assumes unrestricted flow through an open motor, or over the motor casing if it is fully enclosed.

Part IV

Electrical Controls

Chapter

This is both an instructional and a reference section. The overall intent is to provide you with the basic principles and detailed specifics of manual and automatic electrical control. Most of the examples are taken from the HVAC industry, but are also generally applicable to the broader context of electrical control. Motors are included as required, but this section is not specific to motors and motor control. For additional information on motors and motor control, refer to Part III.

"Control Devices," the first chapter in this section, is devoted to manual and automatic switches, switch ratings, and various types of sensors. This chapter includes a detailed explanation of switch contact arrangements, naming conventions, and electrical ratings. A wide variety of automatic control devices are also covered, including electromechanical and electronic types. Control devices with variable, rather than on/off contact, outputs are also introduced.

Chapter 16, "Low-Voltage Controls," examines the additional control components required for low-voltage control systems. The information provided includes control transformers, relays, and timers. These devices are put into context through application examples.

These two general chapters are followed by two chapters on specific control applications. Chapter 17, "Residential Furnaces," is specific to the control circuits and devices found in typical home heating and heating/cooling systems. Chapter 18, "Packaged Systems," provides additional information on the devices and circuits found in typical prepackaged rooftop heating/cooling units.

15 Control Devices

Objectives

After completing the study of this chapter you should be able to:

- Describe common switch configurations using appropriate terminology
- Draw example switch circuits and explain their operation
- Correctly interpret the markings and data found on typical switches
- Test switches and switch circuits with a multimeter to determine proper operation
- Identify hazards associated with incorrect switch application and placement
- Use appropriate safety measures prior to contact with potentially hazardous circuits
- List examples of common automatic control devices and describe their operation
- Identify common symbols used for automatic controls and explain their meaning
- Explain, in general terms, the differences between on/off, multistage, and proportional control

15.1 Introduction

As mentioned in previous sections, virtually all practical electrical circuits include a control device of some kind. Examples include manually and automatically operated switches, electromagnetically operated relays or contactors, variable resistors, and a variety of electronic circuits. Conventional (nonelectronic) controls use moving mechanical parts to control electrical contacts or adjust resistance elements. Electronic controls switch or regulate current through changes in the conductivity of solid-state materials, primarily silicon-based semiconductors. These changes are, in turn, controlled by electrical signals, but the signals are at very low voltage and current levels. Electronic circuits contain conventional switches, variable resistors, capacitors, inductors, and relays, as well as various solid-state components.

Manual switches are those that are normally operated by a human hand. They can be as simple as a two-position on/off switch controlling a low-voltage lamp, as complex as a multiposition selector switch, or as daunting as a large high-voltage disconnect. Automatically operated switches sense physical changes such as temperature, level, flow, pressure, mechanical position, or light level and use them to operate one or more output contacts. Many such devices use ingenious mechanical linkages; others depend on electronic circuitry to sense the physical variable being monitored and, possibly, to do the actual switching.

Switches that operate automatically may be used as protective devices to prevent dangerous conditions, or as controls to determine normal operating levels. Examples of protective devices include high- and low-temperature limits, pressure limits, and end-of-travel switches. Control examples include thermostats for heating and cooling systems, and pressure switches for pumps or compressors. There are many other examples. Protective devices usually have fixed, factory preset operating points. Control devices usually have adjustable set points to allow the end user to make changes to the operating level. Many different control devices may be integrated into a single control system.

15.2 Manual Switches

Although controls can be very elaborate, in many situations all that is required is a conveniently located, manually operated switch that allows the user to turn the circuit on and off. Manual switches are the most common of all the control devices available. They can be very simple devices or amazingly complex. Switches are made in a tremendous variety of styles, sizes, electrical ratings, and switching patterns. The most common is also the simplest, the two-position on/off switch that makes and breaks a single-circuit line.

In an operating electrical circuit, current circulates continuously from the power supply to the load and back to the supply. For the circuit to function, there must be a complete conductive loop in which the current can flow. For basic on/off control, a switch is inserted into this conductive path, in series with the load. When it is operated, the switch contacts act to make or break the circuit at that point. When the contacts are brought together, a connection is made, current can flow, and the load is energized. When the contacts are separated, the connection is broken, current flow is blocked by the insulating air gap between the contacts, and the load is de-energized.

Switches and Valves Compared

Comparing valves to switches can be quite useful. The roles they play in their respective systems are very much the same. Some caution is advised, however, if confusion is to be avoided. The terminology of switches and valves is essentially reversed in meaning. In a hydraulic system, for example, when a valve is open, the oil can flow through; when the valve is closed, the flow is blocked. When a switch is in the closed position, current can flow; when the switch is opened, the current is blocked.

Hydraulic fluids can only flow through an open space, not through solid materials. A closed valve blocks fluid flow by placing a solid barrier in its path. When the valve is open, the barrier is removed, and the fluid is allowed to pass through with little or no restriction. Electrical current flows through the solid conductive materials that form the circuit path. When the switch is in the closed position, a metal-to-metal contact is made between the contact points of the switch. This completes the conductive path through the switch from one terminal to the other. In the open position the contacts are separated, essentially inserting a barrier of insulating air into the circuit and blocking all current flow. When the switch is closed, it is in the on position. When it is open, the circuit is shut off.

The comparison of switches and valves can be extended to include some of their ratings. Because they perform essentially the same types of functions, valve specifications and switch specifications are also quite similar. Fluid flow is normally confined to a specific path by the solid walls of the pipes and tubes through which it moves. The opening through the piping must be large enough that it will not overly restrict the flow. The inside dimensions of the pipe or tube determine the maximum practical flow rate that can be accommodated. The inside dimension of the valve must meet a similar specification. When it is in the fully open position, the passageway through the valve must be large enough to allow the fluid to pass at the full design flow rate without unnecessary restriction or loss. The valve must also be durable enough to withstand repeated opening and closing without undue wear or failure.

The requirements of an electrical circuit are very similar. The current flows through solid conductors rather than the opening of a pipe, but to an electrical supply, a conductor looks the same as a pipe full of oil does to a hydraulic pump. In place of the oil, there are freely mobile electrons that can flow through the conductor just as easily as the oil can flow through the pipe. Just as the cross section of the pipe determines the maximum practical flow rate of the oil, the cross section of the conductor determines the maximum current flow it can support without excessive energy loss and heat buildup. Likewise for the switch. When it is closed, the area of the metal contact surfaces must be large enough to accommodate the current flow. The contacts must also be made of a material that provides a minimum of contact resistance and is able to withstand repeated use without undue deterioration.

In a hydraulic system, the piping system must be strong enough to withstand the operating pressure of the system. The same is true for the valves used. They must not only contain the pressure, but also operate properly with full pressure applied. In an electrical system, the insulation surrounding the conductors must contain the electrical pressure—the potential difference or voltage—of the supply. Switches must be insulated just as the conductors are, because they are part of the circuit path. They must also be able to break the flow of current with the full electrical pressure of the supply voltage applied. Just as the closed valve must provide a barrier that is strong enough to hold back the physical pressure of the fluid, the open switch must also provide an insulating barrier that blocks the electrical pressure of the circuit voltage. This is usually just an air gap wide enough to prevent the voltage from arcing across. The separation between the contact surfaces can be surprisingly narrow.

In addition to all the similarities, there is one notable difference between the operating characteristics of valves and switches. Most valves can regulate flow, at least to some degree. When it is set a some position between fully closed and fully open, the valve acts like an adjustable restriction in the line. Switches cannot regulate flow because they have no halfway position. They can only be fully closed or fully open. The ability to regulate flow, something that is relatively easy to do in a fluid-based system, is comparatively difficult in an electrical circuit.

15.3 Switch Identification

Switches date back to early electrical experiments. The builders of those circuits needed convenient points of control for the same reasons that switches are included in practical circuits today. An early design was the open-blade knife switch. One end of a copper blade was loosely clamped between two copper supports. The other end could be swung into position between the jaws of a set of stationary copper contacts. An insulated handle was attached to the moving end of the blade to allow operation of the switch without contacting the energized circuit. The knife switch was relatively simple to construct and easy to operate. It was also convenient, because its open or closed status could be observed at a glance. The design provided good metal-to-metal contact under spring pressure, reducing the possibility of high resistance at the contact points.

With only a few exceptions, knife switches have given way to switches with enclosed contacts. These are much safer to use, more compact, and easier to install. Although knife switches are almost a thing of the past, the system of identification that was established for them is still in use today.

Poles and Throws

Not all switches are as simple as a typical on/off switch. Even some relatively basic circuit functions require a switch that does more than just open and close a single circuit line. As switches became more complex, a naming convention was developed to describe their operation and to identify the various combinations of contacts and connection positions that were used. The key terms used in these descriptions are "poles" and "throws."

Note

The naming conventions used for manual switches may also be used to define the operation of the contacts in other switching devices, such as relays, circuit breakers, or contactors.

A pole is a terminal point or line that is part of a single conductive circuit path. A throw is a normal operating position to which the switch "blade" can be "thrown" to make a connection between two switch terminals. The basic on/off switch is the simplest possible design. The knife-switch version has a single blade that can be positioned so that it is either closed—the blade is positioned in the contact jaws to complete the circuit through the switch—or open—the blade is pulled away from the jaws so that the circuit is broken. The moving blade and the stationary jaws form a single contact set that provides only two options, open or closed. A switch like this is classed as having one pole, because a single circuit path is controlled. It also has only one throw, because its single pair of contacts can be closed to complete only one circuit path. This device is called a single-pole single-throw (SPST) switch.

An SPST switch has two terminals and controls a single line. The switch handle has only two positions. In one position the contacts between the two terminals of the switch are closed, and in the other position the contacts are open. The SPST designation is used for all types of switches that perform the same function, regardless of the style or size, or whether the contacts are actually visible (see Figure 15–1).

FIGURE 15–1 The single-pole single-throw (SPST) switch gets its name and symbol from the open-blade-knife version of this switch.

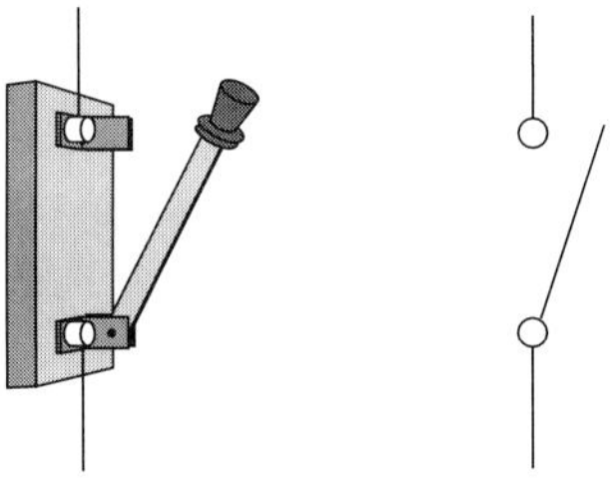

Single-pole single-throw (SPST) switch

When a second connection point, or throw, is added to the switch, it is classed as a single-pole double-throw (SPDT) switch. This switch has three terminals instead of only two. It still controls the connection of just one terminal point, or pole, of the circuit, but it allows that point to be connected to either of two others. The SPDT switch has two "on" positions.

It is important to realize that one terminal of the switch is unique, because it is shared by the other two terminals. In an open-knife-switch, it is always connected to the blade. There is an equivalent terminal on all enclosed SPDT switches. This shared terminal is referred to as the "common" terminal, because although the switch has two possible circuit paths, that terminal is common to both of them.

The switch handle of a SPDT switch may have two positions or three. In the two-position type, there is no true off position. The common is always connected to one of the other two terminals. When the switch has a third handle position, it is in the center of the handle movement and acts as an off position. With the switch off, there is no connection between any of the switch terminals.

Many, but not all, SPDT switches have at least the common terminal identified in some way (see Figure 15–2). Marking methods include one oddly colored terminal screw, or the letters COM or COMMON printed beside the terminal. Sometimes symbols are used instead, showing the contact arrangement and indicating the internal connections to each terminal. Switches or contact sets of relays or contactors that have a "normal" position may be identified with the initials COM, NO, and NC. These stand for common, normally open, and normally closed, respectively. The normal position is the position of the contacts with the device "at rest"—no pressure applied to the switch handle or button, and no power to any operating coils. If the switch is not clearly marked, the contact arrangement can be identified by testing between the terminals with a continuity tester or an ohmmeter.

Multipole single-throw switches operate as multiple SPST switches that are mechanically, but not electrically, interconnected. For example, a double-pole single-throw (DPST) switch is essentially two SPST switches with a common operating handle (see Figure 15–3). A DPST switch has four terminals, two for each of the two contacts. The contacts open and close simultaneously, but there are no electrical connections between the two poles. This type of switch can also have three poles (3PST) or more, all of which behave as mechanically interconnected SPST switches.

FIGURE 15–2 Single-pole double-throw (SPDT) switch operation and typical markings.

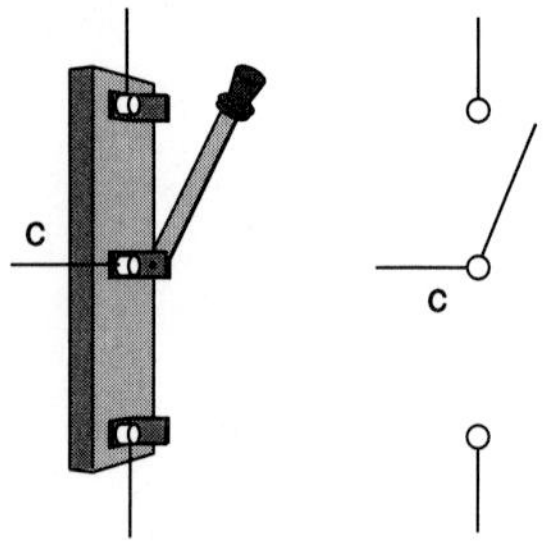

Single-pole double-throw (SPDT) switch operation

Two position With center off

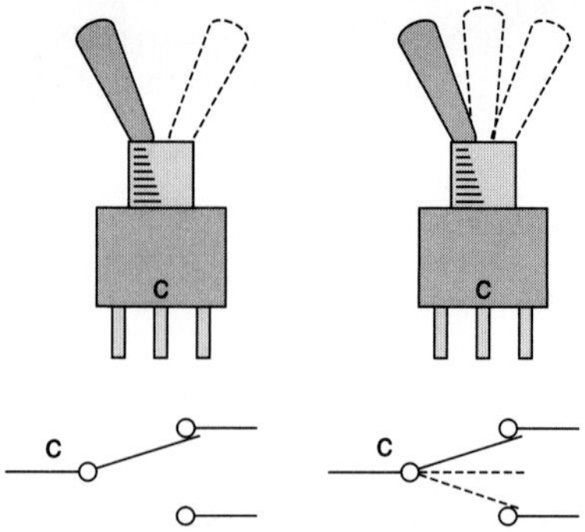

Single-pole double-throw switch examples and their symbols

SPDT microswitch

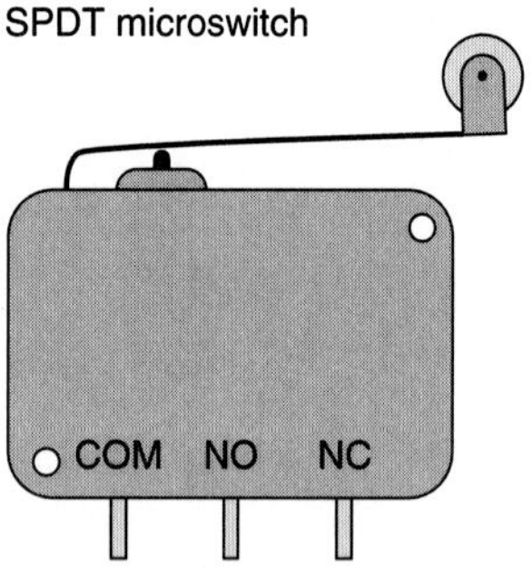

Other SPDT symbols used with switches and relays

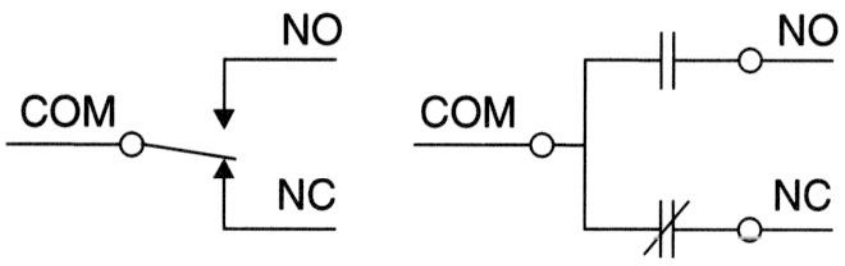

FIGURE 15–3 The double-pole single-throw (DPST) switch controls two lines simultaneously. The dashed line in the symbol indicates a mechanical tie.

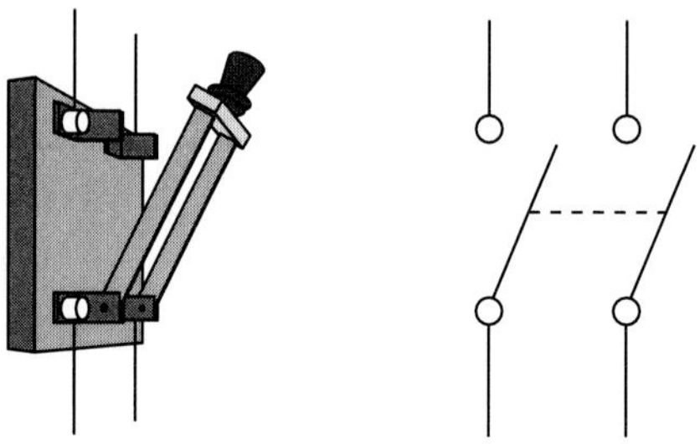

Double-pole single-throw (DPST) switch

Multipole single-throw switches are widely used, particularly in two- and three-pole configurations. They are most often found where more than one line must be simultaneously disconnected. Common applications include two-pole switches for disconnecting 208 or 240 V single-phase circuits, and three-pole switches for disconnecting the three supply lines of three-phase circuits. Electrical codes normally require that a disconnect switch must simultaneously break all the ungrounded lines of the circuit.

Multithrow switches have more than one position in which there is a conductive path through the switch. They always have at least one common terminal, which may be marked in some way. Lever-operated switches, such as those with toggle handles, usually have either two or three positions. The third position is typically a "center off." Multithrow switches are most often single-, double-, or triple-pole.

Rotary selector switches may have many more "throws," but these are usually identified as positions. Selector switches may have one or more common terminals and many different connection points. A common application for a multiposition selector switch is the combined range and function switch found on many electrical multimeters.

Multiposition, Multipole Switches

A double-pole double-throw (DPDT) or triple-pole double-throw (3PDT) switch or relay is essentially multiple SPDT contacts with a common mechanical linkage (see Figures 15–4 and 15–5). Again, there is normally no electrical connection between the various sets of contacts.

It may be obvious by this time that identifying switches by the number of poles and throws is fine within certain limits. However, when the number of combinations becomes excessive, or the switch pattern becomes too complex, such a system quickly becomes clumsy and inadequate. Complex switch operation cannot be conveniently described with the letter and number designations of poles and throws. The usual method of providing the necessary connection information is a small chart packed with the switch (see Figure 15–6). The chart lists the open or closed status of the contacts for each position of the operating handle. Very complex switches may also be supplied with a diagram that shows all the internal connections.

When the connection chart of a complex switch has been lost, it is critically important to spend the time it takes to recreate the chart from scratch. This requires a continuity tester or ohmmeter and a pencil and paper. The best procedure is to first make sure that each terminal and switch position is clearly identified. Next, check every possible combination of continuity between all terminals for each switch position in turn. Record all closed connections by position and terminal marking. Complete all the continuity tests for one switch position before going on to the next. Once the test is complete, organize the information into a clearly labeled chart and save it for future reference.

Many complex switches do not fall neatly into a poles-and-throws type of description. The switch pattern may have been developed to meet very specific requirements for a custom application. It is foolhardy to jump to conclusions about a switching pattern based on the number of switch positions and the external layout of the terminals. There is no substitute for a thorough, well-organized test.

FIGURE 15–4 DPDT, 3PDT switch combinations, symbols, and markings

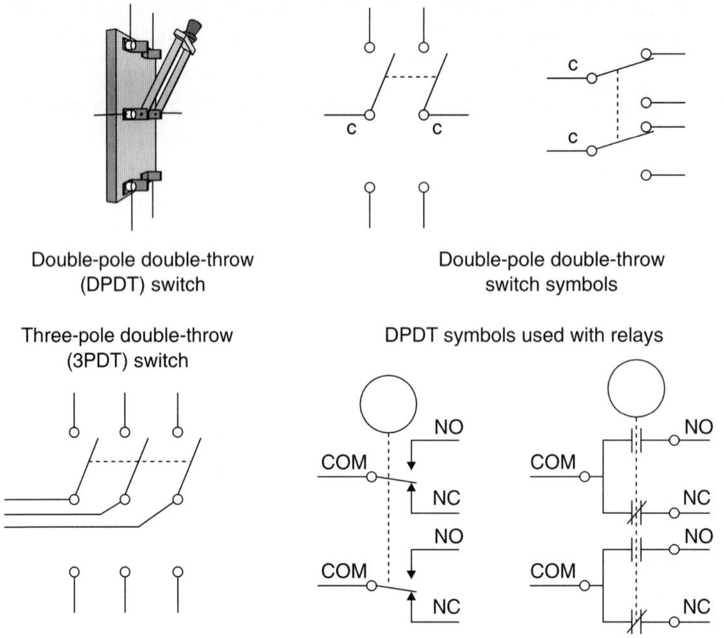

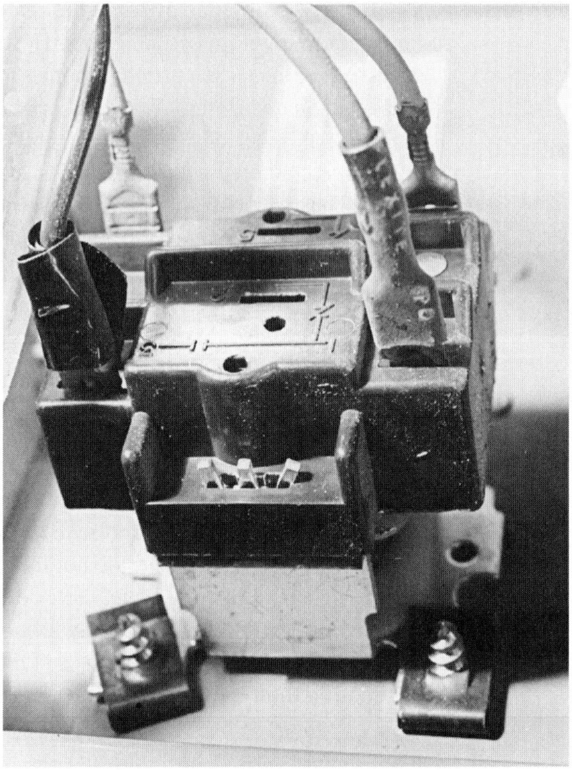

FIGURE 15–5 This relay has two sets of SPST contacts, and could therefore be classed as a DPST relay. The case markings include a normally closed contact that is not supplied with this particular model. With a complete set of contacts, the relay would be a DPDT. The near set of contacts would be accessed through terminals 1, 2, and 3. Terminal 1 would be the common, 2 the normally closed contact, and 3 the normally open contact.

FIGURE 15–6 Complex switches are usually supplied with a chart that indicates the arrangement and operation of the contacts. The Xs designate which terminals are connected for each of the four switch handle positions.

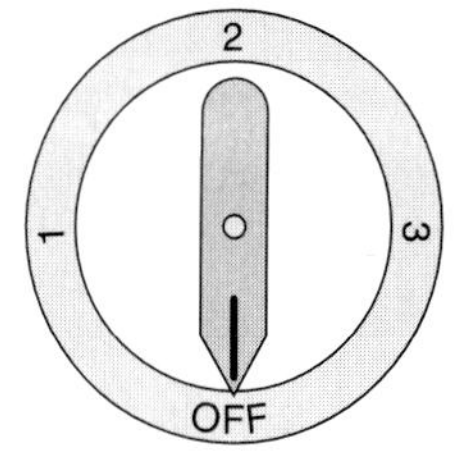

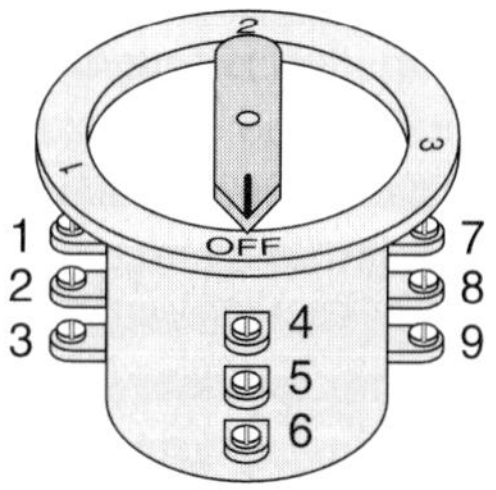

	OFF	1	2	3
1 4		X		
4 7			X	
1 7				X
2 5		X		
5 8			X	
2 8				X
3 6		X		
6 9			X	
3 9				X

Maintained and Momentary Contacts

Switches are available with either maintained or momentary contacts. Maintained contacts, like those in simple on/off switches, hold their last position until the switch handle is physically moved. Momentary-contact switches are held in one position by a spring, and go to a different position only if they are physically held in place. A very common example of momentary-contact operation is the pushbutton switch used to operate a typical doorbell.

Maintained and momentary-contacts are sometimes combined in a single switch. A car ignition switch is a good example. The auxiliary and run positions have maintained contacts; the start position has a momentary-contact.

Momentary-contacts can be either normally open or normally closed. The "normal" condition refers to the state of the contact when the switch is "at rest." For example, the doorbell button has a normally open contact, while the light inside a refrigerator would be controlled by a normally closed contact. In the first case, the doorbell is energized only if the button is pressed. In the second case, the light is on only if the door is opened. Opening the door releases the switch button, allowing its contacts to return to the normally closed position, which turns on the light.

Many pushbutton switches are momentary types, but they may also be maintained. For example, a push-on/push-off switch latches in one position or the other until the next time it is pressed.

Toggle switches usually have maintained contacts; they simply stay in the last position to which they are set. However, momentary types are also available. They have a return spring on one or more throws, so they come back to their "normal" position when they are released. Toggle switches may also have a mix of maintained and momentary-contacts, if there are at least three handle positions. Pushed to one position, the switch handle would spring back to the center when it is released. Pushed to the other position, the handle and the contacts stay in place until the handle is manually moved again.

15.4 Common Switch Circuits

On/Off Control With Single-Pole and Multipole Switches

The simple on/off control provided by SPST switches is adequate only when one line of the circuit is a safe circuit common. Grounded and ungrounded circuit conductors, and the shock hazards they represent, are often misunderstood. The following examples may help explain the problem.

An SPST switch opens only one line of the circuit, leaving the other side of the supply still connected to the load and to the rest of the circuit wiring. The switch breaks the circuit and stops current flow, but it does not completely isolate the load from the supply. When one side of the supply is grounded, only the ungrounded side needs to be disconnected. The grounded line can be left connected to the load. It is safe to touch, so there is no shock hazard when the switch is open. Virtually all 120 V AC circuits are connected this way (see Figure 15–7).

When the conductors between the supply and the load include two or more lines that are not safe to touch, a multipole switch must be used to simultaneously disconnect all of them (see Figure 15–8). Virtually all North American three-phase circuits, and all single-phase AC circuits operating at more than 120 V, are like this. Opening just one side of the circuit path does shut off the load but, unless the switch disconnects every potentially hazardous line, the result can be a very dangerous situation.

FIGURE 15–7 In the circuit on the left, the switch is where it should be, in the ungrounded supply line. If the switch is open, it is safe to touch any terminal or conductor of the circuit from the load side terminal of the switch to the grounded supply conductor. In the circuit on the right, the switch is in the grounded line. With the switch open, the load is de-energized, but the circuit is "hot" to ground from the supply all the way to the load side switch terminal. This can be a very hazardous situation.

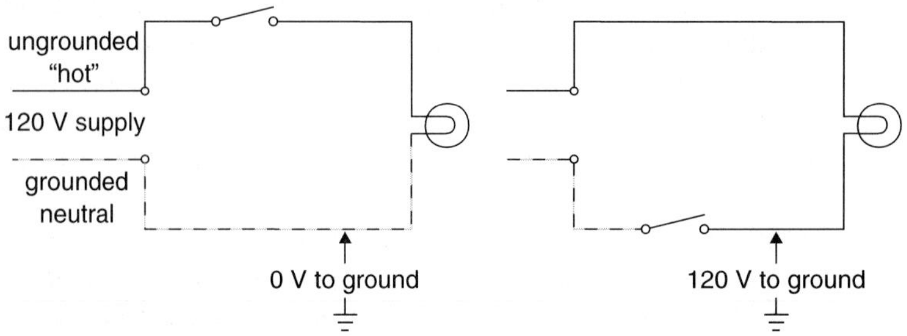

FIGURE 15–8 Both lines of 208 and 240 V circuits have a hazardous voltage to ground. For this reason it is extremely important that disconnect switches break both of the ungrounded lines. If an SPST switch is substituted for the correct SPDT, when the switch is open, the load will appear to be de-energized but will still be a serious shock hazard to anyone making contact with exposed wiring or terminals.

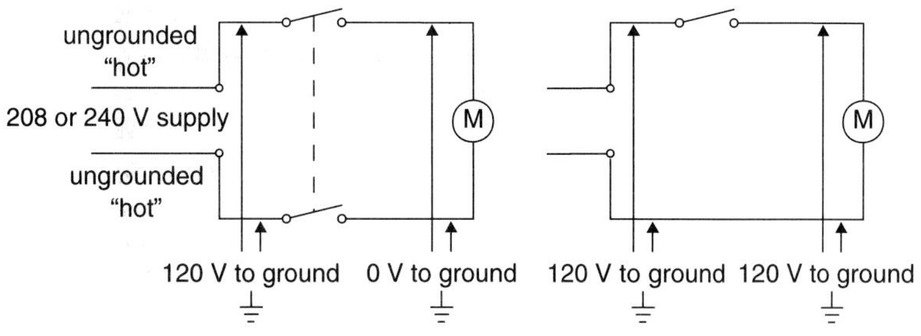

Because the motor or other equipment is shut off, there is an illusion that it is safe to work on. Unfortunately, the reality is that all the internal connection terminals and the associated wiring are still dangerous to touch, because of the shock hazard to ground. This can be a deadly trap for unsuspecting service personnel.

Warning

All electrical equipment operating from a supply of more than 120 V must have a DPST disconnect. Typical examples are single-phase 208 and 240 V motors. These circuits are particularly dangerous because a much less expensive SPST switch can be substituted for the proper double-pole version without attracting any attention. Once the switch cover is in place on the electrical box, a typical single-pole switch looks the same as a double-pole type. Not only do they look the same, they operate the same way, with the same apparent effect. The problem is that one leaves the load energized—as a shock hazard to ground—while the other leaves the load safe to work on.

All three-phase loads must have three-pole disconnects. For smaller motors, these can be box-mounted toggle switches, similar in appearance to the smaller SPST and DPST types. The shock hazard that results from breaking only one line of a single-phase circuit can occur in three-phase systems as well. Occasionally a two-pole control device may be used with a three-phase motor. This creates a similar situation to that of the single-pole switch on a 240 V single-phase circuit. The motor is shut off, but one line is left connected. This energizes the equipment and all the associated wiring at a dangerous voltage to ground.

Switches in Series

On/off control, from one location only, can be provided with an SPST switch, or with a multipole single-throw if two or more circuit conductors must be disconnected. Two single-throw switch contacts, connected in series, can provide interlocking control or a remote

FIGURE 15–9 When switches are in series, they can serve as interlocks or secondary isolation points, because all of them must be closed for the load to be energized.

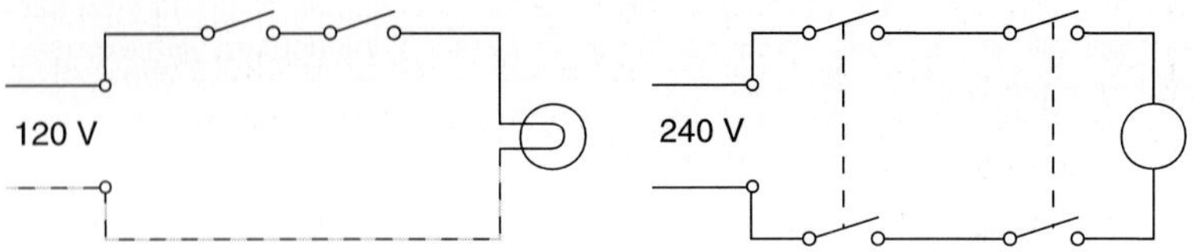

isolation point. With the switches wired in series, both must be closed for the circuit to be operational. If either switch is open, the circuit is broken at that point, and the load is shut off (see Figure 15–9).

When a switch contact is used to ensure that one or more specific conditions are met before a circuit can be energized, it is usually referred to as an interlock. For example, critical equipment enclosure covers or doors are sometimes equipped with interlocks. A typical door interlock has normally open, momentary-contacts. When the door is closed, the contact is closed and the circuit is operational; when the door is opened even a small amount, the contact opens and the equipment is shut off.

Isolation switches are used to provide a safe local disconnect point, primarily for service personnel. The isolation switch is wired in series with the main switch in the supply line to the equipment. It must have enough poles to break all the ungrounded lines of the circuit. Local isolators are mounted close to the equipment, where they can be conveniently operated and visually checked to confirm the switch position. Electrical codes generally require that an isolation switch must have provisions for a padlock. The lock can only be applied with the switch in the off position, and it cannot be turned back on until the lock has been removed.

Switches in Parallel

When two or more switches are connected in parallel, the operation of the circuit is much different than for switches in series. Series-connected switches must all be closed for the circuit to be energized. Parallel-connected switches must all be open for the circuit to be de-energized (see Figure 15–10). With two switches in parallel, there is power to the load if either or both of the switches are in the on position. The load is shut off only when both switches are open.

FIGURE 15–10 When switches are connected in parallel, they must all be open for the circuit to be de-energized.

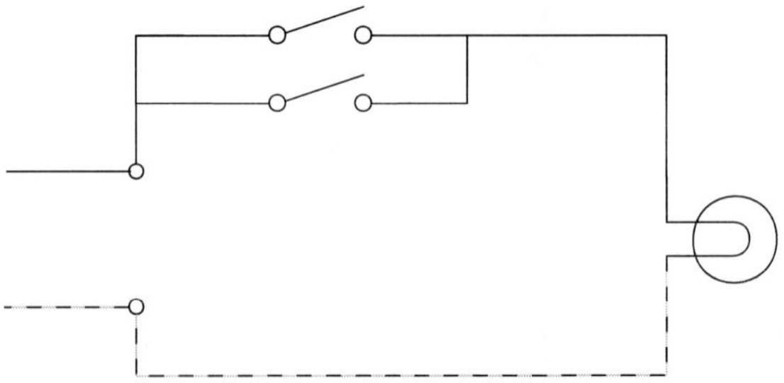

There are fewer application for parallel-connected switches than for the series connection, but one common example is the door switch circuits in a typical automobile. Opening any of the doors turns on the interior lights, and all doors must be closed for the lights to be off. Each of the doors has a momentary-contact, normally closed switch mounted in the door post. The switches are wired in parallel in the ungrounded line of the interior light circuit. An extra maintained-contact switch, also in parallel, is usually provided on the dash or at the dome light.

An example of automatic control contacts that are wired in parallel may be found in the fan control circuits of many combined heating and cooling systems. The circulating air blower motor must operate in both the heating and cooling modes, but these two systems use different blower controls. If the fan is a simple single-speed unit, the contacts of the two controls may be connected in parallel, so either of them can energize the blower motor. Multispeed motor controls are more complex.

Control From Multiple Locations

No combination of SPST switches can provide full on/off control from more than one location. If the switches are connected in series, either can hold the circuit off. If they are connected in parallel, either can hold the circuit on. In both cases, neither of the switches is guaranteed to have on/off control of the load. The only way to provide control from two locations using standard switches is to use a three-way switch circuit.

A three-way circuit is built with two SPDT switches (see Figure 15–11). Both switches must be maintained-contact types with only two positions. They cannot have a center off. As with SPST switch circuits, only one line of the circuit is controlled, while the other is left connected to the load. For this reason, the three-way switch circuit is only suitable for use in 120 V AC, or other systems that have a "safe" common conductor.

For the circuit to operate correctly, the switches must be interconnected in a very specific way. For a normal three-way connection, the power supply line feeds into the common terminal of the first switch. Two conductors are used to interconnect the noncommon terminals of the two switches. Then the common terminal of the second switch is connected to the load. The other side of the load is connected permanently to the circuit common. The two conductors that run between the switches are sometimes called the "travelers"; they are never connected to anything other than the switch terminals.

The three-way circuit operates as follows. The first switch always energizes one of the two travelers, but never both. The second switch always connects the load to one of the two

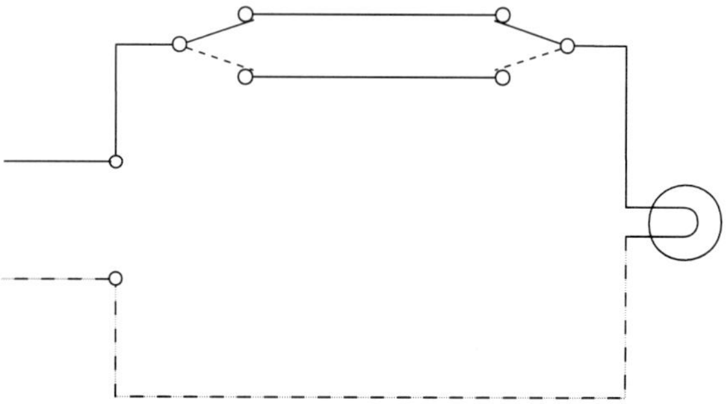

FIGURE 15–11 A three-way switch circuit is made up of two SPDT switches. Either switch has full on/off control of the load, regardless of the position of the other switch.

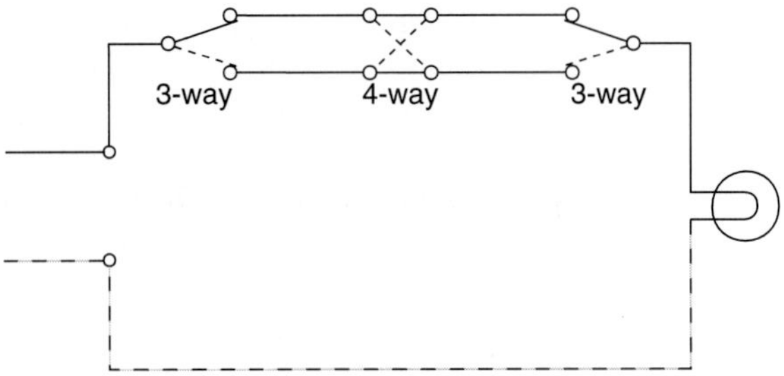

FIGURE 15–12 A four-way switch is specifically designed to cross or uncross the connections of one pair of conductors.

travelers, but never both. The logic is simple enough once understood. Either switch provides full on/off control of the load, regardless of the position of the other switch. The first gets control by deciding which of the two travelers is energized, the second by deciding which of the two travelers connects to the load.

This ingenious circuit is used extensively for lighting controls in hallways, stairways, and rooms with two entrances. Drawn as a schematic, the circuit operation is easy to understand. When it is actually installed, it can be quite confusing. All the conductors shown in the schematic are part of the two- and three-wire cables linking the physically separate switches, the controlled lights, and the power supply. Many different cable layouts are possible, depending on the situations that arise. Connections are further complicated by code rules that must be followed regarding the duties that can be assigned to each of the color-coded conductors.

The three-way circuit can be extended from two control locations to three or more, with the addition of a special switch called a four-way. Four-way switches are manufactured specifically to be connected into the pair of intermediate conductors—the travelers—that link the two three-way switches (see Figure 15–12). These four terminal switches have a unique internal connection pattern. In one position, the incoming conductors from one pair of terminals are connected straight through the switch to the other pair of terminals and the two outgoing conductors. In this position the switch has no effect on the circuit operation. The current passes straight through on each conductor, just as if the switch were not there. In the second switch position, the connections are crossed. Both of the conductors still have continuity through the switch, but this time the connection is to the opposite terminals.

When a four-way switch is installed in a three-way circuit, it adds another control location. Any reasonable number of four-way switches can be used between two three-ways to provide multiple control locations. Three-way and four-way switches are effective and not particularly difficult to install. However, for extensive multilocation control or more sophisticated switching requirements, it is usually less expensive to use relays. Inexpensive relays are available that are specifically designed for lighting control. These are widely used in commercial lighting and in larger homes.

Transfer Switches

A DPDT switch is essentially two SPDT switches with a common operating handle. An SPDT switch has one common terminal and two others, for a total of three. A DPDT switch has six terminals, including two commons, one for each pole of the switch.

The DPDT switch has a variety of interesting uses, including transferring loads between two alternate sources. When there are two sources of power, such as the normal supply utility and a standby generator, a safe and convenient method must be provided to transfer the load from one source to the other. Two separate switches cannot be used, because it is critically important that the two power sources never be interconnected, even temporarily.

From the point of view of the supply utility, an unexpected power source connected to their system will backfeed into the distribution system, energizing lines that are supposed to be dead. This can be an extremely dangerous situation for unsuspecting line service personnel. From the viewpoint of the standby generator, connection to the utility can jeopardize both the generator and its controls. For these and other less crucial reasons, the transfer must take place with no possibility of the two supplies becoming interconnected. A double-throw transfer switch fits the requirements, allowing the loads to be switched from one supply to the other, while keeping the two supplies fully isolated from each other (see Figure 15–13).

For a transfer switch connection in a typical three-wire 120/240 V single-phase system, the grounded neutrals of the two systems are interconnected. The two ungrounded conductors that feed to the loads connect to the common terminals associated with each pole of the switch. The two power supplies are then connected to opposite "ends" of the switch. The two ungrounded utility conductors connect to one pair; the alternate generator supply connects to the others. When the switch is thrown to the utility side, the connection is made for both incoming utility lines through the switch to the ungrounded conductors of the loads. With the switch in this position, the generator side of the switch is isolated from the load and the utility feed. When the switch is moved to the generator position, the utility side is automatically isolated first, and then the switch connects the generator to the load.

For a single-phase three-wire system, a DPDT switch is adequate. For a three-phase system, the transfer switch must be a 3PDT switch to simultaneously switch all three lines. Some manufactured transfer switches are built from pairs of multipole circuit breakers that are mechanically interlocked so that only one can be closed at a time. Although the breakers are only single-throw on/off devices, the resulting switch action is still that of a double-throw switch. Transfer switches must be manufactured and approved as an assembly.

FIGURE 15–13 This DPDT transfer switch connection would be suitable for a single-phase 120/240 V system using a standby generator. The switch can select either supply, but can not interconnect the ungrounded conductors of the two supplies.

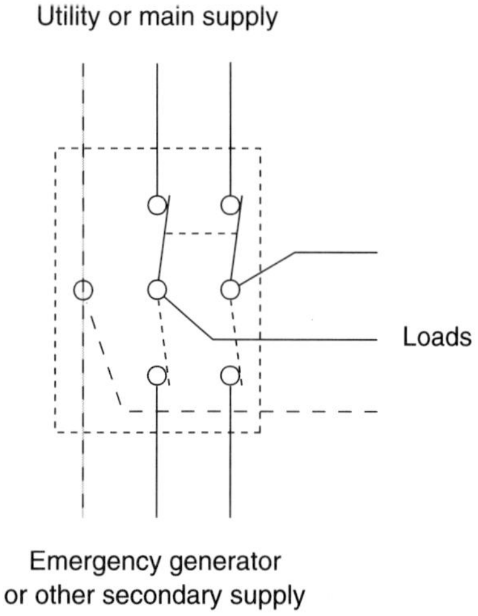

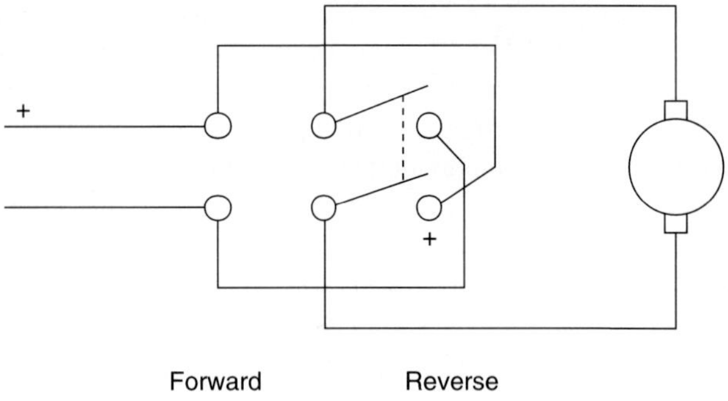

FIGURE 15–14 A DPDT throw switch can be used as a polarity reversing switch. In this example the terminal marked positive will be connected to the top terminal of the motor armature when the switch is thrown to the left, and to the bottom terminal when it is thrown to the right.

Polarity Reversal Switches

Another common use for the DPDT switch is for polarity reversal (see Figure 15–14). For this switch circuit, the noncommon terminal pairs are cross-connected by external jumpers. A pair of input leads connect to either of the jumpered pairs. The output is from the switch commons. This connection is reversible, with the same result.

In one position, the pair of conductors feeding into the switch connects straight through. In the other position, the connections of the incoming and outgoing pair are crossed. This circuit is sometimes used for reversing small motors, such as in a reversible electric drill. The switch is connected to reverse only the armature connections, not the fields. If the switch has a center off position, it can provide forward-off-reverse operation with a single control. If the switch has only two positions, a separate power switch must be provided. In many electric drills, a separate, variable-voltage trigger switch is used to regulate the motor speed with either direction selected.

15.5 Interpreting Switch Markings

The following information may help in identifying switch contact patterns and interpreting the markings on switches that do have identified terminals. Maintained-contact switches, such as general-purpose toggle or rocker types, usually have contact ratings such as voltage and current, but may have little or no terminal identification. Some are fairly obvious, based on the layout of the connection terminals, although looks can be deceiving. Others are a total mystery until they are properly tested.

Switches with more than one throw, such as SPDT or DPDT (both very popular general-purpose layouts), may have the common terminals identified with a letter C or the abbreviation COM. Strap-mounted SPDT (three-way) switches designed for standard electrical boxes usually have the common terminal identified with an oddly colored terminal screw.

Momentary-contact switches, such as typical pushbutton types, often have both normally open and normally closed contacts mounted in a single switch. The contact connections may be brought out to either three or four terminals. If the switch has only three terminals, one is a common. It is typically identified with a C or COM, and the other two terminals are marked NC and NO. The NC identifies the second terminal of the normally closed contact, the first terminal being the common. The NO marking identifies the second

FIGURE 15–15 The switch on the left is a momentary-contact SPDT push-button. In the center is a momentary-contact pushbutton with both normally open and normally closed contacts. On the right, the same switch has an SPDT action when a jumper is added. Note the markings on the switches and their relationship to the symbols.

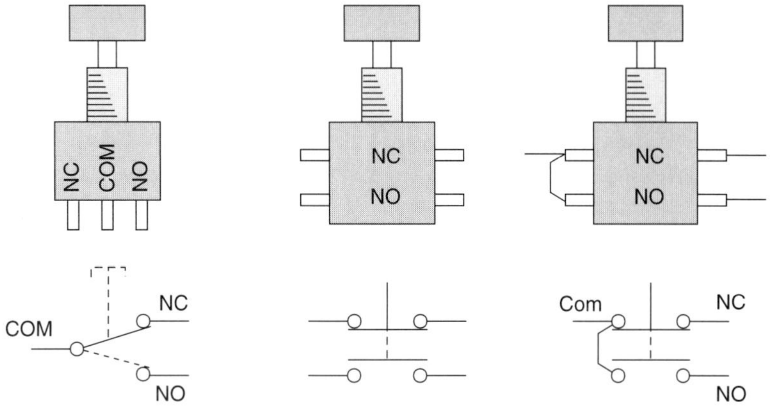

terminal of the normally open contact. When the switch has four terminals, there are separate pairs of terminals for each contact, identified as NO and NC. With two separate terminals for each contact, there is no common. When a four-terminal switch is used in a circuit that requires an SPDT operation, a jumper is added to form a common by joining one terminal from each of the two contacts (see Figure 15–15).

The word "normally," as it is used for normally open or normally closed momentary-contacts, should always be interpreted as the contact status when the switch is at rest. With a pushbutton switch, this would be when the button is not pressed. Assume that the switch is a three terminal pushbutton type with SPDT contacts. When the button is not pressed, there is a connection from the common (COM) to the normally closed contact terminal (NC), and an open circuit from COM to the normally open (NO) terminal. When the button is pressed, the contact status reverses, so there is no connection from COM to NC, but there is a connection from COM to NO.

In switches with more than one contact, all of them change state when the switch is operated. However, unless the switch is designed for some special purpose, the NC and NO contacts operate in sequence with no overlap. The NO contact does not close until after the NC contact is opened. This is true for most contact sets, whether they are in a switch or part of a relay. There are exceptions made for special situations, but such contacts should have special identifying markings, and their operation should be noted in the drawings. Typical examples of special contacts are "late break" NC contacts or "early make" NO contacts. These descriptions indicate contacts that overlap the operation of the "normal" contacts in the same device.

Industrial control panels are often equipped with a variety of oiltight switches that may be made to order with modular operators and stackable contact blocks. Operators can be selected from a wide variety of types, including various styles of push (or pull) buttons, rotary selectors, wobble sticks, and key switches. Any reasonable number of any type of contact can be added to the operator to produce the desired switching patterns. Either the individual contact blocks are marked or the contact can be identified through a clear plastic cover.

FIGURE 15–16 A DPST switch mistakenly installed as a four-way makes the circuit inoperable when the DPST is open. A four-way mistakenly installed as a DPST disconnect is incapable of shutting off the load.

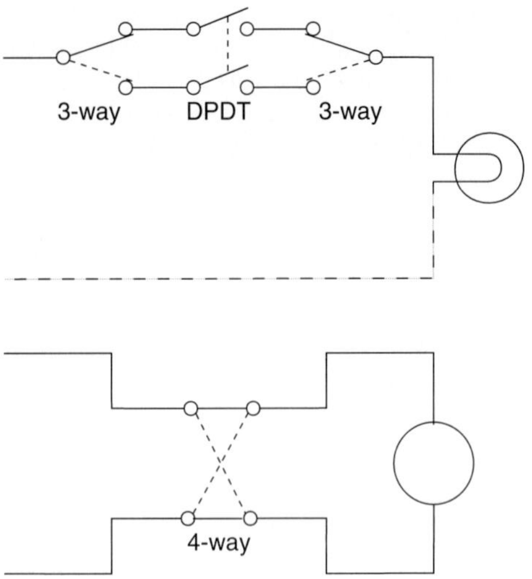

Four-way switches are usually marked in some way to avoid confusing them with DPST switches. Without some identifying marking, there is no externally visible difference between the two. Both have four terminals and two switch handle positions. Both are made in standard strap-mount style, designed to fit a standard electrical box. The terminals of both switches are usually arranged in a pattern that is similar to the way the switch appears in typical symbolic drawings. A pair of conductors enters at one end of the switch, and the second pair exits at the other. The internal circuit usually runs from end to end through the switch, rather than from side to side.

Unless the user is very familiar with the make and markings of the switch, all four terminal switches should be checked out with a continuity tester. Common application errors include incorrectly wired four-ways, or a mixup between a four-way and a DPST (see Figure 15–16). Incorrectly wired four-ways usually cause no damage, but the circuit does not work properly and finding the error can be time-consuming. A DPST switch used in place of a four-way works much like an incorrectly wired four-way: The circuit works with the "four-way" in one position, but not in the other. Correcting the error is a frustrating waste of time, which could have easily been avoided by testing the switch.

When a four-way switch is used in place of a DPST type, the switch will not shut off the load. It is unlikely there will be any damage, but the switch appears to be defective, because the load is on, regardless of the handle position. Check it out before you throw it out. The correct switch will solve the problem.

While most of these errors do not usually cause any damage, they can. With switches such as DPST, extra care should be taken to ensure that the switch pattern is understood. When the switch is installed with the supply conductors connected to the wrong terminals (across one of the contacts instead of one on each contact), the result is a short circuit. This error will blow fuses or trip the circuit breaker, and may also permanently damage the switch contacts (see Figure 15–17).

Larger power switches, and all knife-style switches, have specific line-side and load-side terminals. The line-side terminals connect to the incoming supply. The load-side terminals connect to the loads or other parts of the circuit. Wherever possible, knife-switches

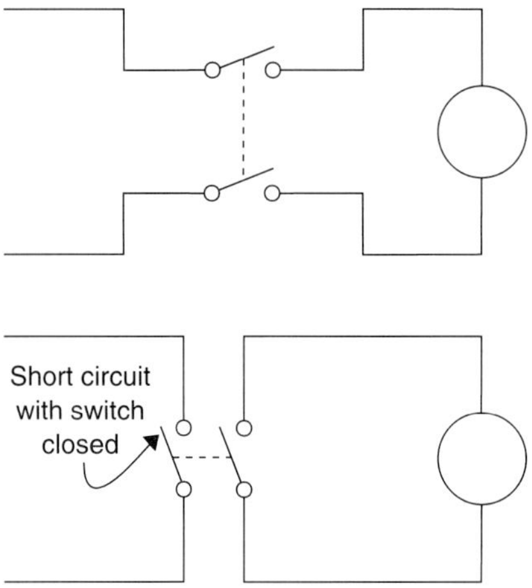

FIGURE 15–17 If a DPST (or any switch) is mistakenly wired with a pair of contacts across the supply, the resulting short circuit may damage the switch contacts.

are connected with the line to the stationary terminals, and the load to the terminals that connect with the moving blades. That way, the switch blades are not energized when the switch is open. Energized blades are unavoidable on double-throw switches, but should be avoided in all other cases. When any switch has markings identifying one side as line and the other as load, these markings should be respected. When there are no markings and the contacts are enclosed, there is no specific line or load terminal, so the switch can be connected either way.

It is conventional to mount vertically oriented switches with the line terminals toward the top. This is a throwback to knife-switches, which are supposed to be mounted that way to prevent an open switch from falling closed. With the line side at the top, the switch blades are pulled down and away from the line terminals when the switch is opened. On large switches and vertically oriented circuit breakers, a good first assumption is that the hot terminals holding the incoming supply lines are at the top. However, it is not a good idea to bet your life on it.

knife-switches helped create the convention that a switch handle should be in the up position when the circuit is on, and down when it is off. This practice is followed with reasonable consistency in North America, but it is not universal around the world. Three- and four-way switches are exempt, because either up or down can be on, depending on the position of the other switches in the circuit. Most control switches are marked in some way to indicate the on position; three- and four-way switches have no markings.

15.6 Practical Exercise: Testing Switches and Circuits

The following information includes practical advice and descriptions of typical switches and switch circuits. If the appropriate materials are on hand, the tests described can be conducted as practical exercises. The materials used include a number of different switches, a DC power supply, a multimeter, three or more low-voltage lightbulbs with sockets, and an assortment of connecting wires. For the complete set of experiments, the following switches are required: two SPST, two SPDT, one DPST, and one DPDT. All of these should have maintained contacts. Two momentary-contact pushbutton switches are also required, one with at least one NC contact, the other with an NO contact.

For the final tests, examining switches in grounded circuits, the power supply must have a grounded common. The test can be done with a low-voltage supply or a standard 120 V outlet. For a 120 V supply, a suitable 120 V lamp and socket and a 120 V rated SPST switch must also be available. The 120 V AC supply must be a normal one, with a grounded neutral. Using an isolated supply defeats the purpose of the test. Many laboratory 120 V supplies have isolation transformers installed for added safety and to avoid problems with oscilloscope connections. This type of supply is not suitable.

Before building any circuits, each switch should be identified and thoroughly tested. It is best to use standard naming conventions where possible, and to sketch the switch layouts in a standard symbolic form. Confirm the operation of each switch by methodically testing it with an ohmmeter or a continuity tester.

Note

A closed contact has a measured resistance of 0 Ω, and an open contact has an infinite resistance, ∞ Ω. The 0 Ω value will be true for all but the most sensitive of ohmmeters. However, a typical digital multimeter indicates a resistance of about 0.1 to 0.4 Ω, with the meter probes shorted together. This is the resistance of the meter leads. Do not confuse this value with the actual resistance of a switch contact.

To measure very low resistance values accurately with a digital meter, the measured meter-lead resistance must be deducted from the reading taken across the device. If the lead resistance is higher than expected or seems to change with each reading, the leads may have broken strands and are due for replacement or repair.

Switch Testing

Modern switches rarely use open blades. The contacts are totally enclosed and, unless the enclosure is clear plastic, it is impossible to see how the switch works. To further complicate the problem, many general-purpose switches have no identifying markings to show the internal arrangement of the contacts. Without conclusive markings, the only sensible alternative is to test the switch with an ohmmeter or a continuity tester. Even if the switch is marked or the terminal layout makes the connections obvious, it is often worth the time it takes to check the switch before it is installed in the circuit.

The key to efficient, error-free switch testing is a methodical check of every possibility. If the switch has more than four terminals, a pencil and paper are almost essential. The best procedure is to choose one switch position and check continuity from each terminal to every other terminal. Then set the handle to the next position and repeat this procedure, testing all possible connections again. When there are more than two handle positions, repeat the same procedure for each one. If the results are recorded, they will be a complete listing of all the internal connections for each of the possible handle positions. For a complex switch, it is usually best to put the information in chart form. When the switch is a relatively common one, it is easier to convert the results to a standard switch symbol.

The Basic Switch Circuit

For this test, an SPST switch is used to control a single lightbulb load (see Figure 15–18).

Note

The single most common beginner's error is connecting both of the supply leads to the switch. This causes a short circuit when the switch is closed. A single-pole switch can break only one line. Therefore, only one side of the supply can pass through the switch. The other side of the circuit is connected directly to the load.

Open-blade switches should have the line (supply) side conductors connected to the stationary contact terminals, not to the terminals that connect directly to the blades. The blade terminal is the load side. This type of switch should also be mounted with the line side to the top. It is a good habit to regard the top terminal of vertically mounted switches as the line side terminal, but typical box-mounted switches used in 120 V AC circuits have unmarked terminals. Power may enter the switch through the top or bottom terminal. In all cases where the rule can be applied, the switch handle position should be up when the switch is closed.

FIGURE 15–18 The pictorial and schematic represent the same switch circuit. A light is turned on and off by an SPST switch.

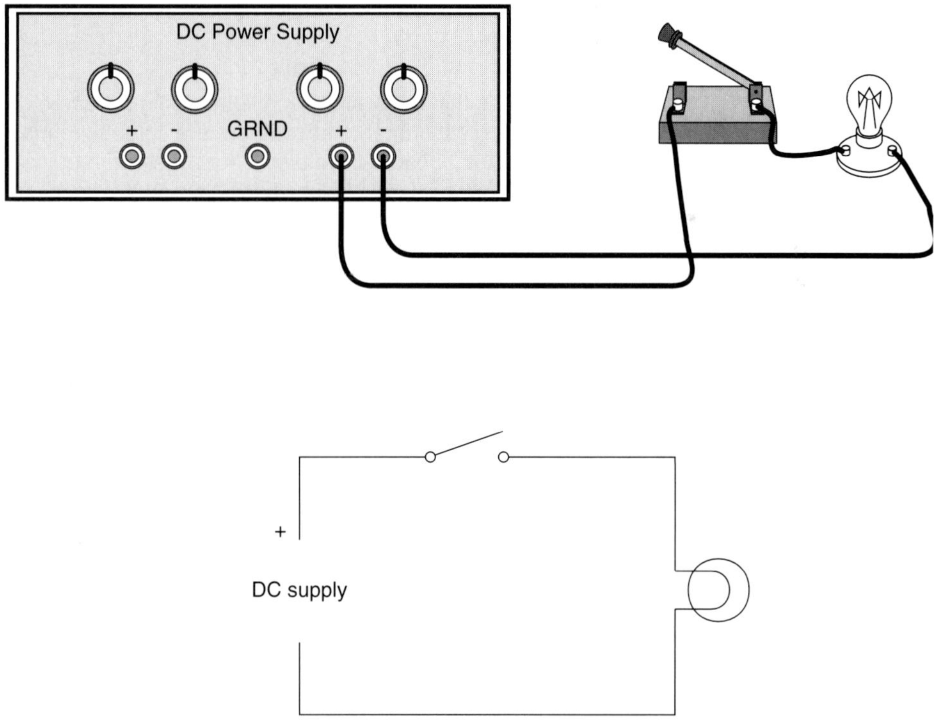

The power supply output must match the rating of the bulb. When the circuit is tested, the bulb goes on and off when the switch is opened and closed. If the test is repeated with a voltmeter across the bulb, the readings are no voltage with the bulb off, and full supply voltage when it is turned on. The results of these tests should not be surprising. When the circuit is shut off at the switch, there is no voltage across the load because there is no connection back to both sides of the supply. When it is on, the load receives full voltage, because the switch connects the supply to the load without adding any resistance to the circuit.

A more informative series of tests can be conducted with one voltmeter lead connected to the circuit common, which is the unswitched side of the circuit (see Figure 15–19). Voltage readings should be taken from the common to each of the following points in the circuit: the supply side of the switch, the load side of the switch, and both the line and common sides of the load. The tests should be repeated for both on and off conditions.

This second test demonstrates how the "hot" or supply side of the circuit can be tracked from point to point around the circuit by referencing all the voltage readings to the circuit common. This is an excellent troubleshooting technique, because an open point in the circuit is identified immediately. The supply voltage disappears when the first opening in the circuit is passed. For example, with the switch open, there will be voltage on the line side terminal, but not on the load side. With the switch closed, there is voltage all the way to the line side of the load. There is no voltage on the common side of the load, because the meter probes are essentially connected together through the circuit common. They are at the same (electrical) point in the circuit; therefore, the meter reads no voltage.

Note

It should be understood that this testing technique works the same on complex circuits. For example, in most burner controls there are several normally closed limit switches wired in series. Checking voltages from the common to both sides of each switch will quickly identify the point where the circuit is open.

Another informative though potentially confusing test can be performed by connecting the voltmeter across the terminals of the switch with power applied. A voltage reading is taken with the switch open, and then again with it closed. A final reading is taken with the meter across the terminals of the open switch and the bulb removed from the lamp socket.

These three tests demonstrate a commonly misinterpreted voltage reading. When the switch is open, the lamp is off, but the meter indicates full circuit voltage. When the switch is closed, the lamp is on, but the meter reads no voltage. When the switch is open but the load is disconnected, the meter again reads no voltage. These results are very easily misunderstood (see Figure 15–20).

With the switch open, the circuit is a series connection through the resistance of the meter and the resistance of the load. The meter has a very high resistance (10 to 20 MΩ for a typical digital multimeter, somewhat less for analog meters) compared to the relatively low resistance of a typical load device. With such an extreme difference, the load is essentially a connecting wire to the meter, and the meter is virtually an open circuit to the load. The result is that the meter reads full circuit voltage, but the load does not operate. When the load is disconnected—the bulb unscrewed—the meter reading drops to zero because the meter no longer has a connection back to the supply.

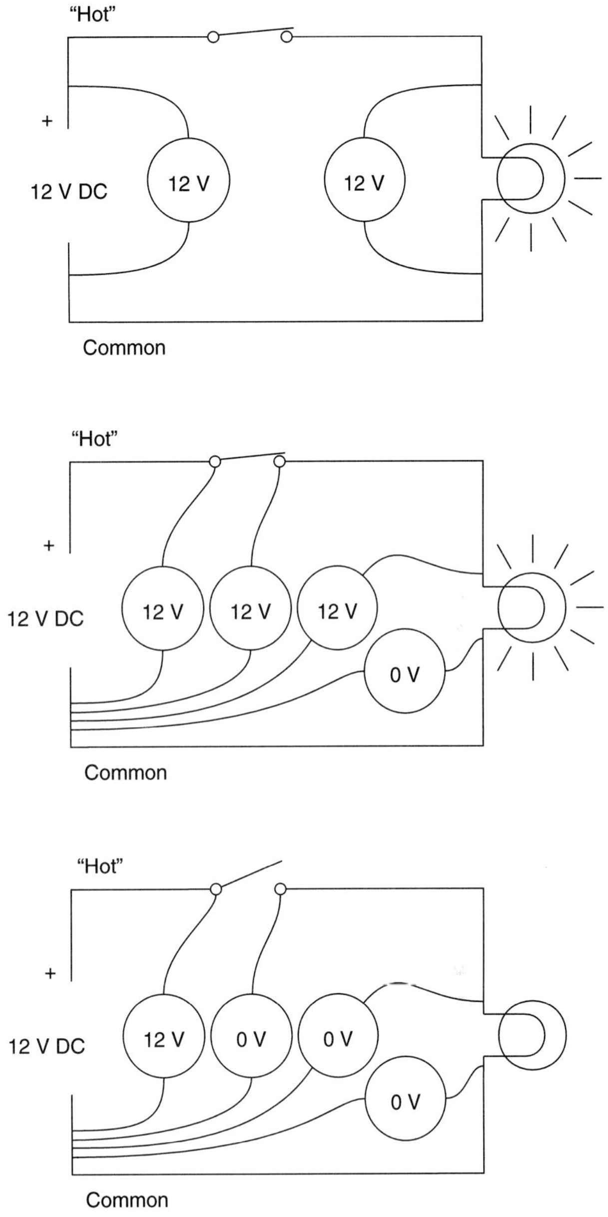

FIGURE 15–19 There are several methods of testing circuits, but voltage measurements from the common are often the most informative and error-free.

There are a number of important lessons in all of this. The best method of tracking through a series of contacts is to make voltage measurements from the circuit common to all the various connection points around the circuit. The fastest way is to check at the last terminal first. If it is live, the circuit is continuous to that point. If it is dead, jump back to the supply terminal and work through each accessible terminal point and wiring connection, all the way back to the line side of the load. The supply voltage will be lost as soon as the first open point is passed.

FIGURE 15–20 These SPST switch voltage readings demonstrate why the results of such tests are often misinterpreted.

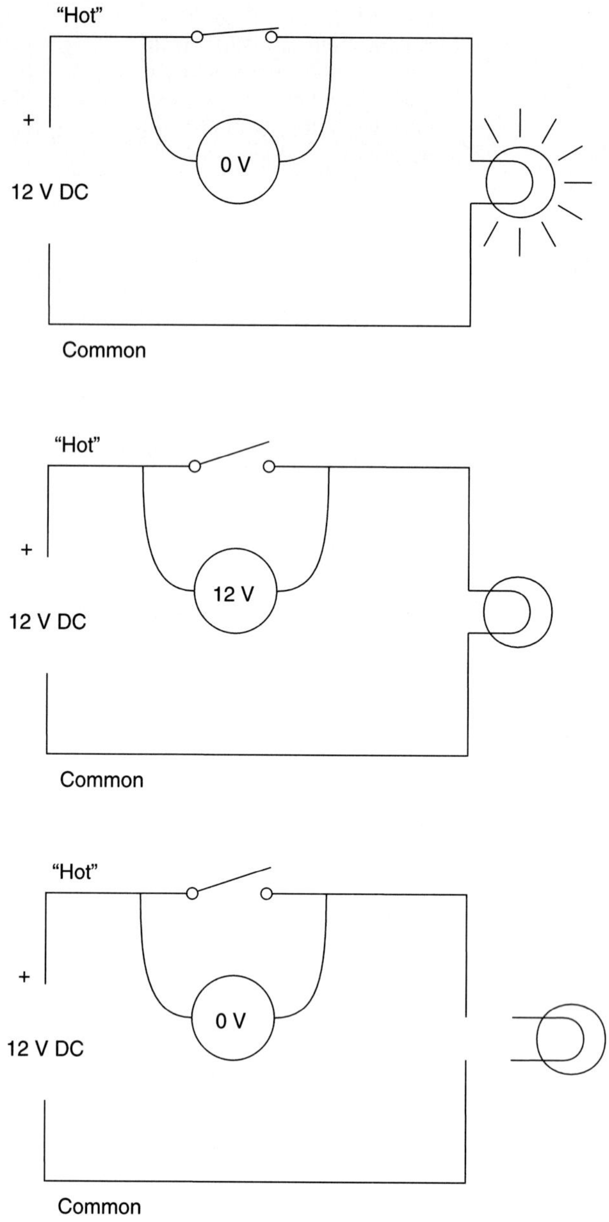

The best method to test for power through a switch or other control is to measure from the circuit common to each side of the control, rather than measuring across the terminals of the device. If there is power on both sides, there is continuity through the control, and its contact is closed. If there is power on the line side only, it is open. If there is no power on either side, the supply has been lost at an earlier point.

When a voltmeter is connected across a switch or other control device, the results are less conclusive. The meter will almost always detect either full supply voltage or no voltage. Full voltage means the switch or contact is open, and the rest of the circuit is probably intact. This tests the continuity of the rest of the circuit (though not very well), but not the switch contacts. When the test indicates no voltage, there are three possible reasons, not just one. First, the circuit may be operating normally, and the switch is closed. Second, there may be no supply voltage to the switch. Finally, there may be an open at some point further into the circuit. This is not a very conclusive test result.

Testing Multiple Controls in Series

There are many instances where several control contacts are connected in series. These circuits are used where a number of conditions must be met before the load is permitted to operate. A test circuit can be built by expanding the basic circuit of the previous test with one or more additional switches. Any reasonable number can be added, but they should all be on the line side of the load, and all in series. The circuit common should not contain any control devices (see Figure 15–21).

Once it is properly connected, the circuit can be tested. The lamp is on only when all the series-connected controls are closed. With the lamp off, a test from the common to both terminals of each switch will identify the first open point in the circuit. The first open switch will have voltage on the line side terminal, but not the load side. Since the rest of the circuit is de-energized, all the terminals on the load side of the first open will show no

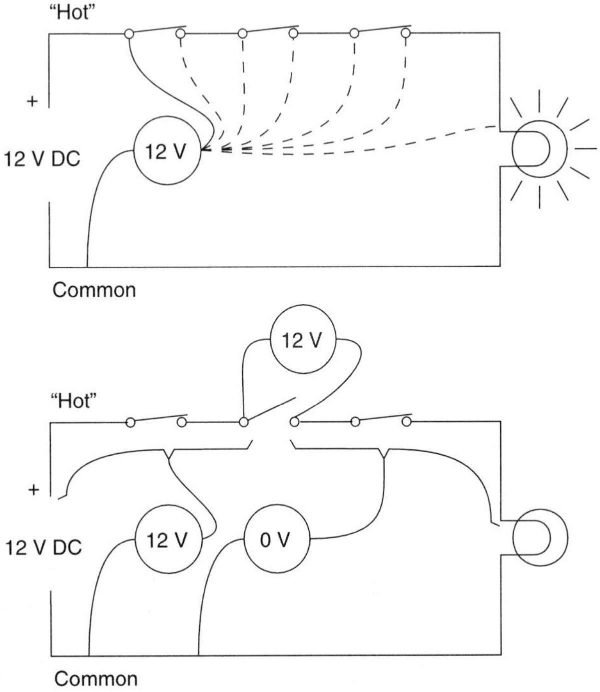

FIGURE 15–21 When there are multiple switch contacts in series, the voltage measured from the common drops to zero when the first open point is crossed. If that is the only open, a test across it will indicate full circuit voltage.

voltage to common. A voltage check across the first open contact tests the continuity of the remainder of the circuit. If full circuit voltage appears across this contact, it is the only one in the circuit that is open; if not, there is at least one more open point.

Series-connected contacts are sometimes used as interlocks. This term generally means a contact that is used to ensure certain conditions are met before the circuit is allowed to operate. For example, a normally open pushbutton door interlock is placed behind the blower chamber access cover on some furnaces. The interlock contact is connected in series with the 120 V AC supply. Under normal circumstances, the cover holds the switch closed. When the cover is removed, the switch opens, shutting off the power supply to the furnace.

Series contacts are also used as limits. Limits function much the same as an interlock, but there is a difference in the intent. An interlock is generally used to prevent a piece of equipment from being started unless specific conditions are met. A limit shuts the circuit down when a condition occurs that is dangerous or otherwise undesirable. They are both series contacts, and in both cases they must be closed for the circuit to function. An interlock is more likely to be open when the equipment is de-energized, while a limit should be closed unless the undesirable condition exists. The following example might help in understanding this concept.

Suppose a conveyor belt system is used to carry material to a storage bin. There are two belts in series, one dropping material onto the next, which then dumps into the bin. The first belt must not operate unless the second is running, because the material would quickly fill the chute between the two belts and spill out on the floor. A contact from the control of the second belt can be used as an interlock in the control circuit of the first belt. This contact is open when the system is idle, forcing the belts to be started in sequence, second belt first. It will also stop the first belt if the second belt shuts off.

An additional control is required to stop the belts when the bin gets full. A limit switch is installed in the bin, near the top. This switch has normally closed contacts wired into the controls of the second belt. When the bin is full, the limit contact opens, shutting off the second belt, which in turn—through the interlock—also shuts off the first belt. The bin limit switch is closed, except when the bin is full. The belt interlock is closed only when the second belt is running.

Testing Multiple Controls in Parallel

When switches are connected in series, any one set of contacts can prevent the load from receiving power. When they are in parallel, any one set of contacts can ensure that the load receives power (see Figure 15–22). Again, a simple test circuit can be constructed with any reasonable number of SPST switches.

When one or more of the switches are closed, the load is energized. Testing from the circuit common shows voltage at both terminals of every switch, regardless of whether it is closed or open. Closing additional switches does not change the operation of the load or the results of the test. The test checks for continuity through the group of switches, but it does not show which of them are open and which are closed. Only when all the switches are open is the load shut off. Then there is no voltage on the load side of any of the switches.

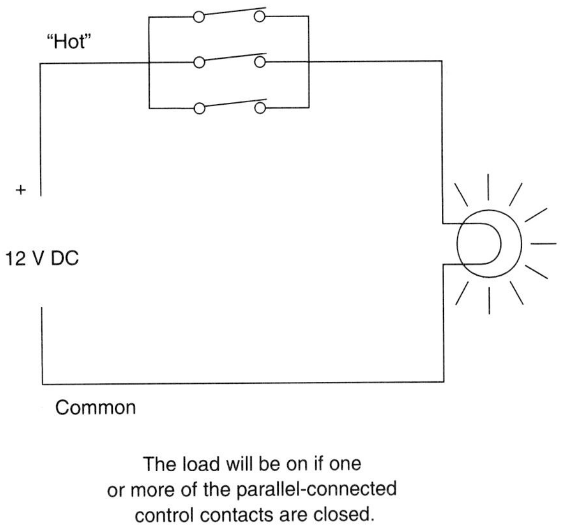

FIGURE 15–22 When there are multiple switches in parallel, any of them can ensure the load is energized.

Testing across any of the switches will produce similar, somewhat inconclusive results. If any of the switches are closed, the voltage across all of the switches is zero. If all of them are open and there is continuity through the rest of the circuit, the meter indicates full voltage across all of them.

If should be obvious from these tests that multiple contacts connected in parallel can be difficult to troubleshoot. When any switch or contact is closed, the others test as if they are also closed (see Figure 15–23). There is no simple solution to this problem. It may be necessary to eliminate the possibilities one at a time by disconnecting the contacts, at least on one side, and testing them for continuity.

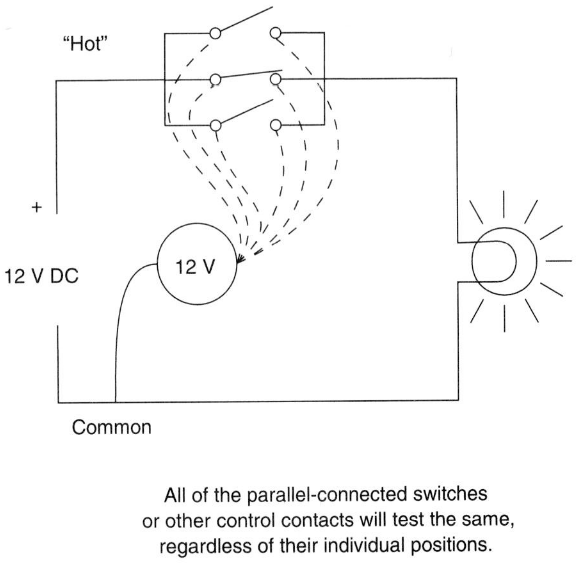

FIGURE 15–23 If one or more switches in a parallel group is closed, all the switches test as if they are also closed.

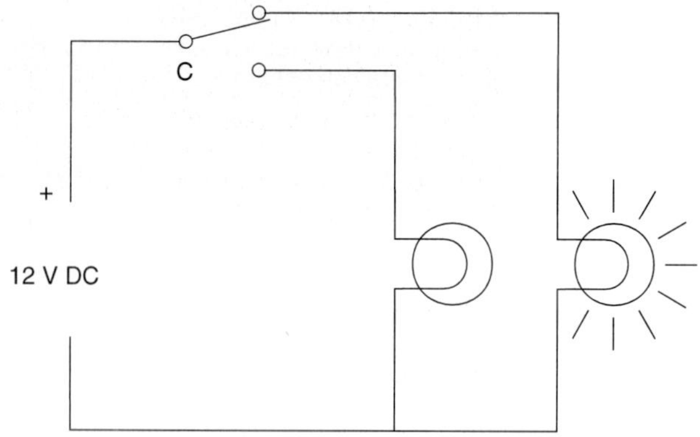

FIGURE 15–24 This SPDT switch can energize only one load at a time.

The SPDT Switch

There are a number of simple experiments that can be used to examine the operation and uses of the SPDT switch. With power to the common terminal, the switch provides an either/or type of control, where only one of two loads can be powered (see Figure 15–24). Testing this circuit is very much like testing one using SPST switches. Measuring from common, there is voltage on only one "output" terminal at a time.

An example of where this is used to advantage is in control circuits for multispeed motors. This type of switch—or an equivalent set of relay contacts—can never energize more than one output at a time. This is important because it protects against simultaneous energization of two motor speeds, a condition that is disastrous for most types of electric motors.

Two SPST controls can be connected in a series cascade connection to extend the motor speed control capability to three speeds (see Figure 15–25). In this arrangement, one output terminal of the first switch or contact set is connected to the common terminal of a second SPDT control. Full protection is provided against simultaneously energizing more than one speed. Voltage tests, taken from the common, can be used to track the current route through the switches. It can get tricky with a motor load, however, because the interconnections inside the motor energize all the motor speed leads whenever the motor is running. It is still possible to follow the circuit, because all the voltages are different. The motor lead that is actually connected to the line will, of course, be at line voltage. All the rest of the speed leads will test higher or lower, depending on the speed they represent. All lower speeds will have progressively higher voltages; high speeds will be progressively lower. A circuit like this is common with multispeed furnace blowers. Several variations are examined in the residential furnace chapter later in this text.

The Three- and Four-Way Switch Circuits

The three-way switch system is based on a special interconnection of two two-position SPDT switches. An experimental three-way circuit can be constructed with a pair of suitable switches, a power supply, and a bulb (see Figure 15–26). In this case, the 12 V lamp and power supply will work nicely. To remember the connections for a three-way switch circuit, note that the supply line connects to the common terminal of the first switch, and the common of the second switch feeds power to the load. A pair of wires connect the

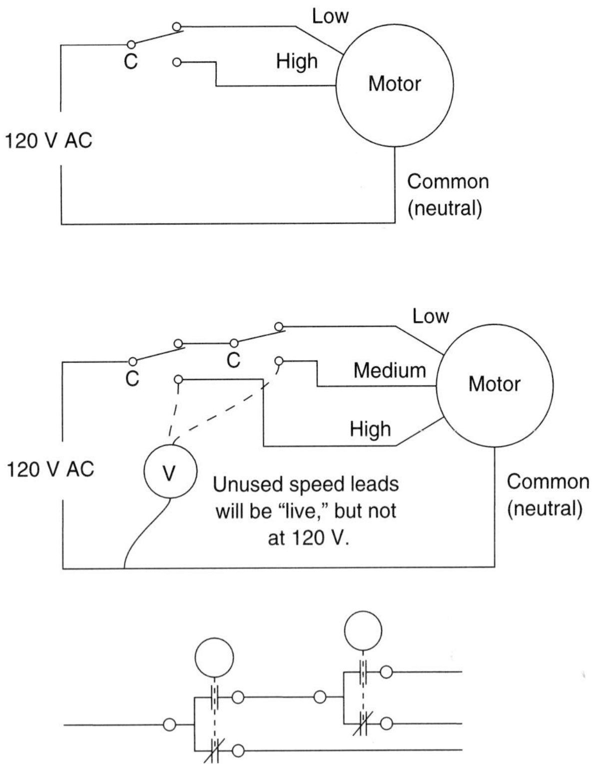

This type of control is usually done with relays.

FIGURE 15–25 Two-speed control with SPDT switches or contacts can be expanded to three speeds, if the second switch is connected in series with the normally closed contact of the first. When the load is a motor, the "disconnected" terminals test "hot" (usually at unexpected voltages) because of feedback from the motor windings.

noncommon terminals of both switches. These intermediate wires only interconnect the two switches; they do not connect directly to the supply or the load.

The three-way switch circuit works much like a single-pole on/off switch. Only one supply line is broken. The circuit must have an unswitched common return conductor from the supply to the load. In a 120 V AC circuit, this common conductor must be the neutral.

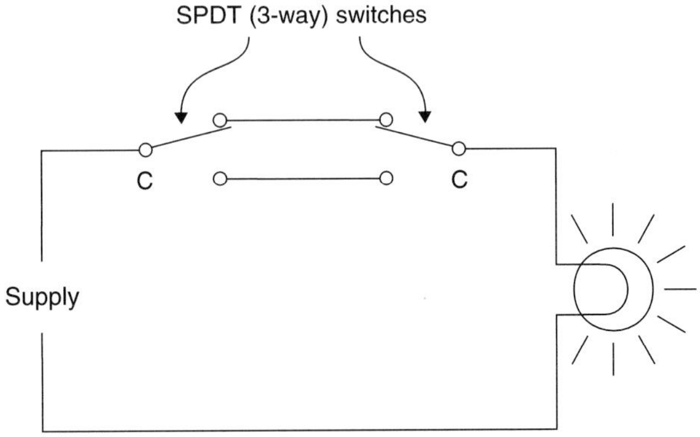

FIGURE 15–26 A three-way switch circuit can be constructed from two SPDT switches. Switches designed for use in three-way switch circuits are identified as "three-way," but they are just two-position SPDT switches.

The operation of the circuit can be examined by taking voltage readings from the circuit common to the switched portions of the wiring. With the load energized, supply voltage appears at the common terminal of the first switch—this is a direct connection to the supply—and at only one of the other two terminals. This is a two-position SPDT switch; therefore, there will always be one, and only one, of the two terminals energized. The energized terminal changes each time the position of the switch handle is changed. At the second switch, one of the incoming pair is live, but the other is not. This switch connects one, and only one, of the incoming pair to its common terminal. Therefore, the common is energized or de-energized, depending on the position of the switch.

The result is, of course, that either switch can turn the load—in this case the lightbulb—on and off. The first switch selects which one of the interconnecting pair is energized. The second switch determines whether the load is connected to the energized line or the one that is de-energized. Regardless of the position of the other switch, either of the two switches can control the load.

The three-way circuit can be extended to three or more switch locations with the addition of four-way switches. A four-way switch is a two-pole polarity reversing switch with no off position. In one position the pair of conductors are connected straight through the switch; in the second position they are crossed internally. Placed between two three-way switches, a four-way provides an additional control point. Just as the first three-way switch alternates which of the two travelers is energized, so does the added four-way switch.

A four-way switch can be adding to the three-way circuit of the previous test (see Figure 15–27). Voltage tests will show the two possible current paths through the switch. If an appropriate four-way switch is not available, a DPDT switch can be used instead. Jumpers must be added to the switch so it reverses polarity (see the polarity reversal experiment that follows).

There is no theoretical limit to the number of four-way switches that can be added to a circuit. However, there are practical limits set by the cost of materials and the amount of

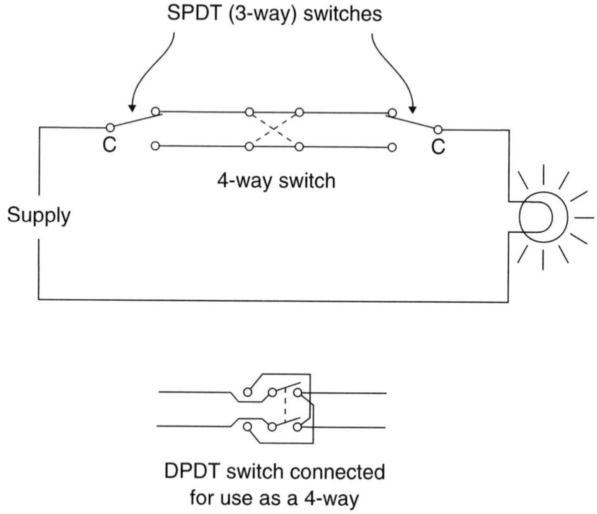

FIGURE 15–27 One or more four-way switches can be added to a three-way circuit to increase the number of control points. If a four-way switch is not available, a DPDT switch can be used, with the appropriate jumpers.

wiring required. When a large number of switch locations is desired, or the distance between the controls is very long, it is usually more efficient to use a relay circuit instead of three- and four-way switches. Conventional relays can be used, but there are relays made specifically for this purpose that are better suited to the task.

Polarity Reversal DPDT Switch

A simple polarity reversing test circuit can be constructed with a DPDT switch and a lamp. Two jumpers are used to cross-connect the terminals at opposite ends of the switch. If the switch has only two positions—no center off—it operates the same as a four-way switch. When the circuit is initially tested, the switch appears to be useless because it has no effect on the lamp; it is on in both positions (see Figure 15–28). The effect of the switch is not apparent until the voltmeter is connected across the terminals of the bulb. The meter indicates full voltage for both switch positions, but the polarity is reversed. The supply must be DC, not AC. With an AC supply, there is no polarity indication on the meter.

Polarity reversal is used to reverse many electric motors. For a permanent magnet DC motor, the switch can be connected into the supply lines. For a wound-field DC motor or series universal motor, it must be connected into the armature circuit. For single-phase AC motors, this type of switch can be used to reverse the connections to the start winding. A

FIGURE 15–28 With no meter connected, this switch circuit appears to do nothing. The light is on with either switch position, but the meter proves the polarity changes.

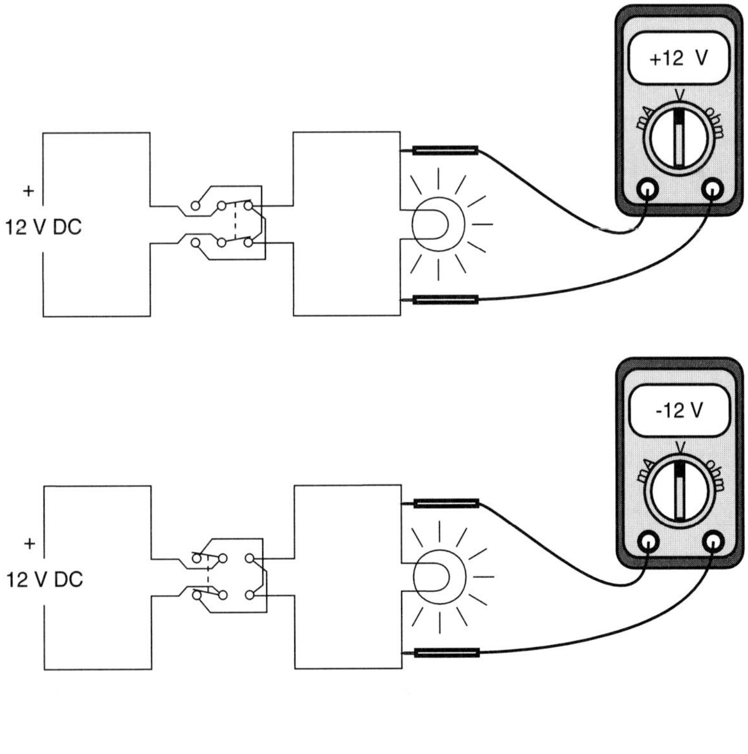

second control will be required to turn the power on and off. There are problems associated with reversing most single-phase motors. Please refer to the single-phase motor control section of this text for a more complete explanation.

There are special motor reversing switches available. They are more expensive than general-purpose switches, but they are also made for motor switching duty and have extra contacts to handle control of the power supply and reversal with only one switch.

Double-Pole Disconnects

In this experiment, a DPST switch is used as a disconnect (see Figure 15–29). The supply feeds into two of the terminals, one from each contact pair. The load connects to the other two. When the switch is closed, the supply current passes through one switch contact, through the load, back through the other switch contact, and finally back to the other side of the supply. When the switch is open, the circuit is broken on both sides, rather than on just one, so the load is totally isolated from the supply.

When a DPST switch is installed, care must be taken to ensure that the correct terminal pairs are used. With a low-current laboratory-type power supply, there is little chance of damage. However, in a real-world application, the switch can be destroyed by a simple wiring error. If two terminals from the same contact are mistakenly connected across the supply, closing the switch will cause a short circuit through the contacts on that side of the switch. A typical AC supply circuit may deliver enough current into the short to damage the contacts.

When a DPST switch is used as a disconnect, it turns the load on and off, exactly the same as if the control were an SPST switch. There appears to be no difference and no obvious advantage to using this more complex—and more expensive—switch in place of the more common SPST. However, the difference between the two is a critical safety concern.

The SPST disconnect is only permitted in circuits where the voltage is too low to be considered a shock hazard, or when one side of the supply is made safe by connecting it to ground. In these situations, a circuit controlled by an SPST switch is not hazardous, as long as the wiring is properly connected. The grounded side must be the circuit common, and it must be permanently connected to each load device. All switches or other controls must be installed in the ungrounded line. With the switch open, one side of the supply is left connected to the load, but it is also connected to ground, and therefore it is safe to touch.

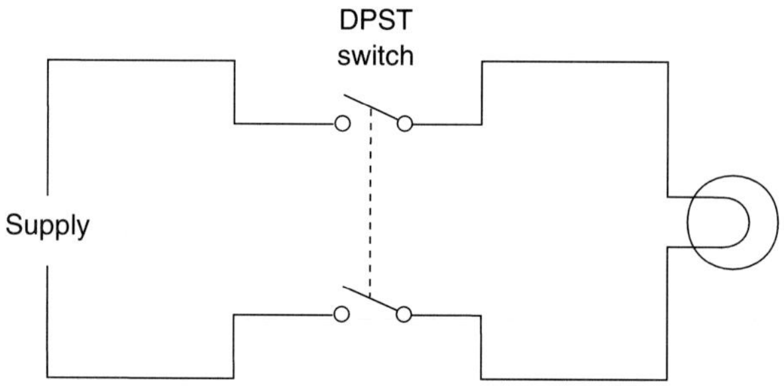

FIGURE 15–29 A DPST switch opens both lines to totally isolate the supply from the load.

All North American AC services voltages—such as the standard 120 V AC that is used almost everywhere—have a grounded common called the neutral. Single-pole controls provide a safe disconnect in such systems. Many lower voltage systems also treat one side of the circuit as a common, and use single-pole controls in the other line. These systems are not always grounded, but this is not a safety hazard because the voltage is so low. However, the situation changes dramatically for AC supplies that are above 120 V.

Almost all North American AC power circuits operating at more than 120 V AC cannot have either side of the circuit connected to ground. Both lines from the power supply represent a shock hazard. In these circuits, contact between any part of the circuit and a grounded surface is extremely dangerous. To make the load truly safe, the disconnect must have one pole for each ungrounded supply conductor. This normally requires two-pole devices for single-phase circuits—typically 208 or 240 V—and three-pole devices for three-phase—which start at 208 V and go up to 600 V or more.

Most of the electrical accidents caused by incorrect switches occur on 208 or 240 V single-phase circuits that feed relatively low current loads, such as motors of two horsepower or less (see Figure 15–30). On these circuits, there is a strong temptation for the installer to substitute a general-purpose SPST switch in place of the required double-pole type. The single-pole switch costs much less than a double-pole switch, and is more readily available. With the switchplate installed on the electrical box, standard strap-mounted SPST and DPST switches look the same. They also have the same apparent effect on the load.

FIGURE 15–30 In either of these circuits, with the switch open, a voltage test across the terminals of the motor will indicate 0 V. However, a test from the neutral or any good ground will indicate a shock hazard of 120 V (for 208 and 240 V circuits) for the SPST switch circuit, and a safe 0 V for the DPST circuit.

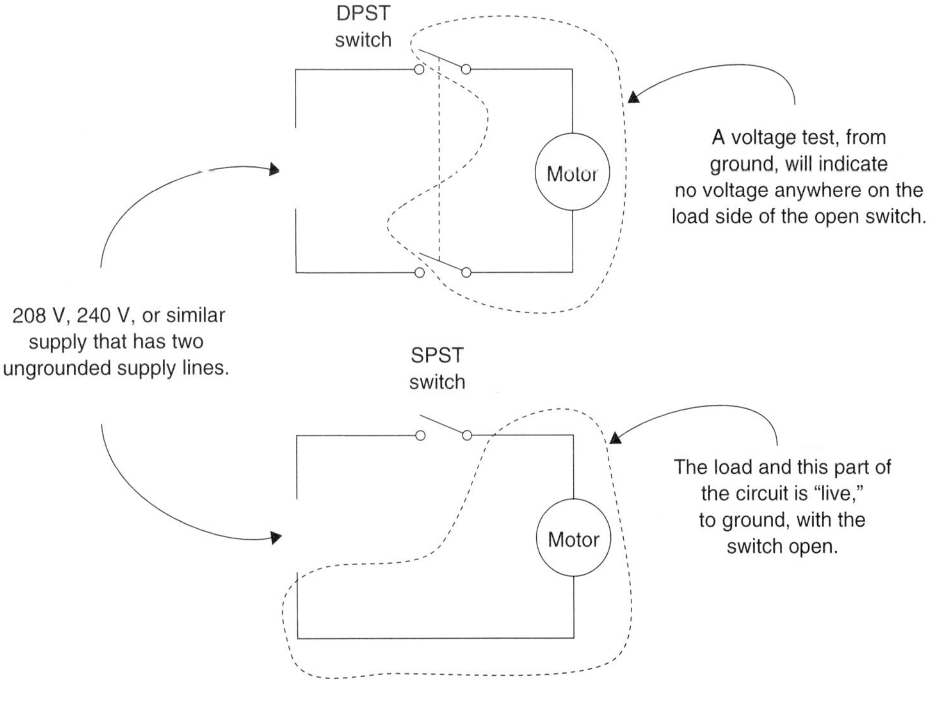

The moral of the story is do not trust the disconnect. Do not touch any electrical terminals until they have been tested for voltage from all the supply terminals to a good ground, such as the frame of the equipment or some metal enclosure in the wiring system. If a voltage is detected, it will not be the full supply value. The normal voltage to ground is either 50% or 58% of the line value, depending on whether the supply transformer is single-phase or three-phase. Any voltage at all should put the disconnect under suspicion. Shut off the supply at the breaker or disconnect and investigate. If there is a problem with the switch or a single-pole type has been used, replace it with a double-pole switch of the correct rating.

This problem can be demonstrated at any power outlet that is rated at 240 V single-phase, or is three-phase. Simply test the voltages from a good ground to each of the supply lines, including the neutral if it is available. The line-to-ground voltage for a 240 or 208 V single-phase circuit is 120 V. For a three-phase system it varies, but almost all three-phase circuits will have voltage from ground to all three lines. All system neutrals, single- or three-phase, have virtually no voltage to ground.

The measured voltage to ground is the line-to-ground shock hazard. Contact with any of the ungrounded lines is dangerous, but the neutral is safe. As long as it is properly connected to the supply, it is also connected to ground.

The nature of this hazard can be demonstrated with a much safer circuit using the same bulbs and switches that were used in previous experiments. The power supply must be a dual DC supply, or a single-phase transformer with either dual 6 V secondary windings or a 12 V secondary that is center tapped. The primary of the transformer must match the available AC supply, typically 120 V.

When a DC supply is used, the outputs must be isolated from each other and from ground—any typical laboratory-type supply should work. For a 12 V lamp, set both of the supplies for 6 V, and then connect them in series to produce a combined output of 12 V. The interconnection point of the two supplies—where the negative of one supply connects to the positive of the other—is then connected to ground. A lead clipped to any grounded metal will do, but many power supplies have a ground terminal for this purpose. This terminal is connected back to earth ground through the ground wire in the 120 V AC power supply cord.

Construct the circuit with a DPST switch between the ungrounded supply lines and the load. With the circuit on, a voltmeter across the lamp indicates 12 V. Meter readings taken from any good ground to the load indicate 6 V on each side (see Figure 15–31). With the switch open, there is no voltage across the load terminals, or from any load terminals to ground. The load is "safe" with the DPST switch in the off position.

When an SPST switch is used instead, one line must go directly to the load. Control of the load appears to be the same, and voltage tests provide identical results when the load is energized. However, when the SPST switch is open, there is 6 V from ground to all parts of the circuit. The SPST disconnect does shut off the load, but it leaves it very unsafe for anyone needing to access the wiring or equipment connections.

The DC power supply, grounded at the midpoint, essentially mimics the three-wire, 120/240 V single-phase AC system that is universal throughout North America. This system has one grounded line connected to the midpoint of a 240 V supply transformer secondary. This results in 120 V, half of the full transformer output, between the ground point and either of the other two lines. The grounded terminal is extended into the system

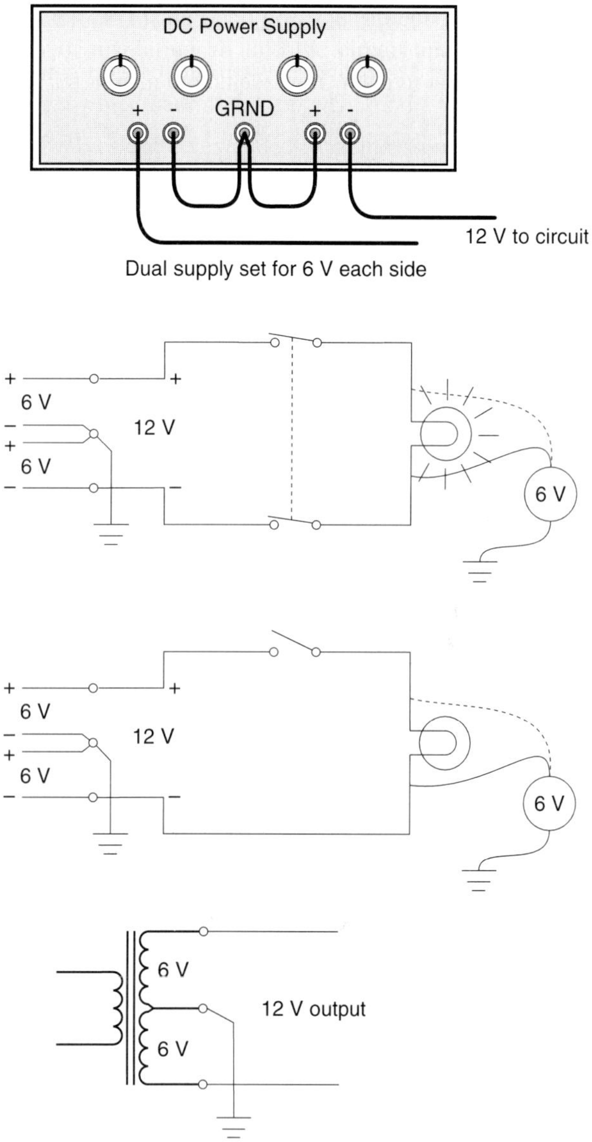

FIGURE 15–31 A dual DC supply or a small center-tapped transformer can be used to build a test circuit that demonstrates the safety hazard of SPST switches in 240 V systems. The test voltage is a much safer 12 V, but the operating characteristics are the same.

as the 120 V neutral. The voltage from line to line is 240 V; from neutral or ground to either line is 120 V.

A small transformer can be used in place of the dual DC supply if it has a center-tapped 12 V secondary or two 6 V secondary coils connected in series. Grounding the center point of the winding duplicates a 120/240 V system at a much safer voltage level. Of course the experiment could be done with other voltages, including a standard 120/240 V system. As long as the supply is a grounded one, and the lightbulbs and other equipment are rated for the voltage applied, the demonstrated results will be the same.

Switches in Circuits With Grounded Commons

The previous experiment investigated one issue relating to circuits with grounded commons. The tests described below examine the reasons for such systems, and demonstrate some other problems caused by incorrect wiring practices and wiring errors.

In most electrical circuits, one side of the circuit is used as a common circuit return. Very often this common return conductor is electrically grounded, which quite literally means connected to the earth. This practice is a source of continuing confusion about how electrical systems operate. A common but erroneous belief is that any electrical connection to the earth will cause current to flow. This is not necessarily true.

Electrical circuits involving the earth—or structures and equipment that are connected to the earth—behave much the same as any other circuit. As long as only one point in the circuit is grounded, no current flows to the ground or through the grounding system. One terminal of a transformer secondary or other supply can be grounded with no negative results, as long as all the other lines from that transformer secondary are kept isolated from any connection to ground. Grounding the supply does not cause problems. The reality is that failing to ground the system causes problems—and the consequences can be serious.

Although lower voltage circuits may also be grounded, it is a universal practice with 120 V AC supplies obtained from a utility. The system is grounded to protect the building and its occupants from the potential hazards of much higher distribution system voltages, and from damage due to lightning strikes on the distribution line. Grounding is initially established by a utility ground connection at the supply transformer. This is supplemented by a second grounding connection at each building's electrical service. This practice is examined more thoroughly in the previous chapter on transformers and services.

In AC systems, the grounded circuit common is called the neutral. Electrical codes stipulate a specific color for the insulation on the neutral conductor, which is white in North America. This color is reserved for neutral identification and cannot be used for any other purpose unless specifically permitted by codes. For both safety and circuit-operation reasons, the neutral is treated as special and is the subject of many rules. In general, these rules are intended to ensure that neutral side of the circuit is never broken. This usually means that no switches can be installed in the neutral line. In situations where switching the neutral is also allowed, the associated hot must be switched at the same time.

The neutral—the grounded circuit conductor—must not be confused with the ground conductor. While the neutral is part of the normal current-carrying circuit, the ground conductor is not. The ground conductors are the bare or green-jacketed wires that interconnect all the metal boxes, the equipment frames, and the ground pins of receptacles. The neutrals and grounds serve different purposes and must not be interchanged or interconnected.

The neutral is grounded, but only at the point where it originates. This is normally at the point where the utility supply connects to the building's electrical service. It may also be at a transformer within the building.

The grounding system interconnects not only the metal electrical boxes and the frames of all the electrical equipment in the building, but also all metal piping, ductwork, concrete floors, structural steel, metal equipment enclosures, and the earth itself.

To conduct the following tests, a circuit must be constructed using a standard 120 V supply that has a true neutral. Any normal 120 V receptacle should work fine. Some 120 V laboratory supplies have isolation transformers. This type of supply will not work for these tests.

Warning

The test circuit described below is potentially dangerous. Appropriate safety precautions are essential.

The circuit is just a simple SPST control circuit, but this time the supply, the load, and the switch are all rated at 120 V AC. Using a standard outlet ensures that the 120 V supply does have a proper grounded neutral. This can be tested by checking the voltage from ground to both sides of the circuit. Any source of ground, such as a metal electrical equipment enclosure, the receptacle box, or the U-shaped ground socket of the outlet, will do. There should be approximately 120 V from the line side to ground, and 0 V from neutral to ground (see Figure 15–32).

The circuit should first be built with the switch in the line side conductor. Testing from common—the neutral—through the circuit gives the same results as in previous tests,

FIGURE 15–32 This 120 V test circuit demonstrates why the switch must be connected in the "hot" line, not the neutral.

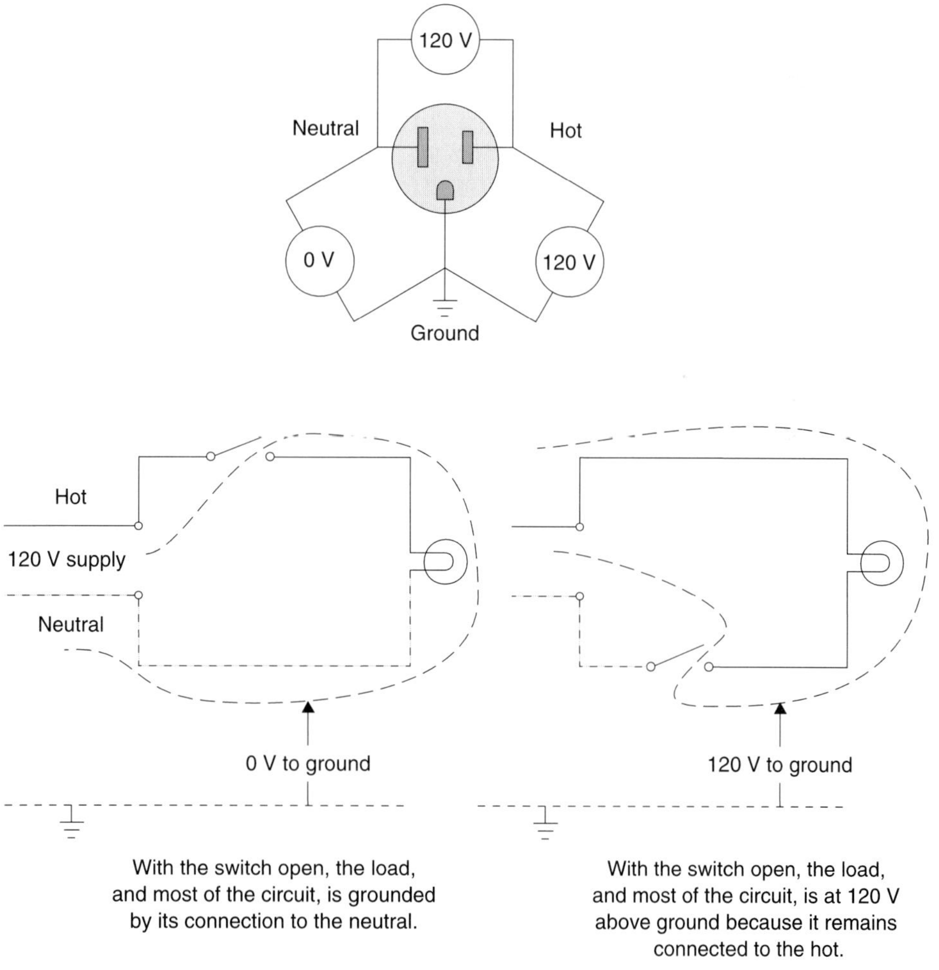

except the voltage is higher. With the switch on, there are 120 V on both terminals of the switch and all the way to the line side of the lamp. There is no voltage on the neutral side of the lamp, because that is essentially at the same point of the circuit as the other voltmeter probe.

With the switch open, there is voltage to the line side terminal, but everything beyond that point is de-energized. The open switch breaks the supply side of the circuit, leaving everything from the load side of the switch and beyond connected only to the grounded neutral.

Note

Because the neutral is grounded, when testing for voltage in 120 V AC circuits, any good source of ground can be substituted for the neutral. This can be very handy when testing inside equipment or at control devices that are remote from the supply. A neutral may not be readily available to use as a reference, but a ground connection is almost always close at hand. Connecting one meter probe to ground is virtually the same as connecting it to the circuit neutral. However, the connection must be a good one. Beware of painted surfaces.

The test circuit can now be modified so the control switch is incorrectly wired into the neutral side. The line side conductor runs directly to the load, as if it were the neutral. This is a simple wiring error with potentially lethal results. A measurement of voltage to ground reveals that any contact with the load, even with the switch off, will result in a 120 V shock. The danger comes from the illusion the load is de-energized. The circuit equipment appears to be safe to touch, but it is a serious safety hazard.

This test not only exposes a common error, it also demonstrates how this error appears when the circuit is tested, and demonstrates the potential hazards that result. There is, however, another hazard that is not so obvious. To understand this second problem, leave the circuit connected incorrectly, with the switch on the neutral side. With the switch off, simulate a ground fault by connecting a jumper wire from any convenient ground point to any exposed part of the circuit between the switch and the load (see Figure 15–33).

FIGURE 15–33 This test circuit demonstrates a not so obvious hazard caused by control switches or contacts installed in the grounded common of the circuit. A simple ground fault in the wiring will bypass the control, instead of tripping the supply breaker as it should.

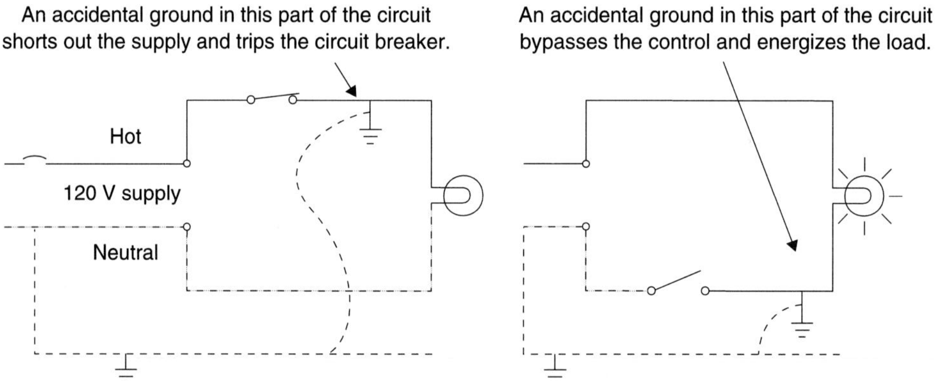

The inadvertent ground condition, simulated by the jumper, could be as simple as a bare ground wire touching a device terminal screw, but the result is that the circuit control device (the switch) is bypassed. The light is on regardless of the position of the control device. On a lighting circuit, this is an obvious error, but not particularly dangerous. In a circuit that controls a motor, a fuel valve, or some similar device, the results could be disastrous. Any control device that is wired into the common side of a circuit creates a potential hazard.

In a properly wired circuit, a ground fault like the one simulated by the jumper would be a short to ground. The circuit would fail, usually tripping the supply breaker or blowing a fuse. This removes power from the circuit, leaving it in a safe condition. This is a much more desirable situation than where the equipment can suddenly start up or fail to shut off properly due to a simple ground fault in the wiring.

15.7 Selecting Switches

A nearly infinite variety of switches are available. Choosing one that is appropriate for a specific task can be difficult. Considerations include the location and its environment, the type of actuator—switch handle—to be used, any markings or labels that may be required, and whether the switch should have an indicator light. Last but not least, its electrical ratings must be adequate for the loads it has to control.

Location and Mounting

The first considerations in choosing a switch are the environmental conditions of the location and the mechanical details of how it will be mounted. The range of possible extremes due to weather, industrial processes, periodic equipment washdowns, or other operational demands must all be considered. The switch must withstand the physical rigors it will encounter. For example, a switch flush-mounted into the wall of a house faces a much less demanding future than one mounted in a control panel on the operating floor of a steel mill.

The switch must physically fit into the allotted space. This can become a serious issue in small equipment, or where many switches must be mounted in the same enclosure. The switch must also match the mounting requirements. For example, the switches used for lighting control in homes and businesses are built onto a mounting strap that is designed for flush mounting in a standard electrical box. This type of switch should only be used in association with a suitable box and cover plate.

Switches that are not intended for electrical box mounting come in a variety of physical sizes with several different mounting provisions. Size categories include standard, miniature, subminiature, and microminiature. Mounting methods include clip-in mounts designed for rectangular holes, switches designed for circuit board mounting, and threaded hubs with lock nuts. Rectangular mounts require special tools to cut the holes. Circuit board-mounted switches have no provision for attachment to the enclosure. They depend on the board's being securely mounted at the right distance from the switch opening. Threaded hub mounts are one of the easiest to field mount.

The diameter and length of the hub vary with the size of the switch and the panel thickness for which it is designed. Hubs may have a notch on one side to mate with a key on a removable indicator plate. This prevents the plate from turning. In some cases, the switch mounting hole is punched with a special tool that leaves a key in the base material. Securing the switch in this manner is particularly important for rotary selector switches.

Switches made for mounting in rectangular- or square-cut holes are generally used in manufactured goods, because the precise rectangular holes required are difficult to cut without specialized equipment. However, some switches with square tops actually mount in round holes; the square face is just for appearance.

When many switches must be mounted in sheet metal, an appropriate chassis punch can save time and provide a more professional fit. These are available for most of the standard hole sizes and shapes. These specialized tools are the only practical method for making several identical holes that would otherwise be difficult to cut. Other methods, such as drilling pilot holes and then filing or nibbling to an accurate pattern line, may be justifiable for one or two holes, but they are too time-consuming for making many holes of the same shape.

Switch Actuators

Switch handles or actuators can take many forms. Examples include different handle shapes—such as toggles, paddles, and rockers—and different switch actions—such as pushbuttons, sliders, and rotary operators. Pushbuttons and toggles are among the most common, but even these have many variations.

Pushbutton actuators are usually used for momentary-contact switches, but they can also control maintained contacts. For maintained-contact operation, the actuator may be a single button with a push-on/push-off action, or two mechanically linked buttons, one for on, and the other for off. Toggle, paddle, and rocker type operators may have two or three positions. These may be maintained or have a spring-return momentary action. A common variation for double-throw center-off switches is a maintained action to one side of center, and spring-return momentary on the other. Rocker-style switch operators perform the same functions as a toggle or paddle, but do not protrude as far as above the surface, making them less prone to accidental operation or breakage.

Rotary switches with two or three positions may provide the same contact options as switches with toggle or pushbutton operators. However, they are also made with many different positions and far more complex contact operations than any other type of switch. A rotary switch can be very simple to use, even with a dozen or more positions. All that is required is a clear identification of all the various selections.

Switch Indications and Labels

Regardless of its style, the switch should be marked, unless its purpose and operation are obvious. Sometimes there is no need for any kind of marking. For example, light switches in homes and businesses are almost never marked, because their intent is usually self-explanatory. At other times, marking is critical; an emergency stop and an alarm button are good examples.

Not only should the purpose of the switch be obvious, but the function of its various positions should be obvious as well. In some cases this is a matter of cultural habit, or regulation, but this is not always consistent. For example, in North America, up is the standard on position for vertically mounted switches, but this standard is not shared by the rest of the world. Switches that are mounted horizontally are subject to much looser interpretations about which way the handle should point. Three- and four-way switches have unpredictable on/off positions, because the operation of any one switch varies with the status of the other switches in the circuit.

When the load and the control switch are some distance apart, it may be necessary to provide some indication of the status of the circuit. This is particularly important for momentary-contact switches or similar controls that give no clue whether the circuit is on or off. Indication lights, sometimes called pilot lights, are a good example. They provide an obvious visual feedback about the condition of the circuit and can be seen at some distance. Switches are available with built-in pilot lights to make mounting easier.

Switch-mounted pilot lights can be confused with locator lights, which serve a different purpose. Locator lights are designed to make a switch easy to find in the dark. They usually cannot be used as pilot lights, even if this might seem like a good idea. Locator lamps are usually wired across the switch contacts. They are high-resistance devices, such as a neon lamp with a series resistor, and are powered by a small amount of current though the connected load. They are only on when the switch is in the off position.

Electrical Ratings

A switch must have the appropriate contact configurations for the control situation. It is possible to use a switch that has more contacts than necessary, but, of course, there must be enough contacts of the right type. The electrical ratings of the contacts must also be appropriate for the voltage, current, and reactive characteristics of the load.

Electrical ratings for switches are usually based on four criteria: the voltage of the circuit, whether it is AC or DC, the circuit current, and the nature of the load. Understanding the effects of each or these can make it much easier to select an appropriate switch or other contact-type device. It also provides a reference for interpreting manufacturers' switch ratings.

When a switch opens, it must break the arc that occurs between the contacts. This is made more difficult by high voltage, by a DC supply instead of AC, and by load reactance. The arc usually has to be cooled down before it will break, and the more quickly this is done, the less damage there is to the contact surfaces. One method is to stretch the arc as far as necessary and as fast as possible. Fast-moving contacts with a long travel are quite effective. This is the reason larger switches, particularly those used for higher voltages, have such stiff actions. The contacts are opened and closed with the aid of springs.

For even higher voltage ratings, the arc may be drawn through a channel surrounded by layers of cooling fins. This is called an arc chute. The fins are usually made of ceramic, or layers of metal in an insulating fiber form. The arc is directed into the chute by its natural tendency to move outward from the contacts. This is sometimes assisted by magnetic effects or even a blast of air. As the arc moves into the arc chute, it is broken into many sections and quickly cooled. Arc chutes are common in circuit breakers and motor contactors.

Warning

Never operate a switch or contactor with the covers or arc chutes removed from the contacts. When the contacts open, there is nothing to confine and extinguish the arc. The contacts may be damaged, or the arc may flash across from pole to pole, an extremely destructive and dangerous occurrence.

AC is generally much easier to switch than DC. In an AC circuit, both the current and voltage pass through zero every half cycle. As the switch contacts open, the arc extinguishes itself the next time the current goes through zero. When a DC switch opens, the current will not stop until the arc is broken by the action of the switch. Many switches are rated for AC only, or have very different ratings for DC than for AC. When a switch designed for AC only is used on a DC circuit, the first operation of the switch is often its last.

High-current circuits require contacts with a large surface area. A high current is also more difficult to make and break. Many types of loads, including incandescent lights, motors, and transformers, draw several times more current when they are first energized than they do during normal operation. The switch must provide this extra "inrush" current the instant the contacts touch. To reduce arcing damage, the closing action of the contacts must be very positive, with very little contact bounce.

Note

As unlikely as it may seem, almost all contacts bounce when they close. Viewed at a microscopic level, typical contacts may bounce open several times, arcing a little on each separation. This behavior can greatly reduce contact life.

Switch bounce causes other problems besides arcing. Digital electronic circuits respond so quickly that a single switch operation may be seen as several separate open-and-close cycles. Extra components and circuit logic may have to be added to "debounce" the switch inputs.

Reactive loads of all kinds are difficult to switch off because the stored energy tries to maintain the voltage and current flow, increasing arcing when the switch opens. Large DC coils and DC motors are particularly difficult, requiring very heavy-duty switching equipment. In AC equipment the results are similar, though less extreme. Reactive loads are hard on contacts, whether they are in a switch or in some other device such as a relay or motor starter.

Switches are often provided with several ratings indicating their limitations for various voltages, AC or DC suitability, and their maximum current ratings for various types of loads. Unfortunately, a full listing of the switch specifications is usually part of the manufacturer's literature, and may not be listed on the switch or in the information provide by the distributor.

Summary

Choosing a suitable switch is not always a simple matter. The correct contact configuration must be determined. The physical demands of environment must be taken into account, the mounting considerations must be satisfied, and the manufacturer's electrical ratings must be interpreted to match the needs of the load. Remember that a switch with a higher than necessary voltage or current rating is perfectly acceptable. These ratings for a switch or any other contact type of device are maximums. Ratings that are somewhat higher than the actual load requirements provide a margin of safety, and may be worth the extra cost.

Maximum voltage ratings are just that—do not exceed the voltage rating of any switch. When there are separate ratings for DC and AC, the appropriate ratings must be used. Typical contact ratings include AC or DC designations, and maximums for both voltage and current. These are sometimes in sets, with a different current rating for AC and for DC, and possibly for each listed voltage.

Very often the words "NONINDUCTIVE" or "RESISTIVE LOAD" appear with the ratings. Either of these means that the ratings are based on switching a resistive load, such as a heating element. These are the easiest loads to switch, so those ratings are the highest. Always assume that the ratings are for resistive loads, unless there is a specific marking indicating otherwise.

When the load has significant inductance, it can be almost impossible to obtain a realistic rating for a general-purpose switch. The readily available ratings are usually for resistive loads, and there is no indication regarding the suitability of the switch to other load types. Additional manufacturer's information may be available, but often only directly from the manufacturer, not the distributor. Any reputable manufacturer can provide incredibly detailed application data for even the simplest of switches. However, the information is not only wonderfully detailed, it is often so technical that it is virtually useless to anyone except an experienced design engineer.

The safest solution is to stick with devices that are specifically rated for the load type, and if they are not available, provide a safe margin by going to a higher rating. For inductive loads, go at least 150% above the normal load current—more if the switch is likely to be operated very frequently. Alternatively, find a switch that has a horsepower rating, instead of just a maximum current. Horsepower-rated switches are specifically designed for motor loads, but are suitable for other inductive loads as well. They have a rating in amps for resistive loads, and in horsepower for motors. As with other types of switches, there are usually several different ratings for various voltages, and possibly for DC supplies. These switches are easy to select because no important ratings are missing. They are the best choice for inductive loads, particularly motor applications.

Tremendous strides have been made in the technology of contact materials and switch design. Although it is difficult to accept, it is very often true that newer switches are as capable as the larger, much more rugged-looking ones they are replacing. This is particularly true for the products of companies with established reputations in the field.

15.8 Automatic Control Switches

The switches discussed in the previous section are manually operated. They allow a human operator to conveniently energize and de-energize an electrical circuit, or make some change in its operation. Many circuits are operated in this way, but many others operate automatically, with very little human intervention.

Automatic controls are particularly attractive for processes that would otherwise require constant attention. A simple example is a storage tank that supplies some liquid to a process. Assume that a pump is used to maintain the level of the tank, drawing the appropriate liquid from a larger supply. If the pump circuit has a manual control switch, a human operator has to start and stop the pump in response to an observed change in the liquid level. When the tank level gets low, the operator turns on the pump and lets it run until the tank is full again.

This manual control system works fine, but the constant attention required is difficult to justify. It also increases the chances of problems. The likelihood of the tank running dry or overflowing is relatively high. It is inevitable that the operator will at least occasionally be distracted by other problems or concerns and forget about the pump. An automatic control system is not only more convenient, it is also more reliable.

Typical automatic controls replace the attentions of a human operator. They sense the physical changes of the process and respond to those changes. Automatic controls for the pump in the example above would have to sense the liquid in the tank and cycle the pump on and off in response to changes in the level. There are a number of ways to do this.

One of the simplest would be a float switch installed in the tank. The float is arranged so that changes in the physical position of the float ball operate the switch contacts inside the control. When the float rises, the contacts open; when it falls, they close. The float-switch contacts start and stop the pump, automatically maintaining the tank level. The automatic control has replaced the operator's attention by doing essentially the same task.

With proper design and a few extra components, the performance and reliability of this simple control can be greatly increased. For example, to protect against unnoticed equipment failures and similar problems, an extra float can be connected to operate an alarm when the tank level becomes too high or too low. If the system is a critical one, a backup pump system could be added in addition to the alarms.

Many automatic controls use the physical change they are designed to sense to directly operate their internal contacts. These devices are essentially switches that are operated by changes in physical variables such as pressure, temperature, flow, position, liquid level, or movement. They are sometimes referred to as pilot devices. The term "pilot," used in this context, means a small element that automatically initiates, guides, or limits the changes that occur during the operation of a larger system.

Note

The term "pilot device" is a general one that applies to a wide array of controls. Unfortunately, there is some potential for confusion. Some devices, such as line-oltage thermostats and pressure controls, have heavy-duty contacts designed for directly switching power loads such as motors or high-wattage heating elements. The contacts in these controls are rated for relatively high voltages and currents, and very often for a specific maximum motor horsepower as well.

Other controls designed to sense the same variables may have only light-duty contacts intended for switching less demanding loads, such as motor starter coils or solenoid valves. The less demanding control circuit use is called pilot-duty. Devices rated for pilot duty are not intended for direct control of power loads.

Despite their lighter contacts, pilot duty devices are often just as expensive as those intended for direct control. It should be noted, however, that they may also respond more accurately to changes in the controlled variable. For a pilot duty device the designer has more opportunity to concentrate on the accuracy of control, rather than on the operation of heavy-duty snap-action contacts.

In almost all automatic control systems, the control devices sense the same physical variables that the controlled equipment is intended to change. In the previous example, the pump affects liquid level, so the controlling pilot device must respond to liquid level. This is often called closed-loop control. The pump changes the level, and the level controls the pump. The control system operates in a feedback loop. When the pump operates, the level rises; when the level rises, the float provides direct feedback that in turn acts to shut off the pump. When the level falls, the controller acts to turn it back on again.

A well-designed control system serves the needs of the process, controlling the physical variables of the system. It also serves the needs of the controlled equipment by avoiding operating conditions that put the equipment unnecessarily at risk. Additional controls are added to protect against hazardous conditions due to human error or equipment failure.

The pump control example above is an oversimplification. For all but the most basic systems, dependable operation usually integrates both manual and automatic devices. Manual controls are provided to allow the human operator to start and stop an entire process, or individual pieces of equipment within the larger system. They may also be used to select optional operating parameters, such as which of two or more pumps or fans are in service.

Automatic controls are the sensing devices that monitor the physical variables of the process. They become the controllers and limits of the system. As controllers, they operate the equipment, attempting to maintain some desired operating condition—such as adequate liquid in a tank. Used as limits, they provide protection against hazards caused by abnormal circumstances—such a tank level that is too high or too low.

Every automatic controller must have a preset operating point. This is the level of pressure, temperature, flow, level, etc., at which the contacts change state. This value is called the set point, and it is the reference that guides automatic operation. If the device is used as a control rather than as a limit, the set point is usually adjustable. If it is a limit device, the set point may be preset at the factory and is not usually field adjustable.

There are several variations in the way that set point adjustments are provided. Besides differences in the physical arrangement of the controls, the effects of the adjustments are not always the same. One of the fundamental differences is the method used to provide an operating parameter called the differential. In almost all control systems, the equipment that is cycled on and off must be allowed to run for a reasonable length of time during each on cycle, and must also be given an acceptable interval of time off between each operating period.

Operating cycles that are too short may reduce system efficiency, and can even cause control problems or equipment failures. Frequent starts cause electric motors to overheat. Sometimes physical parameters such as pressures or flow must be allowed to equalize before the system can be cycled on again. Operating cycles that are too close together can cause restart failures because the system has not stabilized.

To prevent undesirably short cycles, most controls have some degree of differential. Differential, as it is applies here, means the separation between the on and off points of the control. It is an intentional dead spot in the operation of the controller. When the controller is inside the differential band, it ignores small changes in the sensed variable and temporarily leaves the status of the contacts unchanged. The differential setting is critical to proper operation; too wide can be just as problematic as too narrow.

Some controllers provide separate turn-on and turn-off settings, and the differential is the difference between them. Other controllers have only one set point plus a separate differential setting. The differential may be fixed or adjustable, and can be provided in a variety of ways, some of which are quite ingenious.

Along with other control devices, both automatic and manual control systems usually include automatic limits. Almost any type of equipment is subject to improper operation, so limit devices are added to intervene when some controlled condition exceeds allowable maximums or minimums. The devices used as limits are often very similar or even identical to those used as the operating controls of the system, but they fulfill a different role.

Temperature Sensors and Controls

Temperature Units. Temperature is measured with three different scales—Kelvin, Celsius (at one time called Centigrade), and Fahrenheit. The Kelvin degree is the same as the Celsius degree, but the scale starts at absolute zero, which is –273.16° Celsius. The Celsius scale starts with 0° at the freezing point of fresh water at sea level, and is scaled so that 100° is equal to the boiling point. The Fahrenheit scale shares no specific root with the Celsius scale. On the Fahrenheit scale, water freezes at 32° and boils at 212°. The two scales coincide at –40°. Celsius and Kelvin degrees are 1.8 times larger than a degree Fahrenheit. The formulas for converting Celsius to Fahrenheit, and vice versa, are given below.

$$F = \left(\frac{9}{5}C\right) + 32$$

$$C = \frac{5}{9}(F - 32)$$

Bimetallic Devices. Most nonelectronic temperature-operated control devices use one of two methods to convert changes in temperature into the physical action that operates the contacts. For air temperature, the most common sensor element is a bimetallic strip or disk. Bimetallic material is made by tightly bonding two layers of metal into a single sheet. The metals used have different coefficients of thermal expansion. When the temperature of a bimetallic strip changes, it bends, because the metal that forms one surface expands more than the metal on the other side.

The bimetallic strip is an inexpensive and reliable temperature sensor, but it must be matched to the requirements of the control. The amount of force the strip can exert as it moves depends on its physical stiffness. If the contacts require a lot of force to operate, the bimetallic strip must be thick and short. This thickness affects the responsiveness of the control, because the increased mass of the material is slow to change temperature. The length of the strip affects the sensitivity, because a longer strip produces more movement for a given change in temperature.

For maximum sensitivity and responsiveness, the strip must be long and thin. The thin material changes temperature quickly, and the length translates into significant movement for very small changes in temperature. However, a long bimetallic strip is essentially a soft spring. For the control to be useful, the contacts must operate with very little physical force and be free of any tendency to stick or bind. The action of the contacts must be very smooth, or the springiness of the strip will cause them to oscillate. This could cause the contacts to cycle open and closed several times in quick succession, something that is unacceptable to the equipment under control. Most bimetallic controls incorporate some means to combat this tendency.

Several popular designs successfully accommodate all these considerations. For very sensitive sensors, the strip is formed into a helix or spiral to provide maximum length in a compact space. A small gap is maintained between each layer of the spiral to provide good surface exposure between the strip and the surrounding air. This long strip is then coupled to a mercury switch.

A mercury switch consists of a sealed glass bulb containing a small blob of mercury. There are conductors set into the glass at one, or both, ends of the tube. To operate the "contacts," the bulb has to be tilted only slightly one way or the other. When the mercury rolls to a contact end, it electrically connects the imbedded tips of the conductors. Mercury is a very good conductor, and the switch is totally sealed against contamination, so the mercury switch is very reliable. The electrical connections to the bulb are provided with thin flexible wire leads, so almost no physical force is required to operate the switch. This makes it an ideal contact for bimetallic sensors.

The mercury switch, balanced on a long spiral bimetallic strip, can be overly sensitive, not only to temperature change, but also to minor vibrations. To overcome this problem, the glass bulb is given a slight curve, and set so the ends are lower than the center. This slightly decreases sensitivity, because it also increases the amount of tilt required to make the mercury change ends. However, it makes the operation of the contacts very positive, and virtually eliminates any tendency of the switch to cycle on and off due to oscillations of the strip.

The curve in the tube also provides some control differential to help prevent short cycling of the equipment. A well-designed spiral element bimetallic thermostat, using a mercury switch, can provide temperature control to within 0.5° (see Figure 15–34). For the mercury bulb thermostat to be in calibration with the adjustment dial, it must be carefully leveled and fastened securely. If the thermostat is only slightly out of plumb, it will cycle properly, but the temperature setting will not correspond to the operating point.

To be effective and accurate, a temperature control must operate the contacts at exactly the same temperature each time. The contacts must also make and break with a positive action to prevent erratic behavior or burning of the contact surfaces. The bimetallic-operated mercury bulb temperature control meets these needs very well, making it one of the most popular and successful thermostat designs. However, it is not the only type of thermostat.

Open contacts are also used in bimetallic thermostats, although they are less common than the mercury bulb type. One design, common in low-voltage thermostats, uses a small magnet to overcome the problems associated with the springy bimetallic strip. In a typical

FIGURE 15–34 A common automatic temperature-control is a mercury bulb thermostat. The bimetallic strip is long and thin, making it sensitive and relatively quick to respond. The mercury tube is curved to add differential, and control the effects of vibration.

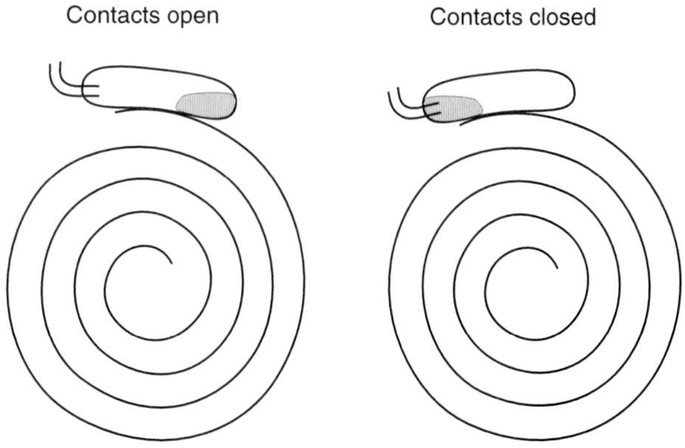

The spiral bimetallic element tips the mercury switch in response to changes in temperature.

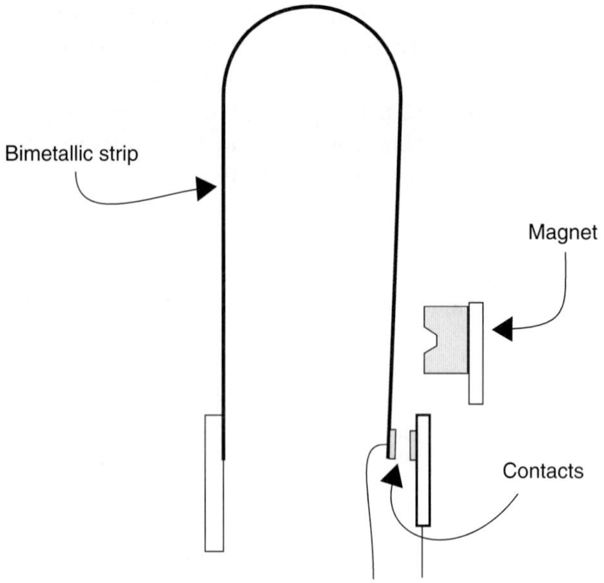

FIGURE 15–35 The addition of a small magnet gives this bimetallic thermostat a positive switch action, that is unaffected by vibration.

example, the moving contact is attached to a relatively short bimetallic strip formed in the shape of an inverted "U." A very small magnet is placed near the moving contact, close enough to affect the movement of the strip, but far enough away that the two can never actually make contact. Starting with the contacts open and a temperature change that is acting to close them, the effect of the magnet is as follows. Just before the contacts touch, the magnetic attraction becomes strong enough to pull the moving contact through the remaining distance. The result is a positive contact closure. The contacts are protected from arcing, and the load device is protected from erratic contact behavior (see Figure 15–35).

Once the contacts are closed, the magnet acts to hold them. To reopen the contacts, the strip has to build up enough stress to overcome the magnetic attraction. This gives the control some differential, because it acts to prevent the contacts from reopening too quickly. Once the temperature change is great enough, the stress overcomes the pull of the magnet, and the contacts spring open. The addition of a tiny magnet gives the contacts a positive action on both opening and closing, and provides the necessary operating differential.

Line-voltage thermostats, such as those used for electric baseboard heating systems, also may use bimetallic elements, but they are typically wide, short, and relatively stiff. The element drives a mechanical linkage that provides a snap-action switch contact. Short cycling is not a problem for electric elements, so the differential can be kept relatively narrow.

Another approach to using bimetallic elements is to shape them into disks. These are used for temperature sensing, but generally not for thermostats. Bimetallic disks operate differently from the strips because they have a natural snap action. The disk is not flat. It is preformed into a slight dish shape. As it is heated, the material used to form the inside surface expands more than the outside. This causes stress to build up in the material until the dish suddenly pops to the opposite curvature. It then stays in that shape until it cools enough to snap back again.

Bimetallic disks are most common where the set point does not have to be adjusted. They are widely used for temperature limits in furnaces, for example. If the limit is a critical one that warrants human attention, the thermal disk can be designed so that is does not snap back until it has been physically reset. These manual-reset temperature limits are used in many modern residential furnaces to provide protection against high flue temperature and flame rollout. (See Chapter 17 for more details.)

Bulb and Capillary Tube Devices. Temperature can also be sensed by thermal expansion in a fluid-filled bulb or tube (see Figure 15–36). This method is used most often when the sensor must be remote from the controller. A thin tube connects the sensing bulb to a bellows inside the control. As the fluid in the bulb expands and contracts, it causes the bellows to extend or shorten. This movement operates one or more contacts through a mechanical linkage. The operating point of the control is changed by an adjustment that affects the linkage. Typical mechanisms use either a threaded rod or a spiral ramp that is turned by the adjustment knob.

Tube-type sensors can be designed to sense the temperature at one specific point, or to provide an average reading taken over a relatively large area. Typical bulb and tube sensors respond to temperature changes at the bulb. For an average reading, the bulb may be eliminated, and the entire length of the capillary tube becomes the sensor. Such controllers are used to measure the average temperature in large air ducts, where the temperatures at various points in the airstream may be quite different. This difference is a normal result of stratification and unpredictable turbulence in the airflow pattern. The averaging sensor improves control, because is gives a more accurate reading than a sensor that monitors only a single point.

An interesting variation of the bulb-and-tube type control is a dual-bulb device that can respond to temperature differential, rather than the temperature of a single sensor location. The sensing bulbs are equipped with extra long capillary tubes to allow placement in quite separate physical locations. Inside the controller, the movements of the two bellows

FIGURE 15–36 Changes in temperature at the sensing bulb cause physical movement of the bellows in the control. Thermostats with remote bulbs and capillary tubes are often found in situations where the temperature must be sensed at some distance from the control.

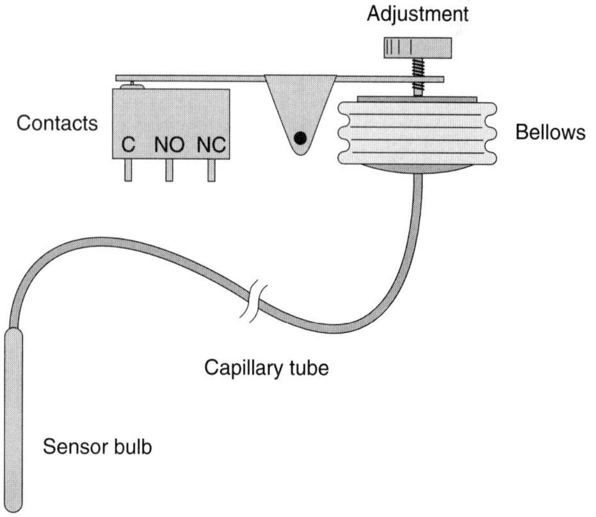

are combined through a shared mechanical linkage. The result is an adjustable control point that is representative of changes in temperature difference, rather than the specific temperature at either point.

An example of differential temperature sensing is the reset control used with a boiler or chilled-water control system. The difference between indoor and outdoor temperatures is proportional to the heating or cooling load. The control adjusts the water temperature set point, thereby adapting the output of the system to the prevailing conditions. In a heating situation, as the outdoor temperature drops the boiler temperature set point is adjusted to a higher value. This provides increased heating capacity, matching the heating system output to the actual requirements of the building. A chilled-water system works essentially the same, but the logic is reversed.

Tube-type temperature sensors are also used to control heating elements or burners in ovens and boilers, and refrigeration equipment in such applications as coolers, display cases, and storage tanks.

Electronic Temperature Controls. Mechanical-type temperature sensors translate changes in temperature into physical movement that is used to operate the contacts. Changes to the controller set points are accomplished through some type of physical adjustment. This can be a change in the position of some part of the mechanical linkage, or an increase or decrease in the tension of an internal spring.

All mechanical controls are prone to wear and loss of calibration. In contrast, typical electronic controllers have few if any moving parts. This is part of the reason that properly designed and field-tested electronic controls can outlast and outperform the mechanical devices they replace. However, this advantage is not without a price. Mechanical controls can be understood by anyone who has some experience with small-scale mechanical devices and an eye for ingenuity in design. There is something inherently appealing about a well-executed mechanical apparatus. It displays a level of human accomplishment that can be difficult to appreciate in a motionless electronic circuit.

A careful examination of an electronic device may reveal very little about how it actually works. Without a considerable amount of knowledge about electronics, it is impossible to understand these controls, and acquiring that knowledge can be time-consuming and expensive. Unfortunately, it also may not provide the level of insight that might be expected. Even to a person with extensive electronics training and experience, the fine details of a specific control often remain a mystery. Even when the basic operating principles are fully understood, it is not always possible to know exactly how it operates.

The workmanship involved in the manufacture of an electronic control is easily appreciated, but a sense of the ingenuity involved can be very elusive. This is not always a bad thing, but it goes against the grain for a lot of installers and service personnel. Many of the best operators, service technicians, tradespeople, and engineers are propelled by an insatiable curiosity. It can be very frustrating to deal with devices that defy all reasonable attempts at understanding their inner workings.

Electronic Temperature Sensors. While mechanical controllers depend on sensors that produce mechanical movement, electronic controllers depend on sensing elements that change their electrical properties. The two basic types of temperature sensors used are thermocouples and resistive temperature devices. Thermocouples generate a voltage that is proportional to temperature. The voltage is produced when two different metals have junctions that are maintained at different temperatures.

Thermocouples are often made up as needed from rolls of thermocouple wire manufactured for the purpose. Thermocouple wire is a thin two-conductor cable, with each wire made of a different metal or alloy. When the wires are joined, one junction point becomes the sensor. The second is made in the measuring instrument. A voltage is generated that is proportional to the temperature difference at the sensing end, compared to the controller end.

There are seven or more standard thermocouples using different metal combinations. They are designated by the code letters B, E, J, K, R, S, and T. Each type has a specific output in millivolts for a specific temperature, and a useful minimum and maximum temperature range. There are also differences in the type of chemical atmosphere that each type can withstand.

Thermocouples have the advantage of small size, low cost, and quick response. However, the voltage produced is generally too low for most low-temperature applications.

There are two categories of resistive temperature sensors: thermistors and RTDs (resistance temperature devices). Thermistors are based on solid-state materials, while RTDs use elements made of materials such as platinum that change their resistance in proportion to changes in temperature. The change in resistance may be inversely proportional to temperature change—the material has a negative temperature coefficient (NTC)—or directly proportional—the material has a positive temperature coefficient (PTC). RTDs have positive temperature coefficients. Thermistors are available in both PTC and NTC types.

RTDs have a number of problems that affect their application. One is that because they are resistance devices, the length and gauge of the conductors between the sensor and the controller can affect the accuracy. A second is that when current passes through the resistance, it generates heat. This self-heating effect will, of course, affect the accuracy of temperature measurements.

Lead resistance is usually not a problem with thermistors, because of their higher resistance. For the lower resistance RTDs, the effect of lead resistance can be reduced by keeping the controller close to the sensor location, or by using four wires between the controller and the sensor instead of only two. One pair of wires carries the sensor current; the second pair carries the sensor voltage drop back to the controller. Because the controller circuitry draws almost no current from the second pair, they can be very long and still have no significant voltage drop. Self-heating is kept to a minimum by using very low currents, or by pulsing the sensor instead of keeping it constantly energized.

Thermistors tend to be nonlinear. This means the change in the resistance, per degree change in temperature, is not consistent throughout the useful temperature range of the device. This can be handled by using them in a relatively narrow temperature range, or by doing scale corrections in the controller. Thermistors are popular because of their low cost and the large change in resistance per degree of temperature change.

RTDs such as the platinum element types are linear and stable over time. They are more expensive than thermistors, and generally more expensive to apply. The change in resistance is relatively small; therefore, a more sophisticated controller is required. RTDs are used where the higher accuracy and long-term stability are worth the extra expense.

Electronic sensing elements can be cost-effective and reliable, but they do require external circuitry and different troubleshooting approaches. An electronic circuit is always required, because the sensing element cannot operate the control contacts directly. The circuit interprets the resistance of the sensor and the user set point, and then uses the difference between them to determine the appropriate status of the output relay contacts. In typical controllers, adjustments such as changes to the set point are provided through a scaled control knob that changes the setting of a variable resistor.

Electronic-based temperature controls tend to be more accurate over time than electromechanical types, because they are unaffected by mechanical wear. They can also be more easily integrated into more complex electronic control systems. This is as true for a self-contained programmable thermostat as it is for a computer-controlled direct digital system used in a large commercial building.

Humidity and Enthalpy Controls

Humidistats and other humidity-controlled switches use hygroscopic elements that expand and contract with changes in relative humidity. A common element material is a nylon strip that is mechanically linked to a set of contacts. The set point is adjusted by a cam or similar arrangement that changes the operating point of the switch relative to the length of the sensing element.

An enthalpy control combines humidity- and temperature-sensing elements to produce a device that responds to total heat content of the air, rather than just the dry bulb temperature. Heat content—enthalpy—varies with both the dry bulb temperature and the moisture content of the air. The controller must sense both of these variables and respond to a combination of their effects. Mechanical enthalpy controls use a temperature-sensing element such as a bulb and bellows, and a humidity element connected through a common mechanical linkage. An adjustment is provided to select the operating point of the control. Because the operating point depends on two variables, the scale is not in specific units. The setting must be matched to a chart or graph provided by the manufacturer.

Electronic elements that are sensitive to relative humidity are also available, although humidity is relatively difficult to sense accurately. There are several different types of sensors, based on different electrical effects. In every case, an electronic circuit is required to convert the variations of the sensor into a usable signal. Electronic humidity sensors are usually used with proportional controls, rather than the simpler contact-type controls discussed in this section. Proportional controls are examined at the end of this chapter.

Simple humidistats are usually electromechanical, but electronic-based enthalpy controls are a common replacement for the more troublesome mechanical versions. As with the mechanical types, electronic enthalpy controls contain both temperature and humidity sensors connected to a single control, but there is no need for a shared mechanical link. The outputs of the two sensors are interpreted and combined by the controller's electronics.

Pressure Switches and Controls

Pressure Units. Pressure is the force exerted by a liquid or gas on the surface of an object with which it comes in contact. The units of measure are based on force per unit area. Examples include the Imperial unit of pounds per square inch (PSI) and the SI unit pascal (Pa). The pascal is based on the metric units of force (in newtons) and area (in square meters). One pascal of pressure is equal to one newton per square meter (N/m^2). The newton is a relatively small unit, and the square meter is relatively large. This makes the pascal a very small base unit, much smaller than the pound per square inch. For this reason, SI pressures are often stated in kilopascals (kPa). One pound per square inch equals 6894.8 Pa or 6.895 kPa.

Note

Pressure can be measured and described in several different ways, depending on what is used as a reference. Absolute pressure is measured relative to a full vacuum. Gauge pressure measurement is referenced to atmospheric pressure. Differential pressure is the difference between the pressures at two different measurement points.

Pressure can be a static value—which is the pressure on the system without adding any of the effects of movement—or it can be a velocity pressure value. Velocity pressure is the pressure or force that results from movement. A stationary volume of air, for example, has only static pressure. A moving stream of air has both velocity and static pressure. They are different values, and both may be present at the same time.

Pressure Switches. Regardless of the type of pressure being measured, it is sensed as a physical force. The pressure is usually applied to a diaphragm, bellows, or similar arrangement, and the force from that device is used to operate contacts or other control elements. The size of the diaphragm generally depends on the amount of pressure being measured. For very low pressures, a relatively large diaphragm is used to provide enough physical force to operate the switch. High pressures apply large forces on a small area, so the diaphragm, bellows, or other system may have a very small area exposed to the force of the fluid.

In most pressure controls, a spring is used to counteract the movement of the sensing element. The set point of the control is changed by turning a bolt or nut, which changes the compression of the spring. A calibrated scale is usually provided as a reference (see Figure 15–37). The operating differential may be fixed or separately adjustable, or it may be the

FIGURE 15–37 The small pressure switch at the front is an encapsulated type with a fixed operating point. The other two have a range of adjustments.

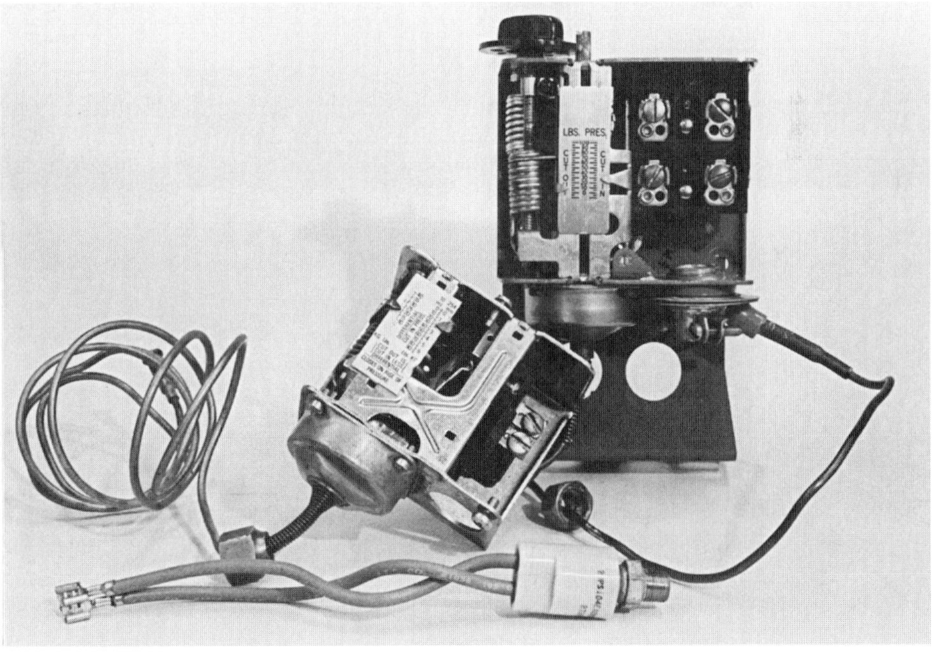

difference between individual cut-in and cut-out settings. The mechanical linkage is usually arranged to provide a snap-action contact operation.

One particularly simple type of low-pressure switch is very common in modern gas-fired heating systems. It monitors the operation of a small blower that is used to draw fresh air into the burner and vent the products of combustion. The blower replaces the natural draft of a traditional vertical vent or chimney. It is critical that combustion airflow is established before the burner is lit, and that it is maintained all the time the burner is operating.

The pressure switch used to monitor combustion airflow is a relatively simple metal box, a few cubic inches in volume. The box is constructed in two halves with a flange joint around the perimeter at the center line. A diaphragm is sandwiched between the flanges, dividing the box into two chambers. The center of the diaphragm connects with the operating plunger of a small microswitch mounted on the outside of the box (see Figure 15–38).

FIGURE 15–38 The small motor shown here drives the combustion air blower of this gas-fired residential furnace. This blower is critical to safe operation of the burner, so it is monitored by the pressure switch mounted on the left. When the blower is running and the vent is not blocked, a slight vacuum is created on the intake side. This low-pressure zone is connected to one side of a diaphragm inside the pressure switch through the hose connected at the top right of the switch chamber. The reduced pressure causes the diaphragm to move to the right and operates the contacts of the microswitch mounted on the left side of the chamber.

The design of this pressure switch allows it to be used for several different functions. Hose fittings can be provided on one or both halves of the box. To sense pressure, the hose is placed so that diaphragm movement away from the pressure source operates the switch contacts. To sense suction or vacuum, the fitting is placed on the opposite half of the box. The diaphragm then moves toward the hose side, pushed by the higher pressure of the atmosphere on the other side. To sense differential pressure, both halves of the diaphragm chamber are equipped with fittings, and hoses are connected to two different points in the system. The diaphragm then moves away from the side that has the highest pressure.

This type of pressure switch is generally made for specific applications and sold in large volume to the equipment manufacturers. The set point and differential are preset at the factory to match the requirements of the equipment. They are not intended for field adjustment. If the switch must be replaced, the new part must match the specifications of the old one. There can be a lot of variation. Two devices that look identical may have different operating points.

Pressure switches for higher operating pressures are smaller and much more ruggedly built. They are made in a wide variety of pressure ranges, in single-pressure and differential styles. Refrigeration and air conditioning systems use pressure switches of this type to monitor or control the operation of the compressor. These controls may be single function units designed for a specific pressure range, or dual function units incorporating both high- and low-pressure switches in the same control. For specific purposes such as limits, where no adjustments are required, an encapsulated pressure switch may also be used. These compact switches have a fixed operating point and differential (see Figure 15–37).

The placement of the control varies with the function it is intended to serve. For example, on the low-pressure or suction side of the compressor, the pressure of the refrigerant varies with the temperature of the evaporator. As the air or other medium cooled by the evaporator gets colder, there is less heat energy available to the evaporating refrigerant. This causes the pressure in the evaporator to drop. An appropriate suction-side pressure control can be used to regulate the temperature of the cooled medium, because evaporator pressure is a function of evaporator temperature. The pressure switch contacts can cycle the compressor or a compressor unloader. In this case, the low-pressure switch is a control. In other systems it has a purely protective or limit function. The compressor is cycled by a thermostat, and the low-pressure switch shuts down the system if the evaporator pressure drops below an acceptable level.

High side-pressure controls for refrigeration compressors may also be used as protective limits or as controls for condenser cooling. Cycling the condenser fan on compressor output pressure helps maintain system operating pressures under changing condenser temperature conditions. If the condenser is overcooled because of a low ambient, the compressor discharge pressure can fall below the desired operating level. By cycling the condenser fan, the pressure can be maintained to a lower ambient temperature than if the fan had been allowed to run continuously.

Another interesting pressure switch is used to monitor lubricating oil pressure in refrigeration compressors. The internal pressure of the compressor can vary over a fairly wide range, so a simple oil-pressure switch is inadequate. For proper lubrication, the oil pressure must always exceed the pressure inside the compressor. It is the difference between these two pressures that represents the useful pressure of the oil. The oil-pressure switch responds to the difference between the compressor crankcase pressure and the pressure of the lubricating oil. Like any differential device, it has two input lines. Inside the control, the two pressure inputs are applied in such a way that their effects oppose one another, and only the difference is passed on to the contact linkage.

Electronic Pressure Controls. In temperature-control applications, an electronic sensing element and its associated electronic circuit is often preferable to a bimetallic or bulb-and-capillary-tube device. The electronic version has no moving parts, making it simpler and potentially more reliable than a mechanical device. This is usually not the case in pressure controls. Electronic versions of pressure controls usually require mechanical parts that perform a similar function to the linkages in electromechanical versions.

Electronic pressure controls monitor the physical movement produced by the pressure, using a variable-resistance strain gauge or capacitance device that changes its value due to flexing under stress. In both cases, the changes produced are very small. The electronic circuit must amplify these effects to produce a useful output signal.

Because of these complications, electronic pressure sensors and controls are not a very popular replacement for conventional pressure switches. However, when the output of the controller is to be a variable signal rather than a change in contact status, the electronic versions become more attractive. Controls with variable output signals, rather than just contacts, are discussed later in this chapter in Section 15.10.

Level Controls

Level controls can be designed to detect liquids, such as the height or depth of a fluid in a tank, or for solids, such as granular material in a bin. Level sensors range from simple electromechanical devices to proprietary systems using jealously guarded electronic technology.

The simplest liquid-level controls are based on floats. The more sophisticated ones use noncontact sensors, such as remote ultrasonic or sonar transmitters and receivers, to detect the surface of the liquid or other material. In between these extremes there are a variety of patented sensors to sense the depth or surface level of a variety of materials.

Indirect sensors are most often used with solid materials. Most liquid-level sensors use more direct approaches. Examples include floats, probes, and a variety of sensing tubes. The best choice depends on variables such as the depth of the tank and the accuracy required. Other considerations include the operating pressure and temperature, and potential hazards such as flammability, corrosiveness, or toxicity.

Float Switches. Many level controls are just floats that operate switches. They consist of one or more sets of contacts that are operated by mechanical attachments to a push rod, weighted chain, or other mechanical link. One common design has a simple switch arm with an open hole in one end. The float is mounted on the end of a rod that passes through this hole and, through a set of guides, down into the tank. Two collars, one above the switch arm and one below, are secured to the rod with set screws. As the liquid level rises, the float rod moves up until the lower collar pushes up on the switch arm. This can control a sump pump, for example, starting it up to pump out the accumulated liquid. As the level falls, the rod moves down until the upper collar pulls the switch arm back down, shutting off the pump. The distance between them creates the differential (see Figure 15–39).

FIGURE 15–39 The operating differential of this float switch is set by the distance between the two collars on the rod.

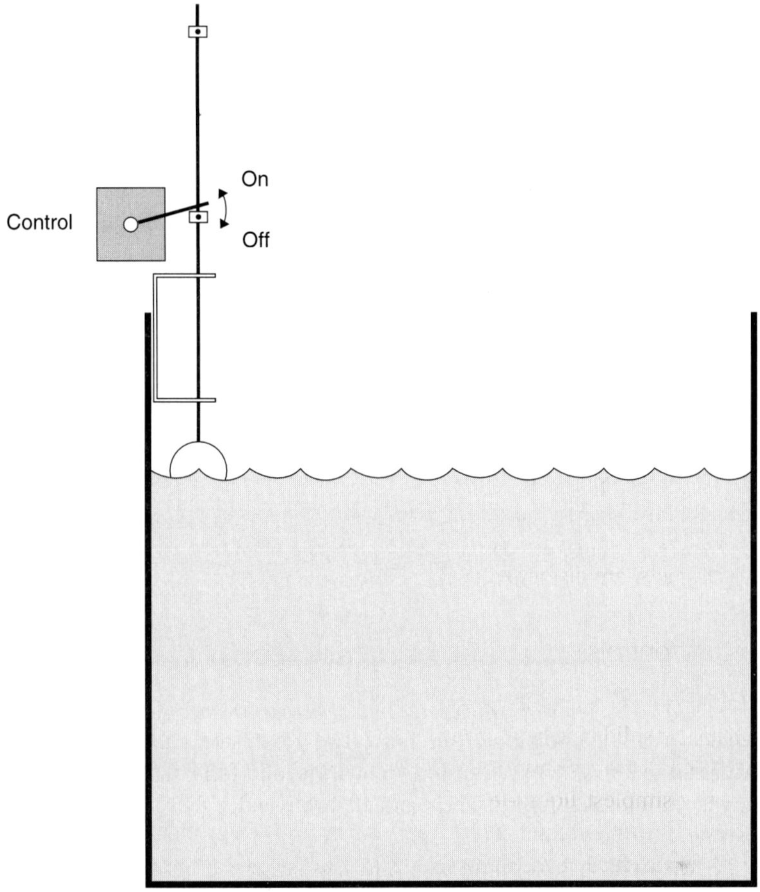

A variation on this theme uses a float with a hole through the middle for the rod. The movement of the float is restricted by collars above and below the float. The operation is similar to the previous system, but much less space is required above the switch arm, because the movement of the rod is quite short. Unless the fluid is clean, the float can become stuck on the rod, and the constant sliding movement may wear through the inside surface of the float, causing it to leak.

When the application requires a perfect seal between the liquid vessel and the outside atmosphere, the movement of the float may be linked to the contacts magnetically rather than by a direct mechanical connection. This eliminates the need for seals, making it particularly useful for situations of extreme pressure or potential contamination. The float is usually installed in a separate float chamber, piped off the main vessel. This allows full access for repairs, because the float chamber can be isolated and, if necessary, completely removed. In one example, a block of magnetic material is mounted on the float rod, which moves inside a tube made of nonmagnetic material. A magnetic yoke is

FIGURE 15–40 This float ball contains a tilt switch. The length of cable between the ball and the anchor point sets the differential.

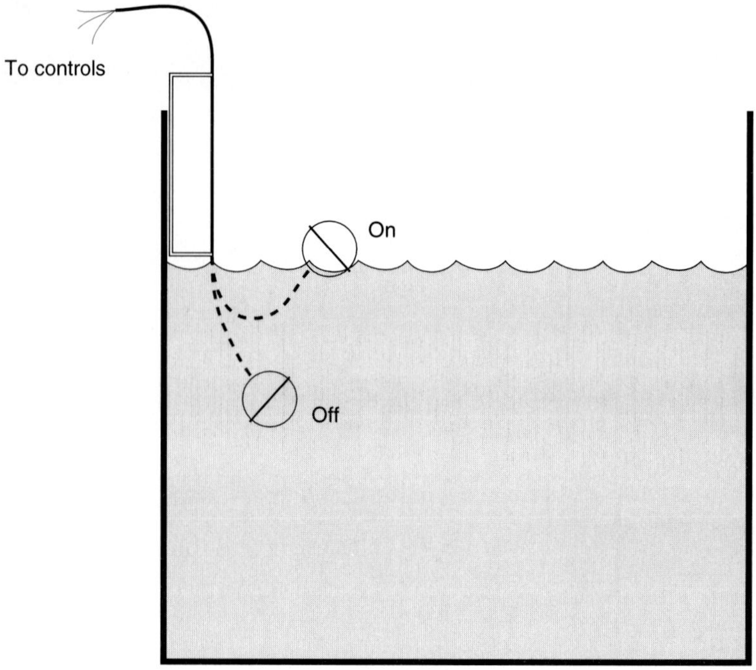

placed around the outside of the tube. The yoke is balanced so it tips forward when the magnetic material is between the yoke arms, and falls back when the action of the float moves the material away. The contacts are usually mercury switches that tilt with the operation of the yoke.

Some float switches have operating contacts, built inside the float ball, that operate by tilting. The float has a durable power cable—sealed into the float ball at one end—that is at least long enough to reach a point completely clear of the tank (see Figure 15–40). The float switch cable is anchored some distance from the ball end. The operating point is set by the position of the anchor in the tank. The differential depends on the amount of free cable between the anchor point and the float ball. When the tank level is very low, the float hangs down, suspended on the power cable. As the liquid rises, the float is lifted until it tips far enough to operate the contacts inside. The contacts do not return to their low-level state until the float ball is nearly vertical again.

Probe-Based Level Controls. When the liquid being monitored is electrically conductive, a probe-type controller can be used. This system uses the liquid to complete an electrical circuit between a set of probes suspended in the tank. Typical probes consist of a heavy brass tip on the end of an insulated conductor. They are suspended in the tank from an open top or hatch. In one common system there are three probes: one at the upper level, one at the lower level, and one slightly below that. The controller contains a transformer and a relay that are wired to the probes. When the liquid level reaches the high-level probe,

FIGURE 15–41 Probe-type controls create an electrical circuit through the liquid. The relay picks up on the high probe and seals it until the level is below the lower probe.

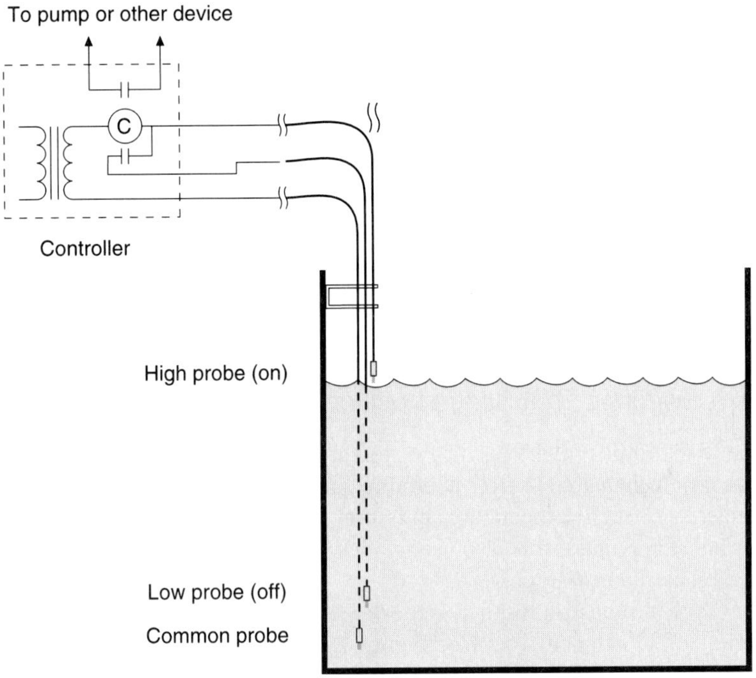

it completes the circuit to a relay coil. The relay picks up and seals itself in through the lower probe (see Figure 15–41). Once energized, the relay will not drop out until the level falls below the low-level probe. The bottom probe acts as a common conductor for both the high- and low-level probe circuits.

An advantage of probe-type controls is that they are simpler to use in very deep tanks, because they permit extreme differentials. There are no moving parts in the tank, just the suspended probes. A disadvantage may be that current must pass through the liquid, making it part of an electrical circuit.

Pressure-Based Level Sensors. The pressure of any fluid varies with the depth. A pressure switch can be mounted at the bottom of the tank or, if this is impractical, it can be connected to a bubbler system. The following description is one example of how this can be done. A rigid tube is installed vertically in the tank, extending nearly to the bottom. The tube is connected to a supply of compressed air that is regulated and restricted so a small amount of air is continually escaping from the bottom of the tube. The pressure in the air line will be directly proportional to the depth of the liquid covering the bubble tube. A pressure switch connected into the air line is set for the desired operating points. This system can be used with a more sophisticated pressure sensor, instead of contacts, to provide a variable level indication on a meter or an electronic display.

Level Sensors for Solid Materials. Sensors for solid materials are often noncontact electronic systems such as an ultrasonic transmitter and receiver mounted above the material. The sensor uses the timing of the reflected signal to determine the distance to the surface of the material. This measurement is then translated by a dedicated electronic circuit into a relative-level signal that can be used to operate contacts, provide an input to a proportional controller, or control a remote level indicator.

Solid materials can also be detected directly. A common example is a switch assembly with a flat pressure plate on one side. The switch is mounted on the bin wall and activated by the physical force of the material against the plate. Plate travel is limited by the interior structure of the switch mechanism, which must be strong enough to withstand the force of the material with the bin full. The actual switch mechanism inside is operated by the movement of the plate, but is isolated from damage by the structure of the housing. Such devices must be very rugged, particularly if the material is in relatively large pieces, has a jagged shape, or is very abrasive.

Limit Switches That Sense Position

The term "limit switch" has two different meanings. It can mean any control device that acts to prevent a system from operating outside some safe maximum or minimum (such as dangerously high temperature, pressure, level, or speed, or mechanical travel beyond normal stop points). It is also used to describe a group of general-purpose switches operated by mechanical force.

Typical industrial limit switches have a short operating shaft that can be turned by a variety of switch actuators (see Figure 15–42). The types of actuators available include

FIGURE 15–42 Industrial limit switches can have a variety of different operators to match the requirements of the situation.

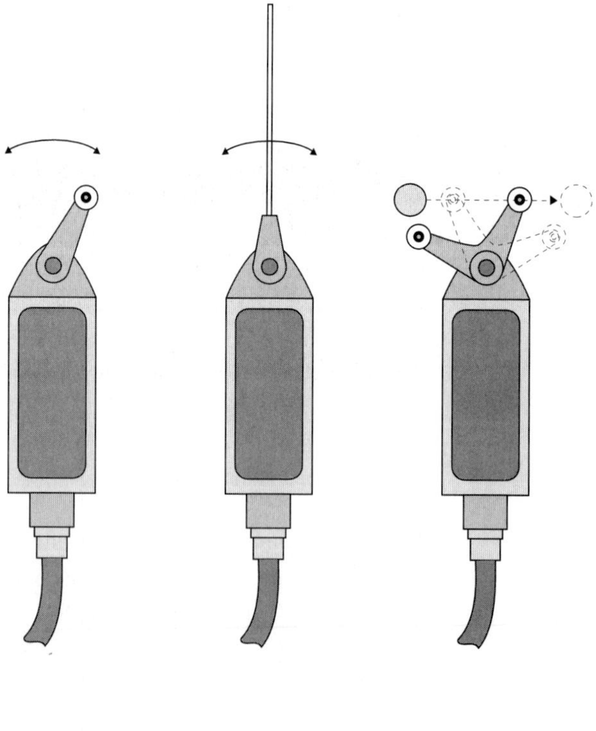

FIGURE 15–43 The symbol used for a limit switch is affected by the type of contact used and the "normal" condition of the monitored device.

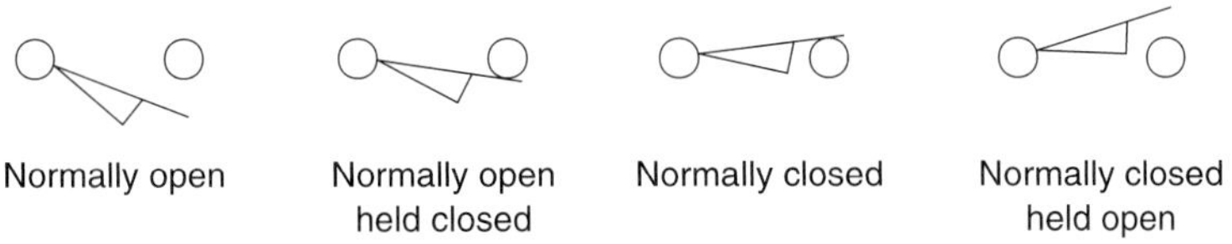

a variety of levers with roller tips, and stiff "whiskers" made of spring wire or nylon. Rocker arms are also available for special situations where the operating device moves back and forth. The rocker arm tips one way or the other, depending on the last direction of travel. The actual switch or contact mechanism inside the limit switch body is protected from excessive force by the nature of the mechanical linkage. It is designed to allow some overtravel of the actuating arm without damage to the arm, the enclosure, or the internal contacts.

Limit switches are widely used to confirm the physical position of some element that is critical to the controlled process. For example, the switch can detect the passage of a box on a material handling conveyor, or the position of a valve, mechanical gate, or damper.

In most applications the limit switch has a momentary action with a spring return. NO or NC contacts may be used, depending on the requirements of the control system. However, it should be noted that the status of these contacts can be very confusing on control drawings. The contacts may be NO, NO held closed, NC, or NC held open. All behave somewhat differently, and all have specific drawing symbols. For a list of these symbols and their interpretations, refer to Figure 15–43.

The following is an example of how these limits could be used, and how they can cause confusion on a drawing. The status of a large service valve, critical to a fire control system, is monitored by a limit switch mounted on the valve. The actuating arm of the switch is depressed only when the valve is in the fully open position. Because the valve is critical to a safety system, the circuit must be fail-safe. Therefore, the contact must be closed when the safety system has not been compromised. In this example, the contact is closed when the valve is in its normal, fully open position. The limit switch contact is a normally open type, held closed when the valve is in the usual operating position. This contact should appear on the drawing as a normally open contact that is held closed. Using the contact in this manner sets off a trouble alarm if the valve is not fully open or there is a wiring problem in the monitoring circuit. This example does not necessarily exist in the real world. However, it does serve to point out how the contact status of a limit switch depends not only on the nature of the contact used, but also on the "normal" condition of the device that engages the switch arm.

Microswitches

Note

"Microswitch" is both a brand name and a common trade term. There is no intention here to confuse the two. The Micro Switch® company is a well-known maker of small switches and other components, but is not the exclusive source of the small general-purpose microswitches described below.

FIGURE 15–44 Microswitches are found in a wide variety of equipment. They can detect mechanical position, and act as contacts for a wide variety of devices.

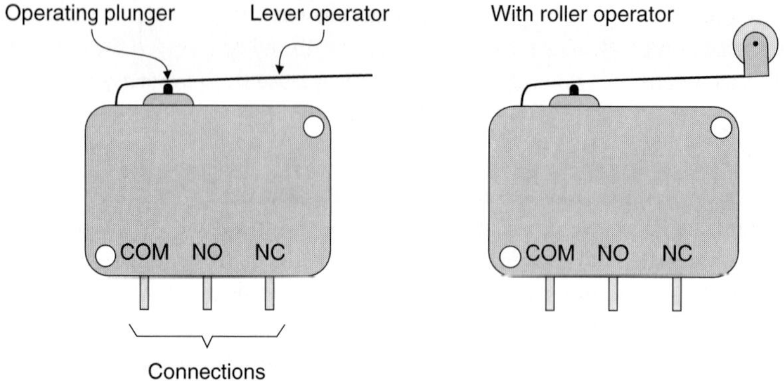

Industrial limit switches have heavy-duty enclosures and threaded hubs to receive a connector for cable or conduit. These switches are designed for exposed mounting in a hostile environment. They are very serviceable, but are also bulky and expensive. Where a similar function must be provided, but conditions are much less demanding, a small general-purpose switch assembly is much more suitable.

Microswitches are usually found inside equipment enclosures, where further protection from the elements is unnecessary. They are also used as contacts in automatic control devices such as motor-driven cam switches, end switches for electrically operated valves, and pressure switches.

Microswitches consist of a contact assembly enclosed in a small, flat, rectangular plastic case. The contact operator is typically a small plunger that protrudes a short distance out of one edge. The switch may be operated directly by the plunger, or with some type of actuating lever. Typical lever actuators are just a narrow strip of metal arranged to engage the contact plunger. A variety of levers are available, including a few with small rollers on the end (see Figure 15–44).

The case of the micro switch has holes in the corners to accommodate mounting bolts. Electrical connections are usually provided by protruding tabs designed for push-on "quick-connect" wire ends. The switch contacts are usually an SPDT configuration, with the terminals identified COM, NO, and NC. Contact ratings are usually marked on the switch.

These generic microswitches are used throughout the electrical manufacturing industry. Small, reliable, inexpensive, and simple to mount, they can be adapted to almost any imaginable situation involving mechanical movement. They have good electrical ratings, although generally pilot duty only, using fully enclosed snap-action contacts. They can be found in almost any type of equipment, including inside other switch devices.

Noncontact Switches

The limit switches and microswitches described above depend on physical contact with the item being sensed. In many situations this is quite acceptable. However, when the item is moving very fast, is very small, or may pass by in a variety of positions, a noncontact type of switch may be a much better solution.

A variety of devices are available, including photoelectric systems using light beams, magnetic sensors that detect metal or the passage of an imbedded magnet, and proximity switches that respond to almost anything within range. They are widely used in high-speed, high-repetition applications such as materials handling and packaging. Noncontact switches are more sophisticated and expensive, but have the advantage of having no mechanical wear.

15.9 Multistage Control

HVAC systems are good examples of equipment that must be capable of a much higher output than is utilized for average operating conditions. In regions with wide climatic variations, the heating and cooling systems must be sized for the peak seasonal loads. This means that the equipment is grossly oversized for most of the year. The equipment must be able to handle worst-case situations, but normally runs at a much lower output. Cycling a high-capacity system to maintain building temperatures within relatively narrow operating limits is often impractical, if not impossible.

Several methods are used to solve this dilemma. One approach is to use a system that can operate at several different output levels. With several stages of output, the equipment capacity can be much more closely matched to the demand. The equipment can then run for longer periods of time without unacceptable swings in the building temperature. This reduces the stress on the equipment and more easily accommodates the necessarily narrow operating differentials.

Multiple stages of heating and cooling can be accomplished in different ways. Where heating is provided by electricity, the elements are put into individually switched groups and switched in and out as required. When the heat source is fuel oil or gas, multiple stages can be difficult or impractical to provide, particularly with small burners.

Burner design is quite complex. It is difficult to run a small burner at different outputs while maintaining safe, efficient operation. For this reason, small gas burners are normally limited to a maximum of two stages, provided by a special two-stage gas valve. The valve provides different rates of gas flow for low and high fire. The combustion airflow may also be adjusted by changing speeds on the induced draft blower. Small oil burners are rarely if ever designed for multistage operation.

Multistage operation of cooling equipment can be accomplished with two-speed compressor motors or multiple compressors. Larger cooling systems may also provide multiple output levels through changes in the mechanical setup of the compressor—by unloading some of the cylinders, for example. Smaller units, typical of residential applications, are usually only single-stage. Heat pumps may provide one or two stages of cooling and, typically, one stage of heating through the refrigerant system. A second stage of heating is provided by an alternate heat source, such as electric heating elements. The second stage of heating provides emergency backup and carries the full heating load when it is too cold for heat pump operation. Because of the emergency heat system, heat pump thermostats must incorporate at least two stages for the heating mode.

No matter what system is used, the benefits of multistage operation are attractive. The equipment can be operated for longer periods of time, yet stay within the narrow temperature range demanded by the building's occupants. These longer operating cycles are possible because the heating or cooling output is closer to the actual load. The equipment can run at high efficiencies with less stress. Multistage controls and control systems are more complex and more expensive than those with only a single stage.

Control Devices for Multiple Stages

When the equipment can operate at more than one output level, the controller must provide control signals for each of the steps or stages of operation. This is often provided by duplicating the control functions several times inside a single multistage control. A typical two-stage heating thermostat, for example, has two mercury switches instead of the usual one. They are arranged so that as the temperature drops, the first-stage contacts close first. Then, if the first-stage of heating is not enough to prevent the temperature from dropping even further, the second-stage will also be energized. In most cases, it is very important that the stages are energized in correct sequence, because the equipment under control usually depends on the sequence for correct operation. For example, a two-stage gas valve cannot be cycled to high fire without first operating on low.

Some multistage systems use multiple separate controls with slightly different settings. For example, a water supply system with dual pumps can be controlled by two pressure switches. A slightly higher pressure setting on one switch starts the first-stage pump. The second pressure switch starts the second-stage pump if the line pressure continues to fall. Time delays may be added to prevent brief line surges from cycling the pumps. An interesting twist to this type of control is a duplex controller with alternating lead and lag pumps. Two equally sized pumps are used, and the controller automatically selects a different pump as the lead after each off cycle. This way the pumps accumulate approximately equal operating hours.

Larger heating and cooling systems often have more than two stages of operation. The controllers are usually more sophisticated multistage devices that provide sequenced outputs for several levels of both heating and cooling. These controllers do not use duplicated electromechanical control switches like many two-stage systems. Most use a sensor that provides a variable signal instead of a contact output. Multiple sensors may also be used, but their signals are usually combined in the controller to produce one value. The controller then stages the heating or cooling equipment based on a preset sequence of slightly different signal levels.

Multistage controls fall in between the single-step on/off controls discussed above and the infinitely variable systems that follow below. All three types of controls may be used in the same system, each applied to the situations where it is most suitable.

15.10 Proportional Control

All the control devices previously described behave as switches, supplying an on/off, contact-type output, not a variable signal. They provide automatic control by cycling the controlled load on and off as required by the process. The cycle rates are kept wide enough to protect the equipment from excessive starts and stops, and narrow enough to satisfy the conditions being regulated. In some cases, such as maintaining the level of a large storage tank, the system differential can be very wide and not cause any problems. In other situations, such as the control of a building heating system, the differential must be kept quite narrow to satisfy the comfort expectations of the occupants.

When the full capacity of the equipment is much higher than the average level of demand, control can be improved by using two or more stages of operation. The equipment is still cycled on and off, but the capacity is adjusted in steps.

The ultimate in control is provided when the equipment output is fully variable, and it is coupled to a controller that can exactly match the output to the demand. Equipment capable of fully proportional output is not limited to just full-on or full-off operation, or even discrete steps—it is truly variable. The output can be adjusted to any point between some reasonably low minimum value and the full rated maximum.

The most basic type of variable system is called proportional control. The controlling signal—and the resulting response of the controlled device—increases and decreases proportionally to the difference between the system set point and the actual operating point. As this difference increases—indicating that the output is not adequate to meet the need—output is increased proportionally. As the difference decreases, the controller acts to reduce output to prevent overshoot.

An example of this type of control could be a throttling valve—equipped with a suitable actuator—that supplies hot water to a heating coil in a hot air supply duct. A temperature sensor is mounted in the duct or the controlled space, and its output is used to control the position of the valve. As the room temperature falls the valve is opened wider, and as it rises the valve is closed down. The sensor supplies a signal to the controller that is proportional to the actual air temperature. The controller compares this temperature signal to the desired room temperature—the system set point—and translates the difference between the two into an appropriate output signal. The valve actuator responds to the output signal, positioning the valve so that the hot water flow rate is proportional to the signal received. The end result is the movement of the valve tracks the heat demand, continuously adjusting the heat added to the room to match the heat lost.

Fully variable control systems can become very sophisticated. What at first looks relatively simple can quickly become very complex. One of the reasons for this is that time lags in response can lead to oscillations. In a heating system, for example, there are delays in sensing temperature change, delays in supplying more heat, and delays in delivering the heat to the area of increased demand. There are also delays in reducing the output. Instead of adjusting smoothly to match the changing demand, a poorly implemented control system may never stabilize. The temperature first overshoots the set point, and then the control overcorrects. The output then goes too low, and the temperature falls back below the set point. As a result, the heat source fluctuates from full-on to full-off, as if it had a simple single-stage control. This problem can only be resolved by matching the response of the controller and the associated equipment to the needs of the controlled space.

Another problem that complicates variable control is that, as the system moves away from an average output, the difference between the set point and the actual operating point becomes wider. This is a normal result of how proportional control works. Just as in a multistage control, increased demand is driven by a greater difference between the set point and the signal from the sensor. This difference is used to provide the increased controller output signal. Unless the control includes what is called an integral function, the operating point will be offset from the set point

The integral function acts to reduce the offset to zero. It is usually combined with proportional control to produce a proportional plus integral (PI) control system. A higher level of sophistication is provided by using proportional, integral, derivative (PID) control. The derivative function responds to the rate of change in the system. PID control systems can respond more aggressively to rapid fluctuations than PI controls. Both PI and PID control are common in large heating, cooling, and ventilation systems. PID control is widely used in industrial processing.

Proportional Sensors

Controls that are continuously variable in operation need sensors that can supply a variable signal. This signal must be proportional to the value being sensed, and must be in a form that can be used by the controller. In pneumatic systems, the signals are variations in air pressure. In electrically operated systems they are variations in resistance, voltage, or current. If the sensor is a variable-resistance device, its output may be converted into a standard voltage or current signal, that is easier to send to the controller and can be interpreted without additional conversions.

Rheostats and Potentiometers. When the current flow in a circuit must be adjusted, instead of just switched on and off, one approach is to vary the circuit resistance (see Figure 15–45). If the variable resistance is inserted as a two-terminal device wired in series, it is called a rheostat. If the resistance is a three-terminal device, forming a voltage divider across some voltage source, it is called a potentiometer.

Adjusting a rheostat changes the total circuit resistance, providing some regulation of the current flow. When the resistance is increased, the current flow is reduced. When it is decreased, the current increases. To be effective, the resistance value has to match the circuit requirements, and must have a wattage rating that is adequate for the application. Unless the resistor is equipped with an auxiliary switch, the rheostat cannot shut off the current completely.

As a potentiometer, the variable resistor is connected in parallel across a voltage. The load is connected between one end of the resistor and the adjustable contact terminal. The potentiometer provides a full range of control, from full voltage to zero. This connection is not suitable in power circuits because of the constant drain of supply energy through the parallel resistor.

In low-powered circuits, variable resistors provide simple, effective control at low cost. Most of the adjustments provided in electrical and electronic devices are variable resistors of some type.

Temperature Sensors. Temperature can be converted to a variable-resistance value either directly, with a temperature-sensitive resistance element, or indirectly, by using a bimetallic- or bulb-type sensor mechanically linked to a variable resistor. The indirect method is widely used in older HVAC controls. These controllers work essentially the same as contact types.

FIGURE 15–45 Variable resistors may be connected as rheostats or as potentiometers.

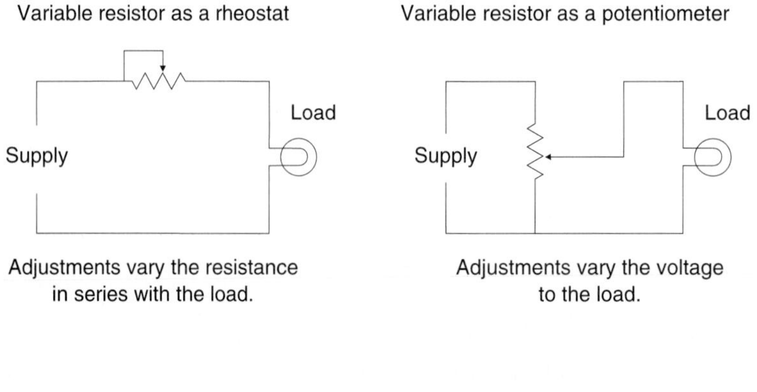

The mechanical action of the sensing element moves the wiper of a variable-resistance element, instead of operating a set of contacts. The result is a resistance that changes in proportion to the monitored temperature.

Thermistors and RTDs make much simpler sensors but require dedicated electronic circuits to convert the signal to voltage or current, because the output of the devices generally cannot be used directly. RTDs are more linear and more accurate, but thermistor technology is constantly improving. Their low price and simple circuit requirements make them cost-effective, and their nonlinear response is essentially irrelevant when the temperature range of the application is relatively narrow. This makes the thermistor popular in many modern HVAC applications.

Humidity Sensors. Resistance-type humidity sensors can be electromechanical. The nylon hygroscopic element is used to move the sliding contact of a variable resistance, in much the same way it would operate the contacts of a humidistat. However, this approach provides very few advantages, because it retains all the mechanical problems of the humidistat. Solid-state humidity sensors may use hygroscopic salts to provide a variable-resistance element. The resistance of a hygroscopic salt drops as moisture is absorbed.

Hygroscopic polymers may be applied in several different ways. A variable-capacitance sensor can be created by using the polymer as the dielectric. The properties of the polymer change with the level of moisture it absorbs, changing the capacitance of the device. The change in capacitance is converted to a variable-voltage or current signal by a dedicated electronic circuit. Polymers can also be used to change the frequency characteristics of a quartz crystal. The crystal is coated with the polymer, which—because of its moisture-absorbing characteristic—changes in mass as the humidity increases. The change in mass alters the oscillation frequency of the crystal, and an electronic circuit is used to sense this change and convert it into a suitable voltage or current output.

Enthalpy Sensors. Humidity sensors are combined with temperature sensors to produce a composite enthalpy signal. This can be done electromechanically, using a mechanical linkage to combine the movement of a bimetallic temperature sensor with that of a ribbon-type humidity sensor. This movement, proportional to enthalpy rather than just temperature or humidity, is linked to a variable resistor.

Enthalpy sensors can also be electronic. A temperature sensor, such as a thermistor, is combined with a humidity sensor, such as a variable-capacitance film device. The two are connected to an electronic circuit that responds to the changes in both temperature and humidity. The circuit supplies a voltage or current signal that varies in proportion to the enthalpy.

Pressure Sensors. As with temperature sensing, changes in pressure can be converted to variable resistance indirectly by using a diaphragm or other device mechanically linked to a variable resistor. These devices are similar in construction to pressure switches, but the mechanical movement of the sensing element changes the position of the wiper on the resistor, instead of operating a switch.

Pressure can also be measured by devices using fewer moving parts, but some electronic circuitry is involved because the changes produced are too small to be used directly. One example is a resistance stain gauge element connected to a diaphragm. As the diaphragm flexes with changes in pressure, the resistance of the strain gauge changes slightly.

This small change is amplified to produce a usable signal. A variable-capacitor element can be used in a similar way. The diaphragm moves one plate of the "capacitor," changing the total capacitance value. This effect is then amplified and converted as required.

Sensing Mechanical Position. Many control systems also require information pertaining to mechanical position. The most common examples in HVAC applications are feedback signals representing the actual position of a damper or valve. For simplicity, this information is usually taken from the actuator rather than from the valve stem or damper arm. The usual method is to use a variable resistor driven directly by the output shaft of the actuator motor. Because the shaft does not rotate continuously, only part of a full turn, the resistor can be mounted on the driveshaft inside the motor enclosure. Such resistors are usually a relatively long wire-wound type, but a carbon strip, or single slide wire fastened to a nonconductive base plate may also be used.

There are other methods used to monitor position. Pneumatic actuators, and electric actuators based on stepper motors, operate without position feedback because they can be depended on to move only the required amount. Pneumatic actuators are positioned by balancing applied pressure against the back pressure of a spring. It is unnecessary to check where the actuator is in its stroke because the reliability is very high.

Stepper motor actuators use an electronic circuit to drive the motor in discrete increments. Stepper motors do not rotate continuously; they are "stepped" by a dedicated control circuit. The controller can keep track of the position by counting the steps the motor is driven each way. This system is reliable as long as the motor is not stalled or the controller memory lost. For better reliability, the motor is periodically run all the way to one end and the counter is reset.

Another method often used with microprocessor controllers is to combine a constant-speed actuator motor with a time record in the controller. The position of the actuator can be predicted on the basis of the total operating time. As with the stepper motor system, reliability is improved by periodically running the actuator through a full stroke to ensure it is correctly zeroed.

Questions

1. Why does current flow stop when a switch is opened?
2. Sketch a diagram for each of the following switches. Indicate the common terminals where applicable: SPST, SPDT, DPST, DPDT, and 3PDT.
3. Explain the difference between maintained contacts and momentary-contacts.
4. Explain why SPDT switches are required for 208 and 240 V motors, when a simpler SPST will control the load?
5. Explain, in general terms, the difference between an interlock and an isolation switch.
6. Draw a three-way switch circuit. Identify the common terminal of both switches.
7. State the switch action required for a switch that transfers selected loads between a utility supply and a standby generator. Draw a diagram that explains how a double-throw switch can be connected to meet these requirements.

8. Compare the sequence of contact operation in a device that has standard normally open and normally closed contacts to that of a device with a standard normally open contact and a late-break normally closed contact.
9. Explain why a zero voltage reading across a switch is not a conclusive indicator of the status of the circuit.
10. In a fully operational circuit with a 120 V AC supply and one SPST control switch, what voltage would be measured from the circuit common to the line and load sides of the switch when it is in the off position?
11. Describe a reliable test procedure that can be used to find the first open point in a circuit containing multiple switches in series.
12. What will a voltmeter read if it is connected across the open switch of an operational 24 V circuit? What will it read if the load is a lightbulb, and it is burned out?
13. A four-way switch is correctly installed in an operational 120 V lighting control circuit. Measured from common, what voltages should appear at the two input terminals and the two output terminals of the switch? Will the results of this test be any different if the light is on or off?
14. Draw a DPDT switch circuit showing the proper connections for polarity reversal.
15. Explain why a test across the terminals of a 240 V motor does not necessarily prove there is no shock hazard. What test procedure should be used, and what readings would indicate the motor terminals are safe to touch?
16. If a 120 V motor has 120 V to ground at the terminals with the motor shut off, what is the most likely wiring error?
17. Explain why it is hazardous to install control contacts in the neutral or other grounded common, instead of in the ungrounded hot.
18. What four different criteria affect the electrical ratings of a switch or other contact device?
19. What is meant by a "pilot duty only" rating on an automatic control device?
20. Describe an example of "closed-loop control."
21. WHat is meant by the terms "set point," "differential" and "feed back," as they are used in reference to an automatic control system.
22. Convert 27°C to Fahrenheit.
23. Convert 175°F to Celsius.
24. Describe the operation of a bimetallic thermodisk.
25. Describe the operation of a differential temperature sensor, and give an example of where it might be useful.
26. State at least one advantage that electronic controls have over electromechanical types.
27. Define NTC and PTC as they apply to resistive temperature sensors.
28. Why are resistive temperature-sensing elements usually operated at very low current?
29. Explain, in general terms, the difference between temperature and enthalpy.
30. Explain the difference between static and velocity pressure.
31. Does the pressure switch used to monitor the combustion air blower of a gas burner operate as a limit, an interlock, or a control?
32. Describe the operation of a probe-type liquid-level control.
33. Explain the difference between "normally open held closed" and "normally closed" as these terms apply to limit switches that sense mechanical position.

34. Explain the difference between a rheostat and a potentiometer.
35. Why are variable-resistance controls not well suited to power circuits?
36. What advantages do multistage or variable-output heating and cooling equipment have over single-stage equipment? Are there any disadvantages?
37. What is the difference in logic between proportional, proportional plus integral (PI), and proportional, integral, derivative (PID) control?
38. Why do control systems with no integral function move further away from the set point as the demand increases?

16 Low-Voltage Controls

Objectives

Chapter 15 examined control devices. This chapter adds the additional components required to interface control devices to the controlled equipment. These components include control transformers, relays, timers, and sequencers. After completing the study of this chapter you should be able to:

- Test a control transformer for common faults, correct phasing, and appropriate ground connections
- Describe the operating characteristics of a control transformer
- Calculate the rated full-load current of a transformer from typical nameplate data
- Describe the operation of a typical electromagnetic relay
- Explain the operation of basic relay circuits
- Describe the operation of typical thermal relays
- Describe the operation of typical electromechanical timers
- Describe the function of sequencers
- Describe the fundamental principles of programmable controllers

16.1 Introduction

Note

The term "low-voltage" is used here in the general, not legal, sense. Electrical codes may define specific voltage ranges using terms such as extra-low-voltage, low-voltage, medium-voltage, etc. The term "low-voltage" as used in this chapter refers to the use of a lower voltage for the control circuit than is supplied to the power loads. The actual voltage level used varies with the situation, and may or may not fall into a category defined as "low-voltage" in electrical codes.

For safety reasons, and to simplify the design of components, control circuits are often operated at a lower voltage than the main supply. The voltage is reduced for the control system, but the main power loads are operated at the full line voltage. This eliminates expensive voltage conversions and keeps the load current as low as possible. As a result of this practice, it is common to have two or more different voltages associated with the same piece of equipment.

Two key elements make this dual-voltage system practical. The first is the transformer, which is used to provide the reduced control circuit voltage. The second is the relay—and its larger cousins the contactor and motor starter—which allows a low-voltage control system to switch loads that operate at a different voltage.

Power transformers and the general theories of transformer operation were covered in Chapter 7. This chapter includes a brief review of transformer basics and an examination of a typical control transformer. This is in the form of a practical exercise demonstrating transformer behavior and the effects of various grounding practices.

Relays, contactors and motor starters, and some specific timers are examined in Part III of this book. They will also be reviewed here briefly, but the focus of this chapter is on low-voltage control systems in a broad sense, not on the specific requirements of motor control.

This chapter completes the examination of the building blocks of electrical control circuits. Manual and automatic control devices allow electrical circuits to respond to nonelectrical events. Transformers provide safe control power. Relays provide an interface between different systems and serve as logic devices. Timers add delays to control logic to ensure correct sequence and timing of automated events. These controls and control circuit components are the basic elements used to create virtually all the nonelectronic control systems now in use. They also play major roles in electronic systems of all types, including those operated by microprocessors or dedicated computers.

16.2 Practical Exercise: Testing a Control Transformer

Warning

The tests described here involve potentially lethal voltages. Appropriate precautions are essential for personal safety.

To conduct this experiment, a small transformer is required. It should have a 120 V primary and a lower secondary voltage. Something in the 12 to 24 V range works well. The transformer should be rated for an output of at least 1 A. If no current rating is provided, it can be found by dividing the transformer's nameplate VA rating by the rated secondary voltage. (See formula and reference below.)

Additional materials include a variable AC supply capable of at least 120 V; a standard, nonisolated 120 V AC outlet; three meters—two for current and one for voltage readings; and suitable power resistors to use as a load. The resistors should be sized to draw at least three different values of current from the transformer. Suggested target current values are approximately half- and full-rated transformer load current, and a 10 to 25% overload. All of these must be based on the transformer's actual rated output voltage and current, so the resistance values used will vary with the transformer available. Adjustable resistors are more adaptable and easier to select. The power rating of the resistors must be adequate for the load they will carry. For full load, the wattage rating should be at least equal to the VA rating of the transformer. Resistors can be combined as necessary, if a single value does not meet the requirements.

Small general-purpose transformers usually lack the standard terminal markings of larger transformers. The leads may be color-coded or identified by their location. If the transformer does not have full nameplate information, consult the manufacturer's data sheet or catalog listing. The important data are the lead identifications, the rated primary and secondary voltages, and either the VA or full-load current rating. Larger transformers have much more complete nameplates, often providing complete specifications and a connection diagram.

The transformer VA rating is a multiple of the rated secondary voltage and current. This relationship can be used with the rated secondary voltage to find current from VA or VA from current.

To find the full-load current rating, use:

$$I_{SEC} = VA \div E_{SEC}$$

To find VA from current, use:

$$VA = E_{SEC} \times I_{SEC}$$

Transformers are generally quite rugged. Modest, short-term overloads should not be a problem, **unless the transformer is Class 2 rated,** in which case the internal fuse link may melt open. Long-term overloads, short circuits, excessive primary voltage, or ambient conditions that cause the transformer to overheat can cause any transformer to fail. Input voltages should be limited to not more than 10 to 15% over the rated.

Testing the Transformer for Faults

Resistance tests can be used to check that the transformer windings are continuous and electrically separate from each other. They can also be used to check for windings that are shorted to the frame. Assuming the transformer is in working order, there will be continuity through both windings—from one lead or terminal to the other—at both the primary and secondary connection points. There should be no continuity from one winding to the other, or from either winding to the transformer frame.

These tests are good quick checks for a suspect transformer. They can detect the most common transformer faults—open windings, shorts between windings, and ground faults from either of the two windings to the core or frame. Large transformers require more sophisticated tests and more elaborate test equipment. Large transformers may also be repairable; small transformers are not.

For winding resistance tests, the meter should be set to the lowest available range. The measured resistance of the low-voltage winding is typically very low, often less than 1 Ω. If a digital meter is used, the readings may have to be corrected to compensate for meter lead resistance. The resistance of the high-voltage winding is typically much higher. Actual values depend on the transformer ratings, size, and manufacturer. Large transformers may have winding resistances that are too low to measure with a conventional multimeter.

For the measurements taken from winding to winding, and from each winding to the frame, the highest available resistance range should be used. All of these readings, from any primary lead to any secondary lead, and from any winding lead to the frame, should be ∞ Ω, which is an open circuit. This proves there are no shorts between the primary and secondary windings, or ground faults from the transformer frame to either winding.

Building up personal experience is the best way to learn how to judge the condition of typical transformers encountered in the field. A few minutes spent taking readings on actual transformers will provide much more useful data than can be found in any book. Remember, the transformer must be completely disconnected before these tests are made.

Transformer Operation at No Load

Once the transformer has passed the initial resistance tests, the experiments can proceed. The first is a no-load voltage test designed to demonstrate the transformer's voltage ratio (see Figure 16–1). The 120 V winding is connected to a variable AC supply, with an ammeter (in series) to measure the primary current. A switch in the supply line can be very convenient. A second meter is used to take voltage readings.

With the supply set to the transformer's rated voltage, the transformer is energized, and the primary and secondary voltages are measured and recorded. The primary voltage will be the rated value as set previously; the secondary voltage should be slightly above the rated value. This is because the nameplate rating is based on full-load conditions. Small transformers are usually designed to produce a slight overvoltage at no load, to compensate for the drop that occurs when the transformer is loaded. The rated no-load or open-circuit voltage is part of the manufacturer's specifications, and may be marked separately on the transformer nameplate.

The primary voltage divided by the no-load secondary voltage is the transformer's no-load voltage ratio. This ratio is a function of the number of turns of coil wire on the primary winding in comparison to the secondary winding. The actual number of turns cannot be found from voltages, but the true numerical turns ratio is very close to the no-load

FIGURE 16–1 This test circuit is used to demonstrate the transformer's voltage ratio.

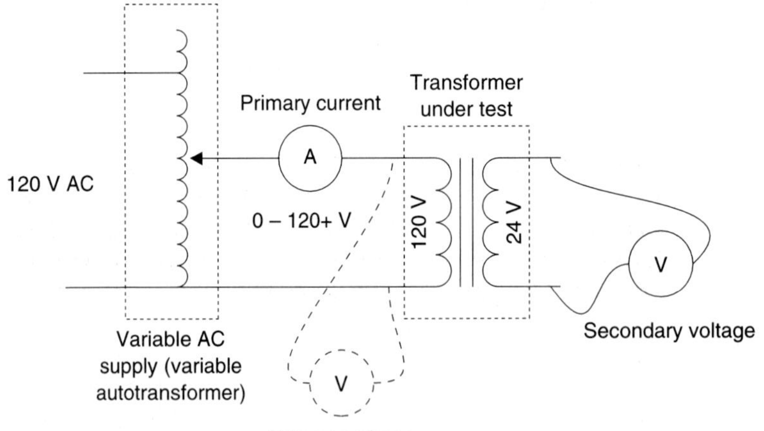

voltage ratio. The relationship between the turns ratio and the voltage ratio is often given as a general-purpose ratio formula:

$$\frac{N_P}{N_S} = \frac{E_P}{E_S}$$

N_P = Primary turns

N_S = Secondary turns

E_P = Primary voltage

E_S = Secondary voltage

The best way to fully examine how a transformer behaves is to use a range of input (primary) voltages. The primary and resulting secondary voltages are recorded and plotted on a line graph (see Figure 16–2). The primary values should start well below the rated voltage and continue at regular intervals to a value that is slightly above this. Most transformers are rated for full-load operation at not more than 105% of rated primary voltage. Brief tests 10 to 15% above that value are unlikely to cause problems.

The line graph should be plotted on standard graph paper. The usual approach is to place the no-load primary voltage on the x axis, and the resulting secondary voltage on the y axis. A separate primary current line can also be plotted. The resulting graph is representative of the transformer's magnetic field strength.

When transformer secondary output is plotted for a broad range of primary values, a predictable pattern emerges. The line is quite straight at first, rising at a fixed angle from the bottom left of the graph up and to the right. This indicates a linear, fixed-ratio relationship between the input and output voltages. Then, as the primary voltage rises through the rated value and above, the rate at which the secondary voltage increases begins to taper off. The line on the graph bends and begins to flatten out. The graph forms a slight "knee," after which each incremental increase in primary voltage produces proportionally less voltage in the secondary.

When the primary current is plotted against the primary voltage, this current curve will show a similar response. The increase in current has a linear relationship to the supply voltage until it approaches the knee of the voltage curve. At that point, the current begins to rise at a rate substantially greater than before.

The cause of this behavior is called magnetic saturation. The magnetizing force producing the field inside the core of the transformer is a product of the primary current and the number of turns in the primary winding. In the transformer's normal operating range, the primary current—and therefore the magnetizing force—rises proportionally with each increase in voltage. This is a reasonable and expected response. To the supply, the transformer is just a load. It is some value of impedance, made up of a lot of inductive reactance and a small amount of resistive losses (a typical coil on an iron core). The primary current should therefore rise in proportion to the voltage that drives it through the circuit.

However, as the voltage, current, and the magnetizing force (MMF) all increase, the strength of the field produced in the core must also increase. At first the increase is linear. Each time the MMF is doubled, the number of magnetic lines of force is also doubled. But the capacity of any core material is limited. At some point the increase in magnetic lines is

FIGURE 16–2 A plot of transformer output voltage is also the transformer's magnetization curve. This example is based on a small 120:24 V control transformer, but the shape of the curve is similar for all transformers.

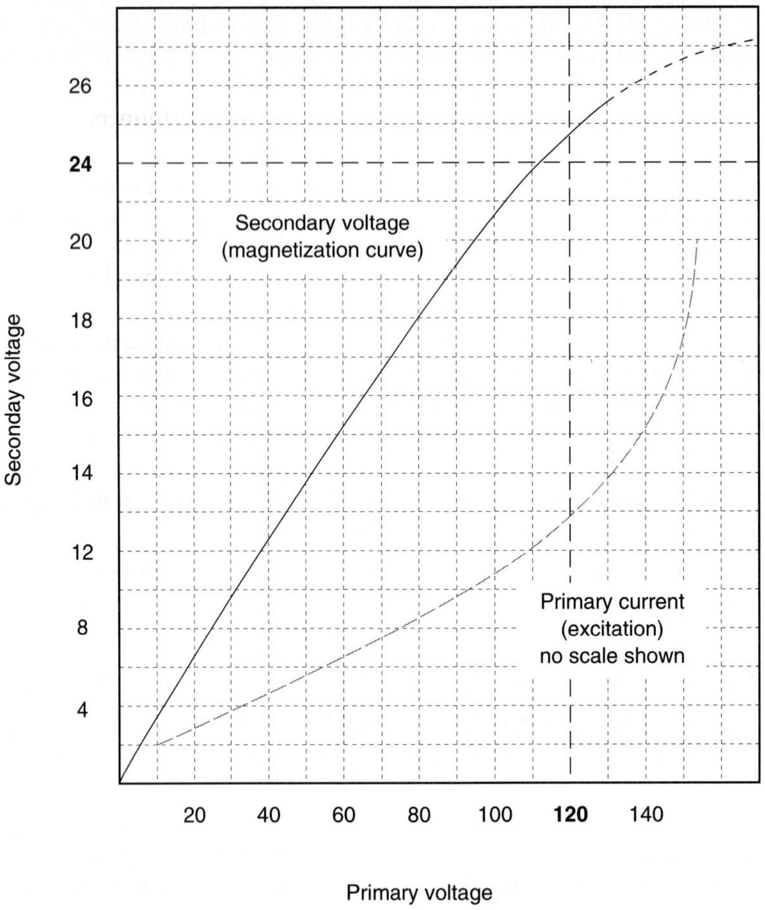

no longer proportional to the increase in magnetizing force. As this is pressed even further, each increase in magnetizing force produces fewer and fewer additional magnetic lines. The core has begun to saturate.

Both the primary back EMF and the transformer's secondary voltage are generated entirely by the changing magnetic field in the core. With no load connected to the transformer, both voltages are directly proportional to the strength of the field. On the linear portion of the curve, they rise in direct proportion to the applied voltage. The primary current also rises steadily, in a reasonably linear fashion, because both the applied voltage and opposing back EMF rise simultaneously. However, at the point where the secondary voltage curve approaches the knee in the graph, the core has begun to saturate. Each increase in applied voltage is no longer balanced by a proportional increase in the back EMF. With less back EMF to oppose the current, each increase in the primary voltage causes an increasingly dramatic rise in the primary current.

At no load, the secondary voltage is directly proportional to the strength of the changing field in the core. There is no current in the secondary, and therefore no losses, so its voltage is an accurate representation of the field strength. A plot of primary voltage vs. no-load secondary voltage is sometimes called a magnetization curve. It has exactly the same shape as a plot of the actual field strength, but can be created from values that are much easier to obtain. The magnetization curve of all transformers has a similar shape, because the relationships are always the same. Most transformers are designed so that the full-load line falls at the beginning of the knee. This represents optimum use of the transformer iron. Selecting a load line that is lower on the curve provides more reserve for overvoltage, while a line higher on the curve results in excessive primary current, which reduces transformer efficiency and can lead to heating problems.

There are a number of lessons in this experiment. Normal transformers are ratio devices, not voltage regulators. When the input voltage changes, so does the output. Transformers are built for a certain maximum voltage. This should not be exceeded for any extended period of time because it causes excessive primary current. A relatively modest overvoltage can burn out a transformer, even with no load on the secondary side. The final lesson is that the no-load input current of a typical transformer is actually quite significant, even though transformer ratio formulas suggest that it is should be zero.

The voltage and current ratios (see Figure 16–3 and the following ratio formulas) are standard fare for any electrical text, including this one. Unfortunately, they give the impression that the transformer primary draws no current from the supply unless there is a load on the secondary side. As the previous test demonstrates, this is not the case.

$$\frac{N_P}{N_S} = \frac{E_P}{E_S} = \frac{I_S}{I_P}$$

N_P, N_S = Primary and secondary turns

E_P, E_S = Primary and secondary volts

I_S, I_P = Secondary and primary current

Voltage is directly proportional to the number of turns, and current is inversely proportional to voltage. Any two of the ratio relationships can be combined to match the known values. Cross-multiplying the voltage and current ratios produces the power in equals power out ($E_P I_P = E_S I_S$) formula that is the underlying principle of transformers.

FIGURE 16–3 Primary and secondary voltage, current, and turns values, and their associated symbols.

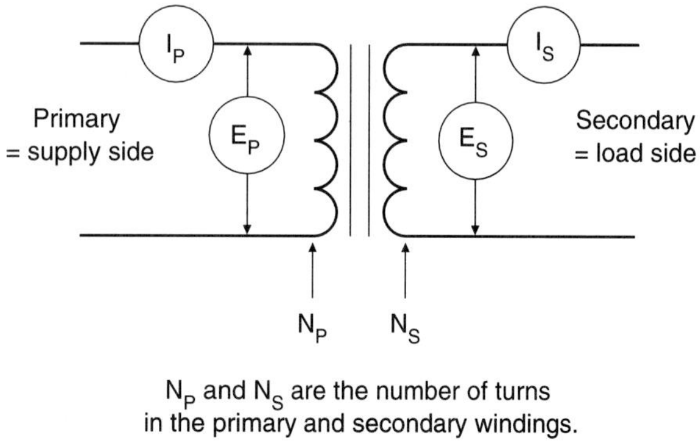

If a peak-reading meter is available, a final no-load test should be conducted. This test helps to explain a problem that occasionally causes mysterious transformer fuse failures. It is a simple test, but the primary current has to be measured with a meter that can record short-duration peaks. This is a common feature on digital clamp-on ammeters. The supply should be set for the nominal voltage of the transformer primary, in this case 120 V. The power to the transformer is simply turned on and off several times, and the current peak is recorded for each trial. The usual result is a random set of current valuesranging from almost nothing to many times the transformer's normal full-load current. The reason for the variation is that the initial "inrush" current depends on where on the AC voltage cycle the switch contacts actually close. This is completely unpredictable, and so is the transformer inrush current.

The lesson here is that the fuses used to supply the transformer should be sized for the transformer, not the probable load. Even if the transformer is going to be operated at less than the rated output, the fuse rating should not be reduced. As the test shows, transformers can draw initial inrush currents as high as 10 times the rated full-load current, even with no secondary load connected. When a transformer is cycled on and off during normal operation, an undersized fuse will occasionally blow for no explainable reason. A caution about this, though: Check for all other causes of failure. Never assume that the problem is an undersized fuse. It is unusual for any transformer, particularly one used for controls, to have a primary fuse that is below the proper rating. However, it does happen occasionally, and when it does it can be a frustrating head-scratcher for those involved.

Transformer Operation with Load

Load tests are done with the input voltage set to the transformer's rated value—in this case 120 V. The most convenient load for test purposes is one or more resistors of appropriate sizes and power ratings (see Figure 16–4). The test should be conducted with at least three

FIGURE 16–4 This test circuit is used to demonstrate transformer behavior under load.

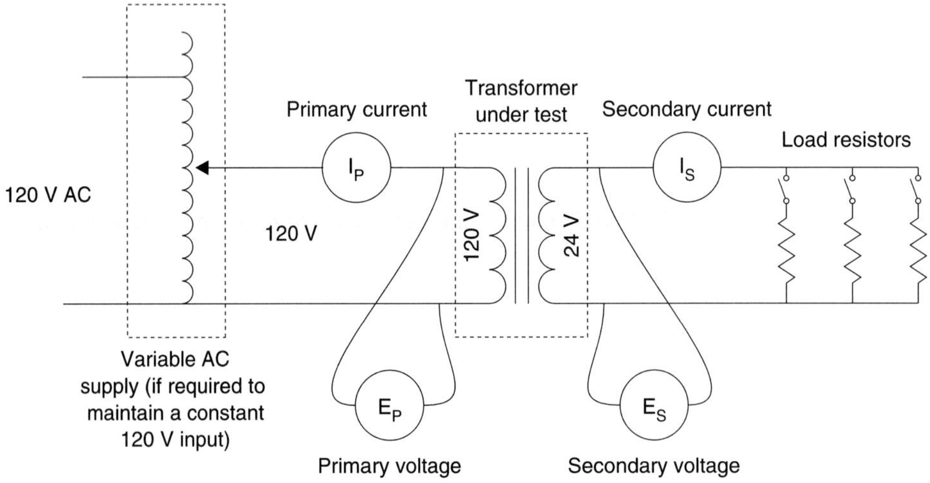

different loads: approximately half the rated load, full-rated load, and an overload of 10 to 25% above the full-load rating. A variable-load box or other adjustable-load device is very convenient. A full range of incremental values may be desirable if a curve is going to be plotted to show the transformer's response to loading.

An ammeter is installed in both the primary and secondary circuits, and a voltmeter is used to take readings on both the primary and secondary sides. A clamp-on ammeter may be substituted for the wired-in ammeters. Primary and secondary current and voltage values are measured and recorded for each different load. The primary voltage should be reasonably constant for each test.

With rated primary voltage applied, and a load that draws full-rated secondary current connected, the transformer output voltage should be at the nameplate rating. At lower loads, the secondary voltage will be slightly higher; above full load, it will be slightly less. The transformer voltage ratio holds up reasonably well from no load to full load, but the current ratios do not fare nearly as well, because primary current is always too high.

It may be tempting to deduct the primary no-load current from the loaded values in an attempt to force-fit the data into the expected outcome. This may even produce a fairly neat result, but it is a misrepresentation of what is actually happening. Transformers always draw "excitation current" to energize the magnetic field. This purely reactive current and a small amount of in-phase current to supply the no-load losses make up the no-load current of the transformer. Excitation current is essentially the same for all loads. The primary current demand that actually supplies energy to the secondary load cannot be calculated by deducting the no-load current from the total supply current, because the two are not in phase. The energy-supplying current has the same phase relationship as the secondary load current, while the no-load primary current is almost entirely reactive.

The main lessons of this experiment are that the transformer output voltage does drop when the transformer is under load—but not very much. The secondary current depends on the impedance of the connected load. The primary current depends on the secondary current and the transformer ratio. Primary current increases proportionally for each increase in secondary current, but the measured current ratio is never really the inverse of the voltage ratio—as ratio formulas indicate it should be—because of the excitation current. The no-load primary current is never zero.

Testing the Effects of Secondary Grounding

The following tests compare the behavior of a transformer with and without a ground connection on the secondary side. The transformer used in the previous tests is connected to a normal 120 V AC supply with a grounded neutral. This can be a typical 120 V outlet or similar source, but it cannot be from an isolated supply that has an ungrounded secondary. No ammeters are required, and no load is connected on the secondary side.

With the primary energized, the voltage readings across both the primary and secondary indicate normal operation. The primary supply voltage is approximately 120 V; the secondary is approximately the transformer's expected no-load output voltage. A voltage reading is then taken from each transformer terminal to ground. Any good ground connection, such as bare metal on a grounded enclosure or a receptacle ground socket, can be used.

The primary should read 120 V to ground on one terminal, and zero or very close to zero on the other terminal. The supply-side "hot" or line terminal is the one with the 120 V reading; the other terminal is the supply neutral. The neutral is grounded at the supply; therefore, it has virtually the same potential as the ground system. Very small neutral-to-ground voltages are possible, particularly in large buildings. Readings from ground to the transformer secondary are impossible to predict (see Figure 16–5). Two otherwise identical transformers may give

FIGURE 16–5 Typical transformers, without a secondary ground connection, have unpredictable secondary voltages to ground. In some cases, these values are dramatically affected by the primary side connections. Only when the secondary is grounded do the voltages to ground become predictable.

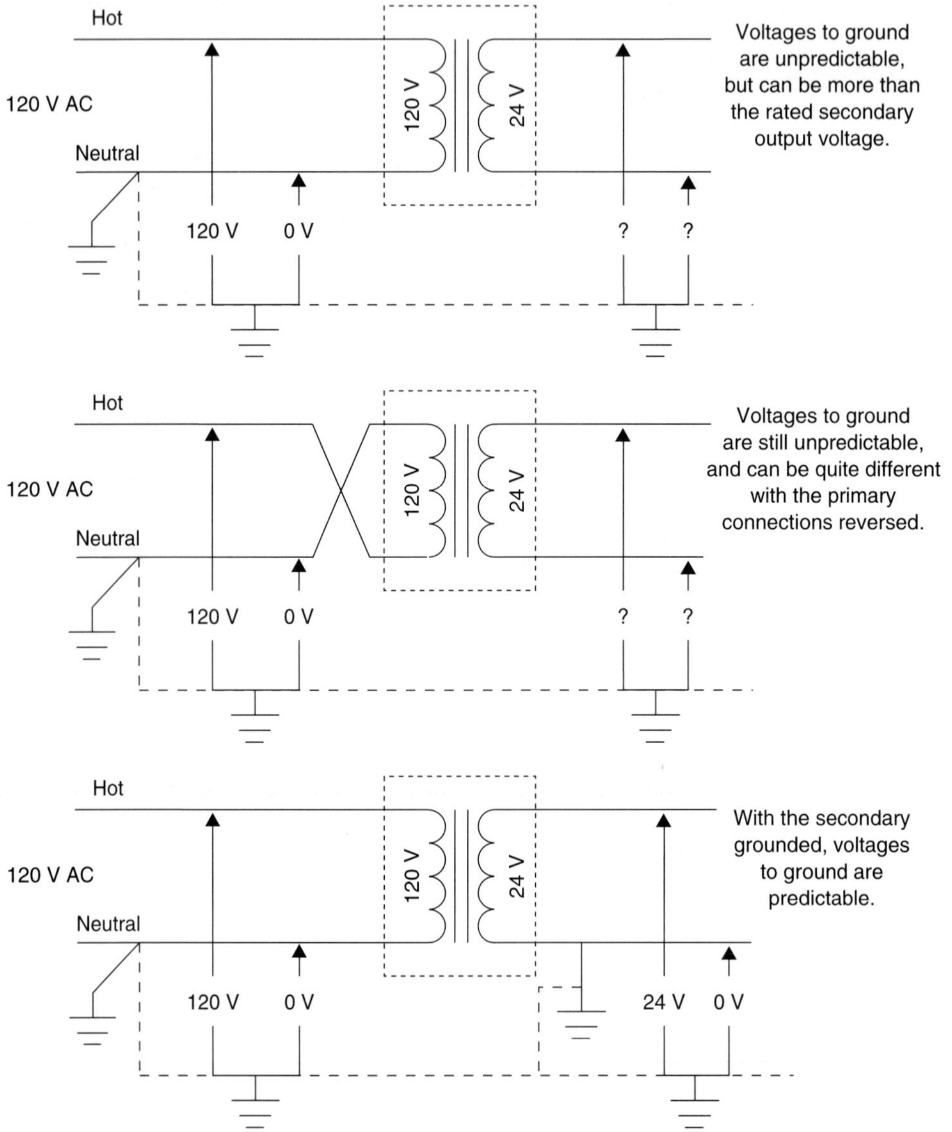

quite different voltage values. Some transformers have relatively high voltages to ground—70 to 80 V or more for transformers of this type—while others have almost none. Reversing the primary side connections can also change the readings, sometimes quite dramatically.

To provide a contrasting view, the secondary of the transformer is connected to ground and the test repeated. The primary terminals read as before; nothing changed on that side. The secondary side voltage from terminal to terminal should also be unchanged, but the voltages to ground are now zero on the grounded terminal, and full secondary voltage on the ungrounded terminal. The random voltages are gone; the secondary now has much the same type of connection as on the primary side.

The primary supply owes its voltage-to-ground readings to the ground connection at the supply transformer and at the building service. With one terminal grounded, the low-voltage transformer secondary behaves in much the same way, and has the same type of relationship to the ground system.

Transformer Phasing

There is some confusion surrounding the issue of transformer phasing, because the term can be applied in two different ways. When transformers are connected in parallel, "phasing" refers to arranging the connections so the transformers can work together. Paralleling transformers can be tricky business. Not only must the transformers be correctly matched, the connections must be correct on both primary and secondary sides if serious problems are to be avoided.

Note

As a general rule, it is unwise, and in most cases a code violation, to connect small transformers in parallel. Paralleling larger transformers can be done, but should not be attempted without a thorough knowledge of the regulations and the problems involved. The requirements for proper parallel connections will not be presented here. If this topic is of interest, there are numerous texts containing the necessary information. Some transformer manufacturers also supply excellent information on this and other aspects of transformer selection and application.

The second use of the term "phasing" refers to selecting the correct terminal when grounding the transformer's secondary winding. This information applies to transformers that have ground connections on both the primary and secondary sides. Correct phasing ensures the lowest possible voltage between the two ungrounded lines—one from the secondary, the other from the primary. This is easier to understand from a demonstration than it is from a theoretical description. It should be noted that some HVAC equipment manufacturers stress the importance of correct phasing. Others ignore the issue completely. At least one make of equipment will not function if the transformer is not properly phased.

To test the phasing of a transformer, the primary must be connected to a grounded supply, such as the normal 120 V circuit used for the previous test (see Figure 16–6). One secondary lead must also be grounded. This is the normal situation for many low-voltage control transformers. To check the phasing, a voltage measurement is taken from the ungrounded primary lead—the incoming hot, not the neutral—and the ungrounded secondary lead. This voltage reading can be only one of two values. It is either the sum of the primary and secondary voltages—about 144 V for a 120:24 V transformer—or the difference—about 96 V.

FIGURE 16–6 When a typical 120:24 V control transformer is tested for correct phasing, there can be only two possible results.

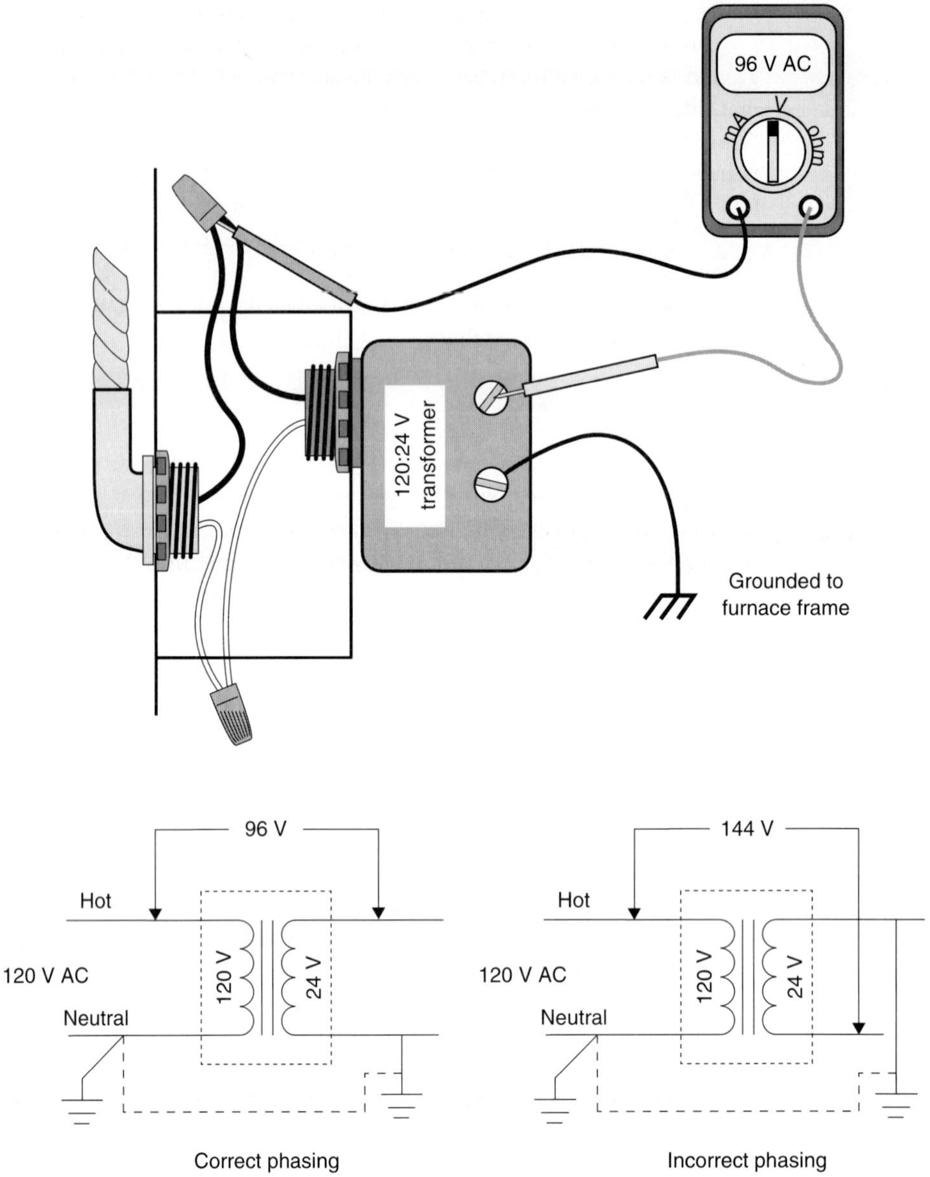

Correct phasing produces the lower of these two readings. When phasing is incorrect, the transformer connections can be changed by reversing the primary leads or connecting the opposite secondary lead to ground. The primary leads should not be reversed if they are color-coded. When one lead is white, it must be connected to the supply neutral. When both leads are black, either lead can be connected to the hot or the neutral. If the ground is moved to the opposite secondary lead, be aware of the following.

Warning

When a control transformer is replaced or the connections are changed in any way, the original grounding should be retained. If the transformer was ungrounded and there are no other reasons to change this, it can be left that way. If the transformer is grounded, the ground connection should always be on the common side of the circuit. Be aware, however, that some manufacturers do not follow this rule. This can cause serious problems, because a ground fault in the control wiring can bypass any control elements that are between the grounded side of the supply and the fault. This can include critical safety limits.

Ground connections for the control transformer in an HVAC system can be at the transformer, on the circuit board, or inside one of the control components—typically the burner ignition module. There should be only one ground, so do not add extra grounds at any other location.

The issue of transformer grounding is not a simple one, especially in HVAC applications. Due to differences in the control equipment and in manufacturers' practices, almost every possible situation presently exists in this field. There are ungrounded and grounded control systems. There are some controls that must be grounded, some that can't be grounded, and some that will work either way. The ground connection may be at the transformer or at a control device. The control itself may require an ungrounded supply because it actually grounds one side of the supply through its own connections.

A simple voltage test to ground determines if the system is grounded. However, it does not indicate if the system is actually supposed to be grounded. An ungrounded control system becomes a grounded one whenever there is an accidental ground connection anywhere in the circuit. This can go undetected for years, but is potentially quite hazardous when the ground happens to be on the wrong side of the control circuit.

16.3 Relays

The second element critical to low-voltage control systems is the relay. The transformer makes it possible to obtain an isolated low-voltage supply to operate the control system. The relay allows one circuit to exercise control over another circuit, with no electrical connection between the two. It can therefore link the control circuit to other circuits while maintaining electrical isolation between them. Relays can be used as an interface between the control and power circuits. They can also be used as a logic device within the control system. The most common types of relays are electromagnetic, thermal or bimetallic, and electronic.

Electromagnetically operated control devices use electromagnetic coils to perform some physical function. This may be opening a valve, releasing a mechanical latch, or, in the case of relays, opening and closing electrical contacts. Electromagnetic contact devices include relays and contactors. Relays are available in a wide variety of open and enclosed styles. Industrial control relays can be as large as some contactors, and may be rated for similar voltages and currents. Subminiature relays, designed for printed circuit board mounting, are so small that they can easily be confused with electronic components.

Relays are generally control circuit related, rather than power circuit related, but there are exceptions. Relays with horsepower-rated contacts are used to control single-phase motors up to about $1^1/_2$ Hp at 240 V. More demanding power loads are switched with contactors. Examples include motors, large electric heating elements, and multiple-fixture lighting circuits such as the area lighting in commercial buildings.

Thermal relays use bimetallic strips or disks in place of coils. Bimetallic materials respond to temperature changes, and are therefore widely used for temperature controls. When they are used to operate relay contacts, a small heating element is used to heat the strip or disk. Bimetallic strips are used in time-delay relays and in place of the solenoid coil in slow-opening gas valves. They are also used in pulsing timer applications. The strip and heater are arranged so that the contacts naturally cycle on and off. This technique is used in light flashers and in variable heat controls for electric range elements.

Bimetallic disks are used in power relays such as those used to control electric furnace elements. Unlike bimetallic strips, disks have a natural snap action. They are made with a slight dish shape, and suddenly pop to the opposite curvature when they are heated. This makes them more suited as actuators for high-current relay applications. When they are used in relays, they have an associated heater, just like bimetallic strip-type relays, and do provide a time delay.

Electronic relays provide much the same functions as electromagnetic and thermal types, but they use solid-state components in place of mechanically operated contacts. They contain no moving parts and are not subject to wear or contact degradation from use.

All types of relays are switching devices. Whether they contain actual contacts, or electronic components that act as contacts, their switching capabilities are a primary specification. The ratings of the "contacts" include the maximum allowable voltage, suitability for AC or DC use, and the continuous-current-carrying capacity. Additional ratings include suitability for specific types of loads, such as coils and motors, that are relatively difficult to switch. The other primary specification for all types of relays is the voltage rating of the operating coil, heater element, or electronic circuit that initiates changes in the status of the contacts.

Electromagnetic Relays

An electromagnetic relay is an electrically operated switch; some examples are pictured in Figure 16–7. The main parts are the operating coil, an iron core, and one or more sets of contacts. The core is in two parts, held a short distance apart by a spring. The coil is placed around the stationary portion of the core. When the coil is energized, a magnetic field is set up in the core and through the air gap that separates the two parts. This field pulls the two core pieces together, against the force of the spring. When the coil is de-energized, the spring pushes them apart again, back to their original position.

Each relay contact set consists of at least one contact surface that moves, and a matching surface that is held stationary. All the moving contacts are connected to the moving portion of the core, so they change position when the coil is energized. The contacts may be classed as normally open (NO) or normally closed (NC), in reference to the position of the contacts when the coil is de-energized. An NO contact pair is open when the coil is de-energized; an NC pair is closed. Contacts may also be arranged in sets consisting of one NO and one NC preconnected to a shared common terminal. These are sometimes called SPDT contacts because they behave the same as an SPDT switch (see Figure 16–8).

FIGURE 16–7 Top left, an industrial control relay with four independent double-break contacts, each of which can be converted to NO or NC operation. Top right, an HVAC "fan center" containing a 120:24 transformer and a SPDT relay. Center, an open-style DPST relay with horsepower-rated contacts. Front, two enclosed- and one open-style relay with SPDT contacts.

FIGURE 16–8 This relay design is a common one. The "clapper"-style magnetic armature operates one or more sets of SPDT contacts. Connections may be brought out to a variety of different terminals or plug pins spaced to fit a relay socket.

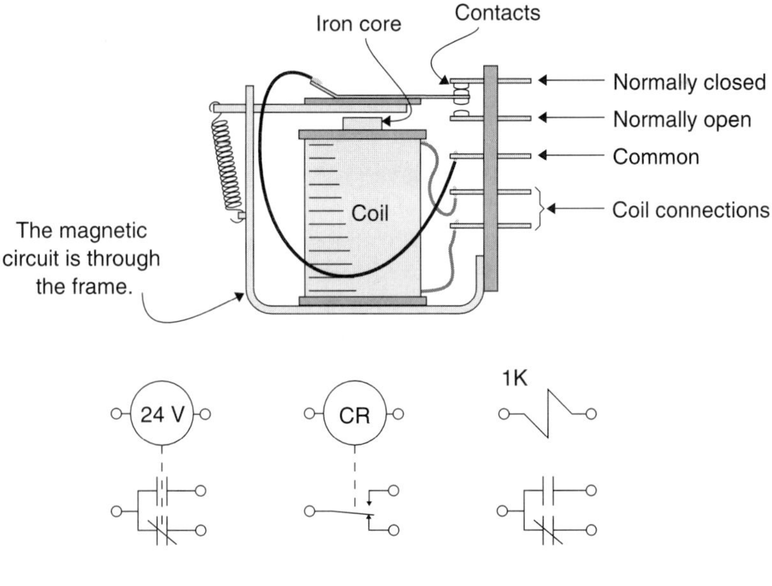

Common relay symbols

There are many possible variations in relay contact arrangement and layout. Two of the most common are SPDT and DPDT. The DPDT relay is the same as an SPDT type, except there are two sets of contacts instead of one. Relays are also made with double-break contacts. The moving portion of a double-break contact set is a short flat strap of conductor with a contact on each end. When the coil is energized, the moving contact pairs bridge across a pair of stationary contact surfaces. The two contacts of each pair are in series, which explains the name double-break. The contacts may be NO or NC, but they are usually independent of each other, rather than in SPDT sets.

Mountings, Connections, and Markings. Small relays may be a plug-in type requiring a separate base socket. All connections are then made to the base terminals, not the relay. These relays are very convenient to remove and replace. Round bases are like old-style radio tube sockets, complete with a small key on the center pin that extends into the hole in the base. They must be aligned properly before they can be inserted. The pins and receptacles of rectangular bases are arranged so that the relay can only be fitted to the base with the correct orientation. Relays may also be self-contained devices with screw terminals, wire leads, or lugs designed for either solder or some type of push-on connector.

Relays mounted on printed circuit boards may have plug-in bases or be directly soldered to the board. Subminiature types can be so small that they look like standard integrated circuit chips. Larger ones are usually cube-shaped. Some have the coil circuit terminals on the bottom and contact connections on the top. The bottom connections plug into a socket or are soldered to the circuit board. The top connections are quick-connect or screw terminals. This allows the manufacturer to isolate the high-voltage power connections from the circuit board traces that operate the low-voltage coil circuit. Replacing a relay that is directly soldered to a circuit board can be very difficult.

Open-style relays usually have only their ratings marked on the base and coil. Enclosed relays have the coil and contact terminals identified by symbols, numbers, or a combination of the two. The terminals for independent contacts may be marked NO or NC, or with an appropriate contact symbol. Because each contact has two terminals, there will be two terminals associated with each symbol. When the contacts are in SPDT pairs, there are only three terminals for each set of contacts. A common marking method is COM, NO, and NC at the appropriate terminals, instead of markings for each terminal pair. When contact symbols are used instead of letter markings, the common may be identified by a line indicating the shared terminal.

On drawings, relay contacts are almost always shown in the de-energized state. The rare exceptions to this include some monitored systems that have a standby mode. Alarms, for example, use monitored circuits that carry current at all times. When relays are used as part of these circuits, they are energized continuously by the monitoring current. The contacts of these relays may be shown in the energized state because it is otherwise quite difficult to follow the normal operation of the circuit.

Basic Relay Circuit. The coil and the contacts of the relay are electrically separate, so the circuit that controls the coil can control loads in other circuits without an electrical connection between the two (see Figure 16–9). The controlled circuit can be at different voltages and use a different type supply (AC or DC); it also may operate at much higher current than the control circuit does. For example, a relay with a 24 V coil can control a load in a 120 V circuit. The 24 V control circuit could have a maximum current rating of 1 amp

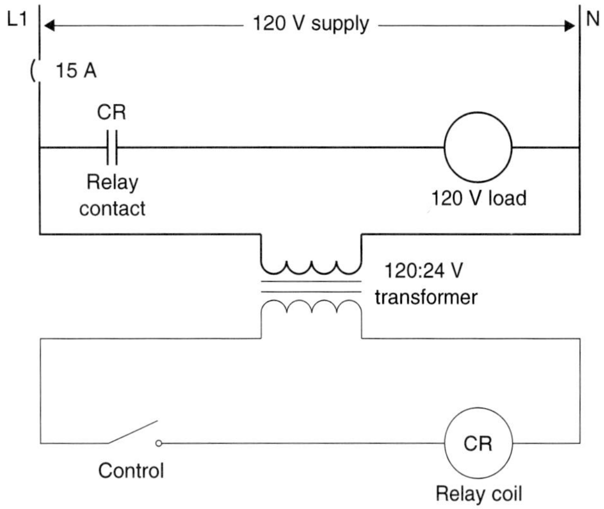

FIGURE 16–9 This basic relay circuit is in ladder schematic form. The relay coil is shown in the 24 V control circuit, while its contact (also marked CR) is shown in its proper electrical location rather than its physical location beside the coil. When the 24 V coil is energized, the normally open contact in the power circuit closes to energize the 120 V load.

or less, while the 120 V power circuit could be fused at 15 A or more. This is a typical role for a control relay. A low-voltage, low-current control circuit operates equipment in a power circuit that is operating at a higher voltage and current.

Thermal Relays

Thermal time-delay relays are very popular in HVAC control. Similar devices are also used in appliances such as electric ranges. Thermal relays do not have an operating coil in the usual sense. In its place, a heating element is used to warm a bimetallic strip or disk. The heater is energized by the control circuit, just like the coil in an electromagnetic relay. When enough heat is generated, the bimetallic material bends and operates the contacts. When the heating element is de-energized, the strip cools and returns to its previous shape, returning the contacts back to their "normal" status.

Thermal relays are generally simple, reliable devices. They are occasionally used as general-purpose relays in place of the electromagnetic types, but most often they are used as inexpensive time delays. Examples of where thermal relays are used for time delay include control of electric heating elements, fan or blower delays, and safety lockouts built into burner controls.

The thermal relays used to control electric heating elements are designed specifically for this purpose. Typical applications include electric furnaces and duct heaters. These heating elements are high-wattage resistive loads. They draw high current but are relatively easy to switch because the supply is AC, and the load is not reactive. However, heating elements do have some initial inrush current, and this, combined with the fact that normal currents are so high, makes it desirable to energize the elements one or two at a time, instead of all at once.

The time delays used are relatively short. To create the desired sequence, relays with different time ratings may be used to energize each element or group. Relays are also available with multiple contacts with different timings, built into a single case. Another alternative is to use relays with extra control contacts. The control contact of the first relay is used to energize the heaters of the second relay, and so on. All of these techniques are used, and many variations are possible.

Thermally operated relays are also used as fan controls in heating circuits. They delay the operation of the circulating air blower until the heat exchanger has had time to warm up. They are most common in larger heating systems with conventional (nonelectronic) controls. Thermal relays are sometimes used in residential heating systems as well, but there are several other more common types of fan controls for those applications. For more information on these circuits, see Chapters 17 and 18.

Safety lockout timers are usually not separate relays, but are built into the burner controller. The lockout heater is energized when the burner ignition sequence is initiated. If flame is detected before the relay contact operates, the lockout heater element is shunted out, and the system runs normally. If flame is not detected, the heater continues to operate until the bimetallic strip trips the lockout contacts. This shuts the system down. Safety lockout relays are built so that the contacts cannot reclose by themselves—a latch mechanism holds them open until a manual reset button is pressed.

Thermal relay contacts are rated for voltage, current, and type of use, just as with conventional relays. Contact ratings can be quite good. Despite the slow response of the bimetallic operator, the contacts can be made quick-acting by using an ingenious mechanical linkage, or a disk instead of a strip.

Solid-State Relays

There are a number of solid-state devices that can be used as replacements for electromechanical contacts. The most common of these for AC use are the silicon-controlled rectifier (SCR) and its bidirectional cousin, the triac. When connected in series between a load and an AC supply, these devices conduct when triggered, and block current when the trigger current is removed. There are no moving parts, and no limitation to the number of operating cycles. Unlike a relay, there is no relationship between the life span of the device and the number of times it is used to switch the load. Failures are not related to the number of operations. Like most solid-state devices, if they fail, it is usually due to heat, voltage spikes, or a high current caused by problems in another part of the system. Failures due to defects in the solid-state components themselves are quite rare, and they usually occur very early in the life of the device.

Most general-purpose solid-state relays are encapsulated in blocks of material with a few terminals and a label. They have a control voltage rating and ratings for the "contacts." They work much like any other relay, but may be more difficult to test. They are generally very reliable, but tend to be very unforgiving when subjected to current or voltage beyond their rated maximums.

Relay Ratings

Virtually all relays have two ratings, one for the coil or its equivalent, and the other for the contacts. These are two separate entities and must be treated as such. The coil, heater, or electronic circuit that triggers the solid-state switching component are load devices within the control circuit. They are rated for a specific voltage and must be operated at that value.

(Solid-state relays sometimes allow a range of voltages and will operate the same for any control voltage within the stated range.)

Coils are also rated for use on AC or DC. Although it is sometimes possible to operate an AC relay coil on DC at a greatly reduced voltage, it is generally not advisable. The opposite is never acceptable: A DC coil cannot be operated directly from an AC supply. Thermal and solid-state relays controls may be suitable for AC, DC, or either one; follow the manufacturer's ratings.

The relay contacts are most often part of a separate power circuit, usually operating at a different voltage, at a higher current, and sometimes with a different type (AC or DC) of supply. Many relays have several electrically separate contacts. Although all the contacts may be used for devices operating from the same supply, that is not always the case. Each contact set is not only electrically isolated from the coil, it is usually isolated from all the other contacts. Therefore, they can be connected into different circuits. It is not unusual for the same relay to have contacts in two or more different circuits, including a mix of both power and control functions. Many variations are possible.

Contact ratings follow the same rules as ratings for any other switching device. Voltages and currents are generally maximums. Lower values are fine, but neither maximum can be exceeded. If the voltage is lower than the rated value, the current cannot be increased proportionally—both maximums apply in all situations. If the type of use is not indicated, just voltage and current values, the current rating is almost certainly based on switching noninductive or mildly inductive loads. Switching motors or other predominantly inductive loads with a relay contact that is not specifically rated for such loads will greatly reduce the life of the relay. The usual cure is to derate the contact, selecting a relay with a current rating that is significantly higher than the current rating of the inductive load. In most cases this will result in an acceptable service life. The manufacturer's recommendations should be followed whenever possible.

When a relay is designed for motor use, the ratings are given in horsepower (see Figure 16–10). A horsepower-rated relay is very much preferred for a motor load. There are usually several different values, one for each operating voltage.

When the relay contacts are rated for both AC and DC use, the DC current ratings are usually much lower than the AC ratings. DC is much more difficult to switch, particularly if the load is a coil. Unless it is stated otherwise, the DC ratings should be interpreted as applicable to resistive loads only. Switching large DC coils is very rugged duty; the contacts must be specially designed for this purpose.

Note

DC coil circuits often have auxiliary components to absorb the energy that is released by the coil field when the circuit is broken. This is usually a reverse polarity diode across the coil, or a solid-state surge suppresser. These components are not added just for the benefit of the relay contacts. They protect electronic devices in other parts of the circuit from the voltage spike generated each time the coil is shut off.

Solid-state relay ratings must be respected, just as for any other electrical device. However, some extra caution may be advised. The surge-current capability of solid-state relays is generally lower than for normal contacts. All types of equipment depend on system fuses or breakers for protection in case of shorts, equipment failures, or similar

FIGURE 16–10 Note the markings on this relay. There are two different maximum motor horsepower ratings and two different noninductive current ratings that depend on the voltage used. The coil on this one is rated for 120 V AC, but it is very likely the same basic relay is also available with several other coil voltages.

problems that can cause extreme currents. Fault currents are usually of very short duration, but unless the fuses respond very quickly, they may not save a solid-state device. For this reason, solid-state components may have very specific fuse requirements. Replacing a fuse rated for protection of solid-state equipment with a less expensive—and easier to find—general-purpose fuse can result in a costly repair when there are problems in other parts of the circuit.

Practical Exercise: Relay Operation

The following experiments describe the physical examination of a typical electromagnetic relay, and demonstrate testing procedures that can be used to check their operation. Several different connections are shown to illustrate typical relay uses and applications. Many different types and styles of relays are suitable, but the relays on which these experiments are based are open-frame general-purpose relays with 24 V AC coils and two sets of SPDT contacts. These relays may also be called a DPDT, because two sets of SPDT contacts make up a DPDT set.

Note

This bit of confusion will always exist, because relays with this contact arrangement may be correctly described either way. However, when there are more than two sets of contacts, the description usually goes something like "4 SPDT contacts," instead of "4PDT." Also, if the NO and NC contacts are separate, instead of part of a SPDT pair, the designation would be "2 NO and 2 NC" contacts instead of "1 DPST (NO) and 1 DPST (NC)."

Two relays are required for the experiments described. Other necessary parts include two momentary-type pushbutton switches, one with an NO contact and the other with an NC contact, two SPST toggle switches, a small 120:24 V control transformer, and three 120 V lamps with sockets. Suitable connecting cords and leads are also required.

Warning

The voltages used in this experiment are dangerous. Caution is advised.

Physical Examination. For the relays used in this experiment, each of the SPDT contact sets is arranged with one NC and one NO contact sharing the same common terminal. The physical construction of the relay makes the contact pairs inseparable, because the moving contacts are mounted back to back on the same strip of conductive material. The conductive strip is part of the electrical circuit, so it is mounted in a plastic base to insulate it from the rest of the relay. The plastic is in turn fastened to a hinged metal plate that is the moving "armature" of the electromagnet. This is a typical "clapper-style" relay, so named because the hinged metal plate operates like the clapper of an electrically operated bell. The relays used in this experiment are DPDT, so there are two SPDT sets. The moving contacts of both sets are mounted side by side in the same insulating base; they operate together, but are not electrically connected to each other (see Figure 16–11).

The metal frame of the relay forms part of the magnetic circuit. The coil core is laminated, as it is for all AC electromagnets. A "shading coil" is embedded in the exposed pole

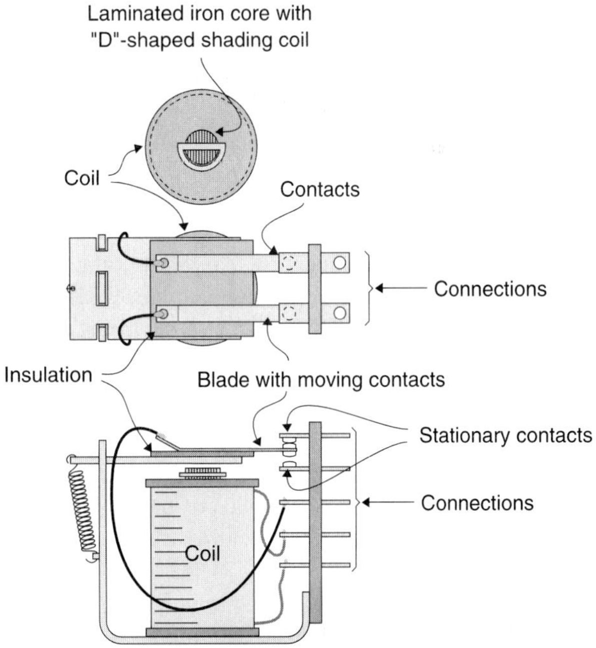

FIGURE 16–11 The details shown in this diagram are quite typical of small relays with AC coils. DC coil relays are similar, but have no shading coil, and may use a solid, rather than a laminated, core.

face of the core at the point where the moving clapper makes contact. This shading coil is a small ring, shaped like a D or flat O, that fits into notches cut in the top of the core material. The shading coil is essential for reducing or eliminating AC relay chatter. It maintains a small amount of field in the core while the AC current passes through zero and the magnetic field reverses polarity. Without it the clapper would chatter continuously, pulling free from the core just slightly at each polarity change.

There are three terminals for each contact set: a common that connects to the moving contacts, and one from each of the two NO and NC contacts. For an open-style relay, the correct connections can be identified by inspection and may not be marked. When the relay is encased, the terminals are identified by one of several methods. There may be an electrical diagram on the case with appropriate terminal numbers—this is common for plug-in relays. There may be symbols and connecting lines indicating where each terminal connects to the relay, or there may be identifying letters such as COM, NC, or NO for each contact set. The coil leads may also be identified, but the markings vary considerably from one make to the next.

If this experiment is actually attempted, the relays used should be closely examined to determine the mechanical layout, the electrical connections, the size and appearance of the contacts, and the contact gap. Any manufacturer's ratings included on the relay or from separate specification sheets should be noted and interpreted. Make particular note of the coil voltage and any contact ratings. For open-style relays, the coil ratings are often written on the coil, rather than on the relay base. This allows the manufacturer to assemble the relay with different coils without changing the other markings. This practice simplifies manufacture, but can lead to problems in the field.

Note

Two relays that look virtually identical may have different coil voltages. ALWAYS look at the relay very carefully to be sure that the coil voltage is correct. Note also that coils are available for both DC and AC voltages, and they are not interchangeable.

Failing to check the coil voltage is a simple and often costly mistake. In the HVAC field, for example, there are several commonly used relays that are available with either 24 or 120 V coils. The two relays are identical in all other respects, so it is alarmingly easy to connect a 24 V relay coil to a 120 V system—which means instant smoke and a ruined relay—or to connect a 120 V relay coil to a 24 V circuit. Connecting the coil to a voltage that is much lower than its rating usually does no harm, but can be very frustrating. More than one relay has been mistakenly discarded as defective when the only error was failure to check the voltage rating of the coil.

Initial Relay Tests. The relay should be given a quick test before it is connected into the circuit. These tests should include measuring the resistance of the coil, and a continuity test of each contact. The contacts should be retested while manually operating the relay—if it is an open type—or after energizing the coil with an appropriate supply. This will provide a ballpark idea of the typical resistances of AC relay coils, and confirms the operation and identification of the contacts. Contact resistance should be virtually 0 Ω; coil resistances vary considerably with relay size, type, and coil voltage rating. AC coils have a lower resistance than DC coils of similar ratings. The DC coil is wound with very fine conductor to intentionally keep the resistance high. The extra resistance is necessary to limit the DC current. In an AC coil, the current is limited by inductive reactance.

Basic Relay Circuit. Once the initial tests are completed, the relay can be connected into a typical control circuit (see Figure 16–12). The transformer is used to provide power to the controls; the 120 V supply is used for the transformer primary and for the power circuit supplying the 120 V lamp used as a load. An SPST toggle switch is used in the coil circuit to control the relay operation. One set of contacts is wired into the line supply of the lamp. The unused contacts are left disconnected.

Circuit operation is best examined in two parts, just as it would be in a real-world troubleshooting situation. The low-voltage control and the power circuit are separate though related systems. The control circuit includes everything that is operated at the control voltage, in this case 24 V. It consists of the 24 V transformer secondary, the switch, and the relay coil. The power circuit consists of the 120 V power source, the connected set of relay contacts, and the 120 V lamp load. In operation, the 24 V circuit has control over the 120 V load, but the interconnection between the two is a magnetic link, not an electrical one.

Voltage measurements within the control circuit are best taken from the 24 V common, which in this example is not grounded. With the switch closed, there will be 24 V from the common to both sides of the switch, and to the line side of the coil. When the switch is open, there is no voltage past the line side terminal of the switch.

In the power circuit with the relay coil energized, there is 120 V from the common (neutral side) to both the line and load sides of the relay contact in use, and the line side of

FIGURE 16–12 A basic relay test circuit shown in wiring diagram and ladder schematic form.

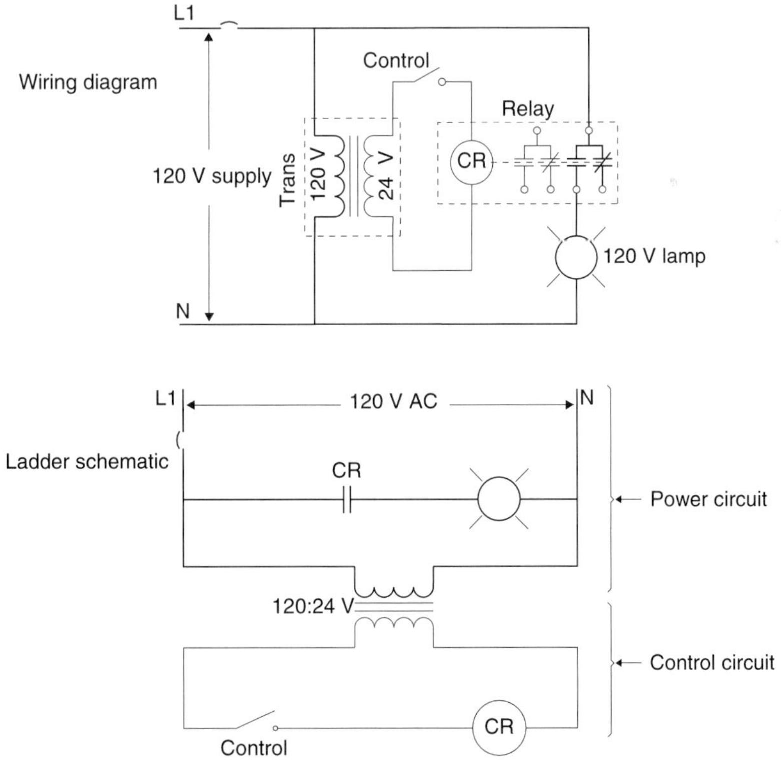

the lamp. The lamp has 120 V across its terminals and should be on. As with the control circuit, when the coil is de-energized, the contacts open, and there is no voltage past the line side of the contact.

This test circuit is as basic as a relay circuit can be made. It uses a single control device to operate the relay coil, and only one relay contact to control a single power load. However, even this simple circuit has some interesting applications. Obviously, it allows low-voltage switching devices to be used in the control of a relatively high-voltage load. A less obvious application is to reduce the cost of remote control. For example, the low-voltage side could be extended for a very long distance quite inexpensively. The control wiring does not need line voltage insulation, nor does it have to be sized for load current. It only has to carry the very low operating current of the relay coil. Such a circuit could provide remote control of a distant load for a small fraction of what it would cost to extend the power circuit.

There are also some interesting variations to this basic circuit. For example, the switching logic can be reversed—switch and relay coil off for load on. This only requires the load to be connected to the normally closed contact instead of the normally open. Another minor change could be the addition of a low-voltage indication light, mounted near the control. This can be very functional feedback, particularly when the power load is not visible. For the example shown in Figure 16–13, the indication light could be an inexpensive 24 V lamp wired across the relay coil or switched with the unused contact. Both methods are used.

Relay Contacts in Series. Multispeed motors cannot tolerate simultaneous energization of more than one speed. An example of multispeed motor control is the fan circuit of many residential furnaces with both heating and air conditioning. Some of these systems use two, or even three, different operating speeds for the fan motor. To avoid motor damage, the fan control circuit must be designed so it is impossible for more than one motor speed lead to be energized at the same time.

The following experimental circuit provides this type of control. The user can select one of three options, but cannot energize two at once. This circuit is not identical to the one used in a typical three-speed furnace fan, but it is very similar (see Figure 16–14). The relays must each provide a set of SPDT contacts. The second set must be wired in series with the normally closed contact of the first.

The first relay selects between low and medium. The second relay shuts off power to the circuit containing both low and medium, and energizes high. In many actual furnaces, the coil of the first relay is rated for 24 V and is energized from the fan control terminal of the thermostat. The 120 V heating mode fan control is connected to the second relay coil, which must have a 120 V coil. With neither relay energized, the low-speed lead is energized for continuous air circulation. When the second relay coil is energized by the heating mode fan control, a medium speed of the motor is energized. When the thermostat calls for cooling, it energizes the coil of the second relay, shutting of the continuous and heating speeds, and switching the motor to a high speed for cooling.

This circuit can be built and thoroughly tested using the same procedures applied in the first experiment, but please note the following warning.

FIGURE 16–13 A pilot lamp provides visual feedback about the status of the load. It could be operated at 24 V (or 120) from an unused contact. A 24 V lamp wired across the coil of the relay can also be used.

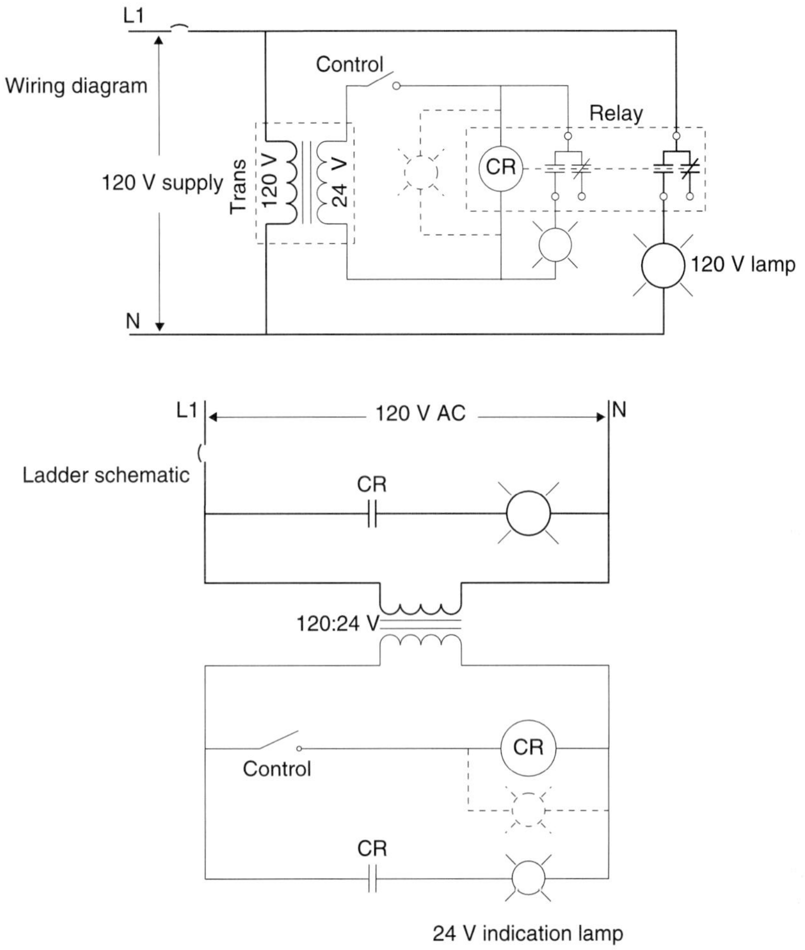

Warning

In a real multispeed fan control circuit like the one described above, each of the different motor speed leads is connected to the same windings inside the motor. When the motor is running, all of these leads are energized, regardless of which one is connected to the supply. However, the voltage on the unused leads is not at the expected 120 V (measured to the neutral). As a result of an autotransformer effect in the motor windings, only the connected speed is at 120 V. All the other leads show significantly more or less voltage than the supply, and no two are the same. For any connection, the voltage is highest on the lead for the lowest speed, and progressively lower on each of the higher speeds.

FIGURE 16–14 This test circuit can be used to simulate a three-speed fan control. The 120 V lamps indicate the three motor speeds. With both control switches open, the low "continuous" speed is energized. Closing the first control switch will shut off the low speed and energize medium. Closing the second switch (with or without opening the first one) will de-energize low or medium, and energize the high speed. It is impossible to energize two lamps (speeds) simultaneously.

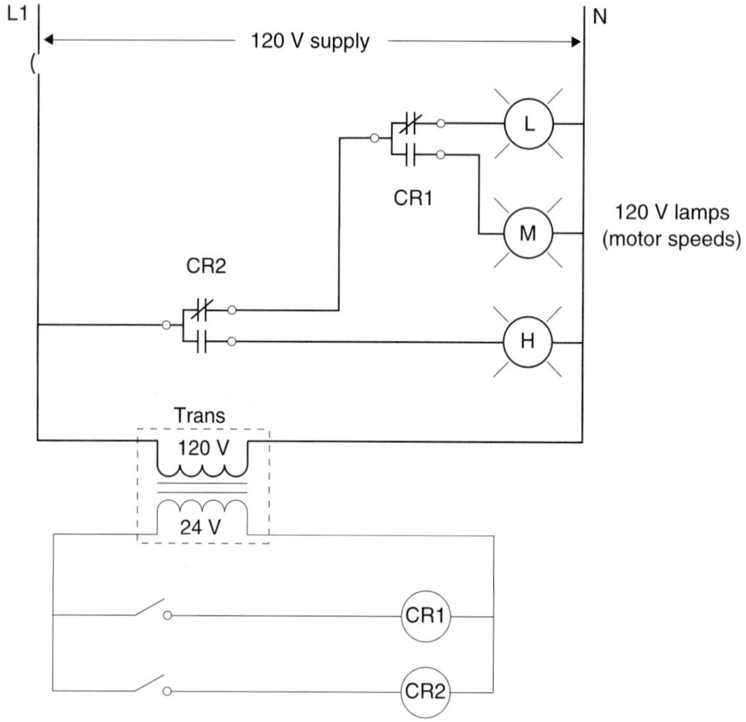

Pushbutton Motor Control. The final experimental relay circuit involves a single relay with two sets of contacts (see Figure 16–15). In this case only the NO contacts are used. The control system contains two momentary-contact pushbutton switches, one NO and one NC. The line side of the 24 V control circuit feeds through the two pushbutton switches in series, and then to the relay coil. One of the relay's NO contacts is connected in parallel with the NO pushbutton contacts. The second NO relay contact is used to control the load.

Note

The relay in Figure 16–15 has contacts operating at two different voltages in two different circuits. This is not uncommon. Never make assumptions about what is connected to a relay contact. It is very possible that it is a completely different circuit, and may be at a much higher voltage.

The NC pushbutton is the stop button, and the NO pushbutton is the start button. Pressing the start button will energize the coil. The NO relay contact in parallel with the start button will act as a "seal-in" or "holding contact" to keep the coil energized after the start button has been released. Once energized, the relay remains picked up until the coil circuit is broken,

FIGURE 16–15 The wiring diagram and ladder schematic show a basic three-wire control circuit, simulated with a low-voltage relay and a lamp load. This circuit is common in motor control.

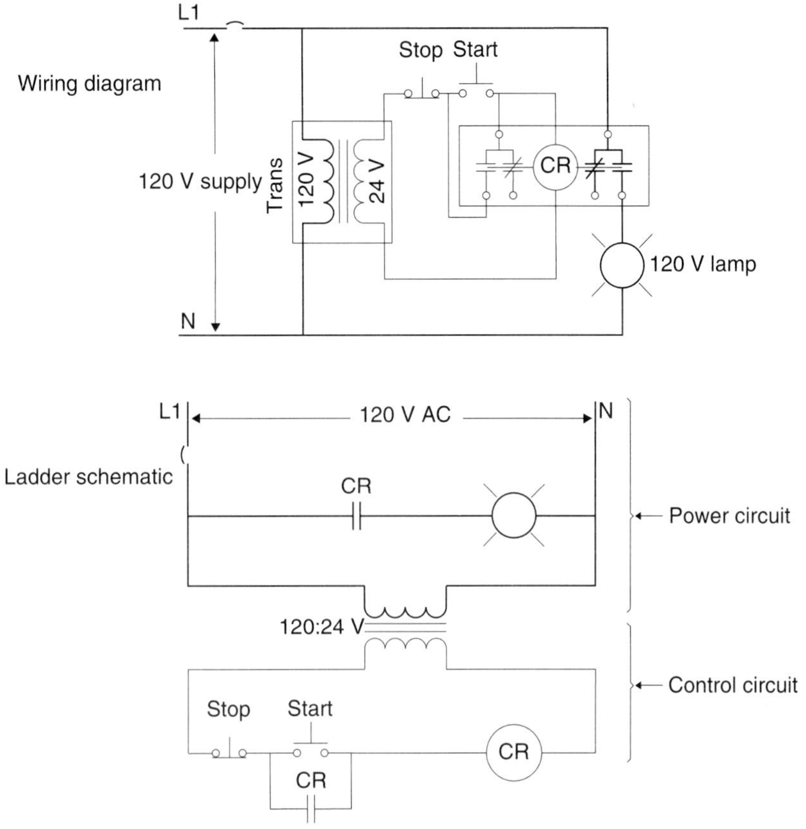

normally by pressing the stop button. The circuit can also be shut off by breaking the control circuit at any point, or by a momentary interruption in power. When the relay is energized, the load is also energized through one or more other contacts on the relay or contactor.

A circuit of this type is widely used in motor control. A motor starter is used in place of the relay, and the switches are usually heavy-duty types suitable for an industrial environment. The starter has two or three main power contacts, plus at least one auxiliary contact for the "seal-in." Motor starters operate much the same as relays, but are designed for higher voltage and current. A typical motor control circuit can be much the same as this relay example, and would operate the same way.

There are two main points about this circuit. First, the circuit seals itself in; therefore, it resets to an off condition on loss of supply power and will not automatically restart. Secondly, any number of control points can be added without complex wiring or special three- and four-way switches. Additional stop buttons can be added in series with the first one. An additional start button can be added in parallel with the original start button and seal-in contact. The control switches must be momentary types, with NC contacts for the stop buttons and NO contacts for the starts.

The fact that it will not automatically restart after a power outage is one of the key features of this circuit. This can be extremely important in motor control applications. It is used where there are specific safety concerns, such as the control of machines with a human operator, where an unexpected startup could be extremely dangerous.

In addition to the safety concerns, there is another reason why this type of control is so popular for motor controls. The electrical service of a typical plant is sized to supply all the motors and other loads that will be operated at the same time. There is some reserve, of course, but during a normal startup, electric motors draw several times their normal running current. Without a circuit like the one used in this experiment, a brief power outage could cause a severe overload when power is restored. Every motor in the plant could attempt to start at the same time. When this three-wire pushbutton control system is used, the motors are restarted in sequence, and the plant is restored to normal operation with no danger of overloading the main supply.

An Unusual DPDT Relay Circuit

This circuit, or some variation of it, is quite often used in residential furnaces with two-speed motors and auxiliary equipment such as electronic air cleaners. It solves yet another control problem that is somewhat unique to furnaces. The auxiliary equipment must operate with either motor speed, but cannot be connected to the same terminals as the motor. Because of the internal connections to the motor windings, connecting any other device to a motor speed lead may burn out the motor or the connected equipment.

This circuit uses a 120 V fan control to operate the heating speed. The cooling speed is controlled by the fan control line of the heating/cooling thermostat. The DPDT relay circuit provides only two motor speeds, different speeds for the heating and cooling modes. It does not provide for a third, continuous low, fan speed.

The commons of the two contact sets are interconnected and used to provide power to the auxiliary loads. One set of contacts is connected with the NO contact directly to the incoming line, and the NC to the line through the 120 V fan control, shown here as a temperature-operated NO contact. The other set goes to the two motor speeds: the heating speed from the NC contact, the cooling speed from the NO (see Figure 16–16).

In the heating mode, the fan control cycles the heating speed and the air cleaner. When the thermostat calls for cooling, it energizes the air conditioning unit and the fan relay coil. The 120 V fan control circuit is disconnected; the auxiliary equipment is powered from the line through one normally open contact, and the high fan speed is powered from the second.

This circuit combines two types of controls: a line voltage switch and a relay operated by a low-voltage control system. There are at least two different variations of this connection in use in the HVAC industry.

The Lockout Relay

Safety limits always act to shut down the equipment or process they are installed to protect. What should happen after the unsafe condition has cleared depends on the circumstances. Serious problems that do not normally resolve themselves are usually protected by manual-reset safety devices. Human intervention is then required right at the equipment before it can be restarted. An example might be a flame rollout protection device in a

FIGURE 16–16 This circuit is found in several makes of residential furnaces. It allows for two motor speeds, plus an auxiliary device that is energized for either motor speed. With the humidifier connected as shown, it can only operate during a heating cycle. It can also be connected in parallel with the air cleaner.

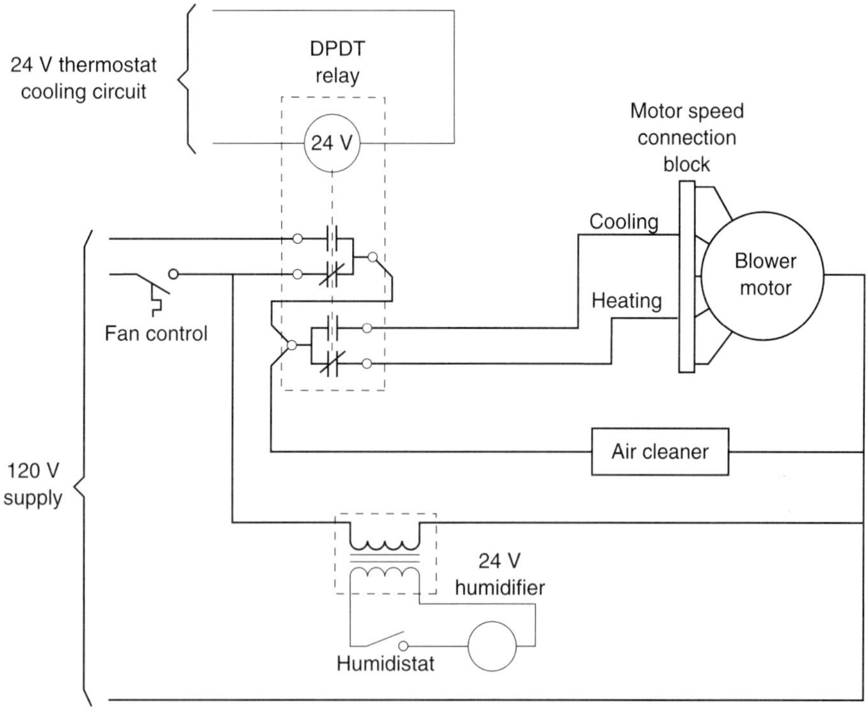

burner system. This limit detects the heat of a burner flame that has "rolled out" to the intake air side of the burner, rather than into the heat exchanger and out of the vent. This is a dangerous condition that should be brought to the attention of the operator.

When the equipment is subject to less dangerous conditions that usually normalize by themselves, a self-resetting limit may be more appropriate. A common example is the high-temperature limit on many heating units. The overtemperature may be caused by blocked registers or a dirty filter that has reduced the heat exchanger airflow. The high-temperature limit can operate repeatedly without danger of fire, so manual reset is not a necessity. Cycling the unit on the high limit is not very desirable, but there is a trade-off. Using a manual reset limit does not appreciably reduce the fire hazard, but it is greatly increases the probability of unnecessary frost damage to the building.

When the problem should be brought to the operator's attention, but it is unlikely to be serious, it may be desirable to allow a remote reset. This avoids the inconvenience of locating an equipment-mounted reset button, while preventing equipment damage due to a problem that went unnoticed. This situation can be handled with a lockout relay (see Figure 16–17). The limits are self-resetting. The relay locks out the equipment each time a limit opens, but allows local reset by cycling the controls. An example of this circuit is found on some air conditioning systems. The limits covered by the lockout relay are the high- and low-pressure switches on the refrigerant loop.

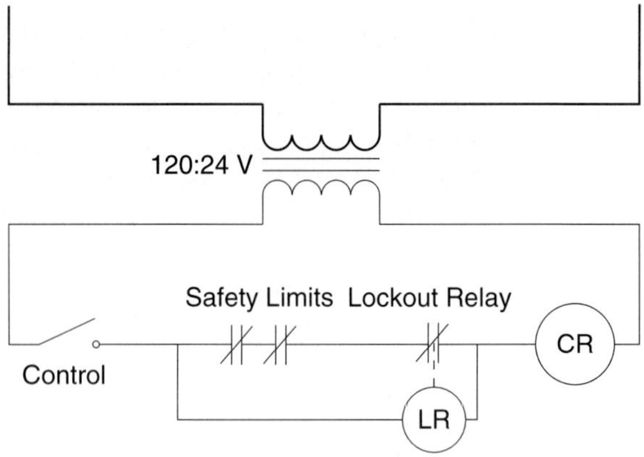

FIGURE 16–17 This partial circuit schematic shows the location of the lockout relay and its contact, as required to add a lockout function to two self-resetting limits. For this circuit to work properly, the impedance of the lockout relay coil must be much higher than the impedance of the main operating coil. Lockout relays are also called *impedance relays* be cause of their high impedance coils.

These limits may operate occasionally without the existence of a serious problem. However, if there is a serious problem, and it is not corrected, costly equipment damage may result. If either pressure switch opens, the lockout relay prevents the equipment from restarting (see Figure 16–17). Once the pressure has normalized, the lockout can be reset by turning the thermostat up, and then back down again, to cycle the control power to the unit.

When an NC limit contact opens, the lockout relay coil is placed in series with the normal operating coil. The lockout coil is made with a high impedance so that it drops most of the circuit voltage and prevents the normal operating coil from picking up. Even after the safety limits reset themselves, the now-open contact of the lockout relay keeps the circuit "locked out" until the circuit is briefly de-energized. The usual method of resetting the circuit is to cycle the operating control.

16.4 Timers

Timers and time-delay relays are used in many control systems. They can add brief delays to accommodate some process characteristic, or set up a sequence of events, forcing them to occur in a specific order. They can also provide periodic scheduling of events that must be repeated at specific intervals. There are a wide variety of timers available. They may be general-purpose devices that mount as individual components, or special-purpose devices made for a specific application. They are also made as add-on accessories to some makes of motor starters, and are built into other controls.

The actual timing can be provided by several methods, including a heated bimetallic element, a pneumatic "dash pot," a clock motor, or an electronic circuit. Bimetallic and pneumatic timers are generally suited to relatively short delay times that do not require split-second accuracy. Most pneumatic timers are adjustable. Bimetallic types are usually fixed. Clock motor-driven timers excel at providing very long delays, multiple scheduled on/off periods, and sequence control, but cannot be set in small increments.

Electronic timers cover the spectrum of timing requirements. They are available for virtually any time period, usually providing a higher degree of accuracy than other timers. They range from simple preset replacements for older designs to full-featured multimode programmable devices.

Programmable timers have the attraction of many different user-selectable timing options, and the ability to change the time base. The timing options allow the user to select one of several modes, such as time-on delay, time-off delay, or pulse timing. The multiple-time-base feature allows the same timer to be set up for time spans that vary from fractions of a second to many hours. Timers with multiple time bases are essentially counters that are coupled to an internal clock pulse. The pulses occur at intervals equal to the time base. Assuming the timer has only three digits, it can count from 000 to 999. With a time base of 0.1 sec, the timer would have a maximum time span of 99.9 sec (1 min 39.9 sec) with 0.1 sec accuracy. A time base of 1 min would extend the timers range to 16 hr 39 min, accurate to 1 min.

Time delays can be provided after the control is energized (time delay on energization) or after it is shut off (time delay on de-energization). These are often abbreviated to "time-on delays," or "time-off delays." The timed contacts may be normally open or normally closed. Some timers also incorporate untimed "instantaneous" contacts that operate just as contacts do in a conventional relay.

The most common type of time delay is a "time delay on energization." The timing period starts after the timer coil (or heater, etc.) is energized, and the timed contacts change state after the period has passed. Most timers of this type are not cumulative; they reset to the starting point each time they are de-energized. If the timer has not completed the preset delay, it will still reset to the starting point when it is shut off. Cumulative timers do not reset until a separate reset terminal is energized by the control system (even if the timer has timed out). The contacts still change state after the preset delay period, but the total time can consist of several shorter periods that have accumulated each time the control was energized.

Certain functions demand time-off delays. In this timer, the contacts change state immediately when the coil or other operator is energized, and stay in the energized state until a specific time after the coil has been de-energized. This is the type of timer used for the light in a typical garage door opener. Each time the opener operates, the light comes on, and stays on until some preset time period after the door stops moving.

Thermal Timers

Virtually all thermal relays have an operating time delay. Bimetallic operators are naturally quite slow, because it takes time for the heating element to affect the disk or strip. The delay can be quite short, but cannot be particularly long. While it is possible to make them adjustable, most thermal timers provide a single, fixed timing period. Delays are usually not more than a few minutes.

Thermal relays are not particularly accurate. The delay time may change slightly if the timer is operated several times in succession, and may also be affected by changes in the ambient temperature. Because they are not adjustable, completely different timers must be used for each different delay time. They also do not instantaneously reset when the heater is de-energized. The bimetallic element must cool down before the contacts can return to their de-energized state. This slow reset can make them difficult to apply in certain circuits. However, for situations where their limitations do not cause problems, thermal timers are a simple, reliable, and comparatively inexpensive option. For more demanding applications, other types of timers may be more suitable.

Pneumatic Timers

In a typical pneumatic timer the delay is produced by a bellows, or by a piston-and-cylinder arrangement that is sometimes called a dash pot. The actual physical layout can vary a lot, but a simple example would be a soft rubber bellows containing a spring. When the relay is de-energized, the spring-and-bellows assembly is held compressed by the plunger of the coil. When it is energized, the coil pulls the plunger away from the bellows, releasing it to the force of the spring. The spring tries to expand, but the airflow into the bellows is restricted by a small, adjustable port. The contacts do not change state until the bellows is almost fully extended. The time delay can be adjusted by varying the size or resistance of the port. This is usually set with a calibrated dial or an uncalibrated set screw. When the timer has no markings, the time is set by trial and error, with a stopwatch if necessary.

These timers are not cumulative. If the solenoid coil is de-energized before the contacts change state, the plunger immediately compresses the bellows and resets the timer. Timing will begin at zero the next time the coil is energized.

Pneumatic timers provide time periods similar to those obtained from thermal types, but have the benefit of being adjustable. They are also more complex and expensive than thermal types. Timing accuracy may be affected by environmental factors such as temperature or contamination. Pneumatic timer technology is often used for add-on auxiliary timers offered as an option on some contactors and motor starters. These are usually simple to install and use, and relatively inexpensive because they do not need their own coil. They are mechanically linked to the moving core of the starter mechanism. These timers and some common applications are examined in Chapter 12.

Clock Motor Timers

When time delays are very long (many minutes, several hours, a day, or even a week) or a number of contact operations must occur in a preset sequence, a mechanical timer driven by a clock motor is a popular choice. Many of these timers are a fairly simple arrangement of one or more contacts operated by lobes or notches on a rotating disk or drum. This mechanism is driven, in one direction only, by a small clock motor. The timer is started by energizing the motor and, in most cases, the controls are designed so the motor continues to operate until the time sequence is complete.

Many of the motorized timers used in HVAC systems are made specifically for the application. A common example is a refrigerator defrost timer, which runs a standard preprogrammed cycle that periodically locks out the compressor and turns on the defrost heater. A separate thermal cutout prevents the heater from operating too long.

Motorized time clocks are used to automatically schedule operations that occur on a daily or weekly basis. These clocks are used to schedule changes, such as day and night operating modes, and to control lighting for business signs, general area illumination, and security. They may also be used to provide some basic energy management by shutting down nonessential loads during peak load times. A common type of time clock has a rotating dial with small on and off "trippers" attached around the rim. The dial is a metal disk marked off in hours. The trippers are held in place by set screws, allowing them to be positioned at any point around the clock dial.

For each on tripper there must be a corresponding off tripper to turn the circuit off again. Standard timers rotate once in 24 hours or once in seven days. A 24-hour timer allows only one daily program, but may be equipped with a skip feature that allows normal operating schedules to be suspended over the weekend. A seven-day timer allows different schedules each day.

A somewhat less common type of motorized timer is offered as an alternative to pneumatic or thermal timers. It has separate connections for the motor, and a magnetic clutch. The motor runs constantly, but the contact control cam turns only when the clutch is engaged. Timing starts when the clutch is energized. When the time period is up, the cam holds the contacts in the "timed out" position. When the clutch is released, the cam immediately returns to its de-energized position. This type of timer is precise and quick to reset, and can have a very long delay if required.

16.5 Sequencers

Multiple timers can be used to provide a sequence of operations, but this is often not the best approach. When the time period is long, or there are many steps that always follow the same pattern, a motor-driven cam timer or sequencer is a popular choice. These timers are reliable and accurate. They can repeat the same sequence over and over for years, with no change in the timing or in the order in which the contacts operate. They are easy to understand, and when they do finally fail, they are often repairable.

The timers in automatic washers and dryers are common examples of specialized motor-driven sequencers. An HVAC equipment example is the timer/sequencer built into the high-capacity burner controls found on boilers with variable firing rates. They are now being replaced by microprocessor systems, but many are still in use. Burners of this type are not cycled; the firing rate is adjusted to match the heating demand, so the burner can stay on indefinitely. When it is shut off or some problem occurs in the system, there is a set sequence of events that must be followed to safely relight the burner flame.

A typical sequence starts with the combustion air blower running and the air dampers wide open to purge any combustible gases from the fire box. After the purge period time has elapsed, the controller closes the dampers to the low-fire position, pretests the flame detector, and energizes the pilot valve and the spark igniter. Once the pilot flame is detected, the main valve is opened, and a few seconds later the pilot valve is closed. Control of the burner firing rate is passed to the main boiler control, and the sequence timer stops.

When the burner shuts down, the timer motor is restarted to provide a short post-purge to clear the fire box. This completes the cycle—the timer has gone once around and is back at the start again. It will either restart immediately and repeat the light-up sequence, or it will stops until the malfunction has been corrected.

Sequence timers, such as this burner control example, often start and stop as other control devices become involved, but the order of events, and the minimum time period for each step are always the same. This type of application is easily handled by an appropriate motorized time control.

There are also a variety of general-purpose timers and sequencers available that may be adjusted to suit various applications. It is time-consuming to set them up, because each cam must be set for the individual start and stop times, but once this is done, they can operate reliably for many years. Motorized timers have the advantage of being almost entirely mechanical. Unlike the newer electronic types, their operation is obvious, the parts are usually open to inspection, and many of them can be repaired if they fail. Most are rugged and reliable devices.

16.6 Solid-State Timers

Solid-state timers are available in a wide variety of types and styles. They may be the most accurate and reliable of all timers, and are certainly available with the most features. There is probably a solid-state timer suitable for every type of timing function that can be imagined. The variety runs from very simple, one-circuit, one-time replacements for thermal or pneumatic timers, all the way to very complex multicircuit programmable devices that may be outweighed by their operating manuals.

Many types of equipment—such as the time clocks and the burner control systems described above—that once contained mechanical timers and sequencers are now entirely electronic. Timing and sequence functions are programmed through a small keypad.

16.7 Programmable Controllers

A modern trend is away from individual control components to an integrated system based on one or more programmable controllers. These may be generic controllers, purchased separately, or custom devices provided with a piece of manufactured equipment. They are microprocessor based, with the operating logic coded into a software program. The program is stored in nonvolatile or battery-backed RAM. Backup copies are kept on disk, on tape, or in electrically erasable memory modules called EEPROMS.

The program is a series of logic instructions that run in a continuously repeating loop. The controller is constantly monitors inputs, checking logic, and updating outputs as required. A programmable controller can provide virtually any type of sequence, timing, or relay function that can be provided by separate components, and usually more advanced functions. They are inexpensive compared to the functions they provide, and extremely flexible to use.

Programmable controllers have two major drawbacks. The first is the time-consuming programming required each time a custom application is created. This is aggravated by the fact that every make of controller, and sometimes even different models in the same product line, uses different programming software.

The second drawback depends on your point of view. The programmable controller, like other electronic devices, makes the operation of the control system virtually inaccessible. All the service person can do is check what happens against what is supposed to happen. If the two don't match, there are no individual devices that can be adjusted or replaced. Programmable controllers can be particularly frustrating because the end user may be locked out of the program.

Programming access can be left open to anyone with a programming terminal or a computer with the required interface and software. But this is rarely done, because it makes it impossible for the designer to protect the system from inappropriate alterations. Good programs for complex systems can be difficult to write. They may also represent a sizable investment in hardware and software.

Simple projects involving only a few input pilot devices and a small number of outputs may be easier to complete with a programmable controller than conventional relays and timers. If there is more than one timer involved, or there are special functions such as sequences or counters, the programmable controller is almost certain to be the least expensive approach. Programming time is relatively short, and alterations are easy.

Complex projects should not be attempted unless the programmer has extensive experience in control system design, and familiarity with programmable controllers. Good documentation is also essential. Without it the program can be nearly impossible to trace. Clear, accurate documentation, often takes longer to write than the program.

Questions

1. Why are control circuits often operated at a lower voltage than the main power loads?
2. If a 120:24 V transformer is rated at 50 VA, what is its full-load current rating?
3. Describe how an ohmmeter can be used to test a transformer for shorts and grounds. List the readings that would indicate the transformer is in working order.
4. What is meant by the term "magnetic saturation" as it applies to transformers?
5. Explain why a transformer cannot be operated at higher than rated voltage, even when the load current is kept within rated values.
6. Why should the supply fuses to a transformer be based on the transformer's rated load, rather than on some reduced value of known load?
7. Can a voltage test usually prove if the supply transformer has a grounded secondary? If so, what is the difference in the test results?
8. Describe a test that can be used to check transformer phasing.
9. If a 120:24 V transformer is properly "phased," what is the voltage from: Ground to the ungrounded primary lead? Ground to the ungrounded secondary lead? Between the ungrounded primary and secondary leads?
10. Explain the fundamental operating principle of a thermal relay.
11. Relay ratings include voltage and type of supply for both the coil and the contacts. Is there a connection between these two sets of ratings, or are they essentially independent of one another?
12. Can DC relay coils be operated directly from an AC supply?
13. Would it be acceptable to use a 120 V 15 A rated relay contact to control a 48 V 10 A load?
14. If a relay has a coil rated for 120 V AC, would it be acceptable to power it from a 24 V AC supply?
15. Contact current ratings with no indication of the load type should generally be interpreted as maximums for what types of loads?
16. What is the purpose of the shading coil on the core of an AC relay?
17. If a replacement relay fails to pick up when it is energized, what mistake may have been made in selecting the relay?
18. Explain the function of the seal-in contact in a three-wire control system.
19. Draw a circuit diagram for a lockout or impedance relay, and explain how it operates.
20. Explain the difference in operation between an on-delay timer that is cumulative and one that is not.
21. Compare the operation of two normally open timer contacts, one operated by an on-delay timer and the other by an off-delay timer.
22. Describe the operation of a pneumatic timer.

17 Residential Furnaces

Objectives

This is primarily a reference chapter. The information is specific to residential HVAC systems. The focus is on gas fired forced air furnaces, but additional information is provided for oil burners, water heating systems, and heat pumps. After completing the study of this chapter you should be able to:

- Summarize the licensing and jurisdictional issues that may affect furnace service workers
- Describe the three different circuits found in a typical standing pilot gas furnace
- Sketch a control drawing for a typical standing pilot furnace with 24 V controls
- Sketch a control drawing for a standing pilot furnace using thermopile-based controls
- Explain the operation of a heat anticipator
- Sketch a schematic drawing showing the internal circuit of a single-stage heat/cool thermostat
- List the types of limits found in a modern gas furnaces and describe their operation
- Describe the operating principles of a flame rectification system
- List, in correct order, the sequence of events that makes up one complete heating cycle of a gas fired furnace with intermittent pilot ignition
- Describe the fundamental differences between intermittent pilot and direct spark ignition
- Describe the operation of a hot surface igniter control system
- Sketch a control drawing for a typical combustion air fan control and explain its operation
- Sketch control drawings for typical single-speed and multispeed blower control circuits
- Explain how a fan center, incorrectly connected into a control system with a combined fan limit control, may inadvertently cause the high limit to be bypassed
- Compare the control circuit of a furnace with a single-circuit-board control system to that of a similar furnace using individual relays and a separate ignition module
- Describe the operation of a typical oil burner control with a CAD cell flame detector
- Describe the operation of a water heating system with multiple-valve-controlled zones and a single circulating pump
- Describe the basic operation of a typical heat pump system

Note

The following information is intended to improve the knowledge and understanding of anyone involved, or simply interested, in the electrical aspects of residential heating systems. A large percentage of all system problems are either directly related to the electrical control system, or can only be identified by examining the status of control related devices. Understanding the operation of the electrical equipment and the control system is essential for efficient troubleshooting and servicing. However, it must also be appreciated that **knowledge is not license.** There are many safety regulations that affect the manufacture, installation, and servicing of HVAC equipment. Understanding the operation of the control system in no way entitles anyone to work on equipment without the applicable qualifications and licenses.

17.1 Introduction

Residential forced air furnaces fall into several broad categories based on the fuel used and the heat exchanger and venting arrangement. Natural gas furnaces are one of the most common. Liquid petroleum gas (propane) fired systems are essentially the same. Basic units have atmospheric burners with natural draft venting. Higher efficiency models have mechanical draft systems using an induced-draft blower.

Fuel oil furnaces are very different from gas systems because the liquid fuel must be atomized into a forced-draft airstream for efficient combustion. Automated coal furnaces use mechanical stokers that feed coal from a hopper. Filling the stoker and removing the ashes are usually manual operations, so the heating unit requires regular attention. Coal fired furnaces are now relatively rare, at least in North America, although in areas where coal is readily available they can be very cost-effective. Wood fired furnaces generally require manual fuel feed, but may have automated draft- and fan-forced air circulation for better performance. Some wood burners are dual-fuel systems to allow extended periods of unattended operation.

Electric furnaces have resistance elements in place of the burner and heat exchanger. These elements are controlled by relays—typically thermally operated time delay types—that energize the elements in stages. The fan controls are similar to those used on other types of furnaces. Electric furnaces have none of the safety concerns associated with flammable fuel, and no exhaust gases to vent from the building. They are 100% efficient in the sense that they convert all of the energy they consume into heat. However, to be fair, electricity is a highly refined form of energy. The energy-conversion process used at the power plant where it was generated is certainly not 100% efficient.

Heat pumps are essentially reversible air conditioning systems. Typical air-to-air heat pumps are split into two parts. The indoor portion is similar to a conventional forced air furnace with a built-in air conditioning coil. In most cases there are electric heating elements to supply heat if the system fails or the outdoor air is too cold for the heat pump to work. The indoor part of the unit also houses most of the controls, either as separate components or on one or more circuit boards.

The outdoor portion of the heat pump is very similar to that of a typical air conditioning unit. It contains the refrigerant compressor, the second heat exchanger coil, the fan for this coil, and the controls that are directly associated with these components. It also contains a solenoid-operated reversing valve. This valve determines the direction of flow for the refrigerant, which switches the system between air conditioning and heating modes.

The heat pump can extract heat from the indoor coil for cooling, or from the outdoor coil for heating. The heat gained is dissipated in the opposite coil—outdoors during air conditioning, indoors when it is heating. When the temperature is too low for efficient operation, the auxiliary heating elements are energized and the compressor unit is shut off. The indoor unit of the heat pump system then becomes an electric furnace.

Heat pumps can exchange heat with media other than air if local conditions make these other options attractive. Groundwater is one alternative. It has a relatively stable temperature year-round, in a range that is acceptable as a heat source for heating or as a heat "sink" for cooling.

Oil burners are usually used with a commercial fuel oil that is pumped from a storage tank into the burner unit. A combustion blower/pressure pump unit forces the fuel oil through a small orifice into an airstream. A powerful high-voltage electric spark is used to ignite the spray, and the resulting flame is directed into a fire box and through a heat exchanger. The circulating air blower is operated in much the same way as for other furnaces.

Boilers may be fired with different fuels, including gas, oil, and electricity. The control systems are somewhat different because the burner is maintaining the temperature of the boiler, not heating the building space directly. Typical residential water heating systems have multiple zones regulated by zone thermostats and motorized zone valves.

17.2 Licensing and Jurisdiction

In a typical natural gas fired residential furnace installation there is a 120 V AC electrical power supply obtained from the building electrical service panel. The furnace is dependent of this electrical circuit, and ultimately on the electrical utility, for continuity of the electrical supply. Natural gas for the burner comes through a gas line and meter from a second utility service connection. If the burner is fired by propane, it typically comes from a bulk storage tank that is periodically refilled by a local supplier.

The electrical utility and the gas supplier (utility or bulk seller) must meet government regulations specific to the installation of their equipment and the operation of their enterprise. They are subject to inspections and are held accountable for failure to comply with the rules. There are serious public safety issues involved, and it is important that there is some control over the operation of the marketplace. However, these issues do not stop with the supplier. It is necessary to regulate not only the distribution and sale of electricity and gas, but also the manufacture and installation of the HVAC equipment to which they connect.

Manufactured items are tested by agencies such as Underwriters Laboratories, the Canadian Standards Association, the American Gas Association, and the Canadian Gas Association. Equipment receives approval as a unit, subject to installation restrictions which are part of the manufacturer's literature, and the applicable installation codes. The installation of the equipment must meet the manufacturer's requirements, plus all the relevant regulating standards and codes. It must also be up to par in terms of accepted trade practice and quality of workmanship.

Installers and service personnel are required to be licensed by appropriate authorities, and their work is subject to inspection and legal responsibility. Part of this regulatory system is the assignment of specific duties to defined trades or similarly recognized groups of skilled workers. Establishing appropriate jurisdictions makes it possible to ensure relevant training, examinations, and licensing.

One result of all this is that installations, such as HVAC, can involve multiple trades. This is unavoidable, but if it is not handled properly, it can result in friction between trade groups and frustration for the equipment owner. Residential furnaces are an interesting case in point. It is inevitable that the work of several trades must converge at the furnace. The exact title and the extent of the involvement of each trade vary somewhat from one geographical area to another, but the following list should be at least identifiable.

A license to install and service gas piping, burners, and appliances is generally part of the gas fitting trade. Duct-work and auxiliary air handling equipment is typically done by sheet-metal workers. Usually only electricians are licensed to install the necessary electrical supply, as well as any field-installed line-voltage auxiliary equipment or controls. Air conditioning or heat pump equipment is installed and serviced by refrigeration and air conditioning mechanics. When the furnace uses a fuel other than natural gas or propane, or there is a boiler involved, other trades such as the steam fitters may be involved, and additional licensing issues will arise.

The jurisdictional issues of HVAC should be at least appreciated by anyone involved in the purchase, installation, or service of this type of equipment. There are strict legal limitations to the "who and what" of work done on all parts of the system. This includes the electrical supply, the gas piping system, the burner and its associated controls, the duct-work and ventilation, and the air conditioning equipment.

This text is all about the electrical side of such equipment, which may fall outside the traditional practice of the electrical trade, or at least inside the traditional practices of other trades. In the case of a gas fired furnace equipped with central air conditioning, the percentage of the installation that is likely to be done by a licensed electrician is relatively small. Typical duties might include the installation of the main furnace supply circuit, the feeder to the outdoor air conditioning unit, and possibly the thermostat cable. That's about it. However, that is only a small part of the total heating and air conditioning equipment electrical system.

Most of the electrical components and the interconnecting conductors are installed at the factory inside the furnace, the outdoor unit of the air conditioning or heat pump system, and any powered ventilation units that make up the complete HVAC installation. These units are installed by the HVAC contractor, and the responsibilities for servicing the equipment, including the electrical portions, are likely to fall to that contractor and his employees. These skilled workers may include electricians, but very often they are from one or more of the HVAC trades.

17.3 Gas Fired Forced Air Furnaces

Natural gas furnaces tend to predominate where gas is available. It is a relatively inexpensive fuel because it requires minimal processing, and it is still widely available. With a proper air mix, it burns cleanly with virtually no residue. Its heat content is high enough that the distribution pressure can be regulated down to a very low value without the need for unreasonably large gas lines. Delivering the gas to the consumer at low pressure improves safety, simplifies the installation, and allows the use of less expensive low-pressure gas controls.

Gas fired burners are quite simple to build; they are compact, low in cost, and require a minimum of maintenance. Although gas is very explosive when mixed with the air in a room or building, the safety shutoff devices in common use have a near-0perfect track record. Gas explosions caused by faulty gas controls are extremely rare.

Gas fired furnaces have evolved from simple standing pilot systems with very basic controls to complex high-efficiency units with electronic ignition, induced-draft systems, and multispeed circulating air blower motors. The initial control circuits presented here are for standing pilot burners, which are the core component of conventional gas furnaces. These are the simplest type of gas fired forced air furnace available. Compared to other types they have the lowest fuel efficiency, but they are also the most reliable and the least expensive to purchase and maintain. Standing pilot furnaces are still in common use—and in demand in many areas.

The controls used in more advanced furnaces are examined later in the chapter. These furnaces have electronically controlled ignition and flame monitoring systems and, usually, more elaborate fan controls than are found in a conventional standing pilot model. The circuits included in this section are typical examples but, of course, do not include every variation. It should be noted, however, that an understanding of the systems described here can be applied to virtually all existing types of residential gas furnaces.

Note

For anyone involved in a study of control systems, it should quickly become clear that the manufacturer's service and installation manuals are of critical importance. Without proper service information and control drawings, even the most skilled technician will have difficulty with new or unusual equipment. In general, the more thorough the knowledge, the more efficient and effective the person is, whether it be an initial installation or a service call.

The 120 V AC Power Supply

The electrical supply to typical residential heating equipment will usually conform to the electrical code standards that were in effect at the time the equipment was installed. The information below is not an explanation of the applicable rules—they are subject to many variations in wording and interpretation, and may be overruled or supplemented by local regulations—but simply an introduction to how things may be done, and some of the reasons why they are usually done in certain ways.

As with most large appliances, the furnace and its associated equipment is usually supplied from a single-branch circuit, typically a separate breaker or fuse in the panelboard (see Figure 17–1). In most cases, this circuit is not only exclusive to the furnace, but the furnace is also exclusive to the circuit—it has no other supplies. In some larger systems, associated equipment such as a boiler circulation pump or an auxiliary ventilation fan that is not required for safe operation of the heating unit, may have its own supply. In heating/cooling systems, the air conditioning unit shares equipment and controls with the furnace, but has its own supply for all the power circuits.

Keep this in mind for your own safety. All electrical equipment associated with the safe operation of the furnace or boiler can normally be de-energized from a single disconnect point, but some types of auxiliary equipment may remain energized.

FIGURE 17–1 A typical residential furnace is supplied with 120 V AC from a dedicated breaker. The system may include a separate disconnect switch. Cable entry into the furnace is often low enough to require mechanical protection.

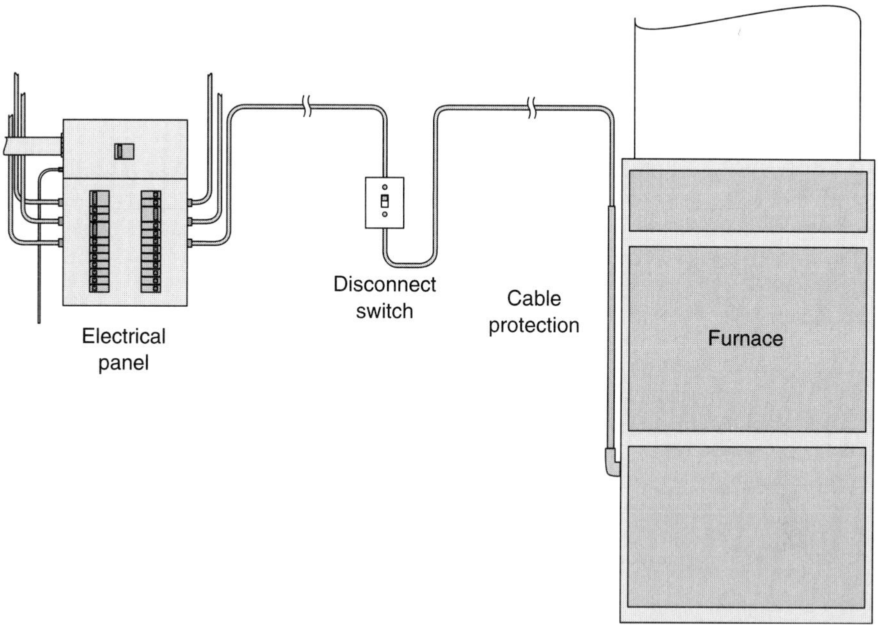

Disconnect Switches

Disconnect switches may be required for different reasons, generally all safety related. However, they may be required to satisfy some quite different intentions. The final installation can vary quite a lot with local codes and regulations. For example, a disconnect may be required because the equipment is gas fired, because it is electrically powered, or because it contains an electric motor that represents a mechanical hazard.

Electrical or gas codes may require that a gas furnace or other fuel-based heating system have a safety disconnect switch that can be accessed at some distance from the furnace. The intention is to provide the homeowner with a shutoff point that can be reached without having to pass close by the furnace. In rare cases, the breaker in the electrical service panel may be acceptable, if the panel is located between the furnace and the point of entry to the area in which it is installed. In most cases a separate switch is installed, and it usually ends up in a hallway or basement stairwell. For a residential furnace, it is usually a standard 120 V wall switch. It should be mounted higher than other switches in the building, so it is unlikely it will be shut off by small children who do not understand its purpose, or by adults who mistake it for a light switch. It is common sense that the switch be marked to identify its purpose; it may also be a code requirement.

Disconnects that are installed for electrical safety reasons are likely to be installed much closer to the furnace, possibly right at the entry point of the power supply. This can be a very convenient service switch for anyone who has to work on the furnace. Disconnects required for mechanical-safety reasons may have to be in sight of the motor. In some cases the codes may require provision for a padlock so the switch can be locked in the off position.

Rules for safety disconnects also apply to large heating equipment installations, but in these cases the disconnect may be a much larger single- or three-phase switch instead of a simple 120 V toggle switch.

Protected Cable

In most jurisdictions, the power supply cable must be protected from damage where it comes down the side of the furnace to the connection box. In residential wiring, most circuits are installed using a plastic-jacketed nonmetallic sheathed cable. Codes require that this type of cable be mounted or covered in such a way that it is unlikely to be damaged. This is why most furnace installations using this type of cable also have a short length of conduit for protection.

As an alternative to nonmetallic sheathed cable, an armored cable can be used without additional protection, as long as it is properly secured. This cable can originate at the supply panel or a nearby junction box. When the armored cable connector enters a hole that is straight through the side of the furnace, the connector should always be a 90° type. This keeps the cable tight to the furnace, greatly reducing the chance that the cable jacket will be bent sharply and damaged.

Grounding

Most modern electrical cables have a separate ground wire to interconnect all the "non-current-carrying" metal parts of electrically powered equipment. This includes the metal enclosure of the furnace and the internal metal parts. Equipment such as electric motors that are not securely bolted to the furnace frame requires additional ground connections. To be fully effective, the electrical ground system must also connect to the nonenergized but conductive materials that are permanently installed in the building. This includes all metal piping such as water and gas lines, ductwork, and structural steel, where applicable. Tying all the conductive parts of the building and its installed equipment to the electrical ground eliminates the possibility of this material becoming electrically live. Accidental energization of any of these conductive materials would be a serious shock and fire hazard.

Grounding is critically important, but bad or missing ground connections are not always noticed. For example, if the ground connection to a furnace is disconnected, it may go undetected indefinitely, even if there are problems in the equipment. If there is an electrical fault to the equipment frame, the current will follow any available path to ground including the gas line, ductwork, metal vents, or water lines. This is not a healthy situation.

Always respect ground connections, even though they make no difference to the operation of the equipment. Grounds may never be tested, so the only assurance of a good ground is good workmanship. This includes the efforts of the installer, and everyone who has any reason to alter the ground connections after the installation is complete.

17.4 Standing Pilot Furnace Controls

The simplest standard-efficiency residential gas furnace has three control circuits (see Figure 17–2). The first is powered directly from the supply panel. It operates at 120 V AC, and includes the fan control, the fan motor and the primary (120 V side) of a 120:24 V

FIGURE 17–2 A typical standing pilot furnace contains three control loops, each operating at a different voltage and performing a different function.

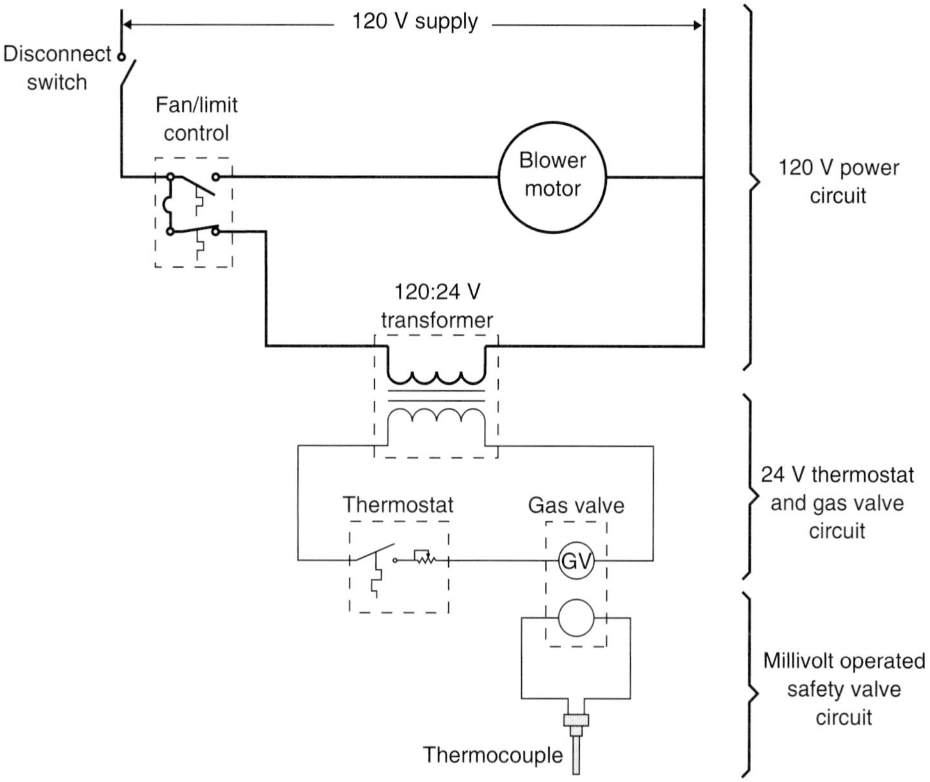

control transformer. The second circuit is powered from the secondary of the control transformer. It operates at 24 V AC, and includes the 24 V secondary winding of the transformer, the thermostat, and the main valve operator inside the gas valve. The third circuit operates from the millivolt output of a thermocouple and includes the thermocouple and the coil of the safety shutoff valve.

The safety valve is usually inside the main gas valve body, but may also be a separate valve piped into the gas line ahead of the main valve. The gas supply to the pilot light is usually piped out after the safety valve, so if the pilot flame goes out, the safety valve will shut off the gas supply to both the main burner and the pilot light.

The 120 V Controls

The 120 V circuit receives power from the building supply panel and is protected against damage from shorts or overloads by the circuit breaker or fuse in the panel. The 120 V supply "hot" (L1) feeds through a high-limit contact to the 120:24 V transformer, and through the fan control contacts to the main air circulation blower. In this example, the fan and limit contacts are both inside the combination fan/limit control. The transformer primary and the second terminal of the motor connect directly to the 120 V neutral (N or L2).

FIGURE 17–3 A typical combination fan/limit control with a spiral bimetallic operator.

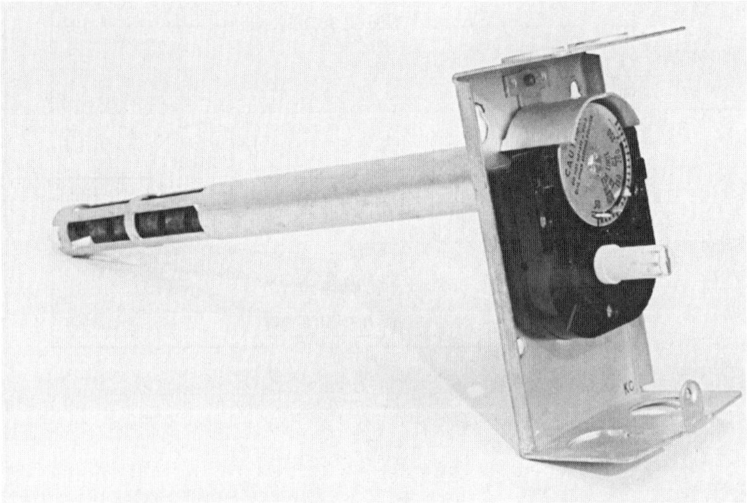

The fan/limit control has a probe containing a bimetallic strip that extends from the back of the control into the air space between sections of the burner heat exchanger. There are several different arrangements, but a common one has a spiral bimetallic strip that rotates a small shaft (see Figure 17–3). The shaft extends into the controller and turns a disk that has cams to operate both the fan and the limit contacts. As a result, both the fan control and limit contacts are operated by the temperature of the heat exchanger.

The fan contact closes on temperature rise and opens on temperature fall. The control is designed to provide an adjustable differential, through separately adjustable "cut in" (fan on) and "cutout" (fan off) temperature settings (see Figure 17–4). Adjustments are made by moving two pointers on the fan control disk. They are identified as off and on, and arranged so the on pointer cannot be moved below the off pointer. They should be set according to the furnace manufacturer's recommendations.

The on setting determines the temperature at which the fan will start, and the off setting the temperature at which it will shut off. The fan cycles on and off according to the temperature of the heat exchanger. The turn-on point should be high enough to allow the exchanger to come up to temperature before the fan starts, to avoid circulating cold air at the start of each heating cycle. The shutoff point should be set low enough that most of the residual heat of the exchanger will be circulated through the building instead of being lost up the vent. The differential (the difference between the on and off settings) must be wide enough to avoid cycling the fan while the burner is running or just after it has shut off.

When the cut-in is set too low, the fan may cycle once or twice at the beginning of each heating period. The cold air blown into the heat exchanger when the fan first comes on briefly lowers the temperature below the cutout point. The fan goes off, and then, because the exchanger is hot, comes back on almost immediately. This can also happen at the end of the cycle if the cut out is set too high. After the airflow stops, there may be enough residual heat in the exchanger to bring the controller back up to the cut-in temperature and run the fan for a few more seconds.

FIGURE 17–4 Combination fan/limit controls are usually located on the back wall of the burner area. The cover can be removed, and the fan off/on temperatures are set by moving the pointers on the rotary dial.

A manual switch is provided on most fan controls to bypass the fan contacts and directly energize the motor for continuous fan operation. This switch may have its own contact, in parallel with the main fan control contact, but usually it is just mechanically connected to the main contact. In the manual on position, the contact is held closed all the time. The manual fan switch may be a rocker, rotary knob, or push-pull operator. It is marked something like "manual fan," "summer fan," or "manual" and "auto" for the two-switch positions. Many more complex variations in fan circuits are discussed later in this section.

The second contact in the fan/limit control is normally closed, opening only if the temperature in the heat exchanger reaches the high-limit temperature setting. The operating point of this contact is preset at the factory and cannot be readjusted. The high limit is one of the primary safety devices of all furnaces. Its function is to shut down the burner if the heat exchanger reaches an unsafe temperature. The high limit should not open unless the furnace blower fails or the heat exchanger airflow is blocked in some way—by a clogged filter or by covered heating registers or cold air returns. This is a very dangerous condition. If the burner is not shut off, the furnace may get hot enough to set fire to the building.

In standing pilot furnaces, particularly older units, the high limit is usually connected in series with the 120 V supply to the 120:24 V control transformer. When the limit contact opens, it cuts off the 120 V supply to the transformer primary, instead of directly breaking the 24 V gas valve circuit (see Figure 17–5, top). However, this makes very little difference in the outcome. With no primary supply, the transformer secondary output is zero, the 24 V circuits are de-energized, and the gas valve closes.

FIGURE 17–5 High-limit contacts may be installed in the 120 or 24 V circuits. If the limit is part of a combination fan/limit control, the jumper must be removed to separate the two contacts when they are installed in separate circuits.

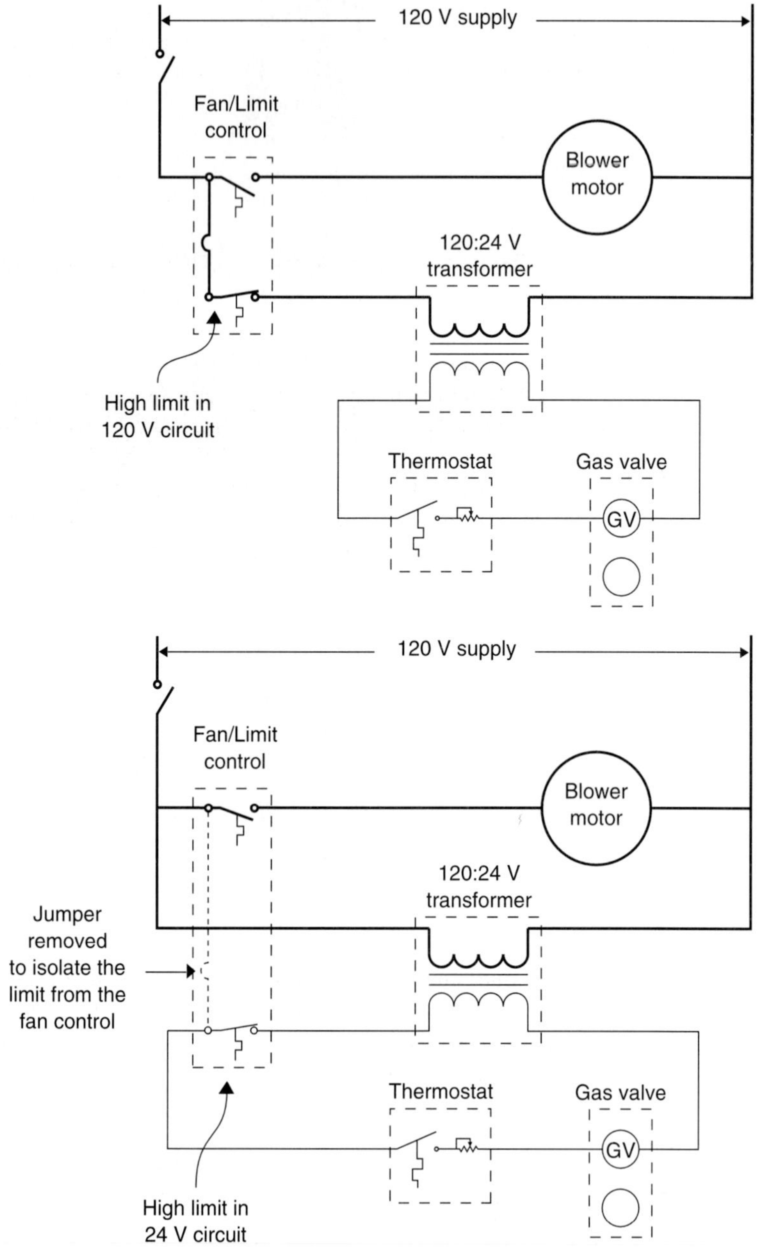

The high limit may also be installed in the 24 V circuit. It has to be in series with the gas valve, and can be before or after the thermostat. It will also work in the common side of the circuit, but this is generally not a good practice. It can be argued that this is an acceptable location because in these furnaces the transformer secondary is usually not grounded. However, this runs counter to the principles of good wiring practice and should be avoided.

A high limit on the 24 V side of the control system has the same effect as in the 120 V side. It cuts off power to the gas valve when the heat exchanger temperature reaches a dangerous level. Almost all newer furnaces have the limits in the 24 V system. Older furnaces usually have 120 V limits, but 24 V connections were also used. Do not make assumptions. Always check the drawings and the circuit itself. When the high limit is connected into the 24 V circuit, the wiring is somewhat different, particularly at the fan/limit control. The jumper that connects the two line terminals must be removed, and the limit contact must be wired independent of the 120 V side of the system (see Figure 17–5, bottom).

The connections on most fan and limit controls are a push-in type designed for solid wire. The wire is stripped to the proper length, and then simply pushed into the appropriate hole. It cannot be removed unless it is released with the aid of a small screwdriver inserted into a slot next to the wire connection hole. The strip length is important, so there is a strip gauge on the face of the control that can be used as a guide.

There are four possible connection points, two on the side marked "Fan" and two more on the side marked "Limit" (see Figure 17–6). Both the fan and limit contacts have a "Line" terminal and a "Load" terminal. In electrical equipment the line side of any type of control always connects to the supply. The load side is always the output side. Therefore, the line terminal should receive power from the appropriate supply, and the load terminal should connect to the controlled equipment.

In a combination fan/limit control, the two line terminals are preconnected at the factory by a removable jumper. This jumper is not replaceable; once it has been removed, it cannot be put back.

Note

Removing the jumper does not render the control unusable, even for situations where the jumper is needed. It can be replaced with an external wire connection between the two line terminals.

The purpose of the jumper is to simplify the wiring for controls with a 120 V limit, while allowing the same control to be used for systems that have the limit connect in a low-voltage circuit. When the limit is on the 120 V side of the transformer, the line side of the fan and limit contacts can be connected together because they come from the same 120 V supply line. The jumper makes this connection in advance. When the limit is wired into a 24 V (or millivolt) circuit, the limit contact must be electrically isolated from the 120 V system. The jumper is removed, and four wires are brought to the fan/limit control instead of three: two for the fan at 120 V, and two at the lower voltage for the limit.

Note

The jumper in a combination fan/limit control is very important. If the control is replaced and the jumper is left in when it should be removed, expensive components can be destroyed.

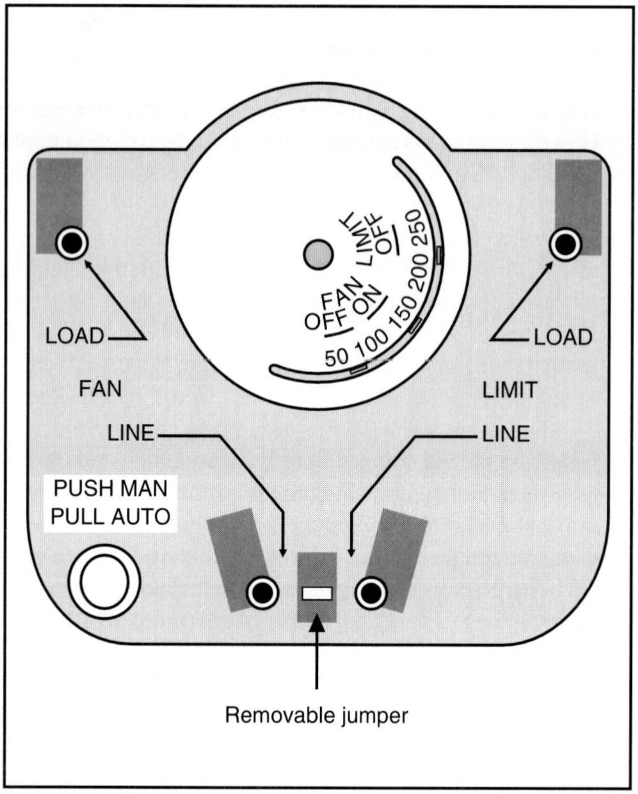

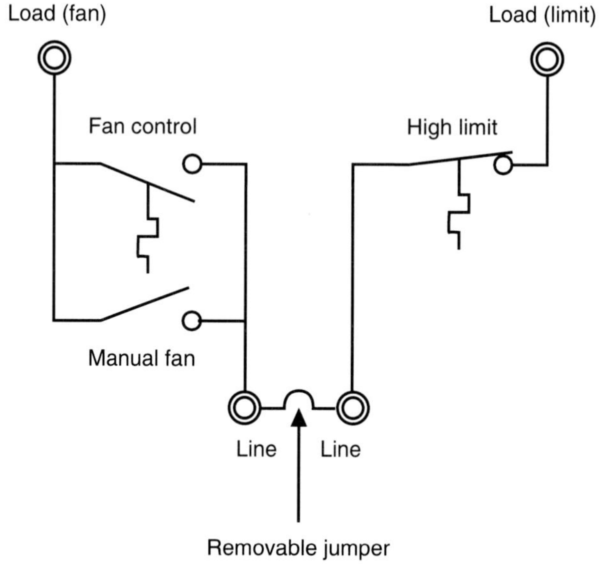

FIGURE 17–6 Compare the markings on this fan/limit control to the markings and internal-contact layout shown in the schematic.

Warning

Misunderstanding limit connections is a common source of problems. Mistakes often result in damage to control components, and can create an extremely serious fire hazard. Remember the purpose of the high limit. It must disconnect power to the gas valve if heat exchanger temperature rises to a dangerous level. It must be in the circuit in such a way that no other control can cause it to be bypassed.

The limit contact can open the 120 V circuit to the control transformer, the 24 V supply to the thermostat, or the 24 V line that goes directly to the gas valve. For the powerpile systems discussed later in this section, it must interrupt the millivolt circuit to the gas valve. No matter how the system actually operates, or at what voltage, an open limit contact must break the circuit to the gas valve.

The 24 V Control Circuit

In most furnaces the 24 V circuit has no fuse or breaker (see Figure 17–7). The equipment is protected from electrical faults by the nature of the 120:24 V control transformer. This transformer has a Class 2 rating, meaning it is a current-limiting supply that is self-protected against shorts and overloads. Unfortunately, most of these small transformers depend on an internal thermal fuse to protect themselves and the circuit. When they are overloaded, the fuse link inside burns open, and the transformer must be replaced. Some manufacturers do add an additional fuse to the circuit, external to the transformer. This fuse can save the transformer if there is a short in the 24 V system.

In a basic standing pilot furnace, the 24 V transformer secondary supplies power through the thermostat to the 24 V main operator in the gas valve. The circuit starts from one 24 V terminal of the transformer, passes through the thermostat to reach the gas valve, and returns to the other 24 V terminal of the transformer. When the thermostat calls for heat, its contact closes, completing the circuit to the gas valve. The thermostat "contact" is usually a mercury switch controlled by a bimetallic strip. In any case, it functions as a normally open contact that closes on a fall in temperature. The gas valve is normally closed, opening

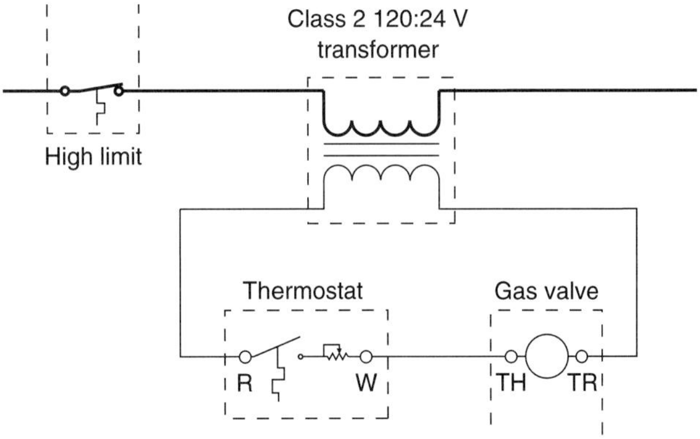

FIGURE 17–7 In a standing pilot furnace, the 24 V circuit may contain only the transformer secondary, the thermostat, and the gas valve. There is no fuse protection in the circuit because a Class 2 transformer is current limiting.

when power is applied and closing automatically when it is removed. It is usually operated by a solenoid, but bimetallic operators are also used when a slower operation is desired.

The Thermocouple Circuit

A standing pilot burner has two gas valves: either separate units, or both in a single-valve body. The main valve is powered by the 24 V AC thermostat circuit, and the safety valve is powered by the low-energy (10 to 30 mV) DC output of the thermocouple. The thermocouple is located in the pilot flame and connects directly to the gas valve by a copper conductor that looks like a small tube. This "tube" is actually a two-conductor cable. One conductor is the outside tube; the second is a smaller inner wire separated from the tube by heat-resistant insulation. The gas-valve end of the thermocouple conductor screws into a threaded hub that is similar to a lightbulb socket. This connects the two conductors of the thermocouple directly to the coil of the safety shutoff valve.

The thermocouple can generate enough current to hold the safety valve open, but not enough to pull it open from a closed position. To light the pilot, this valve has to be held open by a manual pushbutton arrangement on the gas valve. Once the thermocouple is hot, the valve will stay open, and the button can be released. If the pilot light goes out or the thermocouple fails, there is no current in the safety valve coil. The valve is released and springs shut to cut off the gas. For trouble-free operation, the pilot must be properly located and adjusted, free of lint, and protected from drafts that may blow the pilot flame away from the thermocouple.

Thermopile Controls

The thermopile is an alternative to using a 120:24 V transformer for control power. A thermopile contains many thermocouples, connected in series and mounted together inside a common case. It is located in the pilot flame and typically generates about 750 mV. This output is enough to operate a specially designed gas valve.

A thermopile gas-valve system also has two valves. They may be separate or combined into a single valve body, just as in a 24 V control system, but both operate at 750 mV DC. The safety valve operates exactly the same as a thermocouple valve, but it is designed for the higher 750 mV thermopile output (see Figure 17–8). The thermopile output cannot open the valve from a closed position, but it can hold it open. Like the thermocouple-powered safety valve in the 24 V system, this valve has a manual pushbutton arrangement to hold it open while the pilot is lit.

The main valve operator is also designed to operate on 750 mV DC instead of 24 V AC, but this valve opens when power is applied and closes on loss of power, just like a 24 V AC operator type. The 750 mV system can supply very little energy, so the valve is typically a "pilot"-operated diaphragm type. A small 750 mV-operated pilot valve controls the pressure on one side of the diaphragm by trapping or bleeding off the gas on that side. The difference in gas pressure on the two sides of the diaphragm is used to operate the main valve.

In thermopile control systems the high limit must be connected in series in the thermostat-to-gas-valve circuit. This means the high limit is part of a 750 mV circuit, so it must be

FIGURE 17–8 The 750 mV DC powerpile output holds the safety valve open, and operates the main gas valve. The limit must be in the gas-valve circuit.

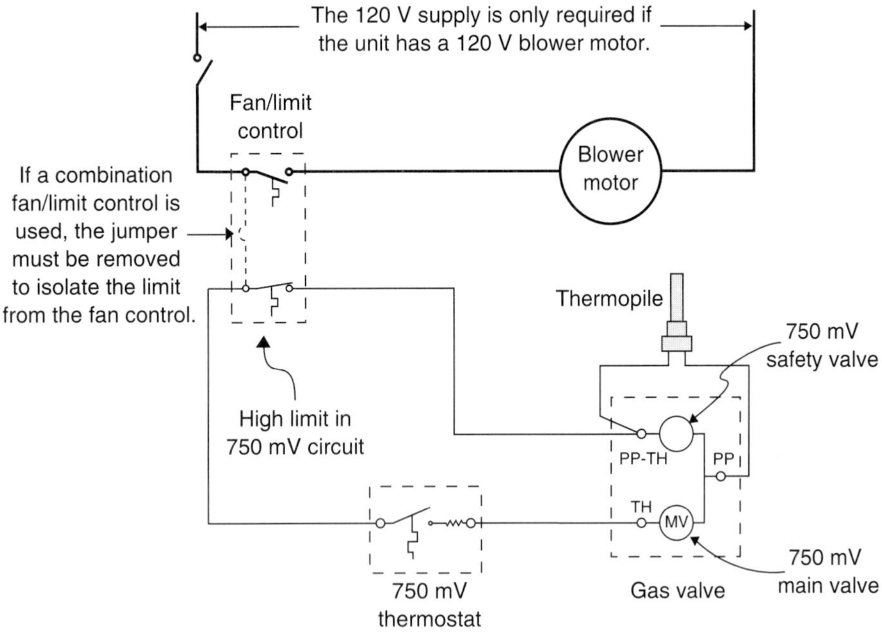

totally separated from the 120 V fan control. A separate limit and fan control or a combination control with the jumper removed may be used.

A 24 V burner control system needs a 120 V supply to energize the control transformer. These systems cannot operate without an external source of power. Thermopile systems are self-powered, making the burner control independent of any external power requirements. When the heating system has no circulating air blower, it can function normally without a utility supply. If it has a 120 V blower, the burner and its controls will function through a power failure, but the blower will not.

With no 120 V supply to the furnace, when the temperature of the building drops, the thermostat calls for heat and the burner cycles on as usual. With no blower to force air through the system, the heat exchanger overheats, and the high-limit contact opens, shutting the burner off again. High limits are self-resetting, so when the heat exchanger cools below the limit setting, it closes and cycles the burner back on again. This process can repeat indefinitely. The thermostat continues to call for heat, and the burner cycles on and off on the high limit. In very cold weather, this may save the building from freezing. As long as there is gas pressure in the supply line, the furnace will continue to supply at least some heat.

Warning

Thermopile systems have specially designed thermostats and gas-valves. Never connect thermopile-rated equipment to a 24 V supply. If the unit has a combination fan/limit control, be sure that the two controls are electrically separate by removing the jumper between the two line side terminals.

17.5 Thermostat Circuits

Heating Thermostats

A simple heating thermostat has only two connection terminals. Inside is a temperature-controlled switch contact—usually a mercury bulb mounted on a spiral bimetallic strip—and an adjustable heat anticipator. The bimetallic strip tilts the mercury bulb switch in response to changes in temperature. The bulb has wires imbedded in the glass at one end and a small blob of mercury sealed inside. Mercury is an excellent electrical conductor, so when the bulb tips the mercury to the contact end, it bridges the embedded wires and closes the "switch."

When temperature rises, the bimetallic strip tilts the bulb so the mercury flows away from the contacts. A drop in temperature moves it the opposite way, causing the mercury contact to close. The glass bulb is arched slightly to provide a "differential" between the on and off positions, resulting in a more positive switch action with less tendency to oscillate. The operation of this, and other types of thermostats, is discussed more thoroughly in Chapter 15, "Electrical Controls."

A standard heating thermostat has no selector switches or extra features. Only two wires are required from the thermostat to the furnace. One connects to the transformer, and the other connects to the gas-valve. Two-terminal thermostats may not have identified terminals, but standard markings are R and W. The R terminal should be interpreted as the "line" terminal from the 24 V supply transformer. The W terminal is the "load" terminal that supplies power to the gas-valve or burner control.

Note

Because it is just a two-wire device, a heating-only thermostat functions properly with the connections reversed. This is not necessarily true for a thermostat with more than two terminals.

Many gas-valves provide a connection point to join the supply wire from the transformer to the line wire feeding the thermostat (see Figure 17–9). This terminal is marked TH-TR and has no connection inside the valve. The TR terminal connects to the opposite side of the transformer; TH comes from the thermostat. Alternatively, the transformer-to-thermostat connection can be made by taking the thermostat wire directly to the transformer terminal, or by joining the wires at some other point.

Heat Anticipators. The thermostat includes a series-connected resistor called a heat anticipator. It is usually adjustable. During a heating cycle, the gas-valve current flows through the anticipator, slightly warming the inside of the thermostat. This causes the thermostat contact to open slightly before the room temperature reaches the thermostat setting. This reduces or eliminates the temperature overshoot that would otherwise occur from the residual heat left in the heat exchanger and duct system after the burner shuts off.

The heat anticipator adjustment is marked in amperes, because it has to be set to the current that the thermostat must conduct (see Figure 17–10). When the only device controlled by the thermostat is the gas-valve, the setting is based on the current rating stamped on the valve. In more complex systems, the thermostat controls more than just the valve. Always check the manufacturer's literature for the required setting. If the anticipator is set too low, the furnace "short cycles"; too high, and the room temperature will overshoot the thermostat setting.

FIGURE 17–9 The TH-TR terminal on standing pilot gas valves is provided as a convenient junction point for the transformer-to-gas valve connection. It does not have to be used.

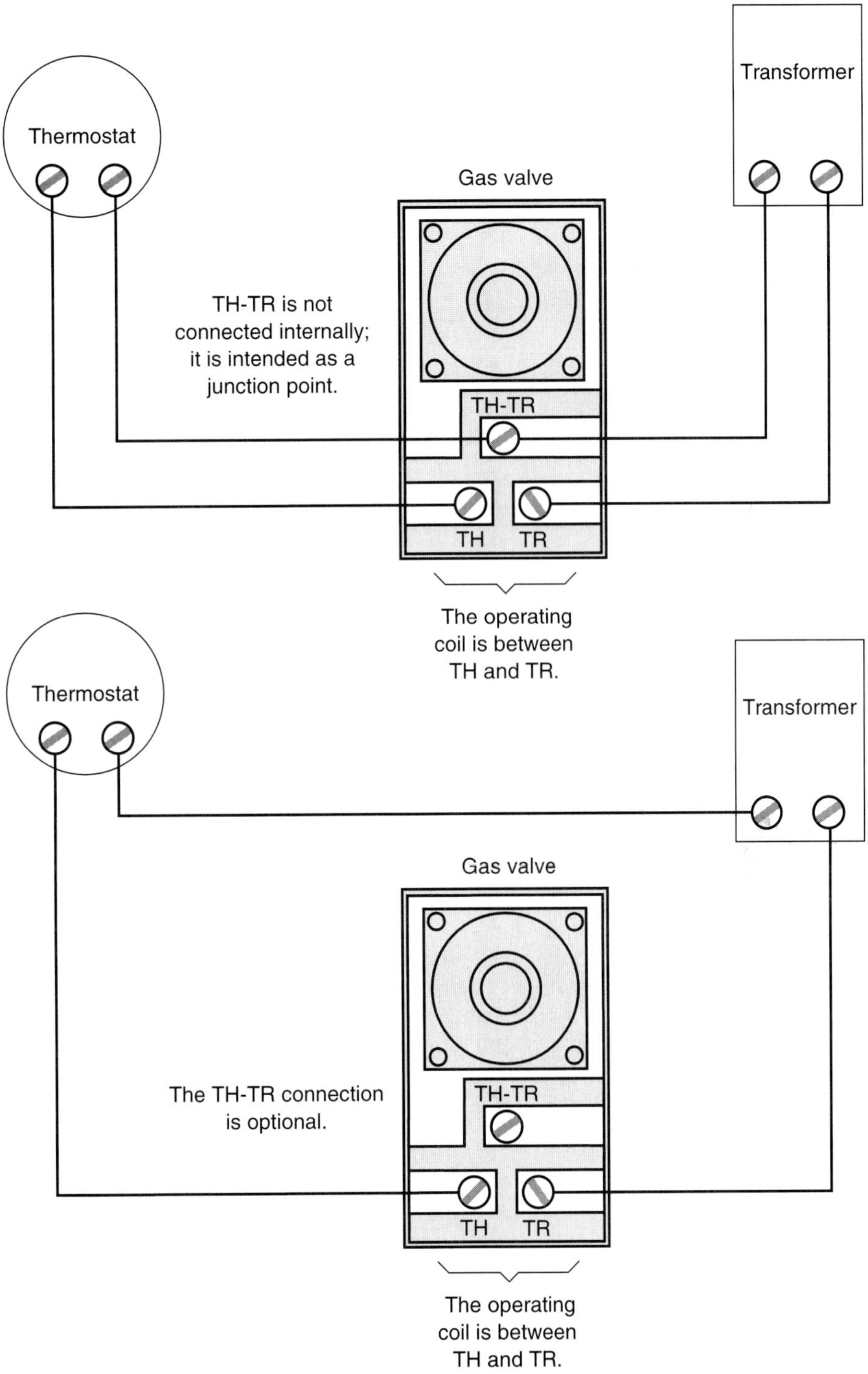

FIGURE 17–10 The thermostat heat anticipator is adjusted by moving the pointer left center) along the scale. The scale markings are in amperes. The wire winding of the anticipator resistance is visible above the scale. The spiral bimetallic strip is also clearly visible.

Note

It is a relatively easy task to measure the actual thermostat current to ensure an accurate setting of the anticipator. To use a typical multimeter, set it for an AC current range of about 1 A, remove one wire from the valve or thermostat, and use the meter to complete the circuit. With the thermostat on, the meter will indicate the current through the thermostat.

A clamp-on ammeter also works well, but, unless the meter has an unusually low current range, extra wraps will have to be taken around the jaws. Each wrap increases the reading by an amount equal to the actual current. Divide the final meter reading by the number of times the conductor passes through the inside of the current jaws. If the conductor is too short to make the extra wraps, use a separate jumper with alligator clips, or make up a 10-wrap accessory coil and keep it with the meter.

Warning

An incorrect connection such as a jumper from TH to TR on the gas-valve can connect the thermostat directly across the transformer. This usually burns out the heat anticipator, destroying the thermostat.

Heating/Cooling Thermostats

When a gas furnace is combined with central air conditioning, the system is usually controlled by a single combination heating and cooling thermostat. Separate thermostats can also be used, but a single unit is more convenient and generally a better solution. A typical residential-type heating/cooling thermostat has temperature-controlled contacts for both heating and cooling. The heating side closes on fall, and opens on rise. The cooling contact is reverse to that, closing on rise and opening on fall. Usually both "contacts" are combined in a single mercury bulb that acts as an SPDT switch. The mercury bulb is balanced on a spiral bimetallic strip, the same as for a heating-only thermostat.

The thermostat also has a three-position selector switch for "heat-off-cool" control of the system. Transfer from one mode to the other is manual, it will not change between heating and cooling automatically. A second switch provides a manual/auto fan control. In most systems this allows the user to turn the fan on for continuous operation, at the cooling speed. Figure 17–11 shows the internal connections of a typical heating/cooling thermostat.

The terminal markings shown in Figure 17–11 are essentially standard throughout the industry. R is the incoming transformer feed, W is for heat, Y is for cooling, and G is for fan control. This thermostat requires a minimum of four conductors from the furnace.

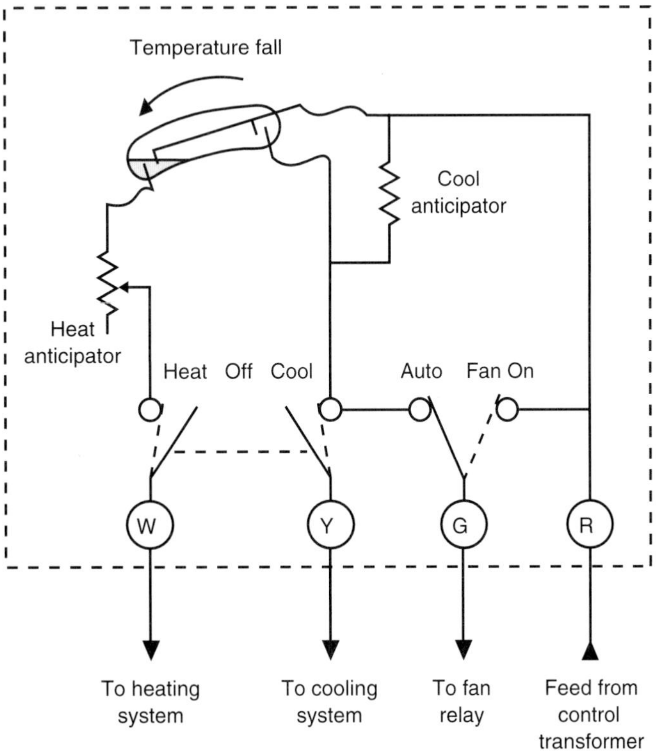

FIGURE 17–11 The internal connections of a typical heating/cooling thermostat.

Note

Although the various controls may be in a slightly different form, most thermostats intended for home heating systems provide the same basic functions. This is true for programmable thermostats as well as the more conventional contact types. However, some of these may require more than four conductors between the thermostat and the furnace.

The thermostat circuit works as follows. For a normal heating mode operation, the heat-off-cool switch is in the heat position and the fan on-auto switch is on auto. The circuit is from the transformer feed at R, through the mercury contact that closes on temperature fall, through the heat anticipator, and out at W. There is no complete circuit through to Y or G.

For normal cooling mode operation, the selector is set to cool, and the fan control is left in auto. The circuit is from the transformer feed at R, through the mercury contact that closes on temperature rise, to terminals Y and G. The circuit through Y is used to energize the cooling system. The circuit from G is used to energize the fan relay, which starts the furnace fan. There is no complete circuit through W.

In a heating mode, the fan is controlled at the furnace, independent of the thermostat, and is not started until after the burner has had time to preheat the heat exchanger. In a cooling mode, the fan is controlled by the thermostat and fan relay, and is started immediately on a call for cooling.

With the heat-off-cool switch in the off position, there is no circuit to either W or Y.

When the fan on-auto switch is in the fan on position, terminal G is energized at all times regardless of any other thermostat setting or function. In most systems this energizes the fan for continuous operation on the cooling speed. Some variations of this depend on the control system in the furnace.

There are two anticipators in the thermostat. The heating anticipator is usually adjustable, while the cooling anticipator is a fixed value, typically a small color-coded resistor. The heat anticipator is in series with the heating side of the mercury switch and works exactly as it does in a heating-only application. The cooling anticipator is in parallel with the cooling side of the mercury switch. With no call for cooling, the mercury "contact" is open, and the anticipator is in series with the control coils for the air conditioning system and the cooling system fan relay. The relatively high resistance of the cooling anticipator limits the current to a low enough value that the control coils will not pick up. A small amount of heat is generated in the anticipator and dissipated inside the thermostat. This has the effect of bringing the cooling system on slightly before the room temperature rises to the set point. When the mercury switch closes on the cooling side, the anticipator is bypassed, and full voltage is supplied to terminal Y and G to start the cooling system and the fan.

The two anticipators work quite differently. The heat anticipator has no effect until the burner is operating, and then it acts to turn it off early. This compensates for overshoot from residual heat in the heat exchanger and the duct system. The cooling anticipator has an effect only when the cooling system is off. It acts to turn the air conditioner on early to compensate for the time it takes for the system to cool the evaporator coil and provide cooled air to the conditioned space.

17.6 Good Wiring Practice

Good wiring practice is essential if the system is going to be reliable and easy to troubleshoot and maintain. All conductors should be identified and installed in a neat, organized manner. Auxiliary equipment such as transformers and relays should be securely mounted. All 120 V connections must be enclosed in an approved box. Allow a 6" free end on each wire, arrange the conductors before making the connections, and fold them in neatly after the connections are done.

Conductors inside the furnace are subjected to high ambient temperature and must be approved for use under those conditions. The wiring is typically rated for 105°C. Lower rated conductors cannot be used as replacements.

Open wiring such as thermostat wires must be secured independent of any terminations. Use suitable staples for thermostat wiring in the building frame; cable ties can be used to anchor the cable to the furnace as required. Arrange the cable in a sensible manner before it is secured and terminated. Pay extra attention to all connections. Low-voltage control systems are less forgiving of bad connections than systems that operate at higher voltages. Connections should be both mechanically and electrically secure, and they should stay that way forever.

Insulated crimp connectors must be appropriate for the wire size and crimped with the proper tool. Twist-on and set-screw wire connectors should be installed only after all the wires have been stripped, twisted, and trimmed to correct length. There must be no insulation caught in the connector, and no bare wire extending beyond the insulated skirt. After the connectors are tightened, each conductor should be vigorously "pull-checked" to be sure it is properly secured. Set-screw and twist-on connectors should then be retightened.

For powerpile (750 mV) systems it is highly recommended that solder be used on all connections that are not made under a terminal screw. Use electronic solder with a nonacid core, or fine solid solder and noncorrosive flux. These connections can then be insulated with electrical tape. The reason for this recommendation is that the voltage is so low in these systems that even the slightest extra resistance in the connections can cause problems. Properly done solder joints are more reliable than other types of connections.

A seal must be provided around the thermostat conductors where they come through the wall. This prevents air movement from the wall cavity through the thermostat. The thermostat must respond to room air, not to the temperature of the air in the wall cavity. Failure to provide a proper seal can lead to poor temperature control.

How Many Thermostat Conductors Are Enough?

Although only two conductors are required for a heating system, it is strongly advised that at least four, and preferably more, be put in place during the initial installation. A standard heating/cooling thermostat requires a minimum of four conductors. Some models need five. Humidistats require an extra two conductors, so installing seven conductors is a sensible allowance for expansion. When the system has multiple stages of heating or cooling, or incorporates special controls such those used with heat pumps, more conductors are going to be necessary.

17.7 Burner Safety and Limit Circuits

Safe burner operation depends on proper installation of gas lines, vents, combustion air supplies, and the safety devices built into the furnace control system. The intent is to protect the building and its occupants from the dangers of fire or explosion, and the ill effects of unvented combustion gases. Installers and service personnel must be aware of the purpose and operation of all such devices and any test procedures that can be used to make sure they are functioning properly.

In standing pilot furnaces, the gas explosion hazard is virtually eliminated by a thermocouple-operated 100% shutoff valve. If the pilot is lit, the burner will ignite. When the pilot is out, the safety valve closes and no gas is allowed into the burner or building space. It is a simple system, but has proven to be very reliable. Venting of the combustion gases has traditionally been left to the natural draft of a chimney or vent. Again, this is a reliable system, although there have been some very tragic incidents. Carbon monoxide gas cannot be sensed by the building occupants, who can easily be overcome before they realize the danger. Visual inspections of vents should be a standard part of annual furnace maintenance. Carbon monoxide detectors provide additional security.

There are, of course, other hazards in addition to the possibility of a gas explosion or unvented gases. An overheated heat exchanger usually caused by insufficient airflow from the circulating air blower, is a serious fire hazard. Blocked air vents, clogged filters, and other, similar problems are beyond the control of the manufacturer or the installer. A broken belt on a belt-drive fan or a failure of the motor or fan controls can happen unpredictably. For these reasons, reliable automatic protective devices are a must.

The traditional primary protection against this danger is provided by the high limit. Tested and preset at the factory, this normally closed contact opens when the exchanger temperature rises to a dangerous level, cutting off power to the gas-valve. In standing pilot furnaces, the limit control is often wired in the 120 V feed to the control transformer. A common trend in newer furnaces is to put all limits, including the high limit, on the 24 V side of the control system. They may be in the common feed to the R terminal of the thermostat, or in the 24 V line from the thermostat to the gas-valve or burner control.

Recent Trends

Most newer furnaces do not use the combination fan/limit controls described previously in this chapter. The high limit is usually separate from the fan control, and there are additional limits added to other parts of the furnace. Newer high limits are quite varied in design. Some look like the older combination unit, but have only the limit side installed (see Figure 17–12). Another popular type of high limit is a bimetallic disk mounted internally or on the surface of the furnace wall. Internally mounted units are typically small, self-resetting thermal disks, on a simple support that extends into the heat exchanger area. The disk and its supporting arms are mounted on the back of a rectangular cover, with two terminals on the front. The disk is not visible until the mounting plate is removed (see Figure 17–13). The second type of thermal-disk high limit is usually a little larger and is mounted directly into a hole in the plenum area above the burner. These are immediately recognizable as thermal disks (see Figure 17–14). They mount directly on the back panel of the burner area, and the wiring goes directly to the terminals of the device.

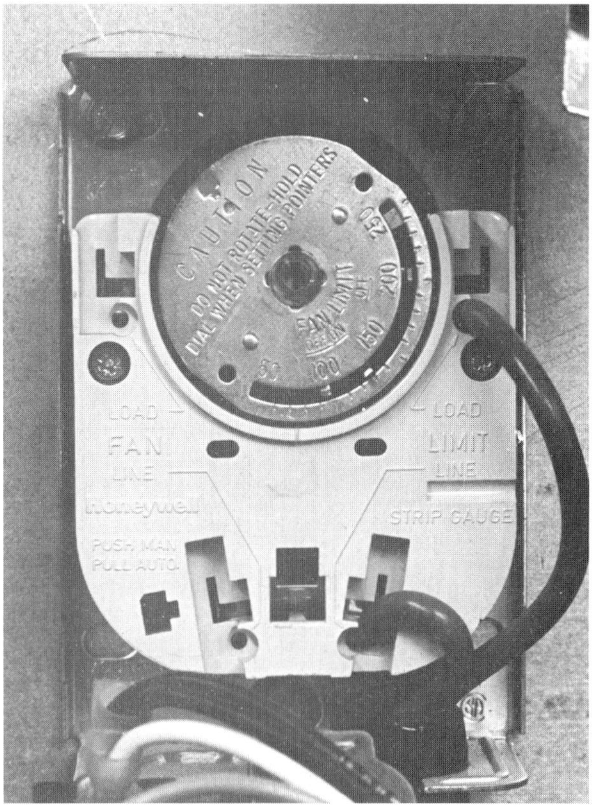

FIGURE 17–12 This high limit looks much like a combination fan/limit control, but has no contact for the fan.

FIGURE 17–13 The high limit of this unit is a thermal disk mounted on two small brackets that extend from the back of this terminal plate.

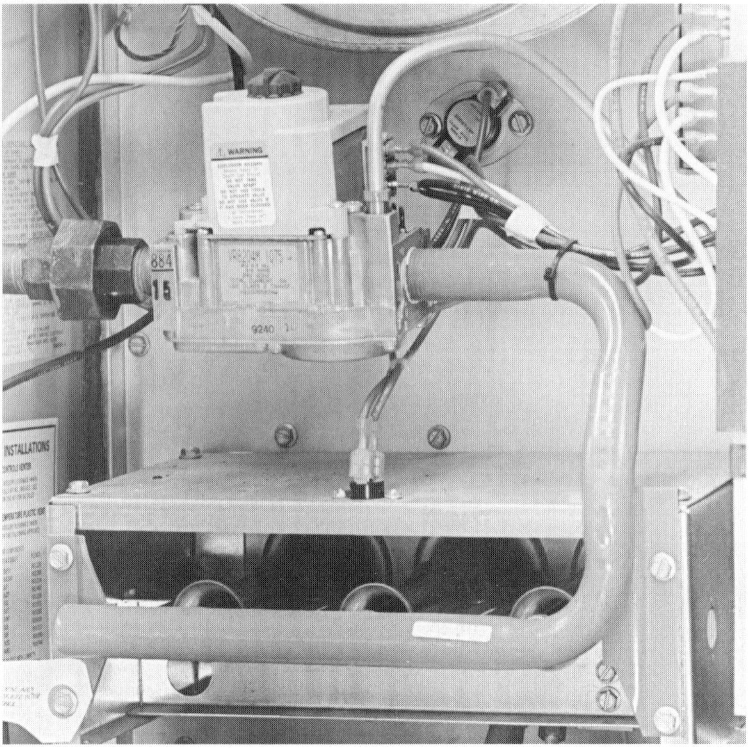

FIGURE 17–14 The thermal disk visible behind the gas valve of this furnace is the high limit.

In addition to the high limit, most newer furnaces also have flame rollout protection and vent temperature limits that protect against hazards caused by burner draft problems (see Figures 17–15, 17–16, and 17–17). They also help prevent the spillage of flue gases into the building. These limits can take several forms, including a sensor with a pilot tube extending across the width of the burner or vent hood, or a small thermal disk mounted near the area of concern. These limits are not always self-resetting. Many must be manually reset, usually by pressing a tiny reset button on the back of the device. Some furnaces also use fusible links that must be replaced after they are overheated.

Limit Circuits Are Fail-Safe

All these limits have normally closed contacts that open when the temperature rises too high. They are always series connected, so when any one opens, the system shuts down. When they are properly installed on the "hot" side of a grounded control circuit normally closed limits are inherently "fail-safe." Opening any of the limit contacts opens the control circuit, but so does a broken or disconnected wire. Accidental ground faults short out the supply transformer. In each case the system fails to a safe condition, which is exactly what is supposed to happen. (If the transformer is not grounded, a single ground fault does nothing, and a second one has less predictable results).

FIGURE 17–15 This small thermal disk is mounted just above the burner to protect against flame roll out. Note the tiny reset button on the back of the device, between the terminals.

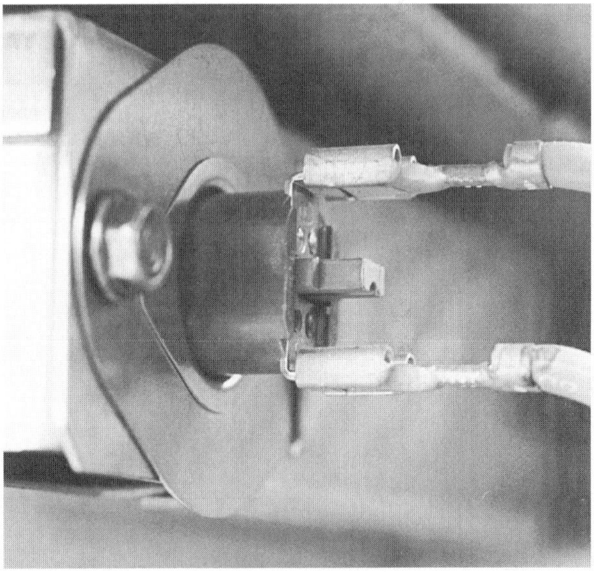

FIGURE 17–16 This small thermal disk is mounted near the vent connection where the exhaust gases exit the furnace. This vent limit is also a manual-reset type.

Note

Multiple limits in furnaces are connected in series, usually in the same circuit line. To test them with an ohmmeter, shut off the power and disconnect the circuit on at least one end, and then check them for continuity, either individually or in groups. When the device has not tripped (or has reset), the resistance will be approximately 0 Ω. If it has tripped, the reading will be an open circuit, or ∞Ω.

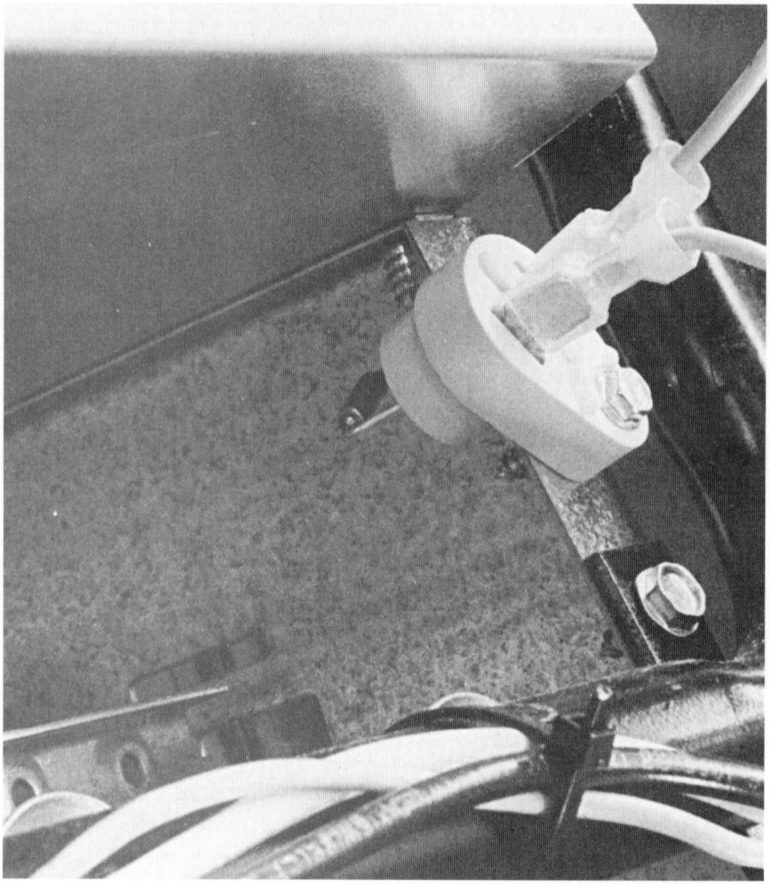

FIGURE 17–17 This furnace has fusible-link flame roll out protection. The link is in the tiny metal cylinder on the back of the ceramic terminal block. There are two of these, one on each side of the burner. If they open, they must be replaced.

As an alternative, it is usually more efficient to test with the circuit power on. Use the voltage function of the multimeter instead of resistance, and connect one lead of the meter to the 24 V circuit common. The supply can then be tracked through the circuit from one device to the next until it shows 0 V—indicating the probe has just passed the open point of the circuit. Once the open limit has been located, check across the terminals of that limit by itself. If the meter shows full voltage across the open limit, the rest of the circuit is intact.

When checking thermal limits, watch for reset buttons; they can be very small or well hidden. On some capillary-tube types, the contact block is mounted through a hole in the furnace frame, with the reset button accessible from outside the furnace. If you are responsible for service on a make of furnace that has fusible links, make sure you have replacements on hand.

17.8 Intermittent Ignition

There are two primary advantages to furnace ignition systems that don't require a continuous pilot. First, the pilot consumes gas even during the seasons when no heat is required. This waste is eliminated, increasing the overall efficiency of the furnace. Second, the continuous heat loss associated with a natural draft vent can also be reduced or eliminated by using vent dampers or a mechanical draft system. With a mechanical draft and a high-efficiency heat exchanger, the vent can be reduced to an inexpensive plastic pipe. This simplifies building construction and frees up the space that would normally be used by the chimney or vent.

The primary disadvantages to these more complex systems are reduced reliability, higher service costs, and possibly lower consumer confidence.

Flame Rectification Principles

The simple thermocouple-based safety valve used on standing pilot furnaces cannot be applied to an intermittent ignition control. Thermocouples of that type are far too slow for a system that must relight the burner on each heating cycle. Although other methods have been tried, virtually all modern residential gas furnaces equipped with intermittent ignition monitor the burner with a flame rectification system. Commercial and industrial burners use a variety of methods, including flame rectification and various scanners sensitive to infrared, the visible light of ultraviolet radiation.

Flame rectification systems are based on the fact that flame conducts electricity. If two conductors spaced a short distance apart are placed in a flame, the ionized gases that make up the flame will complete the circuit. The resistance of the flame is high, so even with a fairly high voltage applied, the current is very low. This is not a serious problem, because sensitive electronic circuits can operate reliably with extremely small current. The flame rectification circuit in a typical residential furnace operates on just a few microamps.

The simple circuit described above detects flame, but does not meet the safety requirements of a burner control, because it is too easily fooled. A bridge of conductive material (such as carbon, moisture, or other contaminants) can form a conductive path across the circuit, allowing current to flow with no flame present. To avoid this dangerous situation, two more techniques are applied. First, most burner controls pretest the sensor circuit before they attempt to light the burner. If the circuit conducts during the pretest, the controller shuts down and locks out. Second, the sensor is constructed so that current passing through the flame will be "rectified," and the detector circuit is designed to tell the difference between a rectified signal and one that is simply conducted (see Figure 17–18).

The term "rectify" means to convert AC to DC. AC continuously reverses direction of flow, while DC always flows the same way. The easiest way to convert AC to DC is to use a diode, which is the electrical equivalent of a one-way valve. A diode conducts current in one direction, but when the polarity reverses, all flow is blocked. The flame rectification circuit uses the flame for this purpose.

Researchers discovered that if the two terminals in the flame are quite different in size, there is less resistance to current flowing in one direction than in the other. The two differently sized terminals, combined with the ionized gas of the flame, become a crude but usable rectifier. In practice the two terminals of a flame rectification system are the "flame rod" or "sensor" and either a larger "ground" terminal or the metal of the burner itself. For reliable operation the size ratio between the two terminals must be at least 4:1.

The controller applies an AC voltage across the gap where the flame is supposed to be and monitors the circuit current. When there is no current, no flame is present. When there is a DC current, the AC voltage has been rectified, and there must be a flame present. When the current is AC, a conductive path other than a flame exists from the sensor to

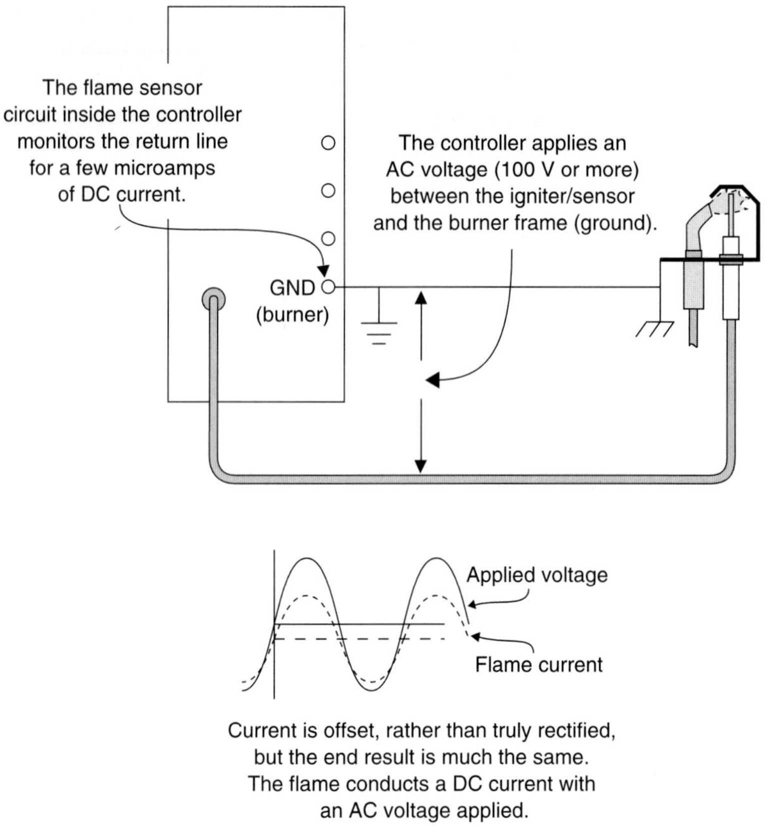

FIGURE 17–18 The burner control may use a combined spark igniter/sensor, as shown here, or a separate flame rod. In either case an AC voltage is applied across the flame to ground. DC current is proof of flame because no other conductive material can "rectify" the current.

ground, and the controller will treat it as a false flame signal. The sensor may serve double duty as the igniter, or a separate sensor rod can be used. Typical ignition controllers are designed for specific arrangements. The type of igniter and the provisions for flame detection must match the requirements of the controller.

Note

The flame rectification system is really a very poor rectifier. Most of the current passed to the controller is still AC, even with a perfectly acceptable flame. Instead of actually cutting off half of the AC voltage wave, like a diode would, the "flame rectifier" causes the AC current to be "offset" from the zero line. This can be reliably detected as DC by a circuit that is designed for that purpose. However, when the current is measured with a suitable microammeter, the AC reading is higher than the DC reading.

Note also that the voltage applied to the flame detection circuit is not the 24 V of the control transformer; there is a transformer in the controller that boosts this voltage to a much higher level.

For intermittent pilot systems, monitoring the pilot flame is sufficient. For a direct ignition system, there is no separate pilot, so the main flame must be monitored instead. In both cases, loss of flame or failure to ignite must result in 100% gas shutoff. All of these systems lock out on flame failure or after a specific number of ignition trials. They can be reset by cycling the thermostat off, and then back on. However, the controller may have to be kept off for a minute or more for the electronic circuits to actually reset.

Intermittent Ignition Systems

There are several ignition methods used as an alternative to standing pilots. Most intermittent pilot systems use an electric spark to light the pilot, which then lights the main burner. The pilot can be eliminated by applying the spark directly to the main burner. A hot surface ignition system uses a high-temperature, silicon carbide resistance element in place of the spark. It can also be used to light the main burner or a separate pilot.

Intermittent Pilot Ignition

For an intermittent pilot system to operate safely, the controller must be able to light the pilot and prove the pilot flame exists before opening the valve for the main burner (see Figure 17–19). The pilot flame must also be monitored continuously while the burner is operating.

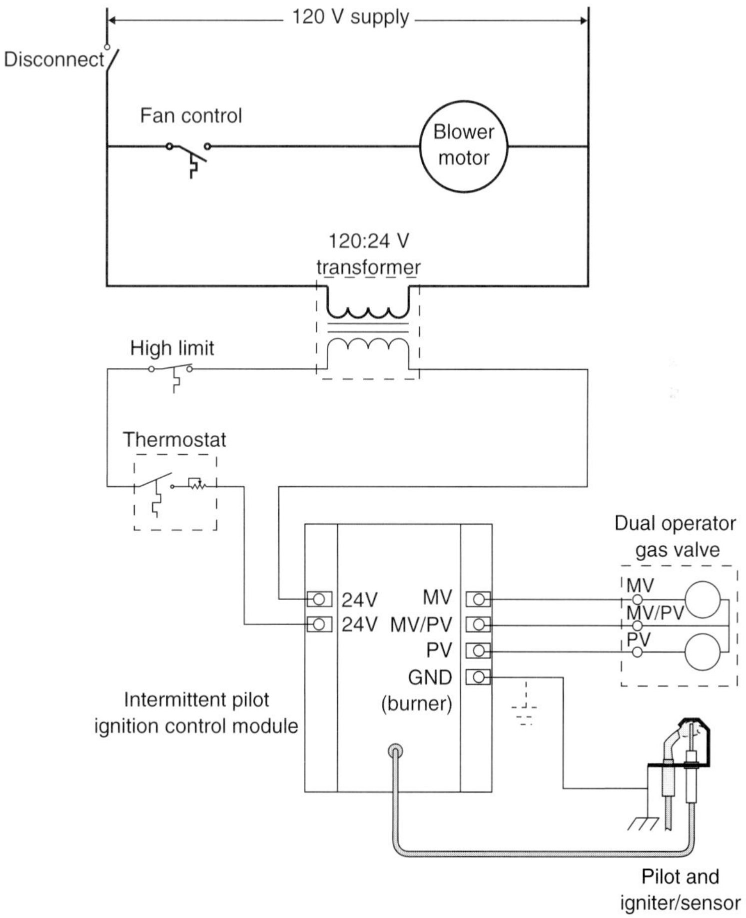

FIGURE 17–19 This diagram shows a typical intermittent pilot ignition system in a basic furnace. The circuit is very similar to a standing pilot system, but the control module is connected in place of the gas valve, and the gas valves are handled by the module, not directly by the thermostat. Two valve operators are required, one for the pilot and one for the main burner.

When the burner is shut down, the gas must be completely shut off to both the main burner and the pilot.

Intermittent pilot systems require a dual gas-valve with separate pilots and main operators. Both valve operators are designed for 24 V AC, and both valves are normally closed, opening with power applied and closing when power is removed. The pilot valve is in series ahead of the main valve in the gas circuit, and acts as a safety shutoff when it is closed. The main valve may be a delayed-action type for a more controlled gas flow when the main burner first ignites.

The burner control may be a single module mounted inside the furnace (see Figure 17–20), or consist of several individual components. It may also be part of a complete furnace control built onto a single circuit board.

Note

Comparing the intermittent pilot control circuit to that of a standing pilot system, the burner control module connects in the thermostat line where the gas-valve would have been. The thermostat cycles the controller, and the controller takes care of the valves.

During idle periods there is no power to the controller and the gas-valves are closed. When the thermostat calls for heat, it connects the controller to the 24 V transformer output. The controller does a circuit pretest, and then opens the pilot valve and, through an internal ignition transformer, generates a 5000 to 15,000 V spark across the igniter to start the pilot flame. When pilot flame is detected by the flame rectification circuit the main valve is energized, lighting the main burner. The pilot is monitored continuously until the thermostat is satisfied.

At the end of the heating cycle, the thermostat opens the circuit, disconnecting the 24 V supply to the controller and closing the gas-valves. When power to the controller is cut off for any reason, it resets to a startup condition, and must go through a complete start up procedure to relight the burner. Some controllers have trial-for-ignition timers that lock out the controller if ignition does not occur within a specific time period. Others make a specific number of attempts before locking out, and some never give up. Some controls also have built-in prepurge time delays on startup. These are used with mechanical draft systems to allow any combustible gases to be purged from the burner before it is re-ignited.

There are many variations—and no doubt more to come. To be knowledgeable and current, you have to read a lot of manufacturers' literature and take advantage of every seminar and training opportunity.

Direct Spark Ignition

Direct spark ignition systems eliminate the pilot by directly lighting and monitoring the main burner (see Figure 17–21). The controller may use a combined igniter/sensor or a separate igniter assembly and sensor rod. The control circuit is virtually identical to the intermittent pilot system, except there are only two conductors to the valve. A dual valve is required to meet safety rules demanding main and redundant valves, but both valve operators operate at the same time from the same supply.

FIGURE 17–20 This intermittent pilot ignition module is mounted (upside down) on the side wall of the furnace. The high-voltage spark igniter serves double duty as a flame rod. Flame rectification current can be measured by removing the burner ground lead (terminal 4) and re-connecting the circuit through a DC microammeter.

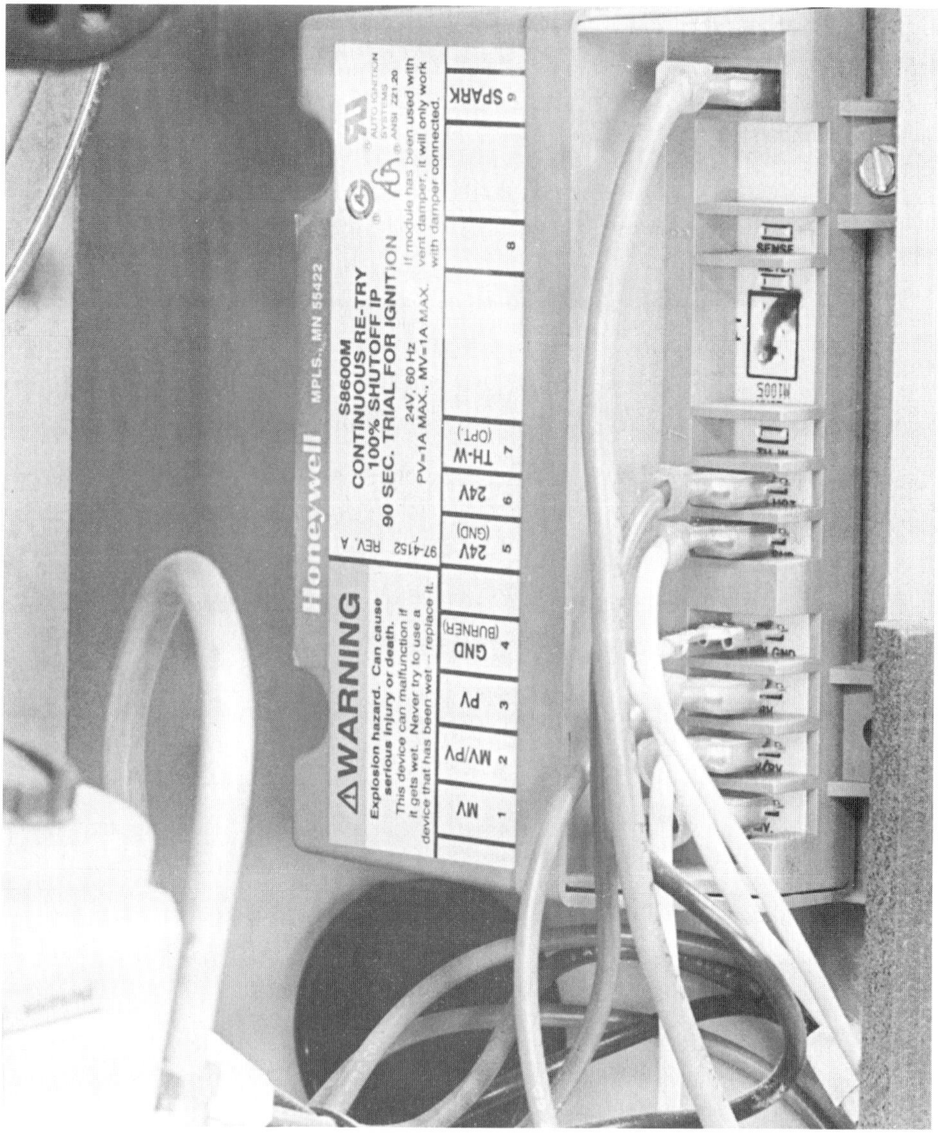

In operation, a direct spark ignition system is very similar to intermittent pilot control. The burner controller is energized by the thermostat circuit and takes care of the valves. It normally incorporates a pretest, and may have a purge time delay prior to trial for ignition. Successful trial for ignition depends on detection of the main flame rather than a pilot

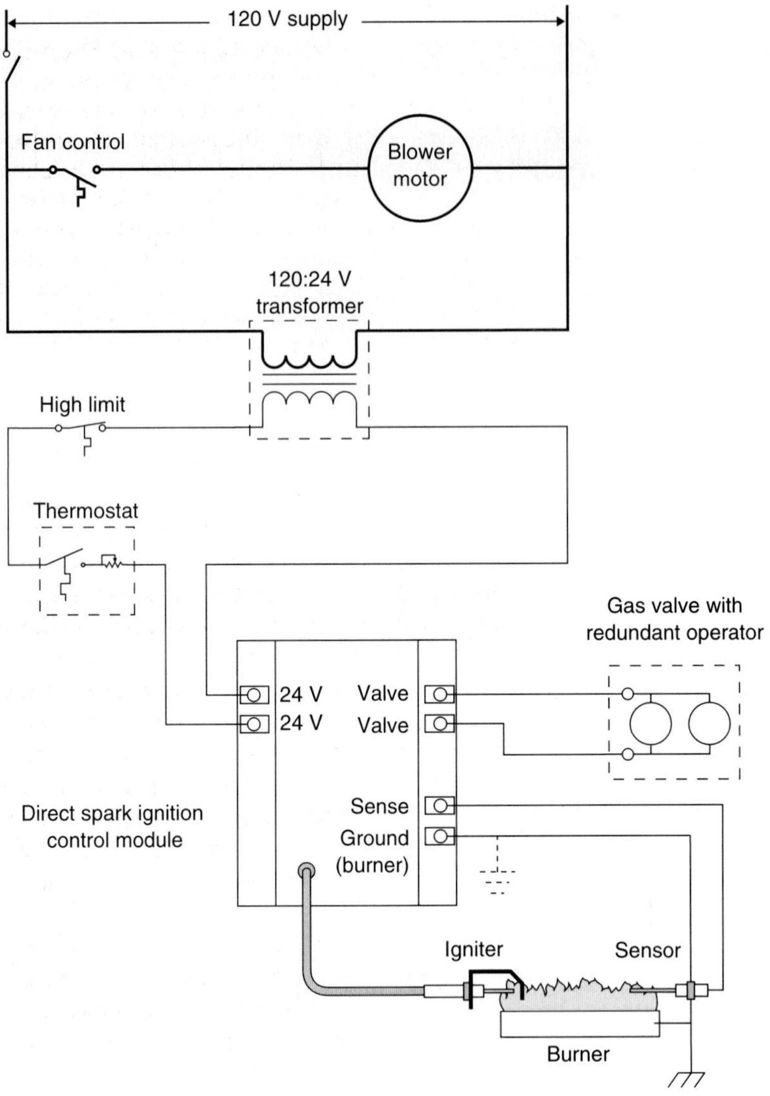

FIGURE 17–21 A direct spark ignition system has no pilot. The spark igniter and flame rod are in the main burner flame. If the igniter and flame rod are separate devices, they should be located at opposite ends of the burner. Dual-operator gas valves are used, but they are not separately controlled.

flame. Once ignited, the main flame is monitored continuously. Ignition failure results in a controller lockout.

As with all these controllers, loss of power, for whatever reason, resets the controller to its startup condition.

Hot Surface Ignition

Direct ignition with a hot surface element is similar to a direct spark system. However, the silicon carbide resistance element requires more voltage and current than can be provided by the 24 V supply. Hot surface systems use 120 V to heat the element (see Figure 17–22).

FIGURE 17–22 This hot surface ignition system uses the igniter as a sensor. Other systems have a separate flame rod. The control module has a 120 V supply for the igniter circuit.

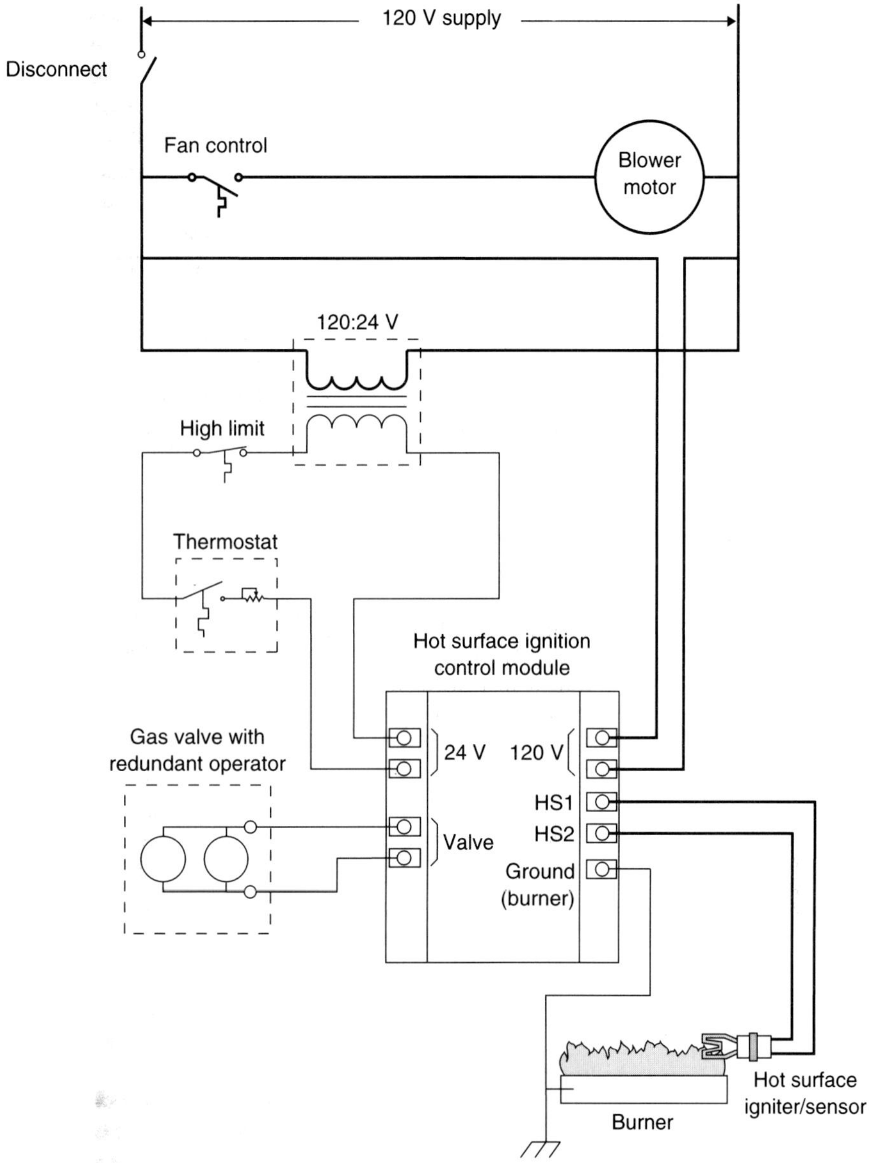

Therefore, the burner control has to have a 120 V feed and output connections in addition to the normal 24 V supply from the thermostat and connections for the valves. These systems can have a separate flame rod or can use the igniter as the sensor.

In operation, the hot surface ignition system is similar to direct spark. It may be a little slower to light, because a delay is built in to allow the igniter to get hot. Igniter current is quite high, a few amps at 120 V, and is easily measured with a clamp-on ammeter. See Figure 17–23 for an example of a gas-valve associated with a hot surface system. Figure 17–24 shows an igniter in place in a furnace. They can be awkward to remove or install.

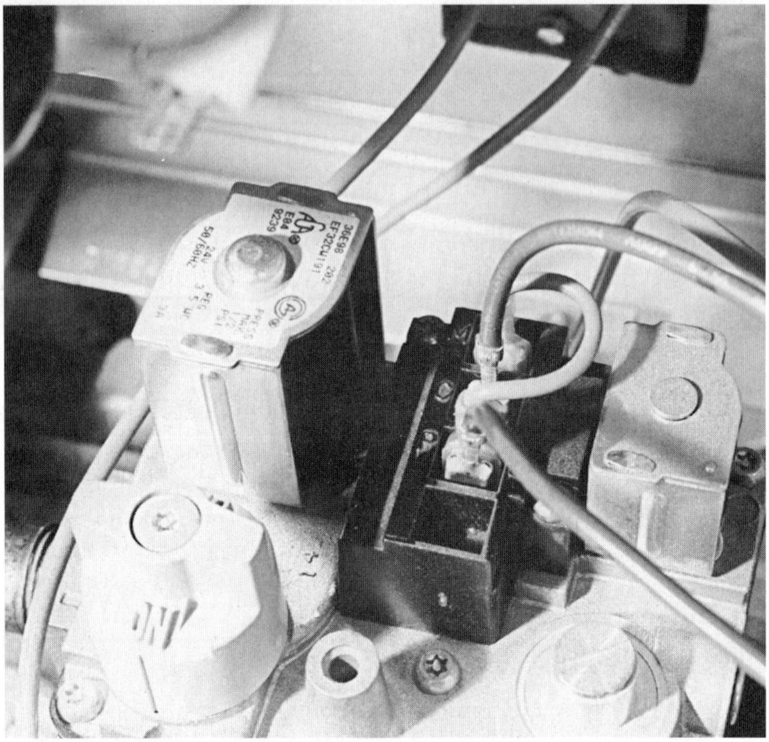

FIGURE 17–23 This dual-operator valve has the markings associated with an intermittent pilot system, but the M (main) and P (pilot) terminals are jumpered together. The valve is being used with a hot surface ignition system that has no pilot.

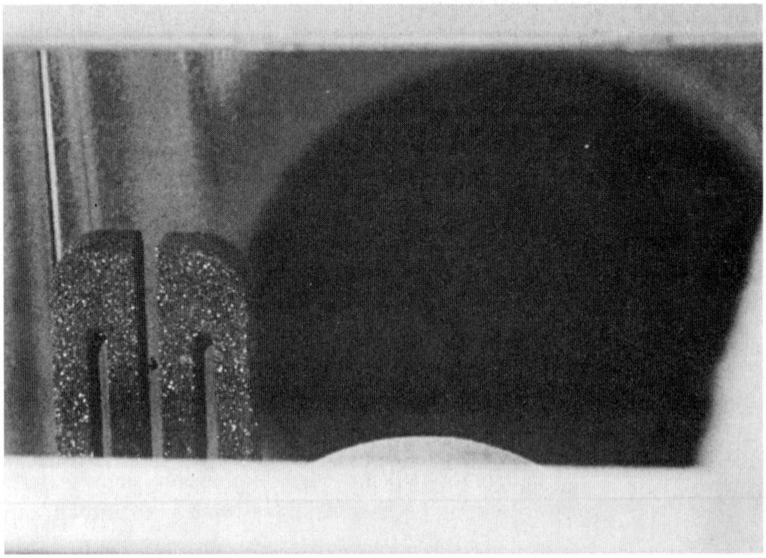

FIGURE 17–24 A silicon carbide igniter is just visible beside the right-hand burner tube.

Note

Silicon carbide igniters are fragile. They are often broken in shipping, or from accidental contact once installed. They are also subject to small cracks that can cause ignition or sensing failures. Sensor failures can also result from contamination buildup on the surface of the element. Do not handle the sensor with bare fingers—oil from the surface of the skin can ruin the element.

17.9 Combustion Air Blower Controls

Furnaces that are designed for mechanical draft require a motor-driven blower to draw in combustion air and exhaust the flue gases from the burner. These fans are relatively small, usually driven with a shaded-pole or PSC motor. The motor is coupled directly to the fan, which is mounted on the exhaust—flue gas—side of the heat exchanger. Typical midefficiency furnaces have the fan mounted high, at the top of the heat exchanger (see Figure 17–25). The fan and housing are metal because the flue gas is quite hot. High-efficiency designs typically have the fan mounted below the burner (see Figure 17–26). These units have a second condensing heat exchanger mounted below the first heat exchanger. The

FIGURE 17–25 This combustion air blower is driven by a shaded-pole motor. It is mounted at the top of this midefficiency heat exchanger. A separate vent over the temperature limit is mounted on the exhaust side just above the blower (out of sight in this picture). The air proving switch is to the right of the blower, with its suction line visible just behind and below the switch. It connects to the flue gas chamber behind the blower.

FIGURE 17–26 This combustion air blower is mounted low in this high-efficiency furnace, at the exhaust side of the condensing heat exchanger. It is made of plastic to resist the mildly corrosive condensate. The motor is a PSC type (note the run capacitor back right). The over-temperature limit is mounted directly on the blower (front right). The tube that passes behind the motor is the air line to the proving switch. It connects to the intake side of the blower. The air proving switch is not visible in this picture.

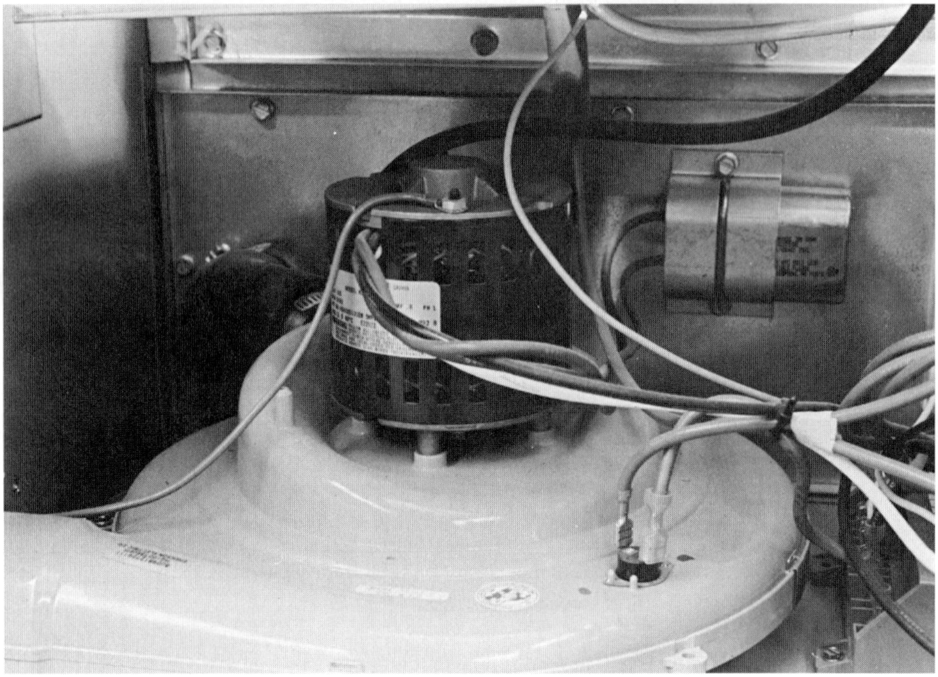

combustion air fan is mounted on the exhaust side of the condensing exchanger. The fan housing is likely to be plastic, because the flue gas is relatively cool but may contain mildly acidic condensate. Both types of fan units are likely to have flue gas temperature limits.

On the control side, mechanical draft presents some special problems. It is not good enough to start the combustion air blower and simply assume there is adequate draft. Blower failure or a blockage in the intake or vent could make it extremely dangerous to operate the burner. Most furnaces use a pressure switch to prove there is sufficient airflow for safe operation. The switch may be a differential type with two hose connections, monitoring the pressure drop across the heat exchanger, or a single-hose low-pressure type connected to the suction side of the combustion air fan. The pressure switch—which may actually operate from a slight vacuum—has a normally open contact that closes when there is adequate airflow (see Figure 17–27).

As an alternative to a pressure switch, some manufacturers use a centrifugal or electronic speed switch on the shaft of the combustion air blower motor. This system assumes that if the motor is up to speed, draft has been established. Failures that would not be detected by this approach are covered by flame rollout or flue over-temperature switches.

FIGURE 17–27 When a combustion air fan is included in the furnace, the control system usually works something like the one shown here. The thermostat energizes the combustion air blower relay, starting the blower. When airflow is proven, the air switch (lower left) closes the circuit to the burner control. The burner control operates through the thermostat contact and the air proving switch. Note that the air switch is symbolized here as a flow switch because it detects airflow, even though it is a pressure switch. There is no clear standard on what symbol is correct for this device.

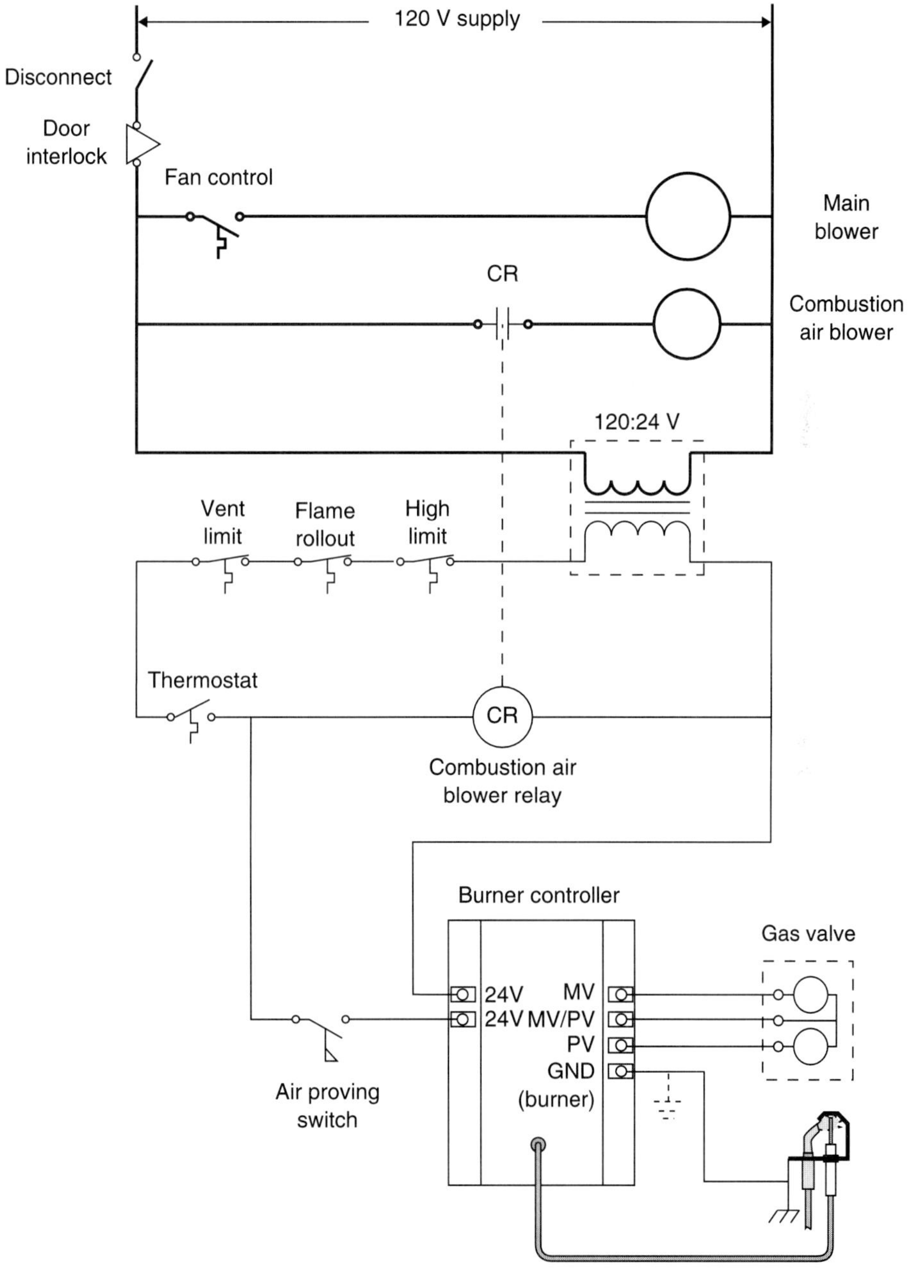

Typical combustion air control systems have the air proving switch (pressure switch) ahead of the burner control. This way it acts as an interlock. Instead of directly energizing the burner control module, the thermostat energizes the combustion air blower relay first. Then the burner control is energized through the air proving contact. The thermostat is still in the circuit ahead of both the relay and the burner control.

A normal operating cycle starts with the thermostat closing. It energizes the 24 V blower relay coil, which closes the 120 V contact in the supply to the combustion air blower motor. When the blower is up to speed and there is sufficient airflow, there is a slight vacuum on the intake side. The air line to the pressure switch is connected at this point, so the pressure on one side of the diaphragm is reduced. The diaphragm moves toward the low-pressure side, operating a microswitch that serves as the contact for the pressure switch. The burner control circuit is wired from the thermostat, through the normally open contact of this microswitch. When the contact closes, the burner control is energized.

At that point the burner control picks up the sequence. It does its prechecks, waits through the prepurge delay, and then lights the pilot or main burner. The burner control monitors the flame, the limits monitor for heat exchanger overtemperature (high limit), flame rollout, and vent overtemperature. There may be additional limits depending on the furnace design and mounting. Cross-flow or downdraft units, for example, have additional high limits. There are many variations.

17.10 Fan Controls

The air circulation fan, its motor, and controls are an integral part of all central heating systems. The fan, or blower, is mounted between the return air intake and the heat exchanger. A replaceable element filter or electronic air cleaner cleans the return air before it passes through the blower, the heat exchanger, and on to the heated space. The blower circulates the conditioned air (heated, filtered, cooled, humidified, etc.) through the building. The blower is either directly mounted on the motor shaft or driven by a speed-reducing V-belt drive. The blower speed should be adjusted to maintain correct heat-exchanger temperature rise. The temperature rise is part of the system specifications supplied on the furnace rating plate.

Blower Characteristics

The blower used in furnaces is typically a radial flow, squirrel-cage type with forward-facing blades. This design has a characteristic operating curve well suited to this application. The power input to this type of fan is inversely proportional to the resistance of the duct system. Restricting airflow reduces the mechanical load on the fan motor. The motor cannot be overloaded by dampering off part or all of the duct. Note, however, that the reverse is also true. If duct resistance is reduced—for example, by running the fan with the access cover removed—the motor current may exceed the full-load rating.

Note

When selecting the blower speed, the motor current should be checked with an ammeter to be sure it is within the motor ratings. This test must be done with the blower operating under normal conditions—all ductwork in place, the blower chamber cover on, and all registers fully open.

Fan Motors

The type of motor used for the blower varies with the drive, the manufacturer, and the style and age of the furnace. With rare exceptions, residential furnaces use 120 V single-phase AC motors. The size range is typically from $^1/_4$ to $^1/_2$ Hp, depending on the capacity of the furnace. The motors may be split-phase (resistance-start/induction-run), PSC (permanent split capacitor), shaded-pole, or capacitor-start/induction-run.

Resistance-start (and capacitor-start) motors are limited by their nature to a maximum of two speeds. All motors with more than two speeds are either permanent split capacitor or shaded-pole. The shaded-pole motor is extremely popular in very small sizes, and less common in this horsepower range. They are available, however, and may be encountered in the field.

Note

Some of the latest furnaces may have a brushless DC motor in place of the more traditional AC types. These motors may use controls that are similar to those discussed below, or something entirely different. For more information see the DC motor chapter of this text; it will not be covered here.

Older furnaces generally use resistance-start motors in combination with a V-belt drive. The pulley on the motor is usually an adjustable type, to allow adjustments to the blower speed. Most furnaces of this vintage have a single-speed, four-pole motor that runs at approximately 1725 rev/min. A slightly more sophisticated unit has a two-speed version that may be wound four-pole/six-pole (for approximately 1725 and 1150 rev/min) or four-pole/eight-pole (for approximately 1725 and 850 rev/min).

The 4/6-pole motor has a dual-run winding, the 4/8-pole may be dual winding or consequent pole. There are both general-purpose versions of these motors as well as some manufactured specifically for the HVAC industry. Blowers have a variable torque characteristic, so the horsepower rating of the lower speed can be dramatically lower than for the higher speed. The blower horsepower is roughly proportional to the cube of the shaft speed.

Note

Some two-speed resistance-start furnace blower motors are specially made so that both the high- and low-speed terminals can be energized simultaneously. General-purpose motors of this type cannot be operated this way without motor overload and possible damage.

Fan controls designed for single-speed motors can also be used with the special-purpose two-speed motors. Fan controls designed for general-purpose two-speed motors are also available. To avoid energizing two speeds at the same time, this fan control has SPDT contacts. A conventional SPST fan control and a relay with SPDT contacts can also be used. Either of these controls—properly connected—will provide two-speed operation with no danger of energizing both speed leads at the same time.

As furnaces have evolved, there have been a number of changes in the types of motor used and the arrangement of the controls. Almost all modern furnaces use a multispeed PSC motor directly connected to the blower. These direct-drive blowers are more compact than the older belt-driven types, and a little simpler in construction. The motor is typically a six-pole design, with extra current-limiting windings added for speed control. The full-load, full-speed rating of a typical motor is approximately 1050 to 1100 rev/min. The multi-speed PSC motor provides greater control flexibility and operates more efficiently than a resistance-start motor and belt drive combination. It also starts much more gently, with a much lower starting current.

Shaded-pole motors are also used for direct-drive blowers. With identical horsepower, mounting dimensions, and similar speed ratings, a shaded-pole motor is physically interchangeable with a PSC and does the same job. However, it is less efficient and has a higher running current. Shaded-pole motors are usually less expensive to purchase than an equivalent-horsepower PSC.

Single-Speed Fan Control—Heating Only

The most basic furnace fan system has a single-speed motor and a fan control contact operated by plenum temperature (see Figure 17–28). The fan control may be a separate unit or part of a combination fan/limit control. Manual fan operation is provided by a manual switch built into the fan control. This system provides acceptable operation of a basic heating cycle. The thermostat cycles the burner on and off based on room temperature. The fan control cycles the blower on and off based on heat exchanger temperature. The blower runs continuously if the manual switch is turned on. This is the type of control built into most older standing pilot furnaces.

Single-Speed Fan With Air Conditioning

On a furnace equipped with air conditioning, a heat/cool thermostat provides outputs for heating, cooling, and fan control. The thermostat fan terminal operates a furnace-mounted

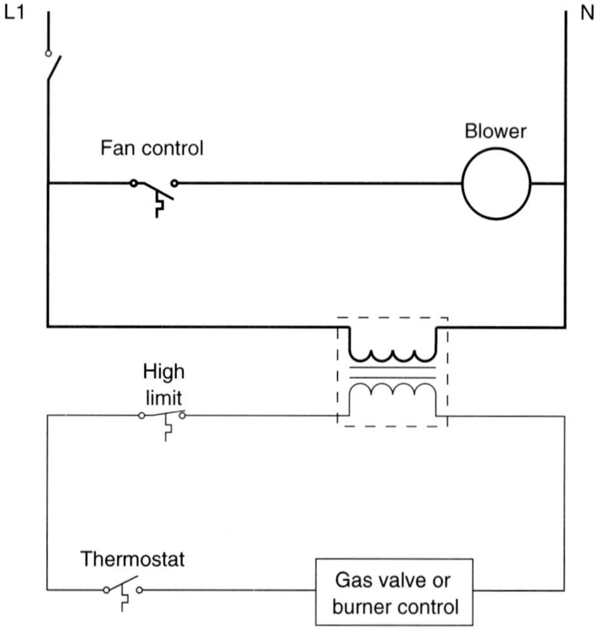

FIGURE 17–28 This basic system controls a single-speed fan according to plenum temperature. There is no provision for air conditioning.

FIGURE 17–29 When air conditioning is added to a basic furnace, a fan relay is used to energize the blower during the cooling cycle.

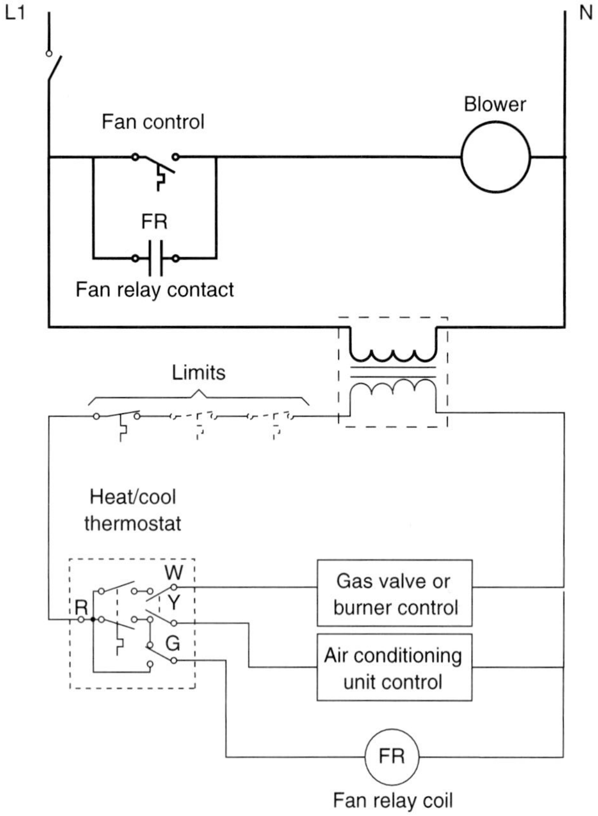

fan relay (see Figure 17–29). The relay contacts are usually in parallel with the heating-mode fan control. They turn the fan on immediately when the thermostat calls for cooling, and off again when the cooling demand is satisfied. Most heat/cool thermostats also provide a manual fan control at the thermostat. This switch energizes the fan relay to provide continuous fan operation. The relay has a 24 V coil, and contacts rated for the 120 V fan motor. These relays are readily available in several styles.

Two-Speed Fans

There are two types of two-speed systems. One provides continuous fan operation at a low speed, and then switches to high for heating (and cooling, if the system also has air conditioning). The second type of two-speed control provides different speeds for the heating and cooling modes.

The first system is generally better than single-speed operation. The low continuous fan speed improves air circulation in the building, and may increase overall efficiency by reducing stratification. It is a relatively simple system that can be used on older furnaces with basic controls. Variations of this control circuit can also be applied to multispeed fan motors.

The second type of two-speed system allows the heating speed to be selected for proper exchanger temperature rise, while the cooling system operates at a more optimal, higher speed. It does not provide continuous fan operation at a reduced speed. The fan may be operated continuously on the cooling speed by turning it on with the thermostat fan control switch. It may also be operated continuously on the heating speed, if the heating mode fan control has a manual fan switch.

One Heat/Cool Speed Plus Continuous Low. The first type of control is similar to single-speed operation and can be provided with standard fan controls when a special two-speed motor is used (see Figure 17–30). The low-speed terminal of the motor can be wired directly to the line or, preferably, through a manual shutoff switch. The high-speed lead connects to the fan control. This type of system is used with special-purpose two-speed resistance-start motors powering the blower through a belt drive. The motor may have six or eight poles for the low speed. The eight-pole version runs the blower at about 50% of the normal speed; the six-pole type runs at about 75%. The difference in the airflow produced by these two different speeds can be quite significant.

When the blower motor is a general-purpose design that cannot tolerate both speed terminals energized at the same time, a SPDT fan control can be used (see Figure 17–31). As an alternative to this special fan control, the standard fan control can be used to switch an SPDT

FIGURE 17–30 This control system provides continuous fan operation at a low speed, plus a higher speed for normal heating (and cooling) operation.

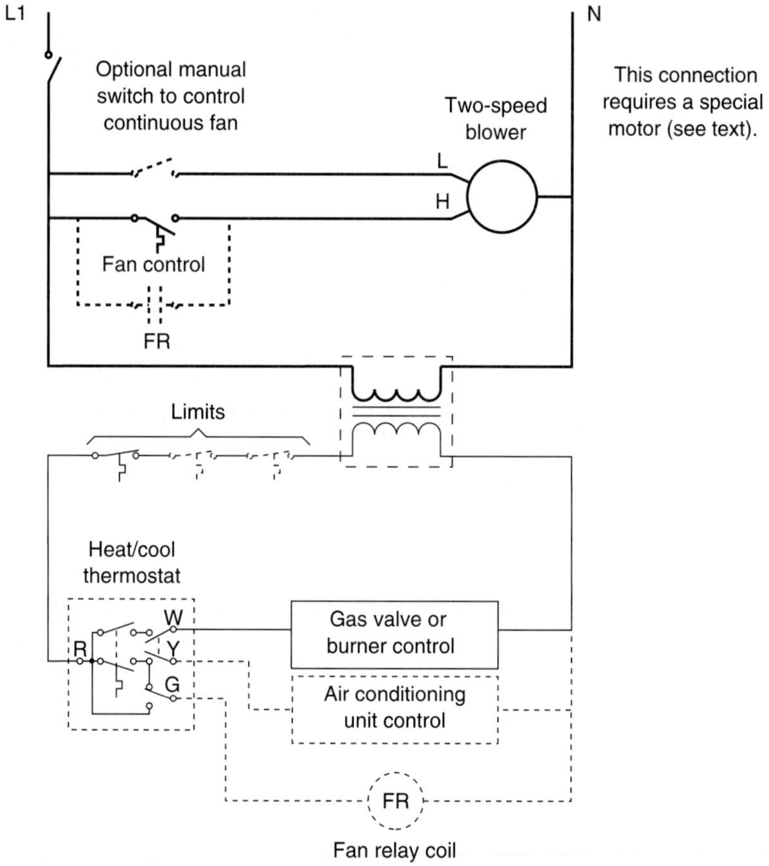

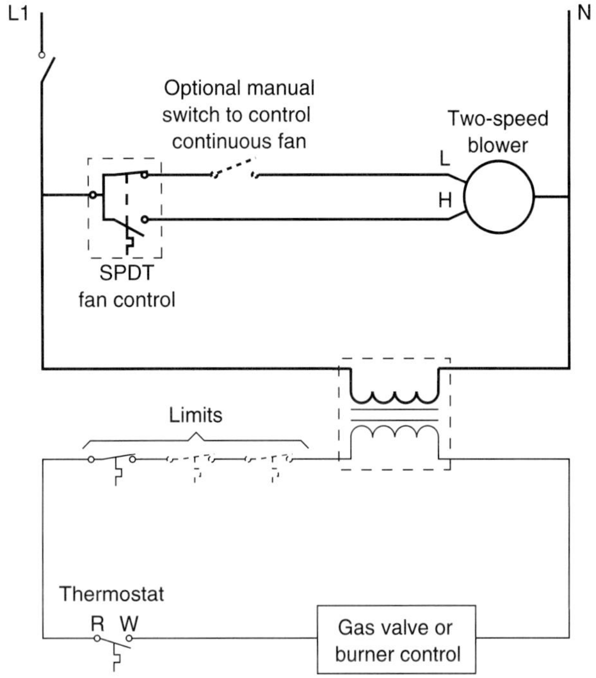

FIGURE 17–31 An SPDT fan control can be used to provide two motor speeds without a special motor.

relay that, in turn, selects the fan speeds. This relay connection is similar to some three-speed controls. Either of these options can also be used to provide a low continuous speed from a multispeed PSC or shaded-pole motor, because only one speed is energized at any one time.

Separate Heating and Cooling Speeds. This second type of two-speed operation—different heating and cooling speeds, but no continuous fan —is very common in newer furnaces. A standard fan relay with SPDT contacts can be used to switch any type of two-speed or multispeed blower motor (see Figure 17–32). Note that the heating-mode fan control contact is in series with the normally closed contact of the relay. It must not be connected in parallel, because the motor could be destroyed if the fan control contact is closed when the relay is energized.

Three-Speed Fan Control

To add a third continuous fan speed to a system with separate heating and cooling speeds, a second fan relay may be added (see Figure 17–33). The circuit can be built with relays that have either 24 V or 120 V coils, but the 120 V coil connection is probably the most common. For a 24 V relay, the fan control and the second relay coil must be connected into the 24 V circuit, instead of the 120 V circuit where it is normally located. The relay must have a least one set of SPDT contacts rated for the motor load, and they must be connected in series as shown in Figure 17–33. A parallel connection will lead to motor damage when both relays are energized.

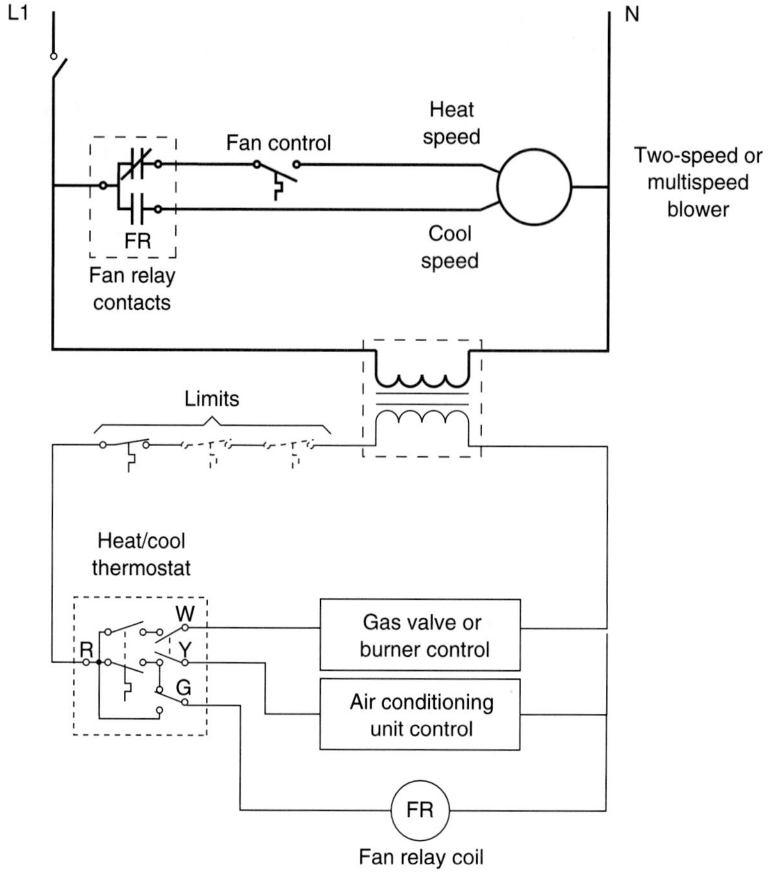

FIGURE 17–32 When a relay with SPDT contacts is connected as shown in this diagram, the system can provide different heating and cooling speeds with any type of two-speed or multispeed motor. There is no provision for continuous operation at a lower speed.

Note

On all systems with a continuous fan speed, it is highly recommended that a clearly identified manual switch be installed in the continuous motor speed line. This allows the fan to be completely shut off without shutting off the entire furnace.

Time-Based Fan Controls

All the control systems discussed so far use temperature-based fan controls to energize the fan motor at its heating speed. There is, however, a strong trend toward using time-based fan controls instead. These fan controls do not monitor plenum temperature; instead, they use a timer that is triggered by the operation of the burner. Some are modified heat-operated controls that use a resistor to heat the bimetallic strip. Newer designs are electronic timing relays with independently adjustable on-delay and off-delay settings.

With a time-based control, the 120 V fan control circuit is essentially the same as it is for a plenum-mounted temperature-operated circuit. However, there are numerous changes to the 24 V control system. Typical time-based fan controls have many connections, including both the line and common sides of the 24 V transformer secondary, the supply side of the main gas-valve, and the load side of the limit circuit (see Figure 17–34). Understanding

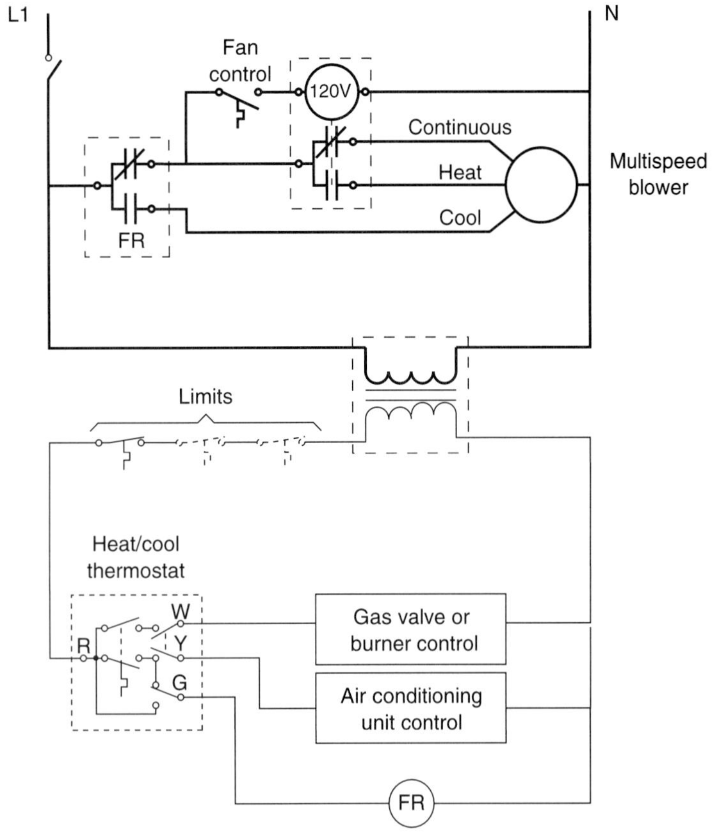

FIGURE 17–33 This three-speed fan control system provides separate heating and cooling speeds, plus continuous fan operation at reduced speed. It is suitable for any type of multispeed blower motor. This circuit is a popular add-on, for furnaces with separate heating and cooling speeds.

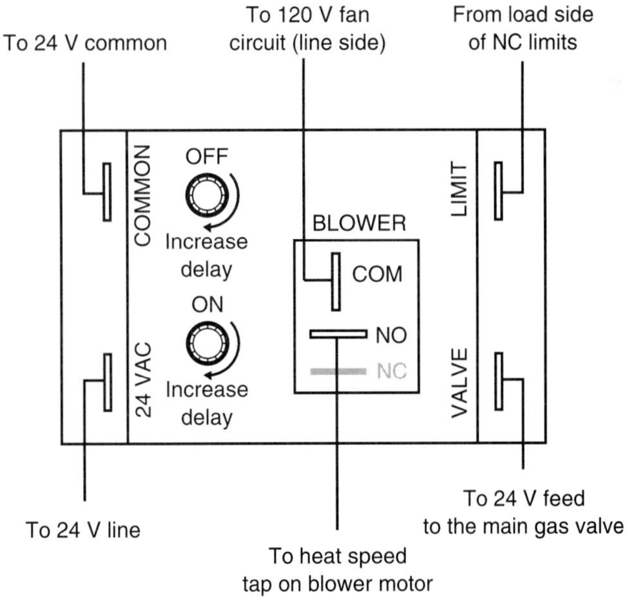

FIGURE 17–34 This pictorial is based on a common time-based fan control. The connection layout, the controls, and all markings are quite typical of these types of controls. The NC contact (terminal shaded in) is an option that is usually not provided.

the reasons for each connection requires an understanding of the control logic and the terminal markings on the control itself.

The fan control is an electronic device that requires a power supply for its own operation. This is the reason both sides of the 24 V transformer output have to be connected. It also has to know when a heating cycle starts and ends. This information is provided by the line from the main gas-valve. The valve connection never supplies power to the valve; it only monitors for voltage, which it uses as a signal to control the timing cycles (see Figure 17–35). When 24 V AC is applied to this terminal, the on-delay timer starts. After the time-delay period has passed, the control energizes an internal relay to start the blower motor. The relay contacts are brought out as the 120 V blower connections on the fan control. When the gas-valve shuts off, the voltage at the valve terminal goes to zero, signaling

FIGURE 17–35 A typical time-based fan control has four 24 V connections, plus a 120 V connection to the contact. It is important to realize that the connection from the control to the gas valve cannot energize the valve. Voltage at this terminal signals the start of a heating cycle.

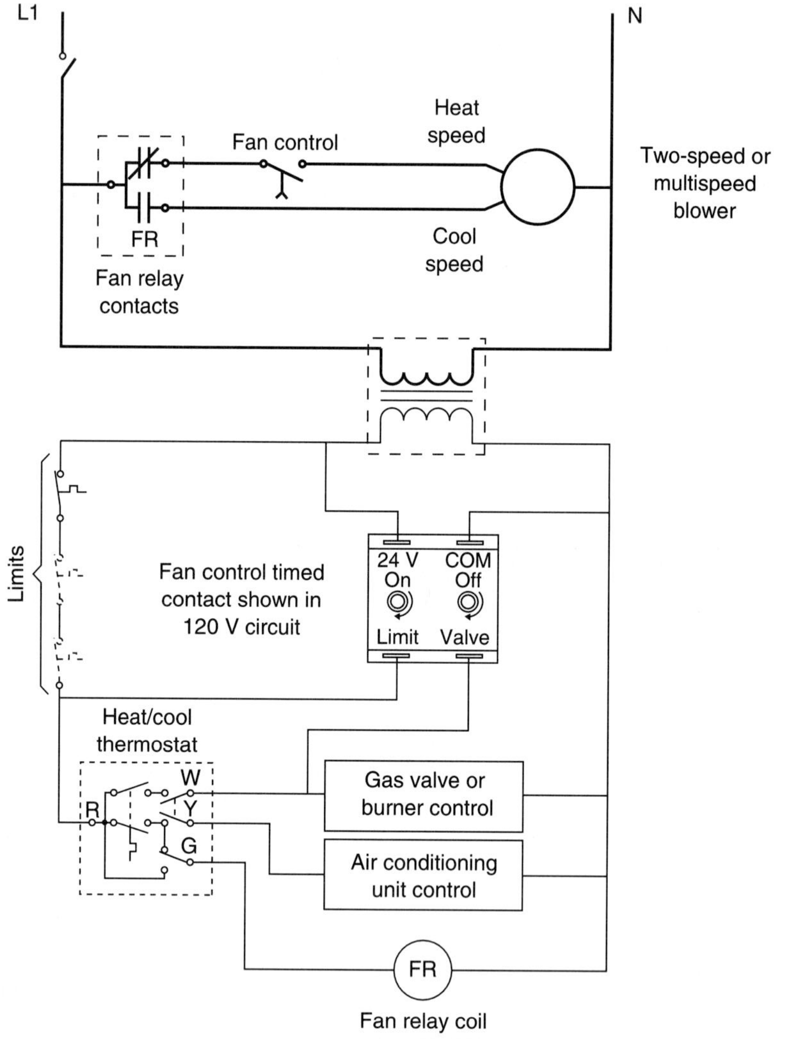

the control to start the off-delay timer. When that time period ends, the relay is de-energized, opening the contacts and shutting off the blower.

If the high limit opens, the fan control should at least attempt to run the blower. This explains the limit terminal on the control. On a traditional fan control, the fan contact is closed if the temperature is high enough to operate the limit. On timed fan controls, plenum temperature is not monitored, so a different approach is required. In the type of control described here, if there is no voltage at the limit terminal, the fan control immediately energizes the blower. The limit terminal of the fan control must be energized at all times for normal operation. The limits perform their normal function of shutting down the burner on any limit operation.

Note

The fan contacts on this particular make of fan control are marked in a somewhat confusing manner (see Figure 17–34). The terminal marked COM is the 120 V "line" terminal of the fan control. This may come direct from the 120 V line (not the neutral, as COM may suggest to some), or possibly from the normally closed fan relay contact or other control. The NO marking on the other 120 V terminal signifies a normally open contact and is the load side of the control. It feeds power to the heating speed of the fan motor. This control may also be available with an SPDT contact set, in which case the markings would be COM, NO, NC.

There are many variations in fan control, and no doubt more will appear in the future. In many furnaces the individual components are combined on one or two control boards. This simplifies internal wiring, but forces replacement of entire boards rather than individual components when parts fail. No matter how a circuit is arranged, the key to successful connection, testing, and troubleshooting of any control system is first knowing what the circuit must do, and then figuring out how it does it.

Electronic Air Cleaners and Other Auxiliaries

When a furnace is equipped with auxiliary equipment—such as an electronic air cleaner, humidifier, or similar device that must be switched on and off with the blower—the circuit required for multispeed fan control becomes more complex. With a single-speed blower motor, this equipment can be connected in parallel with the motor. For most two- or three-speed fan systems, such a connection results in damage to the motor and/or the auxiliary equipment.

One approach used for this type of equipment is to replace the SPDT fan relay with one that has DPDT contacts (two sets of SPDT). The relay connections are unusual, but functional. The auxiliary equipment is energized in the heating or cooling speed, but the two are never connected simultaneously (see Figure 17–36).

An alternative to the DPDT relay system is to field-install a "sail switch" in the airstream of the fan. The switch must have a normally open contact that is rated for 120 V AC, at a current that is at least adequate for the auxiliary equipment. The sail switch operates on airflow, so it can be completely isolated from the rest of the furnace circuits. Sail switches allow an air cleaner to be safely connected to virtually any type of system; some makes have them built in.

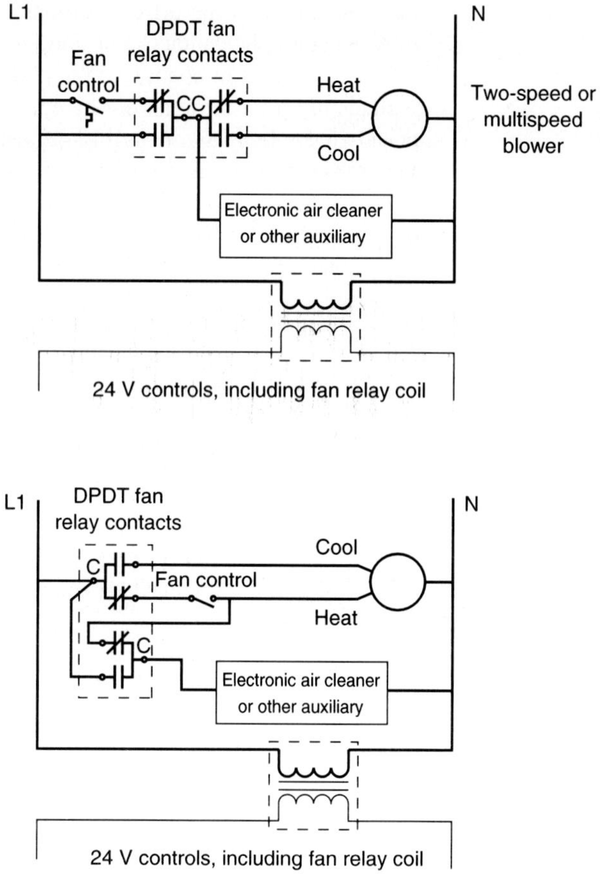

FIGURE 17–36 These diagrams represent two different power circuit connections for DPDT fan relays used to provide power to an electronic air cleaner. In the top circuit, power for a humidifier can be tapped off between the fan control and the NC relay contact, if desired.

Fan Centers

A fan center is often used in place of a separate relay and a control transformer (see Figure 17–37). It has a 120:24 V Class 2 transformer, and a relay mounted in a single enclosure. The relay has a 24 V AC coil and one set of horsepower rated SPDT contacts. Terminals are provided for the thermostat conductors, and leads are provided for the 120 V supply and relay contact outputs. The base of the fan center mounts on a standard electrical box. The 120 V connections are brought out through a bushed hole in the back of the device, where they naturally end up inside the box. This keeps all the 120 V connections inside an approved enclosure, in compliance with electrical codes.

The W and Y terminals on the fan center are for convenience only, to allow all the thermostat conductors to be terminated at one location. They have no internal connections.

On some models, the 120 V line terminal is internally connected to both the common of the relay contacts and the supply side of the internal transformer (see Figure 17–38). Other designs have separate connections for the contact common and the transformer line. This second type is a little more flexible to use.

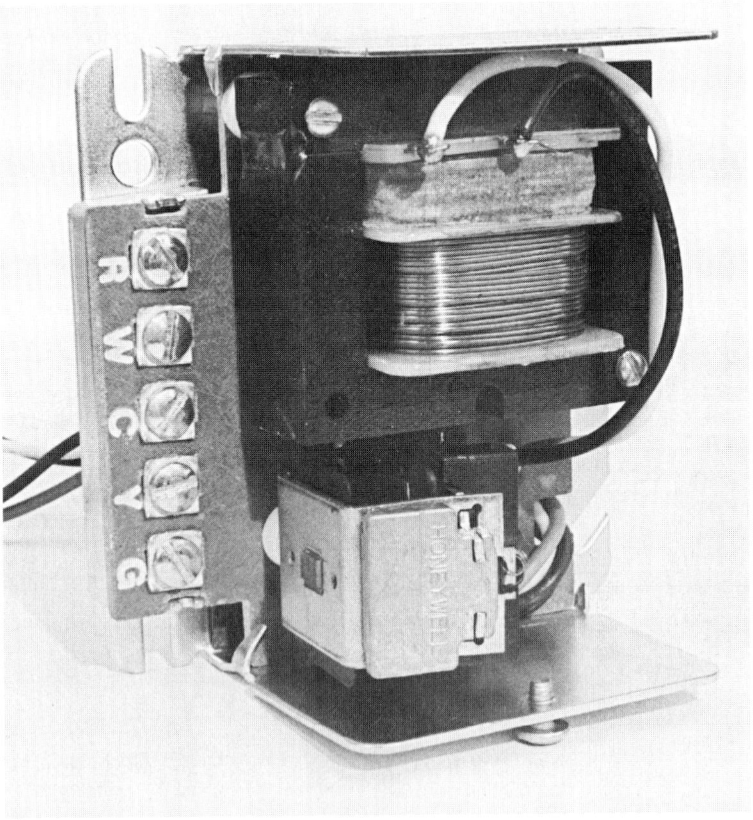

FIGURE 17–37 This is a typical fan center. The 120 and 24 V transformer windings are visible at the top; the relay is at the bottom. The 24 V connections are made to the terminal strip on the left. W and Y terminals are provided as a junction point only; they are not internally connected. The 120 V connections are provided through color-coded leads. The base of the fan center is designed to mount on a standard electrical box. This mounting places all the 120 V connections inside the box, as required by code.

Warning

Incorrectly retrofitting a fan center with the relay contacts internally connected to the transformer creates a serious hazard. Using the existing 120 V high limit and connecting the relay contact to the fan motor creates a circuit that works and looks safe, *but is a serious fire hazard* because it can bypass the high limit.

Examine Figure 17–39 carefully. The problem occurs when the thermostat fan switch is set to manual for continuous air circulation during the heating season. This picks up the fan relay, closing the NO contact. During each heating cycle, the fan contact closes in the fan/limit control. This creates a circuit that bypasses the high limit. There is a connection from the line to the control transformer through the fan control contact and the relay contact. Any problem in the circulating air system will cause the furnace to overheat, but the burner will not shut down. A fire is a very real possibility.

The easiest solution to this problem is to rewire the high limit into the 24 V thermostat circuit. The jumper between the fan contact and the high limit would, of course, have to be removed. Alternately, the fan center could be exchanged for one with separate terminals for the transformer and relay contacts. The 120 V limit can then be wired directly to the transformer, with no other connections to the fan control.

FIGURE 17–38 In the connection shown here, a fan center provides 24 V to the heating and cooling control systems, and a fan relay is used for two-speed fan operation. This fan center has a single 120 V line connection, rather than separate leads for the transformer and the con-

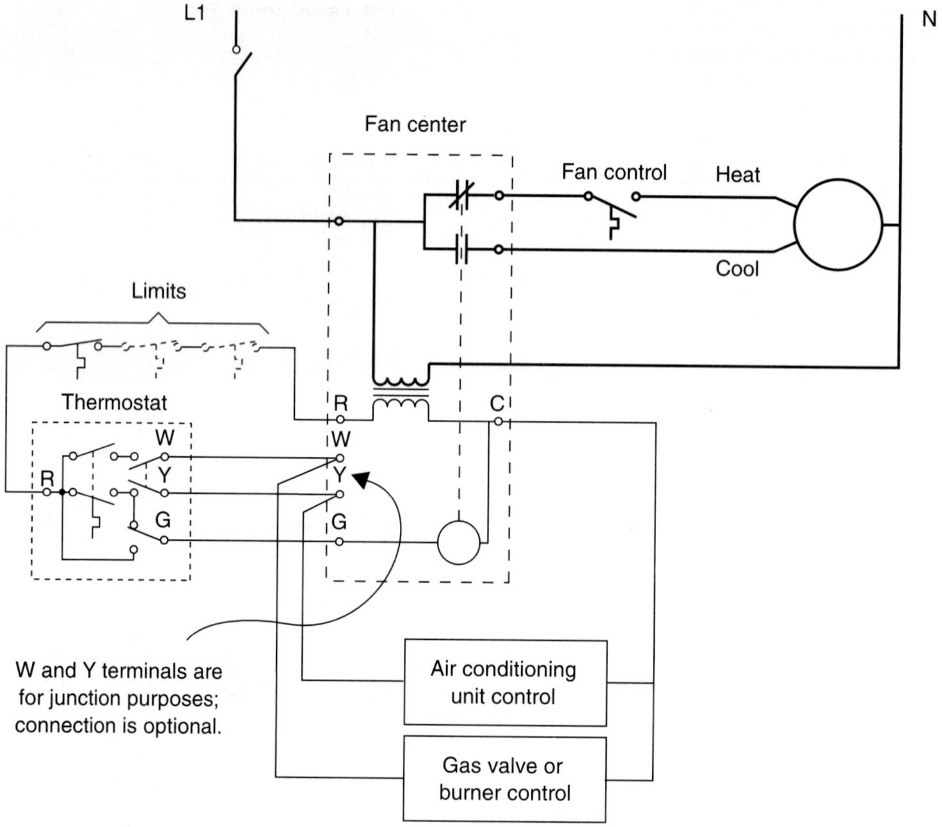

Fan Centers and Separate Transformers

Note

It is possible—but not necessarily a good idea—to use a fan center and a separate control transformer in the same unit. In most cases, it is better to remove the transformer and run the entire control system off the fan center. However, if there is a valid reason to retain both supplies, caution is advised.

When both a fan center and a separate 24 V transformer are used in the same furnace, care must be taken to avoid connecting the secondary of the separate transformer to that of the transformer in the fan center. If the transformer's secondaries do become parallel connected, several different problems can arise, depending on the circumstances. The safety limits may be bypassed, one or both transformers may burn out, or a fire hazard may exist if a fault occurs in the thermostat wiring.

FIGURE 17–39 The connection shown here contains is a very dangerous error that is easily missed. See text for an explanation and suggested methods of solving the problem.

WARNING!
This is not a safe circuit. See text.

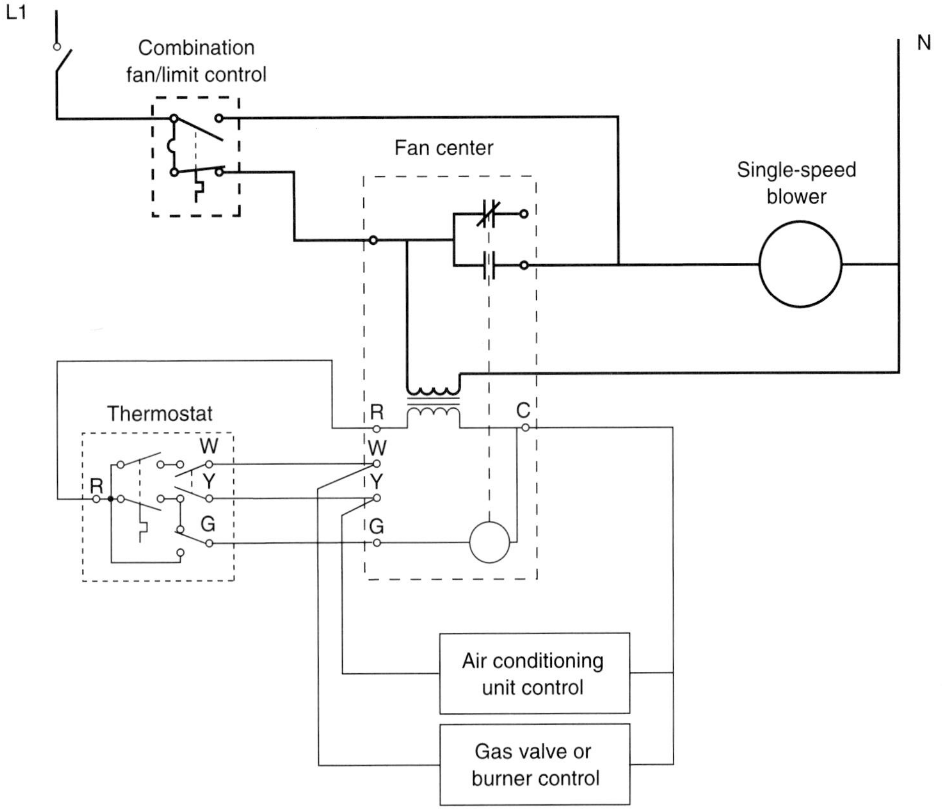

The usual solution is to remove the separate transformer and use the fan center to supply 24 V to both the heating system and the air conditioning control. The VA rating of the fan center transformer should be checked to ensure it can supply enough current.

If both the transformer and the fan center are kept, the best way to make a safe connection is to use a thermostat with separate heating and cooling circuits (see Figure 17–40). This type of thermostat has two R terminals, usually marked R_H and R_C. The units may be shipped with the two R terminals jumpered together to allow the thermostat to be used with a single 24 V supply. For dual supplies, the jumper must be removed to isolate the heating and cooling circuits. A minimum of five wires is required between the thermostat and the furnace.

Properly connected, the fan center supplies the cooling equipment controls, including the fan relay, and the heating system operates from a separate 24 V transformer. There are no electrical interconnections between the two systems.

FIGURE 17–40 A thermostat with separate R terminals for the heating and cooling sections allows two separate power supplies to be used, with no connection between them.

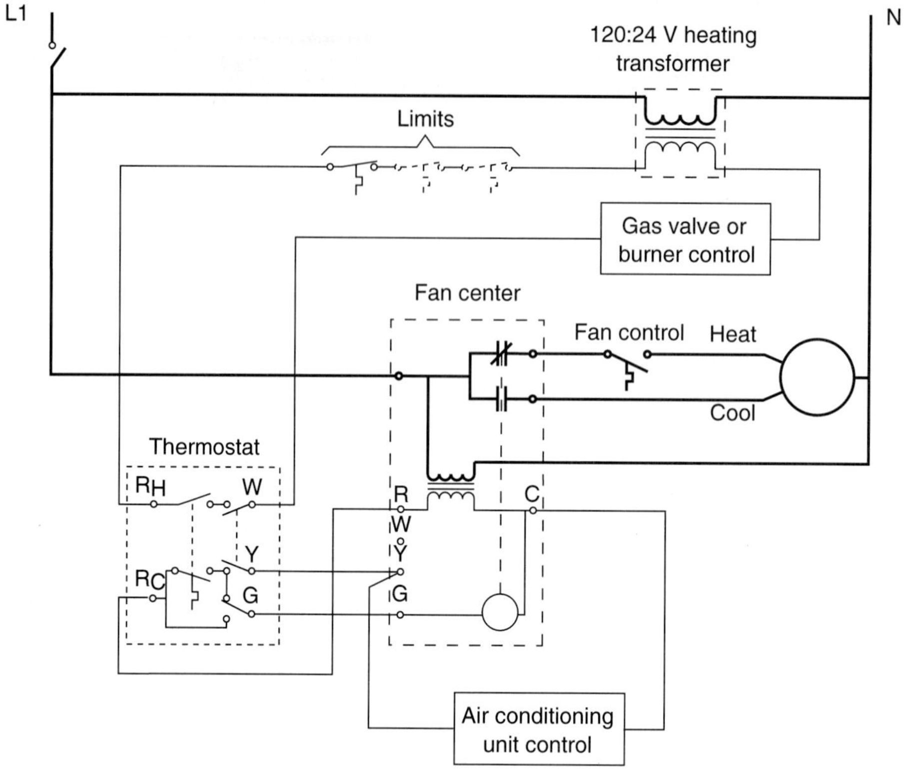

17.11 Computer-Based Controls

Some furnaces are equipped with a single circuit board or control module that uses a microprocessor or integrated logic circuits in place of separate control components. All the control logic, timing, and monitoring functions are handled by a single controller. The advantage is that the manufacturer can provide more advanced features at a lower cost. The disadvantages are unrecognizable control devices, service procedures that may be alien to some service personnel, and the need to swap complete boards or modules to repair minor problems.

Single-module systems are much like oversized ignition controls. They are connected with custom wiring harnesses, which can make the associated control components essentially unique to a single furnace model (see Figure 17–41). Replacement parts must be an exact match. This reduces the chance of errors in field wiring and makes replacement somewhat simpler, but it also creates problems if the parts are not stocked locally or are excessively overpriced.

Open-circuit-board systems (see Figure 17–42) provide more access to the controls and may be easier to test. But the fact that the board is open to inspection does not make it possible, legal, or practical to do board-level component replacement or repairs.

FIGURE 17–41 This single module provides all the control functions for this furnace. Most of the connections are made with plug-in wiring harnesses. Fan delay times are set with a pair of DIP switches on the bottom right. The board has built-in diagnostics to aid in troubleshooting.

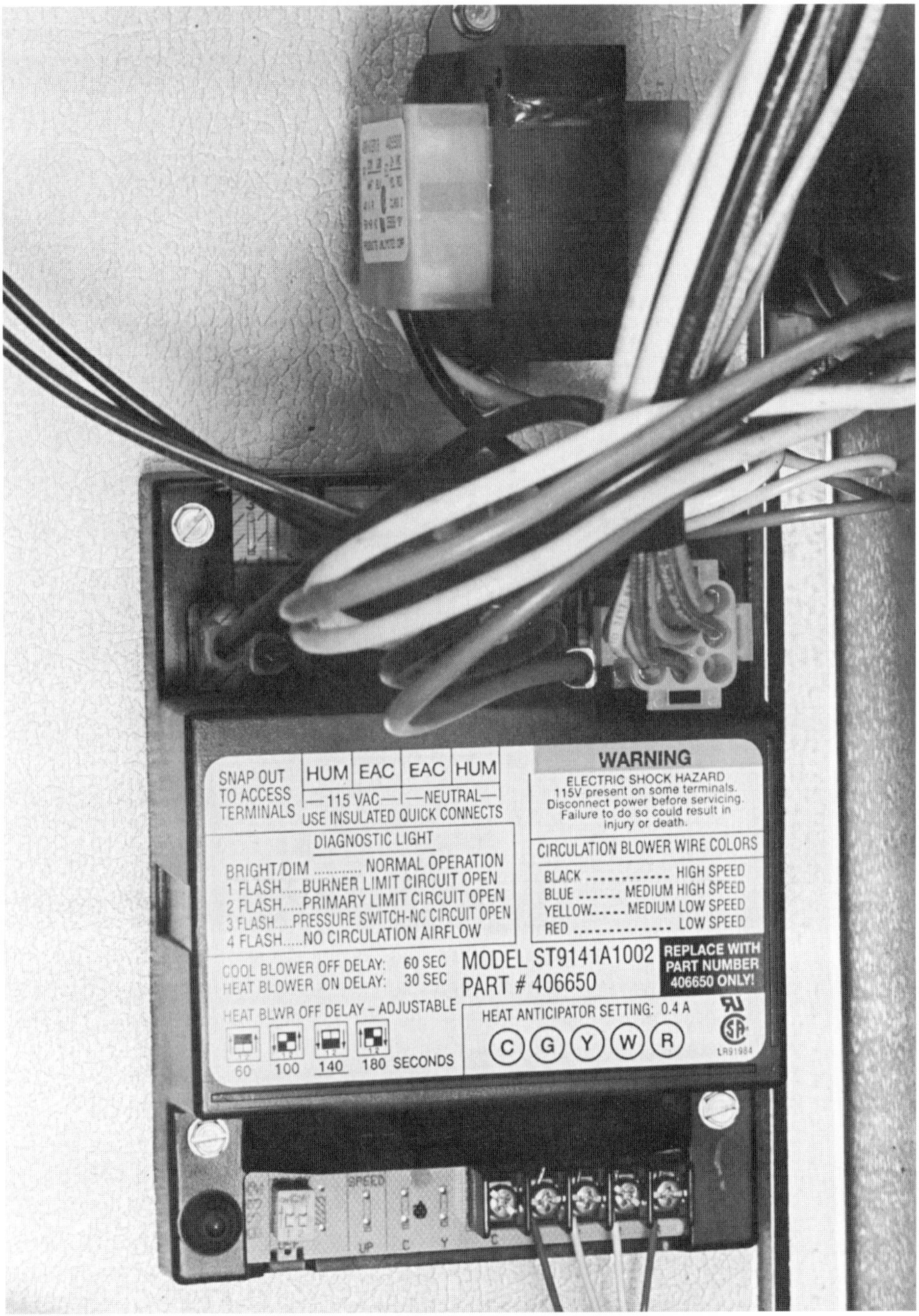

FIGURE 17–42 This circuit board performs the same tasks as the module in Figure 17–41. Note the fan relays (top right) and the automotive-style fuse (top center). The fan delay times are set with a jumper (bottom right). A self-test diagnostic system is initiated through the two terminals in the bottom right corner.

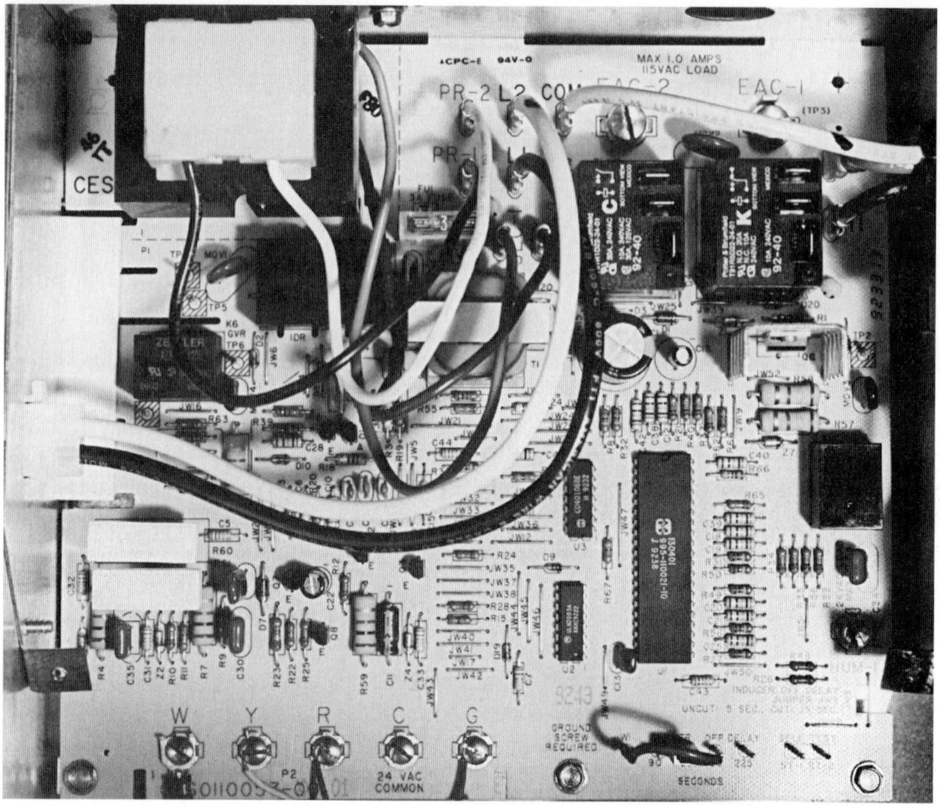

These types of systems require caution when using measuring instruments and careful attention to the manufacturer's literature. The control components external to the circuit board are often quite ordinary, as are some of the devices on the board itself (the relays for example). The controls on the board have to do the same thing as the separate components of a more conventional system, so the external circuits serve very recognizable purposes and are connected in quite similar ways.

The drawing in Figure 17–43 is a ladder-style schematic for the system built around the circuit board shown in Figure 17–42. Note that all the external wiring is shown, but the board wiring includes only enough detail to indicate the connections between the board mounted devices and the external circuits. The relay coils are not shown—only the contacts. The coils are driven by the electronic logic circuits, so there is no simple way to trace how they are energized. This particular board has everything necessary for the complete furnace. All wiring terminates on the board or through an edge connector on one end.

Despite the differences between this system and one with separate components, the individual circuit loops can be followed just as logically. A provision has even been made to test each input line right on to the board itself. Small test points, marked TP1, TP2, etc.,

FIGURE 17–43 This is a ladder schematic for the control system built around the circuit board in Figure 17–42. There are two circulating air blower relays. The system uses an induced-draft fan, hot surface ignition, and a separate flame rod to monitor the main burner. The flame rollouts and the high limit are in the R line ahead of the thermostat. The auxiliary and vent limits and the air proving switch are in the W line after the thermostat. All onboard relays are controlled by the CPU (central processing unit—possibly a microprocessor or a logic IC of some type).

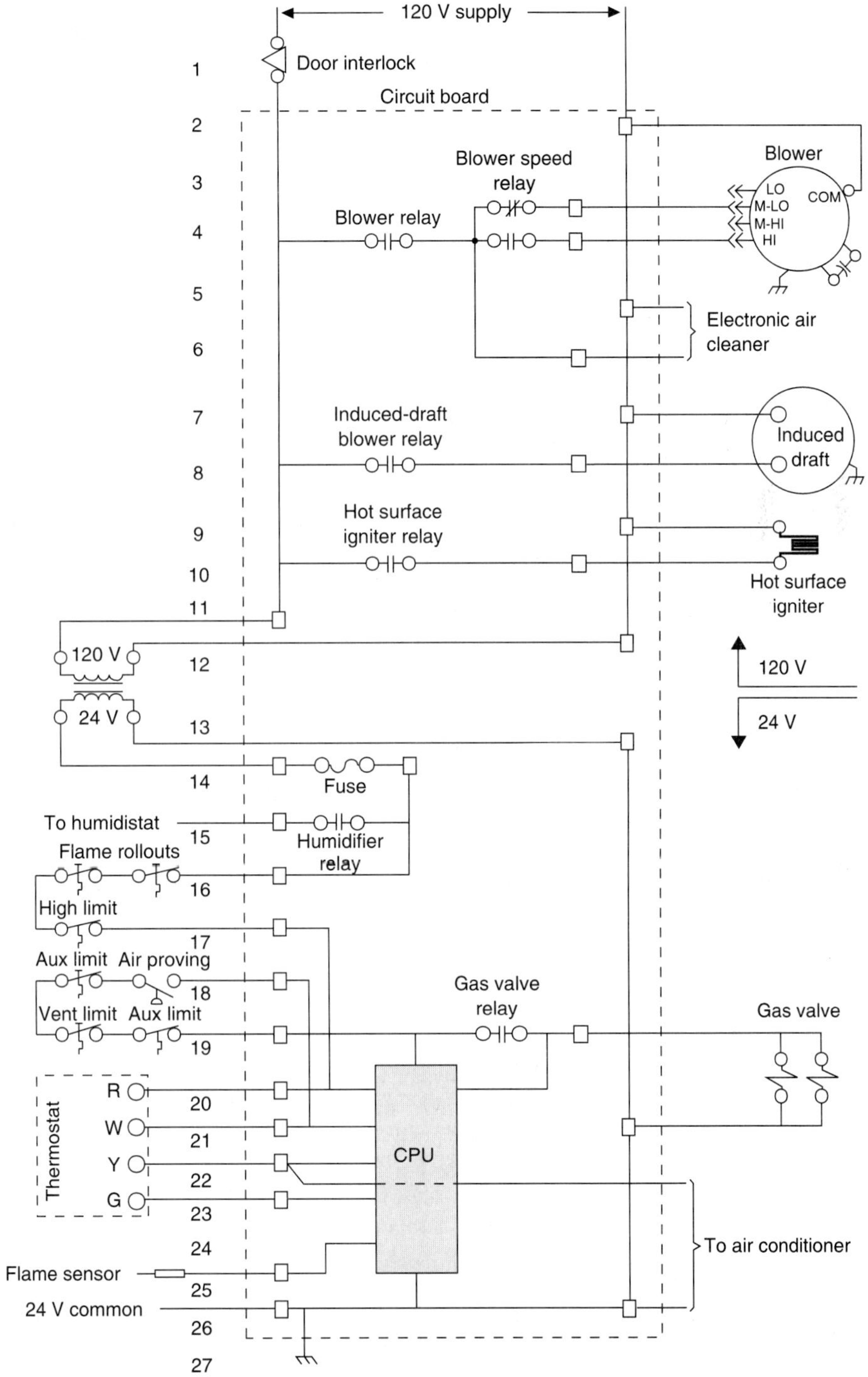

provide access to the board end of critical circuits, making it possible to test them right to traces of the circuit board.

The following are some dos and don'ts that may help with equipment of this type:

Do:

- read the literature. The manufacturer provides detailed explanations of system features, operating sequences, and timings. The literature will also describe test procedures, including how to initiate and interpret built-in self-test routines.
- test for voltage levels with the controls energized, instead of using resistance tests with it off.
- use a good-quality digital meter with an input impedance of 10 MΩ or more.
- test input points at the terminals for proper voltages from auxiliary equipment (such as thermostats, limits, and the transformer) before testing on the circuit board.
- use marked "test points" (and only test points) for tests on the circuit board. They are identified on the drawings and on the board.
- test carefully—working methodically from available drawings and circuit markings, not at random.
- perform "curiosity tests" while the system is in working order. Knowing what is normal is invaluable when a system fails.
- clean all contact points when a board has been removed or is being replaced. Bare copper- or solder-plated edge connections can corrode or become contaminated. The extra resistance at these connections can cause problems with an otherwise healthy board.

Don't:

- assume the fault is on circuit board; most problems will be something simpler.
- poke around with your meter, hoping to get lucky.

17.12 Oil Burner Controls

This section provides examples of the controls used for oil burners. They do not represent all the possible variations, but should provide a reasonable basis for understanding typical oil burner systems. Oil burner controls differ from gas systems in a number of ways. Gas fired burners are relatively simple and clean burning, but gas is potentially more explosive than oil. In a gas fired system, fuel delivery is controlled by an electrically operated valve small enough to be easily be operated from a 24 V supply. The flame is ignited by a standing pilot, an electric spark, or a hot surface igniter. A gas flame is clean enough and at a low enough temperature to be monitored with a conductive probe using a flame rectification circuit.

Fuel oil is much less explosive, but the potential for a fire is much the same. Failure to ignite can cause the burner unit to flood the fire box and, possibly, the surrounding area with fuel oil. Oil burners are more complex, and the flame is hotter and dirtier. This makes indirect flame monitoring a necessity. In either type of furnace, loss of air circulation can result in dangerous heat-exchange temperatures. An oil burner must therefore have a high limit similar to that of a gas furnace.

In a typical oil burner, the fuel is pressurized by a motor-driven pump and sprayed through a small orifice into an airstream. The burner is in the form of a "gun" that incorporates a high-pressure oil pump, a forced-draft blower, and a spark ignition system. The burner unit cannot be operated at 24 V; therefore, although the thermostat circuit may be 24 V, much of the system operates at line-voltage, typically 120 V AC.

Oil Burner Operation

The burner is cycled on by starting the burner motor, opening the fuel valve (if one is used), and turning on the spark igniter. The burner flame is monitored either by a light-sensitive cell that has a line of sight into the fire box, or a bimetallic sensor located in the "stack" that leads to the chimney. Typical systems that sense the light of the flame use a photoconductive device made from cadmium sulfide. This "CAD cell" is located in the cool end of the oil "gun," where it can "see" the flame. The resistance of the cell varies with the amount of light striking the grid of material inside. In darkness, its resistance is very high. In the presence of light, it drops to a much lower value. CAD cells are simple and reliable, but they must be clean to work properly, because a film of soot will block the light.

A common physical layout for this type of burner is shown in Figure 17–44. The ignition transformer is hinged at the back so it can be swung open to allow access to the transformer contacts and the CAD cell. Typical cells plug into a socket to allow easy removal and replacement. The ignition transformer of a typical oil burner is much larger than those used for small gas burners. It has a 120 V primary instead of 24 V, and the secondary is typically rated at 10,000 V. The spark electrodes are also much heavier, and the spark gap is wider.

Note

The spark ignition system in an oil burner must withstand more demanding conditions than for a gas burner, and is therefore more likely to require service. Oil flames are hotter and dirtier than gas, and many igniters run continuously while the burner unit is operating, so the operating hours can be very high. The proper placement of the electrodes relative to the oil nozzle, and the width of the spark gap, are quite critical. Always consult manufacturers' recommendations when these require adjustment.

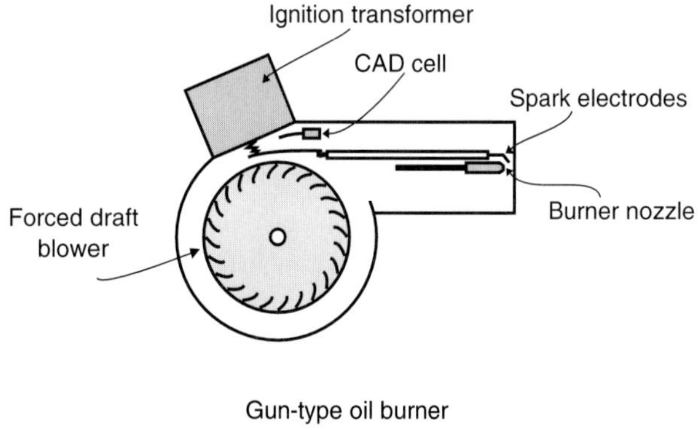

FIGURE 17–44 On a typical gun-type oil burner, the oil pump is connected on the same motor shaft as the forced draft blower. The pressurized oil is atomized by the burner nozzle into the airstream, where it is ignited by spark electrodes close to the nozzle opening. The ignition transformer tips back to allow access to the CAD cell and the back of the spark electrode assembly.

A primary function of any burner control is preventing unsafe operation. All burners have delays between the "trial for ignition" and normal flame detection. The controller has to allow the burner enough, but not too much, time to light. If the burner fails to ignite, unburned fuel accumulates in the system, causing a serious fire hazard. In a typical oil burner control, the safety system is based around a safety lockout timer that starts on a call for heat, and a flame sensor that stops the timer when flame is detected. Each time the burner operates, there is a race between the flame detector and the safety lockout timer. If the flame detector wins, the burner operates normally; if the timer wins, the power to the burner is cut off, and the system will not restart until it has been manually reset.

The safety lockout timer is usually a thermal device using a heater and a bimetallic strip that operates the contacts. When flame is detected, the timer heater is bypassed, stopping the timing cycle before it is complete. Some oil burner controllers have a second timer, or additional relay contacts, that shut off the spark ignition transformer after the flame is detected.

Oil Burner Control Examples

Refer to Figure 17–45. This system has continuous ignition and uses a tapped transformer to provide two different circuits inside the control.

FIGURE 17–45 This oil burner control provides continuous spark ignition and uses a CAD cell to monitor the burner flame. See text for explanation of circuit operation.

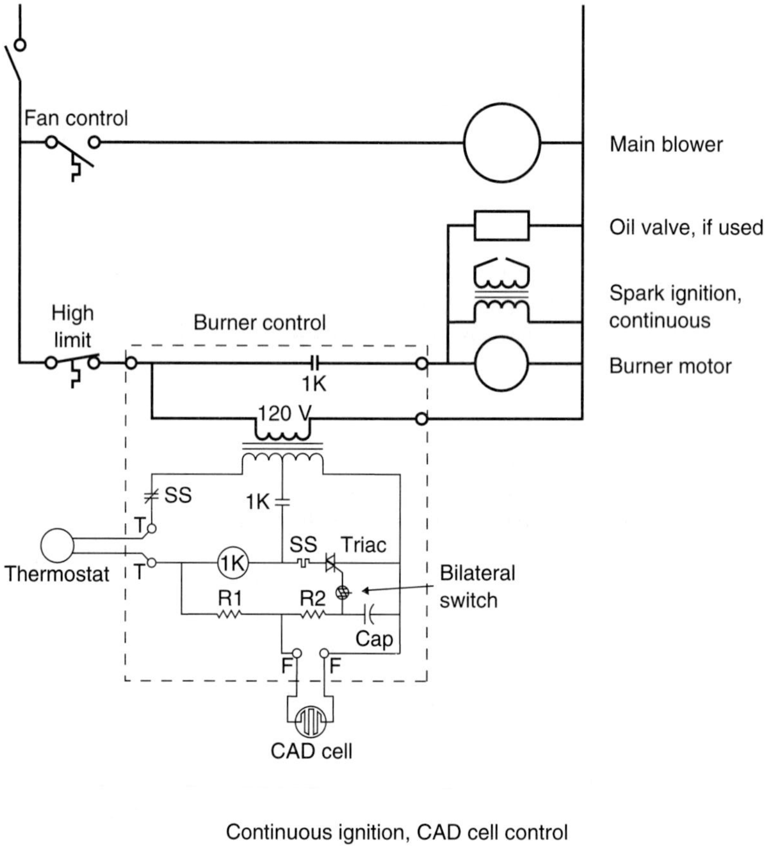

When the thermostat calls for heat, the capacitor charges through R1 and R2, triggering the bilateral switch into conduction. This in turn provides a circuit for the gate of the triac. The triac turns on, essentially acting as a contact in the line containing the 1K relay coil and safety switch heater (SS). The two 1K contacts close to start the burner and ignition, and to connect the 1K coil across one part of the transformer output and the safety switch heater across the other part.

Assuming that the burner lights, the CAD cell will see the light of the flame. When the CAD cell sees light, its resistance drops dramatically, shunting out R2 and the capacitor and shutting off the bilateral switch and triac. The safety switch heater is now out of the circuit, and the burner runs until the thermostat is satisfied.

If the burner fails to light, the safety switch heater will continue to heat until it opens the safety switch contact, SS, and shuts down the burner circuit. It locks out on a manual reset device.

If the burner lights and then fails, the CAD cell resistance will rise, the triac will turn on, and the safety switch heater will be back in the circuit.

In the example in Figure 17–46, ignition is intermittent. The transformer in the controller has two secondaries, one of which is tapped. When the thermostat calls for heat, the initial circuit is through the 1K coil, the safety switch (SS) and T1 NC contacts, and the SS heater and NC 2K contact. This applies full voltage of that transformer secondary across the 1K coil and the SS heater, in series. This picks up 1K, and a 1K contact connects its coil across one part of the transformer and the SS heater across the other part. A second 1K contact starts the burner.

When the CAD cell sees flame, 2K picks up, opening the SS heater circuit and energizing the heater of the second thermal timer, T1. This second timer circuit has its own transformer secondary. When T1 times out, the ignition transformer is de-energized, and the 1K coil is connected in series with the T1 heater circuit across the combined voltages of both transformer secondaries. This keeps both 1K and T1 energized. The burner continues to operate normally, without spark ignition, until the thermostat is satisfied.

If no flame is detected, the SS heater trips the contacts, and the burner locks out. If there is a flame failure after a normal startup, 2K will drop out, putting the 1K coil and the SS heater back on separate parts of their transformer secondary. The T1 heater is also de-energized. Assuming T1 cools before SS times out, the ignition transformer will be energized in an attempt to relight the burner.

The example in Figure 17–47 includes provisions for air conditioning. There are connections for a heat/cool thermostat and a separate fan timer, which is used in place of a conventional fan control operated from plenum temperature. Ignition is continuous, and the control transformer secondary has two taps.

When the thermostat calls for cooling, it energizes the air conditioning control and the fan relay. The fan control circuit is shown with a two-speed fan, but it could also be used with a single-speed unit. The burner is not energized (no power at the W terminal of the thermostat), and the system uses the full secondary voltage (24 V) to operate the air conditioning control and the fan relay.

When the thermostat calls for heat, 1K is energized through the SS NC contact, the SS heater, and the NC 2K contact across the full secondary voltage. The 1K contact in the transformer tap splits the secondary voltage, part to hold in the 1K coil and the rest across the SS heater. When the CAD cell resistance drops due to light from the flame, 2K picks up, disconnecting the SS heater and energizing the fan timer across part of the secondary.

FIGURE 17–46 This oil burner control uses conventional components to provide intermittent spark ignition. A thermal timer is used to open the circuit to the spark igniter after an appropriate delay.

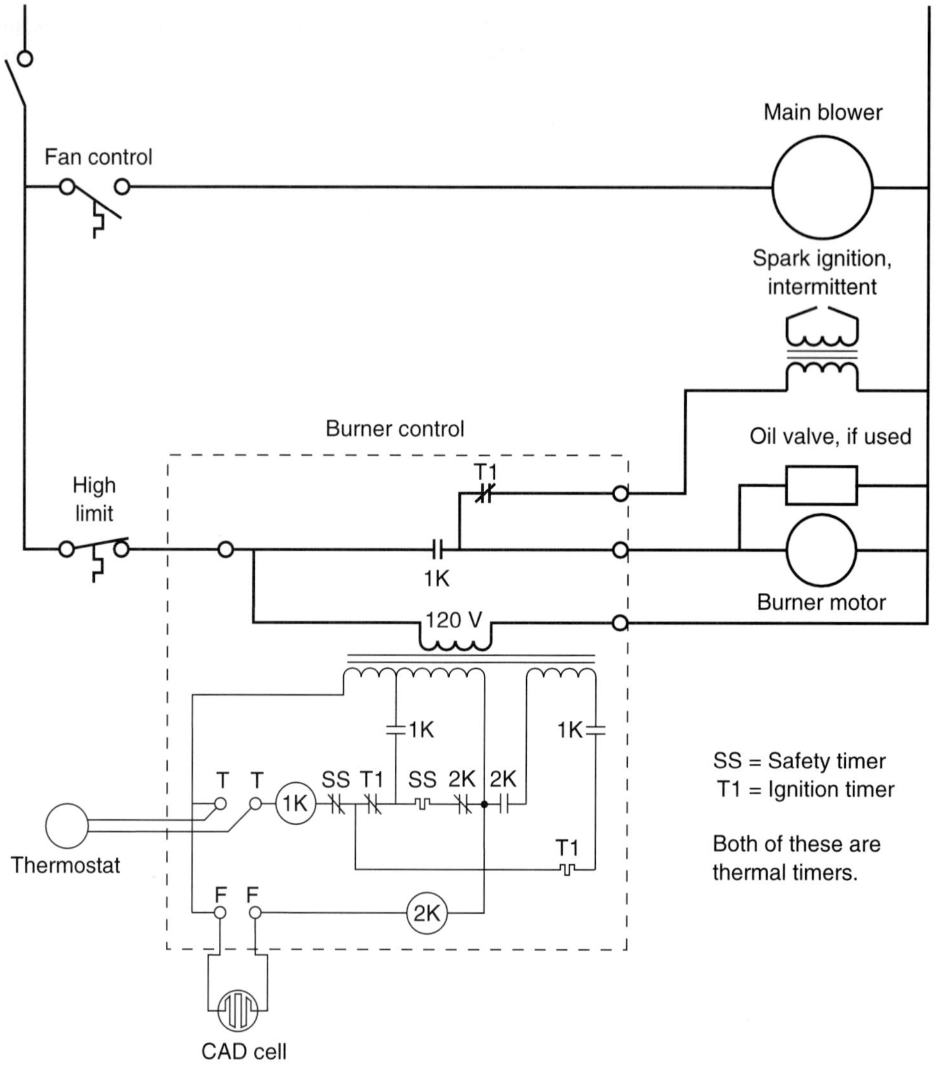

Intermittent ignition, CAD cell control

The fan timer energizes the fan after some delay, and the furnace operates in a normal heating cycle.

If the burner starts and then fails, 2K will drop out, reenergizing the SS heater. The status of the fan will depend on the cooling time required to reset the fan timer contact. This is the normal fan off-delay, so it could be a minute or more.

FIGURE 17–47 This control has provisions for a two-speed blower motor and air conditioning. A separate timer is used to provide the necessary delay for the circulating air blower.

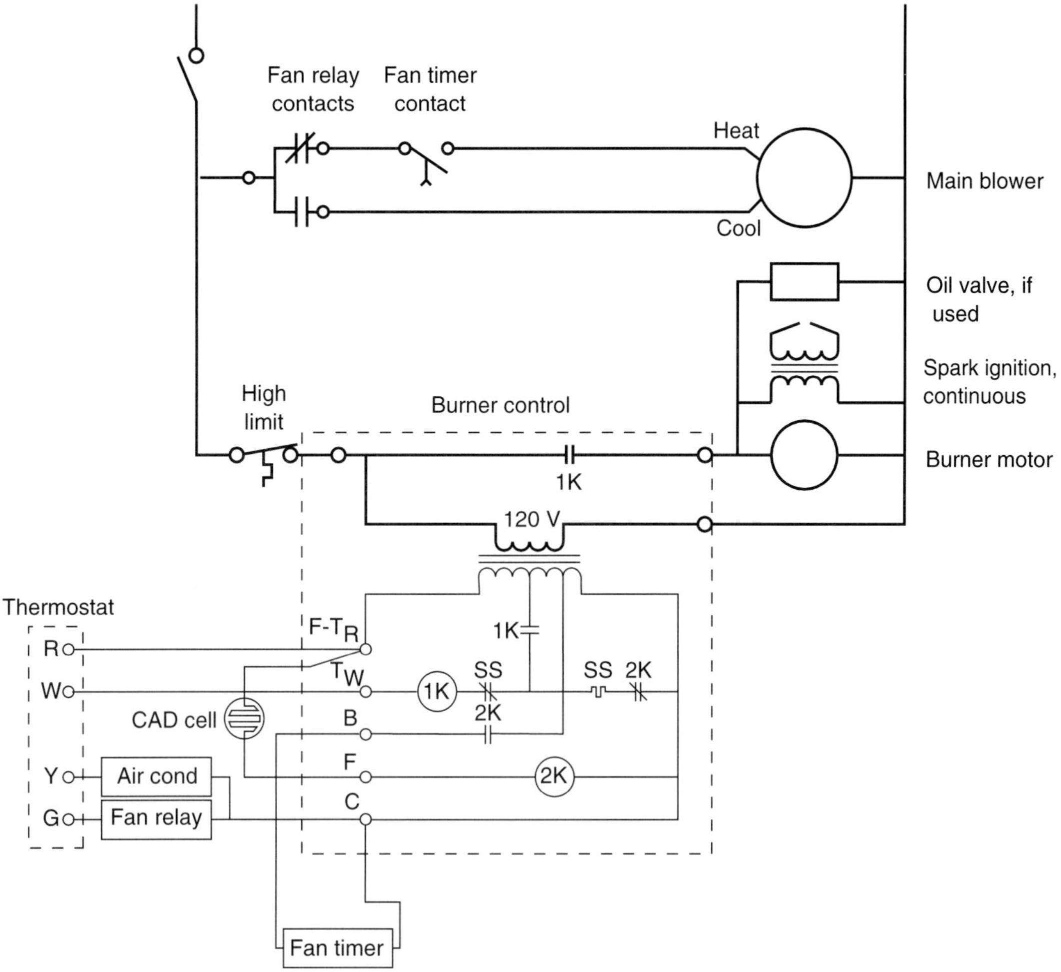

Continuous ignition, CAD cell control with connections for air conditioning

In the example shown in Figure 17–48, the controller uses a remote, stack-mounted temperature control with SPDT contacts. A normal operating sequence starts with a call for heat, energizing 1K through the NC T1 contact, the stack detector contacts (cold position), the SS heater, and, on the back side of the coil, the NC SS contacts. With the 1K contacts closed, the burner and igniter are energized, and the T1 timer heater is connected in parallel with the SS heater. When the temperature of the stack rises to the operating point of the stack switch, the contacts change to the hot position, disconnecting the SS

FIGURE 17–48 This controller senses flame with a stack-mounted temperature sensor. The sensor is an SPDT device that is separate from the main control.

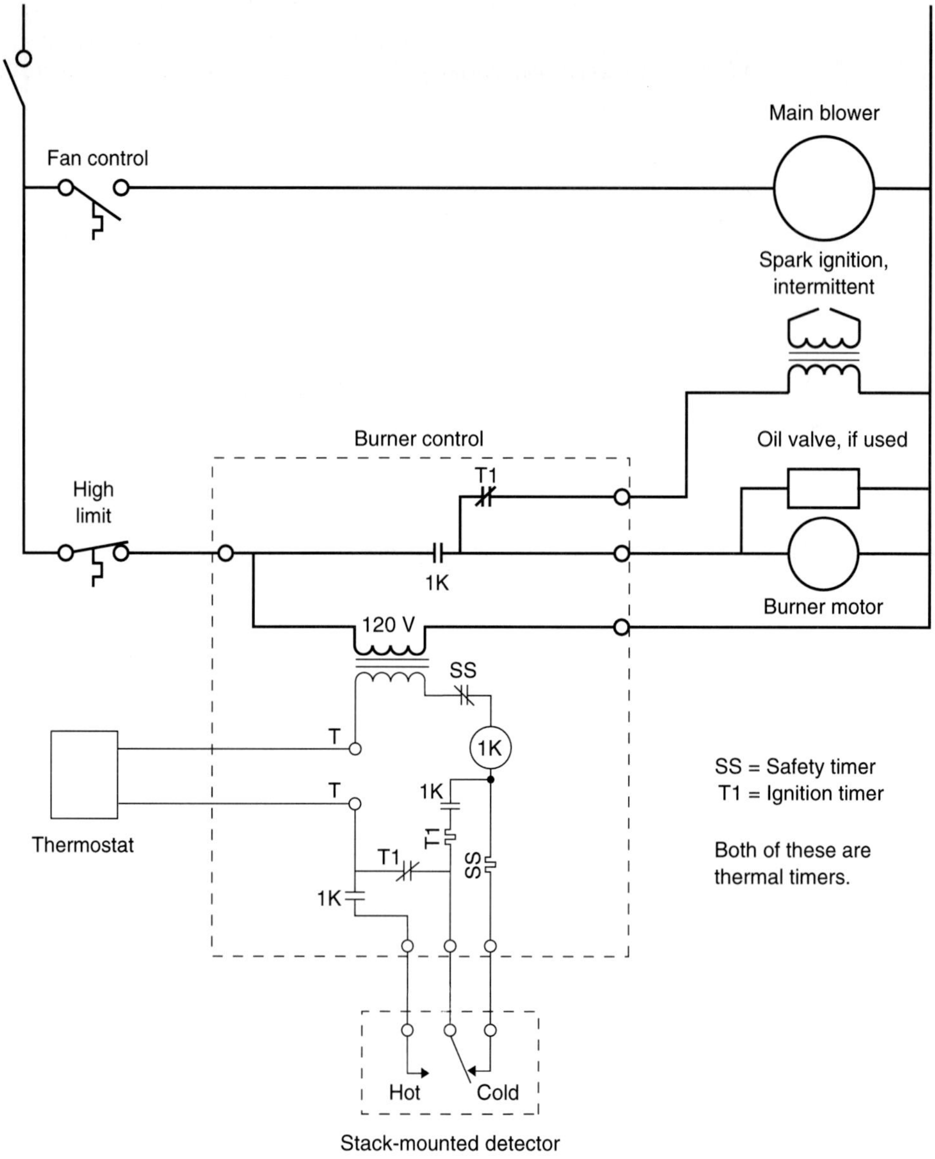

Intermittent ignition, separate stack control

heater. When T1 times out, it shuts down the igniter and opens the NC T1 contact in the 1K coil line. The coil remains energized through the stack switch contact (hot position) and the T1 heater. This keeps the T1 contacts open and the burner on until the thermostat is satisfied.

If there is a flame failure, the stack switch contact will open, dropping out 1K and shutting down the burner. With the cold contacts closed again once T1 cools and resets, the burner will attempt to relight. If it fails, it will lock out on the safety switch.

The final oil burner control example in Figure 17–49 is a one-piece stack control. The contacts operated by stack temperature are mounted inside the controller and operated

FIGURE 17–49 This unit has a one-piece stack-mounted controller that provides intermittent ignition.

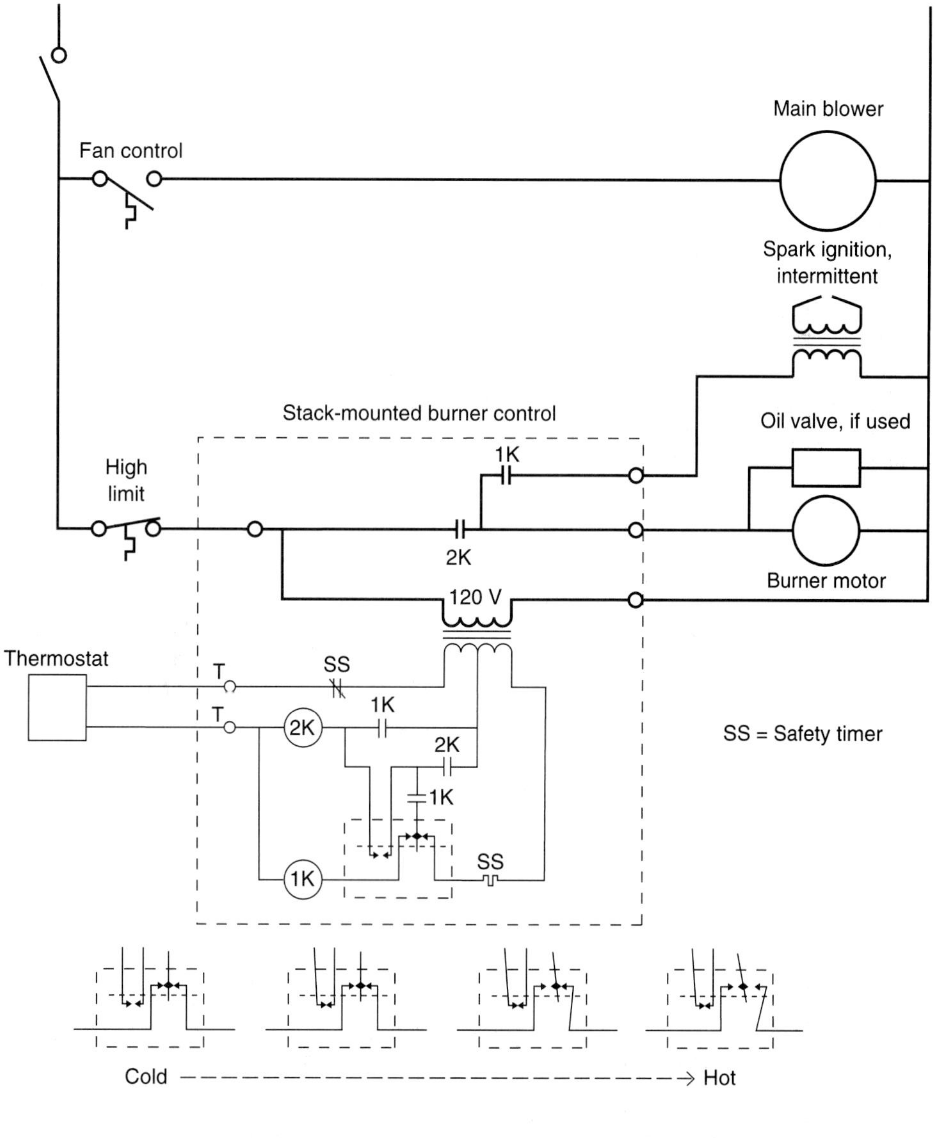

Intermittent ignition, one-piece stack control

mechanically by a rod that comes from the bimetallic sensor. The sensor is inside a tube that extends from the back of the controller into the stack. These contacts have a specific sequence of operation, as shown in the drawing.

The initial thermostat circuit is through the 1K coil, the double-stack switch contact set, and the SS heater. When 1K picks up, it energizes 2K across part of the transformer winding. With 2K picked up, both 1K and 2K are on one part of the transformer through the NC SS contact, and the SS heater is across the other part. The igniter started with 1K, the burner with 2K.

As the bimetallic operator in the stack begins to heat up, the open contact set closes, sealing in 2K, through its own contact, around the NO 1K contact. As the stack continues to heat, one side of the two-contact set opens, de-energizing 1K. This shuts down the igniter and opens the circuit to the SS heater. As heat builds further, the final contact also opens.

If flame fails during the initial attempt, the SS heater will not be de-energized. It will open the NC SS contact, de-energizing the control system and shutting down the burner.

If flame fails after a successful startup, the stack switch will cool, closing the contacts in reverse order. After the last two contacts to open are closed again, 1K will pick up, reenergizing the ignition transformer. The burner attempts a relight, with the SS heater back in the circuit. If it fails, the burner will lock out.

17.13 Boiler Controls

The following information provides control examples for typical residential heating boilers. Boilers may be fired with gas, oil, coal, wood, or electricity. The systems shown here are gas fired.

Boilers are different from forced-air systems in a number of fundamental ways. The burner is heating the water in the boiler, not the room air, so the main burner control is an aquastat set into the boiler, not a thermostat sensing room temperature. The aquastat may serve double duty as the high limit and an operating control.

Thermostats are used to operate zone valves, or individual circulating pumps, to control the flow of hot water through the radiators or heat exchangers associated with each zone. Each zone valve has a contact called an end switch that is used to energize the circulating pump and enable the burner control. The end switches are normally open contacts that close when the valve is fully open.

Boiler Control Examples

Refer to Figure 17–50. Assuming all zones are satisfied (no call for heat), the burner control will be de-energized. When any zone thermostat closes, it energizes the valve motor, opening the valve and closing the end switch contact. The end switch energizes a relay in the boiler control, enabling the burner and turning on the circulating pump. If the water is below the cut-in temperature of the aquastat, the burner will start immediately; if not, it will start when the cool return water from the zone brings the temperature down far enough.

All the end switches are in parallel, so any zone can start the system, and it will continue to operate until all zone valves are closed again. Even if there is a continuous heating

FIGURE 17–50 This boiler control has a built-in 120:24 V transformer and a relay with two NO contacts. One of the contacts is used to operate the 120 V circulator; the other enables the 24 V burner control.

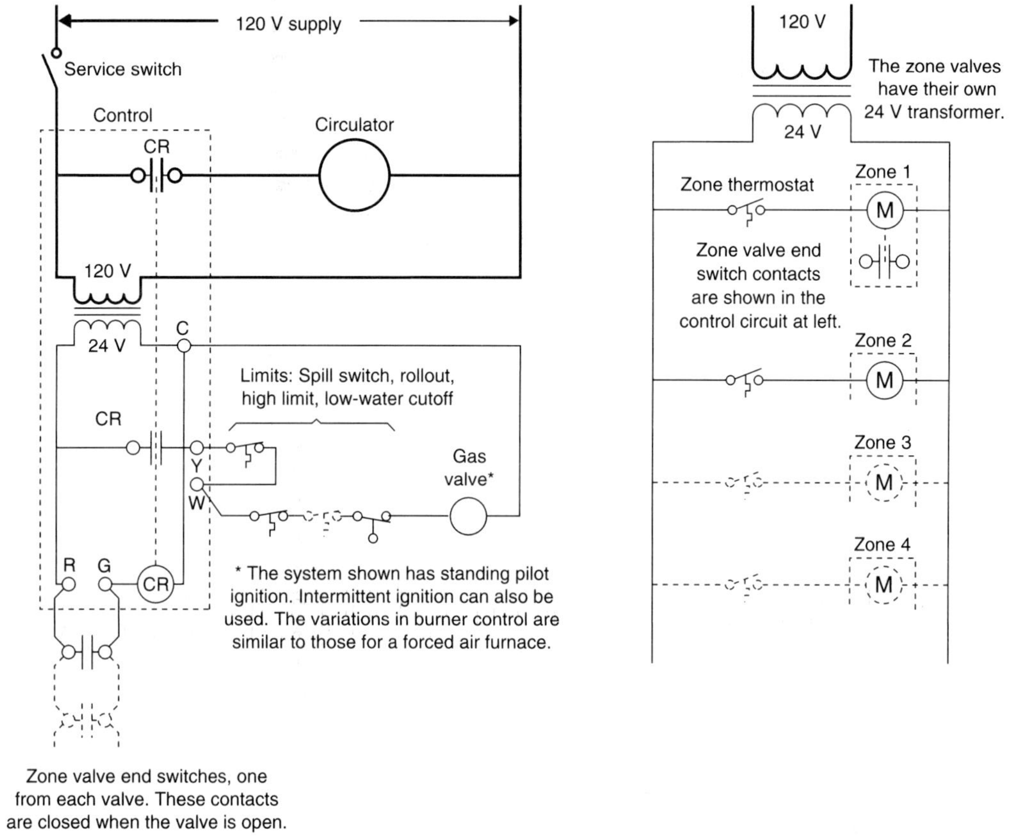

demand, the burner may cycle on and off as required to maintain the water temperature. The relay energized by the end switches starts the circulator, but only serves as an interlock to enable the burner circuit.

The high limit of the boilers may also be the operating control, but it is not the only limit in the system. There will be a low-water cutoff to protect the boiler from operating without the boiler vessel full; there may also be a variety of limits for vent overtemperature or flue gas spillage, flame rollout, etc.

Instead of a common circulator and zone valves, the system may have individual zone circulators. In a system like this, each zone thermostat controls a relay that starts the circulator and provides an end switch contact to the burner control (see Figure 17–51).

Heating boilers are sometimes used to provide domestic hot water through a heat-exchanger coil. When they are used this way, the hot water system functions as an additional zone. In some cases, the domestic hot water "zone" is given priority with a relay that de-energizes all the other zones until the hot water demand is satisfied (see Figure 17–52).

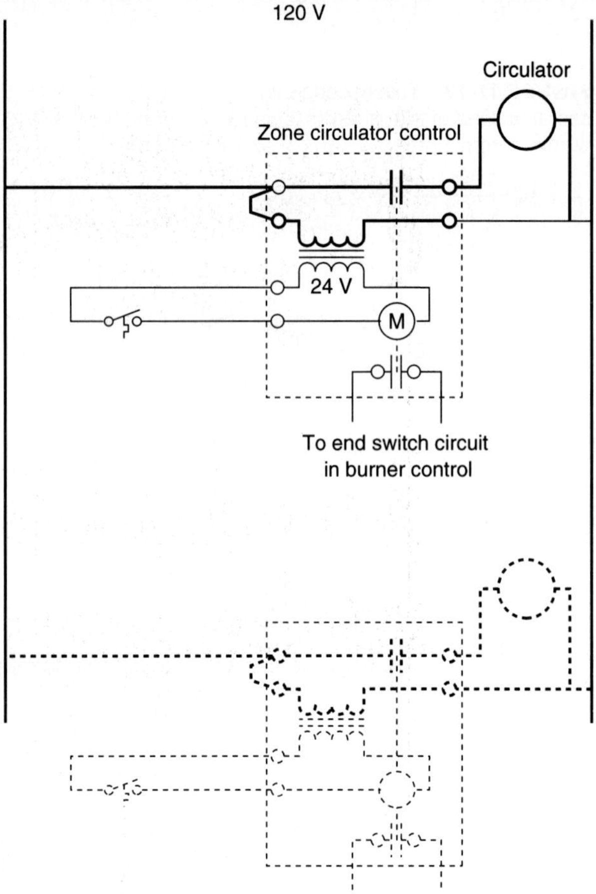

FIGURE 17–51 When each zone has its own circulating pump, a contact from the pump control relay is used as the "end switch."

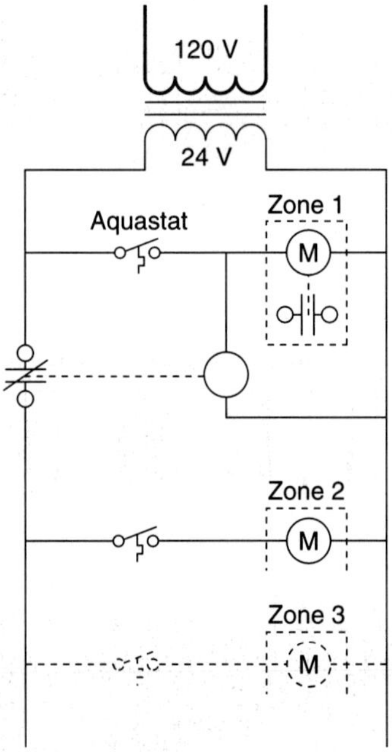

FIGURE 17–52 A priority relay can be used to isolate all other zones when there is a demand for domestic hot water.

Indirect water heater control, with water heater priority relay. Water heater coil is "Zone 1." The priority relay de-energizes the remaining zones when the water heater aquastat calls for heat.

FIGURE 17–53 A boiler control for a system with a high-efficiency gas burner.

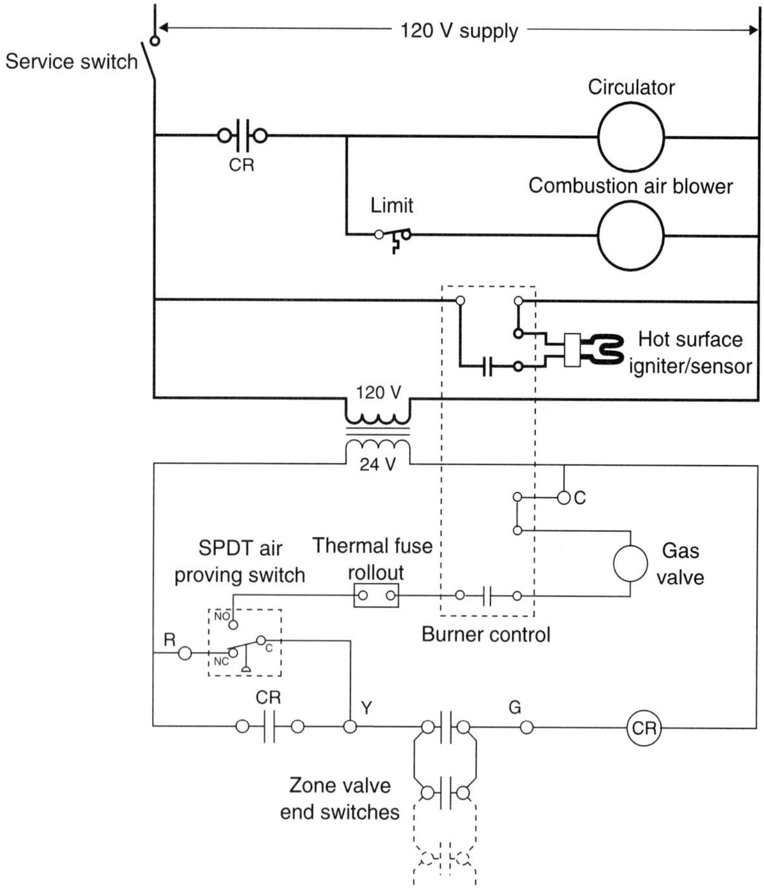

Refer to Figure 17–53. Heating boilers can have high-efficiency burners, just like a forced air furnace. This example has an induced-draft combustion air blower and an electronic control. The details of the burner control are not shown, but it provides similar functions to those used on forced air furnaces. A hot surface igniter is used on this model. It may serve as the flame sensor, or there may be a separate flame rod. The combustion air proving switch is a pressure switch operated by the airflow through the combustion air blower. In this example, the proving switch uses SPDT contacts.

The burner is enabled by one or more end switches that energize relay CR. CR seals in through its own contact around the NC contact of the proving switch. If the proving switch is stuck in the closed position (NC contact open), the unit will not start. A second CR contact in the power circuit starts the circulator and—if the limit is closed, indicating the boiler is not up to temperature—the combustion air blower. Assuming the blower and vent are in working order, the air proving switch will transfer to the NO contact, energizing the burner control.

The burner control will take care of lighting the burner. There may be a purge delay and sensor check, and then the igniter will be preheated and the gas-valve opened. If flame is detected, the system will operate until the limit opens or all zones are satisfied.

17.14 Heat Pumps

A heat pump is essentially an air conditioning system equipped with a reversing valve that can exchange the duties of the two heat-exchanger coils. During a heating cycle, the indoor coil becomes the condenser and the outdoor coil the evaporator. Heat is gained from outdoors and dissipated indoors. When the unit is in cooling mode, it operates as a normal air conditioner. The indoor coil is the evaporator, and the outdoor is the condenser.

The outdoor coil is particularly subject to frost buildup when the unit is operating in the heating mode. A defrost control senses frost buildup by monitoring variables such as refrigerant-liquid line temperature. When defrost is required, the control briefly changes the unit over to cooling mode to heat up the outdoor coil and remove the frost. During this period, auxiliary heating coils may be energized to avoid blowing cold air into the conditioned space.

In areas where heating-season temperatures may drop too low for heat pump operation, the indoor unit may be supplied with one or more stages of heating, usually through electric heat elements mounted in the same blower unit as the indoor coil. The example shown in Figures 17–54 and 17–55 is based on a Lennox unit with a two-speed outdoor heat pump coupled to an indoor unit with four heating elements.

A heat pump/electric heat system requires a more complex thermostat than a conventional heat/cool system. All systems do not necessarily follow the same marking system, but the following is relatively common. Because the compressor supplies both heat and cooling, the Y terminals do not call for cooling specifically, they call for the compressor. The mode is determined by the position of the reversing valve. In the example given here, this valve is normally in the heating mode, and must be energized to switch the unit to cooling. The valve is controlled from terminal O of the thermostat.

On units with two compressor stages, there are two Y terminals, marked Y1 and Y2, for the two stages. The thermostat energizes these in sequence as required to meet the demand. The second stage cannot be energized without Y1 on first, but both can be on when the unit starts. The refrigerant cycle does not allow overly short operating cycles or instant changes in compressor speed. In the system shown in Figure 17–54, these delays are handled by a two-speed compressor control module. This module also incorporates some fault handling that is part of the compressor protection.

The compressor is shown in the "normal" low-speed mode. North/south polarities shown in the motor indicate the relative polarities of the run and auxiliary windings of the motor with the low-speed connection. (The two-pole [high-speed] start winding is dotted to reduce confusion.) The defrost unit is also an electronic unit (sensors not shown) that handles all defrost logic and timing. An outdoor thermostat is used to lock the compressor to high speed when the temperature is too low for effective operation on low speed. Note that terminals O and W1 of the indoor thermostat appear on both Figures 17–54 and 17–55.

The heat relays (contacts upper left, and operating heaters lower right of Figure 17–55) are thermal-delay relays that energize the heating elements in sequence. The blower motor in this particular unit is an electronically commutated, brushless DC motor with an integral controller. The motor is coupled to an external blower-motor controller that determines the appropriate speed for the operating mode. This controller receives signals from the blower relay contact, the thermostat fan terminal (G), the reversing valve control (O), and the thermostat 24 V hot terminal R. This provides the internal logic circuits with indications of the on/off status, continuous fan requirements, and operating mode of the system. Fan speeds

FIGURE 17–54 This is a control drawing for the outdoor unit of a heat pump system; see Figure 17–55 for the indoor portion.

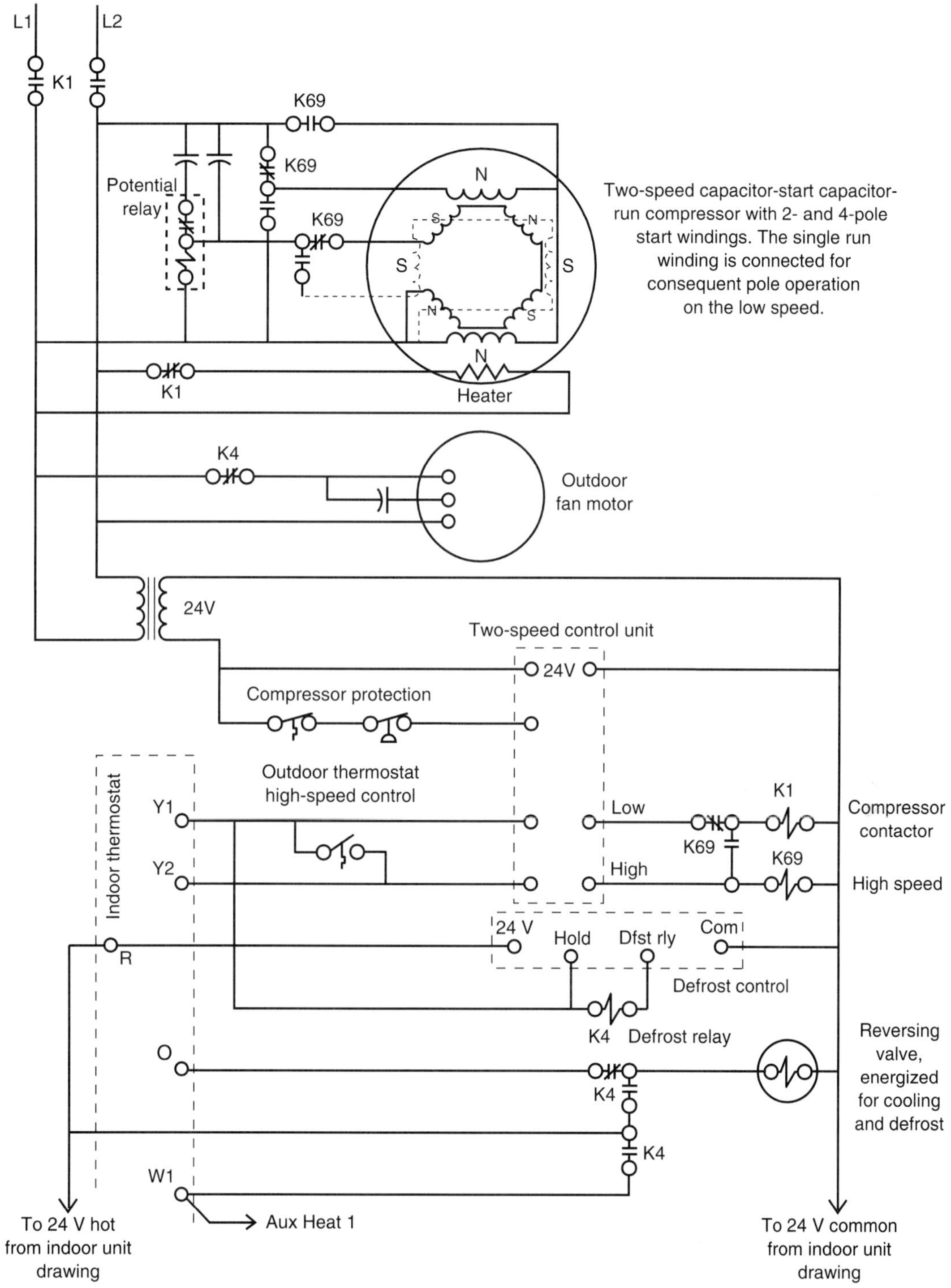

FIGURE 17–55 This control drawing is for the internal blower and electric heat unit that accompanies the outdoor heat pump unit shown in Figure 17–54.

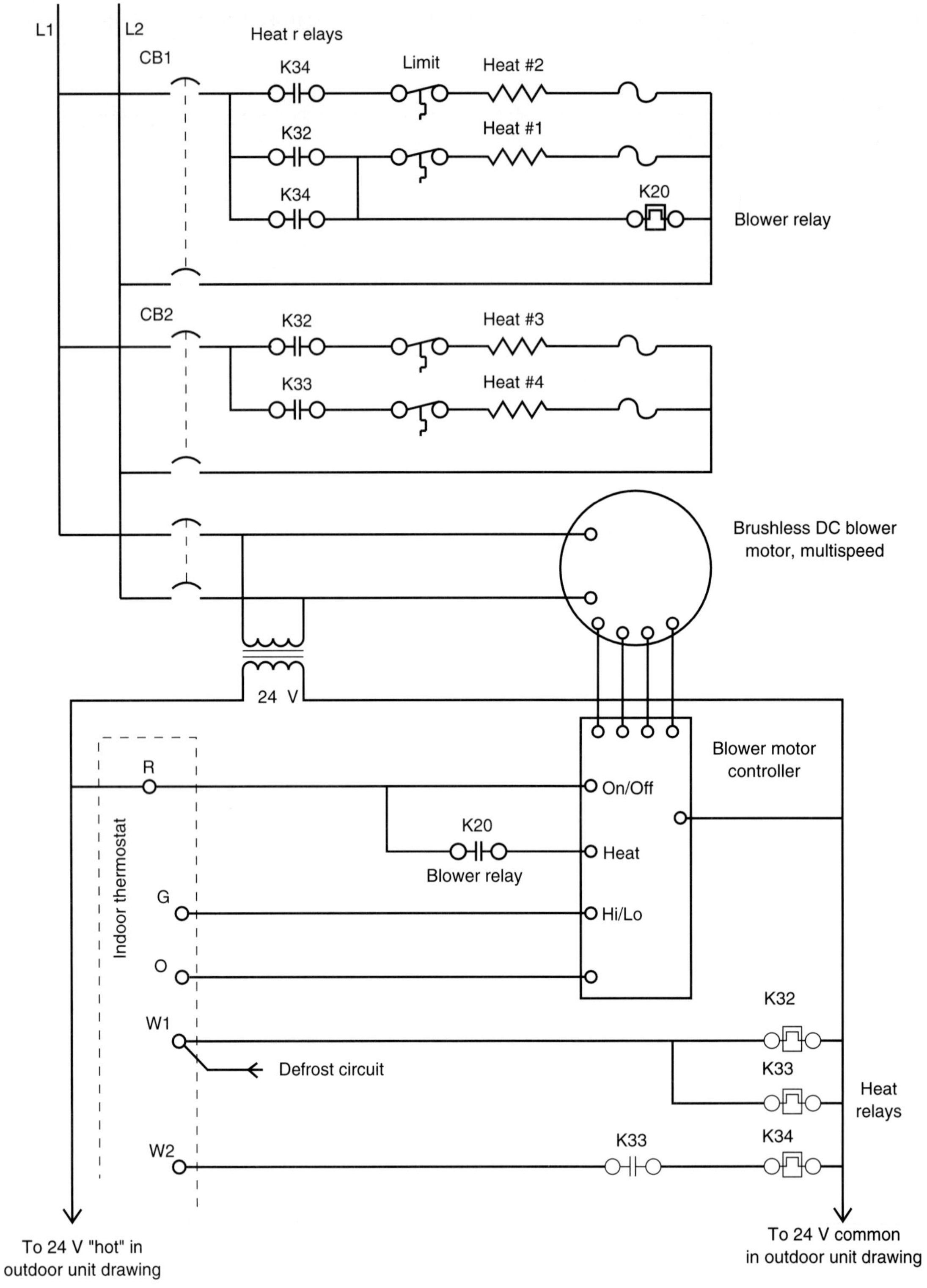

are selected as required, and the appropriate control signal is sent to the electronic control circuit in the blower motor.

Questions

1. What is the function of the thermocouple in a standing pilot gas furnace?
2. How is the circulating air blower (furnace fan) controlled in a typical standing pilot furnace?
3. Why must the jumper of a combination fan/limit control be removed when the limit is installed in the 24 V control circuit?
4. What is meant by the terms “line” and “load” when they are used to mark a control device such as a switch?
5. Once the jumper has been removed from a combination fan/limit control, can the control be used in a system that would normally require the jumper be left in?
6. What is the significance of a Class 2 transformer designation?
7. What is the difference between a thermocouple and a thermopile?
8. In a powerpile control system with a 120 V blower and a combination fan/limit control, how many wires will connect to the fan/limit control?
9. A small 10-turn coil with alligator clips on the lead ends is a handy clamp-on ammeter accessory. Explain how it might be used to measure heat-anticipator current in a thermostat circuit.
10. A standing pilot gas-valve has two wires from the thermostat, terminated on TH and TH-TR, and two wires from the transformer, terminated on TR and TH-TR. What happens if a jumper is connected across each of the following terminals: TH and TH-TR, TR and TH-TR, or TH and TR?
11. Compare the operation of the heat anticipator with that of the cooling anticipator.
12. Why is it important to seal the hole around the thermostat conductors where they come out of the wall into the thermostat?
13. If a limit tests 24 V to common on one side and 0 V to common on the other, is the limit open or closed?
14. Describe the fundamental principles of flame rectification.
15. List the sequence of events for the light-up of a typical intermittent ignition furnace with a combustion air blower. Start with the closing of the thermostat circuit.
16. On a typical residential furnace, will the blower motor draw more current or less current than normal if the duct resistance is reduced?
17. Contrast the operation of a time-based fan control to that of a thermal device that monitors plenum temperature. Explain how the timers are triggered and the reason for a limit connection to the fan control.
18. What problem can be created when a fan center is fitted to a furnace with a combination fan/limit control? What is one solution to this problem?
19. Explain the operation of the CAD cell in a typical oil burner control.
20. In a typical water heating system, the burner may not light immediately when one of the zones calls for heat. Explain how this might happen.
21. Compare the operation of a heat pump system to a conventional split system with an electric furnace and air conditioner.
22. How does the defrost system of a heat pump heat the outdoor coils?

23. Why would the electric heat elements be energized during a defrost cycle of the heat pump?
24. Referring to the heat pump example at the end of this chapter, explain how the compressor windings are connected to provide two-speed operation.

18 Packaged Systems

Objectives

This is primarily a reference chapter. On completing the study of this chapter you should be able to:

- Describe how a packaged HVAC system differs from a typical residential system
- Describe the provisions for fresh air intake in a packaged unit
- Describe the function and operation of an economizer system
- Describe the operation of typical modulating damper actuators

18.1 Introduction

The subject of this chapter is factory-built HVAC packages such as the rooftop units common to many low-rise commercial, industrial, and institutional settings. It should be recognized that much of the operation of these systems can be closely compared to the smaller residential units of the previous chapter. Likewise, an understanding of the packaged units described in this chapter may be very applicable to larger systems using central heating and cooling plants.

Additional information about some of the control devices relevant to this chapter can be found in the sections on multistage and proportional sensors and controls at the end of Chapter 15.

A vast number of medium-sized commercial HVAC systems are based on a factory-built multipurpose unit that is mounted outside the main structure of the building. The unit is often mounted on the roof, but may also be mounted at ground level or on a platform on the side of the building. Sometimes the same type of equipment is mounted inside the building, usually within a dedicated equipment area that is isolated from the normally occupied zones.

A typical packaged unit incorporates all the functions required by a HVAC system for that geographic area. These include ventilation, filtering, heating, dehumidification, cooling, and, in some cases, humidification. The unit contains the circulating air blower, the heating unit—gas, electric, oil, or heat pump—the compressors for the cooling system, all the heat exchangers, and the outdoor air intakes for building ventilation. The controls, with the exception of the thermostats and other sensors, are usually within the unit. Distribution ducts connect to the unit at the roof flange or other point of attachment to the building. If the system has multiple zones, the supply plenum is usually divided within the unit, and individually controlled zone dampers are provided in each division. This approach requires separate duct runs from the unit to each zone. As an alternative, zone dampers may be field-installed remote from the unit, which then supplies conditioned air through a main duct. This type of multizone installation may use a more sophisticated control system that is able to communicate with the remote zone dampers.

Multizone systems have individual thermostats or sensors in each zone. The local thermostat controls the damper for that zone, either directly or through a central controller. The controller for the main unit must accommodate the demands of the various zones. A bypass damper may be required to route some supply air back to the return air duct when most of the zone dampers are closed due to low demand.

Packaged HVAC units are built small enough for use in single-family dwellings. Larger buildings may be serviced from a single large unit or from multiple units located at various points on the building. Multiple units are popular for a variety of reasons. The smaller size is easier to manage during construction and when replacement is required. The duct runs are shorter and smaller, and there is less disruption if a unit fails. Also, individual tenants within a large complex can be supplied from a dedicated HVAC system. This allows the system to be matched to the needs of the tenant, and may allow the building owner to make individual tenants responsible for the operation and maintenance of their portion of the system.

18.2 Residential and Packaged Systems Compared

Typical residential forced air systems are built around a "furnace" installed within the building. The furnace contains the circulating air blower, an air filter, the burner and heat exchanger or other heat source, and, when air conditioning is included, the evaporator coil of the air conditioning system. Ventilation air usually consists of only a combustion air supply for the burner. There are many variations of course, (most of which are discussed in Chapter 17), but this type of forced-air system is very common throughout the U.S. and Canada, and in other parts of the world.

This type of heating and air conditioning equipment is a "split" system, with a separate outdoor enclosure containing the compressor, condenser, condenser fan, and main controls of the cooling system. The connections between this unit and the furnace consist of a single low-voltage control circuit and the two refrigerant lines. The power supply to the outdoor unit is separate from the furnace supply, usually at 208 or 240 V instead of the 120 V used for a typical gas or oil fired furnace. The heating and cooling systems are controlled by a single heat/cool thermostat. When a humidifier is installed, it usually has a separate humidistat. An electronic air filter may be used in place of a conventional filter. It may be powered and controlled by a relay connection within the furnace, or have a built-in sail switch that turns the unit on when there is airflow.

One of the principal weaknesses of the typical residential system is that it has no specific provision for bringing in fresh outside air for general building ventilation. In the past, most homes received adequate fresh air by infiltration through poorly sealed doors, windows, exterior walls, and foundations. Improvements in building techniques and technology have made many modern homes nearly airtight. With little or no natural infiltration into the building, a separate ventilation unit may be required. Fan-forced units that incorporate a heat recovery system are a popular choice. These units bring fresh air in while exhausting stale air. They recover a portion of the energy that would otherwise be lost through the use of heat exchangers that transfer heat between the exhaust and intake airstreams.

A typical rooftop package differs from a comparable residential system in three ways. First, the packaged system is unitary: The main components of both the heating and cooling system are mounted in a single enclosure. Second, the rooftop unit is mounted outside the building; therefore, it is the obvious location of most, if not all, of the building's mechanical ventilation. Fresh outside air can be drawn into the unit, heated or cooled as required, and distributed throughout the building as part of the conditioned air supply. The third difference is in the electrical supply used, and the general size and capacity of the equipment. Typical rooftop packages are available for the common commercial voltages. These include 120/240 V single-phase, and three-phase voltages of 208, 240, 480, and 600 V.

Despite the differences in these systems, the equipment incorporated into the rooftop unit is often very similar to its residential counterparts. The burner and its controls, for example, are often much the same. Typical gas fired versions use electronic ignition, and flame rectification for monitoring the flame. They also have mechanical draft provided by a combustion air blower. With their compact construction and outdoor mounting, they generally use very short exhaust vents. The air proving contact is more likely to be a shaft speed switch on the combustion air blower than a pressure switch monitoring blower airflow (see Figure 18–1). Safety controls such as high limits and flame rollout are applied much the same as for a residential furnace. Larger units are likely to have two-stage gas valves to provide at least two levels of heat output.

When heating is provided by electric elements, electromagnetic contactors may be used in place of the heat relays commonly used in residential systems. The elements may be sequenced by timing relays, by an electromechanical sequencer, or through an electronic control system. Full electronic control of the power to the heating elements is also possible.

Heat pump and/or cooling system components are similar to those used in residential split systems, but the equipment is very closely coupled. The condenser and evaporator are only a short distance apart, mounted in separate compartments of the common enclosure. Multiple cooling stages are very common. Two-speed or dual compressors may be used to provide two-stages. Larger units often have more than two-stages, typically using one compressor for each stage.

In a residential system, the main blower is inside the furnace. The return air duct enters the furnace at one end; the air passes through the filter, through the blower and the heat exchangers of the burner (or directly across the electric heating elements), and then through the evaporator coil of the air conditioner. The only fresh air intake is typically tied directly into the return air duct. It is required by codes to ensure adequate fresh air for combustion, and is not equipped with a damper or other control.

In a rooftop unit, the main blower is also mounted inside the unit. The return air duct enters the unit from the building. The air passes through the filters, the evaporator coil, the blower, and the burner heat exchanger, much as it does in a residential unit. However, the order is not necessarily the same, and heating and cooling exchangers may be in different airstreams. Like the residential unit, fresh outdoor air is added to the return air, but it is not

FIGURE 18–1 This photo shows the gas valve and combustion air blower of a small packaged unit. The motor is a typical PSC blower drive, but note the extra control conductors entering the motor with the power conductors. These lead to an internal sensor mounted near the motor shaft. The sensor is a Hall-effect device that detects the passage of a small magnet bonded to the motor shaft. The output goes to a dedicated circuit board that provides an air proving contact to the main control system.

for combustion. The burner draws its combustion air directly from the outdoors, not from the building. The fresh air drawn into the blower of the unit is ventilation air for the building, and there is usually some provision to exhaust the excess return air so the building does not become pressurized.

The amount of fresh air entering a small system may be controlled by a damper that is set manually or has only two or three operating positions. In larger systems, automatic dampers are used to regulate the mix of outside air and return air that becomes the building supply air. This requires two or more dampers, and one or more actuator motors to adjust their positions. The dampers regulate the return air, outdoor air, and possibly the exhaust air. Smaller systems usually use a single actuator motor mechanically linked to both dampers.

Larger systems may have two or more motors. The "mixed air" input to the system is a combination of fresh outdoor air, and return air that is recycled back into the building. This mixture passes through the filters, the cooling coils, the main blower, and the heat exchanger or electrical heating elements, and into the supply duct of the air distribution system.

Economizers

The physical location of the heating and air conditioning unit and the provision for fresh ventilation air to meet building code requirements have made economizers a natural addition to many unitary HVAC systems. The economizer is a combination of dampers and controls that uses outdoor air to meet the building's air conditioning demand whenever outdoor conditions make this possible. Fresh outdoor air is used to supply "free" cooling whenever the heat content of the outdoor air is low enough to be used in place of the "mechanical" cooling provided by the air conditioning system. This reduces the overall cost of operation by reducing the annual running hours for the compressors. This gives the typical unitary system a major advantage over split systems used in typical residential applications. With no ability to introduce large volumes of outside air, furnace-based residential systems must use expensive mechanical cooling even when the outdoor air is cool.

A typical economizer control system monitors outdoor air temperature (or enthalpy) and selects free cooling or mechanical cooling according to a preset changeover setting. When free cooling is available, the control system will lock out the compressors and modulate the dampers to regulate the amount of outdoor air that is used to replace return air. The positions of the outdoor and return air dampers are continuously adjusted to maintain a target mixed-air temperature into the unit. When the economizer cannot provide the full cooling demand, the system will either switch to mechanical cooling only or to an inbetween stage using a combination of outdoor air plus air conditioning. When outdoor air conditions are such that no free cooling is available, the system reverts to mechanical cooling exclusively.

Minimum Ventilation Air Intake

Regardless of the operating mode—heating, mechanical cooling, or economizer operation—the unit must always provide at least the minimum required ventilation to the building. In typical systems the "minimum air" requirements are satisfied by adjusting the opening of the outdoor air damper with a "minimum air potentiometer" provided for this purpose. This setting overrides other damper controls to ensure the outside air damper is at least partially open whenever the circulating air blower is operating.

During economizer operation, the "minimum air" setting prevents the economizer control from closing the damper past that point. During a normal heating or mechanical cooling cycle, the damper opens to the minimum position. When the system is idle, the blower is off, and the damper is fully closed. When the building has significant unoccupied periods, such as a retail store or office, the control system may have a provision for overriding the minimum air setting and closing the damper during the unoccupied period. This is acceptable because very little fresh air is required when the building is not occupied.

Operating and Control Voltages

Commercial HVAC units are generally much larger than residential systems, and they are usually installed in buildings that are supplied with three-phase power. When it is available, the main unit will be operated from a three-phase supply, very often at higher voltages than those used in smaller equipment. Typical commercial packaged units are available for 208–240, 480, and 600 V operation. As explained previously in this text, the nameplate listings for these nominal voltages are typically 208–280, 460, and 575 V. All of these voltage are in common use.

Please note that in addition to the supply voltage, most packaged units also have other voltage levels within them. It is very common for the manufacturer to install voltage-reducing transformers for some of the less demanding loads. There will also be one or more control transformers, normally rated to provide 24 V AC. In a typical unit the air conditioning compressors and the main circulating air blower operate at the supply voltage. The smaller condensing fans may be three-phase units, typically at line voltage, or single-phase units, also at line voltage or something lower, such as 208–230 V. On a gas fired unit, the combustion air blower is typically very small and is often driven by a single-phase motor operating at 120 V. The modulating motors for the dampers are normally operated as part of the control system at 24 V AC. Some systems use DC motors, which may be operated by dedicated controllers that include a conversion from AC to DC, or they may simple be connected to the AC supply through rectifier diodes.

In addition to the higher operating voltages, there are other safety concerns related to the high output of the burners. Mistakes in connecting safety limits and interlocks, or other errors that cause improper burner operation can have even more devastating consequences. Always be aware of the purpose and operation, of all control components. Never allow any system to operate unmonitored with a bypassed safety device.

A thorough understanding of the controls used in a typical packaged commercial HVAC unit requires an understanding of all the different systems involved. This includes the electrical power and control system, heating and air conditioning components, and the circulating air blower, plus additional knowledge regarding multistage and modulating controls.

18.3 Multistage and Modulating Controls

Note

This information covers only electrical, not pneumatic, control systems. Pneumatic controls are common in commercial and industrial applications. They are inherently proportional and generally somewhat simpler than electrical systems. To take advantage of central computer control, large HVAC installations often have a combination of electronic and electrical devices. Small dampers and valves are driven by low-voltage electric motors, and large ones with pneumatic actuators that have electric-to-pneumatic transducers to regulate the pressure to the piston or diaphragm. Pneumatic positioners are simpler and more powerful than electric types, making them particularly desirable for large dampers and valves.

There are three basic types of control, each of which may be referred to by two or more different names. Two-position or cyclic control is the simplest. Multistage or step control provides multiple stages of operation. Modulating or proportional control regulates smoothly through a range of outputs.

Cyclic or two-position control applies to much of the equipment used in HVAC installations. It is used with equipment that can be either full on or full off, but cannot be throttled or regulated through a range of outputs. The controller has only two positions—on and off—and the turn-on and turn-off points are separated by a differential. The differential is kept as narrow as practical, but must be wide enough for equipment operating cycles to be within acceptable limits. Cycles that are too close together are inefficient and may cause serious problems for the equipment under control. Refrigerant compressors are often protected against this type of operation by a "short-cycle" timer that prevents restarting the compressor until it has been shut down for some preset minimum time.

One alternative to two-position control is some type of staging to provide two or more different output levels. With multistage control, the output is more closely matched to the actual heating or cooling load. It is more accurate, but requires equipment that can operate at different output levels, or multiple smaller units that can be individually controlled to meet the changing demand. A common example of multistage control is a two-stage gas-valve system that provides two different firing rates for the burner. A two-stage thermostat starts the burner on the lower rate, and then moves to the higher output stage if the first stage is unable to meet the demand. When the need for heat is relatively low, the burner runs at a lower output for longer periods of time. This improves efficiency and reduces temperature swings in the system.

Two cooling stages can be provided with a two-speed compressor or dual compressors. The compressors, like the gas-valve in a two-stage heating system, are controlled in sequence by a two-stage thermostat. In systems with economizers, the control system selects the economizer as the first cooling stage. A second mechanical stage may be provided simultaneously with the economizer. A cooling system can, of course, be extended to more than two-stages by using additional units, but there are practical limitations. Multiple-piston compressors can provide several stages from each compressor by "unloading" the cylinders.

When there are more than two-stages, the control system is usually somewhat different. While two-stage control is usually provided with a two-stage thermostat, for three or more stages the conventional contact-type thermostat is usually replaced by one or more sensor modules. The sensor(s) provide a variable signal to a separate controller that in turn converts this into a sequence of three, four, or more stages. Each stage is controlled by a contact in the controller, much like a multistage thermostat, but there are distinct advantages to this system. In multizone systems, for example, the controller can analyze the demand from all of the sensors and provide a more accurate interpretation of the total load. Central control units may also be used to provide more sophisticated control options than a simple thermostat.

Another control alternative is a modulating system that provides fully variable output. The controller adjusts the output proportionally to changes in demand, matching the heating or cooling system to the load. This type of control can only be applied to equipment that is capable of running at a wide range of settings. In the HVAC industry, the most obvious and most common applications for modulating control are valve and damper positioning. A pneumatic or electrically operated actuator positions the valve or damper according to a variable signal from the controller. This type of control is widely used for large systems that have central heating and cooling plants to provide hot water for heating and chilled water for cooling.

The flow of hot and cold water may be regulated by modulating valves, and the movement of air by modulating dampers. Efficiency can be improved by controlling duct pressure or airflow with a variable-pitch fan or a variable-speed motor drive. Pumps and certain types of compressors can also make use of variable-speed drives, reducing the losses associated with throttling the flow with a valve. These systems have become more commonplace in recent years as electronic variable frequency drive technology has improved. Many of these motor controllers can be used with standard single-speed three-phase induction motors to provide an efficient and very effective variable-speed drive.

Modulating control can also be used at the source of heating or cooling, if the equipment is capable of variable output. Larger heating boilers are fired by variable-rate burners that adjust the combustion air and fuel continuously in response to the temperature or pressure of the boiler. Chillers that are based on centrifugal or screw compressors can also be modulated. The centrifugal compressor can be controlled by changing the speed, adjusting the inlet vanes, or both. Screw compressors can be regulated by speed or through a sliding valve arrangement built into the compressor.

Modulating Control Systems

Note

A variety of modulating electric valve and damper actuators are available. Most of those discussed here are either manufactured by Honeywell or follow a similar standard, but there is no intention here to imply that Honeywell equipment is superior or inferior to other makes. Many different manufacturers compete in the same marketplace, with a mix of both compatible and proprietary equipment. Older technology, such as the resistance balance controls and positioners described below, are a de facto standard that is compatible with equipment from a variety of different equipment makers (see Figure 18–2).

The simplest control mode for a modulating controller is called proportional control. Like all modulating control systems, proportional control requires a variable output from the controller and a variable response from the controlled device. As an example, assume a damper has to be positioned in response to changes in temperature. The temperature sensor/controller converts the difference between the actual temperature and the set point into a variable electrical signal that can be "understood" by the actuator. The actuator responds to the signal by driving its internal motor forward or in reverse to position the output arm proportional to the signal it receives. When the controller calls for 25% of full output, the motor would move to a position that represents 25% of its operating range. When the demand rises or falls, the signals change, and the actuator responds accordingly. Although there are delays and some built-in insensitivity—both the controller and the positioner will ignore very small changes—the relationship between the "sensed variable" (the temperature, in this example) and the output of the "final control element" (the damper positioner) can be very close.

A key factor in all modulating controls is the type of signal used for communication between the sensor and controller, and between the controller and the final control element. For electrically operated control systems these signals must, of course, be electrical in nature. They can be either analog—such as variable resistance, voltage, current, or frequency—or digital. Traditional analog systems, which are the most common in the packaged systems of

FIGURE 18–2 These modulating motors are used for damper and valve positioning. The stripped one on the left shows the wire-wound feedback resistance (arc shape above the shaft) and the balancing relay. The coils of the balancing relay are mounted in the metal frame just below the shaft; the contacts are visible just above the coils. The double-sided moving contact sits between the two stationary contacts and is tipped left or right by unequal current in the two coils. The reversible shaded-pole motor (right front) has a single field winding and four wound shading coils. There are two shading coils on each motor pole. Shorting opposing pairs will drive the motor one direction or the other. With all the shading coils open-circuited, the motor acts as a brake.

the type discussed in this chapter, typically depend on a continuously variable signal level that is some combination of resistance, voltage, and current.

Electrical controls are virtually all analog, while electronic controls may be analog, digital, or some combination of the two. Modern microprocessor-based electronic systems may use digital communication between the devices. Digital communication uses electrical signals, but the information is encoded as a string of pulses representing numbers, rather than as a variable level. Digital systems are technologically more complex, but they allow for more direct application of microprocessors and distributed computer control. When communication among various parts of the system is in digital form, troubleshooting takes on a whole new dimension.

Digital technology is widely used, even in many small systems such as residential furnaces. But in these systems, the digital part of the control system is buried within individual devices and proprietary control boards. Communication among the various devices uses traditional electrical signals. Like most electronic equipment, the digital circuits are "nonserviceable" items repaired by replacement. The installer or service person really needs to know only two things: whether the failure is on the circuit board, and how he or she can test the external components safely and efficiently.

In the types of modulating systems examined here, communication among the various components is done with normal analog signals that can be tested with standard instruments. Components that do not modulate—such as relays and motor starters, or two-position dampers and valves—receive their usual on/off, voltage or no-voltage signals from the controller, regardless of whether the control is electronic or electromechanical.

Modulating signals are at relatively low levels of voltage or current. Variable DC voltage signals are usually in the 1 to 5 or 1 to 10 V range, though there are exceptions. DC current signals are typically 0 to 10 mA or 4 to 20 mA. These low-level signals are typical of an electronic system. The electrical controls described below use 24 V AC and a resistance-controlled balance circuit. The current is somewhat higher than in typical DC electronic equipment, but it is still in the milliamp range.

Variable current is generally superior where relatively long control circuits are involved, because accuracy is unaffected by conductor resistance. Current signals can be converted to voltage with an appropriate resistor. For example, a 4 to 20 mA DC current signal passed through a 250 Ω resistor becomes a 1 to 5 V DC voltage signal, because that is the resulting voltage drop across the resistor.

Resistance Balance Circuits

A relatively old modulating control system common enough to be considered an industry standard is based on current balance in a variable-resistance bridge circuit. A good example is the Honeywell Series 90 family of actuators and controls. A full range of these devices is available from Honeywell and other manufacturers. Both the actuator motors and their control circuits operate at 24 V AC.

The Series 90 system uses three wires between the controller and the modulating motor (see Figure 18–3). The control wiring interconnects two potentiometers, one in the controller and one in the motor. The position of the wiper for the "pot" at the controller end is mechanically linked to the sensor. Its starting point is determined by the set point, but its movements follow changes in the sensed variable (typically temperature, enthalpy, or pressure). The pot in the motor acts as a feedback device mechanically linked to the output shaft of the actuator. Motor movement is controlled by the current balance on the two sides of the potentiometer circuit.

The heart of the Series 90 system is the balance relay (or an equivalent electronic circuit) that controls the motor. Electromechanical versions of this relay have two coils, one in each of the outer legs (terminals W1 and B1 at the motor) of the three-wire potentiometer circuit. When the resistance is equal on both sides of the circuit, the currents through the two coils are also equal. The balance relay has a set of SPDT contacts with a center off position. With equal coil currents, the relay contacts are in the off position, and both contacts are open.

FIGURE 18–3 The shaft of a typical Series 90 motor moves through an arc of either 90° or 120°. Through the action of the current balance circuit and feedback potentiometer, the motor shaft automatically moves to track the position of the wiper in the control. Because the wiper is positioned by the sensed variable, the position of the damper or valve is also proportional to the value of the sensed variable.

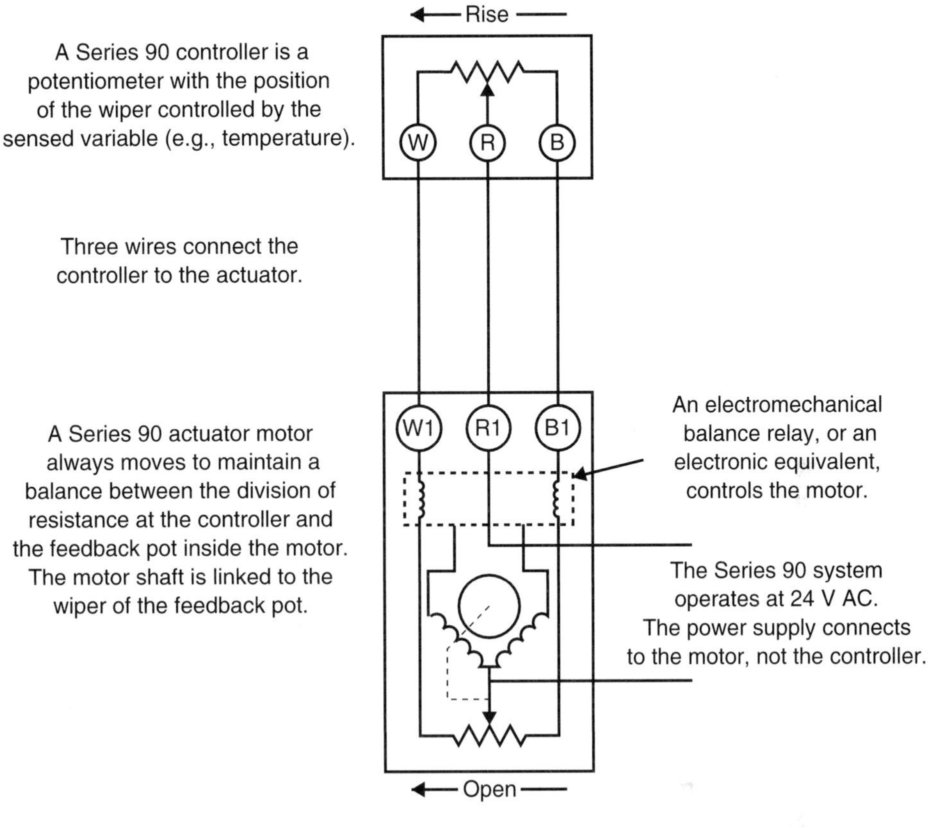

When an electronic circuit is used in place of the relay, the contacts are replaced by a pair of triacs that are switched on and off in the same manner as the contacts of the relay. When the circuit is balanced, the actuator motor is stationary, and the brake is applied to hold the output shaft in position. This "brake" may be the motor, energized but given no directional torque, or it may be a separate device, such as a small mechanical brake released by a solenoid each time the motor is operated. The exact operation depends on the type of motor used and the design of the actuator.

When the resistance of the two sides of the balance circuit is not equal, the two coil currents are different, and the relay tips to one side or the other. This closes either the forward or reverse contact in the motor circuit, driving the actuator in the appropriate direction. Each time there is an imbalance in the system (a normal result of some change in the variable sensed by the controller), the motor starts, and it continues to run until the feedback potentiometer, driven by the output shaft of the damper motor, moves far enough to equalize the circuit resistances and restore the current balance. With the balance restored, the relay returns to the center position, the contact opens, and the motor stops. Again, if an electronic circuit is used in place of the relay, the operation is essentially the same. The imbalance is detected, one of the triacs is turned on, and the motor moves until balance is restored.

The relationship between the potentiometer in the controller and the feedback potentiometer in the actuator controls the motor. The actuator moves quite slowly and may temporarily fall behind changes at the controller, but it always moves until the position of the output shaft matches the position of the controller relative to its set point. The position of the valve or damper that the actuator drives is therefore proportional to the difference between the sensed variable and the set point of the controller.

Electronic versions of this system work much the same as the electromechanical types, but there are fewer moving parts and no contacts. The balance between the two sides of the potentiometer circuit is detected electronically, instead of with a relay. The triacs that replace the relay contacts act as contactless switches to drive the motor forward or in reverse. Triacs are used because the circuit is AC not DC, and the triac is a natural choice for switching this level of AC current. Electronic balance and control circuits are generally superior to electromechanical systems, especially where vibration can cause intermittent operation of the relay contacts. The potentiometer for the electronic circuit can also be simpler and smaller, because less current is required.

For the damper or valve system to work correctly, the direction of movement must always be appropriate. For example, when a temperature controller (a three-wire variable-resistance thermostat) responds to a temperature rise, the direction in which the modulating motor moves must be such that less heat, or more cooling capacity, is directed into the controlled space. The direction in which the motor moves must also act to restore the current balance in the potentiometer circuit, not make it worse. If the wiring is incorrect, or some other problem causes the motor to go the wrong way, it will usually run all the way to one end of its stroke and stay there.

Actuators are often equipped with a spring that forces the output arm to one end when the motor is not under power. This provides a fail-safe control feature that positions the damper or valve to a predictable position during a power failure. The motors can be purchased as normally open—the spring drives the arm to the full-open position—or normally closed, in which case the spring drives the arm toward closed. Normally closed motors are used where the "safe" position is closed; normally open motors are, of course, used where the "safe" position is fully opened. Examples of how these two could be applied would be a normally closed motor for an outdoor air damper—to prevent freeze up—and a normally open motor for the valve in the hot water supply line to a heating coil.

The actuator brake is important, particularly on motors with spring return. Without a brake, any force applied to the operating arm of the motor would cause it to drift off the balance point. It would have to operate again and again, in brief corrective moves, to return the arm to its proper position. This would adversely affect control of the HVAC system, and quickly wear out the motor.

Standards for Series 90 Systems. To assist installers and maintenance personnel, and provide some consistency between applications, there are numerous standards for identification and operation of Series 90 controls. Some of these standards are listed below. Note, however, that you should always refer to the manufacturer's service literature and product manuals. The information given here is gleaned from the same source, but such information is subject to change, and different manufacturers often have different approaches to the same problems. They also may use slightly different terminology, even when their products are compatible with the same control standard.

The power supply for the motor and control is 24 V AC, and it is normally connected at the motor end of the system. Typical actuator motors draw 1 A or less. Despite their relatively high torque (2.8 to 17 Nm, or 25 to 150 in-lb), their electrical demand is quite small. The torque is a product of a very small motor with a high gear ratio. This makes the movement of the actuator arm quite slow; a minute or more to travel full stroke is not unusual.

The control circuit is powered from the motor. The controller is only a three-wire variable-resistance device—a potentiometer. It does not supply power to the motor or the control circuit, it only adjusts the resistance, and therefore the current balance, on the two sides of the resistance bridge circuit. Typical resistance values for Series 90-type control potentiometers are 135 Ω or 270 Ω.

The actuator motor output shaft does not travel a full 360° of rotation; it moves back and forth through an arc (usually either 160° or 90° full travel) that is set by internal limits. The internal limit switches act to open the motor circuit and apply the brake at either end of its stroke. In some motors the limits can be adjusted.

Actuator motors may also have internal auxiliary contacts that can be set to operate at some point of the motor travel. Such contacts are additional to the end limits. They are not part of the motor positioning circuit, but are provided for use in other parts of the overall control system.

Many actuator motors have a specific drive end and an auxiliary end, which are identified on the motor housing. The maximum allowable torque output and the maximum load weight that can be supported are usually much less on the auxiliary end than on the power end of the motor.

Motors have a specific open and closed direction of operation. Viewed from the drive end, the actuator arm is considered to be in the fully open position when it is at the full clockwise end of its stroke. The motor should be mechanically linked to the damper or valve in such a way that the open and closed positions of the motor arm correspond to the open and closed positions of the damper or valve.

The standard color coding for the three-wire potentiometer circuit is red, white, and blue. The connections on the controller and motor are marked R, W, and B (possibly R1, W1, and B1 on the motor). The wipers for both the controller and the internal feedback potentiometers are normally connected to the R terminal (and the red wire, if color-coded wiring is used). The white and blue lines (again, assuming color-coded wiring that matches the terminal markings) are used for the two branches of the bridge.

The terminal markings are standardized, so the controller/motor operation is as follows. When the sensed variable increases, the wiper of the controller potentiometer moves toward the W end of the resistance. This causes the resistance between R and W to decrease, while it adds resistance between R and B. At the motor, when the shaft rotates toward the closed end, the wiper of the feedback pot moves in the opposite direction toward B, adding resistance to the W end. As shown in Figure 18–4, the "direction" of operation for these variable-resistance controllers also applies to the operation of contact-type controls used in some Series 90 control circuits.

The standard heating connection is color to color from the controller to the motor (see Figure 18–5). The W terminal of the controller connects to the motor W1 terminal, R goes

FIGURE 18–4 Series 90 motors can be connected to variable-resistance or contact-type controls. In each case there are standards for the terminal identification and operation of the controller. The variable resistance and floating controller can provide proportional control. The two-position control can only drive the motor all the way to either end of its stroke.

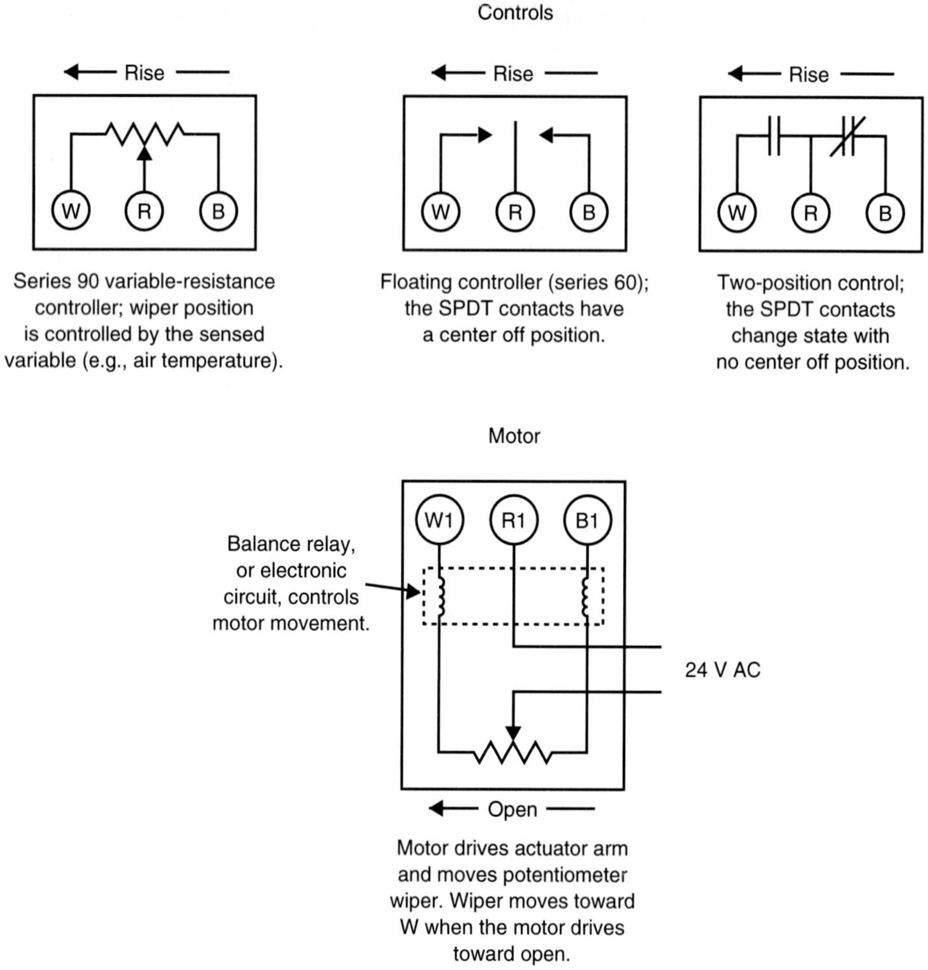

to R1, and B goes to B1. When colored wiring is used, the colors should be applied appropriately. With this connection, an increase in temperature at the controller will drive the motor toward the closed end of its stroke. This is in keeping with the theme that "opening" the damper or valve is equivalent to "adding heat" to the system. As the temperature at the controller rises above the set point, the motor moves to close the damper or valve. When the temperature falls, the operation is, of course, reversed. The controller is always trying to maintain a constant temperature under conditions of changing demand.

The standard "cooling connection" is crossed at the controller end. The two R terminals are connected together as before, but W from the controller connects to B at the motor, and B goes to W. Using a temperature control example, an increase in temperature at the

FIGURE 18–5 The standard heating connection causes the actuator to drive toward open on a fall in temperature (or other sensed variable) and toward closed on a rise. In a standard cooling connection, the wiring between the control and motor is reversed, effectively reversing the response of the motor.

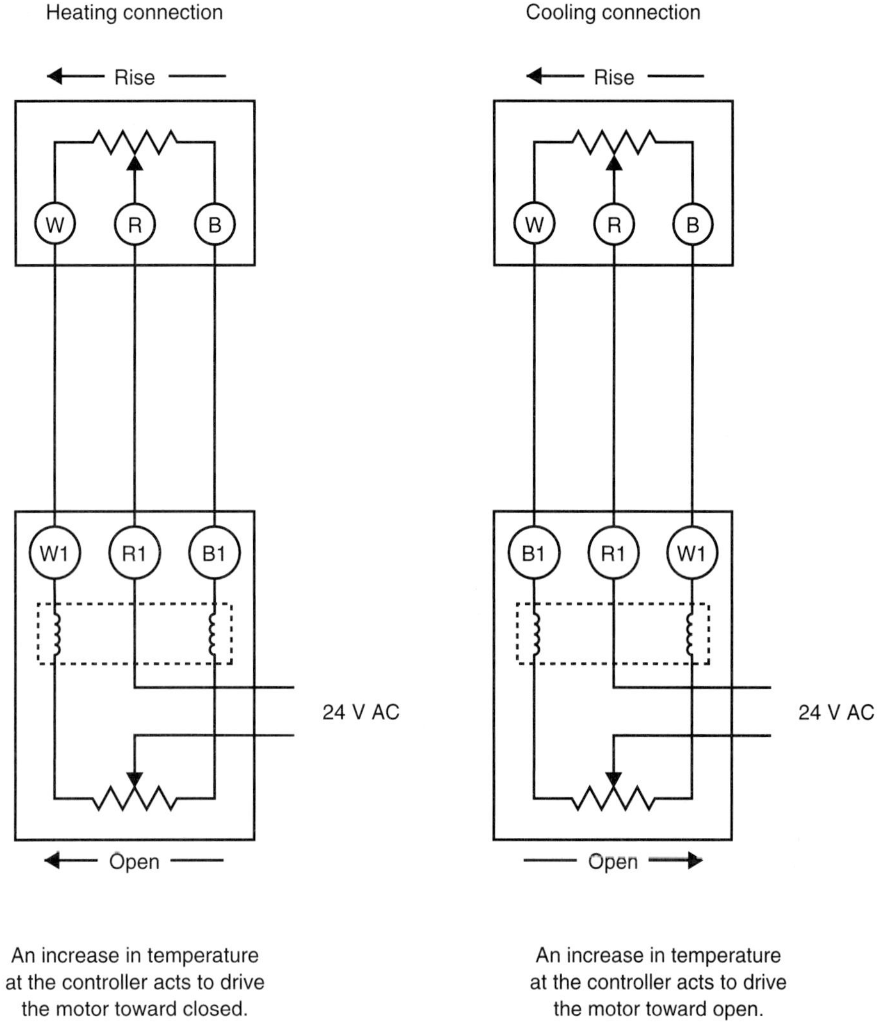

controller now causes the motor to move toward the full-open position. The motor operation has to be reversed, because an increasing temperature at the sensor is equivalent to a demand for more cooling. The correct response at the motor is to adjust the valve or damper to a more open position.

The effects of changing resistance in the control circuit (external to the motor) can be summarized as ARWOBC—Add Resistance to White to Open, Blue to Close. This behavior can be demonstrated by the following tests. With the motor connected to its normal power supply, and the controller disconnected, a jumper can be used to short out one end of the motor potentiometer. This has the effect of reducing the resistance on the jumpered side to 0 Ω, and increasing the resistance on the open side to ∞ Ω. When the jumper is connected from R to W, the motor drives fully closed. Shorting R to B instead drives it fully open. An alternative test is to open the W or B line with the other control lines still connected. This adds an infinite resistance on the open side and forces the motor to the open or closed end of the stroke. An open W line will cause the motor to run fully open; disconnecting B will run it fully closed. A more complete test can be performed with a 135 Ω potentiometer connected in place of the controller. The potentiometer can be used to run the motor through its full stroke in either direction, and to stop it at any point along the way.

Remember, the motor normally moves until it reaches the balance point of the control circuit. Balance occurs when total resistance on the W side (including resistance from the controller, any limits, and the feedback pot) is equal to the total resistance on the B side. Normally, changes in the external control circuit upset the balance, and the motor operates until balance is restored by the feedback pot. In situations where changes to the external circuit are such that even full travel of the motor is not enough to restore balance to the circuit, the motor runs the actuator arm to one end of its stroke and stays there. This can happen when additional external resistance is added to the control circuit, when one side of the control circuit is bypassed by a short or a jumper, or when either the W or B line is open-circuited. The motor will not move off the internal end limit until changes in the control circuit bring the resistance balance back within the operating range.

Series 90 Limits and Minimum Position Controls. A modulating controller and a modulating motor actuator are a complete control system, but they are rarely enough to satisfy all the necessary operating conditions of a typical HVAC application. Other controls are usually added to temporarily override or modify the operation of the actuator. These controls may be manual or automatic, variable-resistance or contact devices, depending on the system requirements.

Contacts are usually used to force the actuator to either the open or closed end of the stroke. The contacts can short R to W to drive the motor closed, or R to B to drive the motor open. Alternatively, a contact can be used to open either the W or B line, again forcing the motor fully open or closed. Sometimes a contact is used to open the R line, but the results depend on the individual motor. Some hold their position, while others may drive fully closed.

Contacts may also be used to disconnect the supply power to the motor. The result of this action depends on whether the motor or the device it drives has a return spring. For most modulating motors, cutting the power releases the brake. If the motor has a return spring, it will return to its normal position. Typical economizer systems, for example, have a normally closed spring return motor. This ensures that the outside air damper will fail to a fully closed position on loss of power. This can be critical in cold climates, because an open outside air damper, with the HVAC system inoperable, can cause serious frost damage. Although the original intention for this was probably as a safety feature, it is commonly exploited as a way for the control system to drive the damper fully closed. A damper or valve with a spring of sufficient weight can be used with a nonspring return motor with the same end result.

Control circuit high or low limits may be contacts (see Figure 18–6) or variable-resistance devices that modify the operation of the control system. Modulating low limits are

FIGURE 18–6 Contact-type limits can be used to drive a Series 90 actuator to one end of the stroke. If the temperature reaches the setting of the high limit, the contacts change state. This connects all the controller resistance into B line of the motor and shorts the W line to R. The motor will drive to the fully closed position. For a cooling connection that controls a fresh air ventilation damper, the same limit could be used to close the damper when the outside air is too hot.

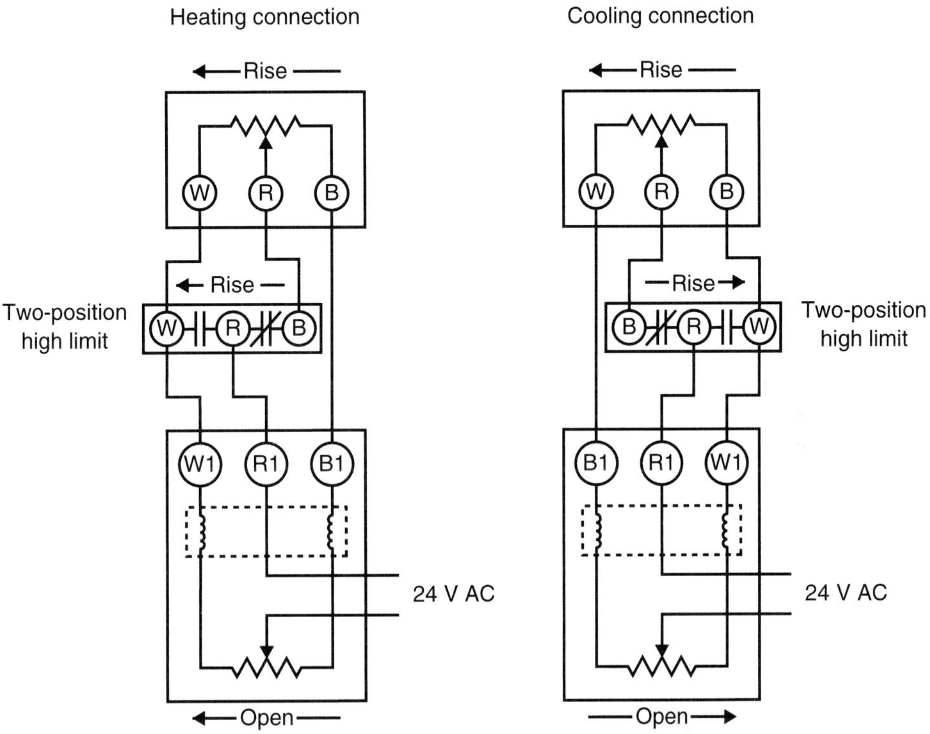

installed in the W line of the controller, and high limits are installed in the B line (see Figure 18–7). A low limit, for example, normally acts to override the controller as it moves toward the upper end of its resistance range. If the controller senses a high temperature and calls for the motor to reduce the heating (or increase the cooling) below the setting of the low limit, the limit will act to prevent the controller from driving the temperature any lower.

A common feature of economizer circuits is a manual minimum air control. This variable-resistance device is connected like a high limit in the B line of the controller, which is the W line of the damper motor, because this is a cooling connection. During economizer operation a modulating controller senses the temperature of the mixed-air and attempts to maintain a set point by modulating the outside air and return air dampers. This adjusts the mix of cool outside air and warm return air that makes up the air supplied to the building. If the mixed-air temperature drops well below the set point of the controller, the wiper of the controller potentiometer goes to all the way to the B end of the resistance element. This applies the full resistance of the controller to the W side of the motor circuit.

FIGURE 18–7 Modulating or variable manual limits may also be used in Series 90 systems. The overall effect depends on the connection of the limit and its position in the control circuit.

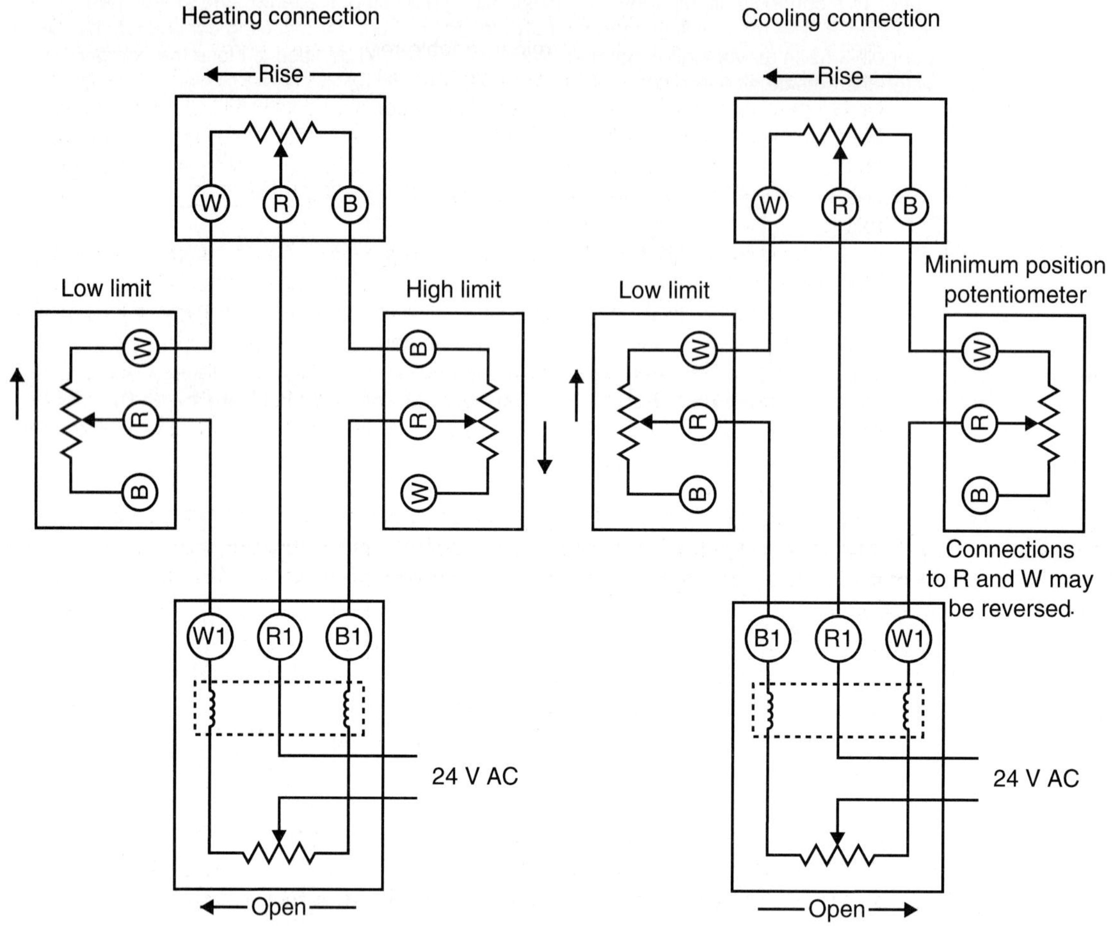

In a heating connection; the low limit acts to open the valve or damper that provides heat.

In a cooling connection, the low limit acts to close the valve or damper that provides cooling.

If the mixed-air control were the only control device in the circuit, each time the mixed-air temperature dropped too low, the damper would go fully closed, and no fresh ventilation air would be drawn into the building. The minimum air control is adjusted to open the damper to the desired minimum air position, with the wiper of the mixed-air control all the way to the B end (135 Ω R to W; 0 Ω from R to W). The extra resistance of the minimum air potentiometer restricts the operating range of the actuator, preventing it from driving the damper past the minimum air setting. In a typical packaged HVAC system with a powered outside air damper, the minimum air adjustment applies to the heating and mechanical cooling modes, as well as to the economizer operation. Examples of these control circuits are provided later in this section.

Note

Although the modulating limits used in Series 90 control systems are potentiometers—and are usually referred to by that name—they are actually connected as rheostats. A rheostat is a two-terminal variable-resistance connected in series with some other device. Because the controls have all three leads available, care must be taken to connect the correct ones. The wiper connection (R) is always used, but selecting W or B determines whether the resistance will increase or decrease for a specific direction of change in the sensed variable. For a manual pot, the connection determines whether the resistance will rise or fall for a specific direction of rotation of the adjustment knob.

Automatic controls must be connected to provide the appropriate response. Manual controls should be connected to operate in accordance with accepted norms. Clockwise rotation should produce an increase in whatever is being adjusted, and counterclockwise rotation should cause a decrease. Once the correct pair of wires or terminals has been selected, they can be connected either way, with identical results. However, if there are standards for the terminal order within the equipment or throughout the manufacturer's product line, following them may minimize the confusion later.

Most Series 90 limits are connected color to color in the W or B line of the controller, with the circuit extending from the R terminal of the limit to the appropriate terminal at the motor. For example, a low limit is essentially "cut into" the W line. The W terminal of the limit connects to the W terminal of the controller, and the R terminal of the limit extends to the motor terminal to which that controller line was originally connected. In this instance, the motor terminal would be W1 for a heating connection, or B1 if it is connected for cooling or ventilation. The unused third terminal of the limit is left disconnected.

Minimum position potentiometers, such as minimum air adjustments for outside air dampers, are an exception. They are usually connected between W and R, in the line between B on the controller and W1 on the motor. With this connection, if the potentiometer is marked OPEN and CLOSED, the behavior of the damper will match these markings.

When additional resistance is added to one side of the balance circuit, the extra resistance does not completely override the normal control: It only modifies the end result. The degree to which the limit controller regulates the outcome is sometimes referred to as its percent authority. The level of authority is determined by the total resistance of the limit, compared to the resistance of the main control. For example, when the motor and controller are typical Series 90 devices with 135 Ω potentiometers, a 135 Ω limit device has only 50% control authority. With the limit adjusted to full resistance, the controller will still be able to modulate the actuator through 50% of its full stroke. When 100% limit authority is required, a 270 Ω device may be used with a 135 Ω motor and controller. The 270 Ω device is capable of completely overriding the controller.

Other Series 90 Devices. The Series 90 system is a complete family of modulating controllers, auxiliary devices, and final control elements. Typical modulating electromechanical controllers include temperature, enthalpy, pressure, and manual potentiometers for applications such as remote ventilation air adjustment. These devices are used where a single controller or a controller/limit combination can provide the necessary features. When the control requirements are more complex, some type of central electronic controller may be used.

Electronic controllers can provide features that are beyond the capabilities of the simpler electromechanical types. In multizone applications, for example, the central controller may accept multiple thermostat/sensor inputs and integrate this information into appropriate outputs for several stages of both heating and cooling. Typical systems also include economizer controls with outputs for one or more actuators. The controller may or may not be compatible with Series 90 motors.

Auxiliary devices for Series 90 motors include motor-mounted auxiliary potentiometers and contact sets. An auxiliary potentiometer is driven by the motor's auxiliary output shaft, and can be used to control one or more slave motors. Multiple motors can also be paralleled from a single controller if its W and B terminals are bridged by the correct value of shunt resistance.

Most Series 90 control circuits are used to position a modulating actuator that is mechanically linked to a damper or valve. In these applications, the final control element of the circuit is the actuator. However, there are other types of final control elements. For example, there are modulating gas-valves with the operating motor integrated into the valve assembly. These valves may be totally self-contained, or they may have an external linkage to modulate a combustion air damper. Although it looks different than other actuators, the valve is a fully compatible Series 90 device designed for use with the same types of variable-resistance controllers. Valves like these are used on larger burners with modulated firing rates.

Another type of Series 90 final control element is a modulating step controller, or sequencer. These units have a small modulating motor linked to a multicontact cam switch that converts the mechanical position of the motor shaft into a sequence of open and closed contacts. This device is used in some electric heating applications. The cam switch contacts are all open with the motor in the fully open position, and close in sequence as the motor moves toward the fully closed position. The contacts are used to energize the coils of the heating element contactors. This modulating controller/sequencer combination responds to changes in heating demand by increasing or decreasing the total number of elements that are energized at one time. The sequencer converts a proportional signal into multiple stages.

The small motor or the sequencer drives the cam switch in exactly the same way that a more powerful actuator opens and closes a damper or valve. The motor is small, because very little torque is required, but it has a feedback potentiometer and is fully compatible with any Series 90 controller.

Floating Control

The Series 90 system is based on balanced currents on two sides of a variable-resistance bridge. In typical installations it operates as a closed-loop control system because it has a feedback device that monitors the physical position of the output shaft of the actuator. Closed-loop control circuits depend on some type of feedback to monitor the operation of the final control element. Open-loop controls drive the final element without a feedback signal.

Some Series 90 motors can also be operated as open-loop devices by connecting them to a suitable floating controller. Motors designed specifically for floating control may be simpler than Series 90 devices, because the internal balance circuit and the feedback potentiometer are not required.

Floating controllers have a three-position SPDT contact-type of output. The controller can provide only three possible output signals to the actuator: close, open, and stop. The control signals do not vary in the usual sense, but because the controller can stop the actuator at any position, it can provide modulating control. Electromechanical floating controllers work best if the actuator is relatively slow, and the resulting effects of its movements are almost immediately apparent to the controller. The controller should also have a significant differential, or "deadband," to reduce the chance of oscillation. An example of where this type of circuit has been used effectively is in the control of static pressure in the supply ducts of multizone variable air volume systems.

In a system of this type, the temperature of each zone is controlled by an independent thermostat/damper circuit that regulates the flow of conditioned air into the zone. For proper operation, the supply duct pressure must be maintained within reasonable tolerances, despite wide variations in demand. The airflow into the supply duct is regulated by a damper, and a floating pressure controller is located in the duct downstream from the damper.

The controller operates the damper actuator in much the same way as human operators would if they were given a three-position manual switch and a pressure gauge. When the static pressure drops below the set point, the controller closes the contact that starts the damper motor moving in the open direction. As the damper slowly opens, the static pressure increases almost immediately, and the controller responds by reopening the contact. The motor stops, holding the damper in this new position. The feedback for this system is the effect the sensed variable has on the controller.

The simplicity of floating control is very attractive, particularly at the motor end. With no need for feedback pots or complex internal controls, the actuator is simpler and less expensive to build. The signal between the controller and the motor is either full off or full on, so it is immune to interference, and the accuracy is essentially unaffected by conductor resistance. Unfortunately, floating control can be difficult to apply with conventional control devices, and sees only limited use in such systems.

It is perhaps fitting that the most technologically complex controllers often allow the simplest form of control. Outputs for floating control are provided on most microprocessor-based controllers. Note, however, this is not necessarily the same type of floating control logic that is described above. A microprocessor control system can operate a simple floating-type actuator motor as a fully proportional device. It can even provide for effective shaft position feedback, even though there is no feedback device installed. In place of direct position feedback, the controller keeps track of the amount of time the actuator has been operated in each direction. To keep this information accurate, the motor is periodically driven through a full stroke, and the recorded "position" is reset to zero.

Actuators With Integrated Control Electronics

The floating control actuators described above may have no internal control components. All the control logic and the switching devices are external to the motor. The emphasis is on "smart" central controllers coupled to simple final control elements. In contrast to this system, some actuators have most of the control components integrated into the motor unit. The actuator contains the "smarts"; the external devices are usually simple variable-resistance sensors or contacts.

For temperature sensing, thermistors are a common choice, but because the resistance range must match the requirements of the internal control circuit, they are often specific to a particular make and model of actuator. Other sensors are likely to follow a similar logic. To extend the capabilities of these actuators, the manufacturer may to adapt them for use with standard control signals. They may also provide add-on modules to enhance the built-in control logic.

Examples of these motors are found in many modestly sized economizer systems. A typical economizer control requires enthalpy or temperature sensors for both outdoor air and mixed-air, and an adjustment for the minimum air position of the outdoor air damper. These features are provided through the motor's internal circuitry or through add-on modules. Like most economizer actuators, the motors are usually equipped with a return spring to close the outside air damper on loss of power.

Actuators and Control Systems Using Standard Signals

Although many manufacturers build proprietary equipment, there is always a demand for controllers and control elements designed to provide or respond to one or more of the accepted standard control signals. Standardization allows the end user to select equipment from multiple manufacturers, and helps ensure a continued supply of replacement parts.

Systems built around standard signals do the same tasks as proprietary equipment. However, the system is built from "standard" components, so more engineering is required, and the components of the system are usually more accurately defined. The control devices may be adapted to a variety of situations, so they are usually defined in terms of their function rather than their specific application. This leads to terminology that can be applied to almost any control application, rather than descriptions that are specific to one field. For example, in HVAC applications, temperature sensors might be referred to as thermostats, aquastats, or mixed-air controls, because they are defined in terms of their use rather than their function. All three of these devices could be the same sensor used for different purposes. The mounting provisions may vary, but the same sensor could be used for each application.

Control systems of all kinds can be categorized into functions. All systems require inputs to provide information, control logic to interpret the information and make decisions on how to respond, and outputs to make these responses into real-world actions.

The inputs can be broken down into two types: analog and discrete. An analog input is a variable signal, such as a voltage or current level, or a standard thermocouple connection. Some input devices are very simple, and an acceptable signal can be taken directly from the sensor. Others require a "signal-conditioning" circuit to convert the changes in the sensor into standard signals. These input devices are sometimes referred to as transducers, because they convert a physical property such as temperature, humidity, or pressure into a standard signal that can be understood by the controller.

A discrete input is an on/off signal from a contact or other switching device. The voltage level used for discrete inputs can be one of several standards, and it is usually provided at the controller. The field devices are just dry contacts that act as switches to the control circuit. That way, the voltage used in the external circuit is not connected to the control system.

Note

Analog signals can also be converted to digital numbers. Systems that use this approach treat both discrete and analog information as digital data. This data may be transmitted and manipulated in digital form. In general, a digital control system performs the same functions as a conventional type, but the method is totally different.

The control logic determines how the control system reacts to changes in the inputs. Discrete logic examples include examining combinations of input conditions to determine the appropriate output, inserting specific time delays, checking the sequence of inputs, and providing sequential outputs. Analog information may require values to be stored, compared, or scaled through mathematical calculations. The end result of all control logic is a set of outputs.

Control logic can be provided with hard-wired relays, timers and sequencers, electronic circuits built for a specific application, or programmable logic circuits based on microprocessors. Combinations of these may also be used. Hard-wired systems use electrical interconnections of the various components to provide the correct logic. These systems are relatively easy to define on a schematic drawing and are often referred to as conventional controls. Many examples are provided throughout this text.

Application-specific electronic circuits are installed as circuit boards or modules. They are usually accompanied by a written description of their functions and features. Charts or logic diagrams may also be used. A circuit diagram is usually not provided, because the circuit is not field repairable, and a diagram would not be self-explanatory. When the circuit board or module is shown on a diagram, it is usually represented by a "black box." Contacts may be shown, but the coils are usually not. The circuits used in the module or on the circuit board may be analog or digital in nature. These two approaches are very different, but the functions provided may be exactly the same.

Programmable control systems may be function-specific devices, running a preset program that cannot be changed, or they may be field-programmable devices such as programmable logic panels, computer-based control systems, or PLCs (programmable logic controllers). Function-specific controls are easier to apply but are not very flexible. They are essentially digital examples of the electronic circuit boards and modules described above.

Field-programmable controls have the advantage that they can be adapted to a wide array of applications, and the disadvantage that they cannot be used until a program has been developed and entered into the controller. Typical programmable controllers designed specifically for HVAC systems provide for a combination of discrete and analog signals on both inputs and outputs. The number and type depend on the application for which the controller is intended.

Note

It is common for programmable HVAC controllers to have outputs for floating control. Floating controls have a three-position SPDT switch action. With programmable control devices, this is provided by a pair of NO output contacts dedicated to each three-wire control circuit. Floating controls are analog in the sense that the resulting output is fully variable, and discrete in the sense that the output signals are only on/off. This same type of function can be provided with other programmable control devices.

General-purpose programmable controllers were originally designed for control of manufacturing processes. They are now used for electrical control applications of all types. In place of hard-wired logic provided by electromechanical or electronic circuits, they use a software program that is stored in the controller. The program is an endlessly repetitive loop of instructions that examines the status of the inputs and determines the appropriate outputs. Programs must be developed and entered into the controller before it can perform any function at all. If the program is at all complex and is going to be maintainable, it must also be properly documented.

The outputs of the control system are the signals sent to the final control elements. Like the inputs, the output signals are either discrete or analog. Discrete outputs include indication lights, solenoid valves, and the coils of relays or contactors. Analog outputs include informational displays such as LCD panels that display text messages or numeric values, variable actuators such as modulating dampers and valves, and other variable systems such as variable-speed motor drives.

Discrete inputs are provided by field contacts, typically isolated from field voltages. Discrete outputs are circuits that are switched by the controller. This switching can be provided in several ways. If the controller has isolated output contacts, each output can be connected directly to an external circuit, within the voltage and current ratings of the contact. The contacts are electrically isolated from one another, so each circuit can be independent from all the others. If the controller outputs are not isolated, the field devices either have to be part of the same circuit, or relays have to be added in between the controller and the load.

Analog output signals must be supplied at a level that can be understood by the final control element. Common standards are 4 to 20 mA and 1 to 5 V, but there are many others, including a variable-resistance circuit or current signal that can be used to control Series 90-type actuators. The final control element converts the variable signal into an actual damper of valve position, a specific shaft speed from a variable-speed drive, and so on. In the case of pneumatic actuators, the signal is converted into a regulated air pressure output by an electric-to-pneumatic interface device.

The actuators for these systems have connections for a power supply and inputs for the control signal. The motor must either be designed for the signals used or be adaptable through some additional components or modules. Usually the flow of signals is one way only. The control system tells the actuator what to do, and then assumes that the actuator did what it was told. The motor may use an internal feedback to keep track of its own shaft position, but usually there is no feedback signal returned to the central controller.

Motor Orientation and Environment

Many modulating actuators have output shafts on both ends of the motor, but one may be specifically designed as the drive end and the other as the accessory end. Both ends may have the same size driveshaft, but the accessory end is intended only for low-torque add-ons such as external potentiometers and limit switches. It cannot physically support or drive a normal load.

Typical motor units must be mounted with the shaft horizontal. This restriction may vary with the type and design of the motor, and is part of the manufacturer's specifications. These limitations are particularly applicable to motors with oil-filled gear cases. Models that are not oil-filled are less likely to be restricted, so the manufacturer may allow almost any mounting orientation.

There are limits to the acceptable ambient temperature and the nature of the atmosphere around the actuator. The ambient temperature greatly affects the ability of the motor unit to cool itself, and must be within the range given in the specifications. This specification typically includes both upper and lower maximums. Extremely cold temperatures may thicken lubricating oils or grease, or cause critical components to become brittle.

Typical motors are designed for operation in a dry atmosphere that is not corrosive or explosive. When the motor is exposed to abnormal conditions, it must be designed for the situation, or be suitable protected and enclosed. Explosive environments are particularly difficult, demanding special wiring methods and procedures, as well as equipment that is approved for the application and the environment. **Never, never, never** install standard electrical components in an environment that may contain flammable material or explosive mixtures of gas or dust.

Damper Linkages

As a rule of thumb, dampers should be considered full-open with the blades at 60° rather 90°. The reason is that a typical damper has almost no effect on the airflow for the last 30° of its travel. This rule is subject to many variables, including the type of damper used, but the basic principles must be taken into account. The controller and its actuator are regulating airflow in the duct system to maintain the temperature or pressure at some set point. If a significant portion of the actuator stroke has virtually no effect, much of the overall control capability is sacrificed. A common way to address this problem is to set the motor and damper linkages so the motor is at the full-open position with the damper blades 60° from the full-closed position.

The linkage for a typical actuator consists of crank arms on the motor and damper shafts, and a rod connecting the two together. The rod connects to the two crank arms with a ball-joint device that bolts into a slot in the arm. The rod slides into a hole in each ball-joint and is secured with a set screw. The placement of the ball-joint sets the effective radius of the crank arm, which affects the force applied by the motor and the length of the stroke. The orientation of the arms determines the relative direction and length of the stroke, and whether the actuator torque is used to best advantage. The linkage must be able to move through the full stroke of the motor with no tendency to stick or bind.

Actuator manufacturers provide various recommendations regarding linkage installation. The examples illustrated in Figure 18–8 are from Honeywell, but they are generally applicable to a variety of actuator installations.

The motor should be mounted between 14″ and 36″ from the center line of the damper shaft. If it is closer than 14″, it is very difficult to adjust the linkage. When the distance is over 36″, the linkage rod may not be stiff enough to withstand the full force of the actuator without buckling.

The motor ball-joint should be set at the bottom of the crank arm slot, as close as possible to the motor shaft. This gives the motor its maximum leverage and highest output force.

With the motor in the fully closed position (fully counterclockwise viewed from the power end), its crank arm should face away from the damper shaft. Using an imaginary reference line drawn through the center of the two shafts, the motor crank arm should be about 10° above the line for a 160° motor, and 45° for a 90° motor. This will place the crank arm perpendicular to this reference line when the motor is at midstroke.

On damper crank arms that are calibrated in degrees, the ball-joint on the damper end should be set to the desired degree of operation, such as 60°. This sets the full stroke of the damper shaft.

With the motor in the fully closed position, the damper blades should, of course, be closed. When the motor arm is at the midpoint, both crank arms should be perpendicular to the imaginary reference line. With the motor ball joint at the bottom of the slot and the damper end at the 60° mark, the damper should open to 60°, with the motor arm in the fully open position.

FIGURE 18–8 The placement of the actuator, and the connection of the operating arms and linkage, are critical to proper operation. The stroke of the motor must be translated into an appropriate arc of travel at the damper end. Placing the crank arms at the correct angle ensures best use of the actuator torque and minimizes any tendency to stick or bind.

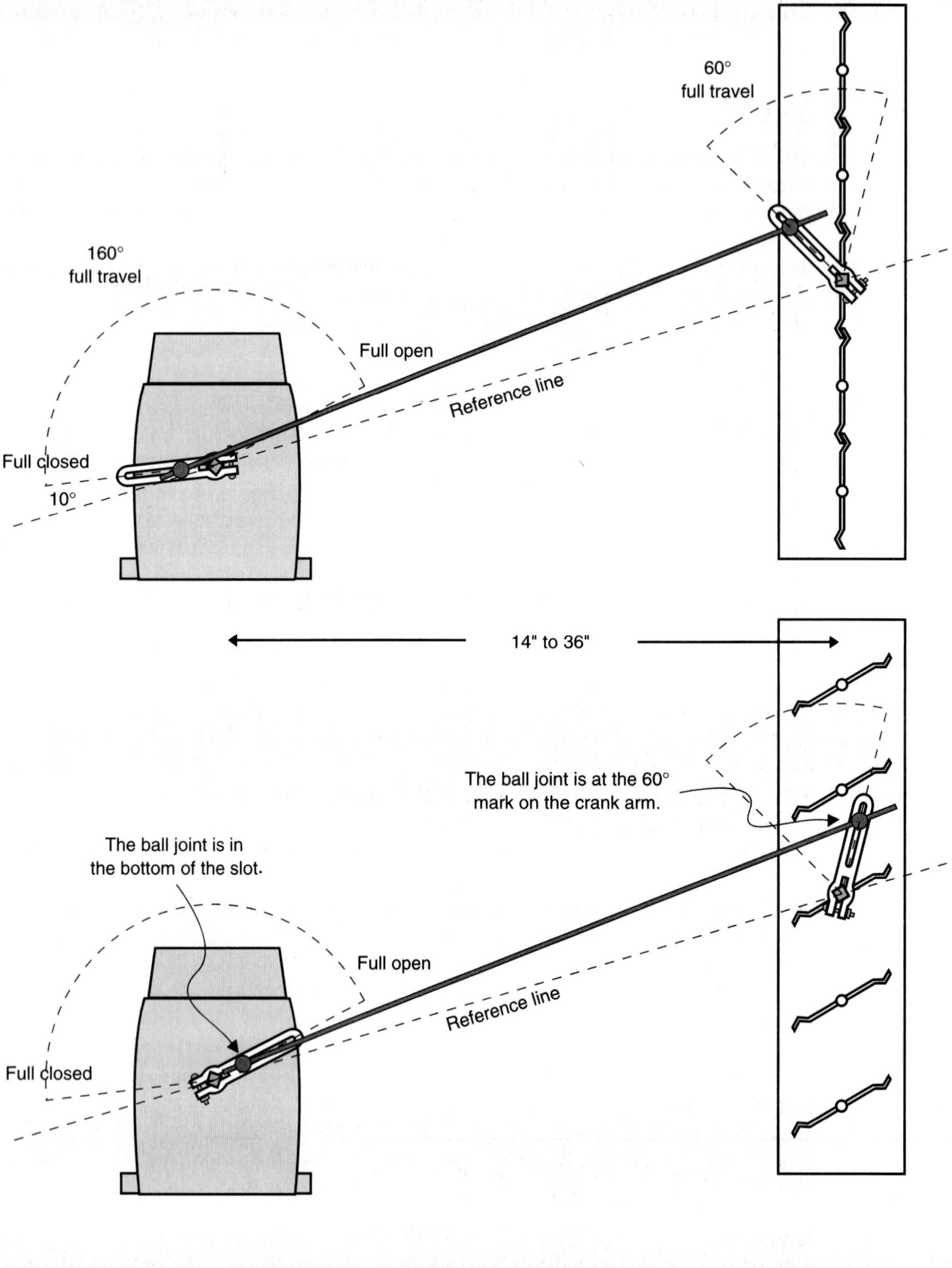

The best way to test the linkage is to connect an appropriate manual potentiometer in place of the controller. The potentiometer can be used to run the motor through its full stroke while providing full control over the motor movement. It is also recommended that initial operations be done with the linkage not fully tight. If there is a problem, it is better to have the linkage slip a little than to cause damage. Care must be taken to ensure that the linkage operates smoothly and that the motor is not stalled at either end of the stroke.

18.4 Control System Examples

The following examples are based on a Lennox-built packaged system with conventional relay controls. The drawings in Figures 18–9 through 18–14 are adapted from the manufacturer's service literature, redrawn and revised as necessary by the author to meet the aims of this text. They should not be considered as replacements for the drawings provided by any manufacturer. Many of the small details that are present in the manufacturer's literature have intentionally been omitted. The main purpose of these examples is to show the operation of this type of control system and some of the various options.

Rooftop Packaged Unit, Gas Heat, No Economizer

To fit the drawing in the available space, it has been separated into two parts—the power circuit in Figure 18–9, and the control circuit in Figure 18–10. This division is a natural one that requires a minimum of references from one drawing to the next. The three-phase power circuit is laid out in a horizontal format that starts with the main disconnect in the top left-hand corner of the drawing. The unit upon which this is based was manufactured in three voltage ratings, 208–230, 460, and 575 V. These different voltage ratings required different motors and some variation in the transformers used. The control circuit is the same in all cases because each system uses 24 V controls.

The control circuit is drawn in vertical ladder schematic form with the power supply at the top. The ungrounded supply line is drawn on the left, and the grounded common on the right. Grounding is a result of a connection at the burner control, not at the transformer. This is not unusual in gas fired systems with electronic ignition. Adding a ground connection at the transformer can cause problems because it bypasses the wiring between the two ground points. Reference numbers are provided on the left side, and cross-reference information is provided on the right. This follows normal conventions for this type of drawing. The reference numbers are sequential; the cross-references indicate the location of the contacts associated with each relay or contactor coil.

The blower and compressors are three-phase units controlled by three-pole contactors. No overloads are shown in the control drawing, and they are not mentioned in the service manual, so a reasonable assumption is that they are line-break units mounted at the motors they protect. The condenser fan operates directly from the three-phase supply, but it is a single-phase, permanent-split capacitor (PSC) motor. This motor operates at the same line voltage as the compressors and blower motor, which is 208/240, 480, or 600 V, depending on the model. The contactor that controls the condenser fan is only a single-pole device, so the motor terminals will be live to ground unless the main disconnect is open.

The combustion air fan motor is a small single-phase unit rated for 230 V. An autotransformer is supplied with 460 and 575 V units to match this motor to the rating of the supply. Note that the terminals of this motor will also be live to ground at all times, unless

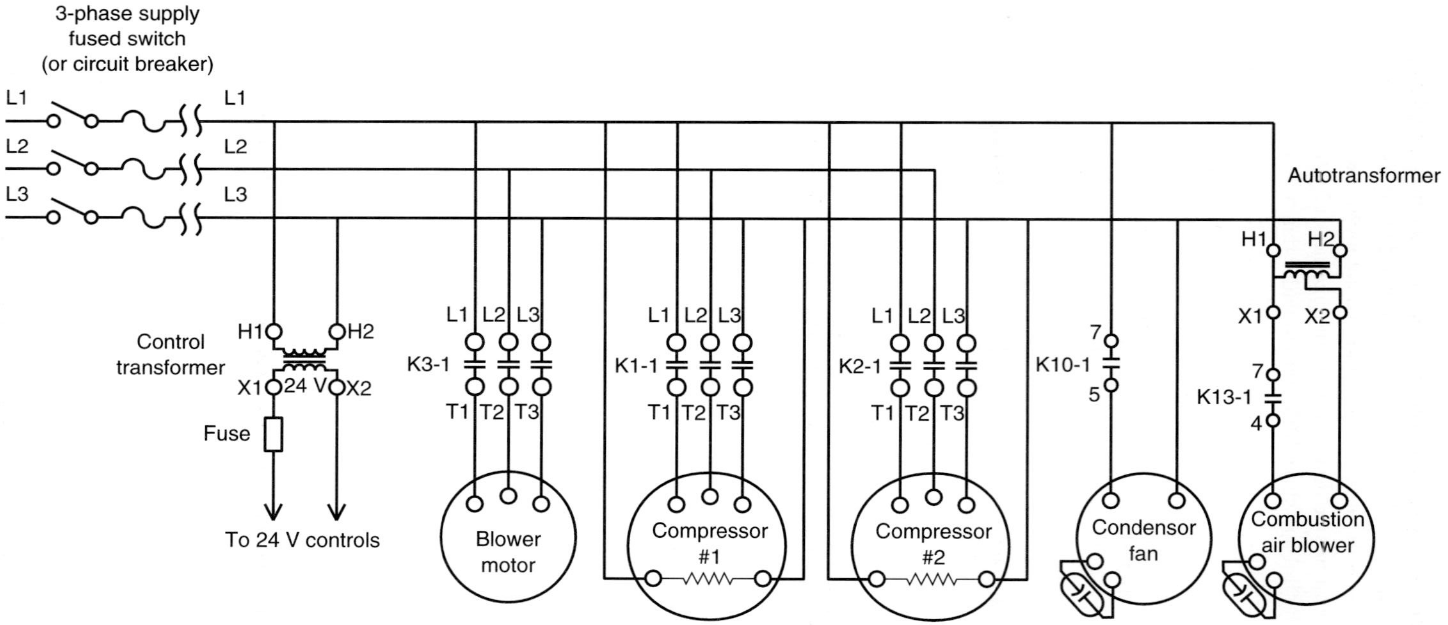

FIGURE 18–9 Power circuit for packaged system example. See text for explanation.

FIGURE 18–10 Control circuit for packaged system example. See text for explanation.

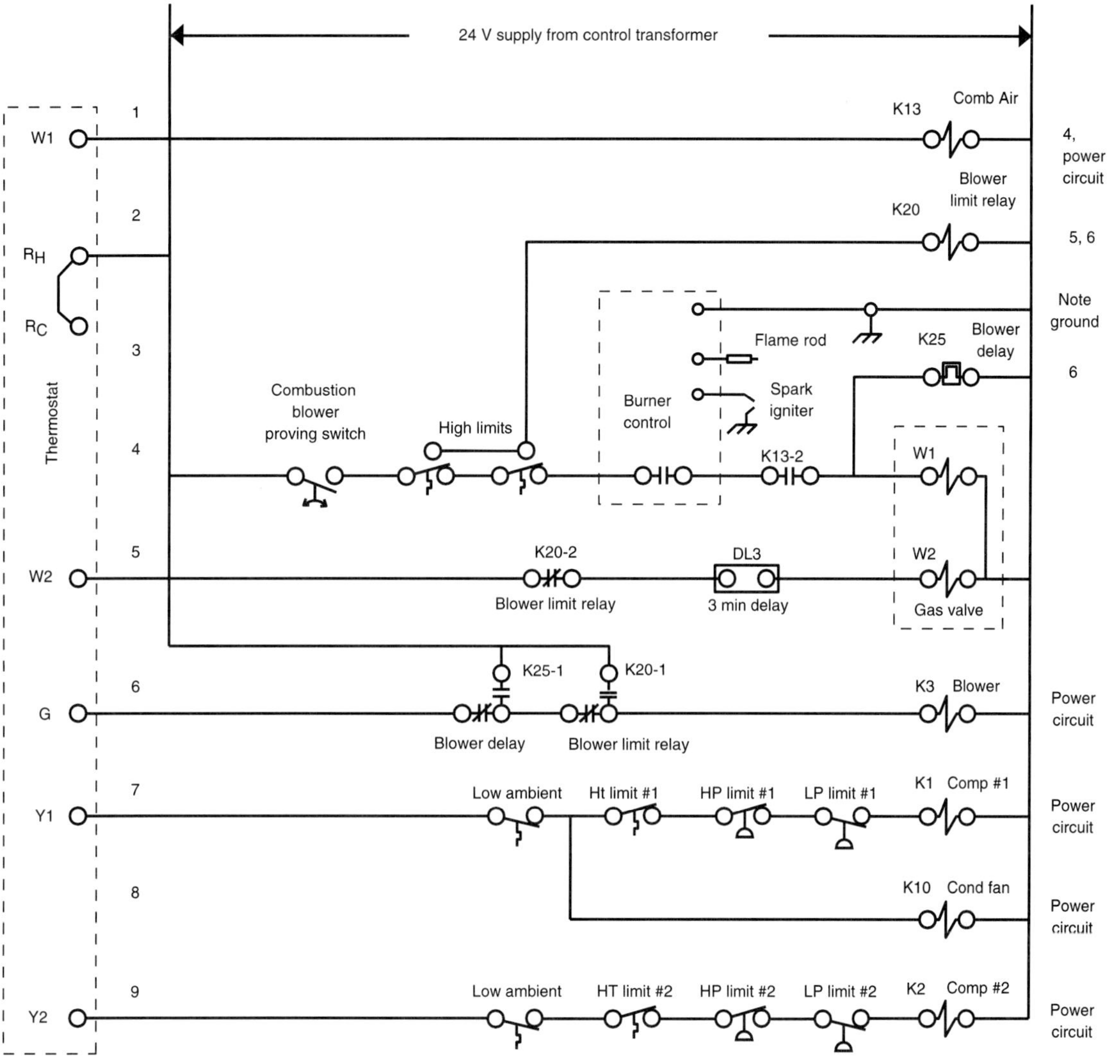

the main disconnect is open. The autotransformer does not isolate the motor from the voltage to ground of the three-phase line. The actual voltage present at the motor terminals may not be the full-rated three-phase value, but it is certainly high enough to be lethal.

The contactors in the power circuit have their coils in the control circuit. Contacts have the same K number as the coil that controls them. The number following the dash—the 1 in K3-1, for example—is a contact number that is unique to that contact or contact set. This feature can help in identification of the contacts on the actual relay or contactor.

In the examples that follow, there are several different "models" of the same packaged unit. All of these have the same power circuit, but there are significant variations in the control circuits. The initial control drawing has two-stages of heating provided by a two-stage gas-valve, and two-stages of cooling provided by two separate compressors. There is no economizer. The primary control device is the two-stage heat/cool thermostat. The system can be best understood by following the operation of each mode through both of its stages. Short diversions are necessary to show how the system responds to typical failures.

Starting in a heating mode, the thermostat calls for first-stage heat by making a connection from the ungrounded side of the 24 V control circuit, supplied to the R_H and R_C terminals, to the W_1 terminal of the thermostat. This energizes K13, the coil of the combustion air blower relay, which closes the motor contact (K13-1) in the power circuit and an interlocking contact (K13-2) on line 4 between the burner control and the gas-valve. Note that the thermostat energizes the combustion blower relay, not the gas burner control. The burner control is energized by the air proving switch, which is a speed switch on the shaft of the combustion air blower. The burner control manages the prepurge timing, the spark igniter, the rectification-type flame detector, and the first stage of the gas-valve.

The K13-2 contact ensures that the gas-valve cannot be energized unless both the combustion relay and the speed switch contacts are closed. Without this contact, the gas-valve would stay energized for a short period of time after the thermostat shuts off the combustion air blower relay. Gas would continue to flow into the burner until the combustion blower motor slows enough for the speed switch contact to reopen. This contact also provides some extra protection in case of a problem with the speed switch.

Once the burner control is energized, it completes its normal preignition routine—typically a prepurge delay and false flame check—and then it fires up the spark igniter and opens the gas-valve. If this trial for ignition succeeds, and flame is detected, it shuts off the igniter and keeps the gas-valve energized for normal burner operation. The flame is monitored continuously. On loss of flame, the burner control shuts off the gas-valve.

When the burner control energizes the gas-valve, it also energizes the thermal blower delay relay (K25) in line 3. After an appropriate delay, the SPDT contacts of this relay (K25-1, line 5) open the line to G, and close to connect the 24 V supply line to the blower contactor coil (K3). This starts the main circulating air blower. The SPDT contacts are necessary because G and Y1 are usually connected together inside the thermostat. (In a typical thermostat, the G terminal is connected to Y when the fan control switch is set to auto, and to the 24 V line when this switch is set to manual fan. Refer to the thermostat information in Chapter 17.) If the blower delay relay did not isolate the G terminal when it energized the K3 coil, terminal Y1 could also be energized, possibly starting the first stage of the cooling system.

Continuing with a normal heating cycle, if the first stage of heating is not adequate the temperature at the thermostat will continue to fall, and the thermostat will energize the second-stage W2 terminal. A series-connected time-delay relay holds the second stage off for a minimum of three minutes. This delay ensures that when there is a simultaneous call for both first- and second-stage heating, the burner must operate for at least three minutes on the first-stage before the second stage brings the burner up to maximum output.

The high limits are both SPDT devices. If either of these operates, the burner control and first stage gas-valve coil are de-energized, shutting down the burner, and the blower limit relay (K20, line 2) is energized. The blower limit relay de-energizes the second stage gas-valve coil and, through an SPDT set of contacts in line 5, energizes the blower contactor coil K3. As in all systems of this type, if the high limit opens, the control tries to run the main blower to cool things down. If the limit tripped because of a blower failure, this will not solve the problem, but there are other reasons for limit operation, and it is important that the blower operate if possible.

The thermal relay used for the blower delay (K20) has a delay on both closing and opening. The delay on closing allows the heat exchanger to warm up before the blower starts. The delay on opening keeps the blower running for some time after the burner is shut down. This improves efficiency and burner safety by removing the residual heat from the heat exchanger.

The cooling controls depend on the thermostat to energize the blower through terminal G. Each cooling stage is energized through its own protective limits. The first stage energizes the single condenser fan. The second stage will not be energized without the first stage, because the thermostat operates the two-stages in sequence.

This system can be equipped with a manual or automatic outside air damper to provide fresh ventilation air into the building. The automatic damper (not shown) is driven by a three-position actuator motor with a spring return to close the damper when the system is shut down. This damper control is for fresh air only. It can be adjusted, but does modulate, and does not function as an economizer.

Economizer Option, Series 90

The control drawing in Figure 18–11 shows the same packaged unit with a Series 90-style modulating damper motor controlling the outdoor air damper. This system allows electrical adjustment of the ventilation air intake, and provides economizer operation. The heating portion of the circuit is unchanged. An additional auxiliary contact (K3-2) from the blower contactor is added into the supply line to the economizer damper motor (line 11). This contact energizes the damper motor when the blower is running, and shuts it off when the blower is stopped. The motor is equipped with a return spring that closes the outside air damper when the motor is de-energized.

The motor is connected to both the outdoor air and return air dampers and operates both at the same time. As the outdoor air opens, the return air damper closes and vice versa. If it is required, a gravity-controlled exhaust damper is available to release excess return air that has been replaced by the intake of fresh air. On larger systems, exhaust air is controlled by a modulating motor-driven damper; on this small system, the exhaust damper operates on the static pressure of the building. When outside air is drawn into the building without an equal amount of air escaping, the static pressure rises and the damper opens to release the pressure. The exhaust damper closes automatically when the static pressure is low.

The air that reenters the heating/cooling system is a blend of return air from the building and outdoor air that is added to improve air quality or to provide free cooling. These return air and outside air dampers are installed in such a way that the two airstreams are forced toward each other to mix them together. The resulting mixed-airstream is conditioned as required, and then returned to the building via the supply air ducts.

The sequence of the two cooling stages and the operation of the fan during cooling are controlled by the thermostat. The fan control switch in the thermostat connects the fan terminal (G) to Y1 or to the 24 V supply. In the normal "auto" position, the G terminal is energized simultaneously with Y1, the first cooling stage. When the switch is set to manual fan, the blower runs continuously. The thermostat is constructed in such a way that it will not energize the second cooling stage (Y2) without first energizing Y1 (and G).

The SPDT electromechanical enthalpy control (line 7) selects between economizer and mechanical cooling. When outdoor enthalpy is low, free cooling is available, and the enthalpy control contact is closed from 1 to 3, just as it is shown in the drawing. A call for the first-stage of cooling is directed to the enthalpy relay coil (K7, line 7) instead of the first-stage

FIGURE 18–11 Control system for packaged unit example equipped with Series 90-type economizer option. See text for explanation.

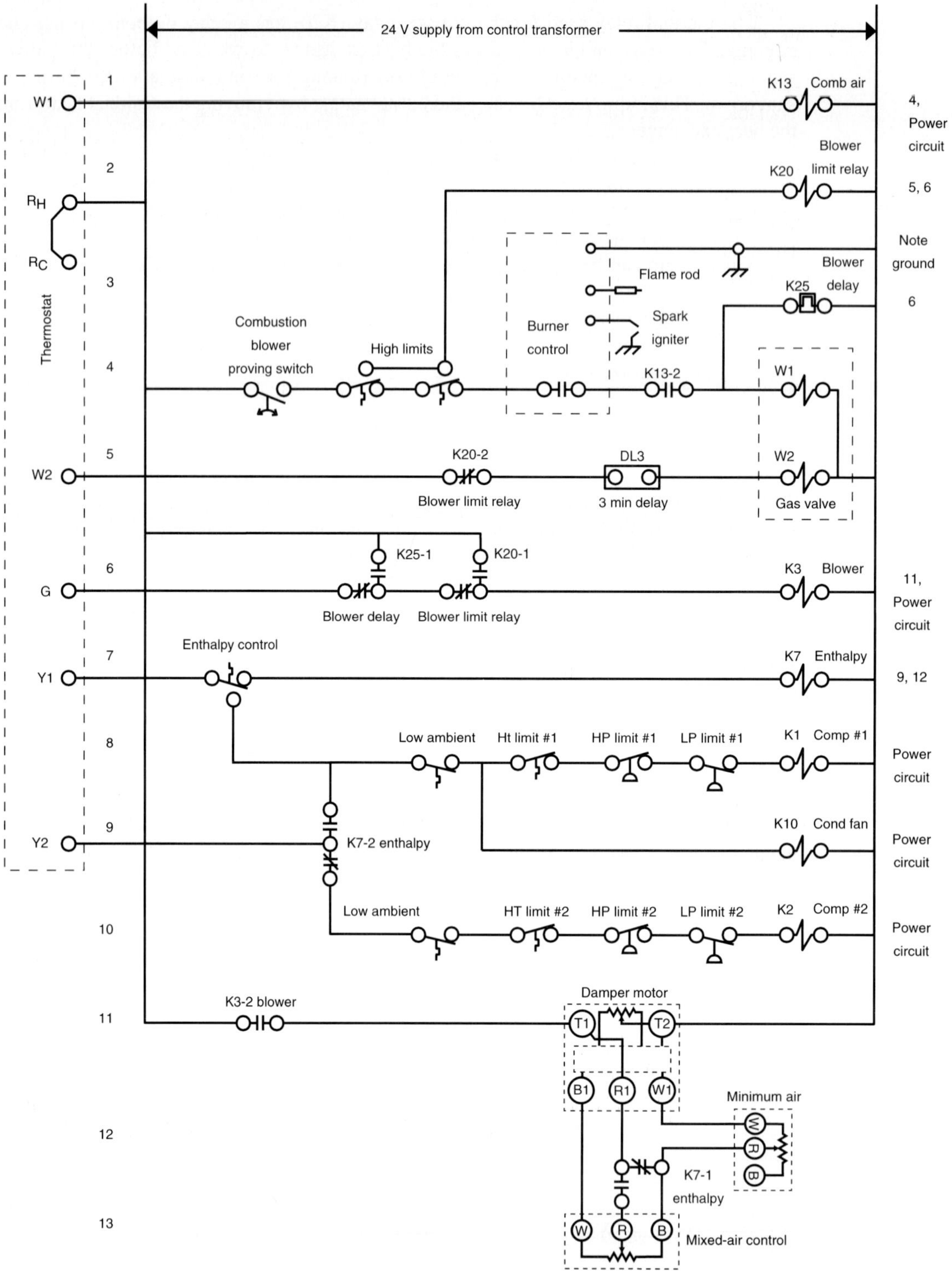

compressor circuit. The enthalpy relay transfers the first-stage mechanical cooling circuit to the second-stage thermostat terminal (Y2) through the SPDT K7-2 contacts in line 9. During economizer operation, the first stage of mechanical cooling is available as the second cooling stage, to assist the economizer if necessary. The second stage of mechanical cooling is not available until enthalpy goes high and shuts off economizer operation.

The enthalpy relay has a second pair of SPDT contacts in the mixed-air control circuit (K7-1, line 12). This is a typical Series 90 economizer connection. When the system is in the economizer mode, the mixed-air temperature controller and the minimum air potentiometer are connected to the damper motor in a standard Series 90 cooling configuration. The controller-to-motor connections are crossed, and the minimum air potentiometer is in series with the W1 line to the motor.

In the economizer mode, the damper position is proportional to the mixed-air temperature. As the mixed-air temperature rises, the wiper of the controller resistance moves toward W, increasing the resistance in the W1 line of the motor and driving the damper toward full-open. The linkage is connected so that this action increases the percentage of outdoor air brought into the system. When the mixed-air temperature drops, this operation is reversed, the proportion of return air is increased, and the outdoor air is reduced. As long as the blower is running (to close K3-2, line 11), the thermostat is calling for cooling, and the enthalpy is low (enthalpy relay energized to switch contacts K7-1), the damper modulates on mixed-air temperature. The economizer control system attempts to maintain the mixed-air temperature at the setting of the control by adjusting the amount of cool outdoor air that is mixed into the warm return air and recirculated back into the building.

When there is a demand for cooling, but the outdoor enthalpy is too high for economizer operation, the enthalpy relay coil is de-energized, putting its contacts in the de-energized state, just as they are shown on the drawing. A first-stage cooling demand energizes Y1, but now it starts compressor #1 and the condenser fan instead of energizing the enthalpy relay. The second-stage control, Y2, is connected to the coil of the contactor for compressor #2. It operates if the first stage of cooling cannot prevent the building temperature from rising to the second-stage cut-in point. The main blower is started on the initial cooling call, because the thermostat energizes the G terminal.

Note the connection at the modulating motor. With enthalpy relay contacts K7-1 in the de-energized state, the resistance balance of the Series 90 control circuit consists of the full resistance of the mixed-air control (135 Ω) between B1 and R1, and the adjusted resistance of the minimum air potentiometer between R1 and W1. The damper will be at the minimum air setting, just as it should be during a mechanical cooling cycle. The damper will open to this same position during a heating cycle, or when the fan control is set for continuous fan setting, because the blower is energized, but the enthalpy relay is not.

Economizer Option, Electronic Damper Motor

The control drawing in Figure 18–12 is very similar to the previous example. The heating portion of the circuit is unchanged. The K3-2 contact in line 11 enables the damper motor, the same as in the previous circuit. This damper actuator is also an NC spring-return motor that closes the damper when it loses power. The enthalpy control circuit in line 7 is also the same. It selects between economizer or mechanical cooling, and transfers the first stage of mechanical cooling to the second-stage Y2 circuit when enthalpy is low.

FIGURE 18–12 Control circuit for packaged unit example, with an electronic economizer actuator and an electromechanical enthalpy control. See text for explanation.

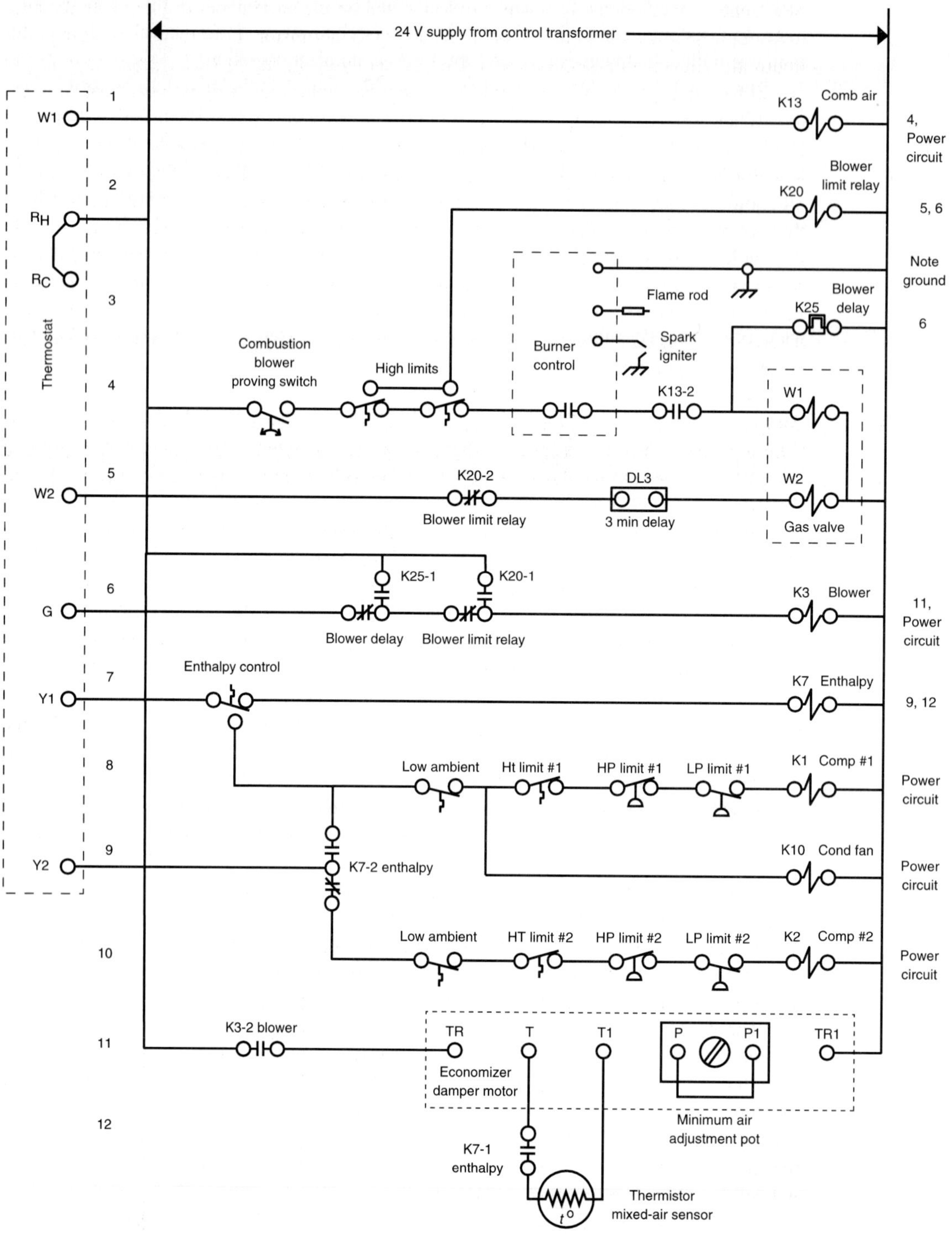

As in the Series 90 circuit, there is an enthalpy relay contact in the damper motor circuit (K7-1, line 12), but the connection is different. When free cooling is available, this SPST enthalpy relay contact closes to connect the mixed-air temperature sensor to the control circuit integrated into the motor. The sensor is a thermistor. The thermistor and the minimum air adjustment potentiometer connect directly to the actuator. The minimum air pot circuit has a jumper from P to P1 in place of an optional remote ventilation air adjustment.

When this actuator motor is de-energized, it goes full closed on the internal spring, just like a Series 90 motor would. However, other behaviors are somewhat different. If the actuator is energized but the thermistor sensor circuit is open (K7-1 is open, for example), the motor will drive the damper to the position set by the minimum air potentiometer. With the thermistor connected, the motor modulates to maintain a specific preset temperature. In this typical economizer circuit, the thermistor sensor is monitoring the temperature of the mixed-air. The motor will drive the outdoor air damper toward open as the temperature rises, and toward closed as it falls. The set point of the control is determined by the characteristics of the thermistor and the design of the internal circuit of the damper motor. This set point is typically 56°F (13°C).

When enthalpy goes high, the cooling system will operate as before, except the damper motor automatically opens to the minimum air position whenever the blower is running. This is also the damper position for a heating cycle. When the unit is supplying heat, the blower is running, but the cooling control system receives no power, so the enthalpy relay is de-energized.

Figure 18–13 shows the previous unit with additional modifications. A night-setback clock and thermostat have been added, with additional controls to provide a morning-warmup cycle. The clock puts the HVAC system under the control of the night thermostat during clock-controlled unoccupied periods. These are typically at night, but can include weekends or any other time period as required. The morning-warmup feature holds the outdoor air damper closed for the first heating cycle that occurs after the clock switches over to normal operation. This allows the heating system to bring the building back up to temperature as quickly and efficiently as possible. After the initial warmup cycle, the system runs with the normally minimum air setting of the damper, to provide fresh ventilation air while the building is occupied.

Ignoring the warmup feature for the moment, note the logic of the night-setback. The clock contact, shown as a NO limit switch in line 13, is open during the day (occupied period) and closed at night. It controls relay coils K11, the night relay, and K42, the warmup-clock relay. The night-setback relay has NC contacts across the night thermostat (K11-1, line 3), and in the minimum air potentiometer circuit of the economizer damper motor (K11-2, line 12). When the clock contact is open, the night thermostat is bypassed, and the minimum air setting is enabled. When the clock contact is closed (the unoccupied period), the night thermostat is in series with the main thermostat, and the minimum air setting is disabled.

The night thermostat is set to a lower temperature than the main thermostat, so during the unoccupied period it effectively takes control of the system. After the clock and its relay put the night thermostat into the circuit, the building temperature has to drop below the night thermostat setting before there will be a call for heat. At this temperature both the first- and second-stage contacts of the normal thermostat will be closed, so when the night thermostat finally closes, both W1 and W2 will be energized simultaneously. The burner will start in first stage and progress to second stage after 3 minutes, when delay timer DL3 (line 5) has timed out. The minimum air setting is disabled by the open K11-2 contact in

FIGURE 18–13 Control circuit for packaged system example, with night-setback and morning warmup options.

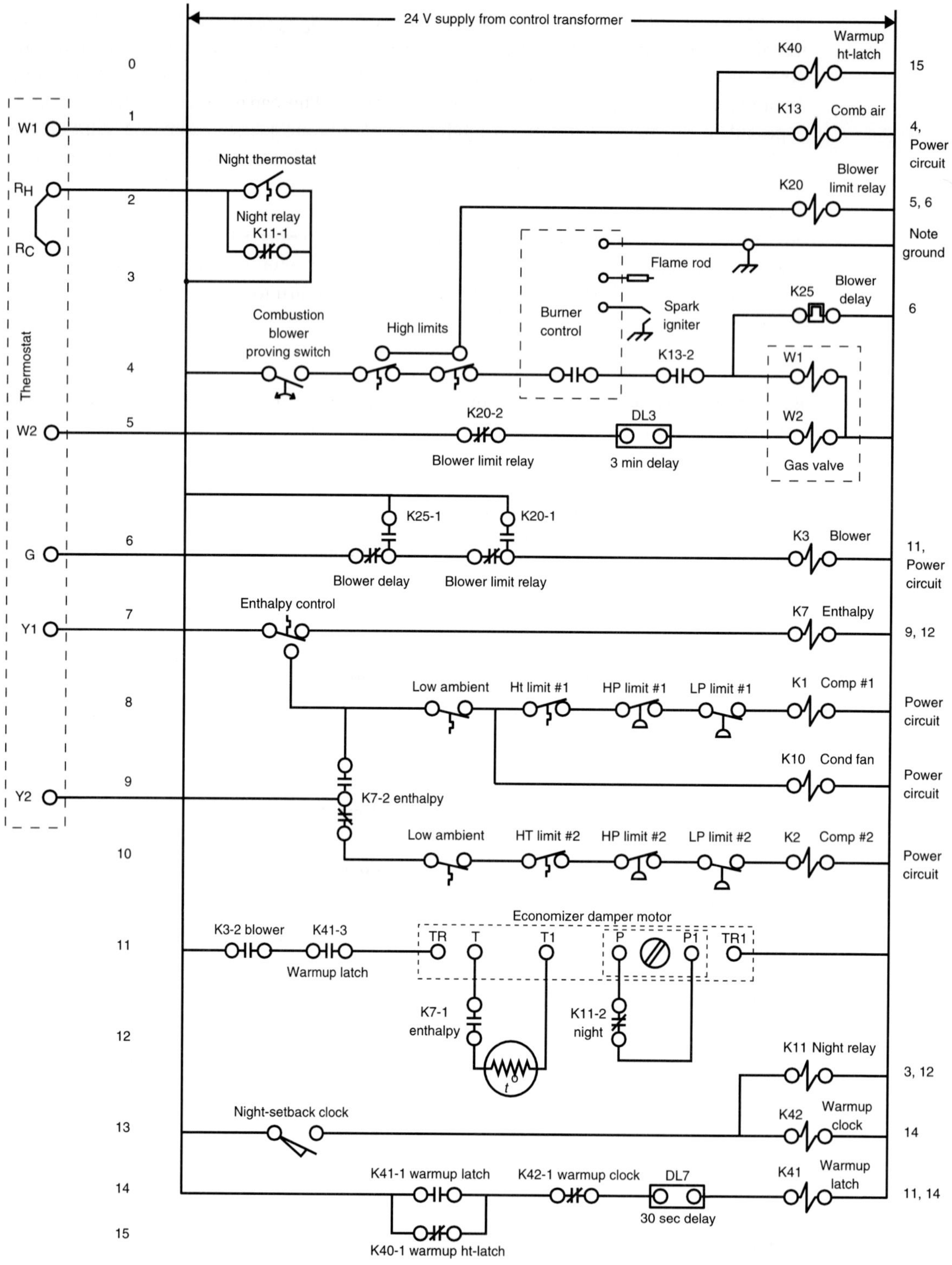

line 12, so the outdoor air damper stays shut. The cooling system is also disabled, because the night thermostat is only a heating thermostat. Even if the building gets hot enough to close the cooling contacts in the main thermostat, the cooling system will stay off because the night thermostat will not close on high temperature. This can easily happen if the system is put into the unoccupied mode during daytime periods such as weekends or holidays.

The intention of the morning-warmup circuit is to hold the damper closed for the first heating cycle following the reduced setting of the unoccupied period. When the unoccupied period ends, the clock-controlled night relay bypasses the night thermostat, returning control to the main thermostat. The building has been kept at a lower temperature by the night thermostat, so when the transfer occurs there is an immediate call for heat. Without the morning-warmup feature, the damper would also return to normal operation, opening to the minimum air setting as soon as the blower starts. With the warmup option, the damper circuit is de-energized to hold the damper closed for the first heating cycle. This brings the building up to normal temperature and, 30 seconds after this first heating call ends, the damper circuit is enabled for normal operation. Normal operation depends on the warmup latch relay, which latches in and stays picked up until the system is switched back to nighttime operation.

Note

A power failure or other event that cycles the power supply can put the system back into the warmup mode, but only for one heating cycle.

Unless there is a call for heating within 30 seconds after the night-setback clock returns the system to normal operation, the warmup latch will pick up and seal in to enable the damper circuit. If the building starts the occupied period at a higher temperature than the setting of the main thermostat, the system will go to an economizer or other cooling mode, with normal damper operation.

The sequence of events that makes the morning-warmup feature work can be difficult to follow. At the end of a normal night-operation period, the clock contact opens, de-energizing the night relay K11 and the warmup-clock relay K42. The system transfers from the night thermostat setting to the higher setting of the daytime thermostat, and will normally call for heat immediately. The first-stage heat line (W1) energizes the combustion blower relay (leading to a normal heating cycle) and the warmup heat latch relay K40. The K40 contact in the warmup latch relay coil circuit (K40-1, line 15) is now open. Although the K42-1 warmup-clock contact in line 14 closed when the clock contact first opened, and K40-1 was also closed for a brief time, the K41 the warmup latch relay coil will not have become energized because of the 30 second delay (DL7).

The warmup latch cannot pick up until the K40-1 contact is closed for 30 seconds. The K40 relay is on during the first stage of heating, so it will release at the end of the first heating cycle, closing the K40-1 contact and allowing DL7 to time out. When K41 is energized, it seals itself in through its K41-1 contact in line 14. Once it has sealed in, K41 will remain energized until the control power is cut off or the warmup-clock contact, K42-1, line 14, is opened by the next clock operation. The K41-3 contact in line 11 puts the economizer damper motor back under the control of the blower motor contactor. The system is now in a normal daytime mode.

The drawing in Figure 18–14 shows the same system with a different economizer control option. The electromechanical enthalpy control has been replaced by a solid-state unit

FIGURE 18–14 Control drawing for packaged system example, with electronic enthalpy control option.

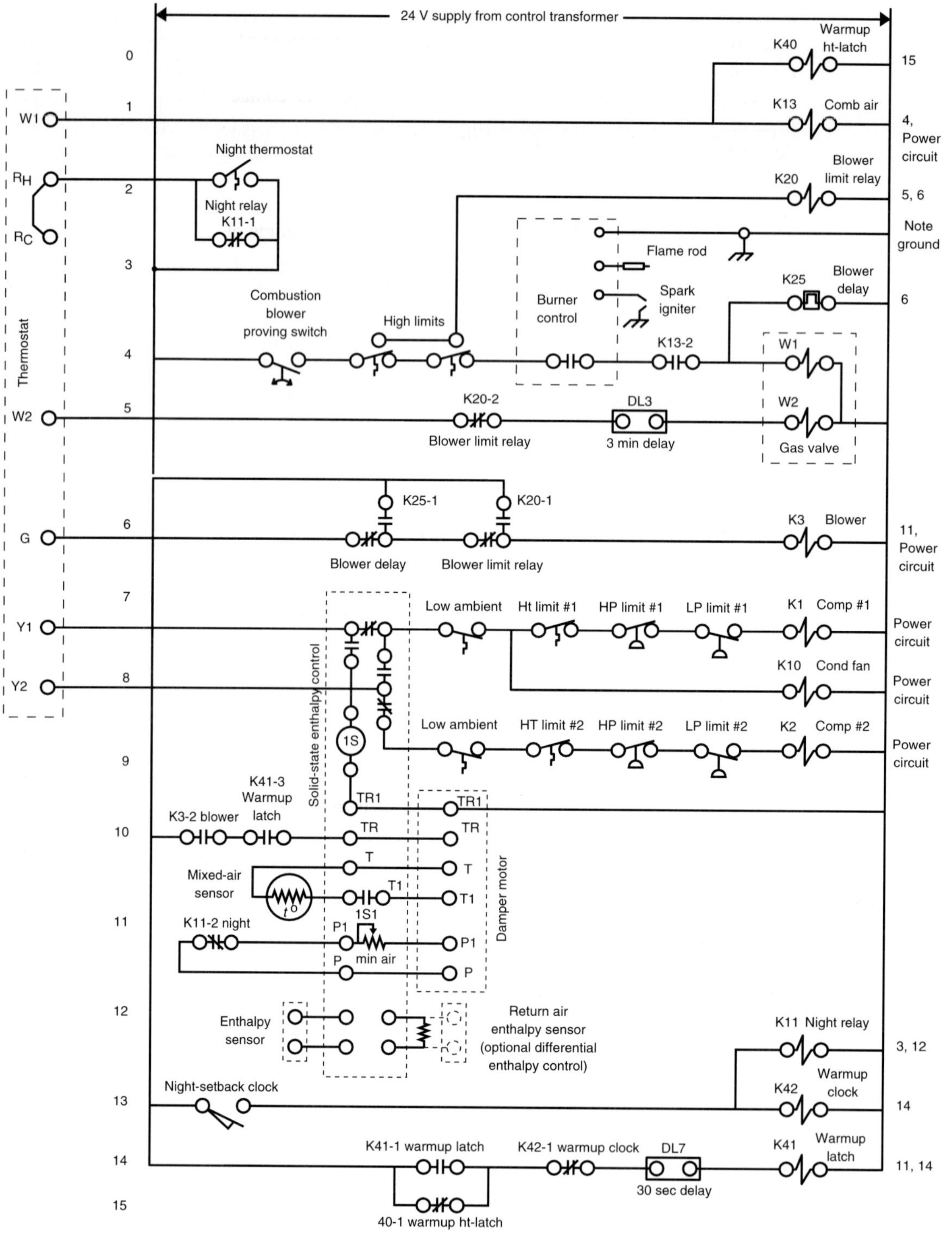

that, for this model, plugs directly onto the terminals of the damper motor. This control has built-in relays that replace the contacts of the enthalpy control and enthalpy relay of the previous circuit. In the electromechanical enthalpy control, the humidity and temperature sensors are built into the control, where they operate contacts through a mechanical linkage. In the solid-state system, the enthalpy sensor is a separate variable-resistance device mounted remotely from the control. The sensor resistance is interpreted by an electronic circuit inside the enthalpy control. This circuit, in turn, controls the relays that select between the operation of the economizer and the mechanical cooling system.

This particular enthalpy control can be used with a single outdoor sensor or two sensors—one outdoors and one in the return air. A two-sensor system responds to differential enthalpy, selecting economizer or mechanical cooling on the basis of the difference between outdoor and return air enthalpy, rather than on outdoor enthalpy alone. When the second sensor is not used, a specific value of resistance must be connected in its place. This system operates essentially the same as the previous one, but more components and functions are integrated into a single control device.

18.5 Integrated Electronic and Microprocessor Systems

As shown in Chapter 17, in residential furnaces, many manufacturers have opted to integrate almost all of the controls onto one or more circuit boards. These systems are electronic and may incorporate digital logic components such as a microprocessor. There are a number of fundamental differences between an electronic control and a system that uses conventional relay logic. These differences require different approaches to reading the drawings, and to troubleshooting in general.

Virtually all burner controls and many refrigeration systems have incorporated electronic devices for many years. In burners, electronics were first incorporated into the flame detection system. Following that, a variety of electronic sensors and control devices were incorporated into what remained a relay-based control system. Many modern systems still follow this approach. Electronics is used in all manner of devices, but the control is basically relay logic.

An alternative is to put the control logic on a circuit board. The external devices may still be a mix of conventional electromechanical and electronic devices, but they all connect to a central electronic control. This has a number of advantages. Electronic controls, particularly those using digital logic, can provide a whole range of logic and timing features very inexpensively. Many individual components can be eliminated, and assembly costs can be reduced by using wiring harnesses and plug-in connectors.

For the service person, circuit board control systems present some different challenges. First, there is a change in the organization of the control drawings and the information they can provide. For a conventional relay control system, virtually all the components are shown, and the operating logic can be traced directly on the drawing. A knowledgeable service technician can usually identify the problem, or isolate the problem component, by relating the symptoms to the control sequence that the drawing shows.

In contrast to relay logic controls, drawings for electronic circuit board systems rarely show the onboard circuit. This is particularly true for microprocessor systems, because even an electronics expert cannot follow the control logic on the drawing. What a typical electronic control drawing shows is the connections between the circuit board and the field devices. These field devices are often the same as the ones used in a conventional control. They include inputs such as the control contacts, limits, and sensors, and outputs such as the valves, motors, and actuators that do the work.

In a conventional control the relay coils are driven directly by wiring connections to the input devices and control contacts. A circuit that operates a relay coil or other control device can be traced directly on the drawing and, with a meter, right in the equipment. Typical electronic control systems also use relays, usually mounted right on the circuit board, to switch the valves, motors, and other output devices. However, the coil side of the circuit, which was traceable logic in the conventional system, is virtually untraceable on a circuit board. (Even if it can be traced, the exercise is usually only worth whatever curiosity it satisfies. It rarely serves any useful purpose in troubleshooting.)

A conventional circuit can usually be traced all the way from the input devices to the final outputs, both on the drawing and in the equipment. The ability to perform this type of troubleshooting procedure accurately and efficiently is the mark of a skilled technician. In an electronic system, the input wiring can be traced to the board terminal or connector, and sometimes to an identified test point on the board itself. The output wiring can also be traced to the board—often to a particular relay contact. However, the circuit that drives the relay coil is usually invisible, and so are the sequence and logic of the control system's operation.

Given a conventional control drawing, an experienced technician will be able to identify the normal sequence of operation, and probably will be able to suggest how the system will respond to typical failures. Given the drawing for an electronic control system, that same technician will ask for the manufacturer's literature describing the logic and features of the system. Although educated guesses are quite often correct—because the electronic system must perform the same kinds of tasks as the conventional system it replaces—there is no way to be certain about the normal operating sequence without the manufacturer's descriptive service literature.

Most modern electronic control boards incorporate a self-test function. For boards that do not have this feature, the manufacturer's literature often describes a test procedure that can be used to check the board. These tests allow the service technician to identify a bad board, or to eliminate the board as the source of the problem. This avoids unwarranted board replacement and allows the technician to continue troubleshooting with reasonable confidence that the problem is in the external field devices or wiring. Most failures are not on the circuit board, but sometimes it is difficult to be sure unless the possibility of an on-board fault can be eliminated.

Questions

1. Describe the layout of a residential split system compared to a unitary packaged system.
2. What is the function of an economizer, and why are they not used in typical residential furnace installations?
3. What changes if the setting of the minimum air potentiometer is altered?

4. What extra safety precautions are applicable to commercial packaged systems, compared to a residential unit?
5. Why does a two-stage gas-valve generally offer better control than a single-stage version?
6. Two-stage systems are often controlled by two-stage thermostats. What is used for systems with more than two-stages?
7. Define the terms "set point," "sensed variable," and "final control element" as they apply to a proportional control system.
8. Why is a variable current signal preferable to a variable voltage signal, particularly in long control circuits?
9. Sketch a circuit representing the controller, balance relay coils, and feedback potentiometer of a Series 90 modulating motor. Explain the basic operation of the circuit.
10. What advantages do electronic balance circuits have over electromechanical balance relays in Series 90 modulating motors?
11. Describe what is meant by normally open and normally closed as they apply to actuator motors equipped with return springs. Give an example of how each could be applied.
12. When an actuator has a marked drive end and auxiliary end, what limitations may apply to its application?
13. What is the standard open position of a Series 90 actuator motor?
14. What is the difference between a standard heating connection and a standard cooling connection for a Series 90 control?
15. If a Series 90 motor is connected to its normal power supply, but not to a controller, what is the effect of shorting terminal R1 to B1?
16. Explain the term "percent authority" as it applies to Series 90 control limits.
17. Describe briefly how a Series 90 sequencer operates.
18. Define "input," "control logic," and "output" in terms of the functions they perform in typical control systems.
19. What are the two common types of inputs, and how are they different?
20. What is meant by the term "signal-conditioning"?
21. Explain the difference between outputs that are isolated and those that are not isolated.
22. Why do diagrams that include electronic controls rarely show the internal connections of the electronic parts of the circuit? What troubleshooting information can be used in place of this untraceable portion of the control system?

Index

Page numbers set in boldface indicate figures.

F

G